Abstract Volterra Integro-Differential Equations

Abstract Volterra Integro-Differential Equations

Marko Kostić
University of Novi Sad
Faculty of Technical Sciences
Novi Sad, Serbia

CRC Press
Taylor & Francis Group
Boca Raton London New York

CRC Press is an imprint of the
Taylor & Francis Group, an **informa** business
A SCIENCE PUBLISHERS BOOK

CRC Press
Taylor & Francis Group
6000 Broken Sound Parkway NW, Suite 300
Boca Raton, FL 33487-2742

First issued in paperback 2019

ISBN-13: 978-1-4822-5430-3 (hbk)
ISBN-13: 978-0-367-37767-0 (pbk)

Library of Congress Cataloging-in-Publication Data

Kostic, Marko, 1977-
Abstract Volterra integro-differential equations / Marko Kostić.
pages cm
"A CRC title."
Includes bibliographical references and index.
ISBN 978-1-4822-5430-3 (hardcover : alk. paper) 1. Volterra equations. 2. Integro-differential equations. I. Title.

QA431.K727 2015
515'.45--dc23 2014044888

Visit the Taylor & Francis Web site at
http://www.taylorandfrancis.com

and the CRC Press Web site at
http://www.crcpress.com

PREFACE

The theory of linear Volterra integro-differential equations has developed rapidly in the last three decades. This book provides an easy to read concise introduction to the theory of abstract Volterra integro-differential equations. A major part of the book is devoted to the study of various types of abstract (multi-term) fractional differential equations with Caputo fractional derivatives, primarily for their invaluable importance in modeling of various phenomena appearing in physics, chemistry, engineering, biology and other sciences. The book also examines the theories of abstract first and second order differential equations, as well as the theories of higher order abstract differential equations and incomplete abstract Cauchy problems, which may be viewed as a part of the theory of abstract Volterra integro-differential equations in a broad sense.

Divided into three individual chapters, this book is a logical continuation of previously published monographs on the subject [20], [141], [463], [531] and [292]. It is not written as a traditional text, but rather as a guidebook suitable as an introduction for advanced graduate students in mathematics or engineering science, researchers in abstract partial differential equations and experts from other areas. The subject matter is intended for readers with an understanding of functions of one complex variable, integration theory and the basic theory of locally convex spaces.

Each chapter is divided in sections and subsections and, with the exception of the introductory one, contains plenty of examples and open problems. The numbering of theorems, propositions, lemmas, corollaries, definitions, etc., are done by chapter and section. The bibliography is by author in alphabetic order, and a reference to an item is of the form [303]. In order to avoid plagiarism, we have tried to make a representative list of accurate, important and consistent references. No attempt has been made to include references to all the research papers valuable for the development of the theory, so that we can freely say that the reference and citation lists are far from being complete.

The book is not intended to be exhaustive and we would like to mention, for now, degenerate Volterra equations, the solvability and asymptotic behaviour of Volterra equations on the line, almost periodic and positive solutions of Volterra equations, semilinear and quasilinear problems, as some of many topics that will not be considered here.

The author would like to express deep appreciation to all his colleagues with whom he worked for many years and without whom this book could not have been written. Special thanks go to my family and closest friends who have offered support, inspiration and encouragement throughout this research effort.

Loznica/Novi Sad
January, 2015

Marko Kostić

NOTATIONS

$\mathbb{N}, \mathbb{Z}, \mathbb{Q}, \mathbb{R}, \mathbb{C}$: the natural numbers, integers, rationals, reals, complexes.

For any $s \in \mathbb{R}$, we denote $\lfloor s \rfloor = \sup\{l \in \mathbb{Z} : s \geqslant l\}$ and $\lceil s \rceil = \inf\{l \in \mathbb{Z} : s \leqslant l\}$.

$\operatorname{Re} z$, $\operatorname{Im} z$: the real and imaginary part of a complex number $z \in \mathbb{C}$; $|z|$: the module of z, $\arg(z)$: the argument of a complex number $z \in \mathbb{C}\setminus\{0\}$.

$\mathbb{C}_+ = \{z \in \mathbb{C} : \operatorname{Re} z > 0\}$.

$B(z_0, r) = \{z \in \mathbb{C} : |z - z_0| \leqslant r\}$ $(z_0 \in \mathbb{C}, r > 0)$.

$\Sigma_\alpha = \{z \in \mathbb{C}\setminus\{0\} : |\arg(z)| < \alpha\}$, $\alpha \in (0, \pi]$.

$\operatorname{card}(G)$: the cardinality of G.

$\mathbb{N}_0 = \mathbb{N} \cup \{0\}$.

$\mathbb{N}_n = \{1, \cdots, n\}$.

$\mathbb{N}_n^0 = \{0, 1, \cdots, n\}$.

$\mathbb{R}^n$: the real Euclidean space, $n \geqslant 2$.

The Euclidean norm of a point $x = (x_1, \cdots, x_n) \in \mathbb{R}^n$ is denoted by $|x| = (x_1^2 + \cdots + x_n^2)^{1/2}$ if no specified otherwise

If $\alpha = (\alpha_1, \cdots, \alpha_n) \in \mathbb{N}_0^n$ is a multi-index, then we denote $|\alpha| = \alpha_1 + \cdots + \alpha_n$.

$x^\alpha = x_1^{\alpha_1} \ldots x_n^{\alpha_n}$ for $x = (x_1, \cdots, x_n) \in \mathbb{R}^n$ and $\alpha = (\alpha_1, \cdots, \alpha_n) \in \mathbb{N}_0^n$.

$f^{(\alpha)} := \partial^{|\alpha|} f / \partial x_1^{\alpha_1} \cdots \partial x_n^{\alpha_n}$; $D^\alpha f := (-i)^{|\alpha|} f^{(\alpha)}$.

In (X, τ) is a topological space and $F \subseteq X$, then the interior, the closure, the boundary, and the complement of F with respect to X are denoted by $\operatorname{int}(F)$ (or F°), $\overline{F}$, ∂F and F^c, respectively.

If X is a vector space over the field $\mathbb{F} \in \{\mathbb{R}, \mathbb{C}\}$, then for each non-empty subset F of X we denote by *span*(F) the smallest linear subspace of X which contains F.

$\circledast$: the abbreviation for the fundamental system of seminorms which defines the topology of a sequentially complete locally convex space E.

SCLCS: shorthand used to denote a sequentially complete locally convex space.

$L(E, X)$: the space of all continuous linear mappings from E into another SCLCS X, $L(E) = L(E, E)$.

$\mathcal{B}$: the family of bounded subsets of E.

E^* : the dual space of E.

E^{**} : the bidual of E.

A : a closed linear operator on E.

C : an injective bounded linear operator on E.

If F is a subspace of E, then we denote by $A_{|F}$ the part of A in F.

A^* : the adjoint operator of A.

$D(A)$, $R(A)$, $\rho(A)$, $\sigma(A)$: the domain, range, resolvent set and spectrum of A.

Kern(A) : the null space of A.

$n(A)$: the stationarity of A.

$\sigma_p(A)$, $\sigma_c(A)$, $\sigma_r(A)$: the point, continuous and residual spectrum of A.

$[D(A)]$: the sequentially complete locally convex space $D(A)$ equipped with the following system of seminorms $p_A(x) = p(x) + p(Ax)$, $x \in D(A)$, $p \in \circledast$.

$D_\infty(A) = \bigcap_{n \geqslant 1} D(A^n)$.

$\rho_C(A)$: the C-resolvent set of A.

$\chi_\Omega(\cdot)$: the characteristic function, defined to be identically one on Ω and zero elsewhere.

$\Gamma(\cdot)$: the Gamma function.

If $\alpha > 0$, then $g_\alpha(t) = t^{\alpha-1}/\Gamma(\alpha)$, $t > 0$; $g_0(t) \equiv$ the Dirac delta distribution.

$\mathcal{D} = C_0^\infty(\mathbb{R})$, $\mathcal{E} = C^\infty(\mathbb{R})$: the Schwartz spaces of test functions.

$\mathcal{S}(\mathbb{R}^n)$: the Schwartz space of rapidly decreasing functions ($n \in \mathbb{N}$); $\mathcal{S} \equiv \mathcal{S}(\mathbb{R})$.

$\mathcal{D}_0$: the subspace of $\mathcal{D}$ which consists of those functions whose support is contained in $[0, \infty)$.

$\mathcal{D}'(E) := L(\mathcal{D}, E)$, $\mathcal{E}'(E) := L(\mathcal{E}, E)$, $\mathcal{S}'(E) := L(\mathcal{S}, E)$: the spaces of continuous linear functions $\mathcal{D} \to E$, $\mathcal{E} \to E$ and $\mathcal{S} \to E$, respectively.

$\mathcal{D}_0'(E)$, $\mathcal{E}_0'(E)$, $\mathcal{S}_0'(E)$: the subspaces of $\mathcal{D}'(E)$, $\mathcal{E}'(E)$ and $\mathcal{S}'(E)$, respectively, containing the elements whose support is contained in $[0, \infty)$.

If $1 \leqslant p < \infty$, $(X, \|\cdot\|)$ is a complex Banach space, and $(\Omega, \mathcal{R}, \mu)$ is a measure space, then $L^p(\Omega, X, \mu)$ denotes the space which consists of those strongly μ-measurable functions $f : \Omega \to X$ such that $\|f\|_p := (\int_\Omega \|f(\cdot)\|^p d\mu)^{1/p}$ is finite;

$L^p(\Omega, \mu) \equiv L^p(\Omega, \mathbb{C}, \mu)$.

$L^\infty(\Omega, X, \mu)$: the space which consists of all strongly μ-measurable, essentially bounded functions.

$\|f\|_\infty = ess\ \sup_{t \in \Omega} \|f(t)\|$, the norm of a function $f \in L^\infty(\Omega, X, \mu)$.

$L^p(\Omega : X) \equiv L^p(\Omega, X) \equiv L^p(\Omega, X, \mu)$, if $p \in [1, \infty]$ and $\mu = m$ is the Lebesgue measure; $L^p(\Omega) \equiv L^p(\Omega : \mathbb{C})$.

$L^p_{loc}(\Omega : X)$: the space which consists of those Lebesgue measurable functions $u(\cdot)$ such that, for every bounded open subset Ω' of Ω, one has $u_{|\Omega'} \in L^p(\Omega' : X)$;

$L^p_{loc}(\Omega) \equiv L^p_{loc}(\Omega : \mathbb{C})$ $(1 \leqslant p \leqslant \infty)$.

$C_0(\mathbb{R}^n)$: the space consisted of those functions $f \in C(\mathbb{R}^n)$ for which $\lim_{|x| \to \infty} |f(x)| = 0$, topologized by the norm $\|f\| := \sup_{x \in \mathbb{R}^n} |f(x)|$.

$C_b(\mathbb{R}^n)$ $(BUC(\mathbb{R}^n))$: the space of bounded continuous functions (bounded uniformly continuous functions) on $\mathbb{R}^n$, topologized by the norm $\|f\| := \sup_{x \in \mathbb{R}^n} |f(x)|$.

$C^\sigma(\mathbb{R}^n)$: the space of bounded Hölder continuous functions on $\mathbb{R}^n$, topologized by the norm $\|f\|_\sigma := \sup_{x \in \mathbb{R}^n} |f(x)| + \sup_{x,y \in \mathbb{R}^n, x \neq y} \frac{|f(x)-f(y)|}{|x-y|^\sigma}$ $(0 < \sigma < 1)$.

If X is a Banach space, then the abbreviation $AC_{loc}([0, \infty) : X)$ $(BV_{loc}([0, \infty) : X))$ stands for the space of all X-valued functions that are absolutely continuous (of bounded variation) on any closed subinterval of $[0, \infty)$.

$AC_{loc}([0, \infty)) \equiv AC_{loc}([0, \infty) : \mathbb{C})$, $BV_{loc}([0, \infty)) \equiv BV_{loc}([0, \infty) : \mathbb{C})$.

$BV\,[0, T]$, $BV_{loc}([0, \tau))$, $BV_{loc}([0, \tau) : X)$: the spaces of functions of bounded variation.

$C^k(\Omega : E)$: the space of k-times continuously differentiable functions ($k \in \mathbb{N}_0$) from a non-empty subset $\Omega \subseteq \mathbb{C}$ into E; $C(\Omega : E) \equiv C^0(\Omega : E)$.

If $k \in \mathbb{N}$, $p \in [1, \infty]$ and Ω is an open non-empty subset of $\mathbb{R}^n$, then we denote by $W^{k,p}(\Omega : X)$ the Sobolev space consisting of those X-valued distributions $u \in \mathcal{D}'(\Omega : X)$ such that, for every $i \in \mathbb{N}^0_k$ and for every $\alpha \in \mathbb{N}^n_0$ with $|\alpha| \leqslant k$, one has $D^\alpha u \in L^p(\Omega : X)$; $H^k(\mathbb{R}^n : X) \equiv W^{k,2}(\mathbb{R}^n : X)$.

$W^{k,p}_{loc}(\Omega : X)$: the space of those X-valued distributions $u \in \mathcal{D}'(\Omega : X)$ such that, for every bounded open subset Ω' of Ω, one has $u_{|\Omega'} \in W^{k,p}(\Omega' : X)$.

$S^{\alpha,p}(\mathbb{R}^n)$: the fractional Sobolev space of order $\alpha > 0$ $(p \in [1, \infty])$.

$\mathcal{L}$, $\mathcal{L}^{-1}$: the Laplace transform and its inverse transform; $\tilde{f}(\lambda) \equiv \mathcal{L}f(\lambda)$.

$\mathcal{F}$, $\mathcal{F}^{-1}$: the Fourier transform and its inverse transform.

$LT - E$: we say that a function h: $(a, \infty) \to E$ belongs to the class $LT - E$ if there exists a function $f \in C([0, \infty) : E)$ such that for each $p \in ⊛$ there exists $M_p > 0$ satisfying $p(f(t)) \leqslant M_p e^{at}$, $t \geqslant 0$ and $h(\lambda) = (\mathcal{L}f)(\lambda)$, $\lambda > a$.

If a function $K(t)$ satisfies the condition (P1) stated in Section 1.2, then we denote $\mathrm{abs}(K) = \inf\{\mathrm{Re}\,\lambda : \tilde{K}(\lambda) \text{ exists}\}$.

$L^1_{loc}([0, \infty))$, resp. $L^1_{loc}([0, \tau))$: the space of scalar valued locally integrable functions on $[0, \infty)$, resp. $[0, \tau)$.

J^α_t: the Riemann-Liouville fractional integral of order $\alpha > 0$.

D^α_t: the Riemann-Liouville fractional derivative of order $\alpha > 0$.

$\mathbf{D}^\alpha_t$: the Caputo fractional derivative of order $\alpha > 0$.

$E_{\alpha,\beta}(z)$: the Mittag-Leffler function $(\alpha > 0, \beta \in \mathbb{R})$; $E_\alpha(z) \equiv E_{\alpha,1}(z)$.

$\Psi_\gamma(t)$: the Wright function ($0 < \gamma < 1$).

$\wp(R)$: the set which consists of all subgenerators of an (a, k)-regularized C-resolvent family $(R(t))_{t\in[0,\tau)}$.

$a^{*,n}(t)$: the n-th convolution power of function $a(t)$.

$\delta_{j,l}$: the Kronecker's delta.

supp(f) : the support of function $f(t)$.

If E and F are non-empty sets, denote by E^F the set which consists of all functions from F into E.

INTRODUCTION

This monograph concentrates on abstract Volterra integro-differential equations and abstract (multi-term) fractional differential equations with Caputo derivatives. We shall work in the setting of sequentially complete locally convex spaces (not necessarily finite-dimensional); the operators we examine in our analyses need not be densely defined and may have empty resolvent sets.

One of the main subjects considered in this book is the following abstract Cauchy problem:

$$(1) \qquad u(t) = f(t) + \int_0^t a(t-s)Au(s)\, ds,\ t \in [0, \tau),$$

where $t \mapsto f(t)$, $t \in [0, \tau)$ is a continuous mapping with values in a sequentially complete locally convex space E, $a \in L^1_{loc}([0, \tau))$ and A is a closed linear operator with domain and range contained in E. Fairly complete information on the general theory of well-posed abstract Volterra equations in Banach spaces, i.e., the theory of resolvent (sometimes also called solution) families for (1), can be obtained by consulting the monograph [463] of J. Prüss. Notice that the problem (1) intervenes, in an unavoidable manner, when we study the motion of viscoelastic materials (cf. [113]-[114], [129]-[130], [162], [178], [194], [200], [203], [228], [234], [278], [376], [398], [400], [418], [442] and [468] for further information on the continuum mechanics for materials with memory). A large part of the monograph [463] is devoted to applications of the abstract theory to problems appearing in the theories of viscoelastic materials behaviour, heat conduction in materials with memory, and electrodynamics with memory. We present some applications of our abstract results in the above-mentioned fields, mainly by making use of the generalized subordination principles clarified in Theorem 2.1.8, and take under consideration some equations that are valuable only from the mathematical point of view and do not have any physical significance.

In the second chapter of this monograph, we shall follow the method which is based on the use of (a, k)-regularized C-resolvent families and which suggests a very general way of approaching Volterra problems of the kind (1). Readers will probably need some time to reflect over this new approach. The notion of resolvent families, which now plays a central role in the theory of well-posed abstract Volterra equations, was introduced by G. Da Prato and M. Iannelli [134] in 1980. Notice that a resolvent family defined in this paper is nothing but an (a, k)-regularized C-resolvent family with $k(t) \equiv 1$ and $C \equiv I$. The notion appearing in [134] has been generalized by several authors who have considered integrated solution operator families (W. Arendt, H. Kellermann [14], H. Oka [445]; $k(t) \equiv t^n/n!$, $C \equiv I$), regularized resolvent families (M. Li, Q. Zheng, J. Zhang [363]; $k(t) \equiv 1$, $C \in L(E)$ injective) or convoluted solutions of abstract Volterra equations (M. Kim [272]-[273], C. Lizama [382]; $k(t)$ kernel, $C \equiv I$). The notion of a (local) (a, k)-regularized C-resolvent family in the Banach space setting was introduced by the author [302] in 2009, whereas the definition in a general sequentially complete locally convex space appeared three years later [303].

The organization and main ideas of this book can be briefly described as follows. A diverse range of tools and materials is collected in the opening chapter of this book, which is necessary for understanding anything that follows. The primary concerns are the integration and Laplace transform in locally convex spaces, the operators of fractional differentiation, as well as the Mittag-Leffler and Wright functions. The second chapter is broken down into ten separate sections, of which the first seven concern in detail the class of (a, k)-regularized C-resolvent families. Material presented in Section 2.1 and Section 2.2, collecting various contributions in the field of ill-posed abstract Volterra equations and their applications, is taken from [302]-[303], [306] and [316]; special accent is on the wellposedness of various types of abstract Cauchy problems closely connected with (1), and on the convoluted C-semigroups and cosine functions in locally convex spaces (Subsection 2.1.1 and Subsection 2.1.2). In the third and fifth section we focus our attention on the analysis of (g_α, C)-regularized resolvent families generated by abstract differential operators ([321]), as well as on the systems of abstract time-fractional equations ([307]); concerning the abstract time-fractional equations in Banach spaces, we can recommend reading the doctoral dissertation of E. Bazhlekova [49] and using it as a manual to guide the reader through the first steps. The proofs of our generation results from Section 2.3 and Section 2.5 rely on the use of functional calculus for commuting generators of bounded C_0-groups [141]; cf. also [214] and the references cited there for further information about functional calculi. A word of caution is also necessary regarding the possibility of applications of integrated solution families to the (systems of) abstract time-fractional PDEs. First of all, it seems that the notion of a $(g_\alpha, g_{\alpha r+1})$-regularized resolvent family is somewhat misleading in the case that $0 < \alpha < 1$ and $r > 0$, because the function $g_\alpha(t)$ is not locally of bounded variation and Theorem 2.1.29(ii) cannot be applied then. Furthermore, it should be emphasized that we could not find in the existing literature an appropriate reference which systematically treats the generation of $(g_\alpha, g_{\alpha r+1})$-regularized resolvent families by coercive differential operators ($\alpha \in (0, 2)\setminus\{1\}$, $r \geqslant 0$), and that the question of transferring of Theorem 2.9.64 and Theorem 2.9.66 to $(g_\alpha, g_{\alpha r+1})$-regularized C-resolvent families is very non-trivial provided that $\alpha \in (0, 2)\setminus\{1\}$ and $\alpha r + 1 \notin \mathbb{N}$. From the theory of abstract differential equations of first order it is well known that C-regularized semigroups provide an efficient tool for dealing with abstract non-coercive differential operators and are far superior to integrated semigroups in this case ([550]); the results established in Section 2.5 show that similar conclusions hold for abstract time-fractional differential equations. Observe, however, that integrated semigroups can be applied successfully to a large class of coercive differential operators, as well as to some special systems of mathematical physics, for example to Maxwell equations. Due to time limitations, we have not been able to consider in Section 2.3 the problem of generation of $(g_\alpha, g_{\alpha r+1})$-regularized resolvent families by homogeneous differential operators on $L^p(\mathbb{R}^n)^N$ type spaces (cf. [20], [139], [141]-[149], [153], [167]-[168], [243], [260] and [548]-[556], among many other research papers not cited here, for more details about applications of integrated semigroups and C-regularized semigroups; concerning various applications of strongly continuous semigroups, we refer to [80], [138], [175], [179], [185], [198], [229], [252], [396] and [450]).

A principal new feature of this book in comparasion with other monographs and papers on abstract Volterra integro-differential equations is, undoubtedly, the consideration of solutions in locally convex spaces. In connection with this still-undeveloped subject, we would like to mention our analysis of q-exponentially equicontinuous (a, k)-regularized C-resolvent families, carried out in the fourth section of the second chapter ([304]), a large number of equations considered in E_l-type spaces ([531]), and the analysis of abstract time-fractional equations associated with (not necessarily coercive) second order differential equations on the n-torus (cf. the paragraph following Theorem 2.2.10). Although the theory of abstract differential equations with solutions in Banach spaces is much more popular, we may conclude from the above that there exist several examples of abstract differential equations which can be analysed more effectively on locally convex spaces as on Banach spaces. We shall quote two more examples in support of this fact. Consider first the space $E := \{f \in C^\infty([0, \infty)) : \lim_{x\to+\infty} f^{(k)}(x) = 0$ for all $k \in \mathbb{N}_0\}$. Then the following increasing family of seminorms $\|f\|_k := \Sigma_{j=0}^{k} \sup_{x\geq 0} |f^{(j)}(x)|$ $(f \in E,\ k \in \mathbb{N}_0)$ makes E a Fréchet space. If $c_0 > 0$, $\beta > 0$, $s > 1$, $l > 0$ and the operator A is defined by $D(A) := \{u \in E : c_0 u'(0) = \beta u(0)\}$ and $Au := c_0 u''$, $u \in D(A)$, then we will prove in Example 2.4.6(ii) that A cannot be the generator of an exponentially equicontinuous fractionally integrated semigroup on E. On the other hand, there exists a suitable chosen continuous kernel $k(t)$ such that A is the integral generator of an equicontinuous analytic k-convoluted semigroup of angle $\pi/2$, which implies that the qualitative properties of abstract second order differential equations in some classes of Fréchet function spaces have not been fully explored. Now we would like to remind the reader of the definition of a bi-continuous semigroup $(T(t))_{t\geq 0}$ on a Banach space $(E, \|\cdot\|)$, which was introduced for the first time by F. Kühnemund in her doctoral dissertation [341]. This definition is based upon the introduction of a new locally convex Hausdorff topology τ on E, which is coarser than the norm topology of space and satisfies that the topological dual $(E, \tau)'$ is norming for $(E, \|\cdot\|)$. For some genuine applications of bi-continuous semigroups, we may refer to [341], [5] and [176]. In the forthcoming paper [326], we will try to clarify the basic structural properties of bi-continuous (a, k)-regularized C-resolvent families, at least to formulate subordination principles.

Perturbation results for abstract Volterra equations are thoroughly discussed in Section 2.6, which is further divided into three separate subsections ([314], [305]). A fairly detailed structure of this section is described as follows. A new line of approach to bounded commuting perturbations of abstract time-fractional equations is developed in Theorem 2.6.3. Our analysis is inspired, on the one side, by the incompleteness of study of bounded perturbations of integrated C-cosine functions and, on the other side, by the possibilities of extension of [292, Theorem 2.5.3] to fractional operator families. We consider an exponentially equicontinuous (g_α, k)-regularized C-resolvent family $(R(t))_{t\geq 0}$ with a subgenerator A $(\alpha > 0)$, a function $k_1(t)$ satisfying certain properties and an A-bounded perturbation B such that $BA \subseteq AB$ and $BC = CB$. In order to prove the existence of perturbed (g_α, k_1)-regularized C-resolvent family $(R_B(t))_{t\geq 0}$ with a subgenerator $A+B$, we employ the method that involves only direct computations and differs from those established in many other papers in that we do not consider $(R_B(t))_{t\geq 0}$ as the unique solution of a corresponding integral equation. The main objective in Theorem 2.6.5 is to show that, under some additional conditions, the perturbed (g_α, k_1)-regularized C-resolvent family $(R_B(t))_{t\geq 0}$ inherits analytical properties from $(R(t))_{t\geq 0}$. In case $\alpha = 2$ and B satisfies the aforementioned conditions, Corollary 2.6.6 produces significantly better results compared with [549, Theorem 3.1]. This is important since M. Hieber [226] proved that the Laplacian Δ with maximal distributional domain generates an exponentially bounded α-times integrated cosine function on $L^p(\mathbb{R}^n)$ $(1 \leqslant p < \infty,\ n \in \mathbb{N})$ for any $\alpha \geqslant (n-1)|\frac{1}{2}-\frac{1}{p}|$. Notice also that V. Keyantuo and M. Warma proved in [266] a similar result for the Laplacian Δ on $L^p([0, \pi]^n)$, with the Dirichlet or Neumann boundary conditions; cf. also [258]. In Corollary 2.6.9, we focus our attention to the case $k(t) = \mathcal{L}^{-1}(\lambda^{-\alpha} e^{-\varrho\lambda^\sigma})(t)$, $t \geqslant 0$ $(\alpha > 1, \varrho > 0, \sigma \in (0, 1))$, which is important in the theory of ultradistribution semigroups of Gevrey type. As a special case of Corollary 2.6.9, we obtain that the class of tempered ultradistribution sines of $(p!^s)$-class $(\{p!^s\}$-class) is stable under bounded commuting perturbations $(s > 1)$; cf. [286], [298, Definition 13, Remark 15], [292, Section 3.5], [412] for more details on the subject. It is worthwhile to mention here that the proof of Theorem 2.6.3 works only in the case that the considered (g_α, k)-regularized C-resolvent family $(R(t))_{t\geq 0}$ is exponentially equicontinuous. It seems to be really difficult to prove an analogue of Theorem 2.6.3 in the context of local (g_α, k)-regularized C-resolvent families (cf. [338], [516] and [292, Section 2.5, Theorem 3.5.17] for further information in this direction), which implies, however, that it is not clear whether the class of ultradistribution sines of $(p!^s)$-class $(\{p!^s\}$-class) retains the property stated above. In Theorem 2.6.12, Remark 2.6.13 and Corollary 2.6.15, we continue the researches of W. Arendt, H. Kellermann [14], C. Lizama, J. Sánchez [384] and A. Rhandi [470]. The local Hölder continuity with exponent $\sigma \in (0, 1]$ is the property stable under perturbations considered in these assertions, as explained in Remark 2.6.14. In a separate subsection, we investigate unbounded perturbation theorems. The main purpose of Theorem 2.6.18-Theorem 2.6.19 is to generalize perturbation results of C. Kaiser and L. Weis [241]. The loss of regularity appearing in Theorem 2.6.18 is slightly reduced in Theorem 2.6.19 by assuming that the underlying Banach space E has certain geometrical properties. As an application, we consider (g_{α}, g_{r+1})-regularized resolvent families generated by higher order differential operators ($0 < \alpha \leqslant 2$, $r \geqslant 0$). Perturbations of subgenerators of analytic (a, k)-regularized C-resolvent families are also analyzed in Theorem 2.6.22, which might be a little surprising in the case $C \neq I$. The above result is applied to differential operators in the spaces of Hölder continuous functions (W. von Wahl [511]). Possible applications of Corollary 2.6.6 and Theorem 2.6.5 can be also made to coercive differential operators considered in the previous section. In the remaining part of this section, we reconsider and slightly improve results of W. Arendt, C. J. K. Batty [19] and W. Desch, G. Schappacher, W. Schappacher [160] on rank–1 perturbations. Moreover, we shall slightly improve results of T.-J. Xiao, J. Liang and J. van Casteren [536] on time-dependent perturbations of abstract Volterra equations (cf. also [49], [370], [482] and [305] for more details).

The study of approximation of strongly continuous semigroups of operators has been initiated by H. F. Trotter ([505], 1958) and T. Kato ([254], 1959). Together with a great number of their variants, the Trotter-Kato approximation theorem and the Chernoff product formula ([97], 1970) play an important role in the mathematical analysis of approximations schemes for abstract differential equations. The reader may consult [20], [73], [80], [85], [199], [213], [337]-[338], [359]-[360], [366], [381], [383], [441], [485], [528], [553] and [556] for further information concerning approximation and convergence of integrated C-semigroups and cosine functions. Approximation and convergence of (a, k)-regularized resolvent families, fractional resolvent families, and propagators of higher-order abstract differential equations, have been analyzed in [49], [273], [319], [364], [383], [386], [429], [463] and [535]. In Section 2.8, which is taken from [322], we shall take a closer look at approximation and convergence of various types of resolvent operator families appearing in the theory of abstract Volterra integro-differential equations. The results on hyperbolic Volterra equations of non-scalar type are also included in this book, though maybe insufficiently. In Section 2.8, we shall also consider the aproximation and convergence of (A, k)-regularized C-pseudoresolvent families, known to be crucially important in the analysis of non-scalar equations ([463], [317], [292]).

The concept of (a, k)-regularized (C_1, C_2)-existence and uniqueness families is introduced in Section 2.8 following the ideas of R. deLaubenfels [153]. It is worth noting that the classes of (a, k)-regularized C-resolvent families and (a, k)-regularized (C_1, C_2)-existence and uniqueness families are defined in this section by using some purely algebraic identities (see, e.g., the papers [89] by C. Chen,

M. Li, and [390] by C. Lizama, F. Poblete). The subordination principles for the introduced class are formulated and after that applied to fractional diffusion equations ([308]).

Section 2.9 consists of eight subsections and its purpose is to analyze the fractional powers of (a, b, C)-nonnegative operators and semigroups generated by them ([90]-[93], [324]). We consider, as an application, incomplete abstract Cauchy problems with modified Liuoville right-sided time-fractional derivatives. An almost complete description of the structure of Section 2.9 is given as follows. In Subsection 2.9.1 we shall outline the main points of the complex analytical method used in the construction of powers of a class of C-nonnegative operators ([90]). We shall briefly describe the Balakrishnan's method for C-nonnegative operators in Subsection 2.9.2. With the notations explained later, we shall prove in Proposition 2.9.20 that the operator A_{∞} is C_{∞}-nonnegative in the space $D_{\infty}(A)$, provided that the operator A belongs to the class $\mathcal{M}_{C,m}$ for some $m \geqslant -1$. In order to construct the power A_{α} $(\alpha \in \mathbb{C}_{+})$ of such an operator A, we can follow two slightly different approaches; in both of them, the central point is the construction of powers of the operator $A_{\infty} \in L(D_{\infty}(A))$. The first approach is based, more or less, on the strict following of the method proposed by C. Martinez, M. Sanz and A. Redondo in [404]. The second approach produces the same powers as the first one and is based on the fact that, for every operator A belonging to the class $\mathcal{M}_{C,m}$ for some $m \geqslant -1$, one can find another regularizing operator C' such that the operator A belongs to the class $\mathcal{M}_{C',-1}$; this approach will be partially used in the construction of powers of (a, b, C)-nonnegative operators (Subsection 2.9.6), and its essence is briefly described in Remark 2.9.37. Several improvements of results established in [404] will be presented; bearing in mind that we would also like to have a complete picture on the available methods for construction of powers of almost C-nonnegative operators, the first approach will be almost always used in Section 2.9. We introduce in Definition 2.9.24 the complex power A_{α} $(\alpha \in \mathbb{C}_{+})$ of a closed linear operator A belonging the class $\mathcal{M}_{C,m}$ for some $m \geqslant -1$. In Lemma 2.9.26 and Proposition 2.9.29, we shall prove the C-maximality of introduced powers (i.e., the equality $A_{\alpha} = C^{-1}A_{\alpha}C$) and the coinciding of powers with the usual ones in the case that $\alpha = n \in \mathbb{N}$ (i.e., the equality $A_{n} = C^{-1}A^{n}C$). The main objective in Lemma 2.9.30 is to prove that the power A_{α} is injective provided that A is. In Definition 2.9.31 (Definition 2.9.32), we essentially use this fact in the construction of complex powers of injective operators belonging the class $\mathcal{M}_{C,m}$ for some $m \geqslant -1$ $(m \in \mathbb{R})$. As indicated in Remark 2.9.33, the introduced powers coincide with those of [90], [404] and [145, Section 5], which have been defined under the stronger conditions. Some basic properties of powers with exponents of negative or purely imaginary parts are clarified in Proposition 2.9.34-Proposition 2.9.35, whereas the additivity of powers is discussed in Theorem 2.9.36. Powers of C-nonnegative operators will be considered in Subsection 2.9.4 in more detail. We apply the abstract results obtained in Theorem 2.9.39-Theorem 2.9.40 in the study of incomplete higher order differential equations involving the Schrödinger-type evolution operators in $L^p(\Omega)$-spaces (cf. Example 2.9.41 and [532]), where Ω is a bounded domain in $\mathbb{R}^n$ with sufficiently smooth boundary. As a special case of Proposition 2.9.42, the operator A_α is C-sectorial with C-spectral angle less than or equal to $\alpha\pi$, provided that A is C-nonnegative and $0 < \alpha < 1$. In Proposition 2.9.43, we transmit some results used in the Balakrishnan's proof of spectral mapping theorem to C-nonnegative operators. In the remaining part of Subsection 2.9.4, we generalize the assertions of [365, Theorem 3.1(a),(c)] and [90, Theorem 3.1(ii)] (Theorem 2.9.44), and further analyze the semigroup and continuity properties of powers (Theorem 2.9.46). The first thing we need to do in Subsection 2.9.5 is to introduce the notion of an almost C-sectorial operator. The incomplete abstract Cauchy problems, in general with time fractional order derivatives, are further considered in Theorem 2.9.48, Remark 2.9.49, Theorem 2.9.51 and Remark 2.9.52. Making use of arguments given in the proof of Theorem 2.9.48(vi), we shall give a short and elegant proof of multiplicative identity for powers of C-nonnegative operators (Theorem 2.9.50). In Subsection 2.9.6, we shall briefly explain how one can reformulate a great number of our results for the class of (a, b, C)-nonnegative operators. In an illustrative example, we shall present some applications of Theorems 2.9.48, 2.9.51, 2.9.58 and 2.9.60 to abstract incomplete problems involving the generators of integrated or C-regularized semigroups. Subsection 2.9.7 is devoted to the study of existence and growth of mild solutions of abstract Cauchy problems of the first (second) order, associated with generators of α-times integrated C-semigroups (cosine functions) in locally convex spaces ($\alpha > 0$). In the last subsection, we investigate the representation of the complex powers of the operator A in the case that its negative generates an equicontinuous (g_α, C)-regularized resolvent family ($\alpha \in (0, 2]$). Observe, finally, that it would be very tempting to examine the further possible applications of (g_α, C)-regularized resolvent families in the theory of sectorial operators ([215], [403]).

In Section 2.10, consisting of four separate subsections, we present an extensive survey of recent results on abstract multi-term fractional differential equations with Caputo fractional derivatives ([318]-[319], [323], [325]). Of concern is the following multi-term problem:

$$(2)\quad \begin{aligned} &\mathbf{D}_t^{\alpha_n} u(t) + \sum_{i=1}^{n-1} A_i \mathbf{D}_t^{\alpha_i} u(t) = A\mathbf{D}_t^{\alpha} u(t) + f(t), \quad t > 0, \\ &u^{(k)}(0) = u_k, \quad k = 0, \cdots, \lceil \alpha_n \rceil - 1, \end{aligned}$$

where $n \in \mathbb{N}\setminus\{1\}$, A and $A_1, \cdots, A_{n-1}$ are closed linear operators on E, $0 \leqslant \alpha_1 < \cdots < \alpha_n$, $0 \leqslant \alpha < \alpha_n$, $f(t)$ is an E-valued function, and $\mathbf{D}_t^\alpha$ denotes the Caputo fractional derivative of order α ([49], [315], [313]).

We start by quoting some special cases of (2). Unquestionably, the most important subcase of (2) is the abstract Cauchy problem

$$(ACP_n): \begin{cases} u^{(n)}(t) + A_{n-1}u^{(n-1)}(t) + \cdots + A_1 u'(t) + A_0 u(t) = 0, \ t \geqslant 0, \\ u^{(k)}(0) = u_k, \ k = 0, \cdots, n-1. \end{cases}$$

It would take too long to consider the qualitative properties of problem (ACP_n) in more details; for further information in this direction, we refer the reader to [30], [147], [180], [182], [186], [323], [428], [435]-[436], [438]-[440], [444], [479], [529]-[530], [533]-[535], [538]-[539], and especially, to the monograph [531] by T.-J. Xiao and J. Liang.

The study of qualitative properties of the abstract Basset-Boussinesq-Oseen equation

$$(3) \qquad u'(t) - A\mathbf{D}_t^{\alpha} u(t) + u(t) = f(t),\ t \geqslant 0,\ u(0) = 0,$$

describing the unsteady motion of a particle accelerating in a viscous fluid under the action of the gravity, has been initiated by C. Lizama and H. Prado in [392]. In 1991, S. Westerlund suggested using fractional derivatives for the description of propagation of plane electromagnetic waves in an isotropic and homogeneous material, lossy dielectric. In the abstract form, the equation suggested by S. Westerlund takes the following form (cf. [458, (10.107)] and (214)):

$$u''(t) + cA\mathbf{D}_t^{\alpha} u(t) + u(t) = f(t),\ \ t \geqslant 0;\ u(0) = x,\ u'(0) = y,$$

where $c \in \mathbb{R}$, $A = \Delta$ and $1 < \alpha < 2$.

In [342]-[345], T. A. M. Langlands, B. I. Henry and S. L. Wearne have considered various types of fractional cable equation models describing electrodiffusion of ions in neurons for the case of anomalous subdiffusion. Notice that in some of these models abstract multi-term fractional equations with the Riemann-Liouville fractional derivatives have occurred, and that it is not clear whether these derivatives can be replaced by Caputo fractional derivatives or some combination of Caputo and Riemann-Liouville fractional derivatives, without losing some physical meaning. In a joint research with C.-G. Li and M. Li, the author has recently analyzed abstract multi-term fractional differential equations with Riemann-Liouville fractional derivatives; with the exception of this, it is quite questionable whether there exists any other significant reference which treats the abstract multi-term fractional differential equations with fractional derivatives that are not of Caputo's type.

I. Podlubny [458] and K. Diethelm [163, Chapter 8] have analyzed scalar-valued multi-term Caputo fractional differential equations. Consider, for illustration purposes, the following abstract time-fractional equation:

$$\mathbf{D}_t^{\alpha} u(t) + \mathbf{D}_t^{\beta} u(t) = au(t),\ t > 0;\ u(0) = u_0,\ u'(0) = 0,$$

where $1 < \alpha < 2$, $0 < \beta < \alpha$ and $A = a$ is a certain complex constant. By performing the Laplace transform (see (1.23)), we get:

$$(\lambda^{\alpha} + \lambda^{\beta})\tilde{u}(\lambda) - (\lambda^{\alpha-1} + \lambda^{\beta-1})u_0 = a\tilde{u}(\lambda).$$

Therefore,

$$\tilde{u}(\lambda) = \frac{\lambda^{\alpha-1} + \lambda^{\beta-1}}{\lambda^{\alpha} + \lambda^{\beta} - a} u_0.$$

By (24) and (26) in [480], it readily follows that:

$$(4) \qquad u(t)=\sum_{n=0}^{\infty}(-1)^n t^{(\alpha-\beta)n}\left[E^{n+1}_{\alpha,(\alpha-\beta)n+1}(at^{\alpha})+t^{\alpha-\beta}E^{n+1}_{\alpha,(\alpha-\beta)(n+1)+1}(at^{\alpha})\right]u_0,$$

where

$$E^{\gamma}_{\alpha,\beta}(z)=\sum_{n=0}^{\infty}\frac{(\gamma)_n z^n}{\Gamma(n\alpha+\beta)n!}$$

is the generalized Mittag-Leffler function. Here $(\gamma)n = \gamma(\gamma + 1) \cdots (\gamma + n - 1)$ $(n \in \mathbb{N})$ and $(\gamma)_0 = 1$. The formula (4) shows that it is quite complicated to apply Fourier multiplier theorems to the abstract time-fractional equations of the kind (2); for some basic references in this direction, the reader may consult [321] and [357]. More generally, we know that there is one and only one solution $u(t)$ of the equation (2) with $A \equiv 0$, $f(t) \equiv 0$ and $A_j = c_j I$ $(c_j \in \mathbb{C}, j \in \mathbb{N}_{n-1})$ and that the solution $u(t)$ can be expressed in terms of the Mittag-Leffler functions and their derivatives (see [458] and [319, Example 8.1]).

We continue by observing that T. M. Atanacković, S. Pilipović and D. Zorica considered in [22], among many other authors, the following fractional generalization of the telegraph equation:

$$\tau\mathbf{D}^{\alpha}_t u(t) + \mathbf{D}^{\beta}_t u(t) = Du_{xx},\ x \in (0, l),\ t > 0,$$

where $0 < \beta < \alpha < 2$, $\tau > 0$ and $D > 0$. In the afore-mentioned paper, solutions to signaling and Cauchy problems in terms of a series and integral representation are given.

Concerning periodic solutions of (2), one may refer to the paper [264] by V. Keyantuo and C. Lizama. The main results of this paper are given as follows. Let $\alpha > 0$, let $1 < p < \infty$, and let X be a complex Banach space; we shall identify the functions defined on $[0, 2\pi]$ with their periodic extensions to $\mathbb{R}$. Define the Riemann difference

$$\Delta^{\alpha}_t f(x) := \sum_{j=0}^{\infty}(-1)^j\binom{\alpha}{j}f(x-tj), \quad x\in[0,2\pi],\ t\in\mathbb{R}.$$

If for $f \in L^p([0, 2\pi] : X)$ there exists $g \in L^p([0, 2\pi] : X)$ such that $\lim_{t\to 0+} t^{-\alpha}\Delta^{\alpha}_t f = g$ in the norm of $L^p([0, 2\pi] : X)$, then $g(\cdot)$ is called the αth Liouville-Grünwald fractional derivative of $f(\cdot)$ in mean of order p, $g = D^{\alpha} f$ for short. The periodic solutions for the so called composite fractional relaxation-oscillation equation

$$(5) \qquad D^{\alpha}u(t) + BD^{\beta}u(t) + Au(t) = f(t),\ \ t \in [0, 2\pi],$$

where A and B are closed linear operators defined on the complex Banach space X, $0 < \beta < \alpha < 2$ and $f \in C([0, 2\pi] : X)$, have been studied in the aforementioned paper; see the reference [200] by R. Gorenflo and F. Mainardi for a detailed explanation of physical meaning of fractional differential equations that are special cases of (5). With the notion introduced in [264], it has been proved that the following assertions are equivalent as long as the inclusion $D(A) \subseteq D(B)$ holds:

(i) The problem (5) is strongly L^p-wellposed.
(ii) $H(ik) := ((ik)^\alpha + (ik)^\beta B + A)^{-1}$ exists in $L(X)$ for each $k \in \mathbb{Z}$ and the sequences $((ik)^\alpha H(ik))_{k\in\mathbb{Z}}$ and $((ik)^\beta BH(ik))_{k\in\mathbb{Z}}$ are (L^p, L^p)-multipliers.

If the underlying Banach space X has the UMD property (cf. [8], [20], [72] and [216] for the notion and further properties), the condition (ii) has been further discussed in terms of R-boundedness of sequences $((ik)^\alpha H(ik))_{k\in\mathbb{Z}}$ and $((ik)^\beta BH(ik))_{k\in\mathbb{Z}}$. An application is made to the equation

$$D^\alpha u(t) + aA^{1/2}D^{\alpha/2}u(t) + Au(t) = f(t),\ t \in [0, 2\pi],$$

where $0 < \alpha \leqslant 4$, $a > 0$ and A is a sectorial operator on X.

Our recent paper [319] deals with the case that $A_j = c_j I$ for some complex constants $c_j \in \mathbb{C}$ $(1 \leqslant j \leqslant n-1)$. In a more general setting, cf. Subsection 2.10.1, we shall introduce various types of k-regularized (C_1, C_2)-existence and uniqueness propagation families for (2), and clarify after that their basic structural properties. This is probably the best concept for the investigation of integral solutions of the abstract time-fractional equation (2) with $A_j \in L(E)$, $1 \leqslant j \leqslant n-1$. If there exists an index $j \in \mathbb{N}_{n-1}$ such that $A_j \notin L(E)$, then the vector-valued Laplace transform cannot be so easily applied (cf. Theorem 2.10.9-Theorem 2.10.10), which implies, however, that there exist some limitations to the introduced classes of propagation families. The notion of a strong solution of the equation (2) is introduced in Definition 2.10.1, and the notions of strong and mild solutions of inhomogeneous equations of the kind (406) below are introduced in Definition 2.10.6. The generalized variation of parameters formula is proved in Theorem 2.10.7.

On the other hand, the notions of C_1-existence families and C_2-uniqueness families for the higher order abstract Cauchy problem (ACP_n) were introduced by T.-J. Xiao and J. Liang in [529, Definition 2.1]. We shall introduce in Subsection 2.10.2 more general classes of (local) k-regularized C_1-existence families for (2), k-regularized C_2-uniqueness families for (2), and k-regularized C-resolvent families for (2). Our intention is to prove the generalizations of results obtained in [529] for abstract time-fractional equations. In addition, various adjoint type theorems for k-regularized C-resolvent families are considered in Theorem 2.10.19. In Subsection 2.10.3 we consider many other questions concerning abstract multi-term fractional differential equations, while in Subsection 2.10.4 we consider some applications of (a, k)-regularized C-resolvent families in the analysis of problem (2).

As already announced, one of the most striking peculiarities of this monograph lies in the fact that we consider solutions of abstract Volterra integro-differential equations in locally convex spaces. The second peculiarity is that we thoroughly analyze hypercyclic and topologically mixing properties of some classes of abstract Volterra integro-differential equations in separable infinite-dimensional

Fréchet spaces (cf. the monograph [211] by K.-G. Grosse-Erdmann and A. Peris for a detailed exposition of results on dynamical properties of single operators and strongly continuous semigroups). The third chapter is devoted to the study of hypercyclic Volterra equations. Even though it is very difficult to state some satisfactory and noteworthy facts about the hypercyclic properties of the abstract Volterra equation (1), when considered in its most general form, this chapter occupies nearly a quarter of the book. In order to motivate our researchers, and to briefly explain the essence of problems from this chapter we are interested in, suppose that X is a separable infinite-dimensional Fréchet space over the field $\mathbb{K} \in \{\mathbb{R}, \mathbb{C}\}$. We assume that the topology of X is induced by the fundamental system $(p_n)_{n\in\mathbb{N}}$ of increasing seminorms. Then the translation invariant metric $d : X \times X \to [0,\infty)$, defined by

$$d(x,y) := \sum_{n=1}^{\infty} \frac{1}{2^n} \frac{p_n(x-y)}{1+p_n(x-y)}, \ x, \ y \in X,$$

satisfies, among many other properties, the following: $d(x+u, y+v) \leqslant d(x, y) + d(u, v)$ and $d(cx, cy) \leqslant (|c|+1)d(x, y)$, $c \in \mathbb{K}$, $x, y, u, v \in X$. Now we are able to remind ourselves of some well known definitions from the theory of hypercyclic single operators. A linear mapping $T : X \to X$ is said to be hypercyclic, resp. cyclic, if there exists an element $x \in X$ whose orbit $\mathrm{Orb}(x, T) := \{T^n x : n \in \mathbb{N}_0\}$ is dense in X, resp. if there exists an element $x \in X$ such that the linear span of $\mathrm{Orb}(x, T)$ is dense in X, while T is said to be topologically transitive if for any pair of open non-empty subsets U, V of X there exists $n \in \mathbb{N}$ such that $T^n(U) \cap V \neq \emptyset$. In our framework, T is hypercyclic iff T is topologically transitive. Furthermore, a linear mapping $T : X \to X$ is said to be chaotic (supercyclic, resp. positively supercyclic) if T is hypercyclic and the set of periodic points of T, defined by $\{x \in X :$ there exists $n \in \mathbb{N}$ such that $T^n x = x\}$, is dense in X (if there exists an element $x \in X$ whose projective orbit $\{cT^n x : n \in \mathbb{N}_0, c \in \mathbb{K}\}$, resp. positive projective orbit $\{cT^n x : n \in \mathbb{N}_0, c \geqslant 0\}$, is dense in X), while T is said to be weakly supercyclic if T is supercyclic in the space X equipped with the weak topology. In the case that X is a Banach space, then we say that T is antisupercyclic if for any $x \in X$ either $T^n x = 0$ for some $n \in \mathbb{N}$ or the sequence $(T^n x/\|T^n x\|)$ weakly converges to zero.

Chronologically, the first examples of hypercyclic operators were given on the space $H(\mathbb{C})$ of entire functions equipped with the topology of uniform convergence on compact subsets of $\mathbb{C}$. In 1929, G. D. Birkhoff proved that the translation operator $f \mapsto f(\cdot + a)$, $f \in H(\mathbb{C})$, $a \in \mathbb{C}\setminus\{0\}$ is hypercyclic in $H(\mathbb{C})$. The hypercyclicity of the derivative operator $f \mapsto f'$, $f \in H(\mathbb{C})$ was proved by G. R. MacLane in 1952. In the setting of Banach spaces, S. Rolewicz [471] was the first who presented examples of a hypercyclic operator and a hypercyclic strongly continuous semigroup (1969); the state space in his analysis is chosen to be $l^2(\mathbb{N})$. The first examples of chaotic semigroups were given by C. R. MacCluer [399] and V. Protopopescu, Y. Azmy [462] in 1992. Some known assertions concerning the existence of hypercyclic and chaotic semigroups are collected in what follows. J. A. Conejero has proved in [116] that every separable infinite-dimensional Fréchet space, except the space $\omega := \Pi_{n\in\mathbb{N}} \mathbb{K}$, admits a hypercyclic semigroup. An interesting result of T. Bermúdez, A. Bonilla, J. A. Conejero and A. Peris [54] says that every separable infinite-dimensional Banach space admits a topologically mixing, analytic semigroup of angle $\pi/2$. Further on, L. Bernal-Gonzáles and K.-G. Grosse Erdmann [61] have proved that every separable infinite-dimensional Banach space admits a norm continuous, weakly mixing semigroup. The existence of chaotic and supercyclic semigroups on locally convex spaces is more delicate: T. Bermúdez, A. Bonilla and A. Martinón [57] have proved that there exists a separable infinite-dimensional Banach space which does not admit a chaotic semigroup and, due to L. Bernal-Gonzáles and K.-G. Grosse Erdmann [61], we know that the space $\varphi := \oplus_{n\in\mathbb{N}} \mathbb{K}$ does not admit a supercyclic semigroup.

Several illustrative examples of bounded linear operators on Banach spaces, possessing certain (super-)cyclic behaviour, are provided. Let $p \in [1, \infty)$, $X := L^p[0, 1]$ and let $r > 0$. Then the Riemann-Liouville integral operator V_p^r, defined by $V_p^r f(x) := \int_0^x g_r(x-t)f(t)\,dt$, $x \in [0, 1]$, $f \in X$ is a bounded linear operator on X. It is well known that the norm of operator V_2^1 equals $2/\pi$ and that its adjoint is given by $(V_2^1)^* f(x) := \int_x^1 f(t)\,dt$, $x \in [0, 1]$, $f \in X$; F. León-Saavedra and A. Piqueras-Lerena have proved in [352] that the operators V_2^1 and $I + V_2^1$ are both cyclic and non-supercyclic. Furthermore, S. Shkarin has proved in [488] that any operator $T \in L(X)$ commuting with V_p^1 cannot be weakly supercyclic. In [58], S. Bermudo, A. Montes-Rodríguez and S. Shkarin have analyzed the asymptotic estimates of the norms of orbits of certain operators that commute with the operator V_p^r. Their results are then applied in the study of cyclic properties of the operator $\phi(V_p^r)$, where $\phi(\cdot)$ is an analytic scalar valued function at 0; for example, it has been shown that for any bounded quasinilpotent operator W on X, the operators $V_p^r(I + W)$ and $I + V_p^r(I + W)$ are both not weakly supercyclic for $1 \leq p \leq \infty$ and antisupercyclic for $1 < p < \infty$. Some results on spectral and cyclic properties of the Riemann-Liouville operator V_p^z with $z \in \mathbb{C}_+$ can be found in the paper [490]. Volterra composition operators on $L^p[0, 1]$-spaces, i.e., the operators of the form $V_p^\varphi f(x) := \int_0^{\varphi(x)} f(t)\,dt$, $x \in [0, 1]$, $f \in L^p[0, 1]$, where the function $\varphi : [0, 1] \to [0, 1]$ satisfies certain properties, have been analyzed in [424]-[425], [473] and [489].

In the third chapter, we will restrict ourselves to the analysis of hypercyclic properties of some very specific classes of abstract multi-term fractional differential equations. In Section 3.1 and Section 3.2 we analyze the hypercyclic properties of first and second order abstract differential equations ([309]-[311]), while in Section 3.3 and Section 3.4 we analyze the hypercyclic properties of the equation (1) with $a(t) = g_\alpha(t)$ for some $\alpha \in (0, 2)\setminus\{1\}$, and hypercyclic properties of the equation (2) with $A_j = c_j I$ for certain complex constants c_j $(1 \leq j \leq n-1)$, respectively. In Section 3.5, we are interested in the analysis of topological dynamics of abstract (multi-term) fractional PDEs with unilateral backwards shifts ([320]). Each separate section of the third chapter will be organized with its main purposes and ideas explained within itself.

CONTENTS

1

PRELIMINARIES

1.1 Vector-valued functions, closed operators and integration in sequentially complete locally convex spaces

Unless specified otherwise, we shall always assume that E is a Hausdorff sequentially complete locally convex space over the field of complex numbers, SCLCS for short; the abbreviation $\circledast$ stands for the fundamental system of seminorms which defines the topology of E. If X is also an SCLCS, then we denote it by $L(E, X)$ the space of all continuous linear mappings from E into X; $L(E) \equiv L(E, E)$. Let $\mathcal{B}$ be the family of bounded subsets of E and let $p_B(T) := \sup_{x\in B} p(Tx)$, $p \in \circledast$, $B \in \mathcal{B}$, $T \in L(E)$. Then $p_B(\cdot)$ is a seminorm on $L(E)$ and the system $(p_B)_{(p,B)\in\circledast\times\mathcal{B}}$ induces the Hausdorff locally convex topology on $L(E)$. The Hausdorff locally convex topology on E^*, the dual space of E, defines the system $(|\cdot|_B)_{B\in\mathcal{B}}$ of seminorms on E^*, where and in the sequel $|x^*|_B := \sup_{x\in B} |\langle x^*, x\rangle|$, $x^* \in E^*$, $B \in \mathcal{B}$. Here $\langle , \rangle$ denotes the duality bracket between E and E^*, sometimes we also write $\langle x, x^*\rangle$ or $x^*(x)$ to denote the value of $\langle x^*, x\rangle$. Let us recall that the spaces $L(E)$ and E^* are sequentially complete provided that E is barreled ([411]). The bidual of E will be denoted by E^{**}.

A linear operator $A : D(A) \to E$ is said to be closed if the graph of the operator A, defined by $G_A := \{(x, Ax) : x \in D(A)\}$, is a closed subset of $E \times E$; since no confusion seems likely, we will identify A with its graph. The resolvent set, spectrum and range of a linear operator A on E are denoted by $\rho(A)$, $\sigma(A)$ and $R(A)$, respectively. A necessary and sufficient condition for a linear operator $A : D(A) \to E$ to be closed is that for every net $(x_\tau)_{\tau\in I}$ in $D(A)$ such that $\lim_{\tau\to\infty} x_\tau = x$ and $\lim_{\tau\to\infty} Ax_\tau = y$, the following holds: $x \in D(A)$ and $Ax = y$; cf. [411] for the notion. If E is a Banach space and A is a linear operator on E, then we introduce the graph norm on $D(A)$ by $\|x\|_{[D(A)]} := \|x\| + \|Ax\|$, $x \in D(A)$. Then $(D(A), \|\cdot\|_{[D(A)]})$ is a Banach space iff A is closed. A subspace $Y \subseteq D(A)$ is called a core for A if Y is dense in $D(A)$ with respect to the graph norm. Generally, the Hausdorff sequentially complete locally convex topology on $D(A)$ $(\overline{D(A)})$ can be introduced by the following system of seminorms: $p_A(x) =: p(x) + p(Ax)$, $x \in D(A)$, $p \in \circledast$ $((p_{|\overline{D(A)}})_{p\in\circledast})$. We shall denote the first of above spaces simply by $[D(A)]$. If $C \in L(E)$ is injective, then we define the C-resolvent set of A, $\rho_C(A)$ for short, by

$$\rho_C(A) := \{\lambda \in \mathbb{C} : \lambda - A \text{ is injective and } (\lambda - A)^{-1} C \in L(E)\}.$$

By The Closed Graph Theorem [411], the following holds: If E is a webbed bornological space (this, in particular, holds if E is a Fréchet space), then the C-resolvent set of A consists of those complex numbers λ for which the operator $\lambda - A$ is injective and $R(C) \subseteq R(\lambda - A)$. The generalized resolvent equation states that

$$(z - A_{n-1})^{-1} C (\lambda - A_{n-1})^{-k} C^k x$$
$$= \frac{(-1)^k}{(z-\lambda)^k} (z - A_{n-1})^{-1} C^{k+1} x + \sum_{i=1}^{k} \frac{(-1)^{k-i} (\lambda - A_{n-1})^{-i} C^{k+1} x}{(z-\lambda)^{k+1-i}}, \tag{6}$$

for any $x \in X$, $k \in \mathbb{N}_0$ and $\lambda, z \in \rho_C(A)$ with $z \neq \lambda$ (cf. also the equality [90, (18)]). Suppose F is a linear subspace of E. Then the part of A in F, denoted by $A_{|F}$, is a linear operator defined by $D(A_{|F}) := \{x \in D(A) \cap F : Ax \in F\}$ and $A_{|F} x := Ax$, $x \in D(A_{|F})$. Further on, a linear operator A is closable iff there exists a closed linear operator B such that $A \subseteq B$. It can be simply proved that a linear operator A is closable if for every net $(x_\tau)_{\tau \in I}$ in $D(A)$ such that $\lim_{\tau\to\infty} x_\tau = 0$ and $\lim_{\tau\to\infty} Ax_\tau = y$, we have $y = 0$. Suppose that A is a closable linear operator. The closure of A, denoted by $\bar{A}$, is defined as the set of all elements $(x, y) \in E \times E$ such that there exists a net (x_τ) in $D(A)$ with $\lim_{\tau\to\infty} x_\tau = x$ and $\lim_{\tau\to\infty} Ax_\tau = y$; then $\bar{A}$ is a closed linear operator and, for every other closed linear operator B which contains A, one has $\bar{A} \subseteq B$. Suppose $A : D(A) \to E$ is a linear operator. We define the powers of A recursively by setting: $A^0 =: I$, $D(A^n) := \{x \in D(A^{n-1}) : A^{n-1}x \in D(A)\}$ and $A^n x := A(A^{n-1}x)$, $x \in D(A^n)$, $n \in \mathbb{N}$. Then $D(A^n) = D((A - \lambda)^n)$, $n \in \mathbb{N}$, $\lambda \in \mathbb{C}$. Put $D_\infty(A) := \bigcap_{n\geq 1} D(A^n)$. For a closed linear operator A, we introduce the subset A^* of $E^* \times E^*$ by

$$A^* := \{(x^*, y^*) \in E^* \times E^* : x^*(Ax) = y^*(x) \text{ for all } x \in D(A)\}.$$

If A is densely defined, then A^* is also known as the adjoint operator of A and it is a closed linear operator on E^*. It is noteworthy that $D(A^*)$ is weak* dense in E^* even in the case that A is not densely defined in E (see e.g. [302, Lemma 2.4] and [20, Proposition B.10]). Let $\alpha \in \mathbb{C}\setminus\{0\}$, and let A and B be linear operators. Then we define αA, $A + B$ and AB in the following way: $D(\alpha A) =: D(A)$, $D(A + B) := D(A) \cap D(B)$ and $D(AB) := \{x \in D(B) : Bx \in D(A)\}$, $(\alpha A)x := \alpha Ax$, $x \in D(\alpha A)$, $(A + B)x := Ax + Bx$, $x \in D(A + B)$ and $(AB)x := A(Bx)$, $x \in D(AB)$. A family Λ of continuous linear operators on E is said to be equicontinuous if for each $p \in \circledast$ there exist $c_p > 0$ and $q_p \in \circledast$ such that

$$p(Ax) \leqslant c_p q_p(x), \quad x \in E, A \in \Lambda.$$

The Gamma function will be denoted by $\Gamma(\cdot)$ and the principal branch will always be used to take the powers. Set, for every $\alpha > 0$, $g_\alpha(t) := t^{\alpha-1}/\Gamma(\alpha)$, $t > 0$ and $g_0(t) \equiv$

the Dirac delta distribution. Given a number $n \in \mathbb{N}$ in advance, set $\mathbb{N}_n := \{1, \cdots, n\}$ and $\mathbb{N}_n^0 := \{0, 1, \cdots, n\}$. If $(M_p)_{p \in \mathbb{N}_0}$ is a sequence of positive real numbers with $M_0 = 1$, then we use the following conditions from the theory of ultradistributions (cf. [78], [106]-[107] and [283]-[285] for more details):

(M.1) : $M_p^2 \leq M_{p+1} M_{p-1}$, $p \in \mathbb{N}$,

(M.2) : $M_p \leq AH^p \min_{p_1, p_2 \in \mathbb{N}, p_1 + p_2 = p} M_{p_1} M_{p_2}$, $n \in \mathbb{N}$, for some $A > 1$ and $H > 1$,

(M.3)' : $\sum_{p=1}^{\infty} \frac{M_{p-1}}{M_p} < \infty$, and

(M.3) : $\sup_{p \in \mathbb{N}} \sum_{q=p+1}^{\infty} \frac{M_{q-1} M_{p+1}}{p M_p M_q} < \infty$.

Recall that the condition (M.3) is slightly stronger than (M.3)' and that, for every $s > 1$, the Gevrey sequence $(p!^s)$ satisfies the above conditions.

Let us recollect, for the sake of convenience of the reader, the main structural properties of vector-valued distribution spaces used henceforward. Suppose, for the time being, that $(E, \|\cdot\|)$ is a complex Banach space. The Schwartz spaces of test functions $\mathcal{D} = C_0^\infty(\mathbb{R})$ and $\mathcal{E} = C^\infty(\mathbb{R})$ carry the usual inductive limit topologies. The topology of the space of rapidly decreasing functions $\mathcal{S}$ is induced by the following system of seminorms: $p_{m,n}(\psi) =: \sup_{x \in \mathbb{R}} |x^m \psi^{(n)}(x)|$, $\psi \in \mathcal{S}$, $m, n \in \mathbb{N}_0$. By $\mathcal{D}_0$ we denote the subspace of $\mathcal{D}$ which consists of the elements supported by $[0, \infty)$. Further on, $\mathcal{D}'(E) := L(\mathcal{D}, E)$, $\mathcal{E}'(E) := L(\mathcal{E}, E)$ and $\mathcal{S}'(E) := L(\mathcal{S}, E)$ are the spaces of continuous linear functions $\mathcal{D} \to E$, $\mathcal{E} \to E$ and $\mathcal{S} \to E$, respectively; $\mathcal{D}_0'(E)$, $\mathcal{E}_0'(E)$ and $\mathcal{S}_0'(E)$ denote the subspaces of $\mathcal{D}'(E)$, $\mathcal{E}'(E)$ and $\mathcal{S}'(E)$, respectively, containing the elements supported by $[0, \infty)$. Denote by $\mathbf{B}$ the family of all bounded subsets of $\mathcal{D}$. Put $p_B(f) := \sup_{\varphi \in B} \|f(\varphi)\|$, $f \in \mathcal{D}'(E)$, $B \in \mathbf{B}$. Then p_B, $B \in \mathbf{B}$ is a seminorm on $\mathcal{D}'(E)$ and the system $(p_B)_{B \in \mathbf{B}}$ induces the topology on $\mathcal{D}'(E)$. The topology on $\mathcal{E}'(E)$, resp., $\mathcal{S}'(E)$, is defined following a similar line of reasoning; cf [481].

In the remaining part of this section, we shall recall the basic facts and definitions from the theory of integration of functions with values in locally convex spaces. By Ω we denote a locally compact and separable metric space (for example, Ω can be chosen to be a (bounded or unbounded) segment I in $\mathbb{R}^n$, where $n \in \{1, 2\}$) and by μ we denote a locally finite Borel measure defined on Ω.

Definition 1.1.1. (i) It is said that a function $f : \Omega \to E$ is simple if there exist $k \in \mathbb{N}$, elements $z_i \in E$, $1 \leq i \leq k$ and pairwise disjoint Borel measurable subsets Ω_k, $1 \leq i \leq k$ of Ω, such that $\mu(\Omega_i) < \infty$, $1 \leq i \leq k$ and

(7)
$$f(t) = \sum_{i=1}^{k} z_i \chi_{\Omega_i}(t), \quad t \in \Omega.$$

(ii) It is said that a function $f : \Omega \to E$ is (strongly) μ-measurable, (strongly) measurable for short, if there exists a sequence (f_n) in E^Ω such that, for every $n \in \mathbb{N}$, $f_n(\cdot)$ is a simple function and $\lim_{n \to \infty} f_n(t) = f(t)$ for a.e. $t \in \Omega$.

(iii) A function $f : \Omega \to E$ is said to be weakly μ-measurable, weakly measurable for short, if for every $x^* \in E^*$, the function $t \mapsto x^* (f(t))$, $t \in \Omega$ is measurable.

(iv) A function $f : \Omega \to E$ is said to be μ-measurable by seminorms, measurable by seminorms for short, if for every $p \in \circledast$ there exists a sequence (f_n^p) in E^Ω such that $\lim_{n\to\infty} p(f_n^p(t) - f(t)) = 0$ a.e. $t \in \Omega$.

It is clear that every strongly measurable function is also weakly measurable and that the converse statement is not true in general. In the case that E is a Fréchet space, the necessary and sufficient conditions under an E-valued weakly measurable function is strongly measurable can be found in [270].

The Pettis's Measurability Theorem in locally convex spaces reads as follows (cf. [20], [81], [165] and [510] for further information).

Theorem 1.1.2. *A function $f : \Omega \to E$ is measurable by seminorms if $f(\cdot)$ is weakly measurable and for every $p \in \circledast$ there exists a set N_p of μ measure zero such that $f(\Omega \setminus N_p)$ is separable with respect to p.*

The concept of Bochner integrability of a strongly measurable function $f : \Omega \to E$ can be simply extended from Banach spaces ([20, Chapter 1]) to the Fréchet spaces; in general case, the situation is a little complicated and we must make some non-trivial adaptation of the method used in Banach spaces. First of all, we define the Bochner integral of a simple function $f : \Omega \to E$, given by (7), as follows $\int_\Omega f\, d\mu := \sum_{i=1}^k z_i \mu(\Omega_i)$. Then one can simply prove that the definition of Bochner integral does not depend on the representation (7).

Let $1 \leqslant p < \infty$, let $(X, \|\cdot\|)$ be a complex Banach space, and let $(\Omega, \mathcal{R}, \mu)$ be a measure space. Then the space $L^p(\Omega, X, \mu)$ consists of all strongly μ-measurable functions $f : \Omega \to X$ such that $\|f\|_p := (\int_\Omega \|f(\cdot)\|^p d\mu)^{1/p}$ is finite. We also use the notation $L^p(\Omega, \mu)$ in the case that $X = \mathbb{C}$. The space $L^\infty(\Omega, X, \mu)$ consists of all strongly μ-measurable, essentially bounded functions, and is equipped with the norm $\|f\|_\infty := ess\ \sup_{t\in\Omega} \|f(t)\|$, $f \in L^\infty(\Omega, X, \mu)$. Herein we identify functions that are equal μ-almost everywhere on Ω; if μ is the Lebesgue measure on the real line, then, for every $p \in [1, \infty]$, the space $L^p(\Omega, X, \mu)$ will also be denoted by $L^p(\Omega : X)$. By Riesz–Fischer theorem, $(L^p(\Omega, X, \mu), \|\cdot\|_p)$ is a Banach space for all $p \in [1, \infty]$, and furthermore, we know that $(L^2(\Omega, X, \mu), \|\cdot\|_2)$ is a Hilbert space. If $\lim_{n\to\infty} f_n = f$ in $L^p(\Omega, X, \mu)$, then there exists a subsequence (f_{n_k}) of (f_n) such that $\lim_{k\to\infty} f_{n_k}(t) = f(t)$ μ-almost everywhere. If the Banach space X is reflexive, then $L^p(\Omega, X, \mu)$ is reflexive for all $p \in (1, \infty)$ and its dual is isometrically isomorphic to $L^{p/(p-1)}(\Omega, X, \mu)$.

The notions of Dunford integrability and Pettis integrability of a function $f : \Omega \to E$ are introduced as follows.

Definition 1.1.3. A function $f : \Omega \to E$ is said to be Dunford integrable if for each $x^* \in E^*$ the function $t \mapsto \langle x^*, f(t)\rangle$, $t \in \Omega$ belongs to the space $L^1(\Omega, \mu)$ and if for each Borel measurable set $\Lambda \subseteq \Omega$ one can find an element $x_\Lambda^{**} \in E^{**}$ such that

$$\langle x^{**}_\Lambda, x^* \rangle = \int_\Lambda \langle x^*, f \rangle \, d\mu, \quad x^* \in E^*. \tag{8}$$

If for each $\Lambda \subseteq \Omega$, the values of x^{**}_Λ are essentially contained in E then we say that E is Pettis integrable.

Although not further used in the remaining part of the book, the following fact on the existence of Dunford integral should be stated. Suppose that the space E^* is barreled as well as that $f: \Omega \to E$ is weakly measurable and satisfies that the function $t \mapsto \langle x^*, f(t) \rangle$, $t \in \Omega$ belongs to the space $L^1(\Omega, \mu)$. Then for each Borel measurable set $\Lambda \subseteq \Omega$ one can find an element $x^{**}_\Lambda \in E^{**}$ such that (8) holds. For further information concerning the Pettis integrability, the reader may consult [81] and references cited there.

In the sequel, we shall use the following notion of μ-integrability, appearing in the monograph by C. Martinez and M. Sanz [403, pp. 99-102]; among many other concepts, we would like to mention the concepts of Gelfand integrability and integrability by seminorms.

Definition 1.1.4. Let $K \subseteq \Omega$ be a compact set, and let a function $f: K \to E$ be strongly measurable. Then it is said that $f(\cdot)$ is (μ-)integrable if there is a sequence $(f_n)_{n \in \mathbb{N}}$ of simple functions such that $\lim_{n\to\infty} f_n(t) = f(t)$ a.e. $t \in K$ and for all $\varepsilon > 0$ and each $p \in ⊛$ there is a number $n_0 = n_0(\varepsilon, p)$ such that

$$\int_K p(f_n - f_m) \, d\mu < \varepsilon \;\; (m, n \geqslant n_0). \tag{9}$$

In this case we define

$$\int_K f \, d\mu := \lim_{n\to\infty} \int_K f_n \, d\mu \tag{10}$$

The equation (9) shows that $(p(f_n))_{n \in \mathbb{N}}$ is a Cauchy sequence in the space $L^1(K, \mu)$, so that the limit $p(f) = \lim_{n\to\infty} p(f_n)$ is μ-integrable. One obtains similarly that each function $p(f_n - f)$ is μ-integrable and that the sequence of its corresponding integrals converges to zero. Notice that the definition (10) makes sense since E is sequentially complete. It can be simply proved that every continuous function f: $K \to E$ is μ-integrable.

Definition 1.1.5. (i) A function $f: \Omega \to E$ is said to be locally μ-integrable if, for every compact set $K \subseteq \Omega$, the restriction $f_{|K}: K \to E$ is μ-integrable.
(ii) A function $f: \Omega \to E$ is said to be μ-integrable if it is locally integrable and if additionally

$$\int_\Omega p(f) \, d\mu < \infty, \; p \in ⊛. \tag{11}$$

If this is the case, we define

$$\int_{\Omega} f \, d\mu := \lim_{n\to\infty} \int_{K_n} f \, d\mu,$$

with $(K_n)_{n\in\mathbb{N}}$ being an expansive sequence of compact subsets of E with the property that $\bigcup_{n\in\mathbb{N}} K_n = \Omega$.

The above definition is meaningful and does not depend on the choice of sequence $(K_n)_{n\in\mathbb{N}}$. Moreover,

$$p\left(\int_{\Omega} f \, d\mu\right) \leq \int_{\Omega} p(f) \, d\mu, \quad p \in \circledast.$$

If the function $f : K \to E$, resp. $f : \Omega \to E$, is μ-integrable, then for each $x^* \in E^*$ we have that

$$\left\langle x^*, \int_K f \, d\mu \right\rangle = \int_K \langle x^*, f \rangle \, d\mu, \text{ resp. } \left\langle x^*, \int_{\Omega} f \, d\mu \right\rangle = \int_{\Omega} \langle x^*, f \rangle \, d\mu.$$

Before we go any further, it could be of importance to stress that Definition 1.1.5 is equivalent with the definition of Bochner integral, provided that E is a Banach space. Any continuous function $f : \Omega \to E$ satisfying (11) is μ-integrable and the following holds.

Theorem 1.1.6. (i) *(The Dominated Convergence Theorem) Suppose that* (f_n) *is a sequence of* μ*-integrable functions from* E^{Ω} *and* (f_n) *converges pointwisely to a function* $f : \Omega \to E$. *Assume that, for every* $p \in \circledast$, *there exists a* μ*-integrable function* $F_p : \Omega \to [0, \infty)$ *such that* $p(f_n) \leq F_p$, $n \in \mathbb{N}$. *Then* $f(\cdot)$ *is a* μ*-integrable function and* $\lim_{n\to\infty} \int_{\Omega} f_n \, d\mu = \int_{\Omega} f \, d\mu$.

(ii) *Let* Y *be an SCLCS, and let* $T : X \to Y$ *be a continuous linear mapping. If* $f : \Omega \to X$ *is* μ*-integrable, then* $Tf : \Omega \to Y$ *is likewise* μ*-integrable and*

$$T \int_{\Omega} f \, d\mu = \int_{\Omega} T f \, d\mu. \tag{12}$$

(iii) *Let* Y *be an SCLCS, and let* $T : D(T) \subseteq X \to Y$ *be a closed linear mapping. If* $f : \Omega \to D(T)$ *is* μ*-integrable and* $Tf : \Omega \to Y$ *is likewise* μ*-integrable, then* $\int_{\Omega} f \, d\mu \in D(T)$ *and (12) holds.*

Further information on integrability of functions in locally convex spaces can be obtained by consulting the references [218]-[219], [279], [402], [466], [472], [474] and [502]. For the basic properties of Banach space valued absolutely continuous functions (functions of bounded variation), we refer the reader to [20, Chapter 1] and [292, Chapter 1]. If X is a Banach space, then the space of all X-valued functions that are absolutely continuous (of bounded variation) on any closed subinterval of $[0, \infty)$ will be denoted by $AC_{loc}([0, \infty) : X)$ $(BV_{loc}([0, \infty) : X))$, while the space of k-times continuously differentiable functions ($k \in \mathbb{N}_0$) from a non-empty subset $\Omega \subseteq \mathbb{C}$ into a general sequentially complete locally convex space E will be denoted by $C^k(\Omega : E)$, $C(\Omega : E) \equiv C^0(\Omega : E)$. If $X = \mathbb{C}$, then we also write $AC_{loc}([0, \infty))$ $(BV_{loc}([0,\infty)))$ instead of $AC_{loc}([0, \infty) : X)$ $(BV_{loc}([0, \infty) : X))$; the spaces $BV[0, T]$, $BV_{loc}([0, \tau))$, $BV_{loc}([0, \tau) : X)$, as well as the space $L^p_{loc}(\Omega : X)$ for $1 \leqslant p \leqslant \infty$ are defined in a very similar way $(T, \tau > 0)$; $L^p_{loc}(\Omega) \equiv L^p_{loc}(\Omega : \mathbb{C})$ and there is no difference between the spaces $L^p_{loc}([0, \tau))$ and $L^p_{loc}((0, \tau))$, for any $\tau > 0$ and $1 \leqslant p \leqslant \infty$. Let $0 < \tau \leqslant \infty$ and $a \in L^1_{loc}([0, \tau))$. Then we say that the function $a(t)$ is a kernel on $[0, \tau)$ iff for each $f \in C([0, \tau))$ the assumption $\int_0^t a(t-s)f(s)\, ds = 0$, $t \in [0, \tau)$ implies $f(t) = 0$, $t \in [0, \tau)$. If $\tau = \infty$ and $a \neq 0$ in $L^1_{loc}([0, \infty))$, then the famous Titchmarsh theorem (cf. [50]) implies that the function $a(t)$ is automatically a kernel on $[0, \infty)$; the situation is quite different in the case that $\tau < \infty$, then we can apply the Titchmarsh–Foiaş theorem (see e.g. [292, Theorem 3.4.40]) in order to see that the function $a(t)$ is a kernel on $[0, \tau)$ iff $0 \in \mathrm{supp}(a)$.

Suppose $k \in \mathbb{N}$, $p \in [1, \infty]$ and Ω is an open non-empty subset of $\mathbb{R}^n$. Then the Sobolev space $W^{k,p}(\Omega : X)$ consists of those X-valued distributions $u \in \mathcal{D}'(\Omega : X)$ (cf. [292, Section 1.3] and Section 3.2) such that, for every $i \in \{0, \cdots, k\}$ and for every multi-index $\alpha \in \mathbb{N}_0^n$ with $|\alpha| \leqslant k$, one has $D^\alpha u \in L^p(\Omega, X)$. Here, the derivative D^α is taken in the sense of distributions. Notice that the space $W^{k,p}((0, \tau) : X)$, where $\tau \in (0, \infty)$, can be characterized by means of corresponding spaces of absolutely continuous functions (cf. for example [38, Chapter I, Section 2.2]). By $W^{k,p}_{loc}(\Omega : X)$ we denote the space of those X-valued distributions $u \in \mathcal{D}'(\Omega : X)$ such that, for every bounded open subset Ω' of Ω, one has $u_{|\Omega'} \in W^{k,p}(\Omega' : X)$.

A Banach space E is said to possess the Radon–Nikodym property if every absolutely continuous function $F : [0, 1] \to E$ is differentiable a.e. It is well known that every reflexive Banach space possesses the Radon–Nikodym property and that the space $L^1[0, 1]$ does not possess the Radon–Nikodym property. Notice, however, that the existing literature is somewhat controversial about the question whether a general locally convex space E possess the Radon–Nikodym property or not. For further information, we refer the reader to [65].

The reader may consult [17], [40], [209], [239] and [290] for the properties of vector-valued analytic functions. As is well known, a function $f : \Omega \to E$, where Ω is an open subset of $\mathbb{C}$, is said to be analytic if it is locally expressible in a neighborhood of any point $z \in \Omega$ by a uniformly convergent power series with coefficients in E. Since E is an SCLCS, and therefore a locally complete space [209], the analyticity of $f(\cdot)$ is equivalent with the weak analyticity of $f(\cdot)$, i.e., the mapping $\lambda \mapsto f(\lambda)$, $\lambda \in \Omega$ is analytic if the mapping $\lambda \mapsto \langle x^*, f(\lambda)\rangle$, $\lambda \in \Omega$ is analytic for every $x^* \in E^*$. Combined with the strong continuity of the mapping $\lambda \mapsto f(\lambda)$, $\lambda \in \Omega$, the above ensures that, for any closed contour Γ in Ω such that $\mathrm{Ind}_\Gamma(z) = 0$, $z \in \mathbb{C} \setminus \Omega$, the following holds: $f(z) = \frac{1}{2\pi i} \oint_\Gamma \frac{f(\lambda)}{\lambda - z}\, d\lambda$, $z \in \Omega \setminus \Gamma$, $\mathrm{Ind}_\Gamma(z) = 1$. Using the dominated convergence theorem and the proof of Cauchy integral formula in the scalar-valued case, one obtains that the mapping $\lambda \mapsto f(\lambda)$, $\lambda \in \Omega$ is infinitely differentiable and

(13) $$f^{(n)}(z)=\frac{n!}{2\pi i}\oint_{\Gamma}\frac{f(\lambda)}{(\lambda-z)^{n+1}}\,d\lambda,\ z\in\Omega\setminus\Gamma,\ \mathrm{Ind}_{\Gamma}(z)=1,\ n\in\mathbb{N}_0,$$

which simply implies that

$$f(z)=\sum_{n=0}^{\infty}(z-z_0)^n\,\frac{f^{(n)}(z_0)}{n!}$$

in a neighborhood of point $z_0 \in \Omega$. It is also worthwhile to note that the use of bipolar theorem implies that the identity theorem for analytic functions [20, Proposition A.2, p. 456] remains true in the case that X is a general locally convex space, which will be used in the sequel.

For further information about locally convex and generalized function spaces, the reader may consult [187], [189], [283]-[285], [327], [411]-[412], [455]-[456] and [481].

1.2 Laplace transform in sequentially complete locally convex spaces

Concerning the Laplace transform of Banach space valued functions, mention should be made of the excellently written monograph [20] by W. Arendt, C. J. K. Batty, M. Hieber and F. Neubrander (already cited multiple times in Section 1.1). Compared with the Banach space case, increasingly less facts have been said about the Laplace transform of functions with values in sequentially complete locally convex spaces (cf. T.-J. Xiao, J. Liang [526]-[527] and Y.-C. Li, S.-Y. Shaw [367]). Although not fully general in the theoretical sense, we shall follow the method employed in the paper [526] (cf. also [531, Section 1.1.1]). Throughout the section we shall always assume that $\Omega = [0, \infty)$ and that μ is the Lebesgue's measure on $[0, \infty)$. If $-\infty < a < b < \infty$ and $f \in C([a, b] : E)$, then the integral $\int_a^b f(t)\,dt$, defined by means of Riemann sums in the same way as for numerical functions, coincides with the integral $\int_a^b f(t)\,dt$ introduced in the previous section.

Let $a \in \mathbb{R}$. Following [531, Definition 1.1.3], it will be said that a function $h : (a, \infty) \to E$ belongs to the class $LT-E$ if there exists a function $f \in C([0, \infty) : E)$ such that for each $p \in \circledast$ there exists $M_p > 0$ satisfying $p(f(t)) \leqslant M_p e^{at}$, $t \geqslant 0$ and

(14) $$h(\lambda)=\int_0^{\infty} e^{-\lambda t}f(t)\,dt,\quad \lambda > a.$$

If this is the case, $h(\lambda)$ is called the Laplace transform of $f(t)$ and $f(t)$ is called the determining function of $h(\lambda)$. We denote by $\mathcal{L}$ and $\mathcal{L}^{-1}$ the Laplace transform and its inverse transform, respectively:

$$\mathcal{L}(f(t))(\lambda) = h(\lambda),\quad \lambda > a \text{ and } \mathcal{L}^{-1}(h(\lambda))(t) = f(t),\quad t \geqslant 0.$$

Sometimes, we also write $\tilde{f}(\lambda) := \mathcal{L}(f(t))(\lambda)$. The integral appearing on the right hand side of (14) converges for all $\lambda \in \mathbb{C}$ with $\mathrm{Re}\,\lambda > \omega$ and defines an analytic function in this region, i.e., the function $\lambda \to h(\lambda)$, $\lambda > a$ can be analytically extended to the right half plane $\{\lambda \in \mathbb{C} : \mathrm{Re}\,\lambda > a\}$. It is not difficult to prove that

$$\frac{d^n}{d\lambda^n}h(\lambda)=(-1)^n\int_0^\infty e^{-\lambda t}t^n f(t)\,dt,\quad n\in\mathbb{N},\ \lambda\in\mathbb{C},\ \operatorname{Re}\lambda>a.$$

Unless stated otherwise, the abbreviation $*$ denotes henceforth the finite convolution product (cf. (15)). We shall employ occasionally the following condition on a scalar valued function $K(\cdot)$:

(P1) $K(\cdot)$ is Laplace transformable, i.e., $K\in L^1_{loc}([0,\infty))$ and there exists $\beta\in\mathbb{R}$ such that $\tilde{K}(\lambda):=\mathcal{L}(K(t))(\lambda):=\lim_{b\to\infty}\int_0^b e^{-\lambda t}K(t)\,dt:=\int_0^\infty e^{-\lambda t}K(t)\,dt$ exists for all $\lambda\in\mathbb{C}$ with $\operatorname{Re}\lambda>\beta$.
Put $\operatorname{abs}(K):=\inf\{\operatorname{Re}\lambda : \tilde{K}(\lambda)\text{ exists}\}$.

The Laplace transform has the following properties.

Theorem 1.2.1. *Let $f\in C([0,\infty):E)$ satisfy that for each $p\in \circledast$ there exists $M_p>0$ such that $p(f(t))\leqslant M_pe^{at}$, $t\geqslant 0$, let $z\in\mathbb{C}$ and $s\geqslant 0$, and let (14) hold.*

(i) *Put $g(t):=e^{-zt}f(t)$, $t\geqslant 0$. Then $\tilde{g}(\lambda)=\tilde{f}(\lambda+z)$, $\lambda\in\mathbb{C}$, $\operatorname{Re}\lambda>a-\operatorname{Re}z$.*

(ii) *Put $f_s(t):=f(t+s)$, $t\geqslant 0$. Then $\tilde{f}_s(\lambda)=e^{\lambda s}(\tilde{f}(\lambda)-\int_0^s e^{-\lambda t}f(t)\,dt)$, $\lambda\in\mathbb{C}$, $\operatorname{Re}\lambda>a$.*

(iii) *Suppose $h\in L^1_{loc}([0,\infty))$ is Laplace transformable and there exist constants M, $a'>0$ such that $\int_0^t|h(s)|e^{\omega s}\,ds\leqslant Me^{a't}$, $t\geqslant 0$. Put*

$$(h*f)(t):=\int_0^t h(t-s)f(s)\,ds,\quad t\geqslant 0. \tag{15}$$

Then the mapping $t\mapsto(h*f)(t)$, $t\geqslant 0$ is continuous and for each $p\in\circledast$ there exists $N_p>0$ such that $p((h*f)(t))\leqslant N_pe^{\omega' t}$, $t\geqslant 0$. Furthermore,

$$\widetilde{h*f}(\lambda)=\tilde{h}(\lambda)\tilde{f}(\lambda),\ \lambda\in\mathbb{C},\ \operatorname{Re}\lambda>\max(\operatorname{abs}(|h|),a',a).$$

(iv) *Let $F(t):=\int_0^t f(s)\,ds$, $t\geqslant 0$. Then $\tilde{F}(\lambda)=\frac{\tilde{f}(\lambda)}{\lambda}$, $\lambda\in\mathbb{C}$, $\operatorname{Re}\lambda>\max(0,a)$.*

(v) *Put*

$$j(t):=\int_0^\infty\frac{e^{-s^2/4t}}{\sqrt{\pi t}}f(s)\,ds\ \text{and}\ j(0):=f(0).$$

Then the mapping $t\mapsto j(t)$, $t\geqslant 0$ is continuous and for each $p\in\circledast$ there exists $m_p>0$ such that $p(j(t))\leqslant m_pe^{\max(a,0)^2t}$, $t\geqslant 0$. Furthermore,

$$\tilde{j}(\lambda)=\frac{\tilde{f}(\sqrt{\lambda})}{\sqrt{\lambda}}\ \text{for all}\ \lambda\in\mathbb{C}\ \text{with}\ \operatorname{Re}\lambda>\max(a,0)^2.$$

(vi) (The uniqueness theorem for the Laplace transform) *Suppose $\lambda_0>a$ and $\tilde{f}(\lambda)=0$ for all $\lambda\in(\lambda_0,\infty)$. Then $f(t)=0$, $t\geqslant 0$.*

(vii) *Let $f_1,f_2\in C([0,\infty):E)$ satisfy that for each $p\in\circledast$ there exists $M'_p>0$ such that $p(f_1(t))+p(f_2(t))\leqslant M'_pe^{at}$, $t\geqslant 0$. Suppose that A is a closed linear operator on E satisfying that for $\lambda>a$,*

$$\int_0^\infty e^{-\lambda t} f_1(t)\, dt \in D(A),$$

and

$$A\int_0^\infty e^{-\lambda t} f_1(t)\, dt = \int_0^\infty e^{-\lambda t} f_2(t)\, dt, \quad \lambda > a.$$

Then, for every $t \geqslant 0$, *we have* $f_1(t) \in D(A)$ *and* $Af_1(t) = f_2(t)$.

(viii) (Post–Widder inversion) *Suppose* $t > 0$. *Then the following holds:*

$$f(t) = \lim_{n\to\infty} (-1)^n \frac{1}{n!}\left(\frac{n}{t}\right)^{n+1} \tilde{f}^{(n)}\left(\frac{n}{t}\right),$$

uniformly on compacts of $(0, \infty)$.

(ix) *For each* $t \geqslant 0$, *we have that*

$$\int_0^t f(s)\, ds = \lim_{\lambda\to\infty} \sum_{n=1}^{\infty} (-1)^{n-1} n!^{-1} e^{n\lambda t} \int_0^\infty e^{-n\lambda r} f(r)\, dr.$$

For more details about the inversion methods for the Laplace transform of vector-valued functions, the reader may consult [44], [50] and [222]. The complex inversion theorem for the vector-valued Laplace transform reads as follows.

Theorem 1.2.2. *Assume* $a > 0$, $r \in \mathbb{R}$, $q : \{\lambda \in \mathbb{C} : \operatorname{Re} \lambda > a\} \to E$ *is analytic, and for each* $p \in \circledast$ *there exists* $M_p > 0$ *such that*

$$p(q(\lambda)) \leqslant M_p |\lambda|^r, \ \operatorname{Re} \lambda > a.$$

Then for each $\alpha > 1$ *there exists a function* $f_\alpha \in C([0, \infty) : E)$ *with* $f_\alpha(0) = 0$ *and*

$$p(h_\alpha(t)) \leqslant M_\alpha M_p e^{at}, p \in \circledast, t \geqslant 0,$$

$$q(\lambda) = \lambda^{\alpha+r} \int_0^\infty e^{-\lambda t} h_\alpha(t)\, dt, \ \operatorname{Re} \lambda > a,$$

where M_α *is independent of* p *and* $q(\cdot)$.

Notice that the function $h_\alpha(\cdot)$ is given by:

$$h_\alpha(t) = \frac{1}{2\pi i} \int_{\bar{a}-i\infty}^{\bar{a}+i\infty} e^{\lambda t} \lambda^{-r-\alpha} q(\lambda)\, d\lambda, \ t \geqslant 0,$$

for any number $\bar{a} > a$, and that the improper integral above does not depend on the choice of a number $\bar{a} > a$. The celebrated Arendt–Widder theorem [10] has been reconsidered and slightly generalized in a series of recent papers (cf. [292,

Section 1.1] for further information); the following version has been proved by T.-J. Xiao and J. Liang in [526].

Theorem 1.2.3. *Let $a \geqslant 0$, $r \in (0, 1]$, $\omega \in (-\infty, a]$, $M_p > 0$ for each $p \in ⊛$, and let $q : (a, \infty) \to E$ be an infinitely differentiable function. Then we have the equivalence of statements (i) and (ii), where:*

(i) *One has $p((\lambda - \omega)^{k+1}\frac{q^{(k)}(\lambda)}{k!}) \leqslant M_p$, $p \in ⊛$, $\lambda > a$, $k \in \mathbb{N}_0$.*
(ii) *There exists a function $F_r \in C([0, \infty) : E)$ satisfying $F_r(0) = 0$,*

$$q(\lambda) = \lambda^r \int_0^\infty e^{-\lambda t} F_r(t)\, dt,\ \ \lambda > a,$$

$$p\left(\int_0^{t+h} \frac{(t+h-s)^{-r}}{\Gamma(1-r)} F_r(s)\, ds - \int_0^t \frac{(t-s)^{-r}}{\Gamma(1-r)} F_r(s)\, ds\right) \leqslant M_p h e^{\omega t} \max(e^{\omega h}, 1),$$

for any $t \geqslant 0$, $h \geqslant 0$ and $p \in ⊛$, if $r \in (0, 1)$, and

$$p(F_r(t + h) - F_r(t)) \leqslant M_p h e^{\omega t} \max(e^{\omega h}, 1),\ t \geqslant 0,\ h \geqslant 0,\ p \in ⊛,$$

if $r = 1$. Moreover, in this case,

$$p(F_r(t + h) - F_r(t)) \leqslant \frac{2M_p}{r\Gamma(r)} h^r \max(e^{\omega(t+h)}, 1),\ t \geqslant 0,\ h \geqslant 0,\ p \in ⊛.$$

The study of analytical properties of Laplace transform in SCLCSs relies upon the following slight improvement of [20, Theorem 2.6.1, Proposition 2.6.3, Theorem 2.6.4]; we first state the following Vitali's type theorem (cf. W. Arendt, N. Nikolski [17, Theorem 2.1] and E. Jordá [239, Theorem 3, p. 742]) which will be necessary to do so.

Lemma 1.2.4. *Let $\emptyset \neq \Omega \subseteq \mathbb{C}$ be open and connected, and let $f_n : \Omega \to E$ be an analytic function ($n \in \mathbb{N}$). Assume further that, for every $z_0 \in \Omega$, there exists $r > 0$ such that the set $\bigcup_{n\in\mathbb{N}} f_n(B(z_0, r))$ is bounded and the set $\Omega_0 := \{z \in \Omega : \lim_{n\to\infty} f_n(z)$ exists$\}$ has a limit point in Ω. Then there exists an analytic function $f : \Omega \to E$ such that (f_n) converges locally uniformly to f.*

Recall that $\Sigma_\alpha = \{z \in \mathbb{C}\setminus\{0\} : |\arg(z)| < \alpha\}$ $(\alpha \in (0, \pi])$.

Theorem 1.2.5. ([303])

(i) *Let $\alpha \in (0, \frac{\pi}{2}]$, $\omega \in \mathbb{R}$ and $q : (\omega, \infty) \to E$. Then the following assertions are equivalent:*
   (a) *There exists an analytic function $f : \Sigma_\alpha \to E$ such that $q(\lambda) = \tilde{f}(\lambda)$, $\lambda \in (\omega, \infty)$ and the set $\{e^{-\omega z} f(z) : z \in \Sigma_\beta\}$ is bounded for all $\beta \in (0, \alpha)$.*
   (b) *The function $q(\cdot)$ admits an analytic extension $\tilde{q} : \omega + \Sigma_{(\pi/2)+\alpha} \to E$ which satisfies that the set $\{(\lambda - \omega)\tilde{q}(\lambda) : \lambda \in \omega + \Sigma_{(\pi/2)+\alpha}\}$ is bounded for all $\gamma \in (0, \alpha)$.*

If this is the case, then we have that, for every $k \in \mathbb{N}$ and $\beta \in (0, \alpha)$, the set $\{z^k e^{-\omega z} f^{(k)}(z) : z \in \Sigma_\beta\}$ is bounded.

(ii) *Let* $\alpha \in (0, \pi]$ *and let* $f : \Sigma_\alpha \to E$ *be an analytic function which satisfies that, for every* $\beta \in (0, \alpha)$, *the set* $\{f(z) : z \in \Sigma_\beta\}$ *is bounded. Let* $x \in E$. *Then the following holds:*
(a) *The assumption* $\lim_{t\to+\infty} f(t) = x$ *implies* $\lim_{z\to\infty, z\in\Sigma_\beta} f(z) = x$ *for all* $\beta \in (0, \alpha)$.
(b) *The assumption* $\lim_{t\to 0+} f(t) = x$ *implies* $\lim_{z\to 0, z\in\Sigma_\beta} f(z) = x$ *for all* $\beta \in (0, \alpha)$.

(iii) *Assume* $x \in E$, $\alpha \in (0, \frac{\pi}{2}]$, $\omega \in \mathbb{R}$, $q : (\omega, \infty) \to E$ *and let (i)(a) of this theorem hold. Then:*
(a) $\lim_{t\to 0+} f(t) = x$ *iff* $\lim_{\lambda\to\infty} \lambda q(\lambda) = x$.
(b) *Let* $\omega = 0$. *Then* $\lim_{t\to+\infty} f(t) = x$ *iff* $\lim_{\lambda\to 0} \lambda q(\lambda) = x$.

It is well known that the Trotter-Kato type theorems provide an effective method for studying convergence of numerical approximations of solutions to PDEs. The following theorem on approximation of Laplace transform enables one to establish several types of Trotter-Kato theorems for (a, k)-regularized C-resolvent families; for further information in this direction, we refer the reader to Section 2.7, [4]-[5], [273], [381], [383], [386] and [429]. In the case that E is a Banach space, the above-mentioned result has appeared for the first time in the paper [528] by T.-J. Xiao and J. Liang.

Theorem 1.2.6. *(Approximation, [303]) Let* $f_n \in C([0, \infty) : E)$, $n \in \mathbb{N}$, *let the set* $\{e^{-\omega t} f_n(t) : n \in \mathbb{N}, t \geqslant 0\}$ *be bounded for some* $\omega \in \mathbb{R}$ *and let* $\lambda_0 \geqslant \omega$. *Then the following assertions are equivalent:*

(i) *The sequence* $(\tilde{f}_n)$ *converges pointwise on* (λ_0, ∞) *and the sequence* (f_n) *is equicontinuous at each point* $t \geqslant 0$.
(ii) *The sequence* (f_n) *converges uniformly on compact subsets of* $[0, \infty)$.

Assuming (ii) holds and $\lim_{n\to\infty} f_n(t) = f(t)$, $t \geqslant 0$, *one has* $\lim_{n\to\infty} \tilde{f}_n(\lambda) = \tilde{f}(\lambda)$, $\lambda > \lambda_0$.

Proof. In order to prove the implication (i) $\Rightarrow$ (ii), set $l^\infty(E) := \{\langle x_n\rangle : x_n \in E, \langle x_n\rangle$ is bounded$\}$, $p_\infty(\langle x_n\rangle) := \sup_{n\in\mathbb{N}} p(x_n)$, $p \in \circledast$ and $c(E) := \{\langle x_n\rangle : x_n \in E, \lim_{n\to\infty} x_n$ exists$\}$. Equipped with the family $(p_\infty)_{p\in\circledast}$ of seminorms, $l^\infty(E)$ becomes an SCLCS and it is obvious that $c(E)$ is a closed subspace of $l^\infty(E)$. Define $\varphi : [0, \infty) \to l^\infty(E)$ by $\varphi(t) := (f_n(t))_n$, $t \geqslant 0$. Then, for every $p \in \circledast$, there exists $l_p > 0$ such that $p_\infty(\varphi(t)) \leqslant l_p e^{\omega t}$, $t \geqslant 0$, and the equicontinuity of (f_n) at each point $t \geqslant 0$ is equivalent to say that $\varphi \in C([0, \infty) : l^\infty(E))$. Denote by $F(\lambda)$ the Laplace transform of $\varphi(t)$. Then the mapping $\lambda \mapsto F(\lambda)$, Re $\lambda > \lambda_0$ is analytic, $F^{(k)}(\lambda) = (\tilde{f}_n^{(k)}(\lambda))_n$, $\lambda > \lambda_0$, $k \in \mathbb{N}_0$, $F^{(k)}(\lambda) = \int_0^\infty e^{-\lambda t}(-t)^k \varphi(t)\, dt$, Re $\lambda > \lambda_0$, $k \in \mathbb{N}$, and since $F(\lambda) \in c(E)$, $\lambda > \lambda_0$ and $c(E)$ is a closed subspace of $l^\infty(E)$, it follows that $F^{(k)}(\lambda) \in c(E)$, $\lambda > \lambda_0$, $k \in \mathbb{N}_0$. Now one can apply Theorem 1.2.1(viii) in an effort to see that $\varphi(t) = \lim_{k\to\infty} (-1)^k \frac{1}{k!} (\frac{k}{t})^{k+1} F^{(k)}(\frac{k}{t}) \in c(E)$, $t > 0$. Taking into account the continuity of $\varphi(t)$ for $t \geqslant 0$, the above implies that the sequence $(f_n(t))$ is convergent for every $t \geqslant 0$. The equicontinuity of (f_n) indicates that (f_n) converges uniformly on compact subsets of $[0, \infty)$, which completes the proof of implication (i) $\Rightarrow$ (ii). The implication (ii) $\Rightarrow$ (i) follows from the dominated convergence theorem. □

As an almost immediate consequence of the preceding theorem, we have the following corollary (cf. [528, Theorem 2.2, Corollary 2.4]).

Corollary 1.2.7. *Let* $f_n \in C([0, \infty) : E)$, $n \in \mathbb{N}$, *let the set* $\{e^{-\omega t}f_n(t) : n \in \mathbb{N},\ t \geqslant 0\}$ *be bounded for some* $\omega \in \mathbb{R}$ *and let* $\lambda_0 \geqslant \omega$. *Suppose* $\gamma \in (0, 1]$ *and the set* $\{e^{-\omega(t+h)} h^{-\gamma} (f_n(t + h) - f_n(t)) : t \geqslant 0,\ h > 0,\ n \in \mathbb{N}\}$ *is bounded in E. Then the following assertions are equivalent:*

(i) *The sequence* $(\tilde{f}_n)$ *converges pointwise on* (λ_0, ∞).
(ii) *There is a function* $f \in C([0, \infty) : E)$ *such that the set* $\{e^{-\omega(t+h)}h^{-\gamma} (f(t + h) - f(t)) : t \geqslant 0,\ h > 0\}$ *is bounded in E and* $\lim_{n\to\infty} \tilde{f}_n(\lambda) = \tilde{f}(\lambda)$, *uniformly on* $[\omega + \sigma, \infty)$ *for any* $\sigma > 0$.
(iii) *The sequence* (f_n) *converges pointwise on* $[0, \infty)$.
(iv) *There is a function* $f \in C([0, \infty) : E)$ *such that the set* $\{e^{-\omega(t+h)}h^{-\gamma} (f(t + h) - f(t)) : t \geqslant 0,\ h > 0\}$ *is bounded in E and* $\lim_{n\to\infty} f_n(t) = f(t)$, *uniformly on compact subsets of* $[0, \infty)$.

Notice, finally, that one can simply prove a slight extension of Theorem 1.2.6 following the approach of M. Li and Q. Zheng [360, Proposition 2.7].

1.3 Operators of fractional differentiation, Mittag-Leffler and Wright functions

In recent years, considerable interest in fractional calculus and fractional differential equations has been stimulated by their applications in modeling of various problems in engineering, physics, chemistry, biology and other sciences. The Mittag-Leffler and Wright functions are known to play fundamental roles in various applications of the fractional calculus. For further information about the topics mentioned above, the reader may consult the monographs by D. Baleanu, K. Diethelm, E. Scalas, J. Trujillo [33], K. Diethelm [163], A. A. Kilbas, H. M. Srivastava, J. J. Trujillo [271], J. Klafter, S. C. Lim, R. Metzler (Eds.) [278], F. Mainardi [400], K. S. Miller, B. Ross [416], K. B. Oldham, J. Spanier [447], I. Podlubny [458] and S. G. Samko, A. A. Kilbas, O. I. Marichev [478]; we also refer to the references [1], [22], [26]-[28], [34], [89], [235], [264]-[265], [281]-[282], [302]-[308], [312]-[316], [318]-[323], [329], [358], [386], [391], [410], [427], [451], [480], [495] and [513]-[514].

Chronologically, the theory of fractional derivatives goes back to the correspondence of W. Leibnitz and de L'Hospital (1695) in which the meaning of the derivative of one half is discussed; in his later correspondence with J. Bernoulli (1695), the fractional derivaties of "general order" are mentioned. Although L. Euler, P. S. Laplace, J. L. Lagrange, J. B. J. Fourier, J. Wallis and S. F. Lacroix also made mention of fractional derivatives, N. H. Abel (1823) was the first person who used fractional operations. Speaking matter-of-factly, N. H. Abel applied the finite convolution product

$$\int_0^x (x-t)^{1/2} f(t)\, dt$$

in the formulation of the isochrone problem. By the end of the nineteenth century, the theory of fractional derivatives and integrals took an almost complete form; surveys on the history of fractional calculus can be found in [166], [416], [447] and [478].

Let $\alpha > 0$, $m = \lceil \alpha \rceil$ and $I = (0, T)$ for some $T > 0$. The Riemann-Liouville fractional integral of order α is defined by

$$J_t^\alpha f(t) := (g_\alpha * f)(t), f \in L^1(I), t \in I.$$

The Riemann-Liouville fractional derivative of order α is defined for those functions $f \in L^1(I)$ for which $g_{m-\alpha} * f \in W^{1,m}(I)$, by

$$D_t^\alpha f(t) := \frac{d^m}{dt^m} J_t^{m-\alpha} f(t), \quad t \in I.$$

Due to [49, Theorem 1.5], we have that $D_t^\alpha J_t^\alpha f(\cdot) = f(\cdot)$, $f \in L^1(I)$ and, if $f(\cdot)$ additionally satisfies $g_{m-\alpha} * f \in W^{1,m}(I)$, then

$$J_t^\alpha D_t^\alpha f(t) = f(t) - \sum_{k=0}^{m-1} \left(g_{m-\alpha} * f\right)^{(k)}(0) g_{\alpha+1+k-m}(t).$$

In this book, we mainly use the Caputo fractional derivatives (one more point depicting the incompleteness of our study); the only exceptions are Theorem 2.1.24, Example 2.9.53, Theorem 3.2.15 and Subsections 2.9.5-2.9.6. The Caputo fractional derivative $\mathbf{D}_t^\alpha u(t)$ is defined for those functions $u \in C^{m-1}([0, \infty) : E)$ for which $g_{m-\alpha} * (u - \Sigma_{k=0}^{m-1} u_k g_{k+1}) \in C^m([0, \infty) : E)$, by

$$\mathbf{D}_t^\alpha u(t) = \frac{d^m}{dt^m}\left[g_{m-\alpha} * \left(u - \sum_{k=0}^{m-1} u_k g_{k+1}\right)\right].$$

Suppose $\beta > 0$, $\gamma > 0$ and $\mathbf{D}_t^{\beta+\gamma} u(t)$ is defined. Then the Caputo fractional derivative $\mathbf{D}_t^\zeta u(t)$ is defined for any number $\zeta \in (0, \beta+\gamma)$ and the following holds:

$$\mathbf{D}_t^\zeta u(t) = \left(g_{\beta+\gamma-\zeta} * \mathbf{D}_t^{\beta+\gamma} u(\cdot)\right)(t) + \sum_{j=\lceil \zeta \rceil}^{\lceil \beta+\gamma \rceil - 1} u^{(j)}(0) g_{j+1-\zeta}(t), \qquad t \geqslant 0;$$

unfortunately, the equality $\mathbf{D}_t^{\beta+\gamma} u = \mathbf{D}_t^\beta \mathbf{D}_t^\gamma u$ does not hold in general.

If $u \in C([0, \infty) : E)$, resp. $u \in C^{m-1}([0, \infty) : E)$ and $g_{m-\alpha} * (u - \Sigma_{k=0}^{m-1} u_k g_{k+1}) \in C^m([0, \infty) : E)$, then the following equality holds:

$$\mathbf{D}_t^\alpha J_t^\alpha u(t) = u(t), \ t \geqslant 0, \text{ resp. } J_t^\alpha \mathbf{D}_t^\alpha u(t) = u(t) - \sum_{k=0}^{m-1} u^{(k)}(0) g_{k+1}(t), \ t \geqslant 0.$$

Assume that $\omega \geqslant 0$ and for each $p \in \circledast$ there exists $M_p > 0$ such that $p(u(t)) + p(\mathbf{D}_t^\alpha u(t)) \leqslant M_p e^{\omega t}$, $t \geqslant 0$. Then the Laplace transform of function $\mathbf{D}_t^\alpha u(t)$ can be computed by

(16) $$\int_0^\infty e^{-\lambda t}\mathbf{D}_t^\alpha u(t)\,dt = \lambda^\alpha \tilde{u}(\lambda) - \sum_{k=0}^{m-1} u^{(k)}(0)\lambda^{\alpha-1-k},\ \mathrm{Re}\,\lambda > \omega.$$

Before proceeding further, we want to observe that operators of fractional differentiation have their q-analogues. We recommend the monograph by M. H. Annaby and Z. S. Mansour [9] for the introduction to the theory of q-fractional calculus and q-fractional differential equations. A brief overview of basic information basic about the theory of integral transforms can be obtained by consulting [492].

Let $\alpha > 0$ and $\beta \in \mathbb{R}$. Then the Mittag-Leffler function $E_{\alpha,\beta}(z)$ is defined by

$$E_{\alpha,\beta}(z) := \sum_{n=0}^{\infty} \frac{z^n}{\Gamma(\alpha n+\beta)}, \quad z \in \mathbb{C}.$$

Here we assume that $1/\Gamma(\alpha n + \beta) = 0$ if $\alpha n + \beta \in -\mathbb{N}_0$. Set, for short, $E_\alpha(z) := E_{\alpha,1}(z)$, $z \in \mathbb{C}$. Like the function $E_1(z) = e^z$, for which the differential relation $(d/dt)e^{\omega t} = \omega e^{\omega t}$ holds, the function $E_\alpha(z)$ satisfies that

$$\mathbf{D}_t^\alpha E_\alpha(\omega t^\alpha) = \omega E_\alpha(\omega t^\alpha).$$

The asymptotic expansion of the entire function $E_{\alpha,\beta}(z)$ is given in the following important theorem (see e.g. [522, Theorem 1.1]):

Theorem 1.3.1. *Let $0 < \sigma < \frac{1}{2}\pi$. Then, for every $z \in \mathbb{C}\setminus\{0\}$ and $m \in \mathbb{N}\setminus\{1\}$,*

$$E_{\alpha,\beta}(z) = \frac{1}{\alpha}\sum_s Z_s^{1-\beta} e^{Z_s} - \sum_{j=1}^{m-1} \frac{z^{-j}}{\Gamma(\beta-\alpha j)} + O(|z|^{-m}),$$

where Z_s is defined by $Z_s := z^{1/\alpha}e^{2\pi i s/\alpha}$ and the first summation is taken over all those integers s satisfying $|\arg(z) + 2\pi s| < \alpha(\frac{\pi}{2} + \sigma)$.

As a special case of Theorem 1.3.1, we have the following asymptotic formulae appearing in [49]. Let $\alpha \in (0, 2)\setminus\{1\}$, $\beta > 0$ and $N \in \mathbb{N}\setminus\{1\}$. Then the following holds:

(17) $$E_{\alpha,\beta}(z) = \frac{1}{\alpha} z^{(1-\beta)/\alpha} e^{z^{1/\alpha}} + \varepsilon_{\alpha,\beta}(z),\ |\arg(z)| < \alpha\pi/2,$$

and

(18) $$E_{\alpha,\beta}(z) = \varepsilon_{\alpha,\beta}(z),\ |\arg(-z)| < \pi - \alpha\pi/2,$$

where

(19) $$\varepsilon_{\alpha,\beta}(z) = \sum_{n=1}^{N-1} \frac{z^{-n}}{\Gamma(\beta-\alpha n)} + O(|z|^{-N}),\quad |z| \to \infty.$$

The Mittag-Leffler function $E_{\alpha,\beta}(z)$ can be integrally represented by

$$E_{\alpha,\beta}(z) = \frac{1}{2\pi i}\int_G \frac{\lambda^{\alpha-\beta}e^{\lambda}}{\lambda^{\alpha}-z}\,d\lambda, \quad z\in\mathbb{C},$$

where G is a contour (the Hankel path) which starts and ends at $-\infty$ and encircles the disc $|\lambda| \geqslant |z|^{1/\alpha}$ counter-clockwise. One of the most intriguing properties of the Mittag-Leffler functions is associated with the following identity:

(20) $$\int_0^{\infty} e^{-\lambda t}t^{\beta-1}E_{\alpha,\beta}(\omega t^{\alpha})\,dt = \frac{\lambda^{\alpha-\beta}}{\lambda^{\alpha}-\omega}, \quad \operatorname{Re}\lambda > \omega^{1/\alpha}, \ \omega > 0.$$

It is worth noting that the function $t \mapsto E_{\alpha,\beta}(-t)$, $t \geqslant 0$ is completely monotonic (i.e., that $(-1)^n(d^n/dt^n)E_{\alpha,\beta}(-t) \geqslant 0$, $t \geqslant 0$, $n \in \mathbb{N}_0$) provided that $\alpha \in (0, 1]$ or $\beta \geqslant \alpha$ ([400]).

There are many identities for the Mittag-Leffler functions. Only the most important ones are stated here:

(i) $E_{\alpha,\beta}(z) = \Gamma(\beta)^{-1} + zE_{\alpha,\alpha+\beta}(z)$, $z \in \mathbb{C}$.

(ii) $E_{\alpha,\beta}(z) = \beta E_{\alpha,\beta+1}(z) + \alpha z \frac{d}{dz} E_{\alpha,\beta+1}(z)$, $z \in \mathbb{C}$.

(iii) For every $j \in \mathbb{N}$ and $\alpha > 0$, there exist uniquely determined real numbers $c'_{l,j,\alpha}$ $(1 \leqslant l \leqslant j)$ such that:

(21) $$E_{\alpha}^{(j)}(z) = \sum_{l=1}^{j} c'_{l,j,\alpha}\, E_{\alpha,\alpha j-(j-l)}(z), \quad z\in\mathbb{C}.$$

(iv) $\frac{d^p}{dz^p}\left[z^{\beta-1}E_{\alpha,\beta}(z^{\alpha})\right] = z^{\beta-p-1}E_{\alpha,\beta-p}(z^{\alpha})$, $p \in \mathbb{N}$, $z \in \mathbb{C}\setminus(-\infty, 0]$.

(v) $E_{1/q}(z) = e^{z}\left[1+\sum_{k=1}^{q-1}\frac{\gamma(1-k/q,z)}{\Gamma(1-k/q)}\right]$, $z \in \mathbb{C}$, $q = 2, 3, \cdots$, where $\gamma(a, z) = \int_0^z e^{-u}u^{a-1}\,du$ denotes the incomplete Gamma function.

Let $\gamma \in (0, 1)$. Then the Wright function $\Phi_{\gamma}(\cdot)$ is defined by

$$\Phi_{\gamma}(t) := \mathcal{L}^{-1}(E_{\gamma}(-\lambda))(t), \qquad t \geqslant 0.$$

The Wright function $\Phi_{\gamma}(\cdot)$ can be analytically extended to the whole complex plane by the formulae

$$\Phi_{\gamma}(z) = \sum_{n=0}^{\infty} \frac{(-z)^n}{n!\Gamma(1-\gamma-\gamma n)}, \quad z\in\mathbb{C}.$$

Furthermore, we have that:

(i) $\Phi_{\gamma}(t) \geqslant 0$, $t \geqslant 0$,

(ii) $\int_0^{\infty} e^{-\lambda t}\gamma s t^{-1-\gamma}\Phi_{\gamma}(t^{-\gamma}s)\,dt = e^{-\lambda^{\gamma}s}$, $\lambda \in \mathbb{C}_+$, $s > 0$, and

(iii) $\int_0^\infty t^r \Phi_\gamma(t)\, dt = \frac{\Gamma(1+r)}{\Gamma(1+\gamma r)}$, $r > -1$.

The Wright function $\Phi_\gamma(\cdot)$ has the following integral representation:

$$\Phi_\gamma(z) = \frac{1}{2\pi i}\int_G \lambda^{\gamma-1}\exp(\lambda - z\lambda^\gamma)\, d\lambda, \quad z \in \mathbb{C},$$

where G is the Hankel path mentioned above. The asymptotic expansion of the Wright function $\Phi_\gamma(\cdot)$, as $|z| \to \infty$ in the sector $|\arg(z)| \leqslant \min((1-\gamma)3\pi/2, \pi) - \varepsilon$ is given by

$$\Phi_\gamma(z) = Y^{\gamma-1/2}e^{-Y}\left(\sum_{m=0}^{M-1} A_m Y^{-M} + O\left(|Y|^{-M}\right)\right),$$

where $Y = (1-\gamma)(\gamma^\gamma z)^{1/(1-\gamma)}$, $M \in \mathbb{N}$ and A_m are certain real numbers ([524]). For further information about the Mittag-Leffler and Wright functions, see also [201], [221], [274] and [523].

2

(a, k)-REGULARIZED C-RESOLVENT FAMILIES IN LOCALLY CONVEX SPACES

2.1 Definition and main structural properties of (a, k)-regularized C-resolvent families

Let E be a Hausdorff sequentially complete locally convex space over the field of complex numbers, SCLCS for short, and let ⊛ be the abbreviation that stands for the fundamental system of seminorms which defines the topology of E. In this section, we take up the study of various structrural properties of (a, k)-regularized C-resolvent families and their subgenerators, ranging from Hille-Yosida and adjoint type theorems, Abel ergodicity, generalized subordination principles with applications, L^p-stability and parabolicity, the wellposedness of abstract Cauchy problem (1) and its fractional relatives (Subsection 2.1.1), to convoluted C-semigroups and cosine functions in locally convex spaces (Subsection 2.1.2).

Definition 2.1.1. Let $0 < \tau \leqslant \infty$, $k \in C([0, \tau))$, $k \neq 0$ and let $a \in L^1_{loc}([0, \tau))$, $a \neq 0$. Suppose that $C \in L(E)$ is injective, A is a closed linear operator on E, and $CA \subseteq AC$. Then it is said that a strongly continuous operator family $(R(t))_{t\in[0,\tau)}$ is a (local, if $\tau < \infty$) (a, k)-regularized C-resolvent family having A as a subgenerator if the following holds:

(i) $R(t)A \subseteq AR(t)$, $t \in [0, \tau)$ and $R(0) = k(0)C$,
(ii) $R(t)C = CR(t)$, $t \in [0, \tau)$ and
(iii) $R(t)x = k(t)Cx + \int_0^t a(t-s)AR(s)x\,ds$, $t \in [0, \tau)$, $x \in D(A)$.

$(R(t))_{t\in[0,\tau)}$ is said to be non-degenerate if the condition $R(t)x = 0$, $t \in [0, \tau)$ implies $x = 0$, and $(R(t))_{t\in[0,\tau)}$ is said to be locally equicontinuous if, for every $t \in (0, \tau)$, the family $\{R(s) : s \in [0, t]\}$ is equicontinuous. In the case $\tau = \infty$, $(R(t))_{t\geqslant 0}$ is said to be exponentially equicontinuous (equicontinuous) if there exists $\omega \in \mathbb{R}$ ($\omega = 0$) such that the family $\{e^{-\omega t}R(t) : t \geqslant 0\}$ is equicontinuous; the infimum of such numbers is said to be the exponential type of $(R(t))_{t\geqslant 0}$.

If $k(t) \equiv g_{\alpha+1}(t)$, where $\alpha > 0$, then it is also said that $(R(t))_{t\in[0,\tau)}$ is an α-times integrated (a, C)-resolvent family. This way, we unify the notions of (local) α-times integrated C-semigroups ($a(t) \equiv 1$) and cosine functions ($a(t) \equiv t$) in locally convex spaces. Furthermore, in the case $k(t) = \int_0^t K(s)\, ds$, $t \in [0, \tau)$, where $K \in L^1_{loc}([0, \tau))$ and $K \neq 0$, we obtain the unification concept for (local) K-convoluted C-semigroups and cosine functions. More precisely, in the definition of a local K-convoluted C-semigroup (cosine function) and its subgenerator we assume additionally that the condition (22) stated below holds with $a(t) \equiv 1$ ($a(t) \equiv t$). We introduce the integral generator of a K-convoluted C-semigroup (cosine function) as in the Banach space case ([292]). If $k(t) \equiv 1$, then $(R(t))_{t\in[0,\tau)}$ is also said to be an (a, C)-regularized resolvent family with subgenerator A. Set $\Theta(t) := \int_0^t k(s)\, ds$, $t \in [0, \tau)$.

In the sequel of this chapter, our standing hypotheses will be that K is a scalar-valued kernel on $[0, \tau)$, as well as that $k, k_1, k_2,...,$ are scalar-valued continuous kernals on $[0, \tau)$ and that $a \neq 0$ in $L^1_{loc}([0, \tau))$. Unless stated otherwise, we shall always assume that $C \in L(E)$ is injective and satisfies $CA \subseteq AC$.

The following conditions will be used in the continuation:

(H1) : A is densely defined and $(R(t))_{t\in[0,\tau)}$ is locally equicontinuous.
(H2) : $\rho(A) \neq \emptyset$.
(H3) : $\rho_C(A) \neq \emptyset$, $\overline{R(C)} = E$ and $(R(t))_{t\in[0,\tau)}$ is locally equicontinuous.
(H3)' : $\rho_C(A) \neq \emptyset$ and $C^{-1}AC = A$.
(H4) : A is densely defined and $(R(t))_{t\in[0,\tau)}$ is locally equicontinuous, or $\rho_C(A) \neq \emptyset$.
(H5) : (H1) V (H2) V (H3) V (H3)'.

Remark 2.1.2. Assume E is barreled and $\emptyset \neq \Omega \subseteq \mathbb{C}$. Using the uniform boundedness principle [411, Proposition 23.27, p. 273], it follows that the family $\{A(z) : z \in \Omega\}$ in $L(E)$ is equicontinuous if the set $\{A(z) : z \in \Omega\}$ is (pointwise) bounded in $L(E)$. The same argument shows that a strongly continuous operator family $(R(t))_{t\in[0,\tau)}$ in a barreled space is locally equicontinuous (cf. also T. Kōmura [291, Proposition 1.1]).

Proposition 2.1.3. ([302]-[303]) *Let A be a subgenerator of a (local) (a, k)-regularized C-resolvent family* $(R(t))_{t\in[0,\tau)}$. *Then the supposition* $\rho_C(A) \neq \emptyset$ *implies* $A\int_0^t a(t-s)R(s)x\, ds = R(t)x - k(t)Cx$, $t \in [0, \tau)$, $x \in R(C)$. *Assuming additionally (H5), we have*

$$A\int_0^t a(t-s)R(s)x\, ds = R(t)x - k(t)Cx,\ t \in [0, \tau),\ x \in E. \tag{22}$$

Henceforth we will consider only non-degenerate (a, k)-regularized C-resolvent families. Notice that $(R(t))_{t\in[0,\tau)}$ is non-degenerate provided that $k(0) \neq 0$ or that (22) holds. The set which consists of all subgenerators of $(R(t))_{t\in[0,\tau)}$, denoted by $\wp(R)$, need not be finite. One can easily verify that:

(i) $A \in \wp(R)$ implies $C^{-1}AC \in \wp(R)$.

(ii) $R(t)(\lambda - A)^{-1}C = (\lambda - A)^{-1}CR(t)$, $t \in [0, \tau)$, provided that $A \in \wp(R)$ and $\lambda \in \rho_C(A)$.

(iii) Let $a(t)$ be a kernel. Then one can define the integral generator $\hat{A}$ of $(R(t))_{t\in[0,\tau)}$ by setting

$$\hat{A} := \left\{(x, y) \in E \times E : R(t)x - k(t)Cx = \int_0^t a(t-s)R(s)y\, ds,\ t \in [0, \tau)\right\}.$$

The integral generator $\hat{A}$ of $(R(t))_{t\in[0,\tau)}$ is a function which satisfies $C^{-1}\hat{A}\,C = \hat{A}$ and $B \subseteq \hat{A}$, $B \in \wp(R)$; the assumptions $B \in \wp(R)$ and $\rho(B) \neq \emptyset$ imply $\hat{A} = B = C^{-1}BC$. Let $(R(t))_{t\in[0,\tau)}$ be locally equicontinuous. Then:
(a) $\hat{A}$ is a closed linear operator.
(b) $\hat{A} \in \wp(R)$, if $R(t)R(s) = R(s)R(t)$, $0 \leqslant t, s < \tau$.
(c) $\hat{A} = C^{-1}BC$, if $B \in \wp(R)$ and (H5) holds with A replaced by B.

(iv) Let $a(t)$ be a kernel and $\{A,B\} \subseteq \wp(R)$. Then $Ax = Bx$, $x \in D(A) \cap D(B)$, and $A \subseteq B \Leftrightarrow D(A) \subseteq D(B)$. Assume also that (22) holds for A (B) and C. Then:
(a) $C^{-1}AC = C^{-1}BC$ and $C(D(A)) \subseteq D(B)$.
(b) A and B have the same eigenvalues.
(c) $A \subseteq B \Rightarrow \rho_C(A) \subseteq \rho_C(B)$.

(v) Let $a(t)$ be a kernel, let $C = I$ and let (H5) hold for some $A \in \wp(R)$. Then card $(\wp(R)) = 1$.

Proposition 2.1.4. ([382], [302]-[303])

(i) *Let A be a subgenerator of an (a, k_i)-regularized C-resolvent family $(R_i(t))_{t\in[0,\tau)}$, $i = 1, 2$. Then $(k_2 * R_1)(t)x = (k_1 * R_2)(t)x$, $t \in [0, \tau)$, $x \in D(A)$; if (H4) additionally holds, then the above equality is valid for any $t \in [0, \tau)$ and $x \in E$.*

(ii) *Let A be a subgenerator of an (a, k)-regularized C-resolvent family $(R(t))_{t\in[0,\tau)}$. If $k(t)$ is absolutely continuous and $k(0) \neq 0$, then A is a subgenerator of an (a, C)-regularized resolvent family on $[0, \tau)$.*

(iii) *Let $(R(t))_{t\in[0,\tau)}$ be an (a, k)-regularized C-resolvent family with a subgenerator A, and let $L^1_{loc}([0, \tau)) \ni b$ be a kernel. Then A is a subgenerator of an $(a, k * b)$-regularized C-resolvent family $((b * R)(t))_{t\in[0,\tau)}$.*

(iv) *Let $(R(t))_{t\in[0,\tau)}$ be an (a,C)-regularized resolvent family having A as a subgenerator. Then $((k * R)(t))_{t\in[0,\tau)}$ is an (a, Θ)-regularized C-resolvent family with a subgenerator A.*

Theorem 2.1.5. *Let $k(t)$ and $a(t)$ satisfy (P1), and let $(R(t))_{t\geqslant 0}$ be a strongly continuous operator family such that there exists $\omega \geqslant 0$ satisfying that the family $\{e^{-\omega t}R(t) : t \geqslant 0\}$ is equicontinuous. Put $\omega_0 := \max(\omega, abs(a), abs(k))$.*

(i) *Assume that A is a subgenerator of the global (a, k)-regularized C-resolvent family $(R(t))_{t\geqslant 0}$ satisfying (22). Then, for every $\lambda \in \mathbb{C}$ with $\operatorname{Re}\lambda > \omega_0$ and $\tilde{k}(\lambda) \neq 0$, the operator $I - \tilde{a}(\lambda)A$ is injective, $R(C) \subseteq R(I - \tilde{a}(\lambda)A)$,*

$$\tilde{k}(\lambda)(I - \tilde{a}(\lambda)A)^{-1} Cx = \int_0^\infty e^{-\lambda t}R(t)x\,dt,\ x \in E,\ \operatorname{Re}\lambda > \omega_0,\ \tilde{k}(\lambda) \neq 0, \tag{23}$$

$$\left\{\frac{1}{\tilde{a}(\lambda)} : \operatorname{Re}\lambda > \omega_0,\ \tilde{k}(\lambda)\tilde{a}(\lambda) \neq 0\right\} \subseteq \rho_C(A) \tag{24}$$

and $R(s)R(t) = R(t)R(s)$, $t, s \geqslant 0$.

(ii) *Assume (23)-(24). Then A is a subgenerator of the global (a, k)-regularized C-resolvent family $(R(t))_{t\geqslant 0}$ satisfying $R(s)R(t) = R(t)R(s)$, $t, s \geqslant 0$.*

The following Hille-Yosida's type theorem can be deduced with the help of Theorem 1.2.2, the Arendt-Widder theorem in SCLCSs (Theorem 1.2.3), the preceding theorem and the principle of analytical continuation in SCLCSs. The proof is more-or-less standard and therefore omitted.

Theorem 2.1.6. (i) *Let $\omega_0 > \max(0, abs(a), abs(k))$, and let $k(t)$ and $a(t)$ satisfy (P1). Assume that, for every $\lambda \in \mathbb{C}$ with $\operatorname{Re}\lambda > \omega_0$ and $\tilde{k}(\lambda) \neq 0$, the operator $I - \tilde{a}(\lambda)A$ is injective and $R(C) \subseteq R(I - \tilde{a}(\lambda)A)$. If there exists a function $\Upsilon : \{\lambda \in \mathbb{C} : \operatorname{Re}\lambda > \omega_0\} \to L(E)$ which satisfies:*

(a) $\Upsilon(\lambda) = \tilde{k}(\lambda)(I - \tilde{a}(\lambda)A)^{-1}C$, $\operatorname{Re}\lambda > \omega_0$, $\tilde{k}(\lambda) \neq 0$,

(b) *the mapping $\lambda \mapsto \Upsilon(\lambda)x$, $\operatorname{Re}\lambda > \omega_0$ is analytic for every fixed $x \in E$, and*

(c) *there exists $r \geqslant -1$ such that the family $\{\lambda^{-r}\Upsilon(\lambda) : \operatorname{Re}\lambda > \omega_0\}$ is equicontinuous,*

then, for every $\alpha > 1$, A is a subgenerator of a global $(a, k * g_{\alpha+r})$-regularized C-resolvent family $(R_\alpha(t))_{t\geqslant 0}$ which satisfies that the family $\{e^{-\omega_0 t}R_\alpha(t) : t \geqslant 0\}$ is equicontinuous.

(ii) *Assume $\omega \in \mathbb{R}$, $k(t)$ and $a(t)$ satisfy (P1), and (H2) or (H3) holds. Assume further that A is a subgenerator of a global (a, Θ)-regularized C-resolvent family $(R(t))_{t\geqslant 0}$ such that the family*

$$\left\{h^{-1}e^{-\omega t}\min(e^{\omega h}, 1)(R_r(t + h) - R_r(t)) : t \geqslant 0,\ h > 0\right\} \text{ is equicontinuous.} \tag{25}$$

Then there exists $b \geqslant \max(0, \omega, abs(a), abs(k))$ such that:

$$\left\{\frac{1}{\tilde{a}(\lambda)} : \lambda > b,\ \tilde{k}(\lambda)\tilde{a}(\lambda) \neq 0\right\} \subseteq \rho_C(A), \tag{26}$$

the mapping $\lambda \mapsto H(\lambda) := \tilde{k}(\lambda)(I - \tilde{a}(\lambda)A)^{-1}C,\ \lambda > b,\ \tilde{k}(\lambda)\tilde{a}(\lambda) \neq 0$

(27) *is infinitely differentiable in* $L(E)$,

and the family

(28) $$\left\{k!^{-1}(\lambda-\omega)^{k+1}\frac{d^k}{d\lambda^k}H(\lambda) : k \in \mathbb{N}_0,\quad \lambda > b,\quad \tilde{k}(\lambda)\tilde{a}(\lambda) \neq 0\right\}$$ *is equicontinuous.*

(iii) *Suppose* $\omega \in \mathbb{R}$, *the functions* $k(t)$ *and* $a(t)$ *satisfy (P1),* $b \geqslant \max(0, \omega, abs(a), abs(k))$, *(26) holds, the function* $H : D(H) \equiv \{\lambda > b : \tilde{a}(\lambda)\,\tilde{k}(\lambda) \neq 0\} \to L(E)$, *given by* $H(\lambda)x = \tilde{k}(\lambda)(I - \tilde{a}(\lambda)A)^{-1}Cx,\ x \in E,\ \lambda \in D(H)$, *satisfies that the mapping* $\lambda \mapsto H(\lambda)x,\ \lambda \in D(H)$ *is infinitely differentiable for every fixed* $x \in E$ *and, for every* $p \in \circledast$, *there exist* $c_p > 0$ *and* $r_p \in \circledast$ *such that:*

(29) $$p\left(k!^{-1}(\lambda-\omega)^{k+1}\frac{d^k}{d\lambda^k}H(\lambda)x\right) \leqslant c_p r_p(x),\ x \in E,\ \lambda \in D(H),\ k \in \mathbb{N}_0.$$

Then, for every $r \in (0, 1]$, *the operator* A *is a subgenerator of a global* $(a, k * g_r)$*-regularized* C*-resolvent family* $(R_r(t))_{t\geqslant 0}$ *satisfying (22) with* $(R(t))_{t\geqslant 0}$ *replaced by* $(R_r(t))_{t\geqslant 0}$, *as well as that, for every* $p \in \circledast$,

(30) $$p(R_r(t+h)x - R_r(t)x) \leqslant \frac{2c_p r_p(x)}{r\Gamma(r)}\max(e^{\omega(t+h)}, 1)h^r,\ t \geqslant 0,\ h > 0,\ x \in E,$$

and that, for every $p \in \circledast$ *and* $B \in \mathcal{B}$, *the mapping* $t \mapsto p_B(R_r(t)),\ t \geqslant 0$ *is locally Hölder continuous with exponent* r.

(iv) *Suppose* $\omega \in \mathbb{R}$, $k(t)$ *and* $a(t)$ *satisfy (P1), and* A *is densely defined.*
   (a) *Let* A *be a subgenerator of a global* (a, k)*-regularized* C*-resolvent family* $(R(t))_{t\geqslant 0}$ *which satisfies that the family* $\{e^{-\omega t}R(t) : t \geqslant 0\}$ *is equicontinuous. Then (26)-(28) hold.*
   (b) *Suppose* $\omega \in \mathbb{R}$, $k(t)$ *and* $a(t)$ *satisfy (P1), and* A *is densely defined. Let* $b \geqslant \max(0, \omega, abs(a), abs(k))$ *such that (26) holds. Suppose that the mapping* $H : D(H) \to L(E)$, *defined as in (i), satisfies that the mapping* $\lambda \mapsto H(\lambda)x,\ \lambda \in D(H)$ *is infinitely differentiable for every fixed* $x \in E$. *If (29) holds, then* A *is a subgenerator of a global* (a, k)*-regularized* C*-resolvent family* $(R(t))_{t\geqslant 0}$ *satisfying (22) and that the family* $\{e^{-\omega t}R(t) : t \geqslant 0\}$ *is equicontinuous.*

The main objective in the following proposition is to clarify the Abel-ergodic properties of an exponentially equicontinous (a, k)-regularized C-resolvent family (see also [463, Corollary 1.6, p. 45]).

Proposition 2.1.7. *Assume $a(t)$ and $k(t)$ satisfy (P1), there exists $\zeta \in \mathbb{R}$ such that $\int_0^\infty e^{-\zeta t}|a(t)|\,dt < \infty$ and A is a subgenerator of an exponentially equicontinous (a, k)-regularized C-resolvent family $(R(t))_{t\geqslant 0}$.*
Then

$$\lim_{\lambda\to+\infty,\tilde{k}(\lambda)\neq 0} \lambda\tilde{k}(\lambda)(I-\tilde{a}(\lambda)A)^{-1}Cx = k(0)Cx,\ x \in \overline{D(A)}. \tag{31}$$

Proof. The existence of number $\zeta \in \mathbb{R}$ such that $\int_0^\infty e^{-\zeta t}|a(t)|\,dt < \infty$ implies that $\lim_{\lambda\to\infty}\tilde{a}(\lambda) = 0$. Let $\omega \geqslant 0$ be such that the family $\{e^{-\omega t}R(t) : t \geqslant 0\}$ is equicontinuous, and let $\omega_0 = \max(0, \omega, \mathrm{abs}(a), \mathrm{abs}(k))$. By Theorem 2.1.5(i), we have that, for every $\lambda \in \mathbb{C}$ with $\operatorname{Re}\lambda > \omega_0$ and $\tilde{k}(\lambda) \neq 0$, the operator $I - \tilde{a}(\lambda)A$ is injective and $R(C) \subseteq R(I - \tilde{a}(\lambda)A)$. Furthermore, $\tilde{k}(\lambda)(I - \tilde{a}(\lambda)A)^{-1}Cx = \int_0^\infty e^{-\lambda t}R(t)x\,dt$, $x \in E$, $\operatorname{Re}\lambda > \omega_0$, $\tilde{k}(\lambda) \neq 0$, which implies that, for every $p \in \circledast$, there exist $c_p > 0$ and $q_p \in \circledast$ such that

$$p\Big(\tilde{k}(\lambda)(I-\tilde{a}(\lambda)A)^{-1}Cx\Big) \leqslant \frac{c_p}{\operatorname{Re}\lambda-\omega}q_p(x),\ \operatorname{Re}\lambda>\omega_0,\ \ \tilde{k}(\lambda)\neq 0, x\in E. \tag{32}$$

Now the proof of (31) in case $x \in D(A)$ follows from (32), the equality $\lim_{\lambda\to+\infty,\tilde{k}(\lambda)\neq 0}\lambda\tilde{k}(\lambda) = k(0)$ (see [20, Proposition 4.1.3]) and the identity

$$(I - \tilde{a}(\lambda)A)^{-1}\,Cx = \tilde{a}(\lambda)\,(I - \tilde{a}(\lambda)A)^{-1}\,CAx + Cx,\ \operatorname{Re}\lambda > \omega_0,\ \tilde{k}(\lambda) \neq 0.$$

The proof of (31) in case $x \in \overline{D(A)}$ follows from the standard limit procedure. □

Suppose that A is a subgenerator of a locally equicontinuous (a, k)-regularized C-resolvent family $(R(t))_{t\in[0,\tau)}$ satisfying the equality (22) for all $t \in [0, \tau)$ and $x \in E$. Given $s \in [0, \tau)$ and $x \in E$, set $u(t) := R(t)R(s)x - R(s)R(t)x$, $t \in [0, \tau)$. Then it is not difficult to prove that $u \in C([0, \tau) : E)$ and $A\int_0^t a(t-s)u(s)\,ds = u(t)$, $t \in [0, \tau)$. Using the proof of [382, Theorem 2.7], it follows that $\int_0^t k(t-s)Cu(s)\,ds = 0$, $t \in [0, \tau)$. Since $k(t)$ is a kernel and C is injective, we obtain that $R(t)R(s) = R(s)R(t)$, $t, s \in [0, \tau)$.

A nonnegative infinitely differentiable function $\varphi : (0, \infty) \to \mathbb{R}$ is called a Bernstein function if the function $\varphi'(\cdot)$ is completely monotonic, i.e., $(-1)^n \varphi^{(n+1)}(t) \geqslant 0$, $n \in \mathbb{N}_0$, $t > 0$. Following [463, Definition 4.4], a function $a : (0, \infty) \to \mathbb{R}$ is said to be a creep function if $a(t)$ is nonnegative, nondecreasing and concave. A creep function $a(t)$ has the standard form

$$a(t) = a_0 + a_\infty t + \int_0^t a_1(t)\,dt,$$

where $a_0 = a(0+) \geqslant 0$, $a_\infty = \lim_{t\to\infty} a(t)/t = \inf_{t>0} a(t)/t \geqslant 0$, and $a_1(t) = a'(t) - a_\infty$ is nonnegative, nonincreasing and $\lim_{t\to\infty} a_1(t) = 0$.

In the subsequent theorem ([463], [49], [382], [302]-[303]) we analyze subordination principles in sequentially complete locally convex spaces. The n-th convolution power of the kernel $a(t)$ is denoted by $a^{*,n}(t)$.

Theorem 2.1.8. (i) *Let $a(t)$, $b(t)$ and $c(t)$ satisfy (P1), and let $\int_0^\infty e^{-\beta t}|b(t)|\,dt < \infty$ for some $\beta \geqslant 0$. Let*

$$\alpha = \tilde{c}^{-1}\left(\frac{1}{\beta}\right) \text{ if } \int_0^\infty c(t)\,dt > \frac{1}{\beta},\ \alpha = 0 \text{ otherwise,}$$

and let $\tilde{a}(\lambda) = \tilde{b}(\frac{1}{\tilde{c}(\lambda)})$, $\lambda \geqslant \alpha$. Suppose that A is a subgenerator of a (b, k)-regularized C-resolvent family $(R_b(t))_{t\geqslant 0}$ satisfying that the family $\{e^{-\omega_b t}R_b(t) : t \geqslant 0\}$ is equicontinuous for some $\omega_b \geqslant 0$, and let (H2) or (H3) hold. Assume, further, that $c(t)$ is completely positive and that there exists a function $k_1(t)$ satisfying (P1) and

$$\tilde{k}_1(\lambda) = \frac{1}{\lambda\tilde{c}(\lambda)}\tilde{k}\left(\frac{1}{\tilde{c}(\lambda)}\right),\quad \lambda > \omega_0,\ \tilde{k}\left(\frac{1}{\tilde{c}(\lambda)}\right) \neq 0, \qquad \text{for some } \omega_0 > 0.$$

Let

$$\omega_a = \tilde{c}^{-1}\left(\frac{1}{\omega_b}\right) \text{ if } \int_0^\infty c(t)\,dt > \frac{1}{\omega_b},\ \omega_a = 0 \text{ otherwise.}$$

Then, for every $r \in (0, 1]$, A is a subgenerator of a global $(a, k_1 * g_r)$-regularized C-resolvent family $(R_r(t))_{t\geqslant 0}$ such that the family $\{e^{-\omega_a t}R_r(t) : t \geqslant 0\}$ is equicontinuous and the mapping $t \mapsto R_r(t)$, $t \geqslant 0$ is locally Hölder continuous with exponent r, if $\omega_b = 0$ or $\omega_b\tilde{c}(0) \neq 1$, resp., for every $\varepsilon > 0$, there exists $M_\varepsilon \geqslant 1$ such that the family $\{e^{-\varepsilon t}R_r(t) : t \geqslant 0\}$ is equicontinuous and the mapping $t \mapsto R_r(t)$, $t \geqslant 0$ is locally Hölder continuous with exponent r, if $\omega_b > 0$ and $\omega_b\tilde{c}(0) = 1$. Furthermore, if A is densely defined, then A is a subgenerator of a global (a, k_1)-regularized C-resolvent family $(R(t))_{t\geqslant 0}$ such that the family $\{e^{-\omega_a t}R_r(t) : t \geqslant 0\}$ is equicontinuous, resp., for every $\varepsilon > 0$, the family $\{e^{-\varepsilon t}R(t) : t \geqslant 0\}$ is equicontinuous.

(ii) *Suppose $\alpha \geqslant 0$, A is a subgenerator of a global exponentially equicontinuous α-times integrated C-semigroup, $a(t)$ is completely positive and satisfies (P1), $k(t)$ satisfies (P1) and $\tilde{k}(\lambda) = \tilde{a}(\lambda)^\alpha$, λ sufficiently large. Then, for every $r \in (0, 1]$, A is a subgenerator of a locally Hölder continuous (with exponent r), exponentially equicontinuous $(a, k * g_r)$-regularized C-resolvent family ($(a, a^{*,n} * g_r)$-regularized C-resolvent family if $\alpha = n \in \mathbb{N}$, resp. (a, g_r)-regularized C-resolvent family if $\alpha = 0$). If, additionally, A is densely defined, then A is a subgenerator of an exponentially equicontinuous $(a, 1 * k)$-regularized C-resolvent family ($(a, 1 * a^{*,n})$-regularized C-resolvent family if $\alpha = n \in \mathbb{N}$, resp. (a, C)-regularized resolvent family if $\alpha = 0$).*

(iii) *Suppose $\alpha \geqslant 0$ and A is a subgenerator of an exponentially equicontinuous α-times integrated C-cosine function. Let $L^1_{loc}([0,\infty)) \ni c$ be completely positive, and let $a(t) = (c*c)(t)$, $t \geqslant 0$. (Given $L^1_{loc}([0,\infty)) \ni a$ in advance, such a function $c(t)$ always exists provided $a(t)$ is completely positive or $a(t) \neq 0$ is a creep function* *and $a_1(t)$ is log-convex.) Assume $k(t)$ satisfies (P1) and $\tilde{k}(\lambda) = \tilde{c}(\lambda)^{\alpha}/\lambda$, λ sufficiently large. Then, for every $r \in (0, 1]$, A is a subgenerator of a locally Hölder continuous (with exponent r), exponentially equicontinuous $(a, k * g_r)$-regularized C-resolvent family ($(a, c^{*n} * g_r)$-regularized C-resolvent family if $\alpha = n \in \mathbb{N}$, resp. (a, g_{r+1})-regularized C-resolvent family if $\alpha = 0$). If, additionally, A is densely defined, then A is a subgenerator of an exponentially equicontinuous $(a, 1 * k)$-regularized C-resolvent family ($(a, 1 * c^{*,n})$-regularized C-resolvent family if $\alpha = n \in \mathbb{N}$, resp. (a, C)-regularized resolvent family if $\alpha = 0$).*

The main problem in application of subordination principles in locally convex spaces lies in the fact that a great number of differential operators do not generate *exponentially equicontinuous* C-regularized semigroups (cosine functions) in SCLCSs. In the Banach space setting, Theorem 2.1.8 can be applied in the analysis of the problem of heat conduction in materials with memory and the Rayleigh problem of viscoelasticity ([463], [302], [292]).

Example 2.1.9. Denote by A_p the realization of the Laplacian with Dirichlet or Neumann boundary conditions on $L^p([0, \pi]^n)$, $1 \leqslant p < \infty$. By [266, Theorem 4.2], A_p generates an exponentially bounded α-times integrated cosine function for every $\alpha \geqslant (n-1)|\frac{1}{2}-\frac{1}{p}|$. In what follows, we employ the notation given in [463]. Assume $c \in BV_{loc}([0, \infty))$ and $m(t)$ is a bounded creep function with $m_0 = m(0+) > 0$. Thanks to [463, Proposition 4.4, p. 94], we have that there exists a completely positive function $b(t)$ such that $dm * b = 1$. After the usual procedure, the problem [463, (5.34)] describing heat conduction in materials with memory is equivalent to

(33) $$u(t) = (a * A_p)(t) + f(t),\ t \geqslant 0,$$

where $a(t) = (b * dc)(t)$, $t \geqslant 0$ and $f(t)$ contains $r * b$ as well as the temperature history. Assume that:

(i) $p \neq 2$, (ii) $\Gamma_b = \emptyset$ or $\Gamma_f = \emptyset$, and
(iii) there exists a completely positive function $c_1(t)$ such that $a(t) = (c_1 * c_1)(t)$, $t \geqslant 0$.

We refer the reader to [463, pp. 140–141] for the analysis of the problem (33) in the case: $p = 2$ and $m, c \in \mathcal{BF}$. Applying Theorem 2.1.8(iii), one gets that A_p is the integral generator of an exponentially bounded $\left(a,\ 1 * \mathcal{L}^{-1}\left(\frac{1}{\lambda}\tilde{c}_1(\lambda)^{(n-1)|\frac{1}{2}-\frac{1}{p}|}\right)(t)\right)$-regularized resolvent family. Notice also that [463, Lemma 4.3, p. 105] implies that, for every $\beta \in [0, 1]$, the function $\lambda \mapsto \tilde{c}_1(\lambda)^{\beta}/\lambda$ is the Laplace transform of a Bernstein function, and that the function $k(t)$ appearing in the formulations of Theorem 2.1.8(ii)–(iii) always exists (provided $\alpha > 0$ in (ii)). On the other hand, we have that there exists $\omega > 0$ such that A_p is the integral generator of an exponentially bounded $(\omega - A_p)^{-\lceil\frac{1}{2}(n-1)|\frac{1}{2}-\frac{1}{p}|\rceil}$-regularized cosine function. By applying Theorem 2.1.8(iii) once again, we have that A_p is the integral generator of an exponentially bounded $(a, (\omega - A_p)^{-\lceil \frac{1}{2}(n-1)|\frac{1}{2}-\frac{1}{p}| \rceil})$-regularized resolvent family, which implies that [292, Theorem 2.1.27(x)] can be applied. In both approaches, regrettably, we must restrict ourselves to the study of pure Dirichlet or Neumann problem. As mentioned above, Theorem 2.1.8(iii) is applicable in the analysis of the Rayleigh problem of viscoelasticity, which can be generally described in the form of the following abstract Cauchy problem (cf. [463, (5.45), p.136]):

$$(RP):\begin{cases} u_t(t,x) = \int_0^t da(s)u_{xx}(t-s,x) + f(t,x),\ t,\quad x > 0, \\ u(t,0) = g(t),\ t > 0\ ;\ u(0,x) = u_0(x),\ x > 0, \end{cases}$$

about which we assume that $a(t)$ is a creep function with $a_1(t)$ being log-convex. The state Banach space in [463] is chosen to be $C_0([0,\infty))$ or $L^q([0,\infty))$ for some $1 \leqslant q < \infty$. The operator A defined by $D(A) := \{u \in L^\infty([0,\infty)) : u', u'' \in L^\infty([0,\infty)), u(0) = 0\}$ and $Au := \Delta u$, $u \in D(A)$, generates an exponentially bounded α-times integrated cosine function in $L^\infty([0,\infty))$ for all $\alpha > 0$, so that (RP) can be also considered in the space of essentially bounded functions on $[0,\infty)$. Since, for every $\alpha > 0$, the operator $Au(x) := u''(x)$, $x \in [0,1]$,

$$u \in D(A) := \{u \in L^\infty[0,1] : u', u'' \in L^\infty[0,1],\ u(0) = u'(1) = 0\},$$

generates a polynomially bounded α-times integrated cosine function $(C_\alpha(t))_{t\geqslant 0}$ in $L^\infty[0,1]$ (the explicit formula for $(C_\alpha(t))_{t\geqslant 0}$ can be obtained following the lines of the proof of [98, Theorem 3.5]), we are in a position to apply Theorem 2.1.8(iii) in the analysis of motion for the axial extension of a viscoelastic rod [463, (5.49), p.138].

We would like to recommend for the reader the following problems.

Example 2.1.10. (i) In [463, Example 1.2], J. Prüss has considered the following initial-boundary value problem in $E := L^2[0,2\pi]$, with $0 < \alpha < 1$,

$$(34)\qquad \begin{aligned} &u_t(t,x) = u_x(t,x) + \int_0^t g_\alpha(t-s)u_x(s,x)\,ds, \quad t \geqslant 0,\ x \in [0,2\pi], \\ &u(0,x) = u_0(x),\ x \in [0,2\pi]\ ; \qquad u(t,0) = u(t,2\pi),\ t \geqslant 0. \end{aligned}$$

After integration with respect to variable t (reasoning in such a way, one can lose some valuable information on the qualitative properties of (34)), the above problem can be rewritten in the following form:

$$(35)\qquad u(t) = u(0) + \big((1 + g_{\alpha+1}) * Au\big)(t), \quad t \geqslant 0,$$

where $Au := u'$ with $D(A) := \{u \in W^{1,2}[0,2\pi] : u(0) = u(2\pi)\}$. It has been proved in [463] that there does not exist an exponentially bounded $(1 + g_{\alpha+1}, 1)$-regularized resolvent family for (35). Observe, however, that it is not clear whether there exist an injective operator $C \in L(E)$ and a kernel $k(t)$ such that there exists an exponentially bounded $(1 + g_{\alpha+1}, k)$-regularized C-resolvent family for (35).

(ii) The following initial value problem describes the angular displacement of a rod of length $l > 0$ and of uniform circular cross section with radius $r > 0$, cf. [463, p.137] for a more complete discussion:

$$\begin{cases} u_t(t,x) = \int_0^t da(s)u_{xx}(t-s,x) + h(t,x), \quad t > 0, x \in (0,1), \\ u(t,0) = 0\ ;\ u(0,x) = u_0(x), \quad t > 0, x \in (0,1), \\ \beta u_t(t,1) = -\int_0^t da(s)u_x(t-s,1) + g(t), \end{cases}$$

where $\beta \geqslant 0$, $h(t, x)$ denotes the given strain history as well as the present distributed torque, and $g(t)$ similarly corresponds to the given boundary strain history and the torque applied to the tip mass. In the case that $\beta > 0$, the above problem can be formulated as the problem (1) with $E := L^2[0, 1] \times \mathbb{C}$ and the operator A being defined by $D(A) := \{(u, \vartheta) \in E : u \in W^{2,2}[0, 1], u(0) = 0, u(1) = \vartheta\}$ and $A(u, \vartheta) := (u'', -\beta^{-1}u'(1))$, $(u, \vartheta) \in D(A)$. Herein E is a complex Hilbert space with respect to the inner product

$$((u_1, \vartheta_1), (u_2, \vartheta_2)) := \int_0^1 u_1(x)\overline{u_2(x)}\, dx + \beta\vartheta_1\overline{\vartheta_2},$$

and A is both selfadjoint and negative definite in this space. Hence, A generates a bounded cosine function so that Theorem 2.1.8(iii) is susceptible to applications in the case that $a(t)$ is a creep function with $a_1(t)$ log-convex. It could be of interest to know in which classes of non-Hilbert spaces the problem of torsion of a rod can be considered.

In the following theorem we correct some mistakes stated in the formulations of [302, Theorem 2.9, Theorem 2.10 and Theorem 2.11]:

Theorem 2.1.11. (i) *Suppose that the following conditions hold:*

(a) *The mapping* $t \mapsto |k(t)|$, $t \in [0, \tau)$ *is non-decreasing.*

(b) *There exist* $\varepsilon_{a,k} > 0$ *and* $t_{a,k} \in [0, \tau)$ *such that*

$$\left| \int_0^1 a(t-s)k(s)\, ds \right| \geqslant \varepsilon_{a,k} \int_0^t |a(t-s)k(s)|\, ds, \ t \in [0, t_{a,k}).$$

(c) A *is a subgenerator of an* (a, k)*-regularized* C*-resolvent family* $(R(t))_{t\in[0,\tau)}$ *satisfying (22).*

(d) *For every seminorm* $p \in \circledast$, *there exist* $c_p > 0$, $\delta_p \in (0, \tau)$ *and* $q_p \in \circledast$ *such that* $p(R(t)x/k(t)) \leqslant c_p q_p(x)$, $x \in E$, $t \in [0, \delta_p)$.

Then the following holds:

(∇) *Suppose* $x \in D(A_{\overline{D(A)}})$. *Then*

$$ACx = \lim_{t\to 0+} \frac{R(t)x - k(t)Cx}{(a*k)(t)}. \tag{36}$$

(Δ) *Suppose* $x \in \overline{D(A)}$ *and* $\lim_{t\to 0+} \frac{R(t)x-k(t)Cx}{(a*k)(t)}$ *exists. Then* $Cx \in D(A_{\overline{D(A)}})$ *and (36) holds.*

(ii) *Suppose* $\min(a(t), k(t)) > 0$, $t \in (0, \tau)$ *and A is a subgenerator of an* (a, k)*-regularized C-resolvent family* $(R(t))_{t\in[0,\tau)}$ *satisfying that the item (i)(d) of this theorem and (22) hold. Then we have* (∇)-(Δ).

(iii) *Suppose* $\min(a(t), k(t)) > 0$, $t \in (0, \tau)$ *and A is a subgenerator of an* (a, k)*-regularized C-resolvent family* $(R(t))_{t\in[0,\tau)}$ *satisfying (i)(d). Then* $\lim_{t\to 0+} \frac{(a*R)(t)x}{(a*k)(t)} = Cx$, $x \in \overline{D(A)}$.

(iv) *Suppose* $\min(a(t), k(t)) > 0$, $t \in (0, \tau)$ *and A is a subgenerator of an* (a, k)*-regularized C-resolvent family* $(R(t))_{t\in[0,\tau)}$ *satisfying (i)(d) and (22). Let* $x \in \overline{D(A)}$, $y \in E$ *and* $\lim_{t\to 0+} \frac{R(t)x-k(t)Cx}{(a*k)(t)} = y$. *Then* $Cx \in D(A)$ *and* $y = ACx$.

(v) *Suppose* $\min(a(t), k(t)) > 0$, $t \in (0, \tau)$, *E is semi-reflexive, A is a subgenerator of an* (a, k)*-regularized C-resolvent family* $(R(t))_{t\in[0,\tau)}$ *satisfying that* $R(s)R(t) = R(t)R(s)$, $0 \leqslant t, s < \tau$ *and (i)(d) holds. Let (22) hold, let* $x \in \overline{D(A)}$ *and let a zero sequence* (t_n) *be such that the set* $\{\frac{R(t_n)x-k(t_n)Cx}{(a*k)(t_n)} : n \in \mathbb{N}\}$ *is bounded. Then* $Cx \in D(A)$.

(vi) *Suppose* $\alpha > 0$ *and A is a subgenerator of an α-times integrated C-semigroup* $(S_\alpha(t))_{t\in[0,\tau)}$, *resp. an α-times integrated C-cosine function* $(C_\alpha(t))_{t\in[0,\tau)}$, *which satisfies that for every seminorm* $p \in \circledast$, *there exist* $c_p > 0$, $\delta_p \in (0, \tau)$ *and* $q_p \in \circledast$ *such that* $p(S_\alpha(t)x/t^\alpha) \leqslant c_p q_p(x)$, $x \in E$, $t \in (0, \delta_p)$, *resp.* $p(C_\alpha(t)x/t^\alpha) \leqslant c_p q_p(x)$, $x \in E$, $t \in (0, \delta_p)$. *Then, for every* $x \in D(A)$ *such that* $Ax \in \overline{D(A)}$:

$$CAx = \lim_{t\to 0+} \frac{\Gamma(\alpha+2)}{\Gamma(\alpha+1)} \frac{\Gamma(\alpha+1)S_\alpha(t)x - t^\alpha Cx}{t^{\alpha+1}}, \text{ resp.}$$

$$CAx = \lim_{t\to 0+} \frac{\Gamma(\alpha+3)}{\Gamma(\alpha+1)} \frac{\Gamma(\alpha+1)C_\alpha(t)x - t^\alpha Cx}{t^{\alpha+2}}.$$

Sketch of Proof for (v). Set $U(t) := \frac{R(t)-k(t)C}{(a*k)(t)}$, $t \in [0, \tau)$. By the proof of [388, Theorem 2.1], one obtains that

$$U(s)(a*R)(t)x = (R(t) - k(t)C)\frac{(a*R)(s)x}{(a*k)(s)}, \quad 0 \leqslant t, s < \tau. \tag{37}$$

Keeping in mind [411, Proposition 23.18, p. 270] and the prescribed assumptions, we get that the set $\{U(t_n)x : n \in \mathbb{N}\}$ is relatively weakly compact. Therefore, there exist an element $y \in \overline{D(A)}$ and a zero sequence (t'_n) in $[0, \tau)$ such that

$$\lim_{n\to\infty} \langle x^*, U(t'_n)x\rangle = \langle x^*, y\rangle \text{ for every } x^* \in E^*. \tag{38}$$

Connecting (37)-(38) and (iii), we get that $\langle x^*, (a*R)(t)y\rangle = \langle x^*, (R(t)-k(t)C)Cx\rangle$, $x^* \in E^*$, $t \in [0, \tau)$ and

$$\frac{(R(t)-k(t)C)}{(a*k)(t)}Cx = \frac{(a*R)(t)y}{(a*k)(t)}, \quad t \in [0, \tau). \tag{39}$$

Using (iii) again, one gets $\lim_{t\to 0+} \frac{(a*R)(t)y}{(a*k)(t)} = Cy$. By (39), $\lim_{t\to 0+} \frac{R(t)-k(t)C}{(a*k)(t)} Cx = Cy$ and the claimed assertion follows from (iv).

Recall that the Hausdorff locally convex topology on E^* defines the system $(|\cdot|_B)_{B\in\mathcal{B}}$ of seminorms on E^*, where $|x^*|_B = \sup_{x\in B} |\langle x^*, x\rangle|$, $x^* \in E^*$, $B \in \mathcal{B}$. This space is sequentially complete provided that E is barreled.

Theorem 2.1.12. (i) *Suppose E is barreled, A is a subgenerator of a (local, global exponentially equicontinuous) (a, k)-regularized C-resolvent family $(R(t))_{t\in[0,\tau)}$, $D(A)$ and $R(C)$ are dense in E and $\alpha > 0$. Then A^* is a subgenerator of a (local, global exponentially equicontinuous) $(a, k * g_\alpha)$-regularized C^*-resolvent family $(R_\alpha^*(t))_{t\in[0,\tau)}$, which is given by $R_\alpha(t)x^* := \int_0^t g_\alpha(t-s)R(s)^* x^*\, ds$, $t \in [0, \tau)$, $x^* \in E^*$.*

(ii) *Suppose E is barreled, A is a subgenerator of a (local, global exponentially equicontinuous) (a, k)-regularized C-resolvent family $(R(t))_{t\in[0,\tau)}$, and $D(A)$ as well as $R(C)$ are dense in E. Then $A^*_{|\overline{D(A^*)}}$ is a subgenerator of a (local, global exponentially equicontinuous) (a, k)-regularized $C^*_{|\overline{D(A^*)}}$-resolvent family $(R(t)^*_{|\overline{D(A^*)}})_{t\in[0,\tau)}$ in $\overline{D(A^*)}$.*

(iii) *Suppose E is reflexive, $D(A)$ and $R(C)$ are dense in E, and A is a subgenerator of a (local, global exponentially equicontinuous) (a, k)-regularized C-resolvent family $(R(t))_{t\in[0,\tau)}$. Then A^* is a subgenerator of a (local, global exponentially equicontinuous) (a, k)-regularized C^*-resolvent family $(R(t)^*)_{t\in[0,\tau)}$.*

Proof. The assertion (i) can be proved as in the Banach space case [302, Theorem 2.27], the assertion (iii) is an immediate consequence of the assertion (ii), and the only non-trivial thing that should be explained in (ii) is the strong continuity of $(R(t)^*_{|\overline{D(A^*)}})_{t\in[0,\tau)}$. For the sake of simplicity, we shall consider only the global case. Let $B \in \mathcal{B}$ and $x^* \in D(A^*)$. Then

$$|R(t)^*x^* - R(s)^*x^*|_B = \sup_{x\in B}|\langle x^*, R(t)x - R(s)x\rangle|$$

$$= \sup_{x\in B\cap D(A)}\left|\left\langle x^*, (k(t)-k(s))Cx + \int_0^t a(t-r)R(r)Ax\, dr - \int_0^s a(s-r)R(r)Ax\, dr\right\rangle\right|$$

$$\leqslant |k(t) - k(s)| \sup_{x\in B\cap D(A)} |\langle x^*, Cx\rangle|$$

$$+ \sup_{x\in B\cap D(A)} \int_0^s |a(t-r) - a(s-r)||\langle A^*x^*, R(r)x\rangle|\, dr$$

$$+ \sup_{x\in B\cap D(A)} \int_s^t |a(t-r)|\, |\langle A^*x^*, R(r)x\rangle|\, dr,\ t, s \geqslant 0. \tag{40}$$

By the uniform boundedness principle, the set $\{\langle A^*x^*, R(r)x\rangle : x \in B \cap D(A),\ r \in [0, s]\}$ is bounded for every $s \geqslant 0$. Then we obtain by means of (40) that the mapping $s \mapsto R(s)^*x^*$, $s \geqslant 0$ is continuous, as required. Assume now $x^* \in \overline{D(A^*)}$, $t \geqslant 0$ and $\varepsilon > 0$. Then there exists a net $(x^*_\tau)_{\tau\in I}$ in $D(A^*)$ such that $\lim_{\tau\to\infty} x^*_\tau = x^*$ in E^*. The set $B' := \{R(r)x : r \in [0, t+1],\ x \in B\}$ is bounded and there exists $\tau_0 \in I$

such that $|x^* - x_{\tau_0}^*|_{B'} < \varepsilon/3$. By the first part of the proof, we get that there exists $\delta \in (0, 1)$ such that $|(R(t)^* - R(s)^*)x_{\tau_0}^*|_B < \varepsilon/3$ for all $s \geqslant 0$ with $s \in (t-\delta, t+\delta)$. Hence,

$$|R(t)^*x^* - R(s)^*x^*|_B \leqslant 2|x^* - x_{\tau_0}^*|_{B'} + |(R(t)^* - R(s)^*)x_{\tau_0}^*|_B < \varepsilon,$$

for all $s \geqslant 0$ with $s \in (t-\delta, t+\delta)$. This completes the proof. □

Recently, L. Wu and Y. Zhang [525] introduced a new topological concept for the purpose of researches of semigroups on L^∞-type spaces and the L^1-uniqueness of the Fokker-Planck equation (cf. also [348, Theorem 2.1, Theorem 2.2]). Let us explain in more detail the importance of such an approach in our research. One can define on E^* the topology of uniform convergence on compacts of E, denoted by $\mathcal{C}(E^*, E)$; more precisely, given a functional $x_0^* \in E^*$, the basis of open neighborhoods of x_0^* w.r.t. $\mathcal{C}(E^*, E)$ is constituted from the sets $N(x_0^* : \mathbf{K}, \varepsilon) := \{x^* \in E^* : \sup_{x\in\mathbf{K}} |\langle x^* - x_0^*, x\rangle| < \varepsilon\}$, where $\mathbf{K}$ runs over all compacts of E and $\varepsilon > 0$. Then $(E^*, \mathcal{C}(E^*, E))$ is locally convex, complete and the topology $\mathcal{C}(E^*, E)$ is finer than the topology induced by the calibration $(|\cdot|_B)_{B\in\mathcal{B}}$.

Theorem 2.1.13. *Suppose $D(A)$ and $R(C)$ are dense in E, and A is a subgenerator of a locally equicontinuous (a, k)-regularized C-resolvent family $(R(t))_{t\in[0,\tau)}$. Then A^* is a subgenerator of a locally equicontinuous (a, k)-regularized C^*-resolvent family $(R(t)^*)_{t\in[0,\tau)}$ in $(E^*, \mathcal{C}(E^*, E))$. Furthermore, if $\tau = \infty$ and $(R(t))_{t\geqslant 0}$ is exponentially equicontinuous, then $(R(t)^*)_{t\geqslant 0}$ is also exponentially equicontinuous and the exponential type of $(R(t)^*)_{t\geqslant 0}$ is less or equal than that of $(R(t))_{t\geqslant 0}$.*

Proof. By [525, Lemma 1.5] we have that, for every $t \in [0, \tau)$, $R(t)^*$ is a continuous linear operator on E^* for the topology $\mathcal{C}(E^*, E)$. Certainly, $R(0)^* = k(0)C^*$, $R(t)^*A^* \subseteq A^*R(t)^*$, $t \in [0, \tau)$ and $R(t)^*C^* = C^*R(t)^*$, $t \in [0, \tau)$. Since $(R(t))_{t\in[0,\tau)}$ is locally equicontinuous, we get that the mapping $g : [0, \tau) \times E \to E$ given by $g(t, x) := R(t)x$, $t \in [0, \tau)$, $x \in E$ is continuous. This, in turn, implies that for every $t \in [0, \tau)$, $x^* \in E^*$ and for every compact subset $\mathbf{K}$ of E, one has $\lim_{s\to t} \sup_{x\in\mathbf{K}} |\langle x^*, R(t)x - R(s)x\rangle| = 0$. Therefore, $(R(t)^*)_{t\in[0,\tau)}$ is strongly continuous. The equality $\langle x^*, R(t)x\rangle = k(0)\langle x^*, Cx\rangle + \int_0^t a(t-s)\langle x^*, R(s)x\rangle\, ds$, $t \in [0, \tau)$, $x \in D(A)$ implies that $R(t)^*x^* = k(0)C^*x^* + \int_0^t a(t-s)R(s)^*A^*x^*\, ds$, $t \in [0, \tau)$, $x^* \in D(A^*)$ and that $(R(t)^*)_{t\in[0,\tau)}$ is an (a, k)-regularized C^*-resolvent family with a subgenerator A^*. One can simply prove that $(R(t)^*)_{t\in[0,\tau)}$ is locally equicontinuous. Assume now $\mathbf{K}$ is compact, $(R(t))_{t\geqslant 0}$ is exponentially equicontinuous and $\mathbb{R} \ni \omega$ satisfies that the family $\{e^{-\omega t}R(t) : t \geqslant 0\}$ is equicontinuous. Then, for every $\omega' > \omega$, the mapping $g_{\omega'} : [0, +\infty] \times E \to E$ given by $g_{\omega'}(t, x) := e^{-\omega' t}R(t)x$, $t \in [0, +\infty)$, $x \in E$ and $g_{\omega'}(+\infty, x) := 0$, $x \in E$ is continuous ($[0, +\infty]$ is the one point compactification of $[0, +\infty)$, cf. also [525, the proof of Step (2), Theorem 1.4, p. 565]). Now it is clear that the set $\mathbf{K}' := g_{\omega'}([0, +\infty] \times \mathbf{K})$ is compact and that, for every $t \geqslant 0$ and $x^* \in E^*$,

$$\sup_{x\in\mathbf{K}} |\langle e^{-\omega' t}R(t)^*x^*, x\rangle| = \sup_{x\in\mathbf{K}} |\langle x^*, e^{-\omega' t}R(t)x\rangle| \leqslant \sup_{x\in\mathbf{K}'} |\langle x^*, x\rangle|.$$

The proof of theorem is completed through a routine argument. □

The following useful characterization of C-pseudeoresolvents in SCLCSs follows from the proofs of [144, Proposition 2.6, Remark 2.7, Proposition 2.8], a weak criterion for vector-valued analyticity [209, Theorem 1] and the Cauchy integral formula (13).

Proposition 2.1.14. *Let $\Omega \subseteq \rho_C(A)$ be open, and let $x \in E$.*

(i) *The local boundedness of the mapping $\lambda \mapsto (\lambda - A)^{-1}Cx$, $\lambda \in \Omega$, resp. the assumption that E is barreled and the local boundedness of the mapping $\lambda \mapsto (\lambda - A)^{-1}C$, $\lambda \in \Omega$, implies the analyticity of the mapping $\lambda \mapsto (\lambda - A)^{-1}C^3x$, $\lambda \in \Omega$, resp. $\lambda \mapsto (\lambda - A)^{-1}C^3$, $\lambda \in \Omega$. Furthermore, if $R(C)$ is dense in E, resp. if $R(C)$ is dense in E and E is barreled, then the mapping $\lambda \mapsto (\lambda - A)^{-1}Cx$, $\lambda \in \Omega$ is analytic, resp. the mapping $\lambda \mapsto (\lambda - A)^{-1}C$, $\lambda \in \Omega$ is analytic.*

(ii) *Suppose that $R(C)$ is dense in E. Then the local boundedness of the mapping $\lambda \mapsto (\lambda - A)^{-1}Cx$, $\lambda \in \Omega$ implies its analyticity as well as $Cx \in R((\lambda - A)^n)$, $n \in \mathbb{N}$ and*

$$\frac{d^{n-1}}{d\lambda^{n-1}}(\lambda - A)^{-1}Cx = (-1)^{n-1}(n-1)!\,(\lambda - A)^{-n}Cx,\ n \in \mathbb{N}. \tag{41}$$

Furthermore, if E is barreled, then the local boundedness of the mapping $\lambda \mapsto (\lambda - A)^{-1}C$, $\lambda \in \Omega$ implies its analyticity as well as $R(C) \subseteq R((\lambda - A)^n)$, $n \in \mathbb{N}$ and

$$\frac{d^{n-1}}{d\lambda^{n-1}}(\lambda - A)^{-1}C = (-1)^{n-1}(n-1)!\,(\lambda - A)^{-n}C \in L(E),\ n \in \mathbb{N}. \tag{42}$$

(iii) *The continuity of mapping $\lambda \mapsto (\lambda - A)^{-1}Cx$, $\lambda \in \Omega$ implies its analyticity and (41). Furthermore, if E is barreled, then the continuity of mapping $\lambda \mapsto (\lambda - A)^{-1}C$, $\lambda \in \Omega$ implies its analyticity and (42).*

Without any substantial difficulties, one can prove the following analogues of [463, Proposition 0.1, Theorem 0.4, pp. 10-12] in SCLCSs.

Theorem 2.1.15. (i) *Assume $g : \mathbb{C}_+ \to E$ is analytic and satisfies that the sets $\{\lambda g(\lambda) : \lambda \in \mathbb{C}_+\}$ and $\{\lambda^2 g'(\lambda) : \lambda \in \mathbb{C}_+\}$ are bounded. Then the set $\{n!^{-1}\lambda^{n+1}g^{(n)}(\lambda) : \lambda \in \mathbb{C}_+,\ n \in \mathbb{N}_0\}$ is bounded as well.*

(ii) *Assume $k \in \mathbb{N}_0$, $g : \mathbb{C}_+ \to E$ is analytic and satisfies that the set $\{\lambda^{n+1}g^{(n)}(\lambda) : \lambda \in \mathbb{C}_+,\ 0 \leqslant n \leqslant k+1\}$ is bounded. Then there exists a function $u \in C^k((0, \infty) : E)$ such that $g(\lambda) = \tilde{u}(\lambda)$, $\lambda \in \mathbb{C}_+$ and the sets $\{t^n u^{(n)}(t) : t > 0,\ 0 \leqslant n \leqslant k\}$ and $\{(t-s)^{-1}(1 + \ln\frac{t}{t-s})(t^{k+1}u^{(k)}(t) - s^{k+1}u^{(k)}(s)) : 0 \leqslant s < t < \infty\}$ are bounded.*

The proof of the following extension of [49, Proposition 3.8] is given for the sake of completeness.

Proposition 2.1.16. *Suppose $\beta \in (0, 2]$, $\Sigma_{\frac{\beta\pi}{2}} \subseteq \rho_C(A)$, the family $\{|\lambda|^{-1}(\lambda - A)^{-1}C : \lambda \in \Sigma_{\frac{\beta\pi}{2}}\}$ is equicontinuous and, for every $x \in E$, the mapping $\lambda \mapsto (\lambda - A)^{-1}Cx$, $\lambda \in \Sigma_{\frac{\beta\pi}{2}}$ is continuous. Then, for every $r \in (0, 1]$, A is the integral generator of a global*

$(g_\beta, g_{\alpha+r+1})$-regularized C^2-resolvent family $(S_r(t))_{t\geqslant 0}$ which satisfies that, for every $p \in ⊛$, there exist $c_p > 0$ and $q_p \in ⊛$ such that

$$p\big(S_r(t+h)x - S_r(t)x\big) \leqslant \frac{c_p q_p(x)}{r\Gamma(r)} h^r, \; t \geqslant 0, \;\; h > 0, \;\; x \in E,$$

and the mapping $t \mapsto p_B(S_r(t))$, $t \geqslant 0$ is locally Hölder continuous with exponent r for every $p \in ⊛$ and $B \in \mathcal{B}$; if A is densely defined, then A is the integral generator of a global equicontinuous $(g_\beta, g_{\alpha+1})$-regularized C^2-resolvent family $(S(t))_{t\geqslant 0}$.

Proof. Let $x \in E$ and $p \in ⊛$. By Proposition 2.1.14(iii), the mapping $g_x : \mathbb{C}_+ \to E$ given by $g_x(\lambda) := \lambda^{\alpha-\beta-1}(\lambda^\beta - A)^{-1}C^2x$, $\lambda \in \mathbb{C}_+$ is analytic and, for every $\lambda \in \mathbb{C}_+$ and $x \in E$, the following holds:

$$(43)\; \lambda^2 \frac{d}{d\lambda} g_x(\lambda) = (\beta - \alpha - 1)\lambda^{\beta-\alpha}(\lambda^\beta - A)^{-1}C^2x - \beta\lambda^{2\beta-\alpha}\Big(\big(\lambda^\beta - A\big)^{-1}C\Big)^2 x.$$

If we avail ourselves of (43) and the equicontinuity of family $\{|\lambda|^{-1}(\lambda - A)^{-1}C : \lambda \in \Sigma_{\frac{\beta\pi}{2}}\}$, we obtain that there exist $c_p > 0$ and $q_p \in ⊛$, independent of x, such that

$$(44) \qquad p\big(\lambda g_x(\lambda)\big) + p\left(\lambda^2 \frac{d}{d\lambda} g_x(\lambda)\right) \leqslant c_p q_p(x), \; \lambda \in \mathbb{C}_+, \; x \in E.$$

By (44) and the proof of [463, Proposition 0.1], we get that:

$$p\left(\lambda^{n+1} \frac{d^n}{d\lambda^n} g_x(\lambda)\right) \leqslant n! c_p q_p(x) e / 2\pi, \; \lambda \in \mathbb{C}_+, \; x \in E, \; n = 2, 3, \cdots.$$

The previous inequality combined with Theorem 2.1.6(iii)-(iv) completes the proof.

□

The following theorem is an extension of [292, Proposition 2.3.12-Proposition 2.3.13].

Theorem 2.1.17. *Suppose $l \in \mathbb{N}$, $z \in \rho_C(A)$, A is a subgenerator of a (local) (a, k)-regularized $((z - A)^{-1}C)$-resolvent family $(S_{a,k}(t))_{t\in[0,\tau)}$ on E, and:*

$$(45) \qquad (z - A)^{-1}C \in L(E), \cdots, (z - A)^{-l}C \in L(E).$$

Set, for every $x \in E$ and $t \in [0, \tau)$,

$$S_{a,k,l}(t)x := z^l(a^{*,l} * S_{a,k}(\cdot)x)(t) + \sum_{j=1}^{l} (-1)^j \binom{l}{j} z^{(l-j)}$$
$$\times \big[(a^{*,l-j} * S_{a,k}(\cdot)x)(t) - (k * a^{*,l-j})(t)(z - A)^{-l}Cx$$
$$-(k * a^{*,l-j+1})(t)A(z - A)^{-l}Cx - \cdots - (k * a^{*,l-1})(t)A^{j-1}(z - A)^{-l}Cx\big].$$

Then the following holds:

(i) $(S_{a,k,l}(t))_{t\in[0,\tau)}$ *is an* $(a, k * a^{*,l})$*-regularized C-resolvent family with a subgenerator* A.

(ii) *If*

(46) $$A\int_0^t a(t-s)S_{a,k}(s)x\,ds = S_{a,k}(t)x - (z-A)^{-l}\,Cx,\ x\in E,\ t\in[0,\tau),$$

then

$$S_{a,k,l}(t)x = (z-A)^l\int_0^t a^{*,l}(t-s)S_{a,k,l}(s)x\,ds,\ x\in E,\ t\in[0,\tau),$$

and

(47) $$A\int_0^t a(t-s)S_{a,k,l}(s)x\,ds = S_{a,k,l}(t)x - (k * a^{*,l})\,(t)Cx,\ x\in E,\ t\in[0,\tau).$$

(iii) *If* $(S_{a,k}(t))_{t\in[0,\tau)}$ *is locally equicontinuous (globally exponentially equicontinuous), then* $(S_{a,k,l}(t))_{t\in[0,\tau)}$ *is likewise locally equicontinuous (globally exponentially equicontinuous)*.

Proof. Using (45) and the binomial formula, it simply follows that $A^{j-1}(z-A)^{-1}C \in L(E)$, $1 \leqslant j \leqslant l$, which implies that $S_{a,k,l}(t)\in L(E)$, $t\in[0,\tau)$. It is checked at once that $S_{a,k,l}(0) = 0$ as well as that $S_{a,k,l}(t)A \subseteq AS_{a,k,l}(t)$, $t\in[0,\tau)$ and $S_{a,k,l}(t)C = CS_{a,k,l}(t)$, $t\in[0,\tau)$. Furthermore, $(S_{a,k,l}(t))_{t\in[0,\tau)}$ is strongly continuous and, because of that, it suffices to show that, for every fixed $x\in D(A)$,

$$\int_0^t a(t-s)S_{a,k,l}(s)Ax\,ds = S_{a,k,l}(t)x - (k * a^{*,l})\,(t)Cx,\ t\in[0,\tau).$$

Using the standard computation involving the functional equation $\int_0^t a(t-s)S_{a,k}(s)Ax\,ds = S_{a,k}(t)x - (z-A)^{-l}Cx$, $t\in[0,\tau)$, we will only have to prove that:

(48) $$Cx = \sum_{j=1}^{l}(-1)^j\binom{l}{j}z^{l-j}A^{j-1}(z-A)^{-l}C\,Ax + z^l(z-A)^{-l}Cx.$$

By the binomial formula, the above equality holds if x is replaced by $(z-A)^{-l}Cx$. Making use of this fact as well as the equalities $A^{j-1}(z-A)^{-l}CA(z-A)^{-l}Cx = (z-A)^{-l}C(A^{j-1}(z-A)^{-l}CAx)$ and $C(z-A)^{-l}Cx = (z-A)^{-l}C^2x$, we obtain (48). The proof of (i) is thereby completed. The remaining part of the proof is simple and therefore omitted.

Corollary 2.1.18. *Suppose* $\alpha > 0$, $l\in\mathbb{N}$, $z\in\rho_C(A)$, *A is a subgenerator of a (local)* $(g_\alpha, (z-A)^{-l}C)$*-regularized resolvent family* $(S_\alpha(t))_{t\in[0,\tau)}$ *on E, and (45) holds. Set, for every* $x\in E$ *and* $t\in[0,\tau)$,

$$S_{l,\alpha}(t)x := z^l(g_{l\alpha} * S_\alpha(\cdot)x)(t) + \sum_{j=1}^{l}(-1)^j \binom{l}{j} z^{(l-j)}$$
$$\times \Big[(g_{(l-j)\alpha} * S_\alpha(\cdot)x)(t) - g_{(l-j)\alpha+1}(t)(z-A)^{-l}Cx$$
$$- g_{(l-j+1)\alpha+1}(t)A(z-A)^{-l}Cx - \cdots - g_{(l-1)\alpha+1}(t)A^{j-1}(z-A)^{-l}Cx\Big].$$

Then the following holds:

(i) $(S_{l,\alpha}(t))_{t\in[0,\tau)}$ *is a* $(g_\alpha, g_{l\alpha+1})$*-regularized C-resolvent family with a subgenerator* A.

(ii) *If*

$$A\int_0^t g_\alpha(t-s)S_\alpha(s)x\,ds = S_\alpha(t)x - (z-A)^{-l}Cx,\ x\in E,\ t\in[0,\tau), \tag{49}$$

then

$$S_{l,\alpha}(t)x = (z-A)^l \int_0^t g_{l\alpha}(t-s)S_\alpha(s)x\,ds,\ x\in E,\ t\in[0,\tau),$$

and

$$A\int_0^t g_\alpha(t-s)S_{l,\alpha}(s)x\,ds = S_{l,\alpha}(t)x - g_{l\alpha+1}(t)Cx,\ x\in E,\ t\in[0,\tau). \tag{50}$$

(iii) *If* $(S_\alpha(t))_{t\in[0,\tau)}$ *is locally equicontinuous (globally exponentially equicontinuous), then* $(S_{l,\alpha}(t))_{t\in[0,\tau)}$ *is likewise locally equicontinuous (globally exponentially equicontinuous).*

Now we consider the situation in which $(S_{a,k,l}(t))_{t\in[0,\tau)}$ is given in advance.

Theorem 2.1.19. *Suppose* $l\in\mathbb{N}$, $r_0 > \max(0, abs(a), abs(k))$, $z\in\rho_C(A)$, *A is a subgenerator of a (local)* $(a, k*a^{*,l})$*-regularized C-resolvent family* $(S_{a,k,l}(t))_{t\in[0,\tau)}$ *on E, and (45) holds. Let $a(t)$ and $k(t)$ satisfy (P1), let* $\lim_{r\to\infty}\tilde{a}(r) = 0$, $|z\tilde{a}(r)| < 1$, $r > r_0$ *and let the following conditions hold:*

(a) *For every* $j = 1, \cdots, l$, *there exists a continuous function* $t\mapsto F_j(t)$, $t\geqslant 0$ *satisfying* $F_j(0) = \delta_{j,l}$ *and*

$$\tilde{F}_j(r) = \frac{\tilde{k}(r)\tilde{a}(r)^{l-j}}{(1-z\tilde{a}(r))^{l+1-j}},\quad r>r_0,\ \tilde{a}(r)\neq 0, \tag{51}$$

where $\delta_{j,l}$ *denotes the Kronecker's delta.*

(b) *There exists a function $G(t)$ satisfying (P1) and:*

$$\tilde{G}(r) = (-1)^l(1-z\,\tilde{a}(r))^{-l} + (-1)^{l+1},\ r>r_0,\ \tilde{a}(r)\neq 0. \tag{52}$$

Set, for every $x\in E$ *and* $t\in[0,\tau)$,

$$S_{a,k}(t)x := (-1)^l S_{a,k,l}(t)x + (G * S_{a,k,l}(\cdot)x)(t) + \sum_{j=1}^{l}(-1)^{l-j}F_j(t)(z-A)^{-j}Cx.$$

Then the following holds:

(i) $(S_{a,k}(t))_{t\in[0,\tau)}$ *is an (a, k)-regularized* $((z-A)^{-l}C)$*-resolvent family with a subgenerator A.*
(ii) *Assume* $R(C)\subseteq D((z-A)^{-(l+1)})$ *and (47). Then (46) holds.*
(iii) *If* $(S_{a,k,l}(t))_{t\in[0,\tau)}$ *is locally equicontinuous, then* $(S_{a,k}(t))_{t\in[0,\tau)}$ *is likewise locally equicontinuous.*
(iv) *Suppose* $\tau=\infty$, $M\geqslant 1$, $\omega\geqslant 0$,

$$\int_0^t |G(s)|\,ds+\sum_{j=1}^{l}|F_j(t)|\leqslant Me^{\omega t},\ t\geqslant 0,$$

and $(S_{a,k,l}(t))_{t\geqslant 0}$ *is exponentially equicontinuous. Then* $(S_{a,k}(t))_{t\geqslant 0}$ *is likewise exponentially equicontinuous.*

Proof. We will only prove the assertion (i). It is clear that the prescribed assumptions imply that $(S_{a,k}(t))_{t\in[0,\tau)}$ is a strongly continuous operator family as well as that $S_{a,k}(0)=(z-A)^{-l}C$, $S_{a,k}(t)A\subseteq AS_{a,k}(t)$, $t\in[0,\tau)$ and $S_{a,k}(t)((z-A)^{-l}C)=((z-A)^{-l}C)S_{a,k}(t)$, $t\in[0,\tau)$. It remains to be proved that, for every fixed $x\in D(A)$,

$$S_{a,k}(t)x-k(t)\,(z-A)^{-l}Cx=\int_0^t a(t-s)S_{a,k}(s)Ax\,ds,\ t\in[0,\tau).$$

With the help of Laplace transform, a straightforward computation involving the functional equation of $(S_{a,k,l}(t))_{t\in[0,\tau)}$, the resolvent equation and (51)-(52) indicates that the above equality holds if, for any $r>r_0$ with $\tilde{a}(r)\neq 0$:

$$\begin{aligned}\sum_{j=1}^{l}(-1)^{l-j}\frac{\tilde{k}(r)\tilde{a}(r)^{l-j}}{\left(1-z\tilde{a}(r)\right)^{l+1-j}}(z-A)^{-j}Cx-\tilde{k}(r)(z-A)^{-l}Cx\\ =(-1)^{l+1}\left(1-z\tilde{a}(r)\right)^{-l}\tilde{k}(r)\tilde{a}(r)^{l}Cx\\ +\sum_{j=1}^{l}(-1)^{l-j}\frac{\tilde{k}(r)\tilde{a}(r)^{l+1-j}}{\left(1-z\tilde{a}(r)\right)^{l+1-j}}\left[z(z-A)^{-j}Cx-(z-A)^{-(j-1)}Cx\right].\end{aligned}$$

It can be easily seen that the coefficients of $(z-A)^{-j}Cx$ $(0\leqslant j\leqslant l)$ on both sides of the previous equality are equal. The conclusion in the theorem follows from this. □

Due to Theorem 1.3.1 we have that, for every $\alpha>1$, there exist $b_\alpha\geqslant 1$ and $c_\alpha\geqslant 1$ such that:

(53) $$E_{\alpha,\alpha}(t)\leqslant b_\alpha t^{(1-\alpha)/\alpha}\exp(t^{1/\alpha}),\ t>0 \text{ and } E_\alpha(t)\leqslant c_\alpha\exp(t^{1/\alpha}),\ t\geqslant 0.$$

Let us focus now on the situation of Theorem 2.1.17 with $a(t)=g_\alpha(t)$ and $k(t)=1$ $(\alpha>0)$. Noticing that

$$\frac{r^{\alpha j}}{(r^{\alpha}-z)^{l}}=\sum_{n=0}^{j}\binom{j}{n}z^{j-n}\left(r^{\alpha}-z\right)^{n-l},\ 0\leqslant j\leqslant l-1,\ r>|z|^{1/\alpha},$$

the formula (20) implies:

$$\mathcal{L}^{-1}\left(\frac{r^{\alpha j}}{(r^{\alpha}-z)^{l}}\right)(t)=\sum_{n=0}^{j}\binom{j}{n}z^{j-n}\left(\cdot^{\alpha-1}E_{\alpha,\alpha}(z\cdot^{\alpha})\right)^{*,l-n}(t), \tag{54}$$

provided $t\geqslant 0$, $0\leqslant j\leqslant l-1$. By making use of the same formula again, we reveal that:

$$\mathcal{L}^{-1}\left(\frac{r^{\alpha-1}}{(r^{\alpha}-z)^{l+1-j}}\right)(t)=\left(E_{\alpha}\left(\cdot^{\alpha}\right)*\left(\cdot^{\alpha-1}E_{\alpha,\alpha}(z\cdot^{\alpha})\right)^{*,l-n}\right)(t), \tag{55}$$

provided $t\geqslant 0$, $1\leqslant j\leqslant l$. Keeping in mind (53) and (55), we obtain that the mappings $t\mapsto\mathcal{L}^{-1}\left(\frac{r^{\alpha-1}}{(r^{\alpha}-z)^{l+1-j}}\right)(t)$, $t\geqslant 0$ $(1\leqslant j\leqslant l)$ are continuous and exponentially bounded. Using (53) and (54), we get that $|t^{\alpha-1}E_{\alpha,\alpha}(zt^{\alpha})|\leqslant b_{\alpha}(t^{\alpha-1}+|z|^{(1-\alpha)/\alpha})e^{|z|^{1/\alpha}t}$, $t\geqslant 0$ and that the mappings $t\mapsto[\mathcal{L}^{-1}\left(\frac{r^{\alpha j}}{(r^{\alpha}-z)^{k}}\right)(\cdot)*S_{k,\alpha}(\cdot)x](t)$, $t\in[0,\tau)$ $(0\leqslant j\leqslant l-1)$ are continuous for every $x\in E$. With $G(t):=\sum_{j=0}^{l-1}(-1)^{j+1}\binom{l}{j}z^{l-j}\mathcal{L}^{-1}\left(\frac{r^{\alpha j}}{(r^{\alpha}-z)^{l}}\right)(t)$, $t>0$ and $F_{j}(t):=\mathcal{L}^{-1}\left(\frac{r^{\alpha-1}}{(r^{\alpha}-z)^{l+1-j}}\right)(t)$, $t\geqslant 0$ $(1\leqslant j\leqslant l)$, we obtain the following important corollary.

Corollary 2.1.20. *Suppose $\alpha>0$, $l\in\mathbb{N}$, $z\in\rho_{C}(A)$, A is a subgenerator of a (local) $(g_{\alpha}, g_{l\alpha+1})$-regularized C-resolvent family $(S_{l,\alpha}(t))_{t\in[0,\tau)}$ on E, and (45) holds. Set, for every $x\in E$ and $t\in[0,\tau)$,*

$$S_{\alpha}(t)x:=(-1)^{l}S_{l,\alpha}(t)x$$

$$+\sum_{j=0}^{l-1}(-1)^{j+1}\binom{l}{j}z^{l-j}\left[\mathcal{L}^{-1}\left(\frac{r^{\alpha j}}{(r^{\alpha}-z)^{l}}\right)*S_{l,\alpha}(\cdot)x\right](t)$$

$$+\sum_{j=1}^{l}(-1)^{l-j}\mathcal{L}^{-1}\left(\frac{r^{\alpha-1}}{(r^{\alpha}-z)^{l+1-j}}\right)(t)(z-A)^{-j}Cx.$$

Then the following holds:

(i) *$(S_{\alpha}(t))_{t\in[0,\tau)}$ is a $(g_{\alpha}, (z-A)^{-l}C)$-regularized resolvent family with a subgenerator A.*
(ii) *Assume $R(C)\subseteq D((z-A)^{-(l+1)})$ and (50). Then (49) holds.*
(iii) *If $(S_{l,\alpha}(t))_{t\in[0,\tau)}$ is locally equicontinuous (globally exponentially equicontinuous), then $(S_{\alpha}(t))_{t\in[0,\tau)}$ is likewise locally equicontinuous (globally exponentially equicontinuous).*

Remark 2.1.21. (i) Assuming that E is a webbed, bornological space, an induction argument combined with the Closed Graph Theorem shows that (45) holds if $R(C) \subseteq D((z-A)^{-l})$. It is also clear that (45) holds provided that $C = I$ and E is a general SCLCS.

(ii) If $z = 0$ and $k(0) = 1$, then the conditions (a) and (b) of Theorem 2.1.19 hold with $F_j(t) = (k * a^{*,l-j})(t)$, $t \geqslant 0$ $(1 \leqslant j \leqslant l)$ and $G(t) \equiv 0$. This enables one to simply reformulate Theorem 2.1.19 in the case that $z = 0$ and $k(0) \neq 0$.

We leave the reader details concerning inheritance of differential and analytical properties from $(S_{a,k,l}(t))_{t\geqslant 0}$ to $(S_{a,k}(t))_{t\geqslant 0}$ (and vice versa).

Let $C \in L(E)$ be injective. Put $p_C(x) := p(C^{-1}x)$, $p \in \circledast$, $x \in R(C)$. Then $p_C(\cdot)$ is a seminorm on $R(C)$ and the calibration $(p_C)_{p\in\circledast}$ induces a locally convex topology on $R(C)$; we denote the above space by $[R(C)]_{\circledast}$. Notice that $[R(C)]_{\circledast}$ is an SCLCS, and that $[R(C)]_{\circledast}$ is a Fréchet (Banach) space provided that E is. In the case that E is a Banach space, we shall omit the subscript $\circledast$.

In what follows, we introduce and analyze the various types of L^p-stability and parabolicity of the abstract Cauchy problem (1); cf. [463, Section 3, Section 10] and [306].

Definition 2.1.22. Let $p \in [1, \infty]$, $a \in L^1_{loc}([0, \infty))$, $a \neq 0$ and $k \in C([0, \infty))$, $k \neq 0$, and let E be a Banach space. The abstract Volterra equation (1) is said to be:

(i) L^p-stable (CR) if for every $g \in L^p([0, \infty) : [R(C)])$ there exists a unique function $u \in L^p([0, \infty) : E)$ such that $a * u \in C([0, \infty) : [D(A)])$ and $u(t) = (a * g)(t) + A(a * u)(t)$ for a.e. $t \geqslant 0$.

(ii) L^p-stable (CS) if, for every $f \in W^{1,p}_{loc}([0, \infty) : [R(C)])$ such that $f' \in L^p([0, \infty) : [R(C)])$, there exists a unique function $u \in L^p([0, \infty) : E)$ satisfying $a * u \in C([0, \infty) : [D(A)])$ and $u(t) = f(t) + A(a * u)(t)$ for a.e. $t \geqslant 0$.

(iii) C-strongly L^p-stable if for every $g \in L^p([0, \infty) : [R(C)])$ there exists a unique function $u \in L^p([0, \infty) : [D(A)])$ such that $a * u \in C([0, \infty) : [D(A)])$ and $u(t) = (a * g)(t) + (a * Au)(t)$ for a.e. $t \geqslant 0$.

(iv) (kC)-parabolic if (iv.1)-(iv.2) hold, where:

(iv.1) $a(t)$ and $k(t)$ satisfy (P1) and there exist meromorphic extensions of the functions $\tilde{a}(\lambda)$ and $\tilde{k}(\lambda)$ on $\mathbb{C}_+$, denoted by $\hat{a}(\lambda)$ and $\hat{k}(\lambda)$. Let N be the subset of $\mathbb{C}_+$ which consists of all zeroes and possible poles of $\hat{a}(\lambda)$ and $\hat{k}(\lambda)$.

(iv.2) There exists $M \geqslant 1$ such that, for every $\lambda \in \mathbb{C}_+ \setminus N$, $1/\hat{a}(\lambda) \in \rho_C(A)$ and $\|\hat{k}(\lambda)(I - \hat{a}(\lambda)A)^{-1}C\| \leqslant M/|\lambda|$.

If $k(t) \equiv 1$, resp. $C = I$, then it is also said that (1) is C-parabolic, resp. k-parabolic.

Before proceeding further, notice that the definition of (kC)-parabolicity of (1) extends the corresponding one given by J. Prüss [463, Definition 3.1, p. 68]. As an illustrative example of a k-parabolic problem, we quote the backwards heat equation on $L^2[0, \pi]$ ([141], [66], [292]).

Remark 2.1.23. (i) Assume (1) is (kC)-parabolic and there exists an analytic mapping $F : \mathbb{C}_+ \to L(E)$ such that $F(\lambda) = \hat{k}(\lambda)(I - \hat{a}(\lambda)A)^{-1}C$, $\lambda \in \mathbb{C}_+ \setminus N$ and $\sup_{\lambda \in \mathbb{C}_+} \|\lambda^2 F'(\lambda)\| < \infty$. By [463, Theorem 0.4] and [302, Theorem 2.7(iii)-(iv)], we infer that, for every $\alpha \in (0, 1]$, A is a subgenerator of an $(a, k * g_\alpha)$-regularized C-resolvent family $(S_\alpha(t))_{t \geqslant 0}$ which satisfies $\sup_{h>0,t\geqslant 0} h^{-\alpha}\|S_\alpha(t+h) - S_\alpha(t)\| < \infty$; furthermore, if A is densely defined, then A is a subgenerator of a bounded (a, k)-regularized C-resolvent family $(S(t))_{t \geqslant 0}$ that is norm continuous in $t > 0$.
(ii) Assume A is the integral generator of a bounded analytic C-regularized semigroup of angle $\alpha \in (0, \frac{\pi}{2}]$, $a(t)$ satisfies (P1) and admits a meromorphic extension $\hat{a}(\lambda)$ on $\mathbb{C}_+$. Let $\varepsilon \in (0, \alpha)$ and $1/\hat{a}(\lambda) \in \Sigma_{\frac{\pi}{2}+\alpha-\varepsilon}$, $\lambda \in \mathbb{C}_+ \setminus N$. Then (1) is C-parabolic.

Assume $n \in \mathbb{N}$, $a(t)$ satisfies (P1) and $\operatorname{abs}(a) = 0$. Following [463, Definition 3.3, p. 69], $a(t)$ is said to be n-regular if there exists $c > 0$ such that

$$|\lambda^m \hat{a}^{(m)}(\lambda)| \leqslant c|\hat{a}(\lambda)|, \lambda \in \mathbb{C}_+, 1 \leqslant m \leqslant n.$$

Set $a^{(-1)}(t) := \int_0^t a(s)\,ds$, $t \geqslant 0$ and suppose that $a(t)$ and $b(t)$ are n-regular for some $n \in \mathbb{N}$. Then $\hat{a}(\lambda) \neq 0$, $\lambda \in \mathbb{C}_+$, $(a * b)(t)$ and $a^{(-1)}(t)$ are n-regular, and $a'(t)$ is n-regular provided that $\operatorname{abs}(a') = 0$. Furthermore, $a(t)$ is n-regular iff there exists $c' > 0$ such that

$$|(\lambda^m \hat{a}(\lambda))^{(m)}| \leqslant c'|\hat{a}(\lambda)|, \lambda \in \mathbb{C}_+, 1 \leqslant m \leqslant n,$$

and in the case $\arg(\hat{a}(\lambda)) \neq \pi$, $\lambda \in \mathbb{C}_+$, n-regularity of $a(t)$ is also equivalent to the existence of a constant $c'' > 0$ such that

$$|\lambda^m (\ln \hat{a}(\lambda))^{(m)}| \leqslant c'', \lambda \in \mathbb{C}_+, 1 \leqslant m \leqslant n.$$

The next theorem is a generalization of [463, Theorem 3.1, p. 73].

Theorem 2.1.24. *Assume $n \in \mathbb{N}$, $a(t)$ is n-regular, (1) is C-parabolic and the mapping $\lambda \mapsto (I - \tilde{a}(\lambda)A)^{-1}C$, $\lambda \in \mathbb{C}_+$ is continuous. Denote by D_t^ζ the Riemann-Liouville fractional derivative of order $\zeta > 0$. Then, for every $\alpha \in (0, 1]$, A is a subgenerator of an $(a, g_{\alpha+1})$-regularized C^2-resolvent family $(S_\alpha(t))_{t \geqslant 0}$ which satisfies $\sup_{h>0,t\geqslant 0} h^{-\alpha}\|S_\alpha(t+h) - S_\alpha(t)\| < \infty$, $D_t^\alpha S_\alpha(t)C^{k-1} \in C^{k-1}((0,\infty) : L(E))$, $1 \leqslant k \leqslant n$ as well as:*

$$\|t^j D_t^j D_t^\alpha S_\alpha(t)C^{k-1}\| \leqslant M, t \geqslant 0, 1 \leqslant k \leqslant n, 0 \leqslant j \leqslant k-1, \tag{56}$$

$$\|t^k D_t^{k-1} D_t^\alpha S_\alpha(t)C^{k-1} - s^k D_s^{k-1} D_s^\alpha S_\alpha(s)C^{k-1}\|$$

$$\leqslant M|t-s|\left(1 + \ln\frac{t}{t-s}\right), 0 \leqslant s < t < \infty, 1 \leqslant k \leqslant n, \tag{57}$$

and, for every $T > 0$, $\varepsilon > 0$ *and* $k \in \mathbb{N}_{n}$, *there exists* $M^{\varepsilon}_{T,k} > 0$ *such that*

$$\|t^k D_t^{k-1} D_t^{\alpha} S_{\alpha}(t)C^{k-1} - s^k D_s^{k-1} D_s^{\alpha} S_{\alpha}(s)C^{k-1}\|$$

$$\leqslant M^{\varepsilon}_{T,k}(t-s)^{1-\varepsilon},\ 0 \leqslant s < t \leqslant T,\ 1 \leqslant k \leqslant n. \tag{58}$$

Furthermore, if A is densely defined, then A is a subgenerator of a bounded (a, C^2)*-regularized resolvent family* $(S(t))_{t\geqslant 0}$ *which satisfies* $S(t)C^{k-1} \in C^{k-1}((0, \infty) : L(E))$, $1 \leqslant k \leqslant n$ *and (56)-(58) with* $D_t^{\alpha} S_{\alpha}(t)C^{k-1}$ *replaced by* $S(t)C^{k-1}$ $(1 \leqslant k \leqslant n)$ *therein.*

Proof. We will prove the theorem, provided $k \geqslant 2$ and A is nondensely defined. It can be easily seen that the mapping $\lambda \mapsto (I - \tilde{a}(\lambda)A)^{-1}C$, $\lambda \in \mathbb{C}_+$ is analytic. Put $F(\lambda) := (I - \tilde{a}(\lambda)A)^{-1}C/\lambda$, $\lambda \in \mathbb{C}_+$. Then, for every $\lambda \in \mathbb{C}_+$,

$$F'(\lambda)C = -\frac{\left(I - \tilde{a}(\lambda)A\right)^{-1} C^2}{\lambda^2} + \frac{\tilde{a}'(\lambda)}{\lambda \tilde{a}(\lambda)}\left[(I - \tilde{a}(\lambda)A)^{-2}C^2 - (I - \tilde{a}(\lambda)A)^{-1}C^2\right],$$

and

$$F''(\lambda)C^2 = \frac{2}{\lambda^3}(I - \tilde{a}(\lambda)A)^{-1}C^3 - \frac{2}{\lambda^2}\frac{\tilde{a}'(\lambda)}{\tilde{a}(\lambda)}\left[(I - \tilde{a}(\lambda)A)^{-2}C^3 - (I - \tilde{a}(\lambda)A)^{-1}C^3\right]$$

$$+\frac{1}{\lambda}\frac{\tilde{a}''(\lambda)\tilde{a}(\lambda) - \tilde{a}'(\lambda)^2}{\tilde{a}(\lambda)^2}\left[(I - \tilde{a}(\lambda)A)^{-2}C^3 - (I - \tilde{a}(\lambda)A)^{-1}C^3\right]$$

$$+\frac{1}{\lambda}\left(\frac{\tilde{a}'(\lambda)}{\tilde{a}(\lambda)}\right)^2\left[2(I - \tilde{a}(\lambda)A)^{-3}C^3 - 3\,(I - \tilde{a}(\lambda)A)^{-2}C^2 + (I - \tilde{a}(\lambda)A)^{-1}C^3\right].$$

Hence, $\sup_{\lambda\in\mathbb{C}+}(\|\lambda F(\lambda)C\| + \|\lambda^2 F'(\lambda)C\| + \|\lambda^3 F''(\lambda)C^2\|) < \infty$. This inequality, in combination with [463, Proposition 0.1] and [292, Theorem 1.1.13], implies that, for every $\alpha \in (0, 1]$, A is a subgenerator of an $(a, g_{\alpha+1})$-regularized C^2-resolvent family $(S_{\alpha}(t))_{t\geqslant 0}$ which satisfies $\sup_{h\to 0,\ t\geqslant 0} h^{-\alpha}\|S_{\alpha}(t + h) - S_{\alpha}(t)\| < \infty$. Inductively, $\sup_{\lambda\in\mathbb{C}+} \Sigma_{k=0}^{n} \|\lambda^{k+1}F^{(k)}(\lambda)C^k\| < \infty$, and one can apply [463, Theorem 0.4] in order to see that, for every $k \in \mathbb{N}_n$, there exists a function $V_k \in C^{k-1}((0, \infty) : L(E))$ such that (56)-(58) hold with $D_t^{\alpha} S_{\alpha}(t)C^{k-1}$ replaced by $V_k(t)$. By the uniqueness theorem for Laplace transform, one gets $S_{\alpha}(t)C^{k-1}x = \int_0^t g_{\alpha}(t - s)V_k(s)x\, ds$, $x \in E$, $t \geqslant 0$, $1 \leqslant k \leqslant n$. Since $V_k \in L^1((0, T) : L(E))$ for all $T > 0$ and $k \in \mathbb{N}_n$, [49, Theorem 1.5] implies $V_k(t) = D_t^{\alpha} S_{\alpha}(t)C^{k-1}$, $t \geqslant 0$, $1 \leqslant k \leqslant n$. This completes the proof of theorem. □

Keeping in mind Theorem 2.1.24, one can simply transfer the representation formula [463, (3.41), p. 81] and the assertions of [463, Corollary 3.2-Corollary 3.3, pp. 74-75] to exponentially bounded (a, C)-regularized resolvent families. An application can be made to Petrovskii correct matrices of operators ([141], [227], [292]).

We continue our analysis in the framework of Banach spaces, the definition of a (strongly, uniformly) integrable family of operators will be understood in the sense of [463, Definition 10.2, p. 256].

Proposition 2.1.25. (i) *Assume A is a subgenerator of an integrable (a, a)-regularized C-resolvent family $(R(t))_{t\geq 0}$ satisfying (22). Then (1) is L^p-stable (CR) for each $p \in [1, \infty]$.*

(ii) *Let (1) be L^p-stable (CR) for some $p \in [1, \infty]$, and let $a(t)$ satisfy (P1). Put $g^{\mu}(t) := e^{-\mu t}$, $t \geqslant 0$, $\mu \in \mathbb{C}_+$.*

(ii.1) *Then, for every $\mu \in \mathbb{C}_+$, A is a subgenerator of an $(a, a^{(-1)} * g^{\mu})$-regularized C-resolvent family $(U_{\mu}(t))_{t\geq 0}$ and there exists $c(\mu) > 0$ such that $\|U_{\mu}(t)\| \leqslant c(\mu)\, t^{1/p'}$, where $[1, \infty] \ni p'$ satisfies $\frac{1}{p}+\frac{1}{p'} = 1$.*

(ii.2) *Let $D(A^2) \neq \{0\}$. Then $\tilde{a}(\lambda)$ admits a meromorphic extension $\hat{a}(\lambda)$ on $\mathbb{C}_+$. Denote by N the set which consists of all zeroes and possible poles of $\hat{a}(\lambda)$, and assume additionally $\overline{D(A)} = E$ or $\rho(A) \neq \emptyset$. Then $1/\hat{a}(\lambda) \in \rho_C(A)$, $\lambda \in \mathbb{C}_+\setminus N$, and the mapping $\lambda \mapsto K(\lambda) := \hat{a}(\lambda)\,(I - \hat{a}(\lambda)A)^{-1}C$, $\lambda \in \mathbb{C}_+\setminus N$ is uniformly bounded; if $\mathbb{C}_+ \ni \lambda_0$ and $\lim_{\lambda\to\lambda_0} \hat{a}(\lambda) = \infty$, then $0 \in \rho_C(A)$.*

(ii.3) *Let $D(A^2) \neq \{0\}$ and $p = 1$. Assume $\tilde{a}(\lambda)$, $K(\lambda)$ and N possess the same meanings as in (ii.2), and $\overline{D(A)} = E$ or $\rho(A) \neq \emptyset$. Then $1/\hat{a}(\lambda) \in \rho_C(A)$, $\lambda \in \mathbb{C}_+\setminus N$, and the mapping $\lambda \mapsto K(\lambda)$ admits a strongly continuous and uniformly bounded extension on $\overline{\mathbb{C}_+}$. What is more, the mapping $\lambda \mapsto \hat{a}(\lambda)$ admits a continuous extension on $\overline{\mathbb{C}_+}$ which takes values in $\mathbb{C} \cup \{\infty\}$; if $\lim_{\lambda\to\lambda_0} \hat{a}(\lambda) = \infty$ for some $\lambda_0 \in \overline{\mathbb{C}_+}$, then $0 \in \rho_C(A)$.*

(iii) *Let (1) be C-strongly L^p-stable. Then (1) is C-parabolic.*

Proof. We will prove only (ii). Fix $\mu \in \mathbb{C}_+$ and denote by $u_{\mu}(t; x)$ the unique function which satisfies $a * u_{\mu}(\cdot; x) \in C([0, \infty) : [D(A)])$, $u_{\mu}(t; x) = (a * g_{\mu})(t)\, Cx + A(a * g^{\mu}(\cdot; x))\,(t)$ for a.e. $t \geqslant 0$ and $u_{\mu}(\cdot; x) \in L^p([0, \infty) : E)$. By the Closed Graph Theorem, it follows that there exists a constant $c > 0$ such that $\|u_{\mu}(\cdot; x)\|_p \leqslant c\|g^{\mu}\|_p = c(p \operatorname{Re} \mu)^{-1/p}\|x\|$ for all $x \in E$. Define $U_{\mu}(t)x := \int_0^t u_{\mu}(s; x)\, ds$, $t \geqslant 0$, $x \in E$. Then $(U_{\mu}(t))_{t\geq 0}$ is a strongly continuous operator family and there exists $c(\mu) > 0$ such that $\|U_{\mu}(t)\| \leqslant c(\mu)t^{1/p'}$, $t \geqslant 0$. By performing the Laplace transform, we get

$$\left(I - \tilde{a}(\lambda)A\right)\widetilde{U_{\mu}}(\lambda)x = \frac{\tilde{a}(\lambda)}{\lambda(\lambda+\mu)},\ x \in E,\ \operatorname{Re}\lambda > \max(0, \operatorname{abs}(a)).$$

This simply implies that, for every $\lambda \in \mathbb{C}$ with $\operatorname{Re} \lambda > \max(0, \operatorname{abs}(a))$ and $\tilde{a}(\lambda) \neq 0$, $1/\tilde{a}(\lambda) \in \rho_C(A)$ and that A is a subgenerator of an $(a, a^{(-1)} * g^{\mu})$-regularized C-resolvent family $(U_{\mu}(t))_{t\geq 0}$, finishing the proof of (ii.1). Set $f_{\mu}(\lambda) := \lambda(\lambda + \mu)\, \widetilde{U_{\mu}}(\lambda)$, $\operatorname{Re} \lambda > 0$. In order to prove (ii.2), notice that for every $x \in D(A)$, $x^* \in$

E^*, and for every $\lambda \in \mathbb{C}$ with $\operatorname{Re} \lambda > \max(0, \operatorname{abs}(a))$, $\tilde{a}(\lambda) \neq 0$ and $\langle x^*, f_\mu(\lambda)x\rangle \neq 0$, we have $\frac{1}{\tilde{a}(\lambda)} = \frac{\langle x^*, Cx\rangle + \langle x^*, f_\mu(\lambda)Ax\rangle}{\langle x^*, f_\mu(\lambda)x\rangle}$. Assume $\langle x^*, f_\mu(\lambda)x\rangle = 0$ for all $x \in D(A)$, $x^* \in E^*$ and $\lambda \in \mathbb{C}_+$. Then $\langle x^*, U_\mu(t)x\rangle = 0$ for all $x \in D(A)$, $x^* \in E^*$ and $t \geqslant 0$. This yields that, for every $x \in D(A^2)$, $x^* \in E^*$ and $t \geqslant 0$,

$$0 = \langle x^*, Cx\rangle(a^{(-1)} * g^\mu)(t) + \int_0^t a(t-s)\,\langle x^*, U_\mu(s)Ax\rangle\, ds.$$

In view of this, we get that $\langle x^*, Cx\rangle = 0$, $x \in D(A^2)$, $x^* \in E^*$. This is a contradiction to the assumption $D(A^2) \neq \{0\}$. The existence of an element $x \in D(A)$ and a functional $x^* \in E^*$ such that $\langle x^*, f_\mu(\cdot)x\rangle \neq 0$ is clear now. Suppose that for such x and x^* we have $\langle x^*, Cx\rangle + \langle x^*, f_\mu(\lambda)Ax\rangle = 0$, $\lambda \in \mathbb{C}_+$. Then one obtains $\frac{1}{\tilde{a}(\lambda)}\langle x^*, f_\mu(\lambda) x\rangle = \langle x^*, f_\mu(\lambda)x\rangle = 0$, $\operatorname{Re} \lambda > \max(0, \operatorname{abs}(a))$, $\tilde{a}(\lambda) \neq 0$ and $\langle x^*, f_\mu(\lambda)x\rangle = 0$, $\lambda \in \mathbb{C}_+$. This is absurd. Therefore, $\langle x^*, Cx\rangle + \langle x^*, f_\mu(\lambda)Ax\rangle \neq 0$, $\lambda \in \mathbb{C}_+$. Let N be the set which consists of those numbers $\lambda_0 \in \mathbb{C}_+$ such that $\langle x^*, Cx\rangle + \langle x^*, f_\mu(\lambda_0)Ax\rangle = 0$ and $\langle x^*, f_\mu(\lambda_0)x\rangle = 0$. Then $\tilde{a}(\lambda) = \frac{\langle x^*, f_\mu(\lambda)x\rangle}{\langle x^*, Cx\rangle + \langle x^*, f_\mu(\lambda)Ax\rangle}$, $\lambda \in \mathbb{C}_+ \backslash N$, $\operatorname{Re}\lambda > \max(0, \operatorname{abs}(a))$, which shows that $\tilde{a}(\lambda)$ admits a meromorphic extension $\hat{a}(\lambda)$ on $\mathbb{C}_+$. Further on, $A \int_0^\infty e^{-\lambda t}\, U_\mu(t)y\, dt = \int_0^\infty e^{-\lambda t}\, U_\mu(t)Ay\, dt$, $\lambda \in \mathbb{C}_+$, $y \in D(A)$ and $Af_\mu(\lambda)y = f_\mu(\lambda)Ay$, $\lambda \in \mathbb{C}_+$, $y \in D(A)$. Since A is closed and the mapping $\lambda \mapsto Af_\mu(\lambda)y = f_\mu(\lambda)Ay$, $\lambda \in \mathbb{C}_+$ is analytic for every fixed $y \in D(A)$, we obtain:

$$Cy = \left(\frac{1}{\hat{a}(\lambda)} - A\right) f_\mu(\lambda)y,\ \ y \in \overline{D(A)},\ \lambda \in \mathbb{C}_+ \backslash N.$$

This implies $\frac{1}{\hat{a}(\lambda)} \in \rho_C(A)$ if $\overline{D(A)} = E$. Because $R(\cdot : A)$ and $f_\mu(\cdot)$ commutes, the above conclusion still holds if $\rho(A) \neq \emptyset$. The uniform boundedness of the mapping $\lambda \mapsto K(\lambda)$, $\lambda \in \mathbb{C}_+ \setminus N$ follows as in [463]. Let $\lambda_0 \in \mathbb{C}_+$ and $\lim_{\lambda \to \lambda_0} \hat{a}(\lambda) = \infty$. Then $\lim_{n\to\infty} f_\mu(\lambda_n) = f_\mu(\lambda_0)$, $\lim_{n\to\infty} Af_\mu(\lambda_n) = \lim_{n\to\infty} \frac{1}{\hat{a}(\lambda_n)} f_\mu(\lambda_n) - Cy = -Cy$, so that the closedness of A implies $-A\, f_\mu(\lambda_0)y = Cy$, $y \in E$ and $0 \in \rho_C(A)$. The proof of (ii.2) is completed. In the case $p = 1$, the existence of a strongly continuous and uniformly bounded extension of the mapping $\lambda \mapsto K(\lambda)$ on $\overline{\mathbb{C}_+}$ can be proved as in [463]. Assume now $\varrho \in \mathbb{R}$, $(\lambda_n)_n$ is a sequence in $\mathbb{C}_+ \backslash N$ and $\lim_{n\to\infty} \lambda_n = i\varrho$. If $\lim_{n\to\infty} \frac{1}{\hat{a}(\lambda_n)} = z \in \mathbb{C}$, then $z \in \rho_C(A)$ and $\lim_{n\to\infty} K(\lambda_n)C = (z - A)^{-1}C^2$ in $L(E)$; furthermore, if $\lim_{n\to\infty} \frac{1}{\hat{a}(\lambda_n)} = \infty$, then $Cy = \frac{1}{\hat{a}(\lambda_n)} f_\mu(\lambda_n)y - f_\mu(\lambda_n)Ay$, $\lim_{n\to\infty} K(\lambda_n)y = 0$, $y \in D(A)$, and $\lim_{n\to\infty} K(\lambda_n)y = 0$, $y \in E$. This enables one to see that $\hat{a}(\lambda)$ admits a continuous extension on $\overline{\mathbb{C}_+}$ taking values in $\mathbb{C} \cup \{\infty\}$. The remaining part of the proof of (ii.3) is simple. □

In almost the same way, one can prove the following proposition.

Proposition 2.1.26. (i) *Let A be a subgenerator of an (a, C)-regularized resolvent family $(S(t))_{t\geqslant 0}$ that is integrable and bounded. Then (1) is L^p-stable (CS) for each $p \in [1, \infty]$, and $(S(t))_{t\geqslant 0}$ is L^1-stable (CS) if $(S(t))_{t\geqslant 0}$ is strongly integrable.*

(ii) *Let (1) be L^p-stable (CS) for some $p \in [1, \infty]$, and let $a(t)$ satisfy (P1).*

(ii.1) *Then, for every $\mu \in \mathbb{C}_+$, A is a subgenerator of an $(a, 1 * g^{\mu})$-regularized C-resolvent family $(V_\mu(t))_{t\geq 0}$ and there exists $c(\mu) > 0$ such that $\|V_\mu(t)\| \leqslant c(\mu)t^{1/p'}$, where $[1, \infty] \ni p'$ satisfies $\frac{1}{p}+\frac{1}{p'} = 1$.*

(ii.2) *Let $D(A^2) \neq \{0\}$. Then $\tilde{a}(\lambda)$ admits a meromorphic extension $\hat{a}(\lambda)$ on $\mathbb{C}_+$. Denote by N the set which consists of all zeros and possible poles of $\hat{a}(\lambda)$, and assume additionally $\overline{D(A)} = E$ or $\rho(A) \neq \emptyset$. Then $1/\hat{a}(\lambda) \in \rho_C(A)$, $\lambda \in \mathbb{C}_+ \setminus N$, and the mapping $\lambda \mapsto H(\lambda) = (I - \hat{a}(\lambda)A)^{-1}C/\lambda$, $\lambda \in \mathbb{C}_+ \setminus N$ is uniformly bounded. Furthermore, A is invertible provided $C = I$.*

(ii.3) *Let $D(A^2) \neq \{0\}$ and $p = 1$. Assume $\tilde{a}(\lambda)$, $H(\lambda)$ and N possess the same meanings as in (ii.2), and $\overline{D(A)} = E$ or $\rho(A) \neq \emptyset$. Then $1/\hat{a}(\lambda) \in \rho_C(A)$, $\lambda \in \mathbb{C}_+ \setminus N$, and the mapping $\lambda \mapsto H(\lambda)$ admits a strongly continuous and uniformly bounded extension on $\overline{\mathbb{C}_+}$. The mapping $\lambda \mapsto \hat{a}(\lambda)$ admits a continuous extension on $\overline{\mathbb{C}_+}$ taking values in $\mathbb{C} \cup \{\infty\}$, $\lim_{\lambda\to 0}\hat{a}(\lambda) = \infty$, and in the case $0 \in \rho_C(A)$, there exists $\lim_{\lambda\to 0}\lambda\hat{a}(\lambda)$ in $(\mathbb{C} \cup \{\infty\}) \setminus \{0\}$.*

Finally, let us mention that it is not clear how one can prove [463, Theorem 10.1, p. 262] and its consequences in the case of a general (a, k)-regularized C-resolvent family. Nevertheless, in many cases, $(S(t))_{t\geq 0}$ is not integrable in any sense but (70) is solvable provided that the function $t \mapsto \|(C^{-1}g)'(t)\|_{[D(A)]}$ decays polynomially as $t \to -\infty$ (see e.g. [302, Example 2.31(iii)]).

2.1.1. Wellposedness of related abstract Cauchy problems. Let $K_C : C([0, \tau) : E) \to C([0, \tau) : E)$ be defined by $K_C u := k * Cu$, $u \in C([0, \tau) : E)$, and let (τ_n) be a strictly increasing sequence in $[0, \tau)$ with $\lim_{n\to\infty} \tau_n = \tau$. The totality of seminorms $(p_n(f) := \sup_{t\in[0,\tau_n]} p(f(t)))_{p\in\circledast, n\in\mathbb{N}}$ induces a Hausdorff locally convex topology on $C([0, \tau) : E)$. Certainly, $C([0, \tau) : E)$ is sequentially complete and $L(C([0, \tau) : E)) \ni K_C$ is injective.

A function $u \in C([0, \tau) : E)$ is said to be:

(i) a (mild) solution of (1) if $(a * u)(t) \in D(A)$, $t \in [0, \tau)$ and $A(a * u)(t) = u(t) - f(t)$, $t \in [0, \tau)$.

(ii) a strong solution of (1) if the mapping $t \mapsto Au(t)$, $t \in [0, \tau)$ is well defined, continuous and $(a * Au)(t) = u(t) - f(t)$, $t \in [0, \tau)$.

(iii) a weak solution of (1) if for every $(x^*, y^*) \in A^*$ and for every $t \in [0, \tau)$, we have $\langle x^*, u(t)\rangle = \langle x^*, f(t)\rangle + \langle y^*, (a * u)(t)\rangle$, $t \in [0, \tau)$.

Remark 2.1.27. In this book, we shall not analyze the abstract nonlinear Volterra equation

$$u(t) \ni f(t) + \int_0^t a(t-s)Au(s)\,ds,\ t \in [0, \tau),$$

where A is a multivalued m-accretive operator on E. For an account of the theory of abstract nonlinear Volterra equations, we refer the reader to [38], [82], [110]-[112], [123], [126], [206], [234], [248], [378], [398], [419], [437], [442], [463, Subsection 13.7], [467]-[468], [494], [503] and references cited there. Also, a singular perturbation problem for abstract Volterra integro-differential equations is one of themes that will not be considered in this book; for more details, the reader may consult [156]-[157], [177], [180], [188], [203], [369], [387], [375]-[377] and [491].

Observe that every strong solution of (1) is a mild solution; however, in general not every mild solution of (1) is a strong solution. Since [302, Lemma 2.4] continues to hold in SCLCSs, the concepts mild solution and weak solution of (1) actually coincide. Notice that this assertion generalizes the corresponding one stated in [463, Proposition 1.4].

A careful examination of the proof of [382, Theorem 2.7] implies the following theorem appearing in [303].

Theorem 2.1.28. *Assume $f \in C([0, \tau) : E)$, A is a subgenerator of a locally equicontinuous (a, k)-regularized C-resolvent family $(R(t))_{t\in[0,\tau)}$ satisfying (22). Then the solutions of (1) are unique, and the following holds:*

(i) *Let $u(t)$ be a solution of (1). Then*

$$(R * f)(t) = (kC * u)(t),\ t \in [0, \tau) \text{ and } R * f \in R(K_C). \tag{59}$$

(ii) *Let (59) hold for some $u \in C([0, \tau) : E)$. Then $u(t)$ is a unique solution of (1).*

Denote by $C^{n,k}([0, \tau) : E)$ $(n \in \mathbb{N}_0, k \in \mathbb{N})$ the space which consists of those n-times continuously differentiable functions $f(\cdot)$ for which $f^{(j)}(0) = 0, j \in \mathbb{N}^0_{k-1}$. Set $C^n_0([0, \tau) : E) \equiv C^{n,n}([0, \tau) : E)$, $n \in \mathbb{N}$.

The previous theorem yields the conclusions stated in the following theorem ([382], [302]-[303]):

Theorem 2.1.29. (i) *Assume $n \in \mathbb{N}$, $f \in C([0, \tau) : E)$, A is a subgenerator of a locally equicontinuous n-times integrated (a, C)-resolvent family $(R(t))_{t\in[0,\tau)}$ satisfying (22). Then (1) has a unique solution iff $C^{-1}(R * f) \in C^{n+1}_0([0, \tau) : E)$.*
(ii) *Assume $n \in \mathbb{N}$, A is a subgenerator of a locally equicontinuous n-times integrated (a, C)-regularized resolvent family satisfying (22) and $a \in BV_{loc}([0, \tau))$, resp. A is a subgenerator of a locally equicontinuous (a, C)-regularized resolvent family satisfying (22). Assume, further, that $C^{-1}f \in C^{(n+1)}([0, \tau) : E)$, $f^{(k-1)}(0) \in D(A^{n+1-k})$ and $A^{n+1-k}f^{(k-1)}(0) \in R(C)$, $1 \leqslant k \leqslant n+1$, resp. $t \mapsto C^{-1}f(t)$, $t \in [0, \tau)$ is a locally integrable E-valued mapping such that the mapping $t \mapsto (d/dt)C^{-1}f(t)$ is defined for a.e. $t \in [0, \tau)$ and locally integrable on $[0, \tau)$, as well as $f(t) = f(0) + \int_0^t f'(s)\, ds$, $t \in [0, \tau)$. Then (1) has a unique solution.*

Denote by $Z_a(A)$ the space which consists of those elements $x \in E$ for which there exists a unique solution of (1) with $f(t) \equiv x$ and $\tau = \infty$. We shall always assume in the sequel that there exists a unique solution of (1) with $\tau = \infty$ and $f(t) \equiv 0$, so that 0 belongs to the solution space $Z_a(A)$; observe that this condition automatically holds provided that the operator A satisfies the assumptions stated in the formulation of Proposition 2.1.30 below. Therefore, $Z_a(A)$ is a linear subspace of E.

The spaces $Z_a(A)$ with $a(t) = 1$ or $a(t) = t$ have been analyzed in [141] and [292] (cf. [141, Section IV] and [292, Proposition 3.1.28(ii) and p. 259]). In the following proposition, we consider the general case.

Proposition 2.1.30. *Suppose that, for every $\tau > 0$, there exists $n_\tau \in \mathbb{N}_0$ such that A is a subgenerator of a locally equicontinuous n_τ-times integrated (a, C)-resolvent family $(R_{n_\tau}(t))_{t\in[0,\tau)}$ satisfying*

$$(60) \qquad A\int_0^t a(t-s)R_{n_\tau}(s)x\,ds = R_{n_\tau}(t)x - \frac{t^{n_\tau}}{n_\tau!}Cx,\ t \in [0, \tau),\ x \in E.$$

Then $x \in Z_a(A)$ if, for every $\tau > 0$, $R_{n_\tau}(t)x \in R(C)$, $t \in [0, \tau)$ and the mapping $t \mapsto C^{-1}R_{n_\tau}(t)x$, $t \in [0, \tau)$ is n_τ-times continuously differentiable. If this is the case, the unique solution $u(\cdot, x)$ of (1) on $[0, \tau)$ is given by $u(t, x) = \frac{d^{n_\tau}}{dt^{n_\tau}}C^{-1}R_{n_\tau}(t)x$, $t \in [0, \tau)$.

Proof. Assume first $x \in Z_a(A)$. Making use of Theorem 2.1.28(i), we obtain that, for every $\tau > 0$, $\int_0^t R_{n_\tau}(s)x\,ds = (g_{n_\tau+1}(\cdot)C * u)(t)$, $t \in [0, \tau)$. This implies that, for every $\tau > 0$, $R_{n_\tau}(t)x \in R(C)$, $t \in [0, \tau)$ and that the mapping $t \mapsto C^{-1}R_{n_\tau}(t)x$, $t \in [0, \tau)$ is n_τ-times continuously differentiable, proving the necessity. In order to prove the sufficiency, notice that, for every $\tau > 0$, the mappings $t \mapsto (a* \frac{d^j}{d^j} R_{n_\tau}(\cdot)x)(t)$, $t \in [0, \tau)$ are continuous $(0 \leqslant j \leqslant n_\tau)$ and that $(a * \frac{d^j}{d^j} R_{n_\tau}(\cdot)x)(t) = (a * 1 * \frac{d^{j-1}}{d^{j-1}} R_{n_\tau}(\cdot)x)(t)$, $t \in [0, \tau)$ $(1 \leqslant j \leqslant n_\tau)$. Using induction and the closedness of A, we infer that $A(a* \frac{d^j}{d^j} R_{n_\tau}(\cdot)x)(t) = \frac{d^j}{dt^j} R_{n_\tau}(t)x - \frac{t^{n_\tau-j}}{(n_\tau-j)!}Cx$, provided $t \in [0, \tau)$ and $0 \leqslant j \leqslant n_\tau$. The remaining part of the proof can be skipped. □

Assume that A is a subgenerator of a locally equicontinuous (a, C)-regularized resolvent family $(R_0(t))_{t\geqslant 0}$ satisfying (60) with $n_\tau = 0$. If $p \in \circledast$ and $n \in \mathbb{N}$, then $p_n(\cdot) := \sup_{t\in[0,n]} p(u(t, \cdot))$ is a seminorm on $Z_a(A)$. The totality of these seminorms induces a Hausdorff locally convex topology on $Z_a(A)$. It is checked at once that $Z_a(A)$ is an SCLCS, and that $Z_a(A)$ a Fréchet space provided that E is. Notice that $Z_a(A)$ is topologically equivalent to a subspace of $C([0, \infty) : E)$, via the embedding $(\Lambda x)(t) := u(t, x)$, $t \geqslant 0$, $x \in Z_a(A)$, and that the inclusion mapping from $Z_a(A)$ into E is continuous. Moreover, the space $[R(C)]_\circledcirc$ is continuously embedded into $Z_a(A)$. Define now, for every $t \geqslant 0$ and $x \in Z_a(A)$, $R(t)x := u(t, x)$. Then $R(0) = I_{|Z_a(A)}$ and $A\int_0^t a(t-s)R(s)x\,ds = R(t)x - x$, $x \in Z_a(A)$, $t \geqslant 0$, where the last integral is taken with respect to the initial topology of E. Assuming $x \in D(A_{|Z_a(A)})$, it simply follows from the uniqueness of solution that $u(t, x) = \int_0^t a(t-s)u(s, Ax)\,ds + x$, $t \geqslant 0$ and $R(t)A_{|Z_a(A)} \subseteq A_{|Z_a(A)}R(t)$, $t \geqslant 0$. If $a(t) = 1$ $(a(t) = t)$, then $R(t)(Z_a(A)) \subseteq Z_a(A)$, $t \geqslant 0$, so that $(R(t))_{t\geqslant 0}$ is a locally equicontinuous semigroup (cosine function) in $Z_a(A)$ with the generator $A_{|Z_a(A)}$; the above assertion is clear provided $a(t) = 1$ while, in the case $a(t) = t$, it follows from the well-known equality $2R(t)R(s)x = R(t + s)x + R(|t - s|)x$, $t, s \geqslant 0$, $x \in Z_a(A)$. It is not clear whether, in general, $R(t)(Z_a(A)) \subseteq Z_a(A)$, $t \geqslant 0$ (which would immediately imply the continuity of mapping $t \mapsto R(t)x \in Z_a(A)$, $t \geqslant 0$ for every fixed $x \in Z_a(A)$; furthermore, $(R(t))_{t\geqslant 0}$ would be an $(a, 1)$-regularized resolvent family in $Z_a(A)$ with a subgenerator $A_{|Z_a(A)}$ if $R(t) \in L(Z_a(A))$, $t \geqslant 0$). In a similar manner, one can consider exponentially equicontinuous solution spaces (cf. [141, Section V] for more details).

The analysis of interpolation and extrapolation spaces for (a, k)-regularized C-resolvent families is a non-trivial problem which will not be further discussed in the context of this book (see e.g. [20], [141] and [249] for the corresponding results in the case of semigroups). Here we want only to mention in passing that E. Alvarez-Pardo and C. Lizama [7] have recently reconsidered and improved the results of S. Kantorovitz [249] concerning the Hille-Yosida space of an arbitrary closed linear operator A on a complex Banach space E. It has been proved that, under suitable conditions on functions $a(t)$ and $k(t)$, one can always find a Banach space $Z \subseteq E$ such that $A_{|Z}$ generates an exponentially bounded (a, k)-regularized resolvent family on Z. This space is maximal-unique in a certain sense and contains a non-zero vector provided that $\sigma_p(A) \cap (0, \infty) \neq \emptyset$.

The following proposition is a generalization of [445, Theorem 2.5] and some results given in [292, Subsection 2.3]. The proof is simple and therefore omitted.

Proposition 2.1.31. *Consider the following assertions.*

(i) *A is a subgenerator of a locally equicontinuous (a, k)-regularized C-resolvent family $(R(t))_{t\in[0,\tau)}$ satisfying (22).*
(ii) *For every $x \in E$, there exists a unique solution of (1) with $f(t) = k(t)Cx$, $t \in [0, \tau)$.*

Then (i) $\Rightarrow$ (ii). If, in addition, E is a Fréchet space, then the above are equivalent.

Consider now the following abstract fractional Cauchy problem:

$$\mathbf{D}_t^{\alpha}u(t) = Au(t) + f(t),\ t \in (0, \tau);\ u^{(k)}(0) = x_k,\ k = 0, 1, \cdots, \lceil\alpha\rceil - 1, \tag{61}$$

where $x_k \in C(D(A))$, $k = 0, 1, \cdots, \lceil\alpha\rceil - 1$ and $f \in C([0, \tau) : E)$. Let $\alpha \in (0, \infty)\setminus\mathbb{N}$. Following M. Li, C. Chen and F.-B. Li [365], a function $u \in C^{\lceil\alpha\rceil-1}([0, \infty) : E)$ is said to be:

(i) a (strong) solution of (61) if $u \in C^{\lceil\alpha\rceil-1}([0, \tau) : E)$, $Au \in C([0, \tau) : E)$, $\int_0^{\cdot} g_{\lceil\alpha\rceil-\alpha}(\cdot - s)\ [u(s) - \sum_{k=0}^{\lceil\alpha\rceil-1} \frac{s^k}{k!}x_k]\ ds \in C^{\lceil\alpha\rceil}([0, \tau) : E)$ and (61) holds.
(ii) a mild solution of (61) if $u \in C([0, \tau) : E)$, $(g_\alpha * f)(t) \in D(A)$, $t \in [0, \tau)$ and

$$A\,(g_\alpha * u)\,(t) = u(t) - (g_\alpha * f)(t) - \sum_{k=0}^{\lceil\alpha\rceil-1} \frac{t^k}{k!}x_k,\ t \in [0, \tau).$$

Let A be a subgenerator of a locally equicontinuous (g_α, k)-regularized C-resolvent family $(R(t))_{t\in[0,\tau)}$ which satisfies (22) with $a(t) = g_\alpha(t)$. Setting $G(t) := (g_\alpha * f)(t) + \Sigma_{k=0}^{\lceil\alpha\rceil-1} \frac{t^k}{k!}x_k$, $t \in [0, \tau)$, it readily follows from Theorem 2.1.28(i) that every mild solution of (61) satisfies $(R * G)(t) = (kC * u)(t)$, $t \in [0, \tau)$. Furthermore, if the above equality holds for some $u \in C([0, \tau) : E)$, then $u(t)$ is a mild solution of (61); in such a way, we have proved an extension of [365, Proposition 4.2].

It is predictable that every strong solution of (61) is also a mild solution of (61). We will prove this fact. Notice first that the assumption $F \in C^1([0, \tau) : E)$ implies $(g_\alpha * F)'(t) = (g_\alpha * F')(t) + F(0)g_\alpha(t)$, $t \in (0, \tau)$, and $(g_\alpha * F)'(t) = (g_\alpha * F')(t) + F(0)g_\alpha(t)$, $t \in [0, \tau)$, under the additional condition $F(0) = 0$. Assume that $u(t)$ is a strong solution of (61). Then $u \in C^{\lceil\alpha\rceil-1}([0, \tau) : E)$, $(g_\alpha * u)(t) \in D(A)$ and $A(g_\alpha * u)(t) = (g_\alpha * Au)(t)$, $t \in [0, \tau)$. Set $F(t) := \int_0^t g_{\lceil\alpha\rceil-\alpha}(t-s)\,[u(s) - \Sigma_{k=0}^{\lceil\alpha\rceil-1} \frac{s^k}{k!}x_k]\,ds$, $t \in [0, \tau)$. Then $F^{(j)}(t) = \int_0^t g_{\lceil\alpha\rceil-\alpha}(t-s)\,[u^{(j)}(s) - \Sigma_{k=0}^{\lceil\alpha\rceil-1} \frac{s^{k-j}}{(k-j)!}x_k]\,ds$, provided $t \in [0, \tau)$ and $0 \leqslant j \leqslant \lceil\alpha\rceil - 1$. This implies $(g_\alpha * F)^{(m)}(t) = (g_\alpha * F^{(m)})(t)$, $t \in [0, \tau)$ and

$$A\,(g_\alpha * u)(t) = (g_\alpha * Au)(t)$$

$$= -(g_\alpha * f)(t) + \left(g_\alpha * \frac{d^{\lceil\alpha\rceil}}{d\cdot^{\lceil\alpha\rceil}} \int_0^{\cdot} g_{\lceil\alpha\rceil-\alpha}(\cdot - s)\left[u(s) - \sum_{k=0}^{\lceil\alpha\rceil-1} \frac{s^k}{k!}x_k\right]ds\right)(t)$$

$$= -(g_\alpha * f)(t) + \frac{d^{\lceil\alpha\rceil}}{dt^{\lceil\alpha\rceil}}\left(g_{\lceil\alpha\rceil} * \left[u(\cdot) - \sum_{k=0}^{\lceil\alpha\rceil-1} \frac{\cdot^k}{k!}x_k\right]\right)(t)$$

$$= -(g_\alpha * f)(t) + u(t) - \sum_{k=0}^{\lceil\alpha\rceil-1} \frac{t^k}{k!}x_k,$$

for any $t \in [0, \tau)$. Therefore, $u(t)$ is a mild solution of (61).

Assume now that $u(t)$ is a mild solution of (61) and, additionally, $Au \in C([0, \tau) : E)$. Then $u(t) = (g_\alpha * Au)(t) + (g_\alpha * f)(t) + \sum_{k=0}^{\lceil\alpha\rceil-1} \frac{t^k}{k!}x_k$, $t \in [0, \tau)$, and consequently, $u \in C^{\lceil\alpha\rceil-1}([0, \tau) : E)$. Furthermore,

$$\left(g_{\lceil\alpha\rceil} * Au\right)(t) = \left(g_{\lceil\alpha\rceil-\alpha} * \left[u(\cdot) - \sum_{k=0}^{\lceil\alpha\rceil-1} \frac{\cdot^k}{k!}x_k\right]\right)(t) - \left(g_{\lceil\alpha\rceil} * f\right)(t),\ t \in [0, \tau),$$

which implies that $(g_{\lceil\alpha\rceil-\alpha} * [u(\cdot) - \Sigma_{k=0}^{\lceil\alpha\rceil-1} \frac{\cdot^k}{k!}x_k]) \in C^{\lceil\alpha\rceil}([0, \tau) : E)$ and (61) holds. Therefore, $u(t)$ is a strong solution of (61).

Suppose A is a subgenerator of an (a, k)-regularized C-resolvent family and $n \in \mathbb{N}$. Then one obtains inductively that, for every $t \in [0, \tau)$ and $x \in D(A^n)$,

$$R(t)x = k(t)Cx + \sum_{j=1}^{n-1} (a^{*,j} * k)(t)\,CA^jx + (a^{*,n} * R(\cdot)A^nx)\,(t). \tag{62}$$

The following is a strengthening of [365, Proposition 4.7, Proposition 4.8].

Proposition 2.1.32. (i) *Suppose* $\alpha \in (0, \infty)\setminus\mathbb{N}$, $x \in D(A)$ *as well as* $C^{-1}f$, $AC^{-1}f \in C([0, \tau) : E)$ *and* A *is a subgenerator of a* (g_α, C)*-regularized resolvent family* $(R(t))_{t\in[0,\tau)}$. *Set* $v(t) := (g_{\lceil\alpha\rceil-\alpha} * f)(t)$, $t \in [0, \tau)$. *If* $v \in C^{\lceil\alpha\rceil-1}([0, \tau) : E)$ *and* $v^{(k)}(0) = 0$ *for* $1 \leqslant k \leqslant \lceil\alpha\rceil - 2$, *then the function* $u(t) := R(t)x + (R * C^{-1}f)(t)x$, $t \in [0, \tau)$ *is a unique solution of the initial value problem:*

$$\begin{cases} u \in C^{\lceil\alpha\rceil}((0, \tau) : E) \cap C^{\lceil\alpha\rceil-1}([0, \tau) : E), \\ \mathbf{D}_t^\alpha u(t) = Au(t) + \frac{d^{\lceil\alpha\rceil-1}}{dt^{\lceil\alpha\rceil-1}}(g_{\lceil\alpha\rceil-\alpha} * f)(t),\ t \in (0, \tau), \\ u(0) = Cx,\ u^{(k)}(0) = 0,\ 1 \leqslant k \leqslant \lceil\alpha\rceil - 1. \end{cases}$$

(ii) *Suppose* $\alpha \in (0, 1)$, $x \in D(A)$, $C^{-1}f$, $AC^{-1}f \in C([0, \tau) : E)$ *and* A *is a subgenerator of a* (g_α, C)*-regularized resolvent family* $(R(t))_{t\in[0,\tau)}$. *Then the function* $u(t) := R(t)x + (R * C^{-1}f)(t)x$, $t \in [0, \tau)$ *is a unique solution of the initial value problem:*

$$\begin{cases} u \in C^1((0, \tau) : E) \cap C([0, \tau) : E), \\ \mathbf{D}_t^\alpha u(t) = Au(t) + (g_{1-\alpha} * f)(t),\ t \in [0, \tau), \\ u(0) = Cx. \end{cases}$$

(iii) *Suppose* $r \geqslant 0$, $n \in \mathbb{N}\setminus\{1\}$, $x \in D(A^n)$, $A^jC^{-1}f \in C([0, \tau) : E)$ *for* $0 \leqslant j \leqslant n$, *and* A *is a subgenerator of a* $(g_{1/n}, g_{r+1})$*-regularized* C*-resolvent family* $(R(t))_{t\in[0,\tau)}$. *Then the function* $v(t) := R(t)x + (R * C^{-1}f)(t)x$, $t \in [0, \tau)$ *is a unique solution of the initial value problem:*

$$\begin{cases} v \in C^1((0, \tau) : E) \cap C([0, \tau) : E), \\ v'(t) = Av(t) + \sum_{j=1}^{n-1} g_{(j/n)+r}(t)CA^jx \\ \quad + \sum_{j=0}^{n-1}\left(g_{(j/n)+r} * A^j f\right)(t) + \frac{d}{dt} g_{r+1}(t)Cx,\ t \in (0, \tau), \\ v(0) = g_{r+1}(0)Cx. \end{cases}$$

Furthermore, $v \in C^1([0, \tau) : E)$ *provided that* $r \geqslant 1$ *or* $x = 0$ *and* $r \geqslant 0$.

Proof. The proof of assertion (iii) follows from the equality (62) and a simple computation. In order to prove (i), set $u_1(t) := (R * C^{-1}f)(t)$, $t \in [0, \tau)$ and $u_2(t) := (g_{\lceil\alpha\rceil-\alpha} * u_1)(t)$, $t \in [0, \tau)$. The proof follows very easily once we show that the function $u_1(t)$ satisfies $u_1 \in C^{\lceil\alpha\rceil-1}([0, \tau) : E)$, $u_1^{(k)}(0) = 0$, $1 \leqslant k \leqslant \lceil\alpha\rceil - 1$ and the following equation:

$$\mathbf{D}_t^\alpha u_1(t) = Au_1(t) + \frac{d^{\lceil\alpha\rceil-1}}{dt^{\lceil\alpha\rceil-1}}(g_{\lceil\alpha\rceil-\alpha} * f)(t),\ t \in (0, \tau).$$

Clearly,

$$\begin{aligned} u_2(t) &= (g_{\lceil\alpha\rceil-\alpha} * R * C^{-1}f)(t) \\ &= (g_{\lceil\alpha\rceil-\alpha+1} * f)(t) + (g_{\lceil\alpha\rceil} * Au_1)(t),\ t \in [0, \tau) \end{aligned}$$

and

$$u_1(t) = \left(g_{1-(\lceil\alpha\rceil-\alpha)} * \left(g_{\lceil\alpha\rceil-\alpha} * f\right)\right)(t) + \left(g_{1-(\lceil\alpha\rceil-\alpha)} * g_{\lceil\alpha\rceil-1} * Au_1\right)(t),$$

for any $t \in [0, \tau)$. □

The abstract fractional Cauchy problem (61) is said to be C-wellposed (cf. [49] and [312] for some special cases) if:

(i) For every $x_0, \cdots, x_{\lceil\alpha\rceil-1} \in C(D(A))$, there exists a unique solution $u_f(t; x_0, \cdots, x_{\lceil\alpha\rceil-1})$ of (61).

(ii) For every $T \in (0, \tau)$ and $q \in ⊛$, there exist $c > 0$ and $r \in ⊛$ such that, for every $x_0, \cdots, x_{\lceil\alpha\rceil-1} \in C(D(A))$, the following holds:

$$q\left(u_f(t; x_0, \cdots, x_{\lceil\alpha\rceil-1})\right) \leqslant c \sum_{k=0}^{\lceil\alpha\rceil-1} r(C^{-1}x_k),\ t \in [0, T]. \tag{63}$$

Assume that there exists a unique solution of (61) in case $x_0 \in C(D(A))$ and $x_j = 0$, $1 \leqslant j \leqslant \lceil\alpha\rceil - 1$. Then $u_f(t; x_0) \equiv u_f(t; x_0, 0, \cdots, 0)$, $t \in [0, \tau)$ is a mild solution of (61) and $Au_f(\cdot; x_0) \in C([0, \tau) : E)$. By the foregoing, we get that $u_f(\cdot; x_0)$ is a unique function satisfying $u_f(\cdot; x_0)$, $Au_f(\cdot; x_0) \in C([0, \tau) : E)$ and

$$u_f(t; x_0) = x_0 + (g_\alpha * f)(t) + \int_0^t g_\alpha(t-s)Au_f(s; x_0)\, ds,\ t \in [0, \tau). \tag{64}$$

If, additionally, A is densely defined, E is complete and (63) holds provided $x_0 \in C(D(A))$, $x_j = 0$, $1 \leqslant j \leqslant \lceil\alpha\rceil - 1$ and $f \equiv 0$, then A is a subgenerator of a locally equicontinuous (g_α, C)-regularized resolvent family on $[0, \tau)$ ([312]). Assume now A is densely defined, E is complete, $g \in C([0, \tau))$ and, for every $x_0 \in C(D(A))$, there exists a unique function $u(\cdot; x_0) \in C([0, \tau) : E)$ satisfying $Au(\cdot; x_0) \in C([0, \tau) : E)$,

$$u(t; x_0) = x_0 + (g_\alpha * g)(t)x_0 + \int_0^t g_\alpha(t-s)Au(s; x_0)\, ds,\ t \in [0, \tau), \tag{65}$$

as well as (63) with $x_j = 0$, $1 \leqslant j \leqslant \lceil\alpha\rceil - 1$, and $u_f(\cdot; x_0, \cdots, x_{\lceil\alpha\rceil-1})$ replaced by $u(\cdot; x_0)$ therein (cf. (64) with $f(t) = g(t)x_0$, $t \in [0, \tau)$). Put $k(t) := 1 + (g_\alpha * g)(t)$, $t \in [0, \tau)$. Arguing as in the proof of [463, Proposition 1.1, p. 32], we obtain that A is a subgenerator of a locally equicontinuous (g_α, k)-regularized C-resolvent family $(S(t))_{t\in[0,\tau)}$. Since $k \in AC_{loc}([0, \tau))$ and $k(0) = 1 \neq 0$, we infer from [382, Proposition 2.5] and its proof that there exists $b \in L^1_{loc}([0, \tau))$ such that $(R(t) \equiv S(t) + (b*S)(t))_{t\in[0,\tau)}$ is a locally equicontinuous (g_α, C)-regularized resolvent family with a subgenerator A. In other words, the above conclusion does not depend on the choice of continuous function $g(t)$ appearing in (65).

Assume now that, for every $x_0 \in C(D(A))$, there exists a unique function $u_f(t) \equiv u_f(t; x_0)$, $t \in [0, \tau)$ satisfying u_f, $Au_f \in C([0, \tau) : E)$ and (64). Obviously, $u_f(t)$ is a unique solution of (61) with $x_j = 0$, $1 \leqslant j \leqslant \lceil\alpha\rceil - 1$. If A is a subgenerator of a (g_α, C)-regularized resolvent family $(S_\alpha(t))_{t\in[0,\tau)}$, then the unique solution of (61) with $f \equiv 0$ is given by:

$$u(t) = S_\alpha(t)C^{-1}x_0 + \sum_{j=1}^{\lceil\alpha\rceil} \int_0^t \frac{(t-s)^{j-1}}{(j-1)!} S_\alpha(s)C^{-1}x_{j-1}\, ds,\ t \in [0, \tau);$$

in this case, the abstract Cauchy problem (61) is C-wellposed if, additionally, $(S_\alpha(t))_{t\in[0,\tau)}$ is locally equicontinuous.

In the subsequent theorem, we shall generalize the Ljubich uniqueness theorem [397]. For the equations of integer order, we may pass to the theory of first order equations, in the customary manner, by means of the following auxiliary lemma of independent interest.

Lemma 2.1.33. *(cf.* [292, Lemma 2.3.22] *and* [292, Theorem 2.3.23] *for the Banach space case)*

(i) *Let* $\lambda \in \mathbb{C}$, *let* $k \in \mathbb{N}\setminus\{1\}$ *and let* A *be a closed linear operator on* E. *Put* $\mathcal{C}_k(x_1, \cdots, x_k) := (Cx_1, \cdots, Cx_k)$, $(x_1, \cdots, x_k) \in E^k$, $D(\mathcal{A}_k) := D(A) \times E^{k-1}$ *and* $\mathcal{A}_k(x_1, \cdots, x_k) := (x_2, \cdots, x_k, Ax_1)$, $(x_1, \cdots, x_k) \in D(\mathcal{A}_k)$. *Then* $\lambda \in \rho_{\mathcal{C}_k}(\mathcal{A}_k)$ *if* $\lambda^k \in \rho_C(A)$. *If this is the case, then we have the following:*

$$(\lambda^k - A)^{-1}\, Cx = \pi_1\, ((\lambda - \mathcal{A}_k)^{-1}\, \mathcal{C}_k(0, \cdots, x)),\ x \in E,$$

where $\pi_1 : E^k \to E$ *denotes the first projection and, for every* $x_1, \cdots, x_k \in E$,

(66)
$$(\lambda - \mathcal{A}_k)^{-1}\, \mathcal{C}_k(x_1, \cdots, x_k) = (y_1, \cdots, y_k),$$

where

(67)
$$\begin{aligned} y_j = {}& \lambda^{j-1}(\lambda^k - A)^{-1}\, C(\lambda^{k-1}x_1 + \lambda^{k-2}x_2 + \cdots + x_k) \\ & - (Cx_{j-1} + \lambda Cx_{j-2} + \cdots + \lambda^{j-2}Cx_1),\ 1 \leqslant j \leqslant k. \end{aligned}$$

(ii) *Suppose* $\lambda > 0$, $\{n\lambda : n \in \mathbb{N}\} \subseteq \rho_C(A)$ *and, for every* $\sigma > 0$ *and* $x \in E$, $\lim_{n\to\infty} \frac{(n\lambda - A)^{-1}Cx}{e^{n\lambda\sigma}} = 0$. *Then, for every* $x \in E$, *there exists at most one solution of the initial value problem*

$$\begin{cases} u \in C^1((0, \infty) : E) \cap C([0, \infty) : E), \\ u'(t) = Au(t),\ t > 0, \\ u(0) = x. \end{cases}$$

(iii) *Suppose* $T > 0$, $u \in C([0, T] : E)$, $\lambda > 0$ *and the set* $\{\int_0^T e^{n\lambda s}u(s)\, ds : n \in \mathbb{N}\}$ *is bounded. Then* $u(t) = 0$, $t \in [0, T]$.

The following theorem will be reconsidered in Theorem 2.10.44 for abstract multi-term fractional differential equations.

Theorem 2.1.34. *Suppose* $\alpha > 0$, $\lambda > 0$, $\{(n\lambda)^\alpha : n \in \mathbb{N}\} \subseteq \rho_C(A)$ *and, for every* $\sigma > 0$ *and* $x \in E$, $\lim_{n\to\infty} \frac{((n\lambda)^\alpha - A)^{-1}Cx}{e^{n\lambda\sigma}} = 0$. *Then, for every* $x_0, \cdots, x_{\lceil\alpha\rceil-1} \in E$, *there exists at most one solution of the initial value problem*

(68)
$$\begin{cases} u \in C^{\lceil\alpha\rceil}((0, \infty) : E) \cap C^{\lceil\alpha\rceil-1}([0, \infty) : E), \\ \mathbf{D}_t^\alpha u(t) = Au(t),\ t > 0, \\ u^{(j)}(0) = x_j,\ 0 \leqslant j \leqslant \lceil\alpha\rceil - 1. \end{cases}$$

Proof. Suppose first $\alpha = k \in \mathbb{N}\setminus\{1\}$. Put $v(t) := (u(t), u'(t), \cdots, u^{(k-1)}(t))$, $t \geqslant 0$. Then

$$\begin{cases} v \in C^1((0,\infty) : E^k) \cap C([0,\infty) : E^k), \\ v'(t) = \mathcal{A}_k v(t),\ t > 0, \\ v(0) = (x_0, \cdots, x_{k-1}). \end{cases}$$

By Lemma 2.1.33(i), we get that $\{n\lambda : n \in \mathbb{N}\} \subseteq \rho_{C_k}(\mathcal{A}_k)$. Furthermore, the representation formulae (66)-(67) imply that, for every $\sigma > 0$ and $\mathbf{x} \in E^k$, we have $\lim_{n\to\infty} \frac{(n\lambda-\mathcal{A}_k)^{-1}C_k\mathbf{x}}{e^{n\lambda\sigma}} = 0$. Now the claimed assertion follows from an application of Lemma 2.1.33(ii). Suppose now $\alpha \in (0,\infty)\setminus\mathbb{N}$ and $u(t)$ is a solution of initial value problem (68) with $x_j = 0$, $0 \leqslant j \leqslant \lceil\alpha\rceil - 1$. Set $z_n(t) := ((n\lambda)^\alpha - A)^{-1} Cu(t)$, $t \geqslant 0$, $n \in \mathbb{N}$. Then it can be easily verified that $z_n(t)$ is a solution of the initial value problem:

$$\begin{cases} z_n \in C^{\lceil\alpha\rceil}((0,\infty) : E) \cap C^{\lceil\alpha\rceil-1}([0,\infty) : E), \\ \mathbf{D}_t^\alpha z_n(t) = (n\lambda)^\alpha z_n(t) - Cu(t),\ t > 0, \\ z_n^{(j)}(0) = 0,\ 0 \leqslant j \leqslant \lceil\alpha\rceil - 1. \end{cases}$$

This implies $z_n(t) = -(u * \cdot^{\alpha-1}E_{\alpha,\alpha}((n\lambda)^\alpha \cdot^{\alpha-1}))(t)$, $t \geqslant 0$, $n \in \mathbb{N}$. By the given assumptions, it follows that, for every $t > 0$ and $\sigma > 0$,

$$\lim_{n\to\infty} e^{-n\lambda\sigma} \int_0^t s^{\alpha-1}E_{\alpha,\alpha}((n\lambda)^\alpha s^\alpha)Cu(t-s)\,ds = 0. \tag{69}$$

We consider separately two possible cases: $0 < \alpha < 4$ and $\alpha > 4$. Let $p \in \circledast$, $t > 0$ and $0 < \sigma_0 < \sigma < t$. In the first case, one can apply the identity $E_{\alpha,\alpha}(z) = zE_{\alpha,2\alpha}(z) + \frac{1}{\Gamma(\alpha)}$, $z \in \mathbb{C}$, and Theorem 1.3.1 with $N = 2$, to obtain the existence of a positive real polynomial $P(x)$ and positive real numbers $T > 0$ and $M \geqslant 1$ such that $n\lambda\sigma_0 \geqslant T$,

$$\left| s^{\alpha-1}(n\lambda s)^\alpha E_{\alpha,2\alpha}((n\lambda s)^\alpha) - \frac{1}{\alpha}(n\lambda)^{1-\alpha} e^{n\lambda s} \right| \leqslant g_\alpha(s) + MT^{-\alpha}, \text{ if } n\lambda s \geqslant T,$$

and

$$\begin{aligned} &e^{-n\lambda\sigma} \int_0^t \left| s^{\alpha-1} E_{\alpha,\alpha}((n\lambda)^\alpha s^\alpha) - \frac{1}{\alpha}(n\lambda)^{1-\alpha} e^{n\lambda s} \right| p(Cu(t-s))\,ds \\ &\leqslant P(n)e^{n\lambda(\sigma_0-\sigma)} \int_0^{\sigma_0} (1 + s^{\alpha-1})p(Cu(t-s))\,ds \\ &+ e^{-n\lambda\sigma} \int_{\sigma_0}^t (2g_\alpha(s) + MT^{-\sigma})p(C(t-s))\,ds. \end{aligned}$$

Because (69) holds and since p was arbitrary, the above implies $\lim_{n\to\infty} \int_0^t e^{n\lambda(t-s-\sigma)} Cu(s)\,ds = 0$. Keeping in mind that $\lim_{n\to\infty} \int_{t-\sigma}^t e^{n\lambda(t-s-\sigma)}Cu(s)\,ds = 0$, we obtain $\lim_{n\to\infty} \int_0^{t-\sigma} e^{n\lambda(t-s-\sigma)}Cu(s)\,ds = 0$. Making use of the injectiveness of C and Lemma 2.1.33(iii), it follows that $u(s) = 0$, $s \in [0, t-\sigma]$, which completes the proof. In the second case, we assume additionally $\sigma > t\cos(\frac{2\pi}{\alpha})$. Then a similar line of reasoning shows that there exist a positive real polynomial $P(x)$ and two positive real numbers $T > 0$ and $M \geqslant 1$ such that $n\lambda\sigma_0 \geqslant T$,

$$\Bigg| s^{\alpha-1}(n\lambda s)^{\alpha} E_{\alpha,2\alpha}((n\lambda s)^{\alpha}) - \frac{1}{\alpha}(n\lambda)^{1-\alpha} e^{n\lambda s} - \frac{1}{\alpha}(n\lambda)^{1-\alpha} \sum_{k\in\mathbb{Z}\setminus\{0\},|k|\leqslant\lfloor\alpha/4\rfloor} e^{n\lambda s e^{2\pi ik/\alpha}} \Bigg| \leqslant g_{\alpha}(s) + MT^{-\alpha}, \text{ if } n\lambda s \geqslant T,$$

and

$$e^{-n\lambda\sigma} \int_0^t \Bigg| s^{\alpha-1} E_{\alpha,\alpha}((n\lambda)^{\alpha}s^{\alpha}) - \frac{1}{\alpha}(n\lambda)^{1-\alpha} e^{n\lambda s} - \frac{1}{\alpha}(n\lambda)^{1-\alpha} \sum_{k\in\mathbb{Z}\setminus\{0\},|k|\leqslant\lfloor\alpha/4\rfloor} e^{n\lambda s e^{2\pi ik/\alpha}} \Bigg| p(Cu(t-s))\,ds$$
$$\leqslant P(n)e^{n\lambda(\sigma_0-\sigma)} \int_0^{\sigma_0} (1+s^{\alpha-1})p(Cu(t-s))\,ds + e^{-n\lambda\sigma} \int_{\sigma_0}^t (2g_{\alpha}(s) + MT^{-\alpha})p(C(t-s))\,ds.$$

Therefore,

$$\lim_{n\to\infty} e^{-n\lambda\sigma} \int_0^t \Bigg[e^{n\lambda s} + \sum_{k\in\mathbb{Z}\setminus\{0\},|k|\leqslant\lfloor\alpha/4\rfloor} e^{n\lambda s e^{2\pi ik/\alpha}} \Bigg] Cu(t-s)\,ds = 0.$$

Keeping in mind that $\sigma > t\cos(\frac{2\pi}{\alpha})$, we obtain

$$\lim_{n\to\infty} e^{-n\lambda\sigma} \int_0^t \sum_{k\in\mathbb{Z}\setminus\{0\},|k|\leqslant\lfloor\alpha/4\rfloor} e^{n\lambda s e^{2\pi ik/\alpha}}\, Cu(t-s)\,ds = 0,$$

so that $\lim_{n\to\infty} \int_0^t e^{n\lambda(t-s-\sigma)}Cu(s)\,ds = 0$ and $\lim_{n\to\infty} \int_0^{t-\sigma} e^{n\lambda(t-s-\sigma)}Cu(s)\,ds = 0$. Since C is injective, Lemma 2.1.33(iii) implies by letting $\sigma \to t\cos(\frac{2\pi}{\alpha})$ that $u(s) = 0$, $s \in [0, t(1-\cos(\frac{2\pi}{\alpha}))]$. The proof of theorem is thereby completed. □

In the remaining part of this subsection, we shall always assume that E is a non-trivial complex Banach space. We shall basically follow the notation used in the monograph [463].

Of concern are the following abstract Volterra equations on the line:

(70) $$u(t) = \int_0^{\infty} a(s)Au(t-s)\,ds + \int_{-\infty}^t k(t-s)g'(s)\,ds,$$

where $g : \mathbb{R} \to E$, $a \in L^1_{loc}([0,\infty))$, $a \neq 0$, $k \in C([0,\infty))$, $k \neq 0$, and

(71) $$u(t) = f(t) + \int_0^t a(t-s)Au(s)\,ds,\ t \in (-\tau, \tau),$$

where $\tau \in (0, \infty]$, $a \in L^1_{loc}((-\tau, \tau))$ and $f \in C((-\tau, \tau) : E)$. Notice that the equation (70) appears in the study of problem of heat flow with memory ([443]); for some other applications and qualitative properties of abstract Volterra equations on the line, we refer the reader to [99], [128], [261], [263] and [463].

Proposition 2.1.35. *Assume A is a subgenerator of a global (a, k)-regularized C-resolvent family $(S(t))_{t\geq 0}$, $g : \mathbb{R} \to R(C)$, $C^{-1}g(\cdot)$ is differentiable for a.e. $t \in \mathbb{R}$, $C^{-1}g(t) \in D(A)$ for a.e. $t \in \mathbb{R}$,*

(i) *the mapping $s \mapsto S(t-s)(C^{-1}g)'(s)$, $s \in (-\infty, t]$ is an element of the space $L^1((-\infty, t] : [D(A)])$ for a.e. $t \in \mathbb{R}$, and*
(ii) *the mapping $s \mapsto k(t-s)g'(s)$, $s \in (-\infty, t]$ is an element of the space $L^1((-\infty, t] : E)$ for a.e. $t \in \mathbb{R}$.*

Put $u(t) := \int_{-\infty}^{t} S(t-s)(C^{-1}g)'(s)\,ds$, $t \in \mathbb{R}$. Then $C(\mathbb{R} : E) \ni u$ satisfies (70).

Proof. The continuity of $u(t)$ can be proved by using the dominated convergence theorem and the strong continuity of $(S(t))_{t\geq 0}$. The proof of (70) follows from the next computation:

$$\int_0^{\infty} a(s)Au(t-s)\,ds + \int_{-\infty}^{t} k(t-s)g'(s)\,ds$$

$$= \int_0^{\infty} a(s)A \int_{-\infty}^{t-s} S(t-s-r)(C^{-1}g)'(r)\,dr\,ds + \int_{-\infty}^{t} k(t-s)g'(s)\,ds$$

$$= \int_0^{\infty}\int_0^{s'} a(s'-r')AS(r')(C^{-1}g)'(t-s')\,dr'\,ds' + \int_{-\infty}^{t} k(t-s)g'(s)\,ds$$

$$= \int_0^{\infty} (S(s') - k(s')C)(C^{-1}g)'(t-s')\,ds' + \int_{-\infty}^{t} k(t-s)g'(s)\,ds$$

$$= u(t) - \int_0^{\infty} k(s)g'(t-s')\,ds' + \int_{-\infty}^{t} k(t-s)g'(s)\,ds = u(t),\ t \in \mathbb{R}.$$

□

Denote by $AP(E)$, $AA(E)$, $AA_c(E)$ and $AAA(E)$ the spaces which consist of all almost periodic functions, almost automorphic functions, compact almost automorphic functions and asymptotically almost automorphic functions defined on $\mathbb{R}$, respectively, and assume that the function $(C^{-1}g)'(t)$ belongs to one of these spaces ([212]). By [74, Theorem 4.6], the uniform integrability of $(S(t))_{t\geq 0}$ implies that the solution $u(t)$ of (70) belongs to the same space as $(C^{-1}g)'(t)$. The above assertion remains true in nonscalar cases.

A function $u \in C((-\tau, \tau) : E)$ is called a (mild) solution of (71) if $a * u \in C((-\tau, \tau) : [D(A)])$ and $u(t) = f(t) + A\int_0^t a(t-s)u(s)\,ds$, $t \in (-\tau, \tau)$.

Proposition 2.1.36. ([302])

(i) *Assume* $a \in L^1_{loc}((-\tau, \tau))$, $k \in C((-\tau, \tau))$, $a \neq 0$ *and* $k \neq 0$. *Let* $k_+(t) = k(t)$, $a_+(t) = a(t)$, $t \in [0, \tau)$, $k_-(t) = k(-t)$ *and* $a_-(t) = a(-t)$, $t \in (-\tau, 0]$. *If* $\pm A$ *are subgenerators of* $(a_\pm, k_\pm)$*-regularized C-resolvent families* $(S_\pm(t))_{t\in[0,\tau)}$, *then, for every* $x \in D(A)$, *the function* $u : (-\tau, \tau) \to E$ *given by* $u(t) = S_+(t)x$, $t \in [0, \tau)$ *and* $u(t) = S_-(-t)x$, $t \in (-\tau, 0]$ *is a solution of (71) with* $f(t) = k(t)Cx$, $t \in (-\tau, \tau)$. *Furthermore, the solutions of (71) are unique provided that* $k_\pm(t)$ *are kernels (the blank hypothesis).*

(ii) *Assume* $n_\pm \in \mathbb{N}$, $f \in C((-\tau, \tau) : E)$, $a \in L^1_{loc}((-\tau, \tau))$, $a \neq 0$, $f_+(t) = f(t)$, $a_+(t) = a(t)$, $t \in [0, \tau)$, $f_-(t) = f(-t)$, $a_-(t) = a(-t)$, $t \in (-\tau, 0]$, *and* $\pm A$ *are subgenerators of* $(n_\pm-1)$*-times integrated* $(a_\pm, C_\pm)$*-regularized resolvent families. Assume, additionally,* $a_\pm \in BV_{loc}([0, \tau))$ *if* $n_\pm > 1$ *(that is:* $a_+ \in BV_{loc}([0, \tau))$ *if* $n_+ > 1$, *and* $a_- \in BV_{loc}([0, \tau))$ *if* $n_- > 1$*) as well as:*
   (ii.1) $C^{-1}_\pm f_\pm \in C^{n_\pm}([0, \tau) : E)$, $f^{(k-1)}_\pm(0) \in D(A^{n_\pm-k})$ *and* $A^{n_\pm-k}f^{(k-1)}_\pm(0) \in R(C_\pm)$, $1 \leqslant k \leqslant n_\pm$, *if* $n_\pm > 1$, *resp.*
   (ii.2) $C^{-1}_\pm f_\pm \in C([0, \tau) : E) \cap W^{1,1}_{loc}([0, \tau) : E)$ *if* $n_+ = n_- = 1$. *Then there exists a unique solution of (71).*

Example 2.1.37. Let $E := L^2[0, \pi]$, $A := -\Delta$ with the Dirichlet or Neumann boundary conditions, $\tau = \infty$, $\beta \in [\frac{1}{2}, 1)$, $\alpha > 1 + \beta$, $a(t) = g_\beta(|t|)$, $t \in (-\tau, \tau)$ and $f(t) = \mathcal{L}^{-1}(\mathbf{h}_{\alpha,\beta}(\lambda))(|t|)$, $t \in (-\tau, \tau)$, where $\mathbf{h}_{\alpha,\beta}(\lambda)$ is defined through [302, (2.64)]. Then there exists a unique solution $u(t)$ of (71) and $u_{|\mathbb{R}\setminus\{0\}}$ is analytically extendible to the sector $\Sigma_{\frac{\pi}{2}\left(\frac{1}{\beta}-1\right)}$. By Proposition 2.1.36(i) and [302, Example 2.31(iii)], it follows that, for every $n \in \mathbb{N}$, there exists an exponentially bounded kernel $k_n(t)$ such that (71) has a unique solution $u_n(t)$ with A replaced by the polyharmonic operator Δ^{2^n} and $f(t)$ replaced by $k_n(t)$; moreover, $u_{n|\mathbb{R}\setminus\{0\}}$ is analytically extendible to the sector $\Sigma_{\frac{\pi}{2}}$. We refer the reader to [302] for the analysis of the preceding example in the case $\beta \in [1, 2)$.

The theory of (a, k)-regularized C-resolvent families has not been fully exploited in the analysis of abstract differential equations with Riemann-Liouville fractional derivatives ([24], [68], [195]-[196] and [256]-[257]), as the next illustrative example shows.

Example 2.1.38. In [257], L. Kexue, P. Jigen and J. Junxiong have analyzed the well-posedness of the following inhomogeneous fractional Cauchy problem:

$$\begin{cases} D^\alpha_t u(t) = Au(t) + f(t), \ t > 0, \\ (g_{2-\alpha} * u)(0) = 0, \ (g_{2-\alpha} * u)'(0) = x, \end{cases}$$

where A is a closed densely defined operator on a complex Banach space E, D^α_t denotes the operator valued Riemann-Liouville fractional derivative of order $\alpha \in (1, 2)$, $x \in E$ and $f : (0, \infty) \to E$. In order to do that, the authors have introduced in [257, Definition 3.1] the notion of an α-order fractional resolvent $(T(t))_{t>0}$. We would like to observe that [257, Theorem 3.12] in combination with Theorem 2.1.5 immediately implies that $(T(t))_{t\geqslant 0}$ is nothing else but an exponentially bounded (g_α, g_α)-regularized resolvent family on E.

Now we focus our attention towards the analysis of previously stated results for hyperbolic equations of non-scalar type. By X and Y we denote non-trivial complex Banach spaces such that Y is continuously embedded in X; $L(X) \ni C$ is an injective operator and $\tau \in (0, \infty]$. The norm in X, resp. Y, is denoted by $\|\cdot\|_X$, resp. $\|\cdot\|_Y$. Let $A(t)$ be a locally integrable function from $(-\tau, \tau)$ into $L(Y, X)$. In the sequel, we assume that $A(t)$ is not of scalar type, which means that there does not exist a function $a \in L^1_{loc}((-\tau, \tau))$, $a \neq 0$, and a closed linear operator A in X such that $Y = [D(A)]$ and $A(t) = a(t)A$ for a.e. $t \in (-\tau, \tau)$.

Definition 2.1.39. ([317]) Let $\tau \in (0, \infty]$, $k \in C([0, \tau))$, $k \neq 0$ and $A \in L^1_{loc}([0, \tau) : L(Y, X))$, $A \neq 0$. An operator family $(S(t))_{t\in[0,\tau)}$ is called an (A, k)-regularized C-pseudoresolvent family if the following holds:

(S1) The mapping $t \mapsto S(t)x$, $t \in [0, \tau)$ is continuous in X for every fixed $x \in X$ and $S(0) = k(0)C$.

(S2) Put $U(t)x := \int_0^t S(s)x\, ds$, $x \in X$, $t \in [0, \tau)$. Then (S2) is equivalent to say that $U(t)Y \subseteq Y$, $U(t)_{|Y} \in L(Y)$, $t \in [0, \tau)$ and that $(U(t)_{|Y})_{t\in[0,\tau)}$ is locally Lipschitz continuous in $L(Y)$.

(S3) The resolvent equations

$$S(t)y = k(t)Cy + \int_0^t A(t-s)\, dU(s)y\, ds,\ t \in [0, \tau),\ y \in Y, \tag{72}$$

$$S(t)y = k(t)Cy + \int_0^t S(t-s)A(s)y\, ds,\ t \in [0, \tau),\ y \in Y, \tag{73}$$

hold; (72), resp. (73), are called the first resolvent equation, resp. the second resolvent equation.

An (A, k)-regularized C-pseudoresolvent family $(S(t))_{t\in[0,\tau)}$ is said to be an (A, k)-regularized C-resolvent family if additionally:
(S4) For every $y \in Y$, $S(\cdot)y \in L^\infty_{loc}([0, \tau) : Y)$.
A family $(S(t))_{t\in[0,\tau)}$ in $L(X)$ is called a weak (A, k)-regularized C-pseudoresolvent family if (S1) and (73) hold; in the case $\tau = \infty$, $(S(t))_{t\geqslant 0}$ is said to be exponentially bounded if there exist $M \geqslant 1$ and $\omega \geqslant 0$ such that $\|S(t)\|_{L(X)} \leqslant Me^{\omega t}$, $t \geqslant 0$.

A (weak) (A, k)-regularized C-(pseudo)resolvent family with $k(t) \equiv g_{\alpha+1}(t)$ $(\alpha \geqslant 0)$ is also said to be a (weak) α-times integrated A-regularized C-(pseudo) resolvent family; a (weak) 0-times integrated A-regularized C-(pseudo)resolvent family is also said to be a (weak) A-regularized C-(pseudo)resolvent family, and a (weak) (A, k)-regularized C-(pseudo)resolvent family with $C = I$ is also said to be a (weak) (A, k)-regularized (pseudo)resolvent family.

Let us consider the equations

$$(74)\qquad u(t) = \int_0^{\infty} A(s)u(t-s)\,ds + \int_{-\infty}^{t} k(t-s)g'(s)\,ds,$$

where $g : \mathbb{R} \to X$, $A \in L^1_{loc}([0, \infty) : L(Y, X))$, $A \neq 0$, $k \in C([0, \infty))$, $k \neq 0$, and

$$(75)\qquad u(t) = f(t) + \int_0^{t} A(t-s)u(s)\,ds,\ t \in (-\tau, \tau),$$

where $\tau \in (0, \infty]$, $f \in C((-\tau, \tau) : X)$ and $A \in L^1_{loc}((-\tau, \tau) : L(Y, X))$, $A \neq 0$.

The following proposition can be applied to a class of nonscalar parabolic equations considered by A. Friedman and M. Shinbrot in [190].

Proposition 2.1.40. *Assume that there exists an (A, k)-regularized C-resolvent family $(S(t))_{t\geq 0}$, $g : \mathbb{R} \to R(C)$, $C^{-1}g(\cdot)$ is differentiable for a.e. $t \in \mathbb{R}$, $C^{-1}g(t) \in Y$ for a.e. $t \in \mathbb{R}$,*

(i) *the mapping $s \mapsto S(t-s)(C^{-1}g)'(s)$, $s \in (-\infty, t]$ is an element of the space $L^1((-\infty, t] : Y)$ for a.e. $t \in \mathbb{R}$, and*
(ii) *the mapping $s \mapsto k(t-s)g'(s)$, $s \in (-\infty, t]$ is an element of the space $L^1((-\infty, t] : X)$ for a.e. $t \in \mathbb{R}$.*

Let $u(t) = \int_{-\infty}^{t} S(t-s)(C^{-1}g)'(s)\,ds$, $t \in \mathbb{R}$. Then $C(\mathbb{R} : X) \ni u$ satisfies (74).

A function $u \in C((-\tau, \tau) : X)$ is said to be:

(i) a strong solution of (75) if $u \in L^{\infty}_{loc}((-\tau, \tau) : Y)$ and (75) holds on $(-\tau, \tau)$,
(ii) a mild solution of (75) if there exists a sequence (f_n) in $C((-\tau, \tau) : X)$ and a sequence (u_n) in $C([0, \tau) : X)$ such that $u_n(t)$ is a strong solution of (75) with $f(t)$ replaced by $f_n(t)$ as well as that $\lim_{n\to\infty} f_n(t) = f(t)$ and $\lim_{n\to\infty} u_n(t) = u(t)$, uniformly on compact subsets of $(-\tau, \tau)$.

Proposition 2.1.41. ([306])

(i) *Assume $k \in C((-\tau, \tau))$, $k \neq 0$ and $A \in L^1_{loc}((-\tau, \tau) : L(Y, X))$, $A \neq 0$. Let $k_+(t) = k(t)$, $A_+(t) = A(t)$, $t \in [0, \tau)$, $k_-(t) = k(-t)$ and $A_-(t) = -A(-t)$, $t \in (-\tau, 0]$. If there exist $(A_\pm, k_\pm)$-regularized C-resolvent families $(S_\pm(t))_{t\in[0,\tau)}$, then for every $x \in Y$ the function $u : (-\tau, \tau) \to X$ given by $u(t) = S_+(t)x$, $t \in [0, \tau)$ and $u(t) = S_-(-t)x$, $t \in (-\tau, 0]$ is a strong solution of (75) with $f(t) = k(t)Cx$, $t \in (-\tau, \tau)$. Furthermore, strong solutions of (75) are unique provided that $k_\pm(t)$ are kernels.*
(ii) *Assume $n_\pm \in \mathbb{N}$, $f \in C((-\tau, \tau) : X)$, $A \in L^1_{loc}((-\tau, \tau) : L(Y, X))$, $A \neq 0$, $f_+(t) = f(t)$, $A_+(t) = A(t)$, $t \in [0, \tau)$, $f_-(t) = f(-t)$, $A_-(t) = -A(-t)$, $t \in (-\tau, 0]$ and there exist $(n_\pm - 1)$-times integrated $A_\pm$-regularized $C_\pm$-resolvent families. Let $f_\pm \in C^{n_\pm}([0, \tau) : X)$ and $f_\pm^{(i)}(0) = 0$, $0 \leqslant i \leqslant n_\pm - 1$. Then the following holds:*

(ii.1) *Let* $(C_{\pm}^{-1} f_{\pm})^{(n_{\pm}-1)} \in AC_{loc}([0, \tau) : Y)$ *and* $(C_{\pm}^{-1} f_{\pm})^{(n_{\pm})} \in L_{loc}^{1}([0, \tau) : Y)$. *Then there exists a unique strong solution* $u(t)$ *of (75), and moreover* $u \in C((-\tau, \tau) : Y)$.

(ii.2) *Let* $(C_{\pm}^{-1} f_{\pm})^{(n_{\pm})} \in L_{loc}^{1}([0, \tau) : X)$ *and* $\overline{Y}^{X} = X$. *Then there exists a unique mild solution of (75).*

Example 2.1.42. (i) ([317, Example 2.1]) Assume $-\infty < \alpha \leqslant \beta < \infty$, $1 \leqslant p \leqslant \infty$, $0 < \tau \leqslant \infty$, $n \in \mathbb{N}$, $X = L^p(\mathbb{R}^n)$ or $X = C_b(\mathbb{R}^n)$, $P(\cdot)$ is an elliptic polynomial of degree $m \in \mathbb{N}$, $\alpha \leqslant \operatorname{Re}(P(x)) \leqslant \beta$, $x \in \mathbb{R}^n$, $A = P(D)$ and $Y = [D(A)]$. Let $r > |\frac{1}{2} - \frac{1}{p}|$, $C_{\pm} = (\omega \mp A)^{-r}$ and let $a \in L_{loc}^{1}(\mathbb{R})$, $a \neq 0$, be such that the mappings $t \mapsto a_+(t)$, $t \geqslant 0$ and $t \mapsto a_-(t) = a(-t)$, $t \geqslant 0$ are completely positive kernels satisfying (P1); in the case $X = L^{\infty}(\mathbb{R}^n)$ or $X = C_b(\mathbb{R}^n)$, we assume $a(t) \equiv 1$. Suppose, in addition, $(B_{0,\pm}(t))_{t\in[0,\tau)} \subseteq L(Y) \cap L(X, [R(C_{\pm})])$, $(B_{1,\pm}(t))_{t\in[0,\tau)} \subseteq L(Y, [R(C_{\pm})])$,

(i.1) $C_{\pm}^{-1} B_{0,\pm}(\cdot)y \in BV_{loc}([0, \tau) : Y)$ for all $y \in Y$, $C_{\pm}^{-1} B_{0,\pm}(\cdot)x \in BV_{loc}([0, \tau) : X)$ for all $x \in X$,

(i.2) $C_{\pm}^{-1} B_{1,\pm}(\cdot)y \in BV_{loc}([0, \tau) : X)$ for all $y \in Y$,

(i.3) $C_{\pm} B_{\pm}(t)y = B_{\pm}(t)C_{\pm}y$, $y \in Y$, $t \in [0, \tau)$, where $B_{\pm}(t)y = B_{0,\pm}(t)y + (a_{\pm} * B_{1,\pm})(t)y$, $y \in Y$, $t \in [0, \tau)$, and

(i.4) $C_{\pm}^{-1} f_{\pm} \in AC_{loc}([0, \tau) : Y)$ and $(C_{\pm}^{-1} f_{\pm})' \in L_{loc}^{1}([0, \tau) : Y)$.

Set $B(t) := B_+(t)$, $t \in [0, \tau)$ and $B(t) := B_-(-t)$, $t \in (-\tau, 0)$. Then there exists a unique strong solution of (75) with $A(t) = a(t)P(D) + B(t)$, $t \in (-\tau, \tau)$.

(ii) ([317]) Let $1 < p < \infty$, $X = L^p(\mathbb{R})$, $Y = W^{4,p}(\mathbb{R})$,

$$A(t)f = -tf'''' - tf'' - 2if' - tf,\ t \in \mathbb{R}, f \in Y,$$

$s \in (1, 2)$ and $f(t) = k_s(t) = \mathcal{L}^{-1}(e^{-\lambda^{1/s}})(|t|)$, $t \in \mathbb{R}$. Then there exist not exponentially bounded $(\pm A(\pm t), k_s)$-regularized resolvent families, and Proposition 2.1.41(i) implies that there exists a unique strong solution $u(t)$ of (75) on $\mathbb{R}$. Finally, one can simply prove that $u(t)$ is s-hypoanalytic on $\mathbb{R}$, which means that, for every compact set $K \subseteq \mathbb{R}$, there exists $h_K > 0$ such that $\sup_{t\in K, p\in \mathbb{N}_0} \left\| \frac{h_K^p \frac{d^p}{dt^p} u(t)}{p!^s} \right\| < \infty$.

2.1.2. Convoluted *C*-semigroups and convoluted *C*-cosine functions in locally convex spaces. The main purpose of this section is to clarify the most valuable results concerning convoluted C-semigroups and convoluted C-cosine functions in locally convex spaces; for the Banach space case, we refer the reader to [102]-[104], [268], [292], [297], and to the recent papers [269] by V. Keyantuo, P. J. Miana, L. Sánchez-Lajusticia, [339]-[340] by C.-C. Kuo, [355] by F. Li, J. Liang, T.-J. Xiao, J. Zhang, [356] by F. Li, H. Wang, J. Zhang, and [415] by P. J. Miana, V. Poblete.

Definition 2.1.43. Let A be a closed operator, $\tau \in (0, \infty]$ and $K \in L_{loc}^{1}([0, \tau))$, $K \neq 0$. A strongly continuous operator family $(S_K(t))_{t\in(-\tau,\tau)}$ is called a (local, if $\tau < \infty$) K-convoluted C-group with a subgenerator A if:

(i) $(S_{K,+}(t) := S_K(t))_{t\in[0,\tau)}$, resp. $(S_{K,-}(t) := S_K(-t))_{t\in[0,\tau)}$, is a (local) K-convoluted C-semigroup with a subgenerator A, resp. $-A$, and
(ii) for every $t, s \in (-\tau, \tau)$ with $t < 0 < s$ and $x \in E$:

$$S_K(t)S_K(s)x = S_K(s)S_K(t)x$$

$$= \begin{cases} \int_{t+s}^{s} K(r-t-s)S_K(r)Cx\,dr + \int_{t}^{0} K(t+s-r)S_K(r)Cx\,dr, & t+s \geqslant 0, \\ \int_{t}^{t+s} K(t+s-r)S_K(r)Cx\,dr + \int_{0}^{s} K(r-t-s)S_K(r)Cx\,dr, & t+s < 0. \end{cases}$$

It is said that $(S_K(t))_{t\in\mathbb{R}}$ is exponentially equicontinuous (equicontinuous) if there exists $\omega \in \mathbb{R}$ ($\omega = 0$) such that the family $\{e^{-\omega|t|}S_K(t) : t \in \mathbb{R}\}$ is equicontinuous. A closed linear operator $\hat{A}$ is said to be the integral generator of $(S_K(t))_{t\in(-\tau,\tau)}$ if $\hat{A}$ is the integral generator of $(S_K(t))_{t\in[0,\tau)}$.

Plugging $K(t) = g_\alpha(t)$, $t \in [0, \tau)$ in Definition 2.1.43, where $\alpha > 0$, we obtain the definition of an α-times integrated C-group. The notions of a (local) C-regularized semigroup (cosine function) and its subgenerator are understood in the sense of consideration given in [292, Section 1.2]. With the exception of [292, Theorem 2.6.10], the structural characterizations of K-convoluted C-groups established in [292, Section 2.6] continue to hold in our framework.

In the following theorem we clarify some rescaling and perturbation properties of subgenerators of (local) K-convoluted C-semigroups in SCLCSs [292, Theorem 2.5.1-Theorem 2.5.3].

Theorem 2.1.44. (i) (a) *Suppose $z \in \mathbb{C}$, $K(t)$ and $F(t)$ satisfy (P1), there exists a > 0 such that*

$$\frac{\widetilde{K}(\lambda)-\widetilde{K}(\lambda+z)}{\widetilde{K}(\lambda+z)} = \int_0^\infty e^{-\lambda t}F(t)\,dt,\ \mathrm{Re}\,\lambda > a, \widetilde{K}(\lambda+z) \neq 0,$$

and A is a subgenerator, resp. the integral generator, of a (local) K-convoluted C-semigroup $(S_K(t))_{t\in[0,\tau)}$. Then $A-z$ is a subgenerator, resp. the integral generator, of a (local) K-convoluted C-semigroup $(S_{K,z}(t))_{t\in[0,\tau)}$, where:

$$S_{K,z}(t)x := e^{-tz}S_K(t)x + \int_0^t F(t-s)e^{-zs}S_K(s)x\,ds,\ t \in [0,\tau),\ x \in E.$$

Furthermore, in the case $\tau = \infty$, $(S_{K,z}(t))_{t\geqslant 0}$ is exponentially equicontinuous provided that $F(t)$ is exponentially bounded and $(S_K(t))_{t\geqslant 0}$ is exponentially equicontinuous.

(b) *Suppose* $z \in \mathbb{C}$, $\alpha > 0$ *and* A *is a subgenerator, resp. the integral generator, of a (local, global exponentially equicontinuous)* α*-times integrated* C*-semigroup* $(S_\alpha(t))_{t\in[0,\tau)}$. *Then* $A - z$ *is a subgenerator, resp. the integral generator, of a (local, global exponentially equicontinuous)* α*-times integrated* C*-semigroup* $(S_{\alpha,z}(t))_{t\in[0,\tau)}$, *which is given by:*

$$S_{\alpha,z}(t)x = e^{-zt}S_\alpha(t)x + \int_0^t \sum_{n=1}^{\infty} \binom{\alpha}{n} \frac{z^n t^{n-1}}{(n-1)!} e^{-zs} S_\alpha(s)x\, ds,$$

for any $t \in [0, \tau)$ *and* $x \in E$.

(ii) *Suppose* $B \in L(E)$, $K(t)$ *is a kernel and satisfies (P1),* A *is a subgenerator (the integral generator) of a (local)* K*-convoluted* C*-semigroup* $(S_K(t))_{t\in[0,\tau)}$, $BA \subseteq AB$, $BC = CB$, *there exist* $M > 0$ *and* $a > 0$ *such that* $p(Bx) \leqslant Mp(x)$, $x \in E$, $p \in \circledast$ *and that the following conditions hold:*

(a) *For every* $n \in \mathbb{N}$, *there is a function* $K_n(\cdot)$ *satisfying (P1) and*

$$\widetilde{K_n}(\lambda) = \widetilde{K}(\lambda)\left(\frac{1}{\widetilde{K}(\cdot)}\right)^{(n)}(\lambda),\ \lambda > a, \widetilde{K}(\lambda) \neq 0.$$

Put $\Theta_n(t) := \int_0^t |K_n(s)|\, ds$, $t \geqslant 0$, $n \in \mathbb{N}$.

(b) $\sum_{n=1}^{\infty} \Theta_n(t) < \infty$, $t \geqslant 0$.

(c) *The function* $t \mapsto \max_{s\in[0,t]} |\Theta(s)| e^{-at} \sum_{n=1}^{\infty} \Theta_n(t)$, $t \geqslant 0$ *is an element of the space* $L^1([0,\infty) : \mathbb{R})$.

Then $A + B$ *is a subgenerator (the integral generator) of a (local)* K*-convoluted* C*-semigroup* $(S_K^B(t))_{t\in[0,\tau)}$, *given by*

$$S_K^B(t)x := e^{tB}S_K(t)x + \sum_{i=1}^{\infty}\sum_{n=1}^{i} \frac{B^i}{i!}(-1)^n \binom{i}{n} \int_0^t K_n(t-s)s^{i-n}S_K(s)x\, ds,$$

for any $t \in [0, \tau)$ *and* $x \in E$. *Furthermore,* $p(S_K^B(t)x - e^{tB}S_K(t)x) \leqslant e^M \max_{s\in[0,t]} p(S_K(s)x) \Sigma_{n=1}^{\infty} \Theta_n(t)e^{Mt}$, $t \in [0, \tau)$, $x \in E$, $p \in \circledast$, *and the assumption* $\tau = \infty$, *the exponential equicontinuity of* $(S_K(t))_{t\geqslant 0}$ *and the existence of numbers* $M_1 > 0$ *and* $\omega \geqslant 0$ *such that* $\Sigma_{n=1}^{\infty} \Theta_n(t) \leqslant Me^{\omega t}$, $t \geqslant 0$, *imply that* $(S_K^B(t))_{t\in[0,\tau)}$ *is also exponentially equicontinuous.*

(iv) *Suppose* $\alpha > 0$, A *is a subgenerator, resp. the integral generator, of a (local, global exponentially equicontinuous)* α*-times integrated* C*-semigroup* $(S_\alpha(t))_{t\in[0,\tau)}$, $B \in L(E)$, $BA \subseteq AB$, $BC = CB$ *and there exist* $M > 0$ *such that* $p(Bx) \leqslant Mp(x)$, $x \in E$, $p \in \circledast$. *Then* $A + B$ *is a subgenerator, resp. the integral generator, of a (local, global exponentially equicontinuous)* α*-times integrated* C*-semigroup* $(S_\alpha^B(t))_{t\in[0,\tau)}$, *which is given by*

$$S_\alpha^B(t)x := e^{tB}S_\alpha(t)x + \sum_{i=1}^{\infty}\sum_{n=1}^{i} \frac{B^i}{i!}(-1)^n n \binom{i}{n}\binom{\alpha}{n} \int_0^t (t-s)^{n-1}s^{i-n}S_\alpha(s)x\, ds,$$

for any $t \in [0, \tau)$ and $x \in E$. The above formula can also be rewritten in the following form:

$$S^B_\alpha(t)x = e^{tB}S_\alpha(t)x + \sum_{i\geqslant 1}\binom{\alpha}{i}(-B)^i\int_0^t\frac{(t-s)^{i-1}}{(i-1)!}e^{Bs}S(s)x\ ds.$$

Remark 2.1.45. Using [292, Remark 2.5.4(iii)], the conditions (iii)(b)-(c) can be slightly modified so that Theorem 2.1.44(iii) applies to a class of functions of the form $K_{a,c}(t) \equiv \mathcal{L}^{-1}(e^{-a\lambda^c})(t)$ ($a > 0$, $c \in (0, 1)$).

Recall that the exponential region $E(a, b)$ ($a, b > 0$) has been defined in [11] by $E(a, b) := \{\lambda \in \mathbb{C} : \operatorname{Re}\lambda \geqslant b, |\operatorname{Im}\lambda| \leqslant e^{a\operatorname{Re}\lambda}\}$; set $E^2(a, b) := \{\lambda^2 : \lambda \in E(a, b)\}$. Sufficient conditions for the generation of local K-convoluted C-semigroups and cosine functions in SCLCSs [292, Theorem 2.7.3, Theorem 2.7.4, Remark 2.7.5] are stated in the following theorem.

Theorem 2.1.46. (i) *Suppose $K(t)$ satisfies (P1), $r_0 \geqslant \max(0, abs(K))$, $\Phi : [r_0, \infty) \to [0, \infty)$ is strictly increasing and there exists a strictly increasing sequence (n_p) in $[r_0, \infty)$ such that the function $\Phi(t)$ is of class C^1 in $[r_0, \infty)\setminus\{n_p : p \in \mathbb{N}\}$. Suppose, further, $\lim_{t\to\infty}\Phi(t) = +\infty$, $\Phi'(\cdot)$ is bounded on $[r_0, \infty)\setminus\{n_p : p \in \mathbb{N}\}$ and there exist $\alpha > 0$, $\beta > r_0$ and $\gamma > 0$ such that*

$$\Psi_{\alpha,\beta,\gamma} := \left\{\lambda \in \mathbb{C} : \operatorname{Re}\lambda \geqslant \frac{\Phi(\alpha\,|\operatorname{Im}\lambda|)}{\gamma} + \beta\right\} \subseteq \rho_C(A).$$

Designate by $\Gamma_{\alpha,\beta,\gamma}$ the upwards oriented boundary of $\Psi_{\alpha,\beta,\gamma}$ and by $\Omega_{\alpha,\beta,\gamma}$ the open region which lies to the right of $\Gamma_{\alpha,\beta,\gamma}$. Let the following conditions hold.

(a) *For every $x \in E$, the mapping $\lambda \mapsto \tilde{K}(\lambda)(\lambda - A)^{-1}Cx$, $\lambda \in \overline{\Omega_{\alpha,\beta,\gamma}}$ is continuous.*

(b) *There exist $M > 0$ and $\sigma > 0$ such that the family $\{e^{\Phi(\sigma|\lambda|)}\tilde{K}(\lambda)(\lambda - A)^{-1}C : \lambda \in \Omega_{\alpha,\beta,\gamma}\}$ is equicontinuous.*

(c) *There exists a function $m : [0, \infty) \to (0, \infty)$ such that $m(s) = 1$, $s \in [0, 1]$ and that, for every $s > 1$, there exists a number $r_s > r_0$ with*
$$\frac{\Phi(t)}{\Phi(st)} \geqslant m(s),\ t \geqslant r_s.$$

(d) $\lim_{t\to\infty} te^{-\Phi(\sigma t)} = 0$.

(e) $(\exists a \geqslant 0)\,(\exists r'_a > r_0)(\forall t > r'_a)\,\frac{\ln t}{\Phi(t)} \geqslant a$.

Then the operator A is a subgenerator of a local K-convoluted C-semigroup on $[0, a+ m(\frac{a}{\sigma\gamma}))$.

(ii) *Suppose $\alpha > 0$, $a > 0$, $b > 0$, $M > 0$, $E(a, b) \subseteq \rho_C(A)$, the family $\{(1+|\lambda|)^{-\alpha}(\lambda - A)^{-1}C : \lambda \in E(a, b)\}$ is equicontinuous and the mapping $\lambda \mapsto (\lambda - A)^{-1}Cx$, $\lambda \in E(a, b)$ is continuous for every $x \in E$. Then, for every $\beta \in (\alpha + 1, \infty)$, A is a subgenerator of a local β-times integrated C-semigroup $(S_\beta(t))_{t\in[0,a(\beta-\alpha-1))}$.*

(iii) *Suppose $K(t)$ satisfies (P1), $r_0 \geqslant \max(0, abs(K))$, the mapping $\Phi : [r_0, \infty) \to [0, \infty)$ is strictly increasing and there exists a strictly increasing sequence (n_p) in $[r_0, \infty)$ such that the function $\Phi(t)$ is of class C^1 in $[r_0, \infty)\setminus\{n_p : p \in \mathbb{N}\}$.*

Suppose, further, $\lim_{t\to\infty}\Phi(t) = +\infty$, $\Phi'(t)$ *is bounded on* $[r_0, \infty)\setminus\{n_p : p \in \mathbb{N}\}$ *and there exist* $\alpha > 0$, $\beta > r_0$ *and* $\gamma > 0$ *such that*

$$\Psi^2_{\alpha,\beta,\gamma} := \{\lambda^2 : \lambda \in \Psi_{\alpha,\beta,\gamma}\} \subseteq \rho_C(A).$$

Designate by $\Gamma_{\alpha,\beta,\gamma}$ *the upwards oriented boundary of* $\Psi_{\alpha,\beta,\gamma}$ *and by* $\Omega_{\alpha,\beta,\gamma}$ *the open region which lies to the right of* $\Gamma_{\alpha,\beta,\gamma}$*. Let the following conditions hold.*

(i) *For every* $x \in E$, *the mapping* $\lambda \mapsto \widetilde{K}(\lambda)(\lambda^2 - A)^{-1}Cx$, $\lambda \in \overline{\Omega_{\alpha,\beta,\gamma}}$ *is continuous.*

(ii) *There exist* $M > 0$ *and* $\sigma > 0$ *such that the family* $\{e^{\Phi(\sigma|\lambda|)}\widetilde{K}(\lambda)[(\lambda^2 - A)^{-1}C + \lambda^{-1}C] : \lambda \in \overline{\Omega_{\alpha,\beta,\gamma}}\}$ *is equicontinuous.*

(iii) *The conditions (i)(c)-(e) given in the formulation of the item (i) of this theorem hold.*
Then A is a subgenerator of a local K-convoluted C-cosine function on $[0, a+ m(\frac{a}{\sigma\gamma}))$.

(iv) *Suppose* $\alpha > 0$, $a > 0$, $b > 0$, $M > 0$, $E^2(a, b) \subseteq \rho_C(A)$, *the family* $\{(1+|\lambda|)^{-\alpha}(\lambda^2 - A)^{-1}C : \lambda \in E(a, b)\}$ *is equicontinuous and the mapping* $\lambda \mapsto (\lambda^2 - A)^{-1}Cx$, $\lambda \in E(a, b)$ *is continuous for every* $x \in E$. *Then, for every* $\beta \in (\alpha + 2, \infty)$, *A is a subgenerator of a local β-times integrated C-cosine function* $(C_\beta(t))_{t\in[0,a(\beta-\alpha-1))}$.

Example 2.1.47. ([51], [105], [292]) Suppose $x_0 > 0$, $\omega : [0, \infty) \to [0, \infty)$ is a continuous, concave, increasing function which satisfies that $\lim_{t\to\infty}\omega(t) = \infty$, $\lim_{t\to\infty}\omega(t)/t = 0$ and $\int_1^\infty \frac{\omega(t)}{t^2}\,dt < \infty$. Put $\Omega(\omega) := \{\lambda \in \mathbb{C} : \operatorname{Re}\lambda \geqslant \max(x_0, \omega(|\operatorname{Im}\lambda|))\}$, and assume that A is a closed linear operator which satisfies:

(i) $\Omega(\omega) \subseteq \rho_C(A)$ and the family $\{e^{-\omega(\sigma|\lambda|)}(1+|\lambda|)^{-n}(\lambda - A)^{-1}C : \lambda \in \Omega(\omega)\}$ is equicontinuous for some $\sigma > 0$ and $n \in \mathbb{N}$.

(ii) For every $x \in E$, the mapping $\lambda \to (\lambda - A)^{-1}Cx$, $\lambda \in \Omega(\omega)$ is continuous.

Combining [101, Lemma 4.5], Theorem 2.1.46(i) and [292, Remark 2.7.5(ii)], it follows that there exist $\tau > 0$, $l > 0$ and a sequence (M_p) satisfying (M.1), (M.2) and (M.3)' such that A generates a local K_l-convoluted C-semigroup $(S_{K_l}(t))_{t\in[0,\tau)}$ with $K_l(t) = \mathcal{L}^{-1}(\frac{1}{\omega_l(\lambda)})(t)$, $t \in [0, \tau)$ (cf. (85)). Furthermore, the mapping $t \mapsto S_{K_l}(t)$, $t \in [0, \tau)$ is infinitely differentiable in $L(E)$ and l can be chosen such that, for every compact set $K' \subseteq [0, \tau)$, there exists $h_{K'} > 0$ satisfying that the set

$$\left\{\frac{h_K^p \frac{d^p}{dt^p} S_{K_l}(t)}{M_p} : t \in K', p \in \mathbb{N}_0\right\}$$

is bounded in $L(E)$; in the case $\sigma = 0$, one can prove that there exists a family of bounded injective operators $(C_\varepsilon)_{\varepsilon>0}$ such that, for every $\varepsilon > 0$, A is a subgenerator of a global C_ε-regularized semigroup that is differentiable in $t > 0$. If one considers the function $\omega(t) := \sigma t^c$ ($\sigma > 0$, $c \in (0, 1)$) and the region Π_{c,σ,x_0} instead of $\omega(t)$ and $\Omega(\omega)$, then there exists $a > 0$ such that the above conclusion holds with $M_p = p!^{1/c}$ and the function $K_{a,c}(t)$, which provides possible applications of Theorem 2.1.44(iii) and [302, Theorem 2.34,

Remark 2.35(ii)]. A concrete example can be simply constructed. Assuming that $P(D)$ and β possess the same meaning as in Example 2.2.14 below (we use the same terminology, notice however that there exist examples in which the above assumptions are fulfilled with $\rho(P(D)) = \emptyset$; for example, we can extract from the analysis given in [357, Example 4.6(b)] that the operator $P(D)$ in the space $L^p(\mathbb{R}^2)$, where $P(x, y) = \frac{i}{4\omega^2}((1 + x^2)^2(1 + (y - x^m)^2)^2 - 4\omega^4)$, $(x, y) \in \mathbb{R}^2$, $m \in \mathbb{N}$, $m \geqslant 5$, $\omega > 0$, $1 < p < \infty$ and $|\frac{1}{2} - \frac{1}{p}| \geqslant \frac{1}{4} + \frac{1}{l}$, is $\frac{4}{l}$-coercive and satisfies $\rho(P(D)) = \emptyset$), we have that there exists $a > 0$ such that the operator $iP(D)$ is a subgenerator of a local $K_{a,1/2}$-convoluted $\mathbf{T}_l(\langle(1 + |x|^2)^{-\beta}\rangle)$-group in E_l. By Theorem 2.2.13, we get that there exists an injective operator $C \in L(E_l)$ such that A is a subgenerator of a global C-regularized group in E_l.

In the existing literature on abstract integro-differential equations, a great number of papers deal with the properties of solutions of the equation

$$u'(t) = Au(t) + \int_0^t B(t - s)u(s)\,ds + f(t),\ t \geqslant 0,\ u(0) = x, \tag{76}$$

where A generates a strongly continuous semigroup and $B(t)$ is dominated by A in a certain sense ([94]-[95], [136], [205], [420], [506]); cf. [147] and [204] for applications of integrated semigroups in the analysis of (76). A class of second order abstract integro-differential equations of the form

$$u''(t) = Au(t) + \int_0^t B(t - s)u(s)\,ds + f(t),\ t \geqslant 0,\ u(0) = x,\ u'(0) = y, \tag{77}$$

where A generates a strongly continuous cosine function and $B \in BV_{loc}([0, \infty) : L([D(A)], E))$ has also been considered by many authors during the 1970s and 1980s ([129]-[130], [155], [507]). More recently, D. Sforza with her collaborators has investigated in a series of papers the qualitative properties of various types of abstract second-order equations of the form like (77); cf. [3], [75]-[77], [393]-[394] and [483]. We close this subsection with the observation that it could be very interesting to further analyze the equation (76), resp. (77), in the case that A subgenerates a (local) K-convoluted C-semigroup, resp. K-convoluted C-cosine function, on a sequentially complete locally convex space E.

2.2 Differential and analytical properties of (a, k)-regularized C-resolvent families

Now we shall proceed to the study of smoothing properties of (a, k)-regularized C-resolvent families in locally convex spaces.

Definition 2.2.1. (i) Let $\alpha \in (0, \pi]$, and let $(R(t))_{t \geqslant 0}$ be an (a, k)-regularized C-resolvent family. Then it is said that $(R(t))_{t \geqslant 0}$ is an analytic (a, k)-regularized C-resolvent family of angle α, if there exists a function $\mathbf{R} : \Sigma_\alpha \to L(E)$ which satisfies that, for every $x \in E$, the mapping $z \mapsto \mathbf{R}(z)x$, $z \in \Sigma_\alpha$ is analytic as well as that:

(i) $\mathbf{R}(t) = R(t)$, $t > 0$ and

(ii) $\lim_{z \to 0, z \in \Sigma_\gamma} \mathbf{R}(z)x = k(0)Cx$ for all $\gamma \in (0, \alpha)$ and $x \in E$.

It is said that $(R(t))_{t \geqslant 0}$ is an exponentially equicontinuous, analytic (a, k)-regularized C-resolvent family of angle α, resp. equicontinuous analytic (a, k)-regularized C-resolvent family of angle α, if for every $\gamma \in (0, \alpha)$, there exists $\omega_\gamma \geqslant 0$, resp. $\omega_\gamma = 0$, such that the set $\{e^{-\omega_\gamma \operatorname{Re} z}\mathbf{R}(z) : z \in \Sigma_\gamma\}$ is equicontinuous. Since there is no risk of confusion, we will identify $R(\cdot)$ and $\mathbf{R}(\cdot)$ in the sequel.

(ii) An (a, k)-regularized C-resolvent family $(R(t))_{t \geqslant 0}$ is said to be entire if, for every $x \in E$, the mapping $t \mapsto R(t)x$, $t \geqslant 0$ can be analytically extended to the whole complex plane.

Remark 2.2.2. Assume E is barreled and A is a subgenerator of an analytic (a, k)-regularized C-resolvent family $(R(t))_{t \geqslant 0}$ of angle α. Then the mapping $z \mapsto R(z)$, $z \in \Sigma_\alpha$ is analytic in $L(E)$ ([209]), and this implies that Definition 2.2.1 is consistent with [302, Definition 2.14] in the case that $\alpha \in (0, \pi/2]$ and E is a Banach space.

Proposition 2.2.3. *Suppose $k(t)$ and $a(t)$ satisfy (P1), $k(0) \neq 0$, A is densely defined, $A \notin L(E)$ and there exists $\omega_0 \geqslant \max(0, abs(k), abs(a))$ such that $\int_0^\infty e^{-\omega_0 t}|a(t)|\, dt < \infty$. Assume that A is a subgenerator of an exponentially equicontinuous, analytic (a, k)-regularized C-resolvent family $(R(t))_{t \geqslant 0}$ of angle $\alpha \in (0, \pi/2]$ and there exists $\omega \geqslant \omega_0$ such that the family*

$$(78) \qquad \{e^{-\omega z} R(z) : z \in \Sigma_\gamma\} \text{ is equicontinuous for all } \gamma \in (0, \alpha).$$

Then the function $\tilde{a}(\lambda)$ can be extended to a meromorphic function defined on the sector $\omega + \Sigma_{\frac{\pi}{2}+\alpha}$.

Theorem 2.2.4. ([303]) *Suppose $\alpha \in (0, \pi/2]$, $k(t)$ and $a(t)$ satisfy (P1), and $\tilde{k}(\lambda)$ can be analytically continued to a function $g : \omega + \Sigma_{(\pi/2)+\alpha} \to \mathbb{C}$, where $\omega \geqslant \max(0, abs(k), abs(a))$. Suppose, further, that A is a subgenerator of an analytic (a, k)-regularized C-resolvent family $(R(t))_{t \geqslant 0}$ of angle α as well as that (22) and (78) hold. Set*

$$N := \{\lambda \in \omega + \Sigma_{(\pi/2)+\alpha} : g(\lambda) \neq 0\}.$$

Then N is an open connected subset of $\mathbb{C}$. Assume that there exists an analytic function $\hat{a} : N \to \mathbb{C}$ such that $\hat{a}(\lambda) = \tilde{a}(\lambda)$, $\operatorname{Re} \lambda > \omega$. Then the operator $I - \hat{a}(\lambda)A$ is injective for every $\lambda \in N$, $R(C) \subseteq R(I - \hat{a}(\lambda)C^{-1}AC)$ for every $\lambda \in N_1 := \{\lambda \in N : \hat{a}(\lambda) \neq 0\}$, the family

(79) $\{(\lambda - \omega)g(\lambda)\,(I - \hat{a}(\lambda)C^{-1}AC)^{-1}\, C : \lambda \in N_1 \cap (\omega + \Sigma_{\frac{\pi}{2}+\gamma_1})\}$ *is equicontinuous for every $\gamma_1 \in (0, \alpha)$, the mapping*

(80) $\lambda \mapsto (I - \hat{a}(\lambda)C^{-1}AC)^{-1} Cx,\ \lambda \in N_1$ *is analytic for every* $x \in E$,

and

$$\lim_{\lambda \to +\infty, \tilde{k}(\lambda) \neq 0} \lambda \tilde{k}(\lambda)\, (I - \tilde{a}(\lambda)A)^{-1}\, Cx = k(0)Cx,\ x \in E.$$

Theorem 2.2.5. *Assume $k(t)$ and $a(t)$ satisfy (P1), $\omega \geqslant \max(0, abs(k), abs(a))$ and $\alpha \in (0, \pi/2]$. Assume, further, that A is a closed linear operator and that, for every $\lambda \in \mathbb{C}$ with $\operatorname{Re} \lambda > \omega$ and $\tilde{k}(\lambda) \neq 0$, the operator $I - \tilde{a}(\lambda) A$ is injective with $R(C) \subseteq R(I - \tilde{a}(\lambda)A)$. If there exists a function $q : \omega + \Sigma_{\frac{\pi}{2}+\alpha} \to L(E)$ such that, for every $x \in E$, the mapping $\lambda \mapsto q(\lambda)x$, $\lambda \in \omega + \Sigma_{\frac{\pi}{2}+\alpha}$ is analytic as well as that:*

(81) $$q(\lambda)x = \tilde{k}(\lambda)\, (I - \tilde{a}(\lambda)A)^{-1}\, Cx,\ \operatorname{Re} \lambda > \omega,\ \tilde{k}(\lambda) \neq 0,\ x \in E,$$

the family $\{(\lambda - \omega)q(\lambda) : \lambda \in \omega + \Sigma_{\frac{\pi}{2}+\gamma}\}$ is equicontinuous for every $\gamma \in (0, \alpha)$, and

(82) $$\lim_{\lambda \to +\infty} \lambda q(\lambda)x = k(0)Cx,\ x \in E,\ \text{if } \overline{D(A)} \neq E,$$

then A is a subgenerator of an exponentially equicontinuous, analytic (a, k)-regularized C-resolvent family $(S(t))_{t \geqslant 0}$ of angle α and the exponential type of $(S(t))_{t \geqslant 0}$ is less or equal than ω. Moreover, $S(z)A \subseteq AS(z)$, $z \in \Sigma_\alpha$.

Proof. Due to Theorem 1.2.5(i), for every $x \in E$, there exists an analytic function $S_x : \Sigma_\alpha \to E$ such that $\tilde{q}(\lambda)x = \int_0^\infty e^{-\lambda t} S_x(t)\, dt$, $\operatorname{Re} \lambda > \omega$. Define $S(z)x := S_x(z)$, $z \in \Sigma_\alpha$, $x \in E$ and $S(0) := k(0)C$. By the uniqueness theorem for Laplace transform, it follows that $S(z)$ is a linear operator for all $z \in \Sigma_\alpha$. The continuity of single operator $S(z)$ ($z \in \Sigma_\alpha$), the commutation of this operator with A and the equicontinuity of family $\{e^{-\omega z}S(z) : z \in \Sigma_\gamma\}$ follow from the condition (81) and the proof of [20, Theorem 2.6.1]. Fix, for the time being, $x \in E$ and $\gamma \in (0, \alpha)$. We will prove that $\lim_{z \to 0, z \in \Sigma_\gamma} S(z)x = k(0)Cx$. It is clear that the mapping $f(z) := e^{-\omega z}S(z)x$, $z \in \Sigma_\alpha$ is analytic and that the set $\{f(z) : z \in \Sigma_\gamma\}$ is bounded for all $\gamma \in (0, \alpha)$. By Theorem 1.2.5(ii), it suffices to show that $\lim_{t \downarrow 0} S(t)x = k(0)Cx$. But, this equality follows from the assumption $\lim_{\lambda \to +\infty} \lambda \tilde{q}(\lambda)x = 0$ and Theorem 1.2.5(iii). By what has just been shown, $(S(t))_{t \geqslant 0}$ is a strongly continuous, exponentially equicontinuous operator family which satisfies that, for every $x \in E$ and for every $\lambda \in \mathbb{C}$ with $\operatorname{Re} \lambda > \omega$ and $\tilde{k}(\lambda) \neq 0$: $q(\lambda)x = \tilde{k}(\lambda)(I - \tilde{a}(\lambda)A)^{-1}Cx = \int_0^\infty e^{-\lambda t}S(t)x\, dt$. This simply implies that $(S(t))_{t \geqslant 0}$ is an exponentially equicontinuous, analytic (a, k)-regularized C-resolvent family of angle α having A as subgenerator. Suppose now A is densely defined. If $x \in D(A)$, then $\mathcal{L}(\int_0^t a(t - s)S(s)Ax\, ds)(\lambda) = \tilde{a}(\lambda)\tilde{q}(\lambda)Ax = \tilde{a}(\lambda)\tilde{k}(\lambda)(I - \tilde{a}(\lambda)A)^{-1}C\, Ax = \tilde{k}(\lambda)((I - \tilde{a}(\lambda)A)^{-1}Cx - Cx) = \tilde{q}(\lambda) - \tilde{k}(\lambda)Cx = \mathcal{L}(S(t)x - k(t)Cx)(\lambda)$ for all sufficiently large λ with $\tilde{k}(\lambda) \neq 0$. By the uniqueness theorem for Laplace transform, one gets $S(t)x - k(t)Cx = \int_0^t a(t - s)S(s)Ax\, ds$, $t \geqslant 0$, $x \in D(A)$, and consequently, $\lim_{t \downarrow 0} S(t)x = k(0)Cx$, $x \in D(A)$. Combined with the exponential equicontinuity of $S(\cdot)$, the above implies that $\lim_{t \downarrow 0} S(t)x = k(0)Cx$ for every $x \in E$.

Then we obtain $\lim_{\lambda\to+\infty}\lambda q(\lambda)x = k(0)Cx$ by Theorem 1.2.5(iii), which completes the proof of theorem. □

Remark 2.2.6. The assertions of [292, Proposition 2.4.2, Corollary 2.4.3] and Kato's analyticity criteria [292, Theorem 2.4.10, Corollary 2.4.11] remain true for exponentially equicontinuous, analytic K-convoluted C-semigroups in SCLCSs. Assume further that the condition (H_1) quoted in the formulation of [292, Theorem 2.4.4] holds and consider the situation of Theorem 2.2.4 with $a(t) \equiv 1$. Then $R(C) \subseteq R(I - \hat{a}(\lambda)A)$ for every $\lambda \in N$ and (79)-(80) hold with $C^{-1}AC$ replaced by A therein.

The proof of following proposition is simple so we refrain from giving it here.

Proposition 2.2.7. (i) *Let $(E_i)_{i\in I}$ be a family of SCLCSs and let $E := \Pi_{i\in I} E_i$ be its direct product. Assume that, for every $i \in I$, $(S_i(t))_{t\in[0,\tau)}$ is an (a, k)-regularized C_i-resolvent family in E_i having A_i as a subgenerator.*
Let $A_i := \Pi_{i\in I} A_i$, $C := \Pi_{i\in I} C_i$ and $S(t) := \Pi_{i\in I} S_i(t)$, $t \in [0, \tau)$. Then A is a subgenerator of an (a, k)-regularized C-resolvent family $(S(t))_{t\in[0,\tau)}$ in E and the local equicontinuity of $(S_i(t))_{t\in[0,\tau)}$ for all $i \in I$ implies the local equicontinuity of $(S(t))_{t\in[0,\tau)}$. Assume further $\tau = \infty$ and there exists $\omega \in \mathbb{R}$ such that the family $\{e^{-\omega t}S_i(t) : t \geqslant 0\}$ is equicontinuous for all $i \in I$. Then the family $\{e^{-\omega t}S(t) : t \geqslant 0\}$ is also equicontinuous.

(ii) *Assume $(R(t))_{t\in[0,\tau)}$ is a (local) (a, k)-regularized C-resolvent family having A as a subgenerator. Set $D_\infty(A) := \cap_{n\in\mathbb{N}} D(A^n)$, $p_n(x) := \Sigma_{i=0}^{n} p(A^i x)$, $x \in D_\infty(A)$, $p \in \circledast$, $n \in \mathbb{N}$, $R_\infty(t) := R(t)_{|D_\infty(A)}$, $t \in [0, \tau)$ and $C_\infty := C_{|D_\infty(A)}$. Then the system $(p_n)_{p\in\circledast, n\in\mathbb{N}}$ induces a Hausdorff sequentially complete locally convex topology on $D_\infty(A)$, $A \in L(D_\infty(A))$ and $(R_\infty(t))_{t\in[0,\tau)}$ is an (a, k)-regularized C_∞-resolvent family having A_∞ as a subgenerator. Furthermore, the following holds:*

(a) *If $(R(t))_{t\in[0,\tau)}$ is locally equicontinuous (global exponentially equicontinuous), then $(R_\infty(t))_{t\in[0,\tau)}$ is.*

(b) *Assume that $(R(t))_{t\geqslant 0}$ is an (exponentially, equicontinuous) analytic (a, k)-regularized C-resolvent family of angle $\alpha \in (0, \pi]$ and $R(z)A \subseteq AR(z)$, $z \in \Sigma_\alpha$. Then $(R_\infty(t))_{t\geqslant 0}$ is.*

The regularity of strongly continuous semigroups in locally convex spaces has been studied by H. Komatsu [290], T. Ushijima [508], B. Dembart [154] and many other authors. Keeping in mind Theorem 2.2.4, Theorem 2.2.5, Remark 2.2.6 and Proposition 2.1.14, the next profiling of analytic integrated C-semigroups follows immediately.

Theorem 2.2.8. *Suppose $r \geqslant 0$ and $\alpha \in (0, \pi/2]$. Then A is a subgenerator of an exponentially equicontinuous, analytic r-times integrated C-semigroup $(S_r(t))_{t\geqslant 0}$ of angle α iff for every $\gamma \in (0, \alpha)$, there exists $\omega_\gamma \geqslant 0$ such that:*

$$\omega_\gamma + \Sigma_{\frac{\pi}{2}+\gamma} \subseteq \rho_C(A),$$

the family

$\{(1+|\lambda|)^{1-r}(\lambda - A)^{-1}C : \lambda \in \omega_\gamma + \Sigma_{\frac{\pi}{2}+\gamma}\}$ *is equicontinuous, the mapping*

$\lambda \mapsto (\lambda - A)^{-1}Cx,\ \lambda \in \omega_\gamma + \Sigma_{\frac{\pi}{2}+\gamma}$ *is analytic (continuous) for every* $x \in E$, *and*

$$\lim_{\lambda\to+\infty} \frac{(\lambda - A)^{-1}Cx}{\lambda^{r-1}} = \chi_{\{0\}}(r)Cx,\ x \in E,\ \textit{if}\ \overline{D(A)} \neq E.$$

Definition 2.2.9. An entire C-regularized group is any operator family $(T(z))_{z\in\mathbb{C}}$ which satisfies that, for every $x \in E$, the mapping $z \mapsto T(z)x$, $z \in \mathbb{C}$ is entire as well as that $T(0) = C$ and $T(z + \omega)C = T(z)T(\omega)$, $z, \omega \in \mathbb{C}$. The integral generator (subgenerator) of $(T(z))_{z\in\mathbb{C}}$ is said to be the integral generator (subgenerator) of $(T(t))_{t\geqslant 0}$.

The next theorem is an extension of [141, Theorem 8.2] and can be applied to differential operators considered in [131], [141, Section XXIV] and [511].

Theorem 2.2.10. (*cf.* [303] *and* [292, Theorem 2.4.13]) *Suppose* $r \geqslant 0$, $\theta \in (0, \pi/2)$ *and* $-A$ *is a subgenerator of an exponentially equicontinuous, analytic r-times integrated C-semigroup* $(S_r(t))_{t\geqslant 0}$ *of angle* θ. *Then there exists an injective operator* $C_1 \in L(E)$ *such that* A *is a subgenerator of an entire* C_1*-regularized group in E. Furthermore, if A is densely defined, then* C_1 *can be chosen such that* $R(C_1)$ *is dense in E.*

It is worth noting that R. T. Moore considered in [426, Section 3] a class of second order differential equations on the 'n-torus' and that Theorem 2.2.5 can be applied to these operators. To make this precise, notice that R. T. Moore has analyzed in [426] the generation of analytic equicontinuous semigroups in the space Ξ which consists of those smooth functions on $\mathbb{R}^n$ with period 1 along each coordinate axis. Equipped with the topology of L^2-convergence of derivatives, Ξ becomes a Fréchet nuclear space. In the situation of [426, Example 3.2], we easily infer from Theorem 2.2.5 that, for every $\beta \in [1, 2)$, the generalized Laplacian $L = \Sigma_{j=1}^{n} \alpha_j \frac{\partial^2}{\partial x_j^2}$ $(\alpha_j \geqslant 0)$ generates an equicontinuous, analytic $(g_\beta, 1)$-regularized resolvent family of angle $\pi(\frac{1}{\beta} - \frac{1}{2})$; in the case $\beta \in (0, 1)$, it follows from [303, Theorem 3.10] that L generates an equicontinuous, analytic $(g_\beta, 1)$-regularized resolvent family of angle $\min(\pi(\frac{1}{\beta} - \frac{1}{2}, \pi)$. By Theorem 2.2.10, we know that the generalized backwards heat operator $-L$ generates an entire C-regularized group, with $R(C)$ dense.

Up to now, we have only a few papers containing results on the existence and uniqueness of solutions of abstract fractional PDEs which can be analytically extended to the region $\mathbb{C}\setminus(-\infty, 0]$ (cf. [141], [23] and [312]). We quote the following two theorems from the last mentioned paper [312].

Theorem 2.2.11. *Suppose that* $\alpha > 2$ *and there exist* $z_0 \in \mathbb{C}\setminus\{0\}$, $\beta \geqslant -1$, $d \in (0, 1]$, $m \in (0, 1)$, $\varepsilon \in (0, 1]$ *and* $\gamma > -1$ *such that:*

(§) $P_{z_0,\beta,\varepsilon,m} := e^{i\arg(z_0)}(|z_0| + (P_{\beta,\varepsilon,m} \cup B_d)) \subseteq \rho_C(A)$, $(\varepsilon, m(1+\varepsilon)^{-\beta}) \in \partial B_d$,

(§§) the family $\{(1+|\lambda|)^{-\gamma}(\lambda - A)^{-1}C : \lambda \in P_{z_0,\beta,\varepsilon,m}\}$ *is equicontinuous, and*

(§§§) *the mapping* $\lambda \mapsto (\lambda - A)^{-1}Cx$, $\lambda \in P_{z_0,\beta,\varepsilon,m}$ *is continuous for every fixed* $x \in E$.

Given $b \in (0, 1/2)$, *set* $\delta_b := \arctan(\cos \pi b)$. *Let* (M_p) *be a sequence of positive real numbers satisfying*

$$p! \prec M_p,\ \textit{i.e., for every } \sigma > 0,\ \sup_{p \geqslant 0} \frac{p!\sigma^p}{M_p} < \infty. \tag{83}$$

Then, for every $b \in (\frac{1}{\alpha}, \frac{1}{2})$, *there exists an operator family* $(T_b(z))_{z\in\Sigma_{\delta_b}}$ *such that, for every* $x \in E$, *the mapping* $z \mapsto T_b(z)x$, $z \in \Sigma_{\delta_b}$ *is analytic and the following holds:*

(i) *For every* $z \in \Sigma_{\delta_b}$ *and* $p \in ⊛$, $T_b(z)$ *is injective and there exist* $c > 0$ *and* $q \in ⊛$ *such that*

$$p\big(T_b(z)x\big) \leqslant c\left((\tan(\cos\pi b)\operatorname{Re} z - |\operatorname{Im} z|)^{-\frac{\gamma+1}{b}}\right) q(x),\ x \in E.$$

(ii) *If* $\lfloor b + \gamma \rfloor \geqslant 0$, $x \in D(A^{\lfloor b+\gamma\rfloor+2})$ *and* $\delta \in (0, \delta_b)$, *then there exists* $\lim_{z'\in\Sigma_\delta, z'\to 0} \frac{T_b(z')x - Cx}{z'}$, *and particularly,* $\lim_{z'\in\Sigma_\delta, z'\to 0} T_b(z')x = Cx$.

(iii) *For every* $z \in \Sigma_{\delta_b}$, *there exists a unique solution* $u(\cdot; z)$ *of the homogeneous counterpart of (61) with initial data* $x_0, \cdots, x_{\lceil\alpha\rceil-1} \in R(T_b(z))$ *and* $u(\cdot; z)$ *can be extended to the whole complex plane. Furthermore, the mapping* $\omega \mapsto u(\omega; z)$, $\omega \in \mathbb{C}\setminus(-\infty, 0]$ *is analytic. Let* $K \subseteq \mathbb{C}\setminus(-\infty, 0]$ *be a compact set, let* $h > 0$ and let $z \in \Sigma_{\delta_b}$. *Then, for every seminorm* $q \in ⊛$, *there exists a constant* $c_{K,h,z,q} > 0$ *and a seminorm* $r_q \in ⊛$ *such that:*

$$\sum_{l=0}^{\lceil\alpha\rceil-1} \sup_{\omega\in K, p\in\mathbb{N}} \frac{h^p q\left(A^p \frac{d^l}{d\omega^l} u(\omega; z)\right)}{M_{\lfloor\alpha p\rfloor-1+l}} \leqslant c_{K,h,z,q} \sum_{i=0}^{\lceil\alpha\rceil-1} r_q\left(T_b(z)^{-1}x_i\right); \tag{84}$$

if $\alpha \in \mathbb{N}\setminus\{1, 2\}$, *then the mapping* $\omega \mapsto u(\omega; z)$, $\omega \in \mathbb{C}$ *is entire* $(z \in \Sigma_{\delta_b})$ *and (84) holds for any compact set* $K \subseteq \mathbb{C}$, $h > 0$, $z \in \Sigma_{\delta_b}$ *and* $q \in ⊛$.

Theorem 2.2.12. *Suppose that* $\alpha \in (1, 2]$ *and there exist* $z_0 \in \mathbb{C}\setminus\{0\}$, $\theta \in (\frac{\pi}{2}(2-\alpha), \frac{\pi}{2})$, $d \in (0, 1]$ *and* $\gamma > -1$ *such that:*

($§_1$) $\Sigma(z_0, \theta, d) := e^{i\arg(z_0)}(|z_0| + (\Sigma(\theta) \cup B_d)) \subseteq \rho_C(A)$,

($§§_1$) *the family* $\{(1+|\lambda|)^{-\gamma}(\lambda - A)^{-1}C : \lambda \in \Sigma(z_0, \theta, d)\}$ *is equicontinuous, and*

($§§§_1$) *the mapping* $\lambda \mapsto (\lambda - A)^{-1}Cx$, $\lambda \in \Sigma(z_0, \theta, d)$ *is continuous for every fixed* $x \in E$.

Let (M_p) be a sequence of positive real numbers satisfying (83). Then, for every $b \in (\frac{1}{\alpha}, \frac{\pi}{2(\pi-\theta)})$ and $\vartheta \in (0, \arctan(\cos(b(\pi-\theta))))$, there exists an operator family $(T_b(z))_{z\in\Sigma_\vartheta}$ such that, for every $x \in E$, the mapping $z \mapsto T_b(z)x$, $z \in \Sigma_\vartheta$ is analytic and the following holds:

(i) *For every $z \in \Sigma_\vartheta$ and $p \in ⊛$, $T_b(z)$ is injective and there exist $c > 0$ and $q \in ⊛$ such that*

$$p\big(T_b(z)x\big) \leqslant c\left((\tan(\vartheta)\operatorname{Re} z - |\operatorname{Im} z|)^{-\frac{\gamma+1}{b}}\right)q(x),\ x \in E.$$

(ii) *If $\lfloor b+\gamma\rfloor \geqslant 0$, $x \in D(A^{\lfloor b+\gamma\rfloor+2})$ and $\delta \in (0, \vartheta)$, then there exists $\lim_{z'\in\Sigma_\delta, z'\to 0}\frac{T_b(z')x - Cx}{z'}$, and particularly, $\lim_{z'\in\Sigma_\delta, z'\to 0} T_b(z')x = Cx$.*

(iii) *For every $z \in \Sigma_\vartheta$, there exists a unique solution $u(\cdot; z)$ of the homogeneous counterpart of (61) with initial data x_0, $x_1 \in R(T_b(z))$ and $u(\cdot; z)$ can be extended to the whole complex plane. Furthermore, the mapping $\omega \mapsto u(\omega; z)$, $\omega \in \mathbb{C}\setminus(-\infty, 0]$ is analytic. Let $K \subseteq \mathbb{C}\setminus(-\infty, 0]$ be a compact set, let $h > 0$ and let $z \in \Sigma_\vartheta$. Then, for every seminorm $q \in ⊛$, there exist a constant $c_{K,h,z,q} > 0$ and a seminorm $r_q \in ⊛$ such that (84) holds with $\lceil\alpha\rceil = 2$; if $\alpha = 2$, then the mapping $\omega \mapsto u(\omega; z)$, $\omega \in \mathbb{C}$ is entire ($z \in \Sigma_\vartheta$) and (84) holds for any compact set $K \subseteq \mathbb{C}$, $h > 0$, $z \in \Sigma_\vartheta$ and $q \in ⊛$.*

The next improvement of [51, Theorem 1'] is stated in a particular case (cf. also Example 2.1.47).

Theorem 2.2.13. ([303]) *Denote, for every $c \in (0, 1)$, $\sigma > 0$ and $\varsigma \in \mathbb{R}$,*

$$\Pi_{c,\sigma,\varsigma} := \{\lambda \in \mathbb{C} : \operatorname{Re}\lambda \geqslant \sigma|\operatorname{Im}\lambda|^c + \varsigma\}.$$

Assume A is a closed linear operator, $\Pi_{c,\sigma,\varsigma} \subseteq \rho_C(A)$, the mapping $\lambda \mapsto (\lambda - A)^{-1}Cx$, $\lambda \in \Pi_{c,\sigma,\varsigma}$ is continuous for every fixed $x \in E$ and there exists $n \in \mathbb{N}$ such that the family $\{(1+|\lambda|)^{-n}(\lambda - A)^{-1}C : \lambda \in \Pi_{c,\sigma,\varsigma}\}$ is equicontinuous. Then there exists an injective operator $C_1 \in L(E)$ such that A is a subgenerator of a global C_1-regularized semigroup $(S(t))_{t\geqslant 0}$ which satifises that the mapping $t \mapsto S(t)$, $t \geqslant 0$ is infinitely differentiable in the topology of $L(E)$ and that, for every compact set $K \subseteq [0, \infty)$ and for every $h > 0$, the set

$$\left\{\frac{h^p \frac{d^p}{dt^p}S(t)}{p!^{1/c}} : t \in K,\ p \in \mathbb{N}_0\right\} \text{ is bounded.}$$

Furthermore, if A is densely defined, then C_1 can be chosen such that $R(C_1)$ is dense in E.

It is noteworthy that Theorem 2.2.8 and Theorem 2.2.10 can be applied to a class of differential operators considered by T.-J. Xiao and J. Liang in E_l-type spaces, see [531, Theorems 1.5.9-1.5.10], [534, Theorems 2.2-2.4] and [534, Theorems 4.1-4.2]; notice also that the proofs of [141, Theorems 14.1] and Theorem

2.2.10 imply that every partial differential operator with constant coefficients in such a space, no matter whether it is bounded above or not, generates an entire C-regularized group. The above remarks enable one to construct important examples of analytic integrated C-semigroups, entire C-regularized groups and analytic (g_α, g_β)-regularized C-resolvent families $(0 < \alpha \leqslant 2, \beta \geqslant 1)$ in these spaces.

Example 2.2.14. Let E be one of the spaces $L^p(\mathbb{R}^n)$ $(1 \leqslant p \leqslant \infty)$, $C_0(\mathbb{R}^n)$, $C_b(\mathbb{R}^n)$, $BUC(\mathbb{R}^n)$, and let $0 \leqslant l \leqslant n$. Put $\mathbb{N}_0^l := \{\alpha \in \mathbb{N}_0^n : \alpha_{l+1} = \cdots = \alpha_n = 0\}$ (it should not be hard to distinguish this set from the Cartesian product of l copies of $\mathbb{N}_0$) and recall that the space E_l $(0 \leqslant l \leqslant n)$ is defined by $E_l := \{f \in E : f^{(\alpha)} \in E \text{ for all } \alpha \in \mathbb{N}_0^l\}$. The totality of seminorms $(q_\alpha(f) := \|f^{(\alpha)}\|_E, f \in E_l;\ \alpha \in \mathbb{N}_0^l)$ induces a Fréchet topology on E_l. Let $\mathbf{T}_1$ possess the same meaning as in [533] and let $m \in \mathbb{N}$, $a_\alpha \in \mathbb{C}$, $0 \leqslant |\alpha| \leqslant m$ and $P(D)f = \Sigma_{|\alpha| \leqslant m}\, a_\alpha f^{(\alpha)}$, with its maximal distributional domain. Set $P(x) := \Sigma_{|\alpha| \leqslant m}\, a_\alpha i^{|\alpha|} x^\alpha$, $x \in \mathbb{R}^n$, $h_{t,\beta}(x) := (1 + |x|^2)^{-\beta/2} \Sigma_{j=0}^{\infty} \frac{t^{2j} P(x)^j}{(2j)!}$, $x \in \mathbb{R}^n$, $t \geqslant 0$, $\beta \geqslant 0$, $\Omega(\omega) := \{\lambda^2 : \operatorname{Re} \lambda > \omega\}$, if $\omega > 0$ and $\Omega(\omega) := \mathbb{C} \setminus (-\infty, \omega^2]$, if $\omega \leqslant 0$. Assume $r \in [0, m]$ and (W) holds with some $\omega \in \mathbb{R}$, where:

(W): $P(x) \notin \Omega(\omega)$, $x \in \mathbb{R}^n$ and, in the case $r \in (0, m]$, there exist $\sigma > 0$ and $\sigma' > 0$ such that $\operatorname{Re}(P(x)) \leqslant -\sigma|x|^r + \sigma'$, $x \in \mathbb{R}^n$.

Then the proofs of [551, Theorem 2.2] and [534, Theorem 2.2-Theorem 2.4] imply that, for every $l \in \mathbb{N}_n^0$, there exists $M \geqslant 1$ such that, for every $\beta > (m - \frac{r}{2})\frac{n}{4}$, $P(D)$ generates an exponentially equicontinuous $\mathbf{T}_1(\langle(1 + |x|^2)^{-\beta}\rangle)$-regularized cosine function $(C_\beta(t))_{t \geqslant 0}$ in E_l which satisfies $C_\beta(t)f = \mathcal{F}^{-1} h_{t,\beta} * f$, $t \geqslant 0$, $f \in E_l$ and $q_\alpha(C_\beta(t)f) \leqslant M q_\alpha(f) G_{n/2}(t)$, $t \geqslant 0$, $f \in E_l$, $\alpha \in \mathbb{N}_0^l$, where the function $G_{n/2}(t)$ is defined on page 40 of [551] and $\mathcal{F}^{-1}$ denotes the inverse Fourier transform. The previous estimate can be additionally refined in the case that $E = L^p(\mathbb{R}^n)$ $(1 < p < \infty)$ by allowing that β takes the value $\frac{1}{2}(m - \frac{r}{2})n|\frac{1}{p} - \frac{1}{2}|$. Further on, it can be easily seen that $P(D)$ generates an exponentially equicontinuous cosine function in the space $E_n(\mathbb{R}^n)$. Concerning time-fractional diffusion-wave equations in E_l-type spaces $(E \neq L^\infty(\mathbb{R}^n), E \neq C_b(\mathbb{R}^n))$, we obtain the same results on the well-posedness as in the case of the space Ξ. This follows from the fact (see e.g. [20, Example 3.4.6, p. 154]) that the generalized Laplacian $L = \Sigma_{j=1}^n \alpha_j \frac{\partial^2}{\partial x_j^2}$ $(\alpha_j > 0)$ generates an equicontinuous, analytic strongly continuous semigroup $(G_l(t))_{t \geqslant 0}$ of angle $\pi/2$ in E_l, which can be computed according to the formula

$$G_l(t)f = \frac{e^{-\sum_{j=1}^{n} \frac{x_j^2}{4t\alpha_j}} * f}{(4\pi t)^{n/2} \left(\prod_{j=1}^{n} \alpha_j\right)^{1/2}}, \quad t > 0, \ f \in E_l;$$

the above conclusions continue to hold, with suitable modifications, in the case that $E = L^\infty(\mathbb{R}^n)$ or $E = C_b(\mathbb{R}^n)$. Actually, there exists one point different from the analysis in the space Ξ : if $\alpha_j = 0$ for some $j \in \mathbb{N}_n^0$, then the use of C-regularized semigroups is inevitable in E_l-spaces.

In the subsequent theorems we clarify some basic results about differentiability of various classes of (a, k)-regularized C-resolvent families (cf. [38], [43], [45], [52], [301]-[303] and [544]-[545] for more details).

Theorem 2.2.15. *Suppose A is a closed linear operator, $k(t)$ and $a(t)$ satisfy (P1), $r \geqslant -1$ and there exists $\omega \geqslant \max(0, abs(k), abs(a))$ such that, for every $z \in \{\lambda \in \mathbb{C} : \operatorname{Re} \lambda > \omega, \tilde{k}(\lambda) \neq 0\}$, we have that the operator $I - \tilde{a}(z)A$ is injective and $R(C) \subseteq R(I - \tilde{a}(z)A)$. If, additionally, for every $\sigma > 0$, there exist $C_\sigma > 0$ and an open neighborhood $\Omega_{\sigma,\omega}$ of the region*

$$\Lambda_{\sigma,\omega} := \{\lambda \in \mathbb{C} : \operatorname{Re} \lambda \leqslant \omega, \operatorname{Re} \lambda \geqslant -\sigma \operatorname{In} |\operatorname{Im} \lambda| + C_\sigma\} \cup \{\lambda \in \mathbb{C} : \operatorname{Re} \lambda \geqslant \omega\},$$

and a function $h_\sigma : \Omega_{\sigma,\omega} \to L(E)$ such that, for every $x \in E$, the mapping $\lambda \mapsto h_\sigma(\lambda) x$, $\lambda \in \Omega_{\sigma,\omega}$ is analytic as well as that $h_\sigma(\lambda) = \tilde{k}(\lambda)(I - \tilde{a}(\lambda)A)^{-1}C$, $\operatorname{Re} \lambda > \omega$, $\tilde{k}(\lambda) \neq 0$, and that the family $\{|\lambda|^{-r} h_\sigma(\lambda) : \lambda \in \Lambda_{\sigma,\omega}\}$ is equicontinuous, then, for every $\zeta > 1$, A is a subgenerator of an exponentially equicontinuous $(a, k * g_{\zeta+r})$-regularized C-resolvent family $(R_\zeta(t))_{t\geqslant 0}$ satisfying that the mapping $t \mapsto R_\zeta(t)$, $t > 0$ is infinitely differentiable in $L(E)$.

Theorem 2.2.16. *Suppose $k(t)$ and $a(t)$ satisfy (P1), A is a subgenerator of an (a, k)-regularized C-resolvent family $(R(t))_{t\geqslant 0}$ satisfying (22) and that the family $\{e^{-\omega' t}R(t) : t \geqslant 0\}$ is equicontinuous for some $\omega' \geqslant \max(0, abs(k), abs(a))$. If there exists $\omega > \omega'$ such that, for every $\sigma > 0$, there exist $C_\sigma > 0$ and $M_\sigma > 0$ so that:*

(i) *there exist an open neighborhood $\Omega_{\sigma,\omega}$ of the region $\Lambda_{\sigma,\omega}$ and the analytic mappings $f_\sigma : \Omega_{\sigma,\omega} \to \mathbb{C}$ and $g_\sigma : \Omega_{\sigma,\omega} \to \mathbb{C}$ such that $f_\sigma(\lambda) = \tilde{k}(\lambda)$, $\operatorname{Re} \lambda \geqslant \omega$ and $g_\sigma(\lambda) = \tilde{a}(\lambda)$, $\operatorname{Re} \lambda \geqslant \omega$,*
(ii) *for every $\lambda \in \Lambda_{\sigma,\omega}$ with $\operatorname{Re} \lambda \leqslant \omega$, the operator $I - \tilde{a}(\lambda)A$ is injective and $R(C) \subseteq R(I - \tilde{a}(\lambda)A)$,*
(iii) *there exists a function $h_\sigma : \Omega_{\sigma,\omega} \to L(E)$ such that, for every $x \in E$, the mapping $\lambda \mapsto h_\sigma(\lambda)x$, $\lambda \in \Omega_{\sigma,\omega}$ is analytic, $h_\sigma(\lambda) = f_\sigma(\lambda)(I - g_\sigma(\lambda)A)^{-1}C$, $\lambda \in \Lambda_{\sigma,\omega}$,*
(iv) *the family $\{|\operatorname{Im} \lambda|^{-1} h_\sigma(\lambda) : \lambda \in \Lambda_{\sigma,\omega}, \operatorname{Re} \lambda \leqslant \omega\}$ is equicontinuous and $\max(|f_\sigma(\lambda)|, |g_\sigma(\lambda)|) \leqslant M_\sigma$, $\lambda \in \Lambda_{\sigma,\omega}$,*

then the mapping $t \mapsto R(t)x$, $t > 0$ is infinitely differentiable for every fixed $x \in D(A^2)$. Assume, additionally, that $D(A^2)$ is dense in E. Then the mapping $t \mapsto R(t)$, $t > 0$ is infinitely differentiable in $L(E)$.

Suppose $(M_n)_{n\in\mathbb{N}_0}$ satisfies (M.1), (M.2) and (M.3)'. Put $m_n := \frac{M_n}{M_{n-1}}$, $n \in \mathbb{N}$, $M(\lambda) := \sup_{n\in\mathbb{N}_0} \ln \frac{|\lambda|^n}{M_n}$, $\lambda \in \mathbb{C} \setminus \{0\}$, $M(0) := 0$, $\omega_L(t) := \sum_{n=0}^{\infty} \frac{t^n}{M_n}$, $t \geqslant 0$,

(85) $$\omega_l(\lambda) := \prod_{n=1}^{\infty} \left(1 + \frac{l\lambda}{m_n}\right), \lambda \in \mathbb{C}\ (l > 0) \text{ and } K_l(t) := \mathcal{L}^{-1}\left(\frac{1}{\omega_l(\lambda)}\right)(t), t \geqslant 0\ (l > 0).$$

For further information, we refer the reader to [283] and [292].

Theorem 2.2.17. (i) *Suppose $k(t)$ and $a(t)$ satisfy (P1), A is a subgenerator of a (local) (a, k)-regularized C-resolvent family $(R(t))_{t\in[0,\tau)}$,* $\omega > \max(0, abs(k), abs(a))$ *and $m \in \mathbb{N}$. Denote, for every $\varepsilon \in (0, 1)$ and a corresponding $K_\varepsilon > 0$,*

$$F_{\varepsilon,\omega} := \{\lambda \in \mathbb{C} : \operatorname{Re}\lambda \geqslant -\ln \omega_L (K_\varepsilon |\operatorname{Im}\lambda|) + \omega\}.$$

Assume that, for every $\varepsilon \in (0, 1)$, there exist $K_\varepsilon > 0$, an open neighborhood $O_{\varepsilon,\omega}$ of the region $G_{\varepsilon,\omega} := \{\lambda \in \mathbb{C} : \operatorname{Re}\lambda \geqslant \omega, \tilde{k}(\lambda) \neq 0\} \cup \{\lambda \in F_{\varepsilon,\omega} : \operatorname{Re}\lambda \leqslant \omega\}$, a mapping $h_\varepsilon : O_{\varepsilon,\omega} \to L(E)$ and analytic mappings $f_\varepsilon : O_{\varepsilon,\omega} \to \mathbb{C}$, $g_\omega : O_{\varepsilon,\omega} \to \mathbb{C}$ such that:

(a) $f_\varepsilon(\lambda) = \tilde{k}(\lambda)$, $\operatorname{Re}\lambda > \omega$; $g_\varepsilon(\lambda) = \tilde{a}(\lambda)$, $\operatorname{Re}\lambda > \omega$,

(b) *for every $\lambda \in F_{\varepsilon,\omega}$, the operator $I - g_\varepsilon(\lambda) A$ is injective and $R(C) \subseteq R(I - g_\varepsilon(\lambda) A)$,*

(c) *for every $x \in E$, the mapping $\lambda \mapsto h_\varepsilon(\lambda)x$, $\lambda \in G_{\varepsilon,\omega}$ is analytic, $h_\varepsilon(\lambda) = f_\varepsilon(\lambda)(I - g_\varepsilon(\lambda) A)^{-1}C$, $\lambda \in G_{\varepsilon,\omega}$,*

(d) *the family $\{(1 + |\lambda|)^{-m} e^{-\varepsilon|\operatorname{Re}\lambda|} h_\varepsilon(\lambda) : \lambda \in F_{\varepsilon,\omega}, \operatorname{Re}\lambda \leqslant \omega\}$ is equicontinuous and the family $\{(1 + |\lambda|)^{-m} h_\varepsilon(\lambda) : \lambda \in \mathbb{C}, \operatorname{Re}\lambda \geqslant \omega\}$ is equicontinuous.*

Then the mapping $t \mapsto R(t)$, $t \in (0, \tau)$ is infinitely differentiable in $L(E)$ and, for every compact set $K \subseteq (0, \tau)$, there exists $h_K > 0$ such that the set $\left\{\frac{h_K^n \frac{d^n}{dt^n} R(t)}{M_n} : t \in K, n \in \mathbb{N}_0\right\}$ *is equicontinuous.*

(ii) *Suppose $k(t)$ and $a(t)$ satisfy (P1), A is a subgenerator of a (local) (a, k)-regularized C-resolvent family $(R(t))_{t\in[0,\tau)}$,* $\omega > \max(0, abs(k), abs(a))$ *and $m \in \mathbb{N}$. Denote, for every $\varepsilon \in (0, 1)$, $\rho \in [1, \infty)$ and a corresponding $K_\varepsilon > 0$,*

$$F_{\varepsilon,\omega,\rho} := \{\lambda \in \mathbb{C} : \operatorname{Re}\lambda \geqslant -K_\varepsilon |\operatorname{Im}\lambda|^{1/\rho} + \omega\}.$$

Assume that, for every $\varepsilon \in (0, 1)$, there exist $K_\varepsilon > 0$, an open neighborhood $O_{\varepsilon,\omega}$ of the region $G_{\varepsilon,\omega,\rho} := \{\lambda \in \mathbb{C} : \operatorname{Re}\lambda \geqslant \omega, \tilde{k}(\lambda) \neq 0\} \cup \{\lambda \in F_{\varepsilon,\omega,\rho} : \operatorname{Re}\lambda \leqslant \omega\}$, a mapping $h_\varepsilon : O_{\varepsilon,\omega} \to L(E)$ and analytic mappings $f_\varepsilon : O_{\varepsilon,\omega} \to \mathbb{C}$ and $g_\varepsilon : O_{\varepsilon,\omega} \to \mathbb{C}$ such that the conditions (i)(a)-(d) of this theorem hold with $F_{\varepsilon,\omega}$, resp. $G_{\varepsilon,\omega}$, replaced by $F_{\varepsilon,\omega,\rho}$, resp. $G_{\varepsilon,\omega,\rho}$. Then the mapping $t \mapsto R(t)$, $t \in (0, \tau)$ is infinitely differentiable in $L(E)$ and, for every compact set $K \subseteq (0, \tau)$, there exists $h_K > 0$ such that the set $\left\{\frac{h_K^n \frac{d^n}{dt^n} R(t)}{n!^\rho} : t \in K, n \in \mathbb{N}_0\right\}$ *is equicontinuous.*

Proof. We will only prove the first part of the theorem. Combining Theorem 2.1.6(i), the condition (d), Theorem 1.2.2 and Cauchy formula, we get that A is a subgenerator of an exponentially equicontinuous $(a, k * t)$-regularized C-resolvent family $(R_{m+2}(t))_{t\geqslant 0}$ which satisfies, for every $\varepsilon \in (0, 1)$, $x \in E$ and $t \in [0, \tau)$,

$R_{m+2}(t)x = \frac{1}{2\pi i}\int_{\omega-i\infty}^{\omega+i\infty} e^{\lambda t}\frac{h_\varepsilon(\lambda)x}{\lambda^{m+2}}\,d\lambda$ and $R_{m+2}(t)x = \int_0^t (t-s)R(s)x\,ds$. Since (M.3)' holds for (M_n), one has $\lim_{\lambda\to+\infty}\frac{M(\lambda)}{\lambda} = 0$ and $\lim_{n\to\infty}\frac{n}{m_n} = 0$ (cf. [283, (4.5), (4.7), p. 56]), which implies that there exists $c > 0$ such that $M(\lambda) \leqslant c\lambda$, $\lambda \geqslant 0$ and

(86)
$$\frac{\omega'_L(t)}{\omega_L(t)} = \frac{\sum_{n=1}^{\infty}\frac{nt^{n-1}}{M_n}}{\sum_{n=0}^{\infty}\frac{t^n}{M_n}} \leqslant c\frac{\sum_{n=1}^{\infty}\frac{t^{n-1}}{M_{n-1}}}{\sum_{n=0}^{\infty}\frac{t^n}{M_n}} = c,\ t \geqslant 0.$$

Further on, for each $\varepsilon \in (0, 1)$ there exists a unique number $a_\varepsilon > 0$ such that $\omega_L(K_\varepsilon a_\varepsilon) = 1$. Let $\Gamma_\varepsilon := \Gamma_{1,\varepsilon} \cup \Gamma_{2,\varepsilon} \cup \Gamma_{3,\varepsilon}$, where $\Gamma_{1,\varepsilon} := \{-\ln(K_\varepsilon s) + \omega + is : s \in (-\infty, -a_\varepsilon]\}$, $\Gamma_{2,\varepsilon} := \{\omega + is : s \in [-a_\varepsilon, a_\varepsilon]\}$ and $\Gamma_{3,\varepsilon} := \{-\ln(K_\varepsilon s) + \omega + is : s \in [a_\varepsilon, \infty)\}$. Define, for every $\varepsilon \in (0, 1)$ and $x \in E$,

(87)
$$R_{m+2,\varepsilon}(t)x := \frac{1}{2\pi i}\int_{\Gamma_\varepsilon} e^{\lambda t}\frac{h_\varepsilon(\lambda)x}{\lambda^{m+2}}\,d\lambda,\ t > \varepsilon.$$

By the proof of [283, Proposition 4.5, p. 58], we have $\omega_L(s) \leqslant 2e^{M(2s)}$, $t \geqslant 0$ and $\ln \omega_L(K_\varepsilon s) \leqslant \ln 2 + M(2K_\varepsilon s)$, $s \geqslant 0$. Let $p \in ⊛$. Due to (86) and (d), we have that there exist a seminorm $q \in ⊛$ and a number $c_\varepsilon > 0$ such that, for every $x \in E$ and $t > \varepsilon$,

$$p(R_{m+2,\varepsilon}(t)x) \leqslant \frac{1}{2\pi}\left(c_\varepsilon + 2e^{(\omega+\varepsilon)t}\int_{a_\varepsilon}^{\infty}\omega_L(K_\varepsilon s)^{\varepsilon-t}(1+\omega+s+\ln 2+2K_\varepsilon cs)^{-2}\,ds\right)q(x),$$

which implies that the improper integral appearing in (87) is convergent and $R_{m+2,\varepsilon}(t) \in L(E)$ for all $t > \varepsilon$. An elementary application of Cauchy formula yields $R_{m+2}(t) = R_{m+2,\varepsilon}(t)$, $t > \varepsilon$. A similar line of reasoning shows that the mapping $t \mapsto R_{m+2}(t)x$, $t > 0$ is infinitely differentiable with

(88)
$$\frac{d^n}{d\lambda^n}R_{m+2}(t)x = \frac{1}{2\pi i}\int_{\Gamma_\varepsilon} e^{\lambda t}\lambda^{n-m-2}h_\varepsilon(\lambda)x\,d\lambda,\ t > \varepsilon,\ x \in E,\ n \in \mathbb{N}_0.$$

Let $K \subseteq (0, \tau)$ be a compact set, and let $k \in \mathbb{N}$ and $\varepsilon \in (0, 1)$ be such that $\inf K - \varepsilon > k^{-1}$. Then there exists $c'_\varepsilon > 1$ such that $|-\ln \omega_L(K_\varepsilon s) + \omega + is| \leqslant c'_\varepsilon s$, $s \geqslant a_\varepsilon$. Let $h_K \in (0, K_\varepsilon/c'_\varepsilon)$. By (M.2), it follows inductively that

(89)
$$M_{kn} \leqslant A^{k-1}H^{k(k+1)/2}M_n^k,\ n \in \mathbb{N}_0.$$

Now (88)-(89) together imply that there exists $c_K > 0$ such that, for every $n \in \mathbb{N}_0$, $t \in K$ and $x \in E$:

$$
\begin{aligned}
&p\left(\frac{h_K^n \frac{d^n}{d\lambda^n} R_{m+2}(t)x}{M_n}\right) \\
&\leqslant \frac{c_K}{2\pi}\left(\omega_L(h_K(\omega+a_\varepsilon)) + 2e^{(\omega+\varepsilon)t}\int_{a_\varepsilon}^{\infty}\omega_L(K_\varepsilon s)^{1/k}\frac{(c'h_K s)^n}{M_n}s^{-2}\,ds\right)q(x) \\
&\leqslant \frac{c_K}{2\pi}\left(\omega_L(h_K(\omega+a_\varepsilon)) + 2e^{(\omega+\varepsilon)t}\int_{a_\varepsilon}^{\infty}\frac{M_{kn}^{1/k}}{M_n}\frac{(c'h_K s)^n s^{-2}}{(K_\varepsilon s)^n}\,ds\right)q(x) \\
&\leqslant \frac{c_K}{2\pi}\left(\omega_L(h_K(\omega+a_\varepsilon)) + \frac{2}{a_\varepsilon}e^{(\omega+\varepsilon)t}A^{k-1/k}H^{k+1/2}\left(\frac{c'h_K}{K_\varepsilon}\right)^n\right)q(x) \\
&\leqslant \frac{c_K}{2\pi}\left(\omega_L(h_K(\omega+a_\varepsilon)) + \frac{2}{a_\varepsilon}e^{(\omega+\varepsilon)t}A^{k-1/k}H^{k+1/2}\right)q(x).
\end{aligned}
$$

The previous estimate implies that the set $\left\{\frac{h_K^n \frac{d^n}{dt^n}R_{m+2}(t)}{M_n} : t \in K,\ n \in \mathbb{N}_0\right\}$ is equicontinuous. Since (M.2) is assumed, the set $\left\{\frac{h_K^n \frac{d^n}{dt^n}R_m(t)}{M_n} : t \in K,\ n \in \mathbb{N}_0\right\}$ is equicontinuous as well.

☐

Theorem 2.2.18. *(The abstract Weierstrass formula)*

(i) *Assume $k(t)$ and $a(t)$ satisfy (P1), and there exist $M > 0$ and $\omega > 0$ such that $|k(t)| \leqslant Me^{\omega t}$, $t \geqslant 0$. Assume, further, that there exist a number $\omega' \geqslant \omega$ and a function $a_1(t)$ satisfying (P1) and $\tilde{a}_1(\lambda) = \tilde{a}(\sqrt{\lambda})$,* Re $> \omega'$. *(Due to* [20, Lemma 1.6.7], *the above holds if $a(t)$ is exponentially bounded; in this case, $a_1(t) = \int_0^\infty s\frac{e^{-s^2/4t}}{2\sqrt{\pi}t^{3/2}}a(s)\,ds$, $t > 0$.) Let A be a subgenerator of an exponentially equicontinuous (a, k)-regularized C-resolvent family $(C(t))_{t\geqslant 0}$ satisfying (22) with $(R(t))_{t\geqslant 0}$ replaced by $(C(t))_{t\geqslant 0}$. Then A is a subgenerator of an exponentially equicontinuous, analytic (a_1, k_1)-regularized C-resolvent family $(R(t))_{t\geqslant 0}$ of angle $\pi/2$, where:*

(90) $$k_1(t) := \int_0^\infty \frac{e^{-s^2/4t}}{\sqrt{\pi t}}k(s)\,ds,\ t > 0,\ \ k_1(0) := k(0),\ \text{and}$$

(91) $$R(t)x := \int_0^\infty \frac{e^{-s^2/4t}}{\sqrt{\pi t}}C(s)x\,ds,\ t > 0,\ x \in E,\ R(0) := k(0)C.$$

(ii) *Assume $k(t)$ satisfy (P1), $\beta > 0$ and there exist $M > 0$ and $\omega > 0$ such that $|k(t)| \leqslant Me^{\omega t}$, $t \geqslant 0$. Let A be a subgenerator of an exponentially equicontinuous $(g_{2\beta+1}, k)$-regularized C-resolvent family $(C(t))_{t\geqslant 0}$ satisfying (22) with $(R(t))_{t\geqslant 0}$ replaced by $(C(t))_{t\geqslant 0}$. Then A is a subgenerator of an exponentially equicontinuous, analytic (g_β, k_1)-regularized C-resolvent family $(R(t))_{t\geqslant 0}$ of angle $\pi/2$, where $k_1(t)$ and $R(t)$ are defined through (90)-(91).*

We close this section with the observation that the assertions of [306, Theorem 2.3, Proposition 2.6], [292, Theorem 2.1.27(xvi)-(xvii)] and [302, Theorems 2.24-2.26, 2.34, Proposition 2.29, Remark 2.35] can be reformulated for the class of (a, k)-regularized C-resolvent families in sequentially complete locally convex spaces.

2.3 Systems of abstract time-fractional equations

Let E be one of the spaces $L^p(\mathbb{R}^n)$ $(1 \leqslant p \leqslant \infty)$, $C_0(\mathbb{R}^n)$, $C_b(\mathbb{R}^n)$, $BUC(\mathbb{R}^n)$, $C^\sigma(\mathbb{R}^n)$ $(0 < \sigma < 1)$ and let $0 \leqslant l \leqslant n$. Recall that $\mathbb{N}_0^l = \{\eta \in \mathbb{N}_0^n : \eta_{l+1} = \cdots = \eta_n = 0\}$, then the space E_l is defined by $E_l := \{f \in E : f^{(\eta)} \in E \text{ for all } \eta \in \mathbb{N}_0^l\}$. The totality of seminorms $(q_\eta(f) := \|f^{(\eta)}\|_E, f \in E_l;\ \eta \in \mathbb{N}_0^l)$ induces a Fréchet topology on E_l (in the case that $E = C^\sigma(\mathbb{R}^n)$, this follows from the proof of [531, Lemma 5.6, p. 25]). Put $D^{\eta\cdot} := (-i)^{|\eta|\cdot(\eta)}$ $(\eta \in \mathbb{N}_0^n)$.

In the proofs of our main results, we will make use of functional calculus for commuting generators of bounded C_0-groups ([141]). Denote by $\mathcal{F}$ and $\mathcal{F}^{-1}$ the n-dimensional Fourier transform and its inverse transform, respectively. That is

$$(\mathcal{F}f)(\xi) := \int_{\mathbb{R}^n} e^{i(x,\xi)} f(x)\, dx,\ \xi \in \mathbb{R}^n \text{ and } \mathcal{F}^{-1} := (2\pi)^{-n}\hat{\mathcal{F}},$$

where ^ denotes the reflection in 0. Let $(E, \|\cdot\|)$ be a complex Banach space, let $n \in \mathbb{N}$ and let iA_j, $1 \leqslant j \leqslant n$ be commuting generators of bounded C_0-groups on E. Set $A := (A_1, \cdots, A_n)$ and $A^\eta := A_1^{\eta_1} \ldots A_n^{\eta_n}$ for any $\eta = (\eta_1, \cdots, \eta_n) \in \mathbb{N}_0^n$. If $\xi = (\xi_1, \cdots, \xi_n) \in \mathbb{R}^n$ and $u \in \mathcal{A} := \{f \in C_0(\mathbb{R}^n) : \mathcal{F}f \in L^1(\mathbb{R}^n)\}$, put $|\xi| := (\Sigma_{j=1}^n \xi_j^2)^{1/2}$, $(\xi, A) := \Sigma_{j=1}^n \xi_j A_j$ and

$$u(A)x := (2\pi)^{-n} \int_{\mathbb{R}^n} \mathcal{F}u(\xi) e^{-i(\xi, A)} x\, d\xi,\ x \in E.$$

Then $u(A) \in L(E)$, $u \in \mathcal{A}$ and there exists a constant $M < \infty$ such that $\|u(A)\| \leqslant M\|\mathcal{F}u\|_{L^1(\mathbb{R}^n)}$, $u \in \mathcal{A}$.

Put $Z_n := \mathcal{F}\mathcal{D}(\mathbb{R}^n)$ and assume that Z_n is equipped with the topology transported by $\mathcal{F}$ from $\mathcal{D}(\mathbb{R}^n)$. By Z_n' we denote the strong dual of Z_n. It is clear that $Z_n = \mathcal{F}^{-1}\mathcal{D}(\mathbb{R}^n)$ and that the dual mapping of $\mathcal{F}_{|Z_n} : Z_n \to \mathcal{D}(\mathbb{R}^n)$ is an isomorphism of $\mathcal{D}'(\mathbb{R}^n)$ onto Z_n'. We have the following equality:

$$\langle \mathcal{F}T, \mathcal{F}\varphi \rangle = (2\pi)^n \langle T, \hat{\varphi} \rangle,\ T \in \mathcal{D}'(\mathbb{R}^n),\ \varphi \in \mathcal{D}(\mathbb{R}^n).$$

The operator $\partial/\partial x_j : Z_n' \to Z_n'$ is defined as the dual operator of $-\partial/\partial x_j : Z_n \to Z_n$, so that $\partial/\partial x_j\, \mathcal{F}T = \mathcal{F}(i\xi_j T)$, $T \in \mathcal{D}'(\mathbb{R}^n)$, $1 \leqslant j \leqslant n$; the actions of $\mathcal{F}$ on $(\mathcal{D}'(\mathbb{R}^n))^m$ and of $\mathcal{F}^{-1}$ on $(Z_n')^m$ are coordinatewise.

Let $m, n, d \in \mathbb{N}$, and let M_m denote the ring of all complex matrices of format $m \times m$. Define $P(x) := \Sigma_{|\eta| \leqslant d} P_\eta x^\eta$, $x \in \mathbb{R}^n$ ($P_\eta \in M_m$), $P(\partial/\partial x) := \Sigma_{|\eta| \leqslant d} P_\eta (\partial/\partial x)^\eta$, and $\widetilde{P}(\xi) := \Sigma_{|\eta| \leqslant d} i^{|\eta|} P_\eta \xi^\eta$ (here, the use of symbol $\tilde{\cdot}$ is a little bit inappropriate but clear). Denote by $\lambda_1(\xi), \cdots, \lambda_m(\xi)$ the eigenvalues of $\widetilde{P}(\xi)$ ($\xi \in \mathbb{R}^n$).

In the sequel of this section, we will consider the case in which the space E is barreled, so that every considered (g_α, C)-regularized resolvent family $(R_\alpha(t))_{t\geqslant 0}$ in E will be locally equicontinuous. Our intention is to clarify the most important well-posedness results for the following system of abstract time-fractional equations

$$\mathbf{D}_t^\alpha \vec{u}(t) = P(\partial/\partial x)_{|E} \vec{u}(t),\ t > 0;\ \vec{u}^{(k)}(0) = \vec{x_k},\ k = 0, 1, \cdots, \lceil \alpha \rceil - 1, \tag{92}$$

where $\alpha > 0$.

At the outset, let us observe that the formula $G_\alpha(t)\,\vec{f} := E_\alpha(t^\alpha\, \widetilde{P}(\xi))\,\vec{f}$, $t \geqslant 0$, $\vec{f} \in (\mathcal{D}'(\mathbb{R}^n))^m$ determines a global (g_α, I)-regularized resolvent family on $(\mathcal{D}'(\mathbb{R}^n))^m$, and that the integral generator of $(G_\alpha(t))_{t\geqslant 0}$ is the multiplication operator $\widetilde{P}(\xi)_{|(\mathcal{D}'(\mathbb{R}^n))^m} \in L((\mathcal{D}'(\mathbb{R}^n))^m)$. Furthermore, the formula

$$R_\alpha(t)\vec{f} := \mathcal{F}E_\alpha(t^\alpha\, \widetilde{P}(\xi))\mathcal{F}^{-1}\vec{f},\ t \geqslant 0, \vec{f} \in (Z_n')^m,$$

determines a global (g_α, I)-regularized resolvent family $(R_\alpha(t))_{t\geqslant 0}$ on $(Z_n')^m$. The operator $P(\partial/\partial x)_{|(Z_n')^m} \in L((Z_n')^m)$ is the integral generator of $(R_\alpha(t))_{t\geqslant 0}$, and $(G_\alpha(t))_{t\geqslant 0}$ as well as $(R_\alpha(t))_{t\geqslant 0}$ can be extended to the whole complex plane. The following holds:

(i) Let $\alpha \in (0, \infty)\setminus\mathbb{N}$ and $\vec{f} \in (Z_n')^m$.
   - (i.1) The mapping $z \mapsto R_\alpha(z)\vec{f}$, $z \in \mathbb{C}\setminus(-\infty, 0]$ is analytic.
   - (i.2) The mapping $t \mapsto R_\alpha(t)\vec{f}$, $t \geqslant 0$ belongs to the space $C^{\lfloor\alpha\rfloor}([0, \infty) : (Z_n')^m)$.
   - (i.3) For every compact set $K \subseteq \mathbb{C}\setminus(-\infty, 0]$, the family $\{R_\alpha(z) : z \in K\} \subseteq L((Z_n')^m)$ is equicontinuous.

(ii) Let $\alpha \in \mathbb{N}$ and $\vec{f} \in (Z_n')^m$.
   - (ii.1) The mapping $z \mapsto R_\alpha(z)\vec{f}$, $z \in \mathbb{C}$ is entire.
   - (ii.2) For every compact set $K \subseteq \mathbb{C}$, the family $\{R_\alpha(z) : z \in K\} \subseteq L((Z_n')^m)$ is equicontinuous.

Observe also that the above assertions continue to hold for $(G_\alpha(t))_{t\geqslant 0}$ and $\vec{f} \in (\mathcal{D}'(\mathbb{R}^n))^m$, and that, for every $z \in \mathbb{C}$, $R_\alpha(z)Z_n^m \subseteq Z_n^m$ and $R_\alpha(z)(\mathcal{FE}'(\mathbb{R}^n))^m \subseteq (\mathcal{FE}'(\mathbb{R}^n))^m$. This implies that $(R_\alpha(t)_{|Z_n^m})_{t\geqslant 0}$, resp. $(R_\alpha(t)_{|(\mathcal{FE}'(\mathbb{R}^n))^m})_{t\geqslant 0}$, is a locally equicontinuous (g_α, I)-regularized resolvent family generated by $P(\partial/\partial x)_{|Z_n^m}$, resp. $P(\partial/\partial x)_{|(\mathcal{FE}'(\mathbb{R}^n))^m}$.

In the following theorem, we shall extend the assertion of [275, Theorem 1, (a) $\Rightarrow$ (b)] to abstract time-fractional equations.

Theorem 2.3.1. *Suppose $\omega \geqslant 0$, $0 < \alpha \leqslant 2$ and*

$$\sup_{z\in\sigma(\widetilde{P}(\xi))} \operatorname{Re}(z^{1/\alpha}) \leqslant \omega. \tag{93}$$

Let E be one of the spaces listed below:

(i) $E = (\mathcal{S}(\mathbb{R}^n))^m$ *or* $E = (\mathcal{S}'(\mathbb{R}^n))^m$.
(ii) $E = X_n$, *where X is* $L^p(\mathbb{R}^n)$ $(1 \leqslant p \leqslant \infty)$, $C_0(\mathbb{R}^n)$, $C_b(\mathbb{R}^n)$, $BUC(\mathbb{R}^n)$ *or* $C^\sigma(\mathbb{R}^n)$ $(0 < \sigma < 1)$.
(iii) $E = \{\vec{f} \in (L^2(\mathbb{R}^n))^m : (P(\partial/\partial x))^l \vec{f} \in (L^2(\mathbb{R}^n))^m$ *for all* $l \in \mathbb{N}\}$, *with the topology induced by the following family of seminorms:*

$$\|\vec{f}\|_l := \|(P(\partial/\partial x))^l \vec{f}\|_{(L^2(\mathbb{R}^n))^m} \ (\vec{f} \in E,\ l \in \mathbb{N}_0).$$

Then the operator $P(\partial/\partial x)_{|E}$ *is the integral generator of a global* (g_α, I)*-regularized resolvent family* $(S_\alpha(t))_{t\geqslant 0}$ *on E satisfying that, for every* $p \in \circledast$ *and* $\varepsilon > 0$, *there exist* $M \geqslant 1$ *and* $q \in \circledast$ *such that:*

$$p(e^{-(\omega+\varepsilon)t} S_\alpha(t)) \leqslant Mq(x),\ t \geqslant 0,\ x \in E. \tag{94}$$

Proof. We will prove the theorem provided that $\alpha \in (0, 2)\setminus\{1\}$. Let $\varepsilon > 0$ be fixed and let Γ_ε denote the boundary of the region $\{z \in \mathbb{C} : \operatorname{Re}(z^{1/\alpha}) \leqslant \omega + \varepsilon\}$.
I. The case $1 < \alpha < 2$.

Suppose a positively oriented curve C_ξ encircles the spectrum of $\widetilde{P}(\xi)$ and is a subset of $\{z \in \mathbb{C} : \operatorname{Re}(z^{1/\alpha}) \leqslant \omega + \varepsilon/2\}$ $(\xi \in \mathbb{R}^n)$. Notice that, for every $\xi_0 \in \mathbb{R}^n$, there exists an open neighborhood U_{ξ_0} of ξ_0 such that C_{ξ_0} encircles the spectrum of $\widetilde{P}(\xi)$ for all $\xi \in U_{\xi_0}$. Let $z_0 > (\omega + 2\varepsilon)^\alpha$. By [275, Theorem II, (iii)] and the Cauchy integral formula, we obtain that there exists $v \in \mathbb{N}$ such that:

$$\begin{aligned} E_\alpha(t^\alpha \widetilde{P}(\xi)) &= (\widetilde{P}(\xi) - z_0 I)^2 \frac{1}{2\pi i} \oint_{C_\xi} \frac{E_\alpha(t^\alpha z)}{(z - z_0)^2} (zI - \widetilde{P}(\xi))^{-1}\, dz \\ &= (\widetilde{P}(\xi) - z_0 I)^2 \left[a_0(t, \xi) I + \cdots + a_{m-1}(t, \xi) \widetilde{P}(\xi)^{m-1}\right],\ t \geqslant 0,\ \xi \in \mathbb{R}^n, \end{aligned} \tag{95}$$

where $a_j(t, \xi)$ can be written as a finite sum, with coefficients independent of t and ξ, of terms like

$$S_{i_1,\cdots,i_m;l}(t, \xi) = \lambda_1^{i_1}(\xi) \cdots \lambda_m^{i_m}(\xi) \frac{1}{2\pi i} \oint_{C_\xi} \frac{E_\alpha(t^\alpha z)}{(z - z_0)^2} \frac{dz}{(z - \lambda_1(\xi)) \cdots (z - \lambda_l(\xi))},$$

where $t \geqslant 0$, $\xi \in \mathbb{R}^n$, $i_j \in \mathbb{N}_0$ for $1 \leqslant j \leqslant m$ and $i_1 + \cdots + i_m \leqslant v$. By (53) and (93), we get that there exists $M_\varepsilon > 0$ such that the term $|E_\alpha(t^\alpha z)|$ does not exceed $M_\varepsilon e^{(\omega+\varepsilon)t}$, for any $t \geqslant 0$ and $z \in \mathbb{C}$ with $\operatorname{Re}(z^{1/\alpha}) \leqslant \omega + \varepsilon/2$. Since $\operatorname{dist}(\Gamma_{\varepsilon/2}, \Gamma_\varepsilon) := \kappa(\omega, \alpha, \varepsilon) > 0$, the residue theorem implies that, for every $t \geqslant 0$ and $\xi \in \mathbb{R}^n$:

$$S_{i_1,\cdots,i_m;l}(t, \xi) = \lambda_1^{i_1}(\xi) \cdots \lambda_m^{i_m}(\xi) \frac{1}{2\pi i} \int_{\Gamma_\varepsilon} \frac{E_\alpha(t^\alpha z)}{(z - z_0)^2} \frac{dz}{(z - \lambda_1(\xi)) \cdots (z - \lambda_l(\xi))},$$

which yields the existence of a number $\sigma > 0$ such that:

$$|S_{i_1,\cdots,i_m;l}(t,\xi)| \leqslant \frac{M_\varepsilon}{2\pi}(1+|\xi|)^\sigma e^{(\omega+\varepsilon)t}\int_{\Gamma_\varepsilon}\frac{|dz|}{|z-z_0|^2},$$

provided $i_j \in \mathbb{N}_0$ for $1 \leqslant j \leqslant m$ and $i_1 + \cdots + i_m \leqslant v$. Taken together with (95), the above implies that there exist $N_\varepsilon > 0$ and $\sigma_1 > 0$ such that:

(96) $$||E_\alpha(t^\alpha \widetilde{P}(\xi))|| \leqslant N_\varepsilon(1+|\xi|)^{\sigma_1} e^{(\omega+\varepsilon)t},\ t \geqslant 0,\ \xi \in \mathbb{R}^n.$$

Now we will prove that, for every multi-index $\eta \in \mathbb{N}_0^n$ with $|\eta| > 0$, there exist $N_{\varepsilon,\eta} > 0$ and $\sigma_\eta > 0$ such that:

(97) $$||D^\eta(E_\alpha(t^\alpha \widetilde{P}(\xi)))|| \leqslant N_{\varepsilon,\eta}(1+|\xi|)^{\sigma_\eta} e^{(\omega+\varepsilon)t},\ t \geqslant 0,\ \xi \in \mathbb{R}^n.$$

Noticing that $D^{-1} = \mathrm{adj}(D)/\det(D)$ for every regular matrix $D \in M_m$, we obtain that there exist $l_\eta \in \mathbb{N}$ and polynomials $q_{ij}^\eta(\xi, z)$ in $(n+1)$ variables such that, for every $\xi \in \mathbb{R}^n$ and $z \in \rho(\widetilde{P}(\xi))$:

$$D^\eta(zI-\widetilde{P}(\xi))^{-1} = \frac{[q_{ij}^\eta(\xi,z)]_{1\leqslant i,j\leqslant m}}{(z-\lambda_1(\xi))^{l_\eta}\cdots(z-\lambda_m(\xi))^{l_\eta}}.$$

Set $N_\eta := \max\{dg(q_{ij}^\eta(\xi, z)) : 1 \leqslant i, j \leqslant m\} + 2$. By the Cauchy integral formula, one obtains:

$$E_\alpha(t^\alpha \widetilde{P}(\xi)) = (\widetilde{P}(\xi)-z_0I)^{N_\eta}\frac{1}{2\pi i}\oint_{C_\xi}\frac{E_\alpha(t^\alpha z)}{(z-z_0)^{N_\eta}}(zI-\widetilde{P}(\xi))^{-1}dz,\ t \geqslant 0,\ \xi \in \mathbb{R}^n.$$

Further on,

$$D^\eta\frac{1}{2\pi i}\oint_{C_\xi}\frac{E_\alpha(t^\alpha z)}{(z-z_0)^{N_\eta}}(zI-\widetilde{P}(\xi))^{-1}dz$$

$$=\frac{1}{2\pi i}\oint_{C_\xi}\frac{E_\alpha(t^\alpha z)}{(z-z_0)^{N_\eta}}D^\eta(zI-\widetilde{P}(\xi))^{-1}dz$$

(98) $$=\frac{1}{2\pi i}\oint_{C_\xi}\frac{E_\alpha(t^\alpha z)}{(z-z_0)^{N_\eta}}\frac{[q_{ij}^\eta(\xi,z)]_{1\leqslant i,j\leqslant m}\,dz}{(z-\lambda_1(\xi))^{l_\eta}\cdots(z-\lambda_m(\xi))^{l_\eta}}$$

$$=\frac{1}{2\pi i}\int_{\Gamma_\varepsilon}\frac{E_\alpha(t^\alpha z)}{(z-z_0)^{N_\eta}}\frac{[q_{ij}^\eta(\xi,z)]_{1\leqslant i,j\leqslant m}\,dz}{(z-\lambda_1(\xi))^{l_\eta}\cdots(z-\lambda_m(\xi))^{l_\eta}}$$

(99) $$\leqslant M\kappa(\omega,\alpha,\varepsilon)^{-ml_\eta}\frac{M_\varepsilon}{2\pi}e^{(\omega+\varepsilon)t}(1+|\xi|)^{N_\eta-2}\int_{\Gamma_\varepsilon}\frac{(1+|z|)^{N_\eta-2}}{|z-z_0|^{N_\eta}}|dz|,$$

where (98) follows from the residue theorem. Using the matrix differentiation rules and (99), we immediately obtain (97).

(i) Let $E = (\mathcal{S}(\mathbb{R}^n))^m$. By the invariance of E under the Fourier transform, it follows that $S_\alpha(t) := R_\alpha(t)_{|E} \in L(E)$ for all $t \geqslant 0$. Keeping in mind [275, Theorem B] and (96)-(97), we get that, for every $p \in ⊛$, there exist $M \geqslant 1$ and $q \in ⊛$ such that (94) holds. Let $\vec{x} \in E$, let $t \geqslant 0$ and let $(\vec{x_n})$ be a sequence in Z_n^m such that $\lim_{n\to\infty} \vec{x_n} = \vec{x}$ in E. Suppose that p is a continuous seminorm on E. Let $M \geqslant 1$ and $q \in ⊛$ be such that (94) holds. Then $p_{|Z_n^m}$ is a continuous seminorm on Z_n^m, and the strong continuity of $(S_\alpha(t))_{t\geqslant 0}$ simply follows from the following estimate:

$$p(S_\alpha(t)\vec{x} - S_\alpha(s)\vec{x}) \leqslant M(e^{(\omega+\varepsilon)t} + e^{(\omega+\varepsilon)s}) + p_{|Z_n^m}(R_\alpha(t)\vec{x_n} - R_\alpha(s)\vec{x_n}).$$

Therefore, $(S_\alpha(t))_{t\geqslant 0}$ is an exponentially equicontinuous (g_α, I)-regularized resolvent family generated by $P(\partial/\partial x)_{|E}$. The proof is quite similar in the case $E = (\mathcal{S}'(\mathbb{R}^n))^m$.

(ii) Suppose first $X \neq C^\sigma(\mathbb{R}^n)$. Then the estimates (96)-(97), taken together with the product rule and the Bernstein's lemma [20, Lemma 8.2.1], imply that there exists a sufficiently large $v \in \mathbb{N}$ such that, for given $t \geqslant 0$ in advance, every entry of the matrix $f_t(\xi) \equiv [E_\alpha(t^\alpha \widetilde{P}(\xi))(1 + |\xi|^2)^{-v}]$ belongs to $\mathcal{A}$. Then it is not difficult to prove that the expression $(W_\alpha(t) \equiv f_t(-i\partial/\partial x_1, \cdots, -i\partial/\partial x_n))_{t\geqslant 0}$ determines an exponentially bounded $(g_\alpha, (1-\Delta)^{-v})$-regularized resolvent family generated by $P(\partial/\partial x)_{|E}$ (here we do not distinguish the operator $(1-\Delta)^{-v}$, acting on X, from the operator $(1-\Delta)^{-v}I_{m,m}$, acting on $E = X^m$; $I_{m,m}$ is the identity matrix). Furthermore, $\|W_\alpha(t)\|_X = O(e^{(\omega+\varepsilon)t})$, $t \geqslant 0$. By the definition of topology of E, it follows that $(R_\alpha(t) \equiv W_\alpha(t)W_\alpha(0)^{-1})_{t\geqslant 0}$ is an exponentially equicontinuous (g_α, I)-regularized resolvent family generated by $P(\partial/\partial x)_{|E}$, and that, for every $p \in ⊛$, there exist $M \geqslant 1$ and $q \in ⊛$ such that (94) holds. Keeping in mind the assertion [227, b), p. 374], a similar proof works in the case $X = C^\sigma(\mathbb{R}^n)$ $(0 < \sigma < 1)$.

(iii) Let Q be the totality of indexes $q = (j_1, \ldots, j_s)$, where $1 \leqslant s \leqslant m$ and $1 \leqslant j_1 < \cdots < j_s \leqslant m$. By [202, Lemma 3] (cf. also [508, Lemma 10.1]), we obtain that there exist absolute constants $\alpha_{p,q}^k$ $(0 \leqslant p \leqslant m-1, q \in Q, 0 \leqslant k \leqslant m-1)$ such that, for every $t \geqslant 0$ and $\xi \in \mathbb{R}^n$,

$$E_\alpha(t^\alpha \widetilde{P}(\xi)) = (\widetilde{P}(\xi) - z_0 I)^{m+1}$$
$$\times \sum_{k=0}^{m-1}\left(\sum_{0\leqslant p\leqslant m-1, q\in Q} \alpha_{p,q}^k \frac{1}{2\pi i}\oint_{C_\xi} \frac{z^p E_\alpha(t^\alpha z)}{(z-z_0)^{m+1}\prod_{j\in q}(z-\lambda_j(\xi))}\,dz\right)\widetilde{P}(\xi)^k$$
$$= \sum_{k=0}^{m-1}\left(\sum_{0\leqslant p\leqslant m-1, q\in Q} \alpha_{p,q}^k \frac{1}{2\pi i}\oint_{C_\xi} \frac{z^p E_\alpha(t^\alpha z)}{(z-z_0)^{m+1}\prod_{j\in q}(z-\lambda_j(\xi))}\,dz\right)$$
$$\times\ (\widetilde{P}(\xi) - z_0 I)^{m+1}\widetilde{P}(\xi)^k.$$

Then one gets the existence of a number $K_\varepsilon > 0$ such that, for every $t \geqslant 0$, $\xi \in \mathbb{R}^n$, $0 \leqslant p \leqslant m-1$ and $q \in Q$,

$$\left\| \frac{1}{2\pi i} \oint_{C_\xi} \frac{z^p E_\alpha(t^\alpha z)}{(z-z_0)^{m+1} \prod_{j \in q} (z - \lambda_j(\xi))} dz \right\|$$

$$= \left\| \frac{1}{2\pi i} \int_{\Gamma_\varepsilon} \frac{z^p E_\alpha(t^\alpha z)}{(z-z_0)^{m+1} \prod_{j \in q} (z - \lambda_j(\xi))} dz \right\| \leqslant K_\varepsilon e^{(\omega+\varepsilon)t} \int_{\Gamma_\varepsilon} \frac{(1+|z|)^{m-1}}{|z-z_0|^{m+1}} |dz|,$$

and the proof of [275, Theorem 1(v)] can be repeated verbatim.

II. The Case $0 < \alpha < 1$.

Although technically complicated, the proof of theorem in this case is almost the same as the proof of theorem in the case I. The essential change is only the passing from the integration along the curve C_ξ, by using the residue theorem, to the integration along Γ_ε. Put $k_0 := \lceil 1/2\alpha \rceil$ and suppose first that $\omega = 0$. Then $\operatorname{Re}(re^{i\theta}) \leqslant 0$ $(r > 0,\ \theta \in (-\pi, \pi])$ iff

$$\theta \in \left(\bigcup_{k'=0}^{k_0-1} \pm \left[\alpha \frac{(4k'+1)\pi}{2},\ \alpha \frac{(4k'+3)\pi}{2} \right] \right) \bigcap (-\pi, \pi] =: S_\alpha.$$

Once this observation has been made, it becomes apparent that the set $\Phi_\alpha := \{z \in \mathbb{C} \setminus \{0\} : \arg(z) \in (-\pi, \pi] \setminus S_\alpha\}$ has a finite number of connected components. This further implies that $\Gamma_\varepsilon = \{(\varepsilon/\cos(\theta/\alpha))^\alpha e^{i\theta} : \theta \in (-\pi, \pi] \setminus S_\alpha\}$ can be represented as a finite union of smooth curves. Set $\Gamma_{\varepsilon,R} := \Gamma_\varepsilon \cap \{z \in \mathbb{C} : |z| = R\}$ $(R > 0)$. Then there exists $M_\varepsilon > 0$ such that $|E_\alpha(t^\alpha z)| \leqslant M_\varepsilon e^{\varepsilon t}$, $t \geqslant 0$, $z \in \bigcup_{R>0} (\Gamma_{\varepsilon,R})^\circ$. This implies that, for every $\xi \in \mathbb{R}^n$, there exists a sufficiently large $R_\xi > 0$ such that, for every $R \geqslant R_\xi$, the path of integration C_ξ, in any of the integrals considered in the case I, can be deformed into the curve $\Gamma_{\varepsilon,R}$. Now the claimed assertion follows by observing that the distance between $\partial\Phi_\alpha$ and Γ_ε is positive, and that

$$\lim_{R \to \infty} \oint_{z \in \Gamma_{\varepsilon,R}, |z|=R} \frac{E_\alpha(t^\alpha z)}{(z-z_0)^2} dz = 0.$$

If $\omega > 0$, then $\operatorname{Re}(re^{i\theta}) \leqslant \omega$ $(r > 0,\ \theta \in (-\pi, \pi])$ is equivalent to

$$\theta \in S_\alpha \text{ or } (\theta \in (-\pi, \pi] \setminus S_\alpha \text{ and } r \leqslant (\omega/(\cos(\theta/\alpha)))^\alpha),$$

so that the proof follows similarly as in the case $\omega = 0$. □

Remark 2.3.2. (i) Let $(E, \|\cdot\|)$ be a complex Banach space and let iA_j, $1 \leqslant j \leqslant n$ be commuting generators of bounded C_0-groups on E. For a polynomial matrix $P(x) = \Sigma_{|\eta| \leqslant d} P_\eta x^\eta$ $(P_\eta \in M_m)$, we define $P(A) \equiv \Sigma_{|\eta| \leqslant d} P_\eta A^\eta$ with maximal domain. Then it is well known that $P(A)$ is closable. Suppose $\omega \geqslant 0$ and

$$\sup_{x \in \mathbb{R}^n} \left\{ \operatorname{Re}(\lambda(x)^{1/\alpha}) : \lambda(x) \in \sigma(P(x)) \right\} \leqslant \omega.$$

Then the proof of part (I) of preceding theorem implies that there exists a sufficiently large $\sigma > 0$ such that $\overline{P(A)}$ is the integral generator of a global $(g_\alpha, (1 + |A|^2)^{-\sigma})$-regularized resolvent family $(S_\alpha(t))_{t\geqslant 0}$ on E^m satisfying that, for every $\varepsilon > 0$, there exists $M_\varepsilon \geqslant 1$ such that $\|S_\alpha(t)\| \leqslant M_\varepsilon e^{(\omega+\varepsilon)t}$, $t \geqslant 0$. Disappointingly, our method produces a completely imprecise estimate for the lower bound of σ; the additional difficulty is that the equality

$$D^\eta \left(E_\alpha(t^\alpha P(x))\right) = \sum_{j=1}^{|\eta|} t^{\alpha j} E_\alpha^{(j)}\left(t^\alpha P(x)\right) Q_j(x),\ t \geqslant 0, x \in \mathbb{R}^n,\ m = 1,$$

where $Q_j(x)$ are complex polynomials of degree $\leqslant Nj - |\eta|$ $(1 \leqslant j \leqslant |\eta|)$, cannot be so easily interpreted in the matricial case $m > 1$. Distributional techniques show that the above-mentioned result remains true, with suitable modifications, if we move to the spaces $L^\infty(\mathbb{R}^n)$, $C_b(\mathbb{R}^n)$ and $C^\sigma(\mathbb{R}^n)$ $(0 < \sigma < 1)$.

(ii) In contrast to [275, Theorem 1], Theorem 2.1(ii) covers the case $E = X_n$, where X is $L^p(\mathbb{R}^n)$ $(p \in [1, \infty) \setminus \{2\})$, $C_0(\mathbb{R}^n)$ or $C^\sigma(\mathbb{R}^n)$ $(0 < \sigma < 1)$. Notice also that it is not clear how one can transfer the implication [275, Theorem 1, (b) $\Rightarrow$ (a)] to abstract time-fractional equations.

Now we will state and prove the following extension of [141, Theorem 14.1].

Theorem 2.3.3. Let $(E, \|\cdot\|)$ *be a complex Banach space and let* iA_j, $1 \leqslant j \leqslant n$ *be commuting generators of bounded* C_0*-groups on E. Suppose* $\alpha > 0$ *and* $P(x) = \Sigma_{|\eta|\leqslant d} P_\eta x^\eta$ $(P_\eta \in M_m, x \in \mathbb{R}^n)$ *is a polynomial matrix. Then there exists an injective operator* $L(E) \ni C$ *with dense range such that the operator* $\overline{P(A)}$ *is the integral generator of a global* (g_α, C_m)*-regularized resolvent family* $(W_\alpha(t))_{t\geqslant 0}$ *on* E^m, *where* $C_m = CI_{m,m}$. *Furthermore, the mapping* $t \mapsto W_\alpha(t)$, $t \geqslant 0$ *can be extended to the whole complex plane and the following holds:*

(i) $R(W_\alpha(z)) \subseteq D_\infty(P(A))$, $z \in \mathbb{C}$ *and*

$$\overline{P(A)} \int_0^z g_\alpha(z-s) W_\alpha(s)\vec{x}\, ds = W_\alpha(z)\vec{x} - C_m\vec{x},\ z \in \mathbb{C}, \vec{x} \in E^m.$$

(ii) *The mapping* $z \mapsto W_\alpha(z)$, $z \in \mathbb{C} \setminus (-\infty, 0]$ *is analytic.*

(iii) *The mapping* $z \mapsto W_\alpha(z)$, $z \in \mathbb{C}$ *is entire, provided that* $\alpha \in \mathbb{N}$.

Proof. Let $2|k$, $k > 1/\alpha$, $a > 0$ and $C := (e^{-a|x|^{kd}})(A)$. Then $C \in L(E)$, C is injective and $D_\infty(A_1^2 + \cdots + A_n^2) \supseteq R(C)$ is dense in E. Assume that $P(x)^l = [p_{ij;l}(x)]_{1\leqslant i,j\leqslant m}$ for $l \in \mathbb{N}_0$ and $x \in \mathbb{R}^n$. Define

$$W_\alpha(z)\vec{x} := \sum_{l=0}^{\infty} \frac{z^{\alpha l}}{\Gamma(\alpha l + 1)} P(A)^l C_m \vec{x},\ z \in \mathbb{C}, \vec{x} \in E^m.$$

Then

$$W_\alpha(z) := \left[\sum_{l=0}^{\infty} \frac{z^{\alpha l}}{\Gamma(\alpha l + 1)} \left(P_{ij;l}(x) e^{-a|x|^{kd}} \right)(A) \right]_{1 \leqslant i,j \leqslant m}, \quad z \in \mathbb{C}.$$

Let $\varepsilon \in (0, 1)$ be fixed. Then there exists a constant $M_1 < \infty$ such that, for every multi-index $\eta \in \mathbb{N}_0^n$ with $|\eta| \leqslant k_1 \equiv 1 + \lfloor n/2 \rfloor$,

$$|p_{ij;l}^{(\eta)}(x)| \leqslant M_1^l (1 + |x|)^{ld}, \ x \in \mathbb{R}^n, \ 1 \leqslant i, j \leqslant m, \ l \in \mathbb{N}_0 \text{ and}$$

(100) $$|(e^{-a|x|^{kd}})^{(\eta)}(x)| \leqslant M_1 e^{-\varepsilon a|x|^{kd}}, \ x \in \mathbb{R}^n.$$

The asymptotic formula for the Gamma function combined with the choice of k implies that $\lim_{l \to \infty} \Gamma(\frac{2ld+n}{kd})^{1/2l} \Gamma(\alpha l + 1)^{(-1)/l} = 0$ and that the mapping $z \mapsto \Sigma_{l=0}^{\infty} \frac{z^l}{\Gamma(\alpha l+1)} \Gamma(\frac{2ld+n}{kd})^{1/2}$, $z \in \mathbb{C}$ is entire. Furthermore, a direct computation shows that there exists a constant $M_3 < \infty$ such that

(101) $$\left(\int_{\mathbb{R}^n} (1 + |x|)^{2ld} e^{-2\varepsilon a|x|^{kd}} dx \right)^{1/2} \leqslant M_3^l \left[1 + \Gamma\left(\frac{2ld+n}{kd} \right)^{1/2} \right], \quad l \in \mathbb{N}_0.$$

Taking into account Theorem 1.3.1, (100)-(101), an elementary calculus involving Bernstein's lemma, the dominated convergence theorem and the product rule, implies that there exist constants M_2, $M_4 < \infty$ such that, for $1 \leqslant i, j \leqslant m$ and $z \in \mathbb{C}$,

$$\begin{aligned}
&\left\| \sum_{l=0}^{\infty} \frac{z^{\alpha l}}{\Gamma(\alpha l+1)} \left(p_{ij;l}(x) e^{-a|x|^{kd}} \right)(A) \right\| \\
&\leqslant M_2 \sum_{l=0}^{\infty} \frac{|z|^{\alpha l}}{\Gamma(\alpha l+1)} \left\| \mathcal{F}\left(p_{ij;l}(x) e^{-a|x|^{kd}} \right) \right\|_{L^1(\mathbb{R}^n)} \\
&\leqslant M_2 \sum_{|\eta| \leqslant K_1} \sum_{l=0}^{\infty} \frac{|z|^{\alpha l}}{\Gamma(\alpha l+1)} \left\| D^\eta \left(p_{ij;l}(x) e^{-a|x|^{kd}} \right) \right\|_{L^2(\mathbb{R}^n)} \\
&\leqslant M_1 M_2 M_4 \sum_{l=0}^{\infty} \frac{|z|^{\alpha l}}{\Gamma(\alpha l+1)} M_1^l \left(\int_{\mathbb{R}^n} (1 + |x|)^{2ld} e^{-2\varepsilon a|x|^{kd}} dx \right)^{1/2} \\
&\leqslant M_1 M_2 M_4 \sum_{l=0}^{\infty} \frac{|z|^{\alpha l}}{\Gamma(\alpha l+1)} M_1^l M_3^l \left[1 + \Gamma\left(\frac{2ld+n}{kd} \right)^{1/2} \right] \\
&\leqslant M_1 M_2 M_4 c_\alpha e^{|z| M_1^{1/\alpha} M_3^{1/\alpha}} \\
&+ M_1 M_2 M_4 \sum_{l=0}^{\infty} \frac{(|z|^\alpha M_1 M_3)^l}{\Gamma(\alpha l+1)} \Gamma\left(\frac{2ld+n}{kd} \right)^{1/2} < \infty.
\end{aligned}$$

Hence, $W_\alpha(z) \in L(E)$, $z \in \mathbb{C}$. It is clear that $W_\alpha(0) = C_m$ and that the mapping $t \mapsto W_\alpha(t)$, $t \geqslant 0$ is strongly continuous. It is straightforward to prove that $C_m \overline{P(A)}$

$\subseteq \overline{P(A)}\, C_m$, $W_\alpha(z)\overline{P(A)} \subseteq \overline{P(A)}\, W_\alpha(z)$, $z \in \mathbb{C}$ and $W_\alpha(z)\underline{C_m} = C_m W_\alpha(z)$, $z \in \mathbb{C}$. The dominated convergence theorem and the closedness of $\overline{P(A)}$ imply that:

$$\overline{P(A)} \int_0^z g_\alpha(z-s)W_\alpha(s)\vec{x}\, ds$$

$$= \overline{P(A)} \sum_{l=0}^{\infty} \int_0^z g_\alpha(z-s) g_{\alpha l+1}(s) \overline{P(A)}^l C_m \vec{x}\, ds$$

$$= \overline{P(A)} \sum_{l=0}^{\infty} g_{\alpha(1+l)+1}(z) \overline{P(A)}^l C_m \vec{x}$$

$$= \sum_{l=0}^{\infty} g_{\alpha(1+l)+1}(z) \overline{P(A)}^{l+1} C_m \vec{x}$$

$$= W_\alpha(z)\vec{x} - C_m \vec{x},\ z \in \mathbb{C},\ \vec{x} \in E^m.$$

Therefore, $(W_\alpha(t))_{t\geqslant 0}$ is a global (g_α, C_m)-regularized resolvent family which do have $\overline{P(A)}$ as a subgenerator. It can be easily verified that $\overline{P(A)}$ is, in fact, the integral generator of $(W_\alpha(t))_{t\geqslant 0}$, finishing the proof of (i). The proofs of assertions (ii) and (iii) are simple and therefore omitted. □

Remark 2.3.4. (i) If $m = 1$, $p_{11}(x) = \Sigma_{|\alpha|\leqslant d}\, a_\alpha x^\alpha$, $x \in \mathbb{R}^n$ ($a_\alpha \in \mathbb{C}$) and E is a function space on which translations are uniformly bounded and strongly continuous, then $\overline{P(A)}$ is just the operator $\Sigma_{|\alpha|\leqslant d}\, a_\alpha i^{|\alpha|}(\partial/\partial x)^\alpha$ with its maximal distributional domain. By Theorem 2.3.3, we infer that for each $\alpha > 0$ there exists a dense subset $E_{0,\alpha}$ of $L^p(\mathbb{R}^n)$ such that the abstract Cauchy problem:

$$\mathbf{D}_t^\alpha u(t, x) = \sum_{|\alpha|\leqslant d} a_\alpha i^{|\alpha|}(\partial/\partial x)^\alpha u(t, x),\ t > 0,\ x \in \mathbb{R}^n,$$

$$\frac{\partial^l}{\partial t^l} u(t, x)_{|t=0} = f_l(x),\ x \in \mathbb{R}^n,\ l = 0, 1, \cdots, \lceil \alpha \rceil - 1,$$

has a unique solution provided $f_l(\cdot) \in E_{0,\alpha}$, $l = 0, 1, \cdots, \lceil \alpha \rceil - 1$. A similar assertion can be proved in case that E is $L^\infty(\mathbb{R}^n)$, $C_b(\mathbb{R}^n)$ or $C^\sigma(\mathbb{R}^n)$ $(0 < \sigma < 1)$.

(ii) The results stated in Remark 2.3.2(i), Theorem 2.3.3, and the first part of this remark, can be proved for (systems of) abstract time-fractional equations considered in E_l-type spaces.

Theorem 2.3.5. (i) *Suppose $\alpha > 0$ and X is $\mathcal{S}(\mathbb{R}^n)$ or $\mathcal{S}'(\mathbb{R}^n)$. Then there exists an injective operator $C \in L(X)$ with dense range such that the operator $P(\partial/\partial x)_{|E}$ is the integral generator of a global (g_α, C_m)-regularized resolvent family $(W_\alpha(t))_{t\geqslant 0}$ on $E \equiv X^m$. Furthermore, the mapping $t \mapsto W_\alpha(t)$, $t \geqslant 0$ can be extended to the whole complex plane and the properties (i)-(ii) stated directly before Theorem 2.3.1 remain true with $R_\alpha(\cdot)$ and $(Z'_n)^m$ replaced by $W_\alpha(\cdot)$ and E, respectively.*

(ii) *Suppose $\alpha > 0$, X is $L^2(\mathbb{R}^n)$ and $E = \{\vec{f} \in (L^2(\mathbb{R}^n))^m : (P(\partial/\partial x))^l \vec{f} \in (L^2(\mathbb{R}^n))^m$ for all $l \in \mathbb{N}\}$. Then there exists an injective operator $C \in L(X)$ such that the* *operator* $P(\partial/\partial x)_{|E}$ *is the integral generator of a global* $(g_\alpha, C_{m|E})$*-regularized resolvent family* $(W_\alpha(t))_{t\geqslant 0}$ *on E. Furthermore,* $R(C_{m|E})$ *is dense in E, the mapping* $t \mapsto W_\alpha(t)$, $t \geqslant 0$ *can be extended to the whole complex plane and the properties (i)-(ii) stated directly before Theorem 2.3.1 remain true with* $R_\alpha(\cdot)$ *and* $(Z_n')^m$ *replaced by* $W_\alpha(\cdot)$ *and E, respectively.*

Proof. Suppose first that $E = (\mathcal{S}(\mathbb{R}^n))^m$. Let $a > 0$, $2|k$ and $k > 1/\alpha$. Define

$$W_\alpha(z) := \mathcal{F}E_\alpha(z^\alpha \widetilde{P}(\xi))e^{-a|\xi|^{kd}}\mathcal{F}^{-1},\ z \in \mathbb{C},$$

and $Cf := \mathcal{F}e^{-a|\xi|^{kd}}\mathcal{F}^{-1}f$, $f \in \mathcal{S}(\mathbb{R}^n)$. Let $\widetilde{P}(\xi)^l = [p_{ij;l}(\xi)]_{1\leqslant i,j\leqslant m}$ ($l \in \mathbb{N}_0$, $\xi \in \mathbb{R}^n$). Then it is obvious that:

$$W_\alpha(z)\vec{f} = \left[\sum_{l=0}^{\infty}\frac{z^{\alpha l}}{\Gamma(\alpha l+1)}\mathcal{F}\left(p_{ij;l}(\xi)e^{-a|\xi|^{kd}}\right)\mathcal{F}^{-1}\right]_{1\leqslant i,j\leqslant m}\vec{f},\ z \in \mathbb{C},\ \vec{f} \in E.$$

Since $\mathcal{F}$ and $\mathcal{F}^{-1}$ are topological isomorphisms of $\mathcal{S}(\mathbb{R}^n)$, we immediately obtain that $L(X) \ni C$ is injective. Clearly, $R(C)$ is dense in X and $L(E) \ni W_\alpha(0) = C_m$ is injective. In order to prove that $W_\alpha(z) \in L(E)$ for every $z \in \mathbb{C}$, it suffices to show that, for every multi-index $\eta \in \mathbb{N}_0^n$, there exist $M_\eta \geqslant 1$ and $N_\eta \in \mathbb{N}$ such that:

$$\left\|D^\eta[E_\alpha(z^\alpha \widetilde{P}(\xi))e^{-a|\xi|^{kd}}]\right\| \leqslant M_\eta(1+|\xi|)^{N_\eta},\ \xi \in \mathbb{R}^n. \tag{102}$$

It can be easily seen that there exists $M_1 \geqslant 1$ such that, for every $\eta \in \mathbb{N}_0^n$, we have $|p_{ij;l}^{(\eta)}(\xi)| \leqslant M_1^l(1+|\xi|)^{ld}$, $\xi \in \mathbb{R}^n$, $1 \leqslant i, j \leqslant m$, $l \in \mathbb{N}_0$, which implies, along with the asymptotic formula for the Mittag-Leffler functions, that for each $z \in \mathbb{C}$, $\xi \in \mathbb{R}^n$ and $l \in \mathbb{N}_0$:

$$\left\|E_\alpha(z^\alpha \widetilde{P}(\xi))\right\| \leqslant E_\alpha(M_1|z|^\alpha(1+|\xi|)^d) \leqslant c_\alpha e^{M_1^{1/\alpha}|z|(1+|\xi|)^{d/\alpha}}.$$

Further on,

$$\begin{aligned}
&\left\|\frac{\partial}{\partial\xi_j}E_\alpha\left(z^\alpha\widetilde{P}(\xi)\right)\right\| \\
&\quad= \left\|\sum_{l=1}^{\infty}\frac{z^{\alpha l}}{\Gamma(\alpha l+1)}\left[\widetilde{P}(\xi)^{l-1}\left(\frac{\partial}{\partial\xi_j}\widetilde{P}(\xi)\right)+\cdots+\left(\frac{\partial}{\partial\xi_j}\widetilde{P}(\xi)\right)\widetilde{P}(\xi)^{l-1}\right]\right\| \\
&\quad\leqslant \sum_{l=0}^{\infty}\frac{|z|^{\alpha l}}{\Gamma(\alpha l+1)}lM_1^{l-1}(1+|\xi|)^{d(l-1)} \\
&\quad\leqslant \sum_{l=0}^{\infty}\frac{|z|^{\alpha l}}{\Gamma(\alpha l+1)}(2M_1)^l(1+|\xi|)^{dl} \\
&\quad\leqslant c_\alpha e^{|z|(2M_1)^{1/\alpha}(1+|\xi|)^{d/\alpha}},\ z \in \mathbb{C},\ \xi \in \mathbb{R}^n,\ 1 \leqslant j \leqslant n.
\end{aligned}$$

Continuing in this way, we obtain that, for every $\eta \in \mathbb{N}_0^n$, there exists $b_\eta > 1$ such that:

$$\left\| D^\eta E_\alpha(z^\alpha \tilde{P}(\xi)) \right\|$$
$$\leqslant \sum_{l=1}^{\infty} \frac{|z|^{\alpha l}}{\Gamma(\alpha l+1)} (b_\eta M_1)^l (1+|\xi|)^{dl}$$
$$\leqslant c_\alpha e^{|z| b_\eta^{1/\alpha} M_1^{1/\alpha} (1+|\xi|)^{d/\alpha}}, \ z \in \mathbb{C}, \ \xi \in \mathbb{R}^n.$$

Taken together with the product rule and (100), the last estimate immediately implies (102). The strong continuity of $(W_\alpha(t))_{t \geqslant 0}$ follows from the estimate $|t^{\alpha l} - s^{\alpha l}| \leqslant l|t^\alpha - s^\alpha| \max(t, s)^{l-1}$, $t, s \geqslant 0$ and the previously given arguments. Applying twice the Darboux inequality, one obtains that, for every $z \in \mathbb{C} \setminus (-\infty, 0]$, there exists a constant $\kappa(z) \geqslant 1$ such that:

$$\left| \frac{(z+h)^{\alpha l} - z^{\alpha l}}{h} - \alpha l z^{\alpha l - 1} \right| \leqslant |h| \kappa(z)^{\alpha l}, \tag{103}$$

for any $h \in \mathbb{C}$ with $|h| < \operatorname{dist}(z, (-\infty, 0])$. By (103), it readily follows that the mapping $z \mapsto W_\alpha(z)\vec{f}$, $z \in \mathbb{C} \setminus (-\infty, 0]$ is analytic for every fixed $\vec{f} \in E$ and that, for every $z \in \mathbb{C} \setminus (-\infty, 0]$ and $\vec{f} \in E$,

$$\frac{d}{dz} W_\alpha(z)\vec{f} = \left[\sum_{l=1}^{\infty} \frac{\alpha l z^{\alpha l - 1}}{\Gamma(\alpha l+1)} \mathcal{F}\left(p_{ij;l}(\xi) e^{-a|\xi|^{kd}} \right) \mathcal{F}^{-1} \right]_{1 \leqslant i,j \leqslant m} \vec{f}.$$

One obtains similarly that, for every $\vec{f} \in E$, the mapping $z \mapsto W_\alpha(z)\vec{f}$, $z \in \mathbb{C}$ is entire, provided $\alpha \in \mathbb{N}$, and that the mapping $t \mapsto W_\alpha(t)\vec{f}$, $t \geqslant 0$ is in $C^{\lfloor \alpha \rfloor}([0, \infty) : E)$. The remainder of proof in the case $E = (\mathcal{S}(\mathbb{R}^n))^m$ is simple. The proof of theorem in the case $E = (\mathcal{S}'(\mathbb{R}^n))^m$ is akin to the previously considered one. For the proof of (ii), set $W_\alpha(z)\vec{f} := \mathcal{F}E_\alpha(z^\alpha \tilde{P}(\xi)) e^{-a|\xi|^{kd}} \mathcal{F}^{-1}\vec{f}$, $z \in \mathbb{C}$, $\vec{f} \in (L^2(\mathbb{R}^n))^m$ and $Cf := \mathcal{F}e^{-a|\xi|^{kd}} \mathcal{F}^{-1}f$, $f \in L^2(\mathbb{R}^n)$, where $a > 0$, $2|k$ and $k > 1/\alpha$. Certainly, $R(C_{m|E})$ is dense in E and there exists $c > 0$ such that:

$$\max_{1 \leqslant j \leqslant m} |\lambda_j(\xi)| \leqslant c(1+|\xi|)^d, \ \xi \in \mathbb{R}^n.$$

Keeping in mind the proofs of [275, Theorem 1(v)] and Theorem 2.3.1(iii), it suffices to show that there exists a non-negative function $t \mapsto f(t)$, $t \geqslant 0$ such that, for every $p \in \{0, \cdots, m-1\}$, $q \in Q$, $t \geqslant 0$ and $\xi \in \mathbb{R}^n$:

$$\left\| \frac{1}{2\pi i} \oint_{C_\varepsilon} \frac{z^p E_\alpha(t^\alpha z)}{\left(z - 3c(1+|\xi|)^d\right)^{m+1} \prod_{j \in q} \left(z - \lambda_j(\xi)\right)} e^{-a|\xi|^{kd}} \, dz \right\| \leqslant f(t). \tag{104}$$

The Cauchy theorem shows that there exists $\mu > 0$ such that:

$$\left\| \frac{1}{2\pi i} \oint_{C_\varepsilon} \frac{z^p E_\alpha(t^\alpha z)}{\left(z - 3c(1 + |\xi|)^d\right)^{m+1} \prod_{j\in q}\left(z - \lambda_j(\xi)\right)} e^{-a|\xi|^{kd}}\, dz \right\|$$

$$= \left\| \frac{1}{2\pi i} \oint_{|z|=2c(1+|\xi|)^d} \frac{z^p E_\alpha(t^\alpha z)}{\left(z - 3c(1 + |\xi|)^d\right)^{m+1} \prod_{j\in q}\left(z - \lambda_j(\xi)\right)} e^{-a|\xi|^{kd}}\, dz \right\|$$

$$\leqslant \mu e^{\mu t} + e^{\mu t|\xi|^{d/\alpha} - a|\xi|^{kd}}, \quad t \geqslant 0, \quad \xi \in \mathbb{R}^n,$$

which implies by the choice of k the existence of function $t \mapsto f(t)$, $t \geqslant 0$ satisfying (104). □

2.4 *q*-Exponentially equicontinuous (*a*, *k*)-regularized *C*-resolvent families

The class of q-exponentially equicontinuous $(C_0, 1)$-semigroups was introduced by V. A. Babalola in [25]. The purpose of this section is to examine the possibility of extension of the results obtained in this paper to abstract Volterra equations and abstract time-fractional equations; in other words, our intention is to analyze the class of q-exponentially equicontinuous (a, k)-regularized C-resolvent families. Although the restriction $C = I$ seems to be inevitable in our study, it is not clear whether the condition $k(0) = 0$, used only in the proof of non-degeneracy of the (a, k)-regularized resolvent family $(\overline{R_p}(t))_{t\geqslant 0}$ appearing in the formulation of Theorem 2.4.3(i), is superfluous (cf. also Proposition 2.1.4(ii)). Theorem 2.4.3 has several obvious consequences of which we will emphasize the most important perturbation type theorems (Theorem 2.4.5 and Example 2.4.6).

For every $p \in ⊛$, we define the factor space $E_p := E/p^{-1}(0)$. The norm of a class $x + p^{-1}(0)$ is defined by $\|x + p^{-1}(0)\|_{E_p} := p(x)$ $(x \in E)$. Then the canonical mapping $\Psi_p : E \to E_p$ is continuous; the completion of E_p under the norm $\|\cdot\|_{E_p}$ is denoted by $\overline{E_p}$ (the local Banach space with respect to p). Since no confusion seems likely, we also denote the norms on E_p and $L(E_p)$ ($\overline{E_p}$ and $L(\overline{E_p})$) by $\|\cdot\|$; $L_{⊛}(E)$ denotes the subspace of $L(E)$ which consists of those bounded linear operators T on E such that, for every $p \in ⊛$, there exists $c_p > 0$ satisfying $p(Tx) \leqslant c_p p(x)$, $x \in E$. The infimum of such numbers c_p, denoted by $P_p(T)$, satisfies $P_p(T) = \sup_{x\in E, p(x)\leqslant 1} p(Tx)$ $(p \in ⊛)$. It is clear that $P_p(T_1T_2) \leqslant P_p(T_1)P_p(T_2)$, $p \in ⊛$, T_1, $T_2 \in L_{⊛}(E)$ and that $P_p(\cdot)$ is a seminorm on $L_{⊛}(E)$. If $T \in L_{⊛}(E)$ and $p \in ⊛$, then the operator $T_p : E_p \to E_p$, defined by $T_p(\Psi_p(x)) := \Psi_p(Tx)$, $x \in E$ belongs to $L(E_p)$. Moreover, the operator T_p can be uniquely extended to a bounded linear operator $\overline{T_p}$ on $\overline{E_p}$ and the following holds: $\|T_p\| = \|\overline{T_p}\| = P_p(T)$. Define $V_p := \{x \in E : p(x) \leqslant 1\}$ $(p \in ⊛)$ and order $⊛$ by: $p \gg q$ if $V_p \subseteq V_q$ $(p, q \in ⊛)$. The function $\pi_{qp} : E_p \to E_q$, defined by $\pi_{qp}(\Psi_p(x)) := \Psi_q(x)$, $x \in E$ is a continuous homomorphism of E_p onto E_q, and extends therefore, to a continuous linear homomorphism $\overline{\pi_{qp}}$ of $\overline{E_p}$

onto $\overline{E_q}$. The reader may consult [25] for the basic facts about projective limits of Banach spaces (closed linear operators acting on Banach spaces) and their projective limits.

We introduce (analytic) q-exponentially equicontinuous (a, k)-regularized C-resolvent families as follows.

Definition 2.4.1. (i) Let $k \in C([0, \infty))$ and let $a \in L^1_{loc}([0, \infty))$. Suppose that $(R(t))_{t \geqslant 0}$ is a global (a, k)-regularized C-resolvent family having A as a subgenerator. Then it is said that $(R(t))_{t \geqslant 0}$ is a quasiexponentially equicontinuous (q-exponentially equicontinuous, for short) (a, k)-regularized C-resolvent family having A as subgenerator if, for every $p \in \circledast$, there exist $M_p \geqslant 1$, $\omega_p \geqslant 0$ and $q_p \in \circledast$ such that:

(105)
$$p(R(t)x) \leqslant M_p e^{\omega_p t} q_p(x),\ t \geqslant 0,\ x \in E.$$

(ii) Let $\beta \in (0, \pi]$ and let A be a subgenerator of an analytic (a, k)-regularized C-resolvent family $(R(t))_{t \geqslant 0}$ of angle β. Then it is said that $(R(t))_{t \geqslant 0}$ is a q-exponentially equicontinuous, analytic (a, k)-regularized C-resolvent family of angle β, if for every $p \in \circledast$ and $\varepsilon \in (0, \beta)$, there exist $M_{p,\varepsilon} \geqslant 1$, $\omega_{p,\varepsilon} \geqslant 0$ and $q_{p,\varepsilon} \in \circledast$ such that:

$$p(R(z)x) \leqslant M_{p,\varepsilon} e^{\omega_{p,\varepsilon}|z|} q_{p,\varepsilon}(x),\ z \in \Sigma_{\beta-\varepsilon},\ x \in E.$$

It is clear from Definition 2.4.1 that every q-exponentially equicontinuous (a, k)-regularized C-resolvent family $(R(t))_{t \geqslant 0}$ is locally equicontinuous. On the other hand, the following example from [25] shows that $(R(t))_{t \geqslant 0}$ need not be exponentially equicontinuous, in general: Let $a(t) = k(t) = 1$, let $C = I$ and let the Schwartz space of rapidly decreasing functions $\mathcal{S}(\mathbb{R})$ be topologized by the following system of seminorms $p_{m,n}(f) = \|x^m f^{(n)}(x)\|_{L^2(\mathbb{R})}$ $(m, n \in \mathbb{N}_0, f \in \mathcal{S}(\mathbb{R}))$; notice that the usual topology on $\mathcal{S}(\mathbb{R})$, induced by the seminorms $q_{m,n}(f) = \|x^m f^{(n)}(x)\|_{L^\infty(\mathbb{R})}$ $(m, n \in \mathbb{N}_0, f \in \mathcal{S}(\mathbb{R}))$, is equivalent to the topology introduced above. Set $(S(t)f)(x) := f(e^t x)$, $t \geqslant 0$, $x \in \mathbb{R}$, $f \in \mathcal{S}(\mathbb{R})$. Then $(S(t))_{t \geqslant 0}$ is a q-exponentially equicontinuous (a, k)-regularized resolvent family (i.e., q-exponentially equicontinuous $(C_0, 1)$-semigroup) whose integral generator is the bounded linear operator $A \in L(\mathcal{S}(\mathbb{R}))$ given by $(Af)(x) := xf'(x)$, $x \in \mathbb{R}$, $f \in \mathcal{S}(\mathbb{R})$; $(S(t))_{t \geqslant 0}$ is not exponentially equicontinous, $(S(t))_{t \geqslant 0}$ has no Laplace transform in $\mathcal{S}(\mathbb{R})$ and $p_{mn}(S(t)f) = e^{(n-m-(1/2))t} p_{mn}(f)$ $(t \geqslant 0, m, n \in \mathbb{N}_0, f \in \mathcal{S}(\mathbb{R}))$. It can be easily proved that there does not exist an injective operator $C \in L(\mathcal{S}(\mathbb{R}))$ such that A is the integral generator of an exponentially equicontinuous C-regularized semigroup in $\mathcal{S}(\mathbb{R})$.

It is very simple to carry over the assertion of Proposition 2.2.7 to q-exponentially equicontinuous (a, k)-regularized C-resolvent families; in such a way, one can construct some artificial examples of q-exponentially equicontinuous (not exponentially equicontinuous, in general) (a, k)-regularized C-resolvent families, with $C \neq I$ or $k(0) = 0$. It is also worth noting that the assertions of [292, Theorem 2.1.27(xiii)-(xiv), Theorem 2.5.1-Theorem 2.5.3, Remark 2.5.4(iii),

Theorem 2.5.5-Theorem 2.5.6] and [316, Theorem 2.1, Corollary 2.2, Theorem 2.3, Corollary 2.4] can be reformulated for (analytic) q-exponentially equicontinuous (a, k)-regularized C-resolvent families in SCLCSs. This is not the case with the assertions of [303, Theorem 2.14-Theorem 2.15]; even on reflexive spaces, the adjoint of a q-exponentially equicontinuous $(C_0, 1)$-semigroup need not be of the same class ([25]).

Before we state the following extension of [303, Theorem 3.9], it will be necessary to recall that, for every $\alpha > 0$, there exists $c_\alpha > 0$ such that:

$$E_\alpha(t) \leqslant c_\alpha \exp(t^{1/\alpha}),\ t \geqslant 0. \tag{106}$$

Theorem 2.4.2. *Assume $k_\beta(t)$ satisfies (P1), $0 < \alpha < \beta$, $\gamma = \alpha/\beta$ and A is a subgenerator of a q-exponentially equicontinuous (g_β, k_β)-regularized C-resolvent family $(S_\beta(t))_{t\geqslant 0}$ satisfying (105) with $R(\cdot)$ replaced by $S_\beta(\cdot)$ therein. Assume that there exist a continuous function $k_\alpha(t)$ satisfying (P1) and a number $\upsilon > 0$ such that $k_\alpha(0) = k_\beta(0)$ and*

$$\widetilde{k_\alpha}(\lambda) = \lambda^{\gamma-1}\widetilde{k_\beta}(\lambda^\gamma),\ \lambda > \upsilon. \tag{107}$$

Then A is a subgenerator of a q-exponentially equicontinuous (g_α, k_α)-regularized C-resolvent family $(S_\alpha(t))_{t\geqslant 0}$, given by

$$S_\alpha(t)x := \int_0^\infty t^{-\gamma}\,\Phi_\gamma(st^{-\gamma})S_\beta(s)x\,ds,\ x \in E,\ t > 0 \text{ and } S_\alpha(0) := k_\alpha(0)C.$$

Furthermore,

$$p(S_\alpha(t)x) \leqslant c_\gamma M_p \exp(\omega_p^{1/\gamma}t)q_p(x),\ p \in \circledast,\ t \geqslant 0,\ x \in E. \tag{108}$$

Let $p \in \circledast$. Then the condition

$$p(S_\beta(t)x) \leqslant M_p(1 + t^{\xi_p})e^{\omega_p t}q_p(x),\ t \geqslant 0,\ x \in E\ (\xi_p \geqslant 0), \tag{109}$$

resp.,

$$p(S_\beta(t)x) \leqslant M_p t^{\xi_p}e^{\omega_p t}q_p(x),\ t \geqslant 0,\ x \in E, \tag{110}$$

implies that there exists $M'_p \geqslant 1$ such that

$$p(S_\alpha(t)x) \leqslant M'_p(1 + t^{\xi_p\gamma})(1 + \omega_p t^{\xi_p(1-\gamma)})\exp(\omega_p^{1/\gamma}t)q_p(x),\ t \geqslant 0,\ x \in E, \tag{111}$$

resp.,

$$p(S_\alpha(t)x) \leqslant M'_p t^{\xi_p\gamma}(1 + \omega_p t^{\xi_p(1-\gamma)})\exp(\omega_p^{1/\gamma}t)q_p(x),\ t \geqslant 0,\ x \in E. \tag{112}$$

We also have the following:

(i) *The mapping $t \mapsto S_\alpha(t)$, $t > 0$ admits an extension to $\Sigma_{\min((\frac{1}{\gamma}-1)\frac{\pi}{2},\pi)}$ and, for every $x \in E$, the mapping $z \mapsto S_\alpha(z)x$, $z \in \Sigma_{\min((\frac{1}{\gamma}-1)\frac{\pi}{2},\pi)}$ is analytic.*

(ii) *Let* $\varepsilon \in (0, \min((\frac{1}{\gamma}-1)\frac{\pi}{2}, \pi))$. *If, for every* $p \in \circledast$, *one has* $\omega_p = 0$, *then* $(S_\alpha(t))_{t\geqslant 0}$ *is an equicontinuous analytic* (g_α, k_α)*-regularized C-resolvent family of angle* $\min((\frac{1}{\gamma}-1)\frac{\pi}{2}, \pi)$.

(iii) *If* $\omega_p > 0$ *for some* $p \in \circledast$, *then* $(S_\alpha(t))_{t\geqslant 0}$ *is a q-exponentially equicontinuous, analytic* (g_α, k_α)*-regularized C-resolvent family of angle* $\min((\frac{1}{\gamma}-1)\frac{\pi}{2}, \frac{\pi}{2})$.

Proof. By definition of Wright function and (106), we have that (cf. also the proof of [49, Theorem 3.1]):

$$p(S_\alpha(t)x) \leqslant q_p(x)\int_0^\infty t^{-\gamma}\,\Phi_\gamma\,(st^{-\gamma})\,M_p e^{\omega_p s}\,ds$$
$$= M_p q_p(x) E_\gamma(\omega_p t^\gamma) \leqslant M_p c_\gamma \exp(\omega_p^{1/q} t) q_p(x),\, p \in \circledast,\, x \in E,\, t \geqslant 0,$$

which implies (108). By the proof of the above-mentioned theorem, we get that $(S_\alpha(t))_{t\geqslant 0}$ is strongly continuous. It can be easily seen that $S_\alpha(t)A \subseteq AS_\alpha(t)$ and $S_\alpha(t)C = CS_\alpha(t)$ $(t \geqslant 0)$. Let $x \in D(A)$ and $p \in \circledast$ be fixed. Taken together, the identity [49, (3.10)]

$$\int_0^\infty e^{-\lambda t}\, t^{-\gamma}\,\Phi_\gamma\,(st^{-\gamma})\,dt = \lambda^{\gamma-1}\exp(-\lambda^\gamma s),\ s > 0,\ \lambda > 0,$$

the functional equation of $(S_\beta(t))_{t\geqslant 0}$, the Fubini theorem and the elementary properties of vector-valued Laplace transform imply after some patching up that there exists a sufficiently large number $\kappa_p > \upsilon$ such that (the integrals are taken in the sense of convergence in $\overline{E_p}$):

$$\int_0^\infty e^{-\lambda t}\,\Psi_p(S_\alpha(t)x)\,dt$$
$$=\int_0^\infty\int_0^\infty \Psi_p\big(e^{-\lambda t}\,t^{-\gamma}\,\Phi_\gamma(st^{-\gamma})S_\beta(s)x\big)\,ds\,dt$$
$$=\int_0^\infty\int_0^\infty \Psi_p\big(e^{-\lambda t}\,t^{-\gamma}\,\Phi_\gamma(st^{-\gamma})S_\beta(s)x\big)\,dt\,ds \tag{113}$$
$$=\lambda^{\gamma-1}\int_0^\infty e^{-\lambda^\gamma s}\,\Psi_p(S_\beta(s)x)\,ds$$

$$= \lambda^{\gamma-1}\int_0^\infty e^{-\lambda^\gamma s}\,\Psi_p\left(k_\beta(s)Cx + \int_0^s g_\beta(s-r)S_\beta(r)Ax\,dr\right)ds$$

$$= \lambda^{\gamma-1}\widetilde{k_\beta}(\lambda^\gamma)\Psi_p(Cx) + \lambda^{\gamma-1}\lambda^{-\beta\gamma}\int_0^\infty e^{-\lambda^\gamma s}\,\Psi_p(S_\beta(s)Ax)\,ds$$

$$(114)\qquad = \lambda^{\gamma-1}\widetilde{k_\beta}(\lambda^\gamma)\Psi_p(Cx) + \lambda^{-\alpha}\lambda^{\gamma-1}\int_0^\infty e^{-\lambda^\gamma s}\,\Psi_p(S_\beta(s)Ax)\,ds$$

$$= \int_0^\infty e^{-\lambda t}\,\Psi_p(k_\alpha(t)Cx)\,dt + \int_0^\infty e^{-\lambda t}\,\Psi_p\left(\int_0^t g_\alpha(t-s)S_\alpha(s)Ax\,ds\right)dt$$

$$= \int_0^\infty e^{-\lambda t}\,\Psi_p\left(k_\alpha(t)Cx + \int_0^t g_\alpha(t-s)S_\alpha(s)Ax\,ds\right)dt,\ \lambda > \kappa_p,$$

where (114) follows from (107) and (113). Therefore,

$$(115)\qquad \int_0^\infty e^{-\lambda t}\,\Psi_p\left(S_\alpha(t)x - k_\alpha(t)Cx - \int_0^t g_\alpha(t-s)S_\alpha(s)Ax\,ds\right)dt = 0,\ \lambda > \kappa_p.$$

By the uniqueness theorem for the Laplace transform and the fact that E is Hausdorff, we obtain from (115) that $S_\alpha(t)x = k_\alpha(t)Cx + \int_0^t g_\alpha(t-s)S_\alpha(s)Ax\,ds$, $t \geqslant 0$. Suppose now that $S_\alpha(t)x = 0$, $t \geqslant 0$ for some $x \in E$. Then, for every $p \in ⊛$, there exists a sufficiently large $\xi_p > 0$ such that (113) holds for any $\lambda > \xi_p$, which implies by the uniqueness theorem for the Laplace transform that $\Psi_p(S_\beta(t)x) = 0$, $t \geqslant 0$. Therefore, $S_\beta(t)x = 0$, $t \geqslant 0$ and $x = 0$, because $(S_\beta(t))_{t\geqslant 0}$ is non-degenerate. Hence, $(S_\alpha(t))_{t\geqslant 0}$ is a q-exponentially equicontinuous (g_α, k_α)-regularized C-resolvent family with a subgenerator A. Suppose now that (109), resp. (110), holds. By making use of the integral representation of the Wright function, the Fubini theorem and the Laplace transform, it can be simply proved that there exists $M''_p \geqslant 1$ such that:

$$\int_0^\infty e^{\omega_p s t^\gamma}\,\Phi_\gamma(s)s^{\xi_p}\,ds \leqslant M''_p\left(1 + \left(\omega_p t^\gamma\right)^{\frac{\xi_p(1-\gamma)}{\gamma}}\right)\exp\left(\omega_p^{1/\gamma}t\right),$$

provided $\omega_p > 0$ and $t \geqslant \omega_p^{(-1)/\gamma}$. This immediately implies that (111), resp. (112), holds. The proofs of (i)-(iii) essentially follows from Lemma 1.2.4-Theorem 1.2.5 and the proof of [49, Theorem 3.3]; here the only non-trivial part is the continuity of mapping $z \mapsto S_\alpha(z)x$ on closed sectors containing the non-negative real axis ($x \in E$). For the convenience of the reader, we will prove this assertion in the case that $\omega_p > 0$ for some $p \in ⊛$ (cf. (iii)). Put $\kappa_\gamma := \min((\frac{1}{\gamma}-1)\frac{\pi}{2}, \frac{\pi}{2})$. Let $p \in ⊛$, $x \in E$ and $\delta \in (0, \kappa_\gamma)$ be fixed, and let $\delta' \in (\delta, \kappa_\gamma)$. By the proof of [49, Theorem 3.3], we infer that there exist $M_{p,\delta'} \geqslant 1$ and $\omega_{p,\delta'} > 0$ such that $p(S_\alpha(z)x) \leqslant M_{p,\delta'} e^{\omega_{p,\delta'} \operatorname{Re} z} q_p(x)$, $z \in \Sigma_{\delta'}$ and that the mapping $z \mapsto \langle x^*, S_\alpha(z)x \rangle$, $z \in \Sigma_{\kappa_\gamma}$ $(x^* \in E^*)$ is analytic, which implies the analyticity of mapping $z \mapsto S_\alpha(z)x$, $z \in \Sigma_{\kappa_\gamma}$. Let $\xi_{p,\delta'} > \omega_{p,\delta'}$. Then the function $z \mapsto e^{-\xi_{p,\delta'} z}\, \Psi_p(S_\alpha(z)x)$, $z \in \Sigma_{\delta'}$ is analytic and bounded. Since $\lim_{t \downarrow 0+} \Psi_p(S_\alpha(t)x) = \Psi_p(k_\alpha(0)Cx)$, we obtain by Theorem 1.2.5(ii) that $\lim_{z \to 0, z \in \Sigma_\delta} \Psi_p(S_\alpha(z)x) = \Psi_p(k_\alpha(0)Cx)$. The above yields $\lim_{z \to 0, z \in \Sigma_\delta} p(S_\alpha(z)x - k_\alpha(0)Cx) = 0$, and since p was arbitrary, $\lim_{z \to 0, z \in \Sigma_\delta} S_\alpha(z)x = k_\alpha(0)Cx$. □

It is worth noting that the preceding theorem has an obvious analogue in the case that A is a subgenerator of an exponentially equicontinuous (g_β, k_β)-regularized C-resolvent family $(S_\beta(t))_{t \geqslant 0}$, and that the angle of analyticity of the resolvent $(S_\alpha(t))_{t \geqslant 0}$ can be improved provided that $(S_\beta(t))_{t \geqslant 0}$ is an exponentially equicontinuous, analytic (g_β, k_β)-regularized C-resolvent family. For more details, we refer the reader to the assertions of [292, Theorem 2.4.19] and [303, Theorem 3.9-Theorem 3.10].

Combining the proof of Theorem 2.4.2 with [20, Lemma 1.6.7], we have proved in [304, Theorem 2.2] a slight generalization of the abstract Weierstrass formula stated in Theorem 2.2.18. It is clear that the last mentioned theorem as well as Theorem 2.4.2 can be applied to a class of differential operators with variable coefficients on $\mathcal{S}(\mathbb{R}^n)$ (cf. [25, Section 6] and [245]). For example, let $\mathcal{S}(\mathbb{R})$ be topologized as before and let the operator $A \in L(\mathcal{S}(\mathbb{R}))$ be defined by $(Af)(x) := x^2 f''(x) + xf'(x)$, $x \in \mathbb{R}$, $f \in \mathcal{S}(\mathbb{R})$. Then A is the integral generator of a q-exponentially equicontinuous cosine function $(C(t)\Diamond \equiv \frac{1}{2}(\Diamond(e^{t}\cdot) + \Diamond(e^{-t}\cdot)))_{t \geqslant 0}$ in $\mathcal{S}(\mathbb{R})$, which implies by Theorem 2.4.2 that, for every $\alpha \in (0, 2)$, the operator A is the integral generator of a q-exponentially equicontinuous, analytic (g_α, g_1)-regularized resolvent family of angle $\delta_\alpha \equiv \min((\frac{2}{\alpha}-1)\frac{\pi}{2}, \frac{\pi}{2})$. Therefore, for every $\alpha \in (0, 2)$, the abstract Cauchy problem:

$$\mathbf{D}_t^\alpha u(t, x) = x^2 u_{xx}(t, x) + x u_x(t, x),\ t > 0,\ x \in \mathbb{R};$$
$$u(0, x) = f_0(x), \text{ and } u_t(0, x) = f_1(x) \text{ if } \alpha \in (1, 2),$$

has a unique solution for any $f_0, f_1 \in \mathcal{S}(\mathbb{R})$, and the mapping $t \mapsto u(t, \cdot) \in \mathcal{S}(\mathbb{R})$, $t > 0$ is analytically extensible to the sector Σ_{δ_α} ([292]). Furthermore, Theorem 2.4.3 stated below and [292, Theorem 2.4.19] together imply that, for every $\alpha \in (0, 1)$, A is the integral generator of a q-exponentially equicontinuous, analytic (g_α, g_1)-regularized resolvent family of angle $\delta'_\alpha \equiv \min((\frac{2}{\alpha}-1)\frac{\pi}{2}, \pi)$.

Keeping in mind the proof of Arendt-Widder theorem in SCLCSs (Theorem 1.2.3), we obtain the representation formulae for (a, k)-regularized C-resolvent families whose existence has been proved in the subordination principle (Theorem 2.1.8). Here we would like to observe that it is not clear whether the above-mentioned result can be extended to the class of q-exponentially equicontinuous (a, k)-regularized C-resolvent families in SCLCSs by means of these formulae and the method described in the proof of Theorem 2.4.2. Nevertheless, Theorem 2.4.3 enables one to prove a generalization of the subordination principle for a subclass of q-exponentially equicontinuous (a, k)-regularized resolvent families in complete locally convex spaces.

The proofs of generation results given in [25] do not work any longer in the case of a general q-exponentially equicontinuous (a, k)-regularized C-resolvent family $(R(t))_{t\geqslant 0}$. We must restrict ourselves to the case in which $C = I$ and (105) holds with $q_p = p$ (cf. also [25, Theorem 2.8]). In other words, we will consider a q-exponentially equicontinuous (a, k)-regularized resolvent family $(R(t))_{t\geqslant 0}$ which satisfies that, for every $p \in \circledast$, there exist $M_p \geqslant 1$ and $\omega_p \geqslant 0$ such that:

$$p(R(t)x) \leqslant M_p e^{\omega_p t} p(x),\ t \geqslant 0,\ x \in E. \tag{116}$$

In the sequel, the operator $\overline{R(t)_p}$ will also be denoted by $\overline{R_p}(t)$ $(t \geqslant 0)$.

We call a closed linear operator A acting on E compartmentalized (w.r.t. $\circledast$) if, for every $p \in \circledast$, $A_p := \{(\Psi_p(x), \Psi_p(Ax)) : x \in D(A)\}$ is a function ([25]). For example, every operator $T \in L_{\circledast}(E)$ is compartmentalized.

Theorem 2.4.3. (i) *Suppose $a(t)$ satisfies (P1), $k(0) \neq 0$ and A is a subgenerator of a q-exponentially equicontinuous (a, k)-regularized resolvent family $(R(t))_{t\geqslant 0}$ which satisfies that, for every $p \in \circledast$, there exist $M_p \geqslant 1$ and $\omega_p \geqslant 0$ such that (116) holds. Then A is a compartmentalized operator and, for every $p \in \circledast$, $\overline{A_p}$ is a subgenerator of the exponentially bounded (a, k)-regularized resolvent family $(\overline{R_p}(t))_{t\geqslant 0}$ in $\overline{E_p}$ satisfying that:*

$$\|\overline{R_p}(t)\| \leqslant M_p e^{\omega_p t},\ t \geqslant 0. \tag{117}$$

Assume additionally that (22) holds. Then, for every $p \in \circledast$,

$$\overline{A_p} \int_0^t a(t-s)\overline{R_p}(s)\overline{x_p}\, ds = \overline{R_p}(t)\overline{x_p} - k(t)\overline{x_p},\ t \geqslant 0,\ \overline{x_p} \in \overline{E_p}, \tag{118}$$

the integral generator of $(R(t))_{t\geqslant 0}((\overline{R_p}(t))_{t\geqslant 0})$ is A $(\overline{A_p})$, and $(\overline{R_p}(t))_{t\geqslant 0}$ is a q-exponentially equicontinuous, analytic (a, k)-regularized resolvent family of angle $\beta \in (0, \pi]$, provided that $(R(t))_{t\geqslant 0}$ is.

(ii) *Suppose $a(t)$ and $k(t)$ satisfy (P1), E is complete, A is a compartmentalized operator in E and, for every $p \in \circledast$, $\overline{A_p}$ is a subgenerator of an exponentially bounded (a, k)-regularized resolvent family $(\overline{R_p}(t))_{t\geqslant 0}$ in $\overline{E_p}$ satisfying (117)-(118). Then, for every $p \in \circledast$, (116) holds and A is a subgenerator of a q-exponentially equicontinuous (a, k)-regularized resolvent family $(R(t))_{t\geqslant 0}$ satisfying (22). Furthermore, $(R(t))_{t\geqslant 0}$ is a q-exponentially equicontinuous, analytic (a, k)-regularized resolvent family of angle $\beta \in (0, \pi]$ provided that, for every $p \in \circledast$, $(\overline{R_p}(t))_{t\geqslant 0}$ is a q-exponentially bounded, analytic (a, k)-regularized resolvent family of angle β.*

Proof. Suppose $x, y \in D(A)$ and $p(x) = p(y)$ for some $p \in \circledast$. Then $p(R(t)(x-y) + \int_0^t a(t-s)R(s)A(y-x)\, ds) = 0$, $t \geqslant 0$, which implies $p(\int_0^t a(t-s)R(s)A(y-x)\, ds) = 0$, $t \geqslant 0$. Therefore,

$$\int_0^{\infty} e^{-\lambda t}\,\Psi_p\left(\int_0^t a(t-s)R(s)A(y-x)\,ds\right)dt$$
$$=\int_0^{\infty} e^{-\lambda t}\int_0^t a(t-s)\,\Psi_p(R(s)A(y-x))\,ds\,dt=0,\ \operatorname{Re}\lambda>\max\,(abs(a),\omega_p),$$

and by the uniqueness theorem for the Laplace transform, $\Psi_p(R(t)A(x-y))=0$, $t \geqslant 0$. Using the fact that $(R(t))_{t\geqslant 0}$ is non-degenerate, we obtain that $p(A(x-y))=0$ and $p(Ax)=p(Ay)$, so that A_p is a linear operator in E_p. Let (x_n) be a sequence in $D(A)$ with $\lim_{n\to\infty}\Psi_p(x_n)=0$ and $\lim_{n\to\infty}\Psi_p(Ax_n)=y$ in $\overline{E_p}$. Then we have $\lim_{n\to\infty} p(\int_0^t a(t-s)R(s)Ax_n\,ds)=\lim_{n\to\infty}\|\int_0^t a(t-s)\Psi_p(R(s)\,Ax_n)\,ds\|_{\overline{Ep}}=\lim_{n\to\infty}\|\int_0^t a(t-s)\overline{R_p}(s)A_px_n\,ds\|_{\overline{Ep}}=0$, $t\geqslant 0$, which implies $0=\lim_{n\to\infty}\int_0^t a(t-s)\overline{R_p}(s)A_px_n\,ds=\int_0^t a(t-s)\overline{R_p}(s)y\,ds=0$, $t\geqslant 0$. Taking the Laplace transform, one obtains $\overline{R_p}(t)y=0$, $t\geqslant 0$ and, in particular, $y=0$ since $\overline{R_p}(0)=k(0)I$ and $k(0)\neq 0$. The above implies that A_p is a closable linear operator in $\overline{E_p}$ and that A is a compartmentalized operator in E. It is checked at once that $\overline{R_p}(t)\overline{A_p}\subseteq\overline{A_p}\,\overline{R_p}(t)$, $t\geqslant 0$. Furthermore, (117) holds and the mapping $t\mapsto\overline{R_p}(t)x_p$, $t\geqslant 0$ is continuous for any $x_p\in E_p$, which implies by the usual limit procedure that the mapping $t\mapsto\overline{R_p}(t)\overline{x_p}$, $t\geqslant 0$ is continuous for any $\overline{x_p}\in\overline{E_p}$. The functional equation of $(R(t))_{t\geqslant 0}$ implies $\overline{R_p}(t)x_p-k(t)x_p=\int_0^t a(t-s)\overline{R_p}(s)A_px_n\,ds$, $t\geqslant 0$, $x_p\in D(A_p)$, and therefore, $\overline{R_p}(t)\overline{x_p}-k(t)\overline{x_p}=\int_0^t a(t-s)\overline{R_p}(s)\overline{A_p}\,\overline{x_p}\,ds$, $t\geqslant 0$, $\overline{x_p}\in D(\overline{A_p})$. Hence, $\overline{A_p}$ is a subgenerator of the exponentially bounded, non-degenerate (a,k)-regularized resolvent family $(\overline{R_p}(t))_{t\geqslant 0}$ in $\overline{E_p}$. If (22) holds, then $\overline{R_p}(t)x_p-k(t)x_p=A_p\int_0^t a(t-s)\overline{R_p}(s)x_p\,ds$, $t\geqslant 0$, $x_p\in E_p$, which implies (118). It is not difficult to see that the integral generator of $(R(t))_{t\geqslant 0}$ $((\overline{R_p}(t))_{t\geqslant 0})$ is A $(\overline{A_p})$. Suppose now that $(R(t))_{t\geqslant 0}$ is a q-exponentially equicontinuous, analytic (a,k)-regularized resolvent family of angle β. Then the mapping $z\mapsto\overline{R_p}(z)x_p$, $z\in\Sigma_\beta$ is analytic for any $p\in\circledast$ and $x_p\in E_p$, because the mapping $z\mapsto R(z)x$, $z\in\Sigma_\beta$ $(x\in E)$ is infinitely differentiable and $\Psi_p(\cdot)$ is continuous. It is clear that the condition

(119) $$p(R(z)x)\leqslant M_{p,\varepsilon}e^{\omega_{p,\varepsilon}|z|}p(x),\ x\in E,\ z\in\Sigma_{\beta-\varepsilon},\ p\in\circledast$$

for some $M_{p,\varepsilon}\geqslant 1$, $\omega_{p,\varepsilon}\geqslant 0$ and $\varepsilon\in(0,\beta)$ implies the following:

(120) $$\|\overline{R_p}(z)\|\leqslant M_{p,\varepsilon}e^{\omega_{p,\varepsilon}|z|},\ z\in\Sigma_{\beta-\varepsilon}.$$

Now the analyticity of mapping $z\mapsto\overline{R_p}(z)\overline{x_p}$, $z\in\Sigma_\beta$ $(p\in\circledast,\ \overline{x_p}\in\overline{E_p})$ follows from Vitali's theorem [20, Theorem A.5]. Let $\delta\in(0,\beta)$. Then the mapping $z\mapsto\overline{R_p}(z)x_p$, $z\in\overline{\Sigma_\delta}$ $(p\in\circledast,\ x_p\in E_p)$ is continuous, which implies by (120) the continuity of mapping $z\mapsto\overline{R_p}(z)\overline{x_p}$, $z\in\overline{\Sigma_\delta}$ $(p\in\circledast,\ \overline{x_p}\in\overline{E_p})$. The above shows that $(\overline{R_p}(t))_{t\geqslant 0}$ is a q-exponentially equicontinuous, analytic (a,k)-regularized resolvent family of angle β $(p\in\circledast)$. In order to prove (ii), notice first that the projective limit of $\{\overline{A_p}:p\in\circledast\}$ is A and that $(x,y)\in D(A)$ if $(\Psi_p(x),\Psi_p(y))\in\overline{A_p}$ for all $p\in\circledast$. Set, for every $p\in\circledast$, $\omega_p':=\max(abs(a),abs(k),\omega_p)$. By Theorem 2.1.5, for every $p\in\circledast$, the following holds:

$$\widetilde{k}(\lambda)(I-\widetilde{a}(\lambda)\overline{A_p})^{-1}\overline{x_p} = \int_0^\infty e^{-\lambda t}\overline{R_p}(t)\,\overline{x_p}\,dt,\ \overline{x_p}\in\overline{E_p},\ \operatorname{Re}\lambda>\omega_p',\ \widetilde{k}(\lambda)\neq 0.$$

Define $F_p : \{\lambda\in\mathbb{C} : \operatorname{Re}\lambda>\omega_p'\}\to L(\overline{E_p})$ by $F_p(\lambda)\overline{x_p} := \int_0^\infty e^{-\lambda t}\overline{R_p}(t)\overline{x_p}\,dt$, $\lambda\in D(F_p)$, $\overline{x_p}\in\overline{E_p}$ ($p\in\circledast$). Then $F_p(\cdot)$ is analytic and $F_p(\lambda)=\widetilde{k}(\lambda)(I-\widetilde{a}(\lambda)\overline{A_p})^{-1}$, provided $\operatorname{Re}\lambda>\omega_p'$ and $\widetilde{k}(\lambda)\neq 0$. Suppose now $p, q\in\circledast$ and $p\gg q$. Then it is clear that $\pi_{qp}(\overline{R_p}(0)\overline{x_p})=\overline{R_q}(0)\,\pi_{qp}(\overline{x_p})$, $\overline{x_p}\in\overline{E_p}$. Fix for a moment $t>0$. Then, for every $\lambda\in\mathbb{C}$ with $\operatorname{Re}\lambda>\max(\omega_p', \omega_q')$ and $\widetilde{k}(\lambda)\,\widetilde{a}(\lambda)\neq 0$, we have by [25, Lemma 4.1]:

$$\begin{aligned}
&\pi_{qp}\left(\tilde{k}(\lambda)\left(I-\tilde{a}(\lambda)\overline{A_p}\right)^{-1}\overline{x_p}\right)\\
&=\pi_{qp}\left(\frac{\tilde{k}(\lambda)}{\tilde{a}(\lambda)}\left(\frac{1}{\tilde{a}(\lambda)}-\overline{A_p}\right)^{-1}\overline{x_p}\right)\\
&=\frac{\tilde{k}(\lambda)}{\tilde{a}(\lambda)}\left(\frac{1}{\tilde{a}(\lambda)}-\overline{A_q}\right)^{-1}\pi_{qp}\left(\overline{x_p}\right)\\
&=\tilde{k}(\lambda)\left(I-\tilde{a}(\lambda)\overline{A_q}\right)^{-1}\pi_{qp}\left(\overline{x_p}\right),\quad \overline{x_p}\in\overline{E_p}.
\end{aligned}$$

The above implies $\pi_{qp}(F_p(\lambda)\overline{x_p})=F_q(\lambda)\pi_{qp}(\overline{x_p})$, $\operatorname{Re}\lambda>\max(\omega_p', \omega_q')$, $\overline{x_p}\in\overline{E_p}$, and:

(121)

$$\pi_{qp}\left(\frac{d^n}{d\lambda^n}F_p(\lambda)\overline{x_p}\right)=\frac{d^n}{d\lambda^n}F_q(\lambda)\pi_{qp}\left(\overline{x_p}\right),\ \operatorname{Re}\lambda>\max(\omega_p', \omega_q'),\ \overline{x_p}\in\overline{E_p},\ n\in\mathbb{N}.$$

By the Post-Widder inversion formula and (121), we get that:

$$\begin{aligned}
\pi_{qp}\left(\overline{R_p}(t)\overline{x_p}\right)&=\lim_{n\to\infty}\pi_{qp}\left((-1)^n n!^{-1}\left(\frac{n}{t}\right)^{n+1}\left[\frac{d^n}{d\lambda^n}F_p(\lambda)\right]_{\lambda=n/t}\overline{x_p}\right)\\
&=\lim_{n\to\infty}(-1)^n n!^{-1}\left(\frac{n}{t}\right)^{n+1}\left[\frac{d^n}{d\lambda^n}F_q(\lambda)\right]_{\lambda=n/t}\pi_{qp}\left(\overline{x_p}\right)\\
&=\overline{R_q}(t)\pi_{qp}\left(\overline{x_p}\right),\ \overline{x_p}\in\overline{E_p}.
\end{aligned}$$

Hence, $\{\overline{R_p}(t) : p\in\circledast\}$ is a projective family of operators. Denote by $(R(t))_{t\geq 0}\subseteq L(E)$ the projective limit of the above family. Then it can be simply verified that $(R(t))_{t\geq 0}$ is a q-exponentially equicontinuous (a, k)-regularized resolvent family which satisfies the required properties. Suppose now that, for every $p\in\circledast$, $(\overline{R_p}(t))_{t\geq 0}$ is a q-exponentially equicontinuous, analytic (a, k)-regularized resolvent family of angle β and that, for every $\varepsilon\in(0,\beta)$, (120) holds. Using the equality $\pi_{qp}(\overline{R_p}(t)\overline{x_p})=\overline{R_q}(t)\pi_{qp}(\overline{x_p})$, $t>0$, $\overline{x_p}\in\overline{E_p}$ and the fact that $\pi_{qp}(\cdot)$ is a continuous homomorphism from $\overline{E_p}$ onto $\overline{E_q}$, we obtain from the uniqueness theorem for analytic functions that $\pi_{qp}(\overline{R_p}(z)\overline{x_p}) = \overline{R_q}(z)\pi_{qp}(\overline{x_p})$, $z \in \Sigma_\beta$, $\overline{x_p} \in \overline{E_p}$. Therefore, $\{\overline{R_p}(z) : p \in \circledast\}$ is a projective family of operators ($z \in \Sigma_\beta$). Define $R(z)$ as the projective limit of $\{\overline{R_p}(z) : p \in \circledast\}$ ($z \in \Sigma_\beta$). Then the mapping $z \mapsto R(z)x$, $z \in \Sigma_\beta \cup \{0\}$ ($x \in E$) is continuous on any closed subsector of $\Sigma_\beta \cup \{0\}$ and, for every $\varepsilon \in (0, \beta)$, there exist $M_{p,\varepsilon} \geqslant 1$ and $\omega_{p,\varepsilon} \geqslant 0$ such that (119) holds. Let $x \in E$ and let C be an arbitrary closed contour in Σ_β. Then, for every $p \in \circledast$, $\Psi_p(\oint_C R(z)x\, dz) = \oint_C \Psi_p(R(z)x)\, dz = \oint_C \overline{R_p}(z)\, \Psi_p(x)\, dz = 0$, which implies $\oint_C R(z)x\, dz = 0$. Hence, for every $x^* \in E^*$, $\oint_C \langle x^*, R(z)x\rangle\, dz = 0$ and the mapping $z \mapsto \langle x^*, R(z)x\rangle$, $z \in \Sigma_\beta$ is analytic by Morera's theorem. It follows that the mapping $z \mapsto R(z)x$, $z \in \Sigma_\beta$ is analytic, and we complete the proof as a matter of routine. □

Remark 2.4.4. In order for the proof of Theorem 2.4.3(ii) to work, we have to identify the operator A with the projective limit of family $\{\overline{A_p} : p \in \circledast\}$. This can be done only in the case that the space E is complete.

Keeping in mind Theorem 2.4.3, one can simply formulate the Hille-Yosida type theorems for (analytic) q-exponentially equicontinuous (a, k)-regularized resolvent families in complete locally convex spaces, provided that $a(t)$ and $k(t)$ satisfy (P1), and that $k(0) \neq 0$. The interested reader will probably find some light relief in carrying out details concerning this question.

Theorem 2.4.5. *(cf. Theorem 2.4.3 and Section 2.6) Let E be complete.*

(i) *Suppose $z \in \mathbb{C}$, $B \in L_\circledast(E)$, A is densely defined and generates a q-exponentially equicontinuous (a, k)-regularized resolvent family $(R(t))_{t\geqslant 0}$ satisfying (116). Let (P1) hold for $a(t)$, $k(t)$, $b(t)$, let $\tilde{a}(\lambda)/\tilde{k}(\lambda) = \tilde{b}(\lambda) + z$, $\operatorname{Re}\lambda > \omega$, $\tilde{k}(\lambda) \neq 0$, for some $\omega > \max(\operatorname{abs}(a), \operatorname{abs}(k), \operatorname{abs}(b))$ and let $k(0) \neq 0$. Suppose that, for every $p \in \circledast$, there exists a sufficiently large number $\mu_p > 0$ and a number $\gamma_p \in [0, 1)$ such that:*

$$\int_0^\infty e^{-\mu_p t}\, p\left(B\int_0^t b(t-s)R(s)x\, ds + zBR(t)x\right) dt \leqslant \gamma_p\, p(x), \quad x \in D(A).$$

Then the operator $A + B$ is the generator of a q-exponentially equicontinuous (a, k)-regularized resolvent family $(R_B(t))_{t\geqslant 0}$. Furthermore, for every $t \geqslant 0$ and $x \in D(A)$:

$$R_B(t)x = R(t)x + \int_0^t R_B(t-r)\left(B\int_0^r b(r-s)R(s)x\, ds + zBR(r)x\right) dr.$$

(ii) *Suppose $B \in L_\circledast(E)$, $l \in \mathbb{N}$, A is densely defined and generates a q-exponentially equicontinuous (a, k)-regularized resolvent family $(R(t))_{t\geqslant 0}$ satisfying (116). Let $k(0) \neq 0$, let $a(t)$ and $k(t)$ satisfy (P1), and let the following conditions hold:*

(ii.1) $A^jB \in L_\circledast(E)$, $1 \leqslant j \leqslant l$.

(ii.2) *There exists a function $b(t)$ satisfying (P1) and z, $\omega \in \mathbb{C}$ such that:*
$\tilde{a}(\lambda)^{l+1}/\tilde{k}(\lambda) = \tilde{b}(\lambda) + z$, $\operatorname{Re}\lambda > \max(\omega, \operatorname{abs}(a), \operatorname{abs}(k))$, $\tilde{k}(\lambda) \neq 0$.

(ii.3) $\lim_{\lambda\to+\infty} \int_0^\infty e^{-\lambda t}|a(t)|\, dt = 0$ *and* $\lim_{\lambda\to+\infty} \int_0^\infty e^{-\lambda t}|b(t)|\, dt = 0$.

Then $A + B$ is the generator of a q-exponentially equicontinuous (a, k)-regularized resolvent family $(R_B(t))_{t\geqslant 0}$.

(iii) *Suppose $\alpha > 0$, A is densely defined and generates a q-exponentially equicontinuous (g_α, g_1)-regularized resolvent family $(R(t))_{t\geqslant 0}$ satisfying (116). Assume exactly one of the following conditions:*
(iii.1) $\alpha \geqslant 1$ *and* $B \in L_{\circledast}(E)$.
(iii.2) $\alpha < 1$ *and* $A^j B \in L_{\circledast}(E)$, $0 \leqslant j \leqslant \lceil \frac{1-\alpha}{\alpha} \rceil$.
Then the operator $A + B$ is the generator of a q-exponentially equicontinuous (g_α, g_1)-regularized resolvent family $(R_B(t))_{t\geqslant 0}$. Furthermore, if $(R(t))_{t\geqslant 0}$ is a q-exponentially equicontinuous, analytic (g_α, g_1)-regularized resolvent family of angle $\beta \in (0, \pi/2]$, then $(R_B(t))_{t\geqslant 0}$ is.

Concerning Theorem 2.4.5(iii), it is worthwhile to mention that the assertion of [314, Corollary 2.15] (cf. also [292, Theorem 2.5.7-Theorem 2.5.8]) does not admit a satisfactory reformulation for q-exponentially equicontinuous $(g_\alpha, g_{\alpha\beta+1})$-regularized C-resolvent families in Fréchet spaces, unless $C = I$ and $\beta = 0$.

Example 2.4.6. (i) Let $\alpha \in (0, 1)$. Set $a_\alpha(t) := \mathcal{L}^{-1}(\frac{\lambda^\alpha}{\lambda+1})(t)$, $t \geqslant 0$, $k_\alpha(t) := e^{-t}$, $t \geqslant 0$ and $\delta_\alpha := \min(\frac{\pi}{2}, \frac{\pi\alpha}{2(1-\alpha)})$. Suppose E is complete, $f \in L^1_{loc}([0, \infty) : E)$ and A is the integral generator of a q-exponentially equicontinuous $(C_0, 1)$-semigroup $(R(t))_{t\geqslant 0}$ satisfying (116). Then Theorem 2.4.3 combined with the analysis given in [314, Example 3.7] implies that A is the integral generator of a q-exponentially equicontinuous, analytic (a_α, k_α)-regularized resolvent family of angle δ_α, which can be applied in the study of qualitative properties of the abstract Basset-Boussinesq-Oseen equation (3).

(ii) Put $E := \{f \in C^\infty([0, \infty)) : \lim_{x\to+\infty} f^{(k)}(x) = 0$ for all $k \in \mathbb{N}_0\}$ and $\|f\|_k := \sum_{j=0}^{k} \sup_{x\geqslant 0} |f^{(j)}(x)|$, $f \in E$, $k \in \mathbb{N}_0$. Then the topology induced by these norms turns E into a Fréchet space. Suppose $c_0 > 0$, $\beta > 0$, $s > 1$, $l > 0$ and define the operator A by $D(A) := \{u \in E : c_0 u'(0) = \beta u(0)\}$ and $Au := c_0 u''$, $u \in D(A)$. Then A cannot be the generator of a C_0-semigroup since $D(A)$ is not dense in E ([291]). Put $A_1 := A/c_0$, $\omega_{l,s}(\lambda) := \prod_{p=1}^{\infty} (1 + \frac{l\lambda}{p^s})$, $\lambda \in \mathbb{C}$ and $k_{l,s}(t) := \mathcal{L}^{-1}(\frac{1}{\omega_{l,s}(\lambda)})(t)$, $t \geqslant 0$. Making use of the well-known estimates for associated functions ([292]) and [302, (2.36)], we infer that there exists a constant $c_1 > 0$ such that, for every $\varepsilon \in (0, \pi)$,

$$|\omega_{l,s}(\lambda)| \geqslant \exp\big(c_1 (l(1 + \cot \varepsilon)^{-1})|\lambda|^{1/s}\big), \ \lambda \in \Sigma_{\pi-\varepsilon}. \tag{122}$$

Furthermore, $0 \in \operatorname{supp}(k_{l,s})$, $k_{l,s}(0) = 0$ and $k_{l,s}(t)$ is infinitely differentiable in $t \geqslant 0$. We will prove that A is the integral generator of an equicontinuous analytic $k_{l,s}$-convoluted semigroup of angle $\pi/2$ and that there does not exist $n \in \mathbb{N}$ such that A is the integral generator of an exponentially equicontinuous n-times integrated semigroup on E (cf. also the proofs of [175, Theorem 4.1-Theorem 4.2, pp. 384–386]). It is checked at once that the operator $\lambda - A$ is injective for all $\lambda \in \mathbb{C}\setminus(-\infty, 0]$. Let $\lambda = re^{i\theta}$ ($r > 0$, $|\theta| < \pi$), $f \in E$ and $\mu = \lambda^{1/2}$. Then de L'Hospital's rule implies that, for every $k \in \mathbb{N}_0$, the

C^∞-functions $x \mapsto u_{1,k}(x) := \int_0^x e^{-\mu(x-s)} f^{(k)}(s)\, ds = e^{-\mu x}\int_0^x e^{\mu s} f^{(k)}(s)\, ds$, $x \geqslant 0$ and $x \mapsto u_{2,k}(x) := \int_x^\infty e^{\mu(x-s)} f^{(k)}(s)\, ds = e^{\mu x}\int_x^\infty e^{-\mu s} f^{(k)}(s)\, ds$, $x \geqslant 0$ tend to 0 as $x \to +\infty$. Taken together with the computation given in the proof of the estimate (125), the above implies that the function

$$u(x) := \frac{1}{2\mu}\left[\int_0^x e^{-\mu(x-s)} f(s)\, ds + \int_x^\infty e^{\mu(x-s)} f(s)\, ds\right],\ x \geqslant 0,$$

belongs to E. Now it readily follows that the function $\omega(x) := u(x) + [\frac{c_0\mu-\beta}{c_0\mu+\beta}\frac{1}{2\mu}\int_x^\infty e^{-\mu s} f(s)\, ds] e^{-\mu s} := u(x) + \kappa(\mu, f)e^{-\mu x}$, $x \geqslant 0$, belongs to $D(A_1)$ and that $(\lambda - A_1)\omega = f$; therefore, $\lambda \in \rho(A_1)$ and $(\lambda - A_1)^{-1} f = \omega$. Direct computation shows that

$$\sup_{x\geqslant 0}|u(x)| \leqslant \frac{\sup_{x\geqslant 0}|f(x)|}{|\lambda|\cos\frac{\theta}{2}} \text{ and } \sup_{x\geqslant 0}|u_{1,k}(x)| \leqslant \frac{\sup_{x\geqslant 0}|f^{(k)}(x)|}{|\lambda|^{1/2}\cos\frac{\theta}{2}},\ k \in \mathbb{N}_0. \tag{123}$$

Now we obtain that there exists an absolute constant $c > 0$ such that, for every $n \in \mathbb{N}$,

$$\begin{aligned}
\left\|(\lambda - A_1)^{-1} f\right\|_n &= \sum_{j=0}^{n} \sup_{x\geqslant 0}|u^{(j)}(x) + \kappa(\mu, f)(-1)^j \mu^j e^{-\mu x}| \\
&\leqslant \sum_{j=1}^{n}\left\{\sup_{x\geqslant 0}\left|\frac{1}{2\mu}\left[\int_0^x e^{-\mu(x-s)} f^{(j)}(s)\, ds + \sum_{l=0}^{j-1}(-1)^l \mu^l f^{(j-1-l)}(0) e^{-\mu x}\right.\right.\right. \\
&\quad - \sum_{l=1}^{j}\sum_{l_0=0}^{l-1}\binom{j}{l}\mu^{j-1}\binom{l-1}{l_0}(-1)^{l-1-l_0}\mu^{l-1-l_0} f^{(l_0)}(x) \\
&\quad \left.\left.\left.+ u^j\int_x^\infty e^{\mu(x-s)} f(s)\, ds\right] + \kappa(\mu, f)(-1)^j \mu^j e^{-\mu x}\right|\right\} + \frac{c\|f\|_0}{|\lambda|\cos\frac{\theta}{2}} \\
&\leqslant \frac{c\|f\|_0}{|\lambda|\cos\frac{\theta}{2}} + \sum_{j=1}^{n}\left[\frac{\|f\|_j}{2|\lambda|\cos\frac{\theta}{2}} + j(|\mu|^{-1} + |\mu|^{j-1})\|f\|_{j-1}\right] \\
&\quad + \sum_{j=1}^{n}\left[4^{j-1}(|\lambda|^{(-1)/2} + |\lambda|^{(j-1)/2})\|f\|_{j-1}\right. \\
&\quad \left. + \frac{1}{2}|\lambda|^{(j-2)/2}\frac{\|f\|_0}{\cos\frac{\theta}{2}} + \frac{c|\mu|^j\|f\|_0}{|\lambda|\cos\frac{\theta}{2}}\right]
\end{aligned} \tag{124}$$

$$\leqslant n\|f\|_n \frac{1}{2|\lambda|\cos\frac{\theta}{2}} + 2n^2(|\mu|^{-1} + |\mu|^{n-1})\|f\|_{n-1} + \frac{2cn\|f\|_0(1 + |\mu|^n)}{|\lambda|\cos\frac{\theta}{2}}$$

$$+ n4^{n-1}(|\lambda|^{1/2} + |\lambda|^{(n-1)/2})\|f\|_{n-1} + n(1+|\lambda|^{n/2})\frac{\|f\|_0}{2|\lambda|\cos\frac{\theta}{2}}.$$

At this point, we can use the inequality $\exp(-\zeta x^{1/s})x^\eta \leqslant (s\eta/\zeta)^{\eta s}$, $x > 0$, $\zeta > 0$, $\eta > 0$ and (122)-(124) to conclude that, for every $\varepsilon \in (0, \pi)$, the family $\{\lambda\widetilde{k_{l,s}}(\lambda)(\lambda - A)^{-1} : \lambda \in \Sigma_{\pi-\varepsilon}\}$ is equicontinuous. Moreover, $\lim_{\lambda\to+\infty} \lambda\widetilde{k_{l,s}}(\lambda)(\lambda - A)^{-1}f = 0 = k_{l,s}(0)f$, $f \in E$. By Theorem 2.2.5 and its proof, it follows that A is the integral generator of an equicontinuous analytic $k_{l,s}$-convoluted semigroup $(R(t))_{t\geqslant 0}$ of angle $\pi/2$ satisfying additionally that, for every $k \in \mathbb{N}_0$ and $\varepsilon \in (0, \pi)$, there exists $c(k, \varepsilon) > 0$ such that $\|R(z)f\|_k \leqslant c(k, \varepsilon)\|f\|_k$, $z \in \Sigma_{\pi-\varepsilon}$, $f \in E$. Assume that there exists $n \in \mathbb{N}$ such that A is the integral generator of an exponentially equicontinuous n-times integrated semigroup on E. Without loss of generality, we may assume that $2n+3 > 2\beta/c_0$. Then there exists a sufficiently large $\nu > 0$ such that the family $\{\lambda^{-n}(\lambda - A)^{-1} : \lambda > \nu\}$ is equicontinuous, which simply implies that there exist $c_n > 0$ and $n' \in \mathbb{N}$ with:

$$\sup_{x\geqslant 0}\Bigg|\frac{1}{2\lambda^{1/2}}\Bigg[\int_0^x e^{-\lambda^{1/2}(x-s)} f^{(2n+5)}(s)\,ds + \sum_{j=0}^{2n+4}(-1)^j\lambda^{j/2} f^{(2n+4-j)}(0)e^{-\lambda^{1/2}x}\Bigg]$$

$$+ \frac{1}{2}\Bigg[\lambda^{n+2}\int_0^\infty e^{\lambda^{1/2}(x-s)} f(s)\,ds - \sum_{l=1}^{2n+5}\sum_{l_0=0}^{l-1}\binom{2n+5}{l}\lambda^{(2n+3-l_0)/2} \tag{125}$$

$$\times(-1)^{l-1-l_0}\binom{l-1}{l_0} f^{(l_0)}(x)\Bigg] - \Bigg[\frac{c_0\lambda^{1/2}-\beta}{c_0\lambda^{1/2}+\beta}\frac{1}{2\lambda^{1/2}}\int_0^\infty e^{-\lambda^{1/2}s} f(s)\,ds\Bigg]$$

$$\times\lambda^{(2n+5)/2}e^{-\lambda^{1/2}x}\Bigg| \leqslant c_n\lambda^n\|f\|_{n'},\ \lambda > \nu,\ f \in E.$$

Denote by $g_f(x, \lambda)$ the function whose supremum appears in (125). Since $\Sigma_{l=1}^{2n+5}\binom{2n+5}{l}(-1)^{l-1} = 1$ and $\Sigma_{l=1}^{2n+5}\binom{2n+5}{l}(l-1)(-1)^l = 1$, it can be easily seen that there exists a sufficiently large number $a_n > 0$, depending only on n, such that:

$$2\sup_{x\geqslant 0} g_{e_-}(x, \lambda) \geqslant 2g_{e_-}(0, \lambda)$$

$$\geqslant \lambda^n\Bigg|2\lambda + \lambda^{1/2} + \sum_{l=1}^{2n+5}\binom{2n+5}{l}\binom{l-1}{2}(-1)^l\lambda^{1/2}$$

$$+\frac{2\beta\lambda^2}{(c_0\lambda^{1/2}+\beta)(1+\lambda^{1/2})}\Bigg| - a_n\lambda^n,\ \lambda > a_n,$$

which implies $\lim_{\lambda\to+\infty}\lambda^{-n}\sup_{x\geqslant 0} g_{e^-}(x,\lambda)=+\infty$. A contradiction. The existence of an injective operator $C \in L(E)$ such that A is the integral generator of an exponentially equicontinuous C-regularized semigroup on E is a non-trivial matter that we will not pursue here anymore; let us only mention in passing that an affirmative answer to the present question can be given only with the help of results concerning ultradistribution semigroups in locally convex spaces (cf. [292, Subsection 3.6.2] for the Banach space case, and [509]). Now we shall explain how one can use the obtained result in the analysis of a control problem for a one-dimensional heat equation for materials with memory (cf. [463, pp. 146-147]), which closely pertains to the problem of gluing in manufacturing polymeric materials. Let $L^1_{loc}([0,\infty)) \ni a$ satisfy (P1), let $abs(a) = 0$ and let the analytic function $\hat{a} : \mathbb{C}\setminus(-\infty, 0] \to \mathbb{C}\setminus(-\infty, 0]$ satisfy $\hat{a}(\lambda) = \tilde{a}(\lambda)$, $\operatorname{Re}\lambda > 0$; notice that the complete monotonicity of the kernel $a(t)$ has been used in the analysis established in [463]. Suppose that $\sigma \in (0, 1)$ and, for every $\varepsilon \in (0, \pi)$ and $n \in \mathbb{N}$, there exist $c^1_{\varepsilon,n} > 0$ and $c^2_{\varepsilon,n} > 0$ such that $|\hat{a}(\lambda)|^{-n} \leqslant c^1_{\varepsilon,n}\exp(c^2_{\varepsilon,n}|\lambda|^{\sigma})$, $\lambda \in \Sigma_{\pi-\varepsilon}$. By Theorem 2.2.5, we get that, for every $l > 0$ and $s \in (1, 1/\sigma)$, the operator A is the integral generator of an equicontinuous analytic $(a, k_{l,s})$-regularized resolvent family $(S(t))_{t\geqslant 0}$ of angle $\pi/2$ satisfying additionally that, for every $k \in \mathbb{N}_0$ and $\varepsilon \in (0, \pi)$, there exists $c(k, \varepsilon)' > 0$ with $\|S(z)f\|_k \leqslant c(k, \varepsilon)'\|f\|_{k'}$, $z \in \Sigma_{\pi-\varepsilon}$, $f \in E$. This, in particular, implies the existence of regularized solutions to the problem [463, (5.68), p. 147].

(iii) Suppose $E = L^2(\mathbb{R}^n)$, $0 \leqslant l \leqslant n$ and $1 \leqslant \alpha \leqslant 2$. The totality of seminorms $(q_\eta(f) := \Sigma_{\mu\leqslant\eta}\|f^{(\mu)}\|_{L^2(\mathbb{R}^n)}, f \in E_l;\ \eta \in \mathbb{N}_0^l)$ induces a Fréchet topology on E_l. Let $a_\eta \in \mathbb{C}$, $0 \leqslant |\eta| \leqslant N$, let $P(x) = \Sigma_{|\eta|\leqslant N} a_\eta x^\eta$, $x \in \mathbb{R}^n$, and let $\omega \geqslant 0$ satisfy $\sup_{x\in\mathbb{R}^n}\operatorname{Re}(P(x)^{1/\alpha}) \leqslant \omega$. Suppose that the operator $P(D)f := \Sigma_{|\eta|\leqslant N} a_\eta(-i)^{|\eta|}f^{(\eta)}$ acts on E_l with its maximal distributional domain. Then we know that $P(D)$ generates an exponentially equicontinuous (g_α, g_1)-regularized resolvent family $(R_\alpha(t))_{t\geqslant 0}$ in the space E_l and that there exists a constant $M \geqslant 1$ such that:

$$q_\eta(R_\alpha(t)f) \leqslant Me^{\omega t}q_\eta(f),\ t \geqslant 0,\ f \in E_l,\ \eta \in \mathbb{N}_0^l.$$

Let $\varphi \in C^\infty(\mathbb{R}^n)$ possess bounded derivatives of all orders and let $(Bf)(x) := \varphi(x)f(x)$, $f \in E_l$, $x \in \mathbb{R}^n$. Then $B \in L_{\circledast}(E_l)$ and, by Theorem 2.4.5(iii), the operator $P(D) + B$ generates a q-exponentially equicontinuous (g_α, g_1)-regularized resolvent family $(R^B_\alpha(t))_{t\geqslant 0}$ in the space E_l.

2.5 Abstract differential operators generating fractional resolvent families

Let $n \in \mathbb{N}$ and let iA_j, $1 \leqslant j \leqslant n$ be commuting generators of bounded C_0-groups on a Banach space E. Let $k = 1 + \lfloor n/2 \rfloor$, $A = (A_1, \cdots, A_n)$ and $A^\eta = A_1^{\eta_1} \cdots A_n^{\eta_n}$ for any $\eta = (\eta_1, \cdots, \eta_n) \in \mathbb{N}_0^n$. If $\xi = (\xi_1, \cdots, \xi_n) \in \mathbb{R}^n$ and $u \in \mathcal{F}L^1(\mathbb{R}^n) = \{\mathcal{F}f : f \in L^1(\mathbb{R}^n)\}$, put $|\xi| = (\Sigma_{j=1}^n \xi_j^2)^{1/2}$, $(\xi, A) = \Sigma_{j=1}^n \xi_j A_j$ and

$$u(A)x = \int_{\mathbb{R}^n} \mathcal{F}^{-1}u(\xi)e^{-i(\xi,A)}x\, d\xi,\ x \in E.$$

Then $u(A) \in L(E)$, $u \in \mathcal{F}L^1(\mathbb{R}^n)$ and, as announced earlier, we have the existence of a constant $M < \infty$ such that:

$$(126) \qquad \|u(A)\| \leqslant M\|\mathcal{F}^{-1}u\|_{L^1(\mathbb{R}^n)},\ u \in \mathcal{F}L^1(\mathbb{R}^n).$$

Let $N \in \mathbb{N}$. For a complex polynomial $P(x) = \Sigma_{|\eta|\leqslant N}\ a_\eta x^\eta$, $x \in \mathbb{R}^n$ ($|\eta| := \Sigma_{j=1}^n \eta_j$), denote $P(A) = \Sigma_{|\eta|\leqslant N}\ a_\eta A^\eta$ with maximal domain. Set $E_0 := \{\phi(A)x : \phi \in \mathcal{S}(\mathbb{R}^n), x \in E\}$. Then it is well known that $P(A)$ is closable and that the following holds:

$$(\rhd)\ \overline{E_0} = E,\ E_0 \subseteq \bigcap_{\eta\in\mathbb{N}_0^n} D(A^\eta),\ \overline{P(A)|_{E_0}} = \overline{P(A)} \text{ and}$$
$$\phi(A)P(A) \subseteq P(A)\phi(A) = (\phi P)(A),\ \phi \in \mathcal{S}(\mathbb{R}^n).$$

If E is a function space on which translations are uniformly bounded and strongly continuous, then the obvious choice for A_j is $-i\partial/\partial x_j$ (notice also that E can consist of functions defined on some bounded domain [141], [347], [552]). If $P(x) = \Sigma_{|\eta|\leqslant N}\ a_\eta x^\eta$, $x \in \mathbb{R}^n$ and E is such a space (for example, $L^p(\mathbb{R}^n)$ with $p \in [1, \infty)$, $C_0(\mathbb{R}^n)$ or $BUC(\mathbb{R}^n)$), then $\overline{P(A)}$ is the operator $\Sigma_{|\eta|\leqslant N}\ a_\eta(-i)^{|\eta|}\partial^{|\eta|}/\partial x_1^{\eta_1} \cdots \partial x_n^{\eta_n} \equiv \Sigma_{|\eta|\leqslant N}\ a_\eta D^\eta$ acting with its maximal distributional domain. Recall that $P(x)$ is called r-coercive ($0 < r \leqslant N$) if there exist $M, L > 0$ such that $|P(x)| \geqslant M|x|^r$, $|x| \geqslant L$; by a corollary of the Seidenberg-Tarski theorem, the equality $\lim_{|x|\to\infty}|P(x)| = \infty$ implies in particular that $P(x)$ is r-coercive for some $r \in (0, N]$ (cf. [20, Remark 8.2.7]). In the sequel of this section, $M > 0$ denotes a generic constant whose value may change from line to line.

Let $p \in [1, \infty]$. Then a function $u \in L^\infty(\mathbb{R}^n)$ is called a Fourier multiplier on $L^p(\mathbb{R}^n)$ if $\mathcal{F}^{-1}(u\mathcal{F}\phi) \in L^p(\mathbb{R}^n)$ for all $\phi \in \mathcal{S}(\mathbb{R}^n)$ and if

$$\|u\|_{\mathcal{M}_p} := \sup\Big\{\|\mathcal{F}^{-1}(u\mathcal{F}\phi)\|_{L^p(\mathbb{R}^n)} : \phi \in \mathcal{S}(\mathbb{R}^n),\ \|\phi\|_{L^p(\mathbb{R}^n)} \leqslant 1\Big\} < \infty.$$

We use the abbreviation $\mathcal{M}_p$ for the space of all Fourier multipliers on $L^p(\mathbb{R}^n)$; cf. [232] and [496] for more details. Then $\mathcal{M}_p$ is a Banach algebra under pointwise multiplication and $\mathcal{F}L^1(\mathbb{R}^n)$ is continuously embedded in $\mathcal{M}_p$. The following lemma plays an important role in our analysis (cf. [531, Lemma 5.2, Lemma 5.4, pp. 20- 22]).

Lemma 2.5.1. (i) *Let $1 \leqslant p \leqslant \infty$, $j, n \in \mathbb{N}$, $j > n/2$ and $\{f_t\}_{t\geqslant 0}$ be a family of $C^j(\mathbb{R}^n)$-functions. Assume that for each $x \in \mathbb{R}^n$, $\eta \in \mathbb{N}_0^n$ with $|\eta| \leqslant j$, $t \mapsto D^\alpha f_t(x)$, $t \geqslant 0$ is continuous and that there exist $a > 0$, $r > n\left|\frac{1}{p}-\frac{1}{2}\right|$ and $M_t > 0$ (M_t is bounded on compacts of $t \geqslant 0$) such that*

$$|D^\eta f_t(x)| \leqslant M_t^{|\eta|}(1 + |x|)^{(a-1)|\eta|-ar},\ |\eta| \leqslant j, x \in \mathbb{R}^n, t \geqslant 0.$$

Then, for any $t \geqslant 0$, $p = 1, \infty$ (resp. $1 < p < \infty$), we have $f_t \in \mathcal{F}L^1(\mathbb{R}^n)$ (resp. $f_t \in \mathcal{M}_p$), $t \mapsto f_t$, $t \geqslant 0$ is continuous with respect to $\|\cdot\|_{\mathcal{F}L^1(\mathbb{R}^n)}$ (resp. $\|\cdot\|_{\mathcal{M}_p}$) and there exists a constant $M > 0$ independent of $t \geqslant 0$ such that

$$\|f_t\|_{\mathcal{F}L^1(\mathbb{R}^n)} (\textit{resp. } \|f_t\|_{\mathcal{M}_p}) \leqslant MM_t^{n\left|\frac{1}{p}-\frac{1}{2}\right|},\ t \geqslant 0.$$

(ii) *Let* $1 < p < \infty$, $j, n \in \mathbb{N}$, $j > n/2$ *and* $f \in C^j(\mathbb{R}^n)$. *Assume that there exist* $a \geqslant 0$, $r \geqslant n|\frac{1}{p}-\frac{1}{2}|$, $M_f \geqslant 1$ *and* $L_f > 0$ *such that*

$$|D^\eta f(x)| \leqslant L_f M_f^{|\eta|} (1+|x|)^{(a-1)|\eta|-ar}, \ |\eta| \leqslant j, x \in \mathbb{R}^n, t \geqslant 0.$$

Then $f \in \mathcal{M}_p$ *and there exists a constant* $M > 0$ *independent of* $f(\cdot)$ *such that*

$$\|f_t\|_{\mathcal{M}_p} \leqslant M L_f M_f^{n|\frac{1}{p}-\frac{1}{2}|}.$$

In the following theorem, we consider the generation of fractional resolvent families by coercive differential operators.

Theorem 2.5.2. *Suppose* $0 < \alpha < 2$, $\omega \geqslant 0$, $N \in \mathbb{N}$, $r \in (0, N]$, $P(x)$ *is an r-coercive complex polynomial of degree N,* $a \in \mathbb{C} \backslash P(\mathbb{R}^n)$, $\gamma > \frac{nN}{2r\min(1,\alpha)}$ *(resp.* $\gamma = n\,|\frac{1}{p}-\frac{1}{2}|\frac{N}{r\min(1,\alpha)}$, *if* $E = L^p(\mathbb{R}^n)$ *for some* $1 < p < \infty$*) and*

$$\sup_{x \in \mathbb{R}^n} \operatorname{Re}(P(x)^{1/\alpha}) \leqslant \omega. \tag{127}$$

Set

$$R_\alpha(t) := \big(E_\alpha(t^\alpha P(x))(a - P(x))^{-\gamma}\big)(A), \ t \geqslant 0.$$

Then $(R_\alpha(t))_{t\geqslant 0}$ *is a global exponentially bounded* $(g_\alpha, R_\alpha(0))$*-regularized resolvent family with the integral generator* $\overline{P(A)}$, $(R_\alpha(t))_{t\geqslant 0}$ *is norm continuous provided* $\gamma > \frac{nN}{2r\min(1,\alpha)}$, *and the following holds*:

$$\|R_\alpha(t)\| \leqslant M(1 + t^{\max(1,\alpha)n/2})e^{\omega t}, \ t \geqslant 0, \text{ resp.},$$

$$\|R_\alpha(t)\| \leqslant M(1 + t^{\max(1,\alpha)n|\frac{1}{p}-\frac{1}{2}|})e^{\omega t}, \ t \geqslant 0. \tag{128}$$

Proof. Put $C := R_\alpha(0)$. Arguing as in the proof of [357, Theorem 4.1], we get that $L(E) \ni C$ is injective as well as that $C^{-1}\,\overline{P(A)}\,C = \overline{P(A)}$ and that, for every multi-index $\eta \in \mathbb{N}_0^n$ with $|\eta| \leqslant k$, there exist complex polynomials $Q_j(x)$ of degree $\leqslant N_j - |\eta|$ $(1 \leqslant j \leqslant |\eta|)$ such that

$$D^\eta(E_\alpha(t^\alpha P(x))) = \sum_{j=1}^{|\eta|} t^{\alpha j} E_\alpha^{(j)}(t^\alpha P(x)) Q_j(x), \ t \geqslant 0, x \in \mathbb{R}^n. \tag{129}$$

It is clear that there exists $L > 0$ such that $|P(x)| \geqslant M|x|^r$, $|x| \geqslant L$ and $|a - P(x)| \geqslant M|x|^r$, $|x| \geqslant L$. Let us now carefully consider the assertion of Theorem 1.3.1. Choosing a sufficiently small number $\sigma > 0$, and keeping in mind that $0 < \alpha < 2$, we obtain that, for every $m \in \mathbb{N}\backslash\{1\}$ and for every $t \geqslant 0$, $x \in \mathbb{R}^n$ with $|t^\alpha P(x)| \geqslant 1$, the term

$$\left| \begin{array}{l} E_{\alpha,\alpha j-(j-l)}(t^\alpha P(x)) \\ -\dfrac{1}{\alpha}(t^\alpha P(x))^{(1-(\alpha j-(j-l)))/\alpha} e^{(t^\alpha P(x))^{1/\alpha}} - \displaystyle\sum_{j=1}^{m-1} \frac{(t^\alpha P(x))^{-j}}{\Gamma(\alpha j-(j-l)-\alpha j)} \end{array} \right|$$

does not exceed $M|t^\alpha P(x)|^{-m}$. Since the function $E_{\alpha,\alpha j-(j-l)}(\cdot)$ is bounded on compacts of $\mathbb{C}$, and $\mathrm{Re}((t^\alpha P(x))^{1/\alpha}) \leqslant \omega t$, $t \geqslant 0$, $x \in \mathbb{R}^n$ (this follows from (127) and a simple computation), we get that, for every $t \geqslant 0$, $x \in \mathbb{R}^n$ and $1 \leqslant l \leqslant j \leqslant k$,

$$(130) \qquad \left|E_{\alpha,\alpha j-(j-l)}(t^\alpha P(x))\right| \leqslant M\left[1+t^{1-(\alpha j-(j-l))}\left|P(x)\right|^{\frac{1-(\alpha j-(j-l))}{\alpha}} e^{\omega t}\right].$$

By virtue of (129), we obtain that, for every $t \geqslant 0$ and $x \in \mathbb{R}^n$ with $|t^\alpha P(x)| \leqslant L$:

$$(131) \qquad \left|D^\eta(E_\alpha(t^\alpha P(x)))\right| \leqslant M(t^\alpha + t^{\alpha|\eta|})(1+|x|)^{|\eta|(N-1)},\ |\eta| \leqslant k.$$

Suppose now $1 \leqslant l \leqslant j \leqslant |\eta| \leqslant k$, $t \geqslant 0$, $x \in \mathbb{R}^n$ and $|t^\alpha P(x)| \geqslant L$. Then

$$(132) \qquad \left|t^\alpha P(x)\right|^{\frac{1-(\alpha j-(j-l))}{\alpha}}(1+|x|)^{Nj-|\eta|} \leqslant Mt^{1-(\alpha j-(j-l))}(1+|x|)^{|\eta|\left(\frac{N}{\alpha}-1\right)},$$

provided $1-(\alpha j-(j-l)) \geqslant 0$, and

$$(133) \qquad \left|t^\alpha P(x)\right|^{\frac{1-(\alpha j-(j-l))}{\alpha}}(1+|x|)^{Nj-|\eta|} \leqslant M(1+|x|)^{|\eta|(N-1)},$$

provided $1-(\alpha j-(j-l)) \leqslant 0$. A straightforward computation involving (21), (129)-(130) and (132)-(133) shows that:

$$\begin{aligned}
&\left|D^\eta(E_\alpha(t^\alpha P(x)))\right| \\
&\leqslant M\sum_{j=1}^{|\eta|} t^{\alpha j}\sum_{l=1}^{j}\left[1+\left|t^\alpha P(x)\right|^{\frac{1-(\alpha j-(j-l))}{\alpha}} e^{\omega t}\right](1+|x|)^{Nj-|\eta|} \\
&\leqslant M\sum_{j=1}^{|\eta|} t^{\alpha j}\sum_{l=1}^{j}\left[1+e^{\omega t}(1+t^{1-(\alpha j-(j-l))})\right](1+|x|)^{|\eta|\left(\frac{N}{\min(1,\alpha)}-1\right)} \\
&\leqslant M(1+t^{\max(1,\alpha)|\eta|})e^{\omega t}(1+|x|)^{|\eta|\left(\frac{N}{\min(1,\alpha)}-1\right)}.
\end{aligned}$$

Invoking (131), we obtain from the previous estimate that, for every $t \geqslant 0$ and $x \in \mathbb{R}^n$,

$$(134) \qquad |D^\eta(E_\alpha(t^\alpha P(x)))| \leqslant M(1+t^{\max(1,\alpha)|\eta|})e^{\omega t}(1+|x|)^{|\eta|\left(\frac{N}{\min(1,\alpha)}-1\right)},\ |\eta| \leqslant k.$$

Set $f_t(x) := E_\alpha(t^\alpha P(x))(a-P(x))^{-\gamma}$, $t \geqslant 0$, $x \in \mathbb{R}^n$. Using [357, (3.19)], (134) and the product rule, we reveal that, for every $t \geqslant 0$, $x \in \mathbb{R}^n$ and $\eta \in \mathbb{N}_0^n$ with $|\eta| \leqslant k$,

$$(135) |D^\eta(E_\alpha(t^\alpha P(x))(a-P(x))^{-\gamma})| \leqslant M(1+t^{\max(1,\alpha)|\eta|})e^{\omega t}(1+|x|)^{|\eta|\left(\frac{N}{\min(1,\alpha)}-1\right)-r\gamma}.$$

Consider first the case $\gamma > nN/2r\min(1, \alpha)$. By Lemma 2.5.1(i), (126) and (135), we get the boundedness of the operator $R_\alpha(t)$ ($t \geqslant 0$), the estimate (127) and the continuity of mapping $t \mapsto R_\alpha(t)$, $t \geqslant 0$. Making use of (▷) and the fact that the mapping $u \mapsto u(A)$ is an algebra homomorphism from $\mathcal{F}L^1(\mathbb{R}^n)$ into $L(E)$, we infer that $R_\alpha(t)\overline{P(A)} \subseteq \overline{P(A)}R_\alpha(t)$, $t \geqslant 0$ and $R_\alpha(t)C = CR_\alpha(t)$, $t \geqslant 0$. Let $x \in D(\overline{P(A)}) =$

$D(\overline{P(A)|_{E_0}})$. Then there exist a sequence (ϕ_n) in $\mathcal{S}(\mathbb{R}^n)$ and a sequence (x_n) in E such that $\lim_{n\to\infty}\phi_n(A)x_n = x$ and $\lim_{n\to\infty}P(A)\phi_n(A)x_n = \lim_{n\to\infty}(P\phi_n)(A)x_n = \overline{P(A)}x$. Making use of the equalities

$$\int_0^t g_\alpha(t-s)(Pf_s\phi_n)(A)x_n\,ds = \left(\int_0^t g_\alpha(t-s)Pf_s\phi_n\,ds\right)(A)x_n$$
$$= \left(f_t\phi_n - (a-P(\cdot))^{-\gamma}\phi_n\right)(A)x_n,\ n\in\mathbb{N},$$

and the inclusion $R_\alpha(t)\overline{P(A)} \subseteq \overline{P(A)}\,R_\alpha(t)$, $t \geqslant 0$, we obtain that:

$$(136)\qquad R_\alpha(t)\phi_n(A)x_n - C\phi_n(A)x_n = \overline{P(A)}\int_0^t g_\alpha(t-s)R_\alpha(s)\phi_n(A)x_n\,ds,\ t\geqslant 0.$$

Letting $n \to \infty$ in (136), we get that:

$$(137)\qquad R_\alpha(t)x - Cx = \overline{P(A)}\int_0^t g_\alpha(t-s)R_\alpha(s)x\,ds,\ t\geqslant 0,\ x\in E.$$

By the foregoing, $(R_\alpha(t))_{t\geqslant 0}$ is a global exponentially bounded (g_α, C)-regularized resolvent family with a subgenerator $\overline{P(A)}$. Notice that $\overline{P(A)}$ is, in fact, the integral generator of $(R_\alpha(t))_{t\geqslant 0}$ since $C^{-1}\overline{P(A)}C = \overline{P(A)}$ and (137) holds. Suppose now $E = L^p(\mathbb{R}^n)$ for some $1 < p < \infty$, and $\gamma = n\left|\frac{1}{p}-\frac{1}{2}\right|\frac{N}{r\min(1,\alpha)}$. Put $R_\alpha(t)\phi := \mathcal{F}^{-1}(f_t\mathcal{F}(\phi))$, $t \geqslant 0$, $\phi \in \mathcal{S}(\mathbb{R}^n)$. Then Lemma 2.5.1(ii) implies that, for every $t \geqslant 0$, $R_\alpha(t)$ can be extended continuously to the whole space E and that the second inequality in (128) holds. Let $\phi \in \mathcal{S}(\mathbb{R}^n)$ be fixed. Then $\tilde{\phi} := \mathcal{F}^{-1}(P(\cdot)\mathcal{F}\phi(\cdot)) \in \mathcal{S}(\mathbb{R}^n)$, the mapping $t \mapsto R_\alpha(t)\phi$, $t \geqslant 0$ is weakly continuous, and therefore, strongly measurable (cf. also the proofs of [531, Theorem 5.9-Theorem 5.10, pp. 30-33]). Combined with the equality $\int_0^t g_\alpha(t-s)f_s(x)P(x)\,ds = f_t(x) - (a-P(x))^{-\gamma}$, $t \geqslant 0$, $x \in \mathbb{R}^n$ and the definition of $R_\alpha(t)\phi$ for $t \geqslant 0$, the above implies:

$$R_\alpha(t)\phi = C\phi + \int_0^t g_\alpha(t-s)R_\alpha(s)\tilde{\phi}\,ds,\ t\geqslant 0.$$

Taken together, the above equality and the exponential boundedness of $(R_\alpha(t))_{t\geqslant 0}$ immediately imply the continuity of the mapping $t \mapsto R_\alpha(t)\phi$, $t \geqslant 0$, and the strong continuity of $(R_\alpha(t))_{t\geqslant 0}$. The remaining part of proof follows from the arguments given in the case of a general space E. □

The main purpose of the following theorem is to consider the generation of fractional resolvent families by non-coercive differential operators.

Theorem 2.5.3. *Suppose* $0 < \alpha < 2$, $\omega \geqslant 0$, *$P(x)$ is a complex polynomial of degree* $N \in \mathbb{N}$, $\beta > \frac{nN}{2\min(1,\alpha)}$ *(resp.* $\beta = n\left|\frac{1}{p}-\frac{1}{2}\right|\frac{N}{\min(1,\alpha)}$ *if* $E = L^p(\mathbb{R}^n)$ *for some* $1 < p < \infty$*) and (127) holds. Set*

$$R_\alpha(t) := \left(E_\alpha(t^\alpha P(x))(1+|x|^2)^{-\beta/2}\right)(A),\ t\geqslant 0.$$

Then $(R_\alpha(t))_{t\geqslant 0}$ *is a global exponentially bounded* $(g_\alpha, R_\alpha(0))$*-regularized resolvent family with the integral generator* $\overline{P(A)}$, $(R_\alpha(t))_{t\geqslant 0}$ *is norm continuous provided* $\beta > \frac{nN}{2\min(1,\alpha)}$, *and (128) holds.*

Proof. We will only outline a few relevant points. It can be simply proved that the inequality (134) continues to hold for $P(x)$. On the other hand, it is clear that

$$\left|D^\eta((1+|x|^2)^{-\beta/2})\right| \leqslant M(1+|x|)^{-|\eta|-\beta},\ |\eta| \leqslant k,\ x \in \mathbb{R}^n.$$

This implies by the product rule that, for every $t \geqslant 0$ and $x \in \mathbb{R}^n$:

$$\left|D^\eta(E_\alpha(t^\alpha P(x))(1+|x|^2)^{-\beta/2})\right|$$

$$\leqslant M(1+t^{\max(1,\alpha)|\eta|})e^{\omega t}(1+|x|)^{|\eta|\left(\frac{N}{\min(1,\alpha)}-1\right)-\beta},\ |\eta| \leqslant k.$$

Now the assertion follows similarly as in the proof of Theorem 2.5.2. □

Remark 2.5.4. (i) Let $1 < \alpha < \alpha' < 2$ and let $\omega \geqslant 0$. Then the proof of [357, Theorem 4.2] shows that

$$P(\mathbb{R}^n) \subseteq \mathbb{C}\backslash(\omega + \Sigma_{\alpha'\pi/2}) \Rightarrow \sup_{x\in\mathbb{R}^n} \operatorname{Re}(P(x)^{1/\alpha}) \leqslant \omega^{1/\alpha}$$

and, by letting $\alpha \to \alpha'$,

$$P(\mathbb{R}^n) \subseteq \mathbb{C}\backslash(\omega + \Sigma_{\alpha'\pi/2}) \Rightarrow \sup_{x\in\mathbb{R}^n} \operatorname{Re}(P(x)^{1/\alpha'}) \leqslant \omega^{1/\alpha'}. \tag{138}$$

Therefore, it readily follows from the above observation and Theorem 2.4.2 that Theorem 2.5.2 (Theorem 2.5.3) is a generalization of [357, Theorem 4.1-Theorem 4.2] ([357, Theorem 4.3]). We have also proved that the fractional resolvent families appearing in the formulations of [357, Theorem 4.1-Theorem 4.3] are norm continuous at $t = 0$ and that [357, Theorem 4.4(b)] ([357, Theorem 4.5(b)]) can be slightly improved by allowing that the number γ (β) in its formulation takes the value $n_p m/r$ (n_p). Notice, finally, that the estimate (138) does not remain true provided $\alpha' \in (0, 1)$, and that the estimate (127), with α replaced by α', is very restrictive in this case.

(ii) Let the assumptions of [357, Theorem 4.2] hold. Then one can simply prove that there exist $\omega_1 > 0$ and $\omega_2 \in \mathbb{R}$ such that $\operatorname{Re}(P(x)) \leqslant -\omega_1|x|^r + \omega_2$, $x \in \mathbb{R}^n$, which implies by [550, Theorem 2.2] that $e^{\pm i\pi/2(\alpha'-1)}\overline{P(A)}$ are integral generators of exponentially bounded $(a - P(x))^{-\gamma}(A)$-regularized semigroups $(R_{\alpha',\pm}(t))_{t\geqslant 0}$ for any $\gamma > nN/2r$ and that $\|R_{\alpha',\pm}(t)\| \leqslant M(1 + t^{n/2})e^{\omega\cos\pi/2(\alpha'-1)t}$, $t \geqslant 0$. This implies (cf. for example [20, Theorem 3.9.7] and [292, Subsection 2.4]) that $\overline{P(A)}$ is the integral generator of an exponentially bounded, analytic $(a - P(x))^{-\gamma}(A)$-regularized semigroup $(R(t))_{t\geqslant 0}$ of angle $\pi/2(\alpha' - 1)$ ($\gamma > nN/2r$) and that, for every $\varepsilon > 0$, there exists $M_\varepsilon > 0$ such that $\|R(t)\| \leqslant M_\varepsilon e^{(\omega+\varepsilon)\operatorname{Re} t}$, $t \in \Sigma_{\pi/2(\alpha'-1)}$. Now Theorem 2.2.4-Theorem 2.2.5 implies that, for every $\gamma > nN/2r$, $\overline{P(A)}$ is the integral generator of an exponentially bounded, analytic $(g_\alpha, (a - P(x))^{-\gamma}(A))$-regularized resolvent family $(R_\alpha(t))_{t\geqslant 0}$ of angle $\pi/2(\alpha'/\alpha - 1)$, and that

(139) $$\|R_\alpha(t)\| = O(e^{(\omega^{1/\alpha}+\varepsilon)t}),\ t \geqslant 0\ (\varepsilon > 0).$$

Similarly, if the assumptions of [357, Theorem 4.3] hold, then [555, Theorem 1.2] and Theorem 2.2.4-Theorem 2.2.5 taken together imply that, for every $\beta > Nn/2$, $\overline{P(A)}$ is the integral generator of an exponentially bounded, analytic $(g_\alpha, (1 + |x|^2)^{-\beta/2}(A))$-regularized resolvent family $(R_\alpha(t))_{t\geqslant 0}$ of angle $\pi/2(\alpha'/\alpha - 1)$ and that (139) holds. Therefore, the assertions of [357, Theorem 4.1-Theorem 4.3], with the exception of obtained representation formulae and results concerning the growth order of constructed fractional resolvent families, follow immediately from Theorem 2.2.4-Theorem 2.2.5 and corresponding results for analytic C-regularized semigroups.

(iii) Consider again the situation in which the assumptions of [357, Theorem 4.2] hold. Then an application of [555, Theorem 1.1] yields that, for every $\beta > (N - r)n/2$, $\overline{P(A)}$ is the integral generator of an exponentially bounded, analytic $(1 + |x|^2)^{-\beta/2}(A)$-regularized semigroup $(R(t))_{t\geqslant 0}$ of angle $\pi/2(\alpha' - 1)$ and that, for every $\delta \in (0, \pi/2(\alpha' - 1))$, there exists $M_\delta > 0$ such that $\|R(t)\| \leqslant M_\delta(1 + t^{n/2})e^{\omega \operatorname{Re} t}$, $t \in \Sigma_\delta$. Making use of Theorem 2.2.4-Theorem 2.2.5, we get that, for every $\beta > (N - r)n/2$, $\overline{P(A)}$ is the integral generator of an exponentially bounded, analytic $(g_\alpha, (1 + |x|^2)^{-\beta/2}(A))$-regularized resolvent family $(R_\alpha(t))_{t\geqslant 0}$ of angle $\pi/2(\alpha'/\alpha - 1)$, and that (139) holds. If $E = L^p(\mathbb{R}^n)$ for some $1 < p < \infty$, then the above assertion holds with $\beta = (N - r)n|1/p - 1/2|$. In the concrete situation of [357, Example 4.6(b)], we have that $P(D)$ is the integral generator of an exponentially bounded, analytic $(g_\alpha, (1-\Delta)^{-\gamma})$- regularized resolvent family $(R_\alpha(t))_{t\geqslant 0}$ of angle $\pi/2(\alpha'/\alpha - 1)$ with $\gamma = (l + 1 - (1/l))2|1/p - 1/2|$; this is an improvement of the estimate $\gamma > (l + 1)2|1/p - 1/2|$ obtained in the cited example. Furthermore, if $P(x)$ is elliptic, then $2|N$ and [552, Lemma 2.1, Theorem 2.2] in combination with [292, Corollary 2.4.11] and Theorem 2.2.4-Theorem 2.2.5 implies that $\overline{P(A)}$ is the integral generator of an exponentially bounded, analytic (g_α, I)-regularized resolvent family $(R_\alpha(t))_{t\geqslant 0}$ of angle $\pi/2(\alpha'/\alpha - 1)$, and that (139) holds.

(iv) The assertions of Theorem 2.5.2-Theorem 2.5.3 seem to be new even in the case $E = L^2(\mathbb{R}^n)$. It is also worthwhile mentioning here that the generation of fractional resolvent families by elliptic differential operators on $L^2(\Omega)$, where Ω is a bounded domain in $\mathbb{R}^n$ with a sufficiently smooth boundary, has been considered in [363, Section 3].

Remark 2.5.5. In this remark, we would like to explain how one can simply reword the assertions of Theorem 2.5.2-Theorem 2.5.3 in E_l-type spaces. Speaking matter-of-factly, we have the following: Let E be one of the spaces $L^p(\mathbb{R}^n)$ $(1 \leqslant p \leqslant \infty)$, $C_0(\mathbb{R}^n)$, $C_b(\mathbb{R}^n)$, $BUC(\mathbb{R}^n)$ and let $0 \leqslant l \leqslant n$. Let $\mathbf{T}_l\langle\cdot\rangle$ possess the same meaning as in [534], let $a_\eta \in \mathbb{C}$, $0 \leqslant |\eta| \leqslant N$ and let $P(D)f = \Sigma_{|\eta|\leqslant N}\, a_\eta D^\eta f$, with its maximal distributional domain. Let $\omega \geqslant 0$ be such that (127) holds. Then $P(D)$ generates an exponentially equicontinuous (g_α, I)-regularized resolvent family in the space E_n; cf. [321, Remark 2.2]. Let γ (β) have the same value as in the formulation of Theorem 2.5.2 (Theorem 2.5.3). Then the following holds:

(i) Theorem 2.5.2: Set $R_\alpha(t) =: \mathbf{T}_l\langle E_\alpha(t^\alpha P(x))(a - P(x))^{-\gamma}\rangle$, $t \geqslant 0$. Then $(R_\alpha(t))_{t\geqslant 0}$ is an exponentially equicontinuous $(g_\alpha, R_\alpha(0))$-regularized resolvent family with the integral generator $P(D)$, $(R_\alpha(t))_{t\geqslant 0}$ is 'norm continuous' provided $\gamma > nN/2r$ $\min(1, \alpha)$ in the sense that, for every bounded subset B of E_l and for every $\eta \in \mathbb{N}_0^l$, the mapping $t \mapsto \sup_{f\in B} q_\eta(R_\alpha(t)f)$, $t \geqslant 0$ is continuous, and the equality [321, (2)] holds with A, C and E replaced by $P(D)$, $R_\alpha(0)$ and E_l, respectively. The estimate (128) becomes

$$q_\eta(R_\alpha(t)f) \leqslant M(1 + t^{\max(1,\alpha)n/2})e^{\omega t}q_\eta(f),\ t \geqslant 0, f \in E_l,\ \eta \in \mathbb{N}_0^l, \text{ resp.,}$$

$$(140)\quad q_\eta(R_\alpha(t)f) \leqslant M(1 + t^{\max(1,\alpha)n\left|\frac{1}{p}-\frac{1}{2}\right|})e^{\omega t}q_\eta(f),\ t \geqslant 0, f \in E_l,\ \eta \in \mathbb{N}_0^l,$$

with M being independent of $f \in E_l$ and $\eta \in \mathbb{N}_0^l$.

(ii) Theorem 2.5.3: Set $R_\alpha(t) =: \mathbf{T}_l\langle E_\alpha(t^\alpha P(x))(1 + |x|^2)^{-\beta/2}\rangle$, $t \geqslant 0$. Then $(R_\alpha(t))_{t\geqslant 0}$ is an exponentially equicontinuous $(g_\alpha, R_\alpha(0))$-regularized resolvent family with the integral generator $P(D)$, $(R_\alpha(t))_{t\geqslant 0}$ is 'norm continuous' provided $\beta > nN/2 \min(1, \alpha)$, (140) holds, and the equality [321, (2)] holds with A, C and E replaced by $P(D)$, $R_\alpha(0)$ and E_l, respectively.

Example 2.5.6. Let $1 \leqslant p \leqslant \infty$, $0 < \alpha < 2$, $E := L^p(\mathbb{R}^n)$, and $\Delta_\alpha := e^{i(2-\alpha)\frac{\pi}{2}}\Delta$, with its maximal distributional domain (cf. [16], [141], [168]-[169] and [223]-[224] for more details about the case $\alpha = 1$). Let $\gamma > n/2 \min(1, \alpha)$ if $p = 1, \infty$, resp. $\gamma = n\left|\frac{1}{p}-\frac{1}{2}\right|/\min(1, \alpha)$ if $1 < p < \infty$, and $(1 - \Delta)^{-\gamma} := (1 + |x|^2)^{-\gamma}(-i\partial/\partial_{x_1}, \cdots, -i\partial/\partial_{x_n})$ if $p < \infty$, resp. $(1 - \Delta)^{-\gamma} := \mathbf{T}_0\langle(1 + |x|^2)^{-\gamma}\rangle$ if $p = \infty$. By Theorem 2.5.2 and Remark 2.5.5, we infer that Δ_α is the integral generator of a global $(g_\alpha, (1 - \Delta)^{-\gamma})$-regularized resolvent family $(R_\alpha(t))_{t\geqslant 0}$ satisfying the estimate (128) with $\omega = 0$. Notice, finally, that the analysis given in [49, Example 3.7] shows that Δ_α cannot be the generator of a local (g_α, I)-regularized resolvent family on $L^1(\mathbb{R})$.

2.6. Perturbation theory for abstract Volterra equations

In this section, we shall consider additive perturbation theorems for subgenerators of (a, k)-regularized C-resolvent families. We contribute to the abstract theory of certain types of integro-differential evolution equations, including second order equations and fractional equations in the time variable.

Given $f \in L^1_{loc}([0, \infty))$ and $n \in \mathbb{N}$, $f^{*,n}(t)$ denotes, as before, the n-th convolution power of $f(t)$, while $f^{*,0}(t)$ denotes the Dirac δ-distribution. Recall that if $D(A)$ is not dense in E, then $\overline{D(A)}$ is a closed subspace of E and therefore an SCLCS itself; the fundamental system of seminorms which defines the topology of $\overline{D(A)}$ is $(p_{|\overline{D(A)}})_{p\in\circledast}$. The following condition will be used sometimes:

(P2): $k(t)$ satisfies (P1) and $\tilde{k}(\lambda) \neq 0$, $\operatorname{Re}\lambda > \beta$ for some $\beta \geqslant \operatorname{abs}(k)$.

2.6.1. Bounded perturbation theorems. Assume $\alpha > 0$ and $l \in \mathbb{N}$. Set, for any E-valued function $f(t)$ satisfying (P1), $F_{\alpha,f}(z) := \int_0^\infty e^{-z^{1/\alpha}t} f(t)\,dt$, $z > \max(abs(f), 0)^\alpha$. Using induction and elementary operational properties of vector-valued Laplace transform, one can simply prove that there exist uniquely determined real numbers $(c_{l_0,l,\alpha})_{1 \leqslant l_0 \leqslant l}$, independent of E and $f(t)$, such that:

$$\frac{d^l}{dz^l} F_{\alpha,f}(z) = \sum_{l_0=1}^{l} c_{l_0,l,\alpha} z^{\frac{l_0}{\alpha}-l} \int_0^\infty e^{-z^{1/\alpha}t} t^{l_0} f(t)\,dt,\ z > \max(abs(f), 0)^\alpha. \tag{141}$$

Furthermore, $c_{l,l,\alpha} = \frac{(-1)^l}{\alpha^l}$, $l \geqslant 1$, $c_{1,l,\alpha} = \frac{(-1)^l}{\alpha^l}(\frac{1}{\alpha} - 1) \cdot \cdot \cdot (\frac{1}{\alpha} - (l-1))$, $l \geqslant 2$ and the following non-linear recursive formula holds:

$$c_{l_0,l+1,\alpha} = \frac{(-1)}{\alpha} c_{l_0-1,l,\alpha} + \left(\frac{l_0}{\alpha} - l\right) c_{l_0,l,\alpha},\ l_0 = 2, \cdot \cdot \cdot, l. \tag{142}$$

Calculating intrinsic values of coefficients $(c_{l_0,l,\alpha})$ is a non-trivial problem.

Lemma 2.6.1. *There exists* $\zeta \geqslant 1$ *such that*

$$\sum_{l_0=1}^{l} l_0! |c_{l_0,l,\alpha}| \leqslant \zeta^l l! \text{ for all } l \in \mathbb{N}. \tag{143}$$

Proof. Clearly, $L_\alpha := \sup_{n \in \mathbb{N}_0} |\binom{1/\alpha}{n}| < \infty$. Applying (142), one gets:

$$\sum_{l_0=1}^{l+1} l_0! | c_{l_0,l+1,\alpha} |$$

$$\leqslant \left| \frac{1}{\alpha}\left(\frac{1}{\alpha} - 1\right) \cdots \left(\frac{1}{\alpha} - l\right) \right| + \sum_{l_0=2}^{l} \left[\frac{l_0!}{\alpha} | c_{l_0-1,l,\alpha} | + l\left(\frac{1}{\alpha} + 1\right) l_0! | c_{l_0,l,\alpha} | \right] + \frac{(l+1)!}{\alpha^{l+1}}$$

$$\leqslant L_\alpha \left(\frac{1}{\alpha} + l\right) l! + \frac{l}{\alpha} \sum_{l_0=1}^{l-1} l_0! | c_{l_0,l,\alpha} | + l\left(\frac{1}{\alpha} + 1\right) \sum_{l_0=2}^{l} l_0! | c_{l_0,l,\alpha} | + \frac{(l+1)!}{\alpha^{l+1}},\ l \geqslant 2.$$

The preceding inequality implies inductively that (143) holds provided $\zeta \geqslant 4 + \frac{4}{\alpha} + 4L_\alpha(1 + \frac{1}{\alpha})$. □

Set $\zeta_\alpha := \inf\{\zeta \geqslant 1 : \Sigma_{l_0=1}^{l} l_0! |c_{l_0,l,\alpha}| \leqslant \zeta^l l! \text{ for all } l \in \mathbb{N}\}$ and $(\frac{1}{\alpha} + 1) \cdot \cdot \cdot (\frac{1}{\alpha} + (l-1)) := 1$ if $l = 1$. Clearly, $\zeta_1 = 1$, $\zeta_\alpha > 1/\alpha$, $\alpha \in (0, 1)$ and $\Sigma_{l_0=1}^{l} l_0! |c_{l_0,l,\alpha}| \leqslant \zeta_\alpha^l l!$ for all $l \in \mathbb{N}$.

The following lemma will be helpful in the analysis of growth order of perturbed integrated (g_α, C)-regularized resolvent families.

Lemma 2.6.2. *Let* $\alpha > 1$. *Then* $\zeta_\alpha = 1$ *and*

$$\sum_{l_0=1}^{l} l_0! | c_{l_0,l,\alpha} | = \frac{1}{\alpha}\left(\frac{1}{\alpha} + 1\right) \cdots \left(\frac{1}{\alpha} + (l-1)\right) \leqslant \frac{l!}{\alpha} \text{ for all } l \in \mathbb{N}. \tag{144}$$

Proof. Plugging $f(t) \equiv 1$ in (141), we obtain that

$$\sum_{l_0=1}^{l} l_0!c_{l_0,l,\alpha} = (-1)^l \frac{1}{\alpha}\left(\frac{1}{\alpha}+1\right)\cdots\left(\frac{1}{\alpha}+(l-1)\right) \text{ for all } l \in \mathbb{N}. \tag{145}$$

Since $\alpha > 1$, it follows inductively from (142) that $(-1)^l c_{l0,l,\alpha} > 0$, provided $l \geqslant 1$ and $1 \leqslant l_0 \leqslant l$. Combined with (145), the above implies (144) and $\zeta_\alpha = 1$. □

Now we are in a position to state the following important result.

Theorem 2.6.3. *Suppose $\alpha > 0$, $k(t)$ and $k_1(t)$ satisfy (P1), A is a subgenerator of a (g_α, k)-regularized C-resolvent family $(R(t))_{t\geqslant 0}$ satisfying (22) with $a(t) = g_\alpha(t)$, $\omega \geqslant \max(abs(k), 0)$, the family $\{e^{-\omega t}R(t) : t \geqslant 0\}$ is equicontinuous and the following conditions hold:*

(i) *$B \in L(E)$, there exists $|B|_{\circledast} > 0$ such that $p(Bx) \leqslant |B|_{\circledast} p(x)$, $x \in E$, $p \in \circledast$, $BA \subseteq AB$ and $BC = CB$.*
There exist $M \geqslant 1$, $\omega' \geqslant 0$, $\omega'' \geqslant 0$ and $\omega''' \geqslant \max(\omega + \omega', \omega + \omega'', abs(k_1))$ such that

$$\{\lambda \in \mathbb{C} : \operatorname{Re}\lambda > \omega''', \tilde{k}_1(\lambda) \neq 0\} \subseteq \{\lambda \in \mathbb{C} : \operatorname{Re}\lambda > \omega''', \tilde{k}(\lambda) \neq 0\} \tag{146}$$

as well as:

(ii) *For every i, l_0, $l \in \mathbb{N}$ with $1 \leqslant l \leqslant i$ and $1 \leqslant l_0 \leqslant l$, there exists a function $k_{i,l0,l}(t)$ satisfying (P1) and*

$$\mathcal{L}(k_{i,l0,l}(t))(\lambda) = c_{l0,l,\alpha}\lambda^{l0-\alpha(l-1)}\tilde{k}_1(\lambda)\left(\frac{1}{z\tilde{k}(z^{1/\alpha})}\right)^{(i-1)}_{z=\lambda^\alpha},$$

provided $\operatorname{Re}\lambda > \omega'''$ *and* $\tilde{k}_1(\lambda) \neq 0$.

(iii) *For every $i \in \mathbb{N}_0$, there exists a function ${}_ik(t)$ satisfying (P1) and a constant $c_i \in \mathbb{C}$ so that*

$$c_i + {}_i\tilde{k}(\lambda) = \lambda^\alpha \tilde{k}_1(\lambda)\left(\frac{1}{z\tilde{k}(z^{1/\alpha})}\right)^{(i)}_{z=\lambda^\alpha}, \operatorname{Re}\lambda > \omega''', \tilde{k}_1(\lambda) \neq 0.$$

(iv) $\sum_{i=0}^{\infty}|c_i|\frac{|B|^i_{\circledast}}{i!} < \infty$ *and* $\sum_{i=0}^{\infty}\frac{|B|^i_{\circledast}}{i!}\int_0^t|{}_ik(s)|\,ds \leqslant Me^{\omega' t}$, $t \geqslant 0$,

(v) $\sum_{i=1}^{\infty}\sum_{l=1}^{i}\sum_{l_0=1}^{l}\frac{|B|^i_{\circledast}}{i!}\binom{i}{l}\int_0^t(t-s)^{l0}|k_{i,l0,l}(s)|\,ds \leqslant Me^{\omega'' t}$, $t \geqslant 0$, *and*

(vi) $\sum_{i=2}^{\infty}\sum_{l=2}^{i}\sum_{l_0=1}^{l-1}\frac{|B|^i_{\circledast}}{i!}l\binom{i}{l}\int_0^t(t-s)^{l0}|k_{i-1,l0,l-1}(s)|\,ds \leqslant Me^{\omega'' t}$, $t \geqslant 0$.

Then $A+B$ is a subgenerator of an exponentially equicontinuous (g_α, k_1)-regularized C-resolvent family $(R_B(t))_{t\geqslant 0}$, which is given by the following formula:

$$R_B(t)x := \sum_{i=0}^{\infty} \frac{(-B)^i}{i!} [c_i R(t)x + ({}_ik * R(\cdot)x)(t)]$$

(147) $$+\sum_{i=1}^{\infty}\sum_{l=1}^{i}\sum_{l_0=1}^{l} \frac{(-B)^i}{i!}\binom{i}{l}(k_{i,l0,l} * \cdot^{l_0}R(\cdot)x)(t),\ t \geqslant 0,\ x \in E.$$

Furthermore,

(148) $$R_B(t)x = k_1(t)Cx + (A+B)\int_0^t g_\alpha(t-s)R_B(s)x\,ds,\ t \geqslant 0,\ x \in E,$$

and the family $\{e^{-(\omega+\omega')t}R_B(t) : t \geqslant 0\}$ *is equicontinuous.*

Proof. By (iv)-(v), we obtain that the series in (147) converge uniformly on compact subsets of $[0, \infty)$, as well as that $(R_B(t))_{t\geqslant 0}$ is strongly continuous and the family $\{e^{-(\omega+\omega')t}R_B(t) : t \geqslant 0\}$ is equicontinuous. By (i) and Theorem 2.1.5, we have that $(z-A)^{-1}CB = B(z-A)^{-1}C$, $z \in \rho_C(A)$ and that, for every $\lambda \in \mathbb{C}$ with $\operatorname{Re}\lambda > \omega$ and $\tilde{k}(\lambda) \neq 0$: $\lambda^\alpha\tilde{k}(\lambda)(\lambda^\alpha - A)^{-1}CBx = \int_0^\infty e^{-\lambda t}R(t)Bx\,dt$, $x \in E$ and $\lambda^\alpha\tilde{k}(\lambda)B(\lambda^\alpha - A)^{-1}Cx = \int_0^\infty e^{-\lambda t}BR(t)x\,dt$, $x \in E$. By the uniqueness theorem for Laplace transform, one gets $R(t)B = BR(t)$, $t \geqslant 0$. The closedness of A, $R(t)A \subseteq AR(t)$, $t \geqslant 0$ and (iv)-(v) taken together imply $R_B(t)A \subseteq AR_B(t)$, $t \geqslant 0$. Hence, $R_B(t)(A+B) \subseteq (A+B)R_B(t)$, $t \geqslant 0$. By Theorem 2.1.5,

(149) $$z\tilde{k}(z^{1/\alpha})(z-A)^{-1}Cx = \int_0^\infty e^{-z^{1/\alpha}t}R(t)x\,dt,\ x \in E,\ \operatorname{Re}(z^{1/\alpha}) > \omega,\ \tilde{k}(z^{1/\alpha}) \neq 0.$$

Exploiting the closedness of A and the product rule, we easily infer from (149) that, for every $x \in E$, $l \in \mathbb{N}$ and for every $z \in \mathbb{C}$ with $\operatorname{Re}(z^{1/\alpha}) > \omega$ and $\tilde{k}(z^{1/\alpha}) \neq 0$:

$$A\frac{d^l}{dz^l}\int_0^\infty e^{-z^{1/\alpha}t}R(t)x\,dt$$

$$= A\frac{d^l}{dz^l}[z\tilde{k}(z^{1/\alpha})(z-A)^{-1}Cx] = \frac{d^l}{dz^l}[z\tilde{k}(z^{1/\alpha})A(z-A)^{-1}Cx]$$

$$= z\frac{d^l}{dz^l}[z\tilde{k}(z^{1/\alpha})(z-A)^{-1}Cx]$$

(150) $$+\,l\frac{d^{l-1}}{dz^{l-1}}[z\tilde{k}(z^{1/\alpha})(z-A)^{-1}Cx] - \frac{d^l}{dz^l}[z\tilde{k}(z^{1/\alpha})Cx].$$

Fix, for the time being, $x \in E$ and $\lambda \in \mathbb{C}$ with $\operatorname{Re}\lambda > \omega'''$ and $\tilde{k}_1(\lambda) \neq 0$. Then (146) implies $\tilde{k}(\lambda) \neq 0$. By (iv)-(v) and the dominated convergence theorem, it follows that the Laplace transform of power series appearing in (147) can be computed term by term. Making use of this fact as well as (141), (149) and (ii)-(iii), we obtain that:

$$\mathcal{L}(R_B(t)x)(\lambda) = \lambda^{\alpha}\widetilde{k_1}(\lambda)\sum_{i=0}^{\infty}\frac{(-B)^i}{i!}$$

$$\times\sum_{l=0}^{i}\binom{i}{l}\left(\frac{1}{z\tilde{k}(z^{1/\alpha})}\right)^{(i-l)}_{z=\lambda^{\alpha}}\left(\frac{d^l}{dz^l}\left[z\tilde{k}(z^{1/\alpha})(z-A)^{-1}Cx\right]\right)_{z=\lambda^{\alpha}}. \tag{151}$$

Our goal is to prove that:

$$\left(I-\frac{A+B}{\lambda^{\alpha}}\right)\mathcal{L}(R_B(t)x)(\lambda)=\widetilde{k_1}(\lambda)Cx. \tag{152}$$

By the product rule, we get

$$\sum_{i=1}^{\infty}\frac{(-B)^i}{i!}\sum_{l=1}^{i}\binom{i}{l}\left(\frac{1}{z\tilde{k}(z^{1/\alpha})}\right)^{(i-l)}_{z=\lambda^{\alpha}}\left(\frac{d^l}{dz^l}\left[z\tilde{k}(z^{1/\alpha})Cx\right]\right)_{z=\lambda^{\alpha}}$$

$$=-\lambda^{\alpha}\tilde{k}(\lambda)\sum_{i=1}^{\infty}\frac{(-B)^i}{i!}\left(\frac{1}{z\tilde{k}(z^{1/\alpha})}\right)^{(i)}_{z=\lambda^{\alpha}}Cx; \tag{153}$$

notice that the convergence of last series follows from the conditions (iii)-(iv). Taking into account (141), (ii) and (vi), one gets that:

$$\frac{1}{\lambda^{\alpha}\tilde{k}(\lambda)}\int_0^{\infty}e^{-\lambda t}\sum_{i=2}^{\infty}\sum_{l=2}^{i}\sum_{l_0=1}^{l-1}\frac{(-B)^i}{i!}l\binom{i}{l}\int_0^t(t-s)^{l_0}R(t-s)k_{i-1,l_0,l-1}(s)x\,ds\,dt$$

$$=\frac{1}{\lambda^{\alpha}\widetilde{k_1}(\lambda)}\sum_{i=2}^{\infty}\sum_{l=2}^{i}\sum_{l_0=1}^{l-1}\frac{(-B)^i}{i!}$$

$$\times l\binom{i}{l}\int_0^{\infty}e^{-\lambda t}\int_0^t(t-s)^{l_0}R(t-s)k_{i-1,l_0,l-1}(s)x\,ds\,dt \tag{154}$$

$$=\sum_{i=2}^{\infty}\sum_{l=2}^{i}\sum_{l_0=1}^{l-1}\frac{(-B)^i}{i!}l\binom{i}{l}c_{l_0,l-1,\alpha}\lambda^{l_0-\alpha(l-1)}\int_0^{\infty}e^{-\lambda t}t^{l_0}R(t)x\,dt$$

$$=\sum_{i=2}^{\infty}\frac{(-B)^i}{i!}\sum_{l=2}^{i}l\binom{i}{l}\left(\frac{1}{z\tilde{k}(z^{1/\alpha})}\right)^{(i-l)}_{z=\lambda^{\alpha}}\left(\frac{d^{l-1}}{dz^{l-1}}\left[z\tilde{k}(z^{1/\alpha})(z-A)^{-1}Cx\right]\right)_{z=\lambda^{\alpha}},$$

which implies that the series

$$\sum_{i=1}^{\infty}\frac{(-B)^i}{i!}\sum_{l=1}^{i}l\binom{i}{l}\left(\frac{1}{z\tilde{k}(z^{1/\alpha})}\right)^{(i-l)}_{z=\lambda^{\alpha}}\left(\frac{d^{l-1}}{dz^{l-1}}\left[z\tilde{k}(z^{1/\alpha})(z-A)^{-1}Cx\right]\right)_{z=\lambda^{\alpha}}$$

is also convergent. Now we get from (150)-(151) and (153)-(154):

$$\left(I-\frac{A+B}{\lambda^{\alpha}}\right)\mathcal{L}(R_B(t)x)(\lambda)$$

$$=\lambda^{\alpha}\sum_{i=0}^{\infty}\frac{(-B)^i}{i!}\sum_{l=0}^{i}\binom{i}{l}\left(\frac{1}{z\tilde{k}(z^{1/\alpha})}\right)^{(i-l)}_{z=\lambda^{\alpha}}\left(\frac{d^l}{dz^l}[z\tilde{k}(z^{1/\alpha})(z-A)^{-1}Cx]\right)_{z=\lambda^{\alpha}}$$

$$-\sum_{i=1}^{\infty}\frac{(-B)^i}{i!}\sum_{l=1}^{i}\binom{i}{l}\left(\frac{1}{z\tilde{k}(z^{1/\alpha})}\right)^{(i-l)}_{z=\lambda^{\alpha}}\Bigg(z\frac{d^l}{dz^l}[z\tilde{k}(z^{1/\alpha})(z-A)^{-1}Cx]$$

$$+l\frac{d^{l-1}}{dz^{l-1}}[z\tilde{k}(z^{1/\alpha})(z-A)^{-1}Cx]-\frac{d^l}{dz^l}[z\tilde{k}(z^{1/\alpha})]Cx\Bigg)_{z=\lambda^{\alpha}}$$

$$-\lambda^{\alpha}\tilde{k}(\lambda)\sum_{i=0}^{\infty}\frac{(-B)^i}{i!}\left(\frac{1}{z\tilde{k}(z^{1/\alpha})}\right)^{(i)}_{z=\lambda^{\alpha}}A(\lambda^{\alpha}-A)^{-1}Cx$$

$$+\lambda^{\alpha}\sum_{i=0}^{\infty}\frac{(-B)^{i+1}}{i!}\sum_{l=0}^{i}\binom{i}{l}\left(\frac{1}{z\tilde{k}(z^{1/\alpha})}\right)^{(i-l)}_{z=\lambda^{\alpha}}\left(\frac{d^l}{dz^l}[z\tilde{k}(z^{1/\alpha})(z-A)^{-1}Cx]\right)_{z=\lambda^{\alpha}}$$

$$=\lambda^{2\alpha}\tilde{k}(\lambda)\sum_{i=0}^{\infty}\frac{(-B)^i}{i!}\left(\frac{1}{z\tilde{k}(z^{1/\alpha})}\right)^{(i)}_{z=\lambda^{\alpha}}(\lambda^{\alpha}-A)^{-1}Cx$$

$$-\sum_{i=1}^{\infty}\frac{(-B)^i}{i!}\sum_{l=1}^{i}\binom{i}{l}\left(\frac{1}{z\tilde{k}(z^{1/\alpha})}\right)^{(i-l)}_{z=\lambda^{\alpha}}\left(l\frac{d^{l-1}}{dz^{l-1}}[z\tilde{k}(z^{1/\alpha})(z-A)^{-1}Cx]\right)_{z=\lambda^{\alpha}}$$

$$-\lambda^{\alpha}\tilde{k}(\lambda)\sum_{i=1}^{\infty}\frac{(-B)^i}{i!}\left(\frac{1}{z\tilde{k}(z^{1/\alpha})}\right)^{(i)}_{z=\lambda^{\alpha}}Cx$$

$$-\lambda^{\alpha}\tilde{k}(\lambda)\sum_{i=0}^{\infty}\frac{(-B)^i}{i!}\left(\frac{1}{z\tilde{k}(z^{1/\alpha})}\right)^{(i)}_{z=\lambda^{\alpha}}(-Cx+\lambda^{\alpha}(\lambda^{\alpha}-A)^{-1}Cx)$$

$$+\lambda^{\alpha}\sum_{i=0}^{\infty}\frac{(-B)^{i+1}}{i!}\sum_{l=0}^{i}\binom{i}{l}\left(\frac{1}{z\tilde{k}(z^{1/\alpha})}\right)^{(i-l)}_{z=\lambda^{\alpha}}\left(\frac{d^l}{dz^l}[z\tilde{k}(z^{1/\alpha})(z-A)^{-1}Cx]\right)_{z=\lambda^{\alpha}}$$

$$=Cx-\sum_{i=1}^{\infty}\frac{(-B)^i}{i!}\sum_{l=1}^{i}\binom{i}{l}\left(\frac{1}{z\tilde{k}(z^{1/\alpha})}\right)^{(i-l)}_{z=\lambda^{\alpha}}\left(l\frac{d^{l-1}}{dz^{l-1}}[z\tilde{k}(z^{1/\alpha})(z-A)^{-1}Cx]\right)_{z=\lambda^{\alpha}}$$

$$+\lambda^{\alpha}\sum_{i=0}^{\infty}\frac{(-B)^{i+1}}{i!}\sum_{l=0}^{i}\binom{i}{l}\left(\frac{1}{z\tilde{k}(z^{1/\alpha})}\right)^{(i-l)}_{z=\lambda^{\alpha}}\left(\frac{d^l}{dz^l}[z\tilde{k}(z^{1/\alpha})(z-A)^{-1}Cx]\right)_{z=\lambda^{\alpha}}=Cx,$$

because the sum of coefficients of $(-B)^i$ $(i \geqslant 1)$ in the last two series is nothing else but 0; this follows from an elementary calculus involving only the product rule. Assume now $x \in D(A)$, $\operatorname{Re} \lambda > \omega'''$, $\widetilde{k_1}(\lambda) \neq 0$ and $(I - \frac{A+B}{\lambda^\alpha})x = 0$. By (152) and $R_B(t)(A + B) \subseteq (A + B)R_B(t)$, $t \geqslant 0$, we obtain that

$$\widetilde{k_1}(\lambda)Cx = \left(I - \frac{A+B}{\lambda^\alpha}\right)\mathcal{L}(R_B(t)x)(\lambda) = \mathcal{L}\left(R_B(t)\left(I - \frac{A+B}{\lambda^\alpha}\right)x\right)(\lambda) = 0,$$

which implies $Cx = x = 0$. Thus, $\{\lambda^\alpha : \operatorname{Re} \lambda > \omega''', \widetilde{k_1}(\lambda) \neq 0\} \subseteq \rho_C(A + B)$ and

$$\widetilde{k_1}(\lambda)\left(I - \frac{A+B}{\lambda^\alpha}\right)^{-1} Cx = \int_0^\infty e^{-\lambda t} R_B(t)x \, dt, \; x \in E, \; \operatorname{Re} \lambda > \omega''', \widetilde{k_1}(\lambda) \neq 0.$$

The proof of theorem completes an application of Theorem 2.1.5. □

Remark 2.6.4. (i) By Proposition 2.1.4(i), we get that $(R_B(t))_{t \geqslant 0}$ is a unique (g_α, k_1)-regularized C-resolvent family with the properties stated in the formulation of Theorem 2.6.3.

(ii) The following comment is also applicable to Theorem 2.6.5 below. Assume $k(t) = k_1(t)$, $t \geqslant 0$, $n \in \mathbb{N}$ and the conditions (iv)-(vi) of Theorem 2.6.3 hold with $|B|^i_\circledast / i!$ replaced by $|B|^i_\circledast / n^i i!$ therein. Writing $A + B$ as $A + \Sigma^n_{i=1} B/n$ and applying Theorem 2.6.3 successively n times, we obtain that $A + B$ is a subgenerator of a global (g_α, k)-regularized C-resolvent family $(R_B(t))_{t \geqslant 0}$ satisfying (148). Furthermore, the family $\{e^{-(\omega + n\omega')t} R_B(t) : t \geqslant 0\}$ is equicontinuous.

(iii) It is not clear whether there exist functions $k(t)$ and $k_1(t)$ such that the conditions (ii)-(vi) of Theorem 2.6.3 are fulfilled in the case $\alpha \in (0, 1)$.

Theorem 2.6.5. *Consider the situation of Theorem 2.6.3 with $(R(t))_{t \geqslant 0}$ being an exponentially equicontinuous, analytic (g_α, k)-regularized C-resolvent family of angle $\beta \in (0, \pi/2]$. Assume that, for every $\gamma \in (0, \beta)$, there exists $\omega_\gamma \geqslant 0$ such that the set $\{e^{-\omega_\gamma \operatorname{Re} z} R(z) : z \in \Sigma_\gamma\}$ is equicontinuous. Assume, additionally, that there exists $\varepsilon > 0$ such that, for every $\gamma \in (0, \beta)$, there exist $\omega_{\gamma,1} \geqslant \max(\sup\{abs(_i k) : i \geqslant 1\}, \omega_\gamma)$ and $\omega_{\gamma,2} \geqslant \max(\sup\{abs(k_{i,l0,l}) : 1 \leqslant l \leqslant i, 1 \leqslant l_0 \leqslant l\}, \omega_\gamma + \varepsilon)$ with the following properties:*

(i) *For every $i \in \mathbb{N}_0$, the function $\lambda \mapsto \widetilde{_i k}(\lambda)$, $\lambda > \omega_{\gamma,1}$ can be analytically extended to the sector $\omega_{\gamma,1} + \Sigma_{\frac{\pi}{2}+\gamma}$ and the following holds:*

$$\sum_{i=0}^{\infty} \frac{|B|^i_\circledast}{i!} \sup_{\lambda \in \omega_{\gamma,1} + \Sigma_{\frac{\pi}{2}+\gamma}} |\widetilde{_i k}(\lambda)| < \infty. \tag{155}$$

(ii) *For every $i, l_0, l \in \mathbb{N}$ with $1 \leqslant l \leqslant i$ and $1 \leqslant l_0 \leqslant l$, the function $\lambda \mapsto \mathcal{L}(k_{i,l0,l}(t))(\lambda)$, $\lambda > \omega_{\gamma,2}$ can be analytically extended to the sector $\omega_{\gamma,2} + \Sigma_{\frac{\pi}{2}+\gamma}$ and the following holds:*

$$\sum_{i=1}^{\infty} \sum_{l=1}^{i} \sum_{l_0=1}^{l} \frac{|B|^i_\circledast}{i!} \binom{i}{l} \frac{l_0!}{\sqrt{2\pi l_0}(\varepsilon \cos\gamma)^{l_0}} \sup_{\lambda \in \omega_{\gamma,2} + \Sigma_{\frac{\pi}{2}+\gamma}} |\mathcal{L}(k_{i,l_0,l}(t))(\lambda)| < \infty. \tag{156}$$

Then $(R_B(t))_{t\geqslant 0}$ is an exponentially equicontinuous, analytic (g_{α}, k_1)-regularized C-resolvent family of angle β.

Proof. Let $p \in \circledast$, $x \in E$, $\gamma \in (0, \beta)$ and $\varepsilon \in (0, \frac{1}{3}\min(\gamma, \frac{\pi}{2} - \gamma))$. Then Stirling's formula implies that there exists $\kappa \geqslant 1$ such that

$$|z|^{l_0} e^{\omega_\gamma \operatorname{Re} z} \leqslant \frac{(\operatorname{Re} z)^{l_0}}{(\cos\gamma)^{l_0}} e^{\omega_\gamma \operatorname{Re} z} \leqslant \frac{e^{-l_0}}{(\cos\gamma)^{l_0}} \frac{l_0^{l_0}}{\varepsilon^{l_0}} e^{(\omega_\gamma+\varepsilon)\operatorname{Re} z}$$

$$\leqslant \kappa \frac{l_0!}{\sqrt{2\pi l_0}(\varepsilon\cos\gamma)^{l_0}} e^{(\omega_\gamma+\varepsilon)\operatorname{Re} z} \leqslant \kappa \frac{l_0!}{\sqrt{2\pi l_0}(\varepsilon\cos\gamma)^{l_0}} e^{\omega_{\gamma,2}\operatorname{Re} z}$$

for all $z \in \Sigma_\gamma$ and $l_0 \in \mathbb{N}$. By Theorem 1.2.5(i) and the proof of implication (i) $\Rightarrow$ (ii) of [20, Theorem 2.6.1], we obtain that the mapping $\lambda \mapsto \mathcal{L}({}_ik * R(\cdot)x)(\lambda)$, $\lambda > \omega_{\gamma,1}$, resp. $\lambda \mapsto \mathcal{L}(k_{i,l0,l} * \cdot^{l_0} R(\cdot)x)(\lambda)$, $\lambda > \omega_{\gamma,2}$ can be analytically extended to the sector $\omega_{\gamma,1} + \Sigma_{\frac{\pi}{2}+\gamma}$, resp. $\omega_{\gamma,2} + \Sigma_{\frac{\pi}{2}+\gamma}$, as well as that there exist $c_p > 0$ and $q_p \in \circledast$, independent of x, such that

$$(157) \qquad \sup_{\lambda\in\omega_{\gamma,1}+\Sigma_{\frac{\pi}{2}+\gamma-\varepsilon}} p\big((\lambda-\omega_{\gamma,1})\widetilde{{}_ik}(\lambda)\mathcal{L}(R(t)x)(\lambda)\big) \leqslant \frac{c_p q_p(x)}{\sin\varepsilon} \sup_{\lambda\in\omega_{\gamma,1}+\Sigma_{\frac{\pi}{2}+\gamma-\varepsilon}} |\widetilde{{}_ik}(\lambda)|$$

and that, for every i, l_0, $l \in \mathbb{N}$ with $1 \leqslant l \leqslant i$ and $1 \leqslant l_0 \leqslant l$,

$$\sup_{\lambda\in\omega_{\gamma,2}+\Sigma_{\frac{\pi}{2}+\gamma-\varepsilon}} p\big((\lambda-\omega_{\gamma,2})\mathcal{L}(k_{i,l_0,l}(t))(\lambda)\mathcal{L}(t^{l_0}R(t)x)(\lambda)\big)$$

$$(158) \qquad \leqslant \kappa \frac{c_p q_p(x)}{\sin\varepsilon} \frac{l_0!}{\sqrt{2\pi l_0}(\varepsilon\cos\gamma)^{l_0}} \sup_{\lambda\in\omega_{\gamma,2}+\Sigma_{\frac{\pi}{2}+\gamma-\varepsilon}} \left|\mathcal{L}(k_{i,l_0,l}(t))(\lambda)\right|.$$

Using (157)-(158), Theorem 1.2.5(i) and the proof of implication (ii) $\Rightarrow$ (i) of [20, Theorem 2.6.1], it follows that the functions $t \mapsto ({}_ik * R(\cdot)x)(t)$, $t > 0$ and $t \mapsto (k_{i,l0,l} * \cdot^{l_0} R(\cdot)x)(t)$, $t > 0$ can be analytically extended to the sector Σ_γ and that the following estimates hold:

$$p(({}_ik * R(\cdot)x)(z))$$

$$(159) \qquad \leqslant \frac{c_p q_p(x)}{\sin\varepsilon} \sup_{\lambda\in\omega_{\gamma,1}+\Sigma_{\frac{\pi}{2}+\gamma-\varepsilon}} |\widetilde{{}_ik}(\lambda)| \left(e^{1+\omega_{\gamma,1}\operatorname{Re} z} + \frac{e^{\omega_{\gamma,1}\operatorname{Re} z}}{\pi\sin\varepsilon}\right), \; z \in \Sigma_{\gamma-3\varepsilon}$$

and

$$p\big((k_{i,l0,l} * \cdot^{l_0} R(\cdot)x)(z)\big) \leqslant \kappa \frac{c_p q_p(x)}{\sin\varepsilon} \frac{l_0!}{\sqrt{2\pi l_0}(\varepsilon\cos\gamma)^{l_0}}$$

$$(160) \qquad \times \sup_{\lambda\in\omega_{\gamma,2}+\Sigma_{\frac{\pi}{2}+\gamma-\varepsilon}} \left|\mathcal{L}(k_{i,l_0,l}(t))(\lambda)\right| \left(e^{1+\omega_{\gamma,2}\operatorname{Re} z} + \frac{e^{\omega_{\gamma,2}\operatorname{Re} z}}{\pi\sin\varepsilon}\right), \; z \in \Sigma_{\gamma-3\varepsilon}.$$

Since Vitali's theorem holds in our framework, we easily infer from (155)-(156), (159)-(160), and the arbitrariness of γ and ε, that the mapping $t \mapsto R_B(t)x$, $t > 0$ can be analytically extended to the sector Σ_β by the formula (147). Thanks to the proof of Theorem 2.6.3, the series appearing in (147) converge uniformly on compact subsets of $[0, \infty)$, which implies $\lim_{t\to 0+} \sum_{i=0}^{\infty} \frac{(-B)^i}{i!} c_i R(t)x = \sum_{i=0}^{\infty} \frac{(-B)^i}{i!} c_i R(0)x$, $\lim_{t\to 0+} \sum_{i=0}^{\infty} \frac{(-B)^i}{i!} ({}_ik * R(\cdot)x)(t) = 0$ and $\lim_{t\to 0+} \sum_{i=1}^{\infty} \sum_{l=1}^{i} \sum_{l0=1}^{l} \frac{(-B)^i}{i!}\binom{i}{l}(k_{i,l0,l} *^{\cdot l0} R(\cdot)x)(t) = 0.$ Furthermore, the functions $z \mapsto f_1(z) := \sum_{i=0}^{\infty} \frac{(-B)^i}{i!} c_i R(z)x$, $z \in \Sigma_\beta$, $z \mapsto f_2(z) := \sum_{i=0}^{\infty} \frac{(-B)^i}{i!} ({}_ik * R(\cdot)x)(z)$, $z \in \Sigma_\beta$ and $z \mapsto f_3(z) := \sum_{i=1}^{\infty} \sum_{l=1}^{i} \sum_{l0=1}^{l} \frac{(-B)^i}{i!}\binom{i}{l}(k_{i,l0,l} *^{\cdot l0} R(\cdot)x)(z)$, $z \in \Sigma_\beta$ are analytic, and the set $\{e^{-(\omega_\gamma+\omega_{\gamma,1}+\omega_{\gamma,2})\operatorname{Re} z} f_j(z) : 1 \leqslant j \leqslant 3, z \in \Sigma_{\gamma-3\varepsilon}\}$ is bounded. An application of Theorem 1.2.5(ii) gives that the mapping $z \mapsto R_B(z)x$, $z \in \Sigma_\beta \cup \{0\}$ is continuous on any closed subsector of $\Sigma_\beta \cup \{0\}$, which completes the proof of theorem. □

It would take too long to go into details concerning stability of certain differential properties under bounded commuting perturbations described in Theorem 2.6.3.

It is noteworthy that the assumptions of Theorem 2.6.3 and Theorem 2.6.5 hold provided $\alpha > 1$ and $k(t) = k_1(t) = g_{r+1}(t)$, where $r \geqslant 0$. In this case, $c_0 = 1$, $k_0(t) = 0$, $c_i = 0$, $i \geqslant 1$,

$$ {}_ik(t) = \left(\frac{r+1}{\alpha}-1\right)\cdots\left(\frac{r+1}{\alpha}-i\right)g_{\alpha i}(t),\ t \geqslant 0,\ i \geqslant 1 $$

and, for every $i, l_0, l \in \mathbb{N}$ with $1 \leqslant l \leqslant i$ and $1 \leqslant l_0 \leqslant l$,

$$ k_{i,l0,l}(t) = c_{l0,l,\alpha}\left(\frac{r+1}{\alpha}-1\right)\cdots\left(\frac{r+1}{\alpha}-(i-1)\right)g_{\alpha i-l0}(t),\ t > 0. $$

In order to verify (iv)-(vi), notice that there exists a constant $c_{r,\alpha} \geqslant 1$ such that $|(\frac{r+1}{\alpha}-1)\cdots(\frac{r+1}{\alpha}-i)| \leqslant c_{r,\alpha} i!$ for all $i \in \mathbb{N}$. Then we obtain from (53) and Lemma 2.6.1-Lemma 2.6.2 that

$$ \sum_{i=1}^{\infty} \frac{|B|_{\circledast}^i}{i!}\left|\left(\frac{r+1}{\alpha}-1\right)\cdots\left(\frac{r+1}{\alpha}-i\right)\right|\int_0^t g_{\alpha i}(s)\,ds $$

$$ \leqslant c_{r,\alpha}\sum_{i=1}^{\infty}\frac{(t^\alpha |B|_{\circledast})^i}{\Gamma(\alpha i+1)} = c_{r,\alpha}(E_\alpha(t^\alpha |B|_{\circledast})-1) \leqslant c_{r,\alpha}c_\alpha e^{t|B|_{\circledast}^{1/\alpha}},\ t \geqslant 0, $$

$$ \sum_{i=1}^{\infty}\sum_{l=1}^{i}\sum_{l_0=1}^{l}\frac{|B|_{\circledast}^i}{i!}\binom{i}{l}\int_0^t (t-s)^{l_0}\,|k_{i,l_0,l}(s)|\,ds $$

$$ \leqslant \sum_{i=1}^{\infty}\sum_{l=1}^{i}\sum_{l_0=1}^{l}\frac{|B|_{\circledast}^i}{i!}\binom{i}{l}|c_{l_0,l,\alpha}| $$

$$ \times\left|\left(\frac{r+1}{\alpha}-1\right)\cdots\left(\frac{r+1}{\alpha}-(i-l)\right)\right| l_0!\,g_{\alpha i+1}(t) $$

$$\leqslant \frac{c_{r,\alpha}}{\alpha}\sum_{i=1}^{\infty} i\frac{(t^{\alpha}|B|_{\circledast})^{i}}{\Gamma(\alpha i+1)} = \frac{c_{r,\alpha}}{\alpha}t^{\alpha}|B|_{\circledast}\sum_{i=1}^{\infty} i\frac{(t^{\alpha}|B|_{\circledast})^{i-1}}{\Gamma(\alpha i+1)}$$

$$= \frac{c_{r,\alpha}}{\alpha}t^{\alpha}|B|_{\circledast}\left(\sum_{i=1}^{\infty}\frac{z^{i}}{\Gamma(\alpha i+1)}\right)'_{z=t^{\alpha}|B|_{\circledast}} = \frac{c_{r,\alpha}}{\alpha^{2}}t^{\alpha}|B|_{\circledast}E_{\alpha,\alpha}(t^{\alpha}|B|_{\circledast})$$

$$\leqslant \frac{c_{r,\alpha}b_{\alpha}}{\alpha^{2}}t^{\alpha}|B|_{\circledast}(t^{\alpha}|B|_{\circledast})^{(1-\alpha)/\alpha}e^{t|B|_{\circledast}^{1/\alpha}} \leqslant \frac{c_{r,\alpha}b_{\alpha}}{\alpha^{2}}t|B|_{\circledast}^{1/\alpha}e^{t|B|_{\circledast}^{1/\alpha}},\ t\geqslant 0,$$

proving the conditions (iv)-(v) and

$$\sum_{i=2}^{\infty}\sum_{l=2}^{i}\sum_{l_0=1}^{l-1}\frac{|B|_{\circledast}^{i}}{i!}l\binom{i}{l}\int_{0}^{t}(t-s)^{l_0}|k_{i-1,l_0,l-1}(s)|\,ds$$

$$\leqslant c_{r,\alpha}\sum_{i=2}^{\infty}\sum_{l=2}^{i}\sum_{l_0=1}^{l-1}\frac{|B|_{\circledast}^{i}}{(l-1)!}l_0!|c_{l_0,l-1,\alpha}|g_{\alpha(i-1)+1}(t)$$

$$\leqslant \frac{c_{r,\alpha}}{\alpha}|B|_{\circledast}\sum_{i=2}^{\infty}(i-1)\frac{(t^{\alpha}|B|_{\circledast})^{i-1}}{\Gamma(\alpha(i-1)+1)}$$

$$= \frac{c_{r,\alpha}}{\alpha}|B|_{\circledast}\sum_{i=1}^{\infty} i\frac{(t^{\alpha}|B|_{\circledast})^{i}}{\Gamma(\alpha i+1)} \leqslant \frac{c_{r,\alpha}b_{\alpha}}{\alpha^{2}}t|B|_{\circledast}^{(1+\alpha)/\alpha}e^{t|B|_{\circledast}^{1/\alpha}},\ t\geqslant 0,$$

proving the condition (vi). Assume now, with the notation used in the formulation of Theorem 2.6.5, $\gamma\in(0,\beta)$, $\omega_{\gamma,1}\geqslant\omega_{\gamma}$, $\omega_{\gamma,1}>0$, $(\omega_{\gamma,1}\cos\gamma)^{\alpha}>|B|_{\circledast}$, $\varepsilon=1/\cos\gamma$, $\omega_{\gamma,2}\geqslant\omega_{\gamma}+\varepsilon$ and $(1+\omega_{\gamma,2}\cos\gamma)^{\alpha-1}>|B|_{\circledast}$. Then

$$\sum_{i=0}^{\infty}\frac{|B|_{\circledast}^{i}}{i!}\sup_{\lambda\in\omega_{\gamma,1}+\Sigma_{\frac{\pi}{2}+\gamma}}|\tilde{k}_{i}(\lambda)| \leqslant c_{r,\alpha}\sum_{i=0}^{\infty}|B|_{\circledast}^{i}\sup_{\lambda\in\omega_{\gamma,1}+\Sigma_{\frac{\pi}{2}+\gamma}}\frac{1}{|\lambda|^{\alpha i}}$$

$$\leqslant c_{r,\alpha}\sum_{i=0}^{\infty}\frac{|B|_{\circledast}^{i}}{(\omega_{\gamma,1}\cos\gamma)^{\alpha i}}<\infty$$

and

$$\sum_{i=1}^{\infty}\sum_{l=1}^{i}\sum_{l_0=1}^{l}\frac{|B|_{\circledast}^{i}}{i!}\binom{i}{l}\frac{l_0!}{\sqrt{2\pi l_0}(\varepsilon\cos\gamma)^{l_0}}\sup_{\lambda\in\omega_{\gamma,2}+\Sigma_{\frac{\pi}{2}+\gamma}}\left|\mathcal{L}(k_{i,l_0,l}(t))(\lambda)\right|$$

$$\leqslant c_{r,\alpha}\sum_{i=1}^{\infty}\sum_{l=1}^{i}\sum_{l_0=1}^{l}\frac{|B|_{\circledast}^{i}}{l!}l_0!|c_{l_0,l,\alpha}|\sup_{\lambda\in\omega_{\gamma,2}+\Sigma_{\frac{\pi}{2}+\gamma}}\frac{1}{|\lambda|^{\alpha i-l_0}}$$

$$\leqslant \frac{c_{r,\alpha}}{\alpha}\sum_{i=1}^{\infty}\sum_{l=1}^{i}\sum_{l_0=1}^{l}\frac{|B|_{\circledast}^{i}}{(1+\omega_{\gamma,2}\cos\gamma)^{\alpha i-l_0}}$$

$$\leqslant \frac{c_{r,\alpha}}{\alpha}\sum_{i=1}^{\infty}\frac{i^{2}|B|_{\circledast}^{i}}{(1+\omega_{\gamma,2}\cos\gamma)^{(\alpha-1)i}}<\infty,$$

proving the conditions (155)-(156).

Corollary 2.6.6. *Suppose $\alpha > 1$, $\omega \geqslant 0$, $r \geqslant 0$ and A is a subgenerator of a global r-times integrated (g_α, C)-resolvent $(R(t))_{t\geqslant 0}$ satisfying (22) with $a(t) = g_\alpha(t)$ and $k(t) = g_{r+1}(t)$. Let the family $\{e^{-\omega t}R(t) : t \geqslant 0\}$ be equicontinuous, and let $B \in L(E)$ satisfy the condition (i) quoted in the formulation of Theorem 2.6.3. Then $A+B$ is a subgenerator of a global r-times integrated (g_α, C)-resolvent $(R_B(t))_{t\geqslant 0}$ satisfying (148) with $k_1(t) = k(t)$. Furthermore, the family $\{(1+t)^{-1} \exp(-(\omega+|B|_{\odot}^{1/\alpha})t)R_B(t) : t \geqslant 0\}$ is equicontinuous, and $(R_B(t))_{t\geqslant 0}$ is an exponentially equicontinuous, analytic r-times integrated (g_α, C)-resolvent of angle $\beta \in (0, \pi/2]$ provided that $(R(t))_{t\geqslant 0}$ is.*

Remark 2.6.7. It is worthwhile to mention (cf. [292, Theorem 2.5.6]) that Corollary 2.6.6 remains true, with a different upper bound for the growth order of $(R_B(t))_{t\geqslant 0}$, in the case $\alpha = 1$. By Lemma 1.2.4 and the proof of cited theorem, it follows that $(R_B(t))_{t\geqslant 0}$ is entire provided that $\alpha \in \mathbb{N}$ and $(R(t))_{t\geqslant 0}$ is entire.

Example 2.6.8. Corollary 2.6.6 is a proper extension of [295, Lemma 4.7] provided $\alpha = 2$ and $B = zI$ $(z \in \mathbb{C})$, which can be applied in the analysis of the problem

$$u_{tt} + 2\beta u_t - \Delta u + 2\sum_{i=1}^{n} \alpha_i u_{x_i} + \mu u = 0$$

in $L^p([0, \pi]^n)$, with Dirichlet boundary conditions; here we assume $n \in \mathbb{N}$, $1 \leqslant p < \infty$ and $\beta, \alpha_i, \mu \in \mathbb{C}$ (see e.g. [137, pp. 144-145] and [266, Theorem 4.2]). It is clear that Corollary 2.6.6 can be applied to (r-coercive) differential operators generating integrated cosine functions ([14], [169], [223], [267], [554]) or exponentially equicontinuous (g_α, C)-regularized resolvent families (Section 2.5); in what follows, we shall apply Corollary 2.6.6 to abstract differential operators generating C-regularized cosine functions. Let E be one of the spaces $L^p(\mathbb{R}^n)$ $(1 \leqslant p \leqslant \infty)$, $C_0(\mathbb{R}^n)$, $C_b(\mathbb{R}^n)$, $BUC(\mathbb{R}^n)$, let $0 \leqslant l \leqslant n$, and let the Fréchet topology on E_l be defined as before. Let $m \in \mathbb{N}$ and let $a_\alpha \in \mathbb{C}$ for $0 \leqslant |\alpha| \leqslant m$. Consider the operator $P(D)f := \sum_{|\alpha|\leqslant m} a_\alpha f^{(\alpha)}$ with its maximal distributional domain. Set $P(x) := \sum_{|\alpha|\leqslant m} a_\alpha i^{|\alpha|}x^\alpha$, $x \in \mathbb{R}^n$, $h_{t,\beta}(x) := (1 + |x|^2)^{-\beta/2}\sum_{j=0}^{\infty} \frac{t^{2j}P(x)^j}{(2j)!}$, $x \in \mathbb{R}^n$, $t \geqslant 0$, $\beta \geqslant 0$, $\Omega(\omega) := \{\lambda^2 : \operatorname{Re} \lambda > \omega\}$, if $\omega > 0$ and $\Omega(\omega) := \mathbb{C}\setminus(-\infty, \omega^2]$, if $\omega \leqslant 0$. Assume $r \in [0, m]$, $\omega \in \mathbb{R}$ and the condition (W) holds (cf. Example 2.2.14 for more details). Then, for every $l \in \mathbb{N}_{n}^{0}$, there exists $M \geqslant 1$ such that, for every $\beta > (m - \frac{r}{2})\frac{n}{4}$, $P(D)$ generates an exponentially equicontinuous $\mathbf{T}_l(\langle(1+|x|^2)^{-\beta}\rangle)$-regularized cosine function $(C_\beta(t))_{t\geqslant 0}$ in E_l satisfying $C_\beta(t)f = \mathcal{F}^{-1}h_{t,\beta} * f$, $t \geqslant 0$, $f \in E_l$ and $p_\alpha(C_\beta(t)f) \leqslant Mp_\alpha(f)$ $G_{n/2}(t)$, $t \geqslant 0$, $f \in E_l$, $\alpha \in \mathbb{N}_0^l$. If $1 < p < \infty$ and $E = L^p(\mathbb{R}^n)$, then the previous result can be slightly refined by allowing that β takes the value $\frac{1}{2}(m - \frac{r}{2})n|\frac{1}{p} - \frac{1}{2}|$. Given $\varphi \in L^1(\mathbb{R}^n)$, define the bounded linear operator B on E_l by $(Bf)(x) := \int_{\mathbb{R}^n} \varphi(x - t) f(t)\, dt$, $f \in E_l$, $x \in \mathbb{R}^n$. Then $BP(D) \subseteq P(D)B$, $\mathbf{T}_l(\langle(1+|x|^2)^{-\beta}\rangle)B = B\mathbf{T}_l(\langle(1+|x|^2)^{-\beta}\rangle)$

and $p_\alpha(Bf) \leqslant \|\varphi\|_{L^1(\mathbb{R}^n)} p_\alpha(f)$, $f \in E$, $\alpha \in \mathbb{N}_l^0$. Applying Corollary 2.6.6, we get that $P(D)+B$ generates an exponentially equicontinuous $\mathbf{T}_l(\langle(1+|x|^2)^{-\beta}\rangle)$-regularized cosine function $(C_{\beta,B}(t))_{t\geqslant 0}$ in E_l.

Assume now $\alpha > 1$, $\varrho > 0$, $\sigma \in (0, 1)$ and $k(t) = k_1(t) = \mathcal{L}^{-1}(\lambda^{-\alpha}e^{-\varrho\lambda^\sigma})(t)$, $t \geqslant 0$. By the consideration given in [292, Remark 2.5.4(iii)], it follows that, for every $l \in \mathbb{N}$, there exist real numbers $(p_{m,l,\alpha,\varrho,\sigma})_{1\leqslant m\leqslant l}$ such that $p_{1,l,\alpha,\varrho,\sigma} = \varrho\frac{\varrho}{\alpha}(\frac{\varrho}{\alpha}-1)\cdots(\frac{\varrho}{\alpha}-(l-1))$, $p_{l,l,\alpha,\varrho,\sigma} = (\varrho\frac{\varrho}{\alpha})^l$, as well as that the following holds:

$$\frac{d^l}{dz^l}\left(\frac{1}{z\tilde{k}(z^{1/\alpha})}\right) = \frac{d^l}{dz^l}e^{\varrho z^{\sigma/\alpha}} = e^{\varrho z^{\sigma/\alpha}}\sum_{m=1}^{l} p_{m,l,\alpha,\varrho,\sigma} z^{m\frac{\sigma}{\alpha}-l}, \quad z > 0$$

and

$$p_{m,l+1,\alpha,\varrho,\sigma} = \varrho\frac{\sigma}{\alpha}p_{m-1,l,\alpha,\varrho,\sigma} + \left(m\frac{\sigma}{\alpha}-l\right)p_{m,l,\alpha,\varrho,\sigma}, \; 2 \leqslant m \leqslant l. \tag{161}$$

This implies $c_0 = 1$, $k_0(t) = 0$, $c_i = 0$, $i \geqslant 1$,

$$_ik(t) = \sum_{m=1}^{i} p_{m,i,\alpha,\varrho,\sigma} g_{\alpha i - m\sigma}(t), \quad t > 0, \; i \geqslant 1,$$

$$k_{i,l0,l}(t) = c_{l0,l,\alpha}\sum_{m=1}^{i-l} p_{m,i-l,\alpha,\varrho,\sigma} g_{\alpha i - l0 - m\sigma}(t), \quad t > 0, \; 1 \leqslant l < i, \; 1 \leqslant l_0 \leqslant l$$

and

$$k_{i,l0,i}(t) = c_{l0,i,\alpha} g_{\alpha i - l0}(t), \; t > 0, \; 1 \leqslant l_0 \leqslant i.$$

By means of (161) and the proof of Lemma 2.6.1, we obtain the existence of a constant $\zeta_{\alpha,\varrho,\sigma} \geqslant 1$ such that

$$\sum_{m=1}^{l} m!|p_{m,l,\alpha,\varrho,\sigma}| \leqslant \zeta_{\alpha,\varrho,\sigma}^l \, l! \text{ for all } l \in \mathbb{N}. \tag{162}$$

In what follows, we assume that $\zeta_{\alpha,\varrho,\sigma} \geqslant 1$ is minimal with respect to (162); notice that $\zeta_{\alpha,\varrho,\sigma} > \varrho\sigma/\alpha$ and that it is not clear whether Lemma 2.6.2 can be reconsidered in the newly arisen situation. Then

$$\begin{aligned} &\sum_{i=1}^{\infty}\sum_{l_0=1}^{i}\frac{|B|_{\circledast}^i}{i!}\int_0^t (t-s)^{l_0}\,|k_{i,l_0,i}(s)|\,ds \\ &= \sum_{i=1}^{\infty}\sum_{l_0=1}^{i}\frac{|B|_{\circledast}^i}{i!}|c_{l_0,i,\alpha}|\,l_0!\int_0^t \frac{(t-s)^{l_0}}{l_0!}\frac{s^{\alpha i-l_0-1}}{\Gamma(\alpha i-l_0)}ds \\ &\leqslant \sum_{i=1}^{\infty}\frac{(t^\alpha|B|_{\circledast})^i}{\Gamma(\alpha i+1)} \leqslant E_\alpha(t^\alpha|B|_{\circledast}) \leqslant c_\alpha e^{t|B|_{\circledast}^{1/\alpha}}, \; t \geqslant 0. \end{aligned} \tag{163}$$

Since $\Gamma(\cdot)$ is increasing in (ξ, ∞), where $\xi \sim 1.4616...$, we obtain that

$$\frac{\Gamma(\alpha i+1)}{\Gamma(\alpha i-m\sigma+1)m!} \leqslant \frac{\Gamma(\alpha i+1)}{\Gamma(\alpha i-m+1)m!} = \frac{\alpha i(\alpha i-1)\cdots(\alpha i-m+1)}{m!} \leqslant \binom{\lceil \alpha i \rceil}{m},$$

provided $i \geqslant 2$ and $1 \leqslant m \leqslant i-1$. Combining this with (53), Lemma 2.6.1-Lemma 2.6.2 and (162), we get

$$\sum_{i=2}^{\infty}\sum_{l=1}^{i-1}\sum_{l_0=1}^{l}\frac{|B|_{\circledast}^{i}}{i!}\binom{i}{l}\int_0^t (t-s)^{l_0}\,|k_{i,l_0,l}(s)|\,ds$$

$$\leqslant \sum_{i=2}^{\infty}\sum_{l=1}^{i-1}\sum_{l_0=1}^{l}\frac{|B|_{\circledast}^{i}}{i!}\binom{i}{l} l_0!\,|c_{l_0,l,\alpha}\,|\sum_{m=1}^{i-l} m!\,|\,p_{m,i-l,\alpha,\varrho,\sigma}|\frac{t^{\alpha i-m\sigma}}{\Gamma(\alpha i-m\sigma+1)m!}$$

$$\leqslant \frac{1}{\alpha}\sum_{i=2}^{\infty}\sum_{l=1}^{i-1}\frac{|B|_{\circledast}^{i}}{\Gamma(\alpha i+1)}\zeta_{\alpha,\varrho,\sigma}^{i-l}\sum_{m=1}^{i-l}\frac{t^{\alpha i-m\sigma}\Gamma(\alpha i+1)}{\Gamma(\alpha i-m\sigma+1)m!}$$

(164) $$\leqslant \frac{1}{\alpha}\sum_{i=2}^{\infty}\sum_{l=1}^{i-1}\frac{|B|_{\circledast}^{i}}{\Gamma(\alpha i+1)}\zeta_{\alpha,\varrho,\sigma}^{i-l}\sum_{m=1}^{i-l} t^{\alpha i-m\sigma}\binom{\lceil \alpha i \rceil}{m}$$

$$\leqslant \frac{1}{\alpha}(1+t^{-\sigma})\sum_{i=2}^{\infty}\sum_{l=1}^{i-l}\frac{|B|_{\circledast}^{i}}{\Gamma(\alpha i+1)}\zeta_{\alpha,\varrho,\sigma}^{i-l}(t+t^{1-\sigma})^{\alpha i}$$

$$\leqslant \frac{1}{\alpha\zeta_{\alpha,\varrho,\sigma}}(1+t^{-\sigma})\sum_{i=1}^{\infty} i\frac{(|B|_{\circledast}(t+t^{1-\sigma})^{\alpha}\zeta_{\alpha,\varrho,\sigma})^{i+1}}{\Gamma(\alpha i+1)}$$

$$\leqslant \frac{1}{\alpha}(1+t^{-\sigma})\,|B|_{\circledast}(t+t^{1-\sigma})^{\alpha}\left(\sum_{i=1}^{\infty} i\frac{z^i}{\Gamma(\alpha i+1)}\right)_{z=|B|_{\circledast}(t+t^{1-\sigma})^{\alpha}\zeta_{\alpha,\varrho\sigma}}$$

(165) $$\leqslant \frac{b_\alpha}{\alpha^2}(1+t^{-\sigma})\,|B|_{\circledast}^{(1+\alpha)/\alpha}\,(t+t^{1-\sigma})^{1+\alpha}\zeta_{\alpha,\varrho,\sigma}^{1/\alpha}e^{|B|_{\circledast}^{1+/\alpha}(t+t^{1-\sigma})\zeta_{\alpha,\varrho,\sigma}^{1/\alpha}}, t \geqslant 1.$$

Noticing that $t^{\alpha i-m\sigma} \leqslant 1$, $t \in [0, 1)$ $i \geqslant 2$, $1 \leqslant m \leqslant i-1$, we obtain from (164) that there exists $\xi_{\alpha,\varrho,\sigma} \geqslant 1$ such that:

(166) $$\sum_{i=2}^{\infty}\sum_{l=1}^{i-1}\sum_{l_0=1}^{l}\frac{|B|_{\circledast}^{i}}{i!}\binom{i}{l}\int_0^t (t-s)^{l_0}\,|k_{i,l_0,i}(s)|\,ds \leqslant \xi_{\alpha,\varrho,\sigma},\quad t \in [0,1).$$

By (163)-(166), (v) holds for any $\omega' > (|B|_{\circledast}\zeta_{\alpha,\varrho,\sigma})^{1/\alpha}$. In almost the same way, one can prove that (iv) and (vi) hold for any $\omega' > (|B|_{\circledast}\zeta_{\alpha,\varrho,\sigma})^{1/\alpha}$. Assume now that $(R(t))_{t\geqslant 0}$ is an exponentially equicontinuous, analytic (g_α, k)-regularized C-resolvent family of angle $\beta \in (0, \pi/2]$, $\gamma \in (0, \beta)$, $\omega_{\gamma,1} \geqslant \omega_\gamma$, $\omega_{\gamma,1} > 0$, $(\omega_{\gamma,1} \cos \gamma)^{\alpha-\sigma} > |B|_{\circledast}\zeta_{\alpha,\varrho,\sigma}$, $\varepsilon = 1/\cos \gamma$, $\omega_{\gamma,2} \geqslant \omega_\gamma + \varepsilon$ and $(1 + \omega_{\gamma,2} \cos \gamma)^{\alpha-1} > |B|_{\circledast}\zeta_{\alpha,\varrho,\sigma}$. Then

$$\sum_{i=0}^{\infty}\frac{|B|_{\circledast}^{i}}{i!}\sup_{\lambda\in\omega_{\gamma,1}+\Sigma_{\frac{\pi}{2}+\gamma}}|_{i}\widetilde{k}(\lambda)|\leqslant\sum_{i=0}^{\infty}\frac{|B|_{\circledast}^{i}}{i!}\sum_{m=1}^{i}\frac{m!|p_{m,i,\alpha,\varrho,\sigma}|}{(\omega_{\gamma,1}\cos\gamma)^{\alpha i-m\sigma}}$$

$$\leqslant\mu_{\sigma,\gamma}\sum_{i=0}^{\infty}\frac{|B|_{\circledast}^{i}\zeta_{\alpha,\varrho\sigma}^{i}}{(\omega_{\gamma,1}\cos\gamma)^{(\alpha-\sigma)i}}<\infty$$

for an appropriate constant $\mu_{\sigma,\gamma}\geqslant 1$,

$$\sum_{i=1}^{\infty}\sum_{l_0=1}^{i}\frac{|B|_{\circledast}^{i}}{i!}\frac{l_0!}{\sqrt{2\pi l_0}(\varepsilon\cos\gamma)^{l_0}}\sup_{\lambda\in\omega_{\gamma,2}+\Sigma_{\frac{\pi}{2}+\gamma}}\left|\mathcal{L}(k_{i,l_0,i}(t))(\lambda)\right|$$

$$\leqslant\sum_{i=1}^{\infty}i\frac{|B|_{\circledast}^{i}}{(1+\omega_{\gamma,2})^{(\alpha-1)i}}<\infty,$$

and, for every $\varepsilon>0$ with $(1+\omega_{\gamma,2}\cos\gamma)^{\alpha-1}>|B|_{\circledast}(\zeta_{\alpha,\varrho,\sigma}+\varepsilon)$, there exists $\mu_{\varepsilon}>0$ such that

$$\sum_{i=2}^{\infty}\sum_{l=1}^{i-1}\sum_{l_0=1}^{l}\frac{|B|_{\circledast}^{i}}{i!}\binom{i}{l}\frac{l_0!}{\sqrt{2\pi l_0}(\varepsilon\cos\gamma)^{l_0}}\sup_{\lambda\in\omega_{\gamma,2}+\Sigma_{\frac{\pi}{2}+\gamma}}\left|\mathcal{L}(k_{i,l_0,l}(t))(\lambda)\right|$$

$$\leqslant\sum_{i=2}^{\infty}\sum_{l=1}^{i-1}\sum_{l_0=1}^{l}\frac{|B|_{\circledast}^{i}}{(i-l)!l!}l_0!|c_{l_0,l,\alpha}|\sum_{m=1}^{i-l}\frac{|p_{m,i-l,\alpha,\varrho,\sigma}|}{(1+\omega_{\gamma,2}\cos\gamma)^{\alpha i-l_0-m\sigma}}$$

$$\leqslant\sum_{i=2}^{\infty}\sum_{l=1}^{i-1}\sum_{l_0=1}^{l}\frac{|B|_{\circledast}^{i}}{(i-l)!l!}l_0!|c_{l_0,l,\alpha}|\sum_{m=1}^{i-l}\frac{|p_{m,i-l,\alpha,\varrho,\sigma}|}{(1+\omega_{\gamma,2}\cos\gamma)^{(\alpha-1)i}}$$

$$\leqslant\frac{1}{\alpha}\sum_{i=2}^{\infty}\sum_{l=1}^{i-1}\frac{|B|_{\circledast}^{i}\zeta_{\alpha,\varrho\sigma}^{i-l}}{(1+\omega_{\gamma,2}\cos\gamma)^{(\alpha-1)i}}\leqslant\frac{1}{\alpha}\sum_{i=2}^{\infty}(i-1)\frac{(|B|_{\circledast}\zeta_{\alpha,\varrho\sigma}+\varepsilon)^{i}}{(1+\omega_{\gamma,2}\cos\gamma)^{(\alpha-1)i}}<\infty,$$

proving the conditions (155)-(156).

Corollary 2.6.9. *Suppose* $\alpha>1$, $\omega\geqslant 0$, $\varrho>0$, $\sigma\in(0,1)$ *and* $k(t)=\mathcal{L}^{-1}(\lambda^{-\alpha}e^{-\varrho\lambda^{\sigma}})(t)$, $t\geqslant 0$. *Let A be a subgenerator of a global* (g_{α}, k)*-regularized C-resolvent family* $(R(t))_{t\geqslant 0}$ *satisfying (22) with* $a(t)=g_{\alpha}(t)$. *Let* $B\in L(E)$ *satisfy the condition (i) quoted in the formulation of Theorem 2.6.3. Then* $A+B$ *is a subgenerator of a global* (g_{α}, k)*-regularized C-resolvent family* $(R_B(t))_{t\geqslant 0}$ *satisfying (148) with* $k_1(t)=k(t)$. *Furthermore, for every* $\varepsilon>0$, *the family* $\{\exp(-(\omega+(|B|_{\circledast}\zeta_{\alpha,\varrho,\sigma})^{1/\alpha}\varepsilon)t)R_B(t):t\geqslant 0\}$ *is equicontinuous, and* $(R_B(t))_{t\geqslant 0}$ *is an exponentially equicontinuous, analytic* (g_{α}, k)*-regularized C-resolvent family of angle* $\beta\in(0,\pi/2]$ *provided that* $(R(t))_{t\geqslant 0}$ *is.*

Example 2.6.10. ([292, Example 2.8.3(iii)], [314]) Let $s>1$,

$$E:=\left\{f\in C^{\infty}[0,1]\,;\,\|f\|=\sup_{p\geqslant 0}\frac{\|f^{(p)}\|_{\infty}}{p!^{s}}<\infty\right\}$$

and

$$A := -d/ds,\ D(A) := \{f \in E\ ; f' \in E, f(0) = 0\}.$$

Then $\rho(A) = \mathbb{C}$, A generates a tempered ultradistribution semigroup of $(p!^s)$-class and A cannot be the generator of a distribution semigroup since A is not stationary dense (see e.g. [332, Example 1.6] and [301], cf. also [371] and [292] for more details on the subject). If $f \in E$, $t \in [0, 1]$ and $\lambda \in \mathbb{C}$, set $f_\lambda^1(t) := \int_0^t e^{-\lambda(t-s)} f(s)\, ds$ and $f_\lambda^2(t) := \int_0^t e^{-\lambda(t-s)} f(s)\, ds$. Then $f_\lambda^1(\cdot), f_\lambda^2(\cdot) \in E$, $\lambda \in \mathbb{C}$ and there exist $b > 0$ and $M \geqslant 1$, independent of $f(\cdot)$, such that

$$\|f_\lambda^1(\cdot)\| \leqslant M\|f\| e^{b|\lambda|^{1/s}},\ \operatorname{Re}\lambda \geqslant 0, f \in E. \tag{167}$$

It is clear that $\|f_\lambda^2(\cdot)\|_{L^\infty[0,1]} \leqslant e^{|\lambda|}\|f\|$, $\operatorname{Re}\lambda \geqslant 0$ and $\|\frac{d}{dt} f_\lambda^2(\cdot)\|_{L^\infty[0,1]} \leqslant (|\lambda|e^{|\lambda|} + 1)\|f\|$, $\operatorname{Re}\lambda \geqslant 0$. Hence, by iteration, we obtain that, for every $n \geqslant 2$, $t \in [0, 1]$ and $\lambda \in \mathbb{C}$ with $\operatorname{Re}\lambda \geqslant 0$:

$$\frac{d^n}{dt^n} f_\lambda^2(t) = \frac{d^{n-1}}{dt^{n-1}} f(t) + \sum_{k=1}^{n-1} \lambda^k \frac{d^{n-1-k}}{dt^{n-1-k}} f(t) + \lambda^n f_\lambda^2(t). \tag{168}$$

On the other hand, [283, Proposition 4.5] implies that there exists $c > 0$ such that $\Sigma_{p=0}^{\infty} t^p/p!^s = O(\exp(ct^{1/s}))$, $t \geqslant 0$. Combined with (168) and the logarithmic convexity, the last estimate yields:

$$\frac{1}{n!^s}\left\|\frac{d^n}{dt^n} f_\lambda^2(\cdot)\right\|_{L^\infty[0,1]} \leqslant \|f\| + \|f\| e^{c|\lambda|^{1/s}} + \frac{|\lambda|^n}{n!^s} e^{|\lambda|}\|f\|$$

$$\leqslant (1 + e^{c|\lambda|^{1/s}} + e^{c|\lambda|^{1/s}} e^{|\lambda|})\, \|f\|,\ \operatorname{Re}\ \lambda \geqslant 0,\ \lambda \neq 0. \tag{169}$$

In view of (169) we get that, for every $\eta > 1$, there exists $M_\eta \geqslant 1$, independent of $f(\cdot)$, such that

$$\|f_\lambda^2(\cdot)\| \leqslant M_\eta \|f\| e^{\eta|\lambda|},\ \operatorname{Re}\lambda \geqslant 0, f \in E. \tag{170}$$

Consider now the complex polynomial $P(z) = \Sigma_{j=0}^n a_j z^j$, $z \in \mathbb{C}$, $a_n \neq 0$, $n \geqslant 2$. Set, for every $\lambda \in \mathbb{C}$, $P_\lambda(\cdot) := P(\cdot) - \lambda$ and consider the operator $P(A)$ defined by

$$D(P(A)) := D(A^n) \text{ and } P(A)f := \sum_{j=0}^n a_j A^j, f \in D(P(A)).$$

Clearly, $P(A)$ is not stationary dense. Let $r > 0$ and $d > 0$ be such that $P(z) \neq 0$, $|z| \geqslant r$ and $P'(z) \neq 0$, $|z| \geqslant d$. Let $z_{1,\lambda}, \cdots, z_{n,\lambda}$ denote the zeros of the polynomial $z \mapsto P_\lambda(z)$, $z \in \mathbb{C}$ and let $0 < m := \min_{|z| \geqslant d+1} |P'(z)|$. Then an old result of J. L. Walsh [512] says that $|z_{\lambda,j}| \leqslant r + |a_n|^{-1/n}|\lambda|^{1/n}$, $1 \leqslant j \leqslant n$, $\lambda \in \mathbb{C}$. Furthermore, it is checked at once that there exists a sufficiently large $\lambda_0 > 0$ such that $z_{j,\lambda}$ is a simple zero of $P_\lambda(z)$ and that $|z_{j,\lambda}| \geqslant d + 1$, provided $|\lambda| \geqslant \lambda_0$ and $1 \leqslant j \leqslant n$. Therefore, for every $\lambda \in \mathbb{C}$ with $|\lambda| \geqslant \lambda_0$ and for every $i, j \in \mathbb{N}_n$ with $i \neq j$, the following holds:

$$d + 1 \leqslant |z_{j,\lambda}| \leqslant r + |a_n|^{-1/n}|\lambda|^{1/n} \text{ and } |P'(z_{j,\lambda})| \geqslant m,\ z_{i,\lambda} \neq z_{j,\lambda}. \tag{171}$$

It is quite easy to verify that

(172) $\rho(p(A)) = \mathbb{C}$ and $R(\lambda : p(A)) = (-1)^{n+1} a_n^{-1} R(z_{1,\lambda} : A) \cdots R(z_{n,\lambda} : A)$, $\lambda \in \mathbb{C}$.

Assume now $|\lambda| \geqslant \lambda_0$. Then de L'Hospital's rule implies:

$$a_n \prod_{\substack{1 \leqslant i \leqslant n \\ i \neq j}} (z_{i,\lambda} - z_{j,\lambda}) = (-1)^{n+1} P'(z_{j,\lambda}),\ 1 \leqslant j \leqslant n. \tag{173}$$

Using the resolvent equation, (167), (170)-(171) and (173), one can rewrite and evaluate the right hand side of equality appearing in (172) as follows:

$$\begin{aligned} &\| (-1)^{n+1} a_n^{-1} R(z_{1,\lambda} : A) \cdots R(z_{n,\lambda} : A) \| \\ &= \left\| (-1)^{n+1} a_n^{-1} \sum_{j=1}^{n} \frac{R(z_{j,\lambda} : A)}{\prod_{\substack{1 \leqslant i \leqslant n \\ i \neq j}} (z_{i,\lambda} - z_{j,\lambda})} \right\| \\ &= \left\| \sum_{j=1}^{n} \frac{R(z_{j,\lambda} : A)}{P'(z_{j,\lambda})} \right\| \leqslant \frac{1}{m} \sum_{j=1}^{n} \| R(z_{j,\lambda} : A) \| . \end{aligned} \tag{174}$$

By (172) and (174) we finally get that, for every $\eta > 1$,

$$\|R(\lambda : p(A))\| = O\Big(e^{b|a_n|^{-1/n}|\lambda|^{1/ns}} + e^{\eta |a_n|^{-1/n}|\lambda|^{1/n}}\Big),\ \lambda \in \mathbb{C}. \tag{175}$$

Since the preceding estimate holds for any $\lambda \in \mathbb{C}$, it is quite complicated to inscribe here all of its consequences (cf. [286], [292], [302, (2.35)-(2.37)] and [334]); for example, $P(A)$ generates a tempered ultradistribution sine of $(p!^s)$-class provided $n \geqslant 2s$, and $P(A)$ generates an exponentially bounded, $\mathcal{L}^{-1}(e^{-\varrho\lambda^{1/n}})$-convoluted group provided $\varrho > |a_n|^{-1/n}/\cos(\pi/2n)$. Let us also mention that the consideration given in the example following [301, Corollary 3.8] enables one to construct important examples of (pseudo-)differential operators generating ultradistribution sines and that the estimate (175) can be derived, with insignificant technical modifications, in the case that we consider a general sequence (M_p) of positive numbers satisfying $M_0 = 1$ and $M_{p+q} \geqslant M_p M_q$ $(p, q \geqslant 0)$. In what follows, we shall illustrate an application of Corollary 2.6.9. Suppose $n > \alpha \geqslant 1$, $\delta \in (0, \pi/2]$, $(\pi/2 + \delta)\alpha/n < \pi/2$, $\varrho \geqslant 1/\cos((\pi/2 + \delta)\alpha/n)$ and $k(t) = \mathcal{L}^{-1}(\lambda^{-\alpha} e^{-\varrho\lambda^{\alpha/n}})(t)$, $t \geqslant 0$. By Theorem 2.2.5 and (175), $P(A)$ is the integral generator of an exponentially bounded, analytic (g_α, k)-regularized resolvent family of angle δ (cf. also [301, Proposition 3.12]). Let $\varphi \in E$ and $Bf(t) := (\varphi * f)(t)$, $t \in [0, 1]$, $f \in E$. Then $B \in L(E)$, $BP(A) \subseteq P(A)B$ and, by Corollary 2.6.9, $P(A) + B$ is the integral generator of an exponentially bounded, analytic (g_α, k)-regularized resolvent family of angle δ.

Motivated by the idea of A. Rhandi (cf. [470, Theorem 1.1 and Example 1.3]), C. Lizama and J. Sánchez have proved in [384, Theorem 3.1] an additive perturbation result for generators of (a, k)-regularized resolvent families. An extension of the last mentioned theorem is stated as follows.

Theorem 2.6.11. *(cf.* [303, Theorem 2.13]*) Suppose* $M > 0$, $\omega \geqslant 0$ *and A is a densely defined subgenerator of an (a, k)-regularized C-resolvent family* $(R(t))_{t\geqslant 0}$ *which satisfies that, for every seminorm* $p \in \circledast$, $p(R(t)x) \leqslant Me^{\omega t}p(x)$, $t \geqslant 0$, $x \in E$. *Suppose, further,* $C^{-1}B \in L(E)$, $BCx = CBx$, $x \in D(A)$, *there exist a function b(t) satisfying (P1) and a number* $\omega_0 \geqslant \omega$ *such that* $\tilde{b}(\lambda) = \frac{\tilde{a}(\lambda)}{\tilde{k}(\lambda)}$, $\lambda > \omega_0$, $\tilde{k}(\lambda) \neq 0$. *Let* $\mu > \omega_0$ *and* $\gamma \in [0, 1)$ *be such that*

$$\int_0^\infty e^{-\mu t}p\left(C^{-1}B\int_0^t b(t-s)R(s)x\,ds\right)dt \leqslant \gamma p(x),\ \ x \in D(A),\ p \in \circledast. \tag{176}$$

Then the operator A+B is a subgenerator of an (a, k)-regularized C-resolvent family $(R_B(t))_{t\geqslant 0}$ *which satisfies* $p(R_B(t)x) \leqslant \frac{M}{1-\gamma}e^{\mu t}p(x)$, $x \in E$, $t \geqslant 0$, $p \in \circledast$ *and*

$$R_B(t)x = R(t)x + \int_0^t R_B(t-r)C^{-1}B\int_0^r b(r-s)R(s)x\,ds\,dr,\ t \geqslant 0,\ x \in D(A). \tag{177}$$

Concerning Theorem 2.6.11, it is worth noting that, in many cases, we do not have the existence of a function $b(t)$ and a complex number z such that $\tilde{a}(\lambda)/\tilde{k}(\lambda) = \tilde{b}(\lambda) + z$, $\operatorname{Re}\lambda > \omega_1$, $\tilde{k}(\lambda) \neq 0$. The following theorem is an attempt to fill this gap.

Theorem 2.6.12. *Suppose* $M, M_1 > 0$, $\omega \geqslant 0$, $l \in \mathbb{N}$ *and A is a subgenerator of an (a, k)-regularized C-resolvent family* $(R(t))_{t\geqslant 0}$ *such that* $p(R(t)x) \leqslant Me^{\omega t}p(x)$, $x \in E$, $t \geqslant 0$, $p \in \circledast$ *and (22) holds. Let a(t) and k(t) satisfy (P1), and let the following conditions hold:*

(i) $BCx = CBx$, $x \in D(A)$,

$$p(CA^jC^{-1}Bx) \leqslant M_1p(x),\ x \in \overline{D(A)},\ p \in \circledast,\ 0 \leqslant j \leqslant l-1 \text{ and}$$

$$p(A^lC^{-1}Bx) \leqslant M_1p(x),\ x \in \overline{D(A)},\ p \in \circledast. \tag{178}$$

(ii) *There exist a function b(t) satisfying (P1) and a complex number z such that*

$$\tilde{a}(\lambda)^{l+1}/\tilde{k}(\lambda) = \tilde{b}(\lambda) + z,\ \operatorname{Re}\lambda > \max(\omega, abs(a), abs(k)),\ \tilde{k}(\lambda) \neq 0.$$

(iii) $\lim_{\lambda\to+\infty}\int_0^\infty e^{-\lambda t}|a(t)|\,dt = 0$ *and* $\lim_{\lambda\to+\infty}\int_0^\infty e^{-\lambda t}|b(t)|\,dt = 0$.

Define, for every $x \in \overline{D(A)}$ *and* $t \geqslant 0$,

$$S(t)x := \sum_{j=0}^{l-1} a^{*,j+1}(t)CA^jC^{-1}Bx$$
$$+\int_0^t b(t-s)R(s)A^lC^{-1}Bx\,ds + zR(t)A^lC^{-1}Bx.$$

Then, for every $x \in E$, *there exists a unique solution of the integral equation*

$$R_B(t)x = R(t)x + (S * R_B)(t)x,\ t \geqslant 0; \tag{179}$$

furthermore, $(R_B(t))_{t\geqslant 0}$ *is an (a, k)-regularized C-resolvent family with a subgenerator A + B, there exist* $\mu \geqslant \max(\omega, abs(a), abs(k))$ *and* $\gamma \in [0, 1)$ *such that* $p(R_B(t)x) \leqslant \frac{M}{1-\gamma} e^{\mu t}p(x)$, $x \in E$, $t \geqslant 0$, $p \in ⊛$ *and (22) holds with* $k_l(t)$ *and* $g_\alpha(t)$ *replaced by* $k(t)$ *and* $a(t)$ *therein.*

Proof. It is clear that $\int_0^\infty e^{-\lambda t}|a^{*,j}(t)|\,dt \leqslant (\int_0^\infty e^{-\lambda t}|a(t)|\,dt)^j$, $j \in \mathbb{N}$, $\lambda > abs(a)$. By Theorem 1.2.1(ix) it follows that $R(t)x \in \overline{D(A)}$, $t \geqslant 0$, $x \in E$. Making use of this fact and (i), we get that $S(t) \in L(\overline{D(A)})$, $t \geqslant 0$. Keeping in mind the condition (ii), it is not difficult to prove that, for every $x \in \overline{D(A)}$,

$$\mathcal{L}(S(t)x)(\lambda) = \tilde{a}(\lambda)\,(I-\tilde{a}(\lambda)A)^{-1}\,CC^{-1}Bx,\ x \in \overline{D(A)},\ \mathrm{Re}\,\lambda > \omega_1,\ \tilde{k}(\lambda) \neq 0. \tag{180}$$

Using the conditions (i) and (iii), we obtain the existence of numbers $\mu > \max(\omega, abs(a), abs(k))$ and $\gamma \in [0, 1)$ such that

$$\int_0^\infty e^{-\mu t}p(S(t)x)\,dt \leqslant \gamma p(x),\ x \in \overline{D(A)},\ p \in ⊛ \tag{181}$$

and (PH1) holds, where
(PH1) : For every strongly continuous function $f: [0, \infty) \to L(E, \overline{D(A)})$ such that $p(f(t)x) \leqslant Mp(x)$, $x \in E$, $t \geqslant 0$, $p \in ⊛$, the following inequality holds:

$$\int_0^t e^{-\mu(t-s)}p\,(S(t-s)f(s)x)\,ds \leqslant \gamma Mp(x),\ x \in E,\ t \geqslant 0,\ p \in ⊛.$$

Now one can define inductively, for every $t \geqslant 0$, the sequence $(T_n(t))_{n\in\mathbb{N}_0}$ in $L(E, \overline{D(A)})$ by $T_0(t) := R(t)$ and $T_{n+1}(t)x := \int_0^t S(t-s)T_n(s)x\,ds$, $x \in E$, $n \in \mathbb{N}_0$; observe that, for every $n \in \mathbb{N}_0$, $(T_n(t))_{t\geqslant 0}$ is strongly continuous and the family $\{T_n(t) : t \geqslant 0\}$ is locally equicontinuous (with the clear meaning). By (181), (PH1) and the proof of [384, Theorem 3.1], it follows inductively that

$$p(T_n(t))x \leqslant M\gamma^n e^{\mu t}p(x),\ x \in E,\ t \geqslant 0,\ p \in ⊛$$

and that, for every $x \in E$ and $t \geqslant 0$, the sequence $(R_B^n(t)x := \Sigma_{i=0}^n T_i(t)x)_n$ is Cauchy in E and therefore convergent. Set $R_B(t)x := \lim_{n\to\infty} R_B^n(t)x$, $x \in E$, $t \geqslant 0$. It is obvious that the mapping $t \mapsto R_B(t)x$, $t \geqslant 0$ is continuous for every fixed $x \in E$ and (179) holds. Therefore, it suffices to show that

$$\tilde{k}(\lambda)\,(I-\tilde{a}(\lambda)(A+B))^{-1}\,Cx = \int_0^\infty e^{-\lambda t}R_B(t)x\,dt,\ x \in E,\ \mathrm{Re}\,\lambda > \mu,\ \tilde{k}(\lambda) \neq 0. \tag{182}$$

Towards this end, notice that (181) and (PH1) together imply that $I - \tilde{S}(\lambda)$ is invertible for $\mathrm{Re}\,\lambda > \mu$ and $(I-\tilde{S}(\lambda))^{-1} = \Sigma_{n=0}^\infty[\tilde{S}(\lambda)]^n$, $\mathrm{Re}\,\lambda > \mu$. Now we obtain that

$$\widetilde{R_B}(\lambda)x = (I-\tilde{S}(\lambda))^{-1}\tilde{k}(\lambda)\,(I-\tilde{a}(\lambda)A)^{-1}Cx,\ \mathrm{Re}\,\lambda > \mu,\ x \in E, \tag{183}$$

which immediately implies with (180) the validity of (182) in case $\tilde{a}(\lambda) = 0$, $\mathrm{Re}\,\lambda > \mu$ and $\tilde{k}(\lambda) \neq 0$. Assume now $\tilde{a}(\lambda)\,\tilde{k}(\lambda) \neq 0$ and $\mathrm{Re}\,\lambda > \mu$. Then a simple computation including the equality $BCx = CBx$, $x \in D(A)$, as well as (180) and (183), shows that the operator $I-\tilde{a}(\lambda)(A+B)$ is injective and

$$(I-\tilde{S}(\lambda))^{-1}\tilde{k}(\lambda)\,(I-\tilde{a}(\lambda)A)^{-1}C\,(I-\tilde{a}(\lambda)(A+B))x=\tilde{k}(\lambda)Cx,\ x\in D(A).$$

The representation $(I-\tilde{S}(\lambda))^{-1}=\Sigma_{n=0}^{\infty}[(\frac{1}{\tilde{a}(\lambda)}-A)^{-1}CC^{-1}B]^n$ implies

$$\left(\frac{1}{\tilde{a}(\lambda)}-(A+B)\right)(I-\tilde{S}(\lambda))^{-1}\tilde{k}(\lambda)\left(\frac{1}{\tilde{a}(\lambda)}-A\right)^{-1}Cx=\tilde{k}(\lambda)Cx,\quad x\in E,$$

$R(C)\subseteq R(I-\tilde{a}(\lambda)(A+B))$ and (182), finishing the proof of theorem. □

Remark 2.6.13. Now we will explain how one can reword Theorem 2.6.11 in the case that B is not necessarily a bounded operator from $\overline{D(A)}$ into E. Consider the situation of Theorem 2.6.12 with E being complete, and replace in Theorem 2.6.11 the condition $C^{-1}B\in L(E)$ by:

(♮) $C^{-1}B:D(A)\to E$ and, for every $p\in ⊛$, there exist $c_p>0$ and $q\in ⊛$ such that $p(C^{-1}Bx)\leqslant c_p(q(x)+q(Ax)),\ x\in D(A)$.

Denote, with a little abuse of notation, $T_0(t)=R(t),\ t\geqslant 0,\ S(t)x=C^{-1}B(\int_0^t b(t-s)R(s)x\,ds+zR(t)x),\ t\geqslant 0,\ x\in D(A)$ and $T_1(t)x=\int_0^t T_0(t-s)S(s)x\,ds,\ t\geqslant 0,\ x\in D(A)$. Then (♮) implies that the mapping $t\mapsto S(t)x,\ t\geqslant 0$ is continuous for every $x\in D(A)$ and $p(T_1(t)x)\leqslant\gamma Mp(x),\ x\in D(A),\ t\geqslant 0,\ p\in ⊛$. By [411, Lemma 22.19] and the completeness of E, one can extend the operator $T_1(t)$ to the whole space E ($t\geqslant 0$). Proceeding inductively, one can define for each $t\geqslant 0$ the sequence $(T_n(t)=\int_0^t T_{n-1}(t-s)S(s)x\,ds)_{n\in\mathbb{N}_0}$ in $L(E)$ so that $p(T_n(t)x)\leqslant\gamma^n Mp(x),\ x\in E,\ t\geqslant 0,\ p\in ⊛$. The preceding inequality implies that, for every $x\in E$, the sequence $(R_B^n(t)x\equiv\Sigma_{i=0}^n T_i(t)x)_n$ is convergent in E. Put $R_B(t)x=\lim_{n\to\infty}R_B^n(t)x,\ x\in E,\ t\geqslant 0$. As in the proof of Theorem 2.6.12, the mapping $t\mapsto R_B(t)x,\ t\geqslant 0$ is continuous for every fixed $x\in E$ and (177) holds. The closedness of A and the condition (♮) imply that $\int_0^\infty e^{-\lambda t}S(t)x\,dt=C^{-1}B\tilde{a}(\lambda)(I-\tilde{a}(\lambda)A)^{-1}Cx,\ x\in D(A)$, $\operatorname{Re}\lambda>\mu,\ \tilde{k}(\lambda)\neq 0$ and $C^{-1}B\tilde{a}(\lambda)(I-\tilde{a}(\lambda)A)^{-1}C\in L(E)$, $\operatorname{Re}\lambda>\max(\omega_1,\mu),\ \tilde{k}(\lambda)\neq 0$. It can be easily seen that $p(\tilde{S}(\lambda)x)\leqslant\gamma p(x),\ x\in D(A),\ \operatorname{Re}\lambda>\mu,\ p\in ⊛$, by the denseness of $D(A)$ in E, the last estimate holds for all $x\in E$. Hence, the operator $I-\tilde{S}(\lambda)$ is invertible and $(I-\tilde{S}(\lambda))^{-1}x=\Sigma_{n=0}^{\infty}[C^{-1}B\tilde{a}(\lambda)(I-\tilde{a}(\lambda)A)^{-1}C]^n x,\ x\in E,\ \operatorname{Re}\lambda>\mu,\ \tilde{k}(\lambda)\neq 0$. Suppose provisionally $\operatorname{Re}\lambda>\mu$ and $\tilde{a}(\lambda)\tilde{k}(\lambda)\neq 0$. The closedness of the operator $A+B$ can be proved as follows. Let a net $(x_\tau)_{\tau\in T}$ in E satisfy $x_\tau\to x,\ \tau\to\infty$ and $(I-\tilde{a}(\lambda)(A+B))x_\tau\to y,\ \tau\to\infty$. Then a simple computation shows that $(I-\tilde{a}(\lambda)(A+B))x_\tau=(I-\tilde{S}(\lambda))(I-\tilde{a}(\lambda)A)x_\tau$, whence we may conclude that $(I-\tilde{a}(\lambda)A)x_\tau=(I-\tilde{S}(\lambda))^{-1}(I-\tilde{a}(\lambda)(A+B))x_\tau\to(I-\tilde{S}(\lambda))^{-1}y,\ \tau\to\infty$. Since $I-\tilde{a}(\lambda)A$ is closed, we infer that $x\in D(A),\ (I-\tilde{a}(\lambda)A)x=(I-\tilde{S}(\lambda))^{-1}y$ and $(I-\tilde{a}(\lambda)(A+B))x=y$. Therefore, the closedness of $A+B$ follows from that of $I-\tilde{a}(\lambda)(A+B)$. Suppose now $\operatorname{Re}\lambda>\mu$ and $\tilde{k}(\lambda)\neq 0$. Similarly as in the proof of Theorem 2.6.12, we get $\tilde{R}_B(\lambda)x=\tilde{R}(\lambda)(I-\tilde{S}(\lambda))^{-1}x,\ x\in E$, the injectiveness of $I-\tilde{a}(\lambda)(A+B)$, $R(C)\subseteq R(I-\tilde{a}(\lambda)(A+B))$ and $\tilde{k}(\lambda)(I-\tilde{a}(\lambda)(A+B))^{-1}Cx=\tilde{R}_B(\lambda)x$, $x\in E$, which implies that the conclusions of Theorem 2.6.11 continue to hold. We leave to the interested reader details concerning the possibilities of extension of perturbation results clarified in [445] to abstract Volterra equations in SCLCSs.

Remark 2.6.14. The local Hölder continuity with exponent $\sigma \in (0, 1]$ is an example of the property which is stable under perturbations described in Theorem 2.6.11, Theorem 2.6.12 and Remark 2.6.13, as indicated below. Consider the situation of Theorem 2.6.11 in which $D(A)$ is dense in E. Employing the same notation as in Remark 2.6.13, we have $p(S(t)x) \leqslant c_p Mq(x)[\int_0^t |b(t-r)| e^{\omega r}\, dr + |z| e^{\omega t}]$, $p \in ⊛$, $t \geqslant 0$, $x \in D(A)$. Suppose now that, for every $T > 0$ and $p \in ⊛$, there exist $c_{T,p} > 0$ and $h_{T,p} \in ⊛$ such that

$$p(R(t)x - R(s)x) \leqslant c_{T,p}(t-s)^\sigma h_{T,p}(x),\ x \in E,\ 0 \leqslant s < t \leqslant T.$$

Let $T > 0$ and $p \in ⊛$ be fixed. Then, for every $x \in D(A)$ and $0 \leqslant s < t \leqslant T$,

$$\begin{aligned}
&p(S(t)x - S(s)x) \\
&\leqslant c_p q\left(\int_0^s b(r)\,(R(t-r)x - R(s-r)x)\, dr + \int_0^s b(r)R(t-r)x\, dr\right) \\
&+ c_p |z| q(R(t)x - R(s)x) \\
&\leqslant c_p\left[c_{T,q}(t-s)^\sigma h_{T,p}(x)\left(\int_0^T |b(r)|\, dr + |z|\right) + Me^{\omega T} q(x)\int_t^s |b(r)|\, dr\right],
\end{aligned}$$

which implies that:

$$\begin{aligned}
&p\big((R_B * S)(t)x - (R_B * S)(s)x\big) \\
&\leqslant \frac{Me^{\mu^T}}{1-\gamma}\left(\int_0^s p(S(t-r)x - S(s-r)x)\, dr + \int_s^t p(S(t-r)x)\, dr\right) \\
&\leqslant \frac{c_p Me^{\mu^T}}{1-\gamma}\Bigg(c_{T,q}(t-s)^\sigma h_{T,p}(x)\int_0^s \int_0^T |b(v)|\, dv\, dr \\
&+ Me^{\omega T} q(x)\int_0^s \int_{s-r}^{t-r} |b(v)|\, dv\, dr + |z| c_{T,q} T(t-s)^\sigma h_{T,p}(x)\Bigg) \\
&+ \frac{c_p M^2 e^{2\mu^T} q(x)}{1-\gamma}\left(\int_s^t \int_0^{t-r} |b(t-r-v)|\, dv\, dr + |z|(t-s)\right).
\end{aligned}$$

One can simply prove that there exists $c_T > 0$ such that, for $0 \leqslant s < t \leqslant T$,

$$\int_0^s \int_{s-r}^{t-r} |b(v)|\, dv\, dr + \int_s^t \int_0^{t-r} |b(t-r-v)|\, dv\, dr \leqslant c_T (t-s)^\sigma.$$

The previous computation and the denseness of A show that there exists $b_{T,p} > 0$ such that, for every $x \in E$ and $0 \leqslant s < t \leqslant T$:

$$p(R_B(t)x - R_B(s)x) \leqslant b_{T,p}(t-s)^\sigma \max(p(x), h_{T,p}(x), q(x)). \tag{184}$$

In the case of Remark 2.6.13 we obtain that, for every $x \in D(A)$ and $0 \leqslant s < t \leqslant T$,

$$\begin{aligned}
&\frac{p(R_B(t)x - R_B(s)x)}{(t-s)^\sigma} \\
&\leqslant b_{T,p} \max\,(p(x) + p(Ax),\ h_{T,p}(x) + h_{T,p}(Ax),\ q(x) + q(Ax)).
\end{aligned}$$

Assuming additionally

$$\sup_{0\leqslant s<t\leqslant T}\frac{1}{(t-s)^{\sigma}}\sum_{j=0}^{l-1}\left(\int_0^{t-s}\left|a^{*,j+1}(r)\right|dr+\int_0^{s}\left|a^{*,j+1}(t-r)-a^{*,j+1}(s-r)\right|dr\right)<\infty,$$

then an estimate of the kind (184) holds in the case of Theorem 2.6.12.

The following corollary is an immediate consequence of Theorem 2.6.11, Theorem 2.6.12 and Remark 2.6.13.

Corollary 2.6.15. *Suppose* $M, M_1 > 0$, $\omega \geqslant 0$, $\alpha > 0$, $\beta \geqslant 0$, *A is a subgenerator of a* $(g_{\alpha}, g_{\alpha\beta+1})$*-regularized C-resolvent family* $(R(t))_{t\geqslant 0}$ *satisfying* $p(R(t)x) \leqslant Me^{\omega t}p(x)$, $x \in E$, $t \geqslant 0$, $p \in ⊛$ *and (22) with* $a(t) = g_{\alpha}(t)$ *and* $k(t) = g_{\alpha\beta+1}(t)$. *Assume exactly one of the following conditions:*

(i) $\alpha - 1 - \alpha\beta \geqslant 0$, $BCx = CBx$, $x \in D(A)$, *and* (a) $\vee$ (b), *where:*
  (a) $p(C^{-1}Bx) \leqslant M_1(x)$, $x \in \overline{D(A)}$, $p \in ⊛$.
  (b) *E is complete, (176) and (♮).*

(ii) $\alpha - 1 - \alpha\beta < 0$, $BCx = CBx$, $x \in D(A)$, $l = \lceil \frac{\alpha\beta+1-\alpha}{\alpha} \rceil$ *and (178) holds.*

Then there exist $\mu > \omega$ and $\gamma \in [0, 1)$ *such that* $A + B$ *is a subgenerator of a* $(g_{\alpha}, g_{\alpha\beta+1})$*-regularized C-resolvent family* $(R_B(t))_{t\geqslant 0}$ *satisfying* $p(R_B(t)x) \leqslant \frac{M}{1-\gamma}e^{\mu t}p(x)$, $x \in E$, $t \geqslant 0$, $p \in ⊛$, *and (22) with* $k(t)$ *replaced by* $g_{\alpha\beta+1}(t)$ *therein.*

Instructive examples of integrated (a, C)-regularized resolvent families, providing possible applications of Corollary 2.6.15, can be constructed following the analysis given in the proof of [434, Proposition 2.4].

Remark 2.6.16. Let $0 < \alpha < 2$ and let $(R(t))_{t\geqslant 0}$ be an exponentially equicontinuous, analytic $(g_{\alpha}, g_{\alpha\beta+1})$-regularized C-resolvent family of angle $\delta \in (0, \pi/2]$. Suppose additionally that, for every $\zeta \in (0, \delta)$, there exist $M_{\zeta} \geqslant 1$ and $\omega_{\zeta} \geqslant 0$ such that $p(R(z)x) \leqslant M_{\zeta}e^{\omega_{\zeta}\operatorname{Re} z}p(x)$, $x \in E$, $z \in \Sigma_{\zeta}$, $p \in ⊛$. If (ii) or (i)(a) holds, then we obtain from Corollary 2.6.15 and the proofs of Kato's analyticity criteria [89, Theorem 4.3, Theorem 4.6] that $(R_B(t))_{t\geqslant 0}$ is also an exponentially equicontinuous, analytic $(g_{\alpha}, g_{\alpha\beta+1})$-regularized C-resolvent family of angle δ; furthermore, for every $\zeta \in (0, \delta)$, there exist $M'_{\zeta} \geqslant 1$ and $\omega'_{\zeta} \geqslant 0$ such that $p(R_B(z)x) \leqslant M'_{\zeta}e^{\omega'_{\zeta}\operatorname{Re} z}p(x)$, $x \in E$, $z \in \Sigma_{\zeta}$, $p \in ⊛$. If (i)(b) holds, then one has to assume additionally that there exist $\eta > \omega$ and $\gamma \in [0, 1)$ such that, for every $\zeta \in (-\delta, \delta)$, $x \in D(A)$ and $p \in ⊛$, the following holds:

$$\int_0^{\infty} e^{-\mu t}p\left(C^{-1}B\int_0^t b(t-s)R(se^{i\zeta})x\,ds + zC^{-1}BR(te^{i\zeta})x\right)dt \leqslant \gamma p(x).$$

Example 2.6.17. Let $E := l^1$, $0 < \alpha < 1$ and $l := \lceil \frac{1-\alpha}{\alpha} \rceil$. Define a closed densely defined linear operator A_{α} on E by $D(A_{\alpha}) := \{\langle x_n\rangle \in l^1 : \Sigma_{n=1}^{\infty} n|x_n| < \infty\}$ and $A_{\alpha}\langle x_n\rangle := \langle e^{i\alpha\pi/2}nx_n\rangle$, $\langle x_n\rangle \in D(A_{\alpha})$. Then A_{α} is the integral generator of a bounded

$(g_\alpha, 1)$-regularized resolvent family, $A_\alpha + I$ is not the integral generator of an exponentially bounded $(g_\alpha, 1)$-regularized resolvent family and $\sigma(A_\alpha) = \{e^{i\alpha\pi/2}n : n \in \mathbb{N}\}$; see [49, Example 2.24]. Suppose

$$B \in L(E) \text{ and } R(B) \subseteq D(A^l) = \left\{ \langle x_n \rangle \in l^1 : \sum_{n=1}^{\infty} n^l |x_n| < \infty \right\}.$$

Then it follows from Corollary 2.6.15 that $A + B$ is the integral generator of an exponentially bounded $(g_\alpha, 1)$-regularized resolvent family.

Before we switch to the next subsection, it could be of importance to stress that the perturbation properties of (1) have been also analyzed in [463, Theorem 1.2, Theorem 2.3, Theorem 3.2].

2.6.2. Unbounded perturbation theorems. In the subsequent theorems, we shall extend the assertions of [241, Theorem 3.1, Theorem 3.3] and [292, Theorem 2.5.9, Corollary 2.5.10] to abstract Volterra equations.

Theorem 2.6.18. *Suppose E is a Banach space, k(t) and a(t) satisfy (P1)-(P2) and A is the integral generator of an exponentially bounded (a, k)-regularized resolvent family* $(R(t))_{t\geqslant 0}$ *satisfying (22) with* $C = I$. *Let* $M > 0$ *and* $\omega \geqslant 0$ *be such that* $\|R(t)\| \leqslant Me^{\omega t}$, $t \geqslant 0$, *and let* $\lambda_0 > \max(\omega, abs(a), abs(k))$ *satisfy* $\tilde{k}(\lambda)\tilde{a}(\lambda) \neq 0$, $\operatorname{Re}\lambda \geqslant \lambda_0$. *Suppose that, for every* $\varepsilon > 0$, *there exists* $C_\varepsilon > 0$ *such that:*

$$(185) \qquad \frac{1}{|\tilde{k}(\lambda)|} \leqslant C_\varepsilon e^{\varepsilon|\lambda|},\ \operatorname{Re}\lambda \geqslant \lambda_0 \text{ and } \frac{|\tilde{a}(\lambda)|}{|\tilde{a}(\lambda_0 + i\operatorname{Im}\lambda)|} \leqslant C_\varepsilon e^{\varepsilon|\lambda|},\ \operatorname{Re}\lambda \geqslant \lambda_0.$$

(i) *Let B be a linear operator, let* $D(A) \subseteq D(B)$ *and let*

$$\left\| BR\left(\frac{1}{\tilde{a}(\lambda)} : A\right) \right\| \leqslant M_\varrho |\lambda|^{-\varrho},\ \operatorname{Re}\lambda = \lambda_0$$

for some $\varrho > 0$ *and* $M_\varrho > 0$ *(for* $\varrho = 0$ *and some* $M_0 \in (0, 1)$*). Then, for every* $\zeta > 1$, $A + B$ *is the integral generator of an exponentially bounded* $(a, k * g_\zeta)$*-regularized resolvent family* $(R_B(t))_{t\geqslant 0}$ *satisfying (148) with* $k_1(t) = (k * g_\zeta)(t)$, $C = I$, *and* $g_\alpha(t)$ *replaced by* $a(t)$ *therein.*

(ii) *Let B be a densely defined linear operator, and let*

$$\left\| R\left(\frac{1}{\tilde{a}(\lambda)} : A\right) Bx \right\| \leqslant M_\varrho |\lambda|^{-\varrho} \|x\|,\ x \in D(B),\ \operatorname{Re}\lambda = \lambda_0$$

for some $\varrho > 0$ *and* $M_\varrho > 0$ *(for* $\varrho = 0$ *and some* $M_0 \in (0, 1)$*). Then there exists a closed extension D of the operator A+B such that, for every* $\zeta > 1$, *D is the integral generator of an exponentially bounded* $(a, k * g_\zeta)$*-regularized resolvent family* $(R_B(t))_{t\geqslant 0}$ *satisfying (148) with* $k_1(t) = (k * g_\zeta)(t)$, $C = I$, *and* $g_\alpha(t)$ *replaced by* $a(t)$ *therein. Furthermore, if A and A* are densely defined, then D is the part of the operator* $(A^* + B^*)$ *in E.*

Proof. By Theorem 2.1.5, $\{1/\tilde{a}(\lambda) : \operatorname{Re}\lambda > \lambda_0\} \subseteq \rho(A)$ and

(186)
$$\left\| R\left(\frac{1}{\tilde{a}(\lambda)} : A\right) \right\| \leqslant \frac{M\,|a(\lambda)|}{|\tilde{k}(\lambda)|\,(\operatorname{Re}\lambda - \omega)}, \quad \operatorname{Re}\lambda > \lambda_0.$$

Given $z \in \mathbb{C}$ with $\operatorname{Re} z > \lambda_0$, put $\lambda_z := \lambda_0 + i \operatorname{Im} z$. Then the prescribed assumptions combined with (186) imply

(187)
$$\begin{aligned}
\left\| BR\left(\frac{1}{\tilde{a}(z)} : A\right) \right\| &= \left\| BR\left(\frac{1}{\tilde{a}(\lambda_z)} : A\right)\left(I + \left(\frac{1}{\tilde{a}(\lambda_z)} - \frac{1}{\tilde{a}(z)}\right) R\left(\frac{1}{\tilde{a}(z)} : A\right)\right) \right\| \\
&\leqslant \left\| BR\left(\frac{1}{\tilde{a}(\lambda_z)} : A\right) \right\| \left[1 + \left| \frac{1}{\tilde{a}(\lambda_z)} - \frac{1}{\tilde{a}(z)} \right| \left\| R\left(\frac{1}{\tilde{a}(z)} : A\right) \right\| \right] \\
&\leqslant \frac{M_\varrho}{\lambda_0^\varrho}\left[1 + \left| \frac{1}{\tilde{a}(\lambda_z)} - \frac{1}{\tilde{a}(z)} \right| \frac{M\,|\tilde{a}(z)|}{|\tilde{k}(z)|\,(\operatorname{Re} z - \omega)} \right] \\
&\leqslant \frac{M_\varrho}{\lambda_0^\varrho} + \frac{M_\varrho M}{\lambda_0^\varrho(\lambda_0 - \omega)\,|\tilde{k}(z)|} \left| \frac{\tilde{a}(z)}{\tilde{a}(\lambda_z)} - 1 \right|.
\end{aligned}$$

Consider now the function $h : \{z \in \mathbb{C} : \operatorname{Re} z \geqslant 0\} \to L(E)$ defined by $h(z) := z^\varrho BR(\frac{1}{\tilde{a}(\lambda_0+z)} : A)$, $\operatorname{Re} z \geqslant 0$, $z \neq 0$, $h(0) := 0$ if $\varrho > 0$, and $h(0) := BR(\frac{1}{\tilde{a}(\lambda_0)} : A)$ if $\varrho = 0$. Then the function $z \mapsto h(z)$ is continuous for $\operatorname{Re} z \geqslant 0$ and analytic for $\operatorname{Re} z > 0$. Furthermore, $\|h(it)\| \leqslant M_\varrho$, $t \in \mathbb{R}$ and, by (185)–(187), we have that, for every $\varepsilon > 0$, there exists $C'_\varepsilon > 0$ such that $\|h(z)\| \leqslant C'_\varepsilon e^{\varepsilon|z|}$ for all $z \in \mathbb{C}$ with $\operatorname{Re} z \geqslant 0$. By the Phragmén-Lindelöf type theorems (cf. for instance [20, Theorem 3.9.8]), we get that $\|h(z)\| \leqslant M_\varrho$ for all $z \in \mathbb{C}$ with $\operatorname{Re} z \geqslant 0$. This, in turn, implies that there exists $a > \lambda_0$ such that $\|BR(1/\tilde{a}(\lambda) : A)\| < \frac{1}{2}$, $\operatorname{Re}\lambda \geqslant a$ if $\varrho > 0$, and $\|BR(1/\tilde{a}(\lambda) : A)\| < M_0$, $\operatorname{Re}\lambda \geqslant a$ if $\varrho = 0$. Therefore, $1/\tilde{a}(\lambda) \in \rho(A+B)$, $\operatorname{Re} \geqslant a$ and there exists $c_\varrho > 0$ such that for $\operatorname{Re}\lambda \geqslant a$:

(188)
$$\begin{aligned}
&\left\| \frac{1}{\tilde{a}(\lambda)} R\left(\frac{1}{\tilde{a}(\lambda)} : A + B\right) \right\| \\
&= \left\| \frac{1}{\tilde{a}(\lambda)} R\left(\frac{1}{\tilde{a}(\lambda)} : A\right)\left(I - BR\left(\frac{1}{\tilde{a}(\lambda)} : A\right)\right)^{-1} \right\| \leqslant \frac{c_\varrho}{|\tilde{k}(\lambda)|}.
\end{aligned}$$

The proof of (i) follows from [302, Theorem 2.7(i), Remark 2.3(v)]. If we avail ourselves of [241, Lemma 3.2], we obtain similarly the validity of (ii). □

Recall that a Banach space E has Fourier type $p \in [1, 2]$ if the Fourier transform extends to a bounded linear operator from $L^p(\mathbb{R} : E)$ to $L^q(\mathbb{R} : E)$, where $1/p + 1/q = 1$. Each Banach space E has Fourier type 1, and E^* has the same Fourier type as E. A space of the form $L^p(\Omega, \mu)$ has Fourier type $\min(p, p/p{-}1)$ and there exist examples of non-reflexive Banach spaces which do have non-trivial Fourier type.

Theorem 2.6.19. *Let E be a Banach space of Fourier type $p \in (1, 2]$.*

(i) *Let the assumptions of Theorem 2.6.18(i) hold and let $\zeta > 1/p$. Assume that at least one of the following conditions holds:*
(a) *A and A^* are densely defined, there exist $\lambda_0' > \lambda'$ and $\eta > 1$ such that*

$$\tilde{k}(\lambda) = O(|\lambda|^{\zeta-\eta}),\ \mathrm{Re}\,\lambda > \lambda_0' \text{ and } |\tilde{a}(\lambda)| = O(|\lambda|^{\zeta-\eta}),\ \mathrm{Re}\,\lambda > \lambda_0'. \tag{189}$$

(b) *A is densely defined and E is reflexive.*
(c) *$B(D(A^2)) \subseteq D(A)$ and $BAx = ABx$, $x \in D(A^2)$.*
Then $A + B$ is the integral generator of an exponentially bounded, $(a, k * g_\zeta)$-regularized resolvent family $(R_B(t))_{t\geqslant 0}$ satisfying (148) with $k_1(t) = (k * g_\zeta)(t)$, $C = I$, and $g_\alpha(t)$ replaced by $a(t)$ therein.

(ii) *Let the assumptions of Theorem 2.6.18(ii) hold and let $\zeta > 1/p$. Then there exists a closed extension D of the operator $A + B$ such that D is the integral generator of an exponentially bounded, $(a, k * g_\zeta)$-regularized resolvent family $(R_B(t))_{t\geqslant 0}$ satisfying (148) with $k_1(t) = (k*g_\zeta)(t)$, $C = I$, and $g_\alpha(t)$ replaced by $a(t)$ therein. Furthermore, if A and A^* are densely defined, then D is the part of the operator $(A^* + B^*)^*$ in E.*

Proof. Assume that (c) holds. According to (189), $R(\frac{1}{\tilde{a}(\lambda)} : A)(I - BR(\frac{1}{\tilde{a}(\lambda)} : A))^{-1} = (I - BR(\frac{1}{\tilde{a}(\lambda)} : A))^{-1} R(\frac{1}{\tilde{a}(\lambda)} : A)$, $\mathrm{Re}\,\lambda \geqslant a$ and $\frac{1}{\tilde{a}(\lambda)} R(\frac{1}{\tilde{a}(\lambda)} : A + B) = \frac{1}{\tilde{a}(\lambda)}(I - BR(\frac{1}{\tilde{a}(\lambda)} : A))^{-1}R(\frac{1}{\tilde{a}(\lambda)} : A)$, $\lambda \geqslant a$. Define

$$R_B(t)x := \frac{1}{2\pi i}\int_{a-i\infty}^{a+i\infty} e^{\lambda t}\lambda^{-\zeta}\left[\frac{\tilde{k}(\lambda)}{\tilde{a}(\lambda)} R\left(\frac{1}{\tilde{a}(\lambda)} : A + B\right)x\right] d\lambda,\ x \in E,\ t \geqslant 0. \tag{190}$$

By the first part of the proof of [241, Theorem 3.3], $A + B$ is the integral generator of an exponentially bounded, $(a, k * g_\zeta)$-regularized resolvent family $(R_B(t))_{t\geqslant 0}$ satisfying (148) with $k_1(t) = (k * g_\zeta)(t)$, $C = I$, and $g_\alpha(t)$ replaced by $a(t)$ therein. The property (148) holds in any particular case considered below and the assertion (ii) is also an unambiguous consequence of the proof of [241, Theorem 3.3]. Assume now that (b) holds. Then A^* is densely defined and, by Theorem 2.1.12(ii), $(R^*(t))_{t\geqslant 0}$ is an exponentially bounded (a, k)-regularized resolvent family with the integral generator A^*. Let q be such that $1/p + 1/q = 1$, and let $J : E \longrightarrow E^{**}$ denote the canonical embedding of E in its bidual E^{**}. Since E^* has Fourier type p and $\frac{1}{\tilde{a}(\lambda)} R(\frac{1}{\tilde{a}(\lambda)} : A+B)^* = \frac{1}{\tilde{a}(\lambda)}((I - BR(\frac{1}{\tilde{a}(\lambda)} : A))^{-1})^* R(\frac{1}{\tilde{a}(\lambda)} : A)^*$, $\mathrm{Re}\,\lambda \geqslant a$, it follows that there exists $c_1 > 0$ such that, for every $x^* \in E^*$ and $r \geqslant a$,

$$\int_{-\infty}^{\infty}\left\|\frac{\tilde{k}(r+is)}{\tilde{a}(r+is)} R\left(\frac{1}{\tilde{a}(r+is)} : A + B\right)^* x^*\right\|^q ds \leqslant c_1 \|x^*\|^q .$$

Set, for every $x^* \in E^*$ and $t \geqslant 0$,

$$R_{B,*}(t)x^* := \frac{1}{2\pi i}\int_{a-i\infty}^{a+i\infty} e^{\lambda t}\lambda^{-\zeta}\left[\frac{\tilde{k}(\lambda)}{\tilde{a}(\lambda)}R\left(\frac{1}{\tilde{a}(\lambda)} : A+B\right)^* x^*\right]d\lambda.$$

Then $(R_{B,*}(t))_{t\geqslant 0} \subseteq L(E^*)$ is strongly continuous, exponentially bounded and

(191) $$\frac{\tilde{k}(\lambda)}{\tilde{a}(\lambda)}R\left(\frac{1}{\tilde{a}(\lambda)} : A+B\right)^* x^* = \lambda^{\zeta}\int_0^{\infty} e^{-\lambda t}R_{B,*}(t)x^*\,dt,\ \operatorname{Re}\lambda > a,\ x^* \in E^*.$$

By Theorem 2.1.5, $(R_{B,*}(t))_{t\geqslant 0}$ is an $(a, k * g_\zeta)$-regularized resolvent family with the integral generator $(A + B)^*$. By Theorem 2.1.12(iii), it follows that $(R_B(t) \equiv J^{-1}R_{B,*}(t)^*J)_{t\geqslant 0}$ is an $(a, k * g_\zeta)$-regularized resolvent family with the integral generator $A + B = J^{-1}((A + B)^*)^*J$. We continue the proof by assuming that (a) holds. By (188)-(189), we easily infer that the improper integral in (190) converges absolutely for $x \in D(A)$ and

(192) $$\frac{\tilde{k}(\lambda)}{\tilde{a}(\lambda)}R\left(\frac{1}{\tilde{a}(\lambda)} : A+B\right)x = \lambda^{\zeta}\int_0^{\infty} e^{-\lambda t}R_B(t)x\,dt,\ \operatorname{Re}\lambda > a,\ x \in D(A).$$

By (191)-(192) and the uniqueness theorem for Laplace transform, we get

$$\langle R_{B,*}(t)x^*, x\rangle = \langle x^*, R_B(t)x\rangle,\ t \geqslant 0,\ x^* \in E^*,\ x \in D(A),$$

and $R_{B,*}(t)^*J_x = J_{R_B(t)x}$, $t \geqslant 0$, $x \in D(A)$. Now one can simply prove that $((R_{B,*}(t)^*)_{|E})_{t\geqslant 0}$ is an exponentially bounded, $(a, k * g_\zeta)$-regularized resolvent family with the integral generator $A + B$. □

Remark 2.6.20. (i) It is noteworthy that C. Kaiser and L. Weis analyzed in [242, Theorem 3.1] an analogue of Theorem 2.6.19 for operator semigroups in Hilbert spaces. The question whether the perturbed semigroup $(R_B(t))_{t\geqslant 0}$ is strongly continuous at $t = 0$ was answered in the affirmative by C. J. K. Batty [46]; here we would like to underline that it is not clear in which way one can transmit the assertion of [46, Theorem 1] to abstract Volterra equations.

(ii) To the author's knowledge, the denseness of $D(A^*)$ in E^* cannot be simply dropped from the formulation of (a). The main problem is that we do not know whether the mapping $t \mapsto R(t)^*x$, $t \geqslant 0$ is measurable provided $x^* \in E^* \setminus \overline{D(A^*)}$ (cf. [241, (5)-(6), p. 221; l. 7-8, p. 222] and [430, Section 3]). Notice also that the assertion (c), although practically irrelevant, may help one to better understand the proof of [241, Theorem 3.3].

(iii) Let $\alpha > 0$ and $a(t) = g_\alpha(t)$. Then the assumptions of Theorem 2.6.18 and Theorem 2.6.19[(i)(b)-(c), (ii)] hold, while the assumptions of Theorem 2.6.19(i)(a) hold provided $\zeta + \alpha > 1$.

In the following non-trivial example, we will transfer the assertion of [241, Proposition 8.1] to abstract time-fractional equations.

Example 2.6.21. ([292, Example 2.8.5(vi)], [314]) Let $1 < p < \infty$, $1/p+1/q = 1$, $k \in \mathbb{N}_0$, $0 < \beta \leqslant 2$ and $E := L^p(\mathbb{R})$. Define a closed linear operator $A_{\beta,k}$ on E by $D(A_{\beta,k}) := W^{4k+2,p}(\mathbb{R})$ and $A_{\beta,k}f := e^{i(2-\beta)\frac{\pi}{2}} f^{(4k+2)}$, $f \in D(A_{\beta,k})$. Put $Bf(x) := V(x)f^{(l)}(x)$, $x \in \mathbb{R}$ with maximal domain $D(B) := \{f \in E : V \cdot f^{(l)} \in E\}$; here $V(x)$ is a potential and $l \in \mathbb{N}_0$. Assume first that

$$V \in L^p(\mathbb{R}) \text{ and } l \leqslant \frac{1}{p}\left((4k+2)p - 1 - \frac{(4k+2)(p-1)}{\beta}\right). \tag{193}$$

Given $\mathrm{Re}\,\lambda > 0$, denote $\mu_{j,\lambda}$ $(1 \leqslant j \leqslant 2k+1)$ $(2k+1)$ solutions of the equation $\mu_{j,\lambda}^{4k+2} = \lambda^\beta e^{i(\beta\frac{\pi}{2}-\pi)}$ with $\mathrm{Re}\,\mu_{j,\lambda} > 0$. Then $D(A) \subseteq D(B)$,

$$(R(\lambda^\beta : A_{\beta,k})f)(x) = \frac{e^{i\beta\frac{\pi}{2}}}{4k+2}\int_{-\infty}^{\infty}\sum_{j=1}^{2k+1}\frac{e^{-\mu_{j,\lambda}|x-s|}}{(-\mu_{j,\lambda})^{4k+1}}f(s)\,ds,$$

provided $f \in E$, $x \in \mathbb{R}$, $\mathrm{Re}\,\lambda > 0$,

$$(BR(\lambda^\beta : A_{\beta,k})f)(x) = \frac{e^{i\beta\frac{\pi}{2}}}{4k+2}V(x)\sum_{j=1}^{2k+1}\left(\int_{-\infty}^{x}\frac{e^{-\mu_{j,\lambda}(x-s)}}{(-\mu_{j,\lambda})^{4k-l+1}}f(s)\,ds - \int_x^{\infty}\frac{e^{\mu_{j,\lambda}(x-s)}}{\mu_{j,\lambda}^{4k-l+1}}f(s)\,ds\right), \; f \in E,\; x \in \mathbb{R},\; \mathrm{Re}\,\lambda > 0,$$

$$\|R(\lambda^\beta : A_{\beta,k})\| \leqslant \left(|\lambda|^{\beta(1-\frac{1}{4k+2})}\min(\mathrm{Re}\,\mu_{1,\lambda}, \cdots, \mathrm{Re}\,\mu_{2k+1,\lambda})\right)^{-1}, \; \mathrm{Re}\,\lambda > 0, \tag{194}$$

and

$$\|BR(\lambda^\beta : A_{\beta,k})\| \leqslant \|V\|_p\left(|\lambda|^{\beta(1-\frac{l+1}{4k+2})}\min((\mathrm{Re}\,\mu_{1,\lambda})^{1/q}, \cdots, (\mathrm{Re}\,\mu_{2k+1,\lambda})^{1/q})\right)^{-1}, \tag{195}$$

provided $\mathrm{Re}\,\lambda > 0$. Furthermore, $\mathrm{Re}\,\mu_{j,\lambda} = |\lambda|^{\frac{\beta}{4k+2}}\cos(\arg(\mu_{j,\lambda}))$, $\mathrm{Re}\,\lambda > 0$, $1 \leqslant j \leqslant 2k+1$, and

$$\min\{\mathrm{Re}\,\mu_{j,\lambda} : 1 \leqslant j \leqslant 2k+1\} = |\lambda|^{\frac{\beta}{4k+2}}$$
$$\times\min\left(\cos\left(\frac{\arg(\lambda)\beta + (\beta\pi)/(2)}{4k+2} + \frac{(2k-1)\pi}{4k+2}\right), -\cos\left(\frac{\arg(\lambda)\beta + (\beta\pi)/(2)}{4k+2} + \frac{\pi}{2}\right)\right),$$

provided $\mathrm{Re}\,\lambda > 0$. The above implies that there exists a constant $c_{\beta,k} > 0$ such that

$$|\lambda|^{\frac{\beta}{4k+2}}\cos(\arg(\lambda))/\min(\mathrm{Re}\,\mu_{1,\lambda}, \cdots, \mathrm{Re}\,\mu_{2k+1,\lambda}) \leqslant c_{\beta,k}, \; \mathrm{Re}\,\lambda > 0. \tag{196}$$

Looking back at (193)-(196), it is seen that

$$\|R(\lambda^\beta : A_{\beta,k})\| = O(|\lambda|^{1-\beta}(\mathrm{Re}\,\lambda)^{-1}), \; \mathrm{Re}\,\lambda > 0 \tag{197}$$

and

$$\text{(198)} \qquad \|BR(\lambda^\beta : A_{\beta,k})\| = O\Big(\|V\|_p(\text{Re}\,\lambda)^{-\beta(1-\frac{l+1}{4k+2}+\frac{1}{(4k+2)_q})}\Big) = O\Big(\|V\|_p(\text{Re}\,\lambda)^{(-1)/q}\Big),$$

provided Re $\lambda > 0$. Denote by β_k the infimum of all non-negative real numbers $r \geqslant 0$ such that the operator $A_{\beta,k}$ generates an exponentially bounded (g_β, g_{r+1})-regularized resolvent family. The integration rate β_k is very difficult to precisely compute (cf. also the representation formula [49, Example 3.7, (3.15)] and notice that it is not clear whether Theorem 2.6.11 or Remark 2.6.13 can be applied in case $\beta \in (1, 2]$). Clearly, (197) yields the imprecise estimate $\beta_k \leqslant 1$, furthermore, $\beta_k = 0$ provided $p = 2$ ([321]), and $\beta_k \leqslant |\frac{1}{2}-\frac{1}{p}|$ provided $\beta \in \{1, 2\}$ ([223], [554]). Set $\kappa_p := \min(1/p, (p-1)/p)$. By Theorem 2.6.19, $A_{\beta,k} + B$ generates an exponentially bounded $(g_\beta, g_{\sigma_{\beta,k,p}+1})$-regularized resolvent family for any $\sigma_{\beta,k,p} > \beta_k + \kappa_p$. By (197)-(198) and the proof of [241, Proposition 8.1], the above remains true provided $(4k+2)p-1-((4k+2)(p-1)/\beta) > 0$, $l = 0$ and $V \in L^p(\mathbb{R}) + L^\infty(\mathbb{R})$; similarly, one can consider the operators $A^1_{\beta,k}$ ($k \in \mathbb{N}$, $0 < \beta \leqslant 2$) and $A^2_{\beta,k}$ ($k \in \mathbb{N}$, $0 < \beta \leqslant 1$) given by $A^1_{\beta,k}f := e^{-\beta\frac{\pi}{2}} f^{(4k)}$, $f \in W^{4k,p}(\mathbb{R}) := D(A^1_{\beta,k})$ and $A^2_{\beta,k}f := e^{\pm i\frac{\pi}{2}(1-\beta)} f^{(2k+1)}$, $f \in W^{2k+1,p}(\mathbb{R}) := D(A^2_{\beta,k})$.

Notice that *C.* Lizama and H. Prado have recently analyzed in [392] the qualitative properties of the equation (3), where E is a Banach space and $f \in L^1_{loc}([0, \infty) : E)$. By a (strong) solution of (3) we mean any function $u \in C^1([0, \infty) : E)$ such that (3) holds for a.e. $t \geqslant 0$. The following extension of [49, Theorem 2.25] (cf. also [463, p. 65]) will be helpful in the study of perturbation properties of (3).

Theorem 2.6.22. *Let $k(t)$ and $a(t)$ satisfy (P1). Suppose $\delta \in (0, \pi/2]$, $\omega \geqslant \max(0, abs(a), abs(k))$, there exist analytic functions $\hat{k} : \omega + \Sigma_{\frac{\pi}{2}+\delta} \to \mathbb{C}$ and $\hat{a} : \omega + \Sigma_{\frac{\pi}{2}+\delta} \to \mathbb{C}$ such that $\hat{k}(\lambda) = \tilde{k}(\lambda)$, Re $\lambda > \omega$, $\hat{a}(\lambda) = \tilde{a}(\lambda)$, Re $\lambda > \omega$ and $\hat{k}(\lambda)\,\hat{a}(\lambda) \neq 0$, $\lambda \in \omega + \Sigma_{\frac{\pi}{2}+\delta}$. Let A be a subgenerator of an analytic (a, k)-regularized C-resolvent family $(R(t))_{t\geqslant 0}$ of angle δ, and let (22) hold. Suppose that, for every $\eta \in (0, \delta)$, there exists $c_\eta > 0$ such that*

$$\text{(199)} \qquad p(e^{-\omega\,\text{Re}\,z}R(z)x) \leqslant c_\eta p(x),\ x \in E,\ p \in \circledast,\ z \in \Sigma_\eta,$$

as well as b, $c \geqslant 0$, B is a linear operator satisfying $D(C^{-1}AC) \subseteq D(B)$, $BCx = CBx$, $x \in D(C^{-1}AC)$ and

$$\text{(200)} \qquad p(C^{-1}Bx) \leqslant bp(C^{-1}ACx) + cp(x),\ x \in D(C^{-1}AC),\ p \in \circledast.$$

Assume that at least one of the following conditions holds:

(i) *A is densely defined, the numbers b and c are sufficiently small, there exists $|C|_\circledast > 0$ such that $p(Cx) \leqslant |C|_\circledast p(x)$, $x \in E$, $p \in \circledast$ and, for every $\eta \in (0, \delta)$,* *there exists* $\omega_\eta \geqslant \omega$ *such that* $|\hat{k}(\lambda)^{-1}| = O(|\lambda|)$, $\lambda \in \omega_\eta + \Sigma_{\frac{\pi}{2}+\eta}$ *and* $|\hat{a}(\lambda)/\hat{k}(\lambda)| = O(|\lambda|)$, $\lambda \in \omega_\eta + \Sigma_{\frac{\pi}{2}+\eta}$.

(ii) *A is densely defined, the number b is sufficiently small, there exists* $|C|_{\circledast} > 0$ *such that* $p(Cx) \leqslant |C|_{\circledast} p(x)$, $x \in E$, $p \in \circledast$ *and, for every* $\eta \in (0, \delta)$, *there exists* $\omega_\eta \geqslant \omega$ *such that* $|\hat{k}(\lambda)^{-1}| = O(|\lambda|)$, $\lambda \in \omega_\eta + \Sigma_{\frac{\pi}{2}+\eta}$ *and* $\hat{a}(\lambda)/(\lambda\hat{k}(\lambda)) \to 0$, $|\lambda| \to \infty$, $\lambda \in \omega_\eta + \Sigma_{\frac{\pi}{2}+\eta}$.

(iii) *A is densely defined, the number c is sufficiently small,* $b = 0$ *and, for every* $\eta \in (0, \delta)$, *there exists* $\omega_\eta \geqslant \omega$ *such that* $|\hat{a}(\lambda)/\hat{k}(\lambda)| = O(|\lambda|)$, $\lambda \in \omega_\eta + \Sigma_{\frac{\pi}{2}+\eta}$.

(iv) $b = 0$ *and, for every* $\eta \in (0, \delta)$, *there exists* $\omega_\eta \geqslant \omega$ *such that* $\hat{a}(\lambda)/(\lambda\hat{k}(\lambda)) \to 0$, $|\lambda| \to \infty$, $\lambda \in \omega_\eta + \Sigma_{\frac{\pi}{2}+\eta}$.

Then $C^{-1}(C^{-1}AC + B)C$ *is a subgenerator of an exponentially equicontinuous, analytic (a, k)-regularized C-resolvent family* $(R_B(t))_{t\geqslant 0}$ *of angle* δ, *which satisfies* $R_B(z)[C^{-1}(C^{-1}AC+B)C] \subseteq [C^{-1}(C^{-1}AC+B)C]R_B(z)$, $z \in \Sigma_\delta$ *and the following condition:*

$$\forall \eta \in (0, \delta)\ \exists \omega'_\eta > 0\ \exists M_\eta > 0\ \forall p \in \circledast :$$

(201) $$p(R_B(z)x) \leqslant M_\eta e^{\omega'_\eta \operatorname{Re} z} p(x),\ x \in E,\ z \in \Sigma_\eta.$$

Furthermore, in cases (iii) and (iv), the above remains true with the operator $C^{-1}(C^{-1}AC + B)C$ *replaced by* $C^{-1}AC + B$.

Proof. First of all, notice that the closedness of operator $C^{-1}AC + B$ trivially follows in cases (iii) or (iv). Secondly, it is not clear how one can prove that the operator $C^{-1}AC+B$ is closed in cases (i) or (ii). We will only prove the assertion provided that (i) holds and remark the minor modifications in case that (iv) holds. Let $\eta \in (0, \delta)$ and $\sigma \in (0, \delta)$. Clearly, $A \subseteq C^{-1}AC$, $C[C^{-1}AC] \subseteq [C^{-1}AC]C$, $C[C^{-1}AC+B] \subseteq [C^{-1}AC+B]C$, $C[C^{-1}(C^{-1}AC+B)C] \subseteq [C^{-1}(C^{-1}AC+B)C]C$ and $C^{-1}AC + B \subseteq C^{-1}(C^{-1}AC + B)C$. Invoking (199), Theorem 2.2.4 and the proof of [20, Theorem 2.6.1], we obtain that

(202) $$\lim_{\lambda \to +\infty} \lambda \frac{\hat{k}(\lambda)}{\hat{a}(\lambda)} \left(\frac{1}{\hat{a}(\lambda)} - C^{-1}AC \right)^{-1} Cx = k(0)Cx,\ x \in E$$

and that there exists $N_\eta > 0$ such that $\{\frac{1}{\hat{a}(\lambda)} : \lambda \in \omega + \Sigma_{\frac{\pi}{2}+\eta}\} \subseteq \rho_C(C^{-1}AC)$ and

(203) $$\sup_{\lambda \in \omega + \Sigma_{\frac{\pi}{2}+\eta}} p\left((\lambda - \omega) \frac{\hat{k}(\lambda)}{\hat{a}(\lambda)} \left(\frac{1}{\hat{a}(\lambda)} - C^{-1}AC \right)^{-1} Cx \right) \leqslant N_\eta p(x),\ x \in E.$$

By (200) and (203), we infer that, for every $\lambda \in \omega_\eta + \Sigma_{\frac{\pi}{2}+\eta}$, $x \in E$ and $p \in ⊛$:

$$p\left(C^{-1}B\left(\frac{1}{\hat{a}(\lambda)} - C^{-1}AC\right)^{-1} Cx\right)$$

$$\leqslant b|C|_{⊙}p(x) + b\frac{N_\eta c_\eta p(x)}{|\lambda - \omega||\hat{k}(\lambda)|} + cc_\eta \frac{N_\eta p(x)}{|\lambda - \omega|}\frac{|\hat{a}(\lambda)|}{|\hat{k}(\lambda)|},$$

which implies by the given assumption the existence of a number $\omega_\eta' > \omega_\eta$ such that $p(C^{-1}B(\frac{1}{\hat{a}(\lambda)} - C^{-1}AC)^{-1} Cx) \leqslant \sigma p(x)$, $x \in E$, $\lambda \in \omega_\eta' + \Sigma_{\frac{\pi}{2}+\eta}$, $p \in ⊛$, provided that the numbers b and c are sufficiently small; if (iv) holds, then

$$\lim_{\lambda \to +\infty} C^{-1}B\left(\frac{1}{\hat{a}(\lambda)} - C^{-1}AC\right)^{-1} Cx = 0,\ x \in E. \tag{204}$$

By making use of the same argument as in the proof of Theorem 2.6.12, it follows that, for every $\lambda \in \omega_\eta' + \Sigma_{\frac{\pi}{2}+\eta}$, $R(C) \subseteq R(\frac{1}{\hat{a}(\lambda)} - (C^{-1}AC + B)) \subseteq R(\frac{1}{\hat{a}(\lambda)} - C^{-1}(C^{-1}AC + B)C)$ as well as that the operators $\frac{1}{\hat{a}(\lambda)} - (C^{-1}AC + B)$ and $\frac{1}{\hat{a}(\lambda)} - C^{-1}(C^{-1}AC + B)C$ are injective. Moreover, for any $\lambda \in \omega_\eta' + \Sigma_{\frac{\pi}{2}+\eta}$:

$$\left(\frac{1}{\hat{a}(\lambda)} - (C^{-1}AC + B)\right)^{-1} C = \left(\frac{1}{\hat{a}(\lambda)} - C^{-1}(C^{-1}AC + B)C\right)^{-1} C$$

$$= \left(\frac{1}{\hat{a}(\lambda)} - C^{-1}AC\right)^{-1} C\left(I - C^{-1}B\left(\frac{1}{\hat{a}(\lambda)} - C^{-1}AC\right)^{-1} C\right)^{-1}.$$

We contend that the operator $C^{-1}(C^{-1}AC + B)C$ is closed. Indeed, let $(x_\tau)_{\tau\in I}$ be a net in E satisfying $x_\tau \to x$, $\tau \to \infty$ and $C^{-1}(C^{-1}AC+B)Cx_\tau \to y$, $\tau \to \infty$. Then $(\frac{1}{\tilde{a}(\lambda)} - (C^{-1}AC + B))^{-1}CC^{-1}(C^{-1}AC + B)Cx_\tau \to (\frac{1}{\tilde{a}(\lambda)} - (C^{-1}AC + B))^{-1}Cy$, $\tau \to \infty$, i.e., $-Cx_\tau + \frac{1}{\tilde{a}(\lambda)}(\frac{1}{\tilde{a}(\lambda)} - (C^{-1}AC + B))^{-1}Cx_\tau \to (\frac{1}{\tilde{a}(\lambda)} - (C^{-1}AC + B))^{-1}Cy$, $\tau \to \infty$, which simply implies $Cx \in D(C^{-1}AC + B)$ and $(C^{-1}AC + B)Cx = Cy$. Therefore, $x \in D(C^{-1}(C^{-1}AC + B)C)$, $C^{-1}(C^{-1}AC + B)Cx = y$ and $C^{-1}(C^{-1}AC + B)C$ is closed, as required. Notice that, for every $x \in E$, the analyticity of mapping

$$\lambda \mapsto \left(\frac{1}{\hat{a}(\lambda)} - C^{-1}(C^{-1}AC + B)C\right)^{-1} Cx$$

$$= \left(\frac{1}{\hat{a}(\lambda)} - C^{-1}AC\right)^{-1} C\sum_{n=0}^{\infty}\left[C^{-1}B\left(\frac{1}{\hat{a}(\lambda)} - C^{-1}AC\right)^{-1} C\right]^n x,\ \lambda \in \omega_\eta' + \Sigma_{\frac{\pi}{2}+\eta}$$

follows from Lemma 1.2.4 and the fact that an E-valued mapping is analytic if it is weakly analytic. By Theorem 2.2.5, $C^{-1}(C^{-1}AC + B)C$ is a subgenerator of an exponentially equicontinuous, analytic (a, k)-regularized C-resolvent family $(R_B(t))_{t\geqslant 0}$ of angle η and (201) holds; assuming (iv), we obtain from (203):

$$p\left((\lambda-\omega_\eta')\frac{\hat{k}(\lambda)}{\hat{a}(\lambda)}\left(\frac{1}{\hat{a}(\lambda)}-(C^{-1}AC+B)\right)^{-1}Cx\right)$$
$$=p\left(\frac{(\lambda-\omega_\eta')}{(\lambda-\omega_\eta)}(\lambda-\omega)\frac{\hat{k}(\lambda)}{\hat{a}(\lambda)}\left(\frac{1}{\hat{a}(\lambda)}-C^{-1}AC\right)^{-1}C\right.$$
$$\left.\times\left(I-C^{-1}B\left(\frac{1}{\hat{a}(\lambda)}-C^{-1}AC\right)^{-1}C\right)^{-1}x\right)$$
$$\leqslant\left(1+\frac{1}{\cos\eta}\right)\frac{N_\eta}{1-\sigma}p(x),\ x\in E,\ \lambda\in\omega_\eta'+\Sigma_{\frac{\pi}{2}+\eta}.$$

Combined with (202) and (204), the above implies

$$\lim_{\lambda\to+\infty}\lambda\frac{\hat{k}(\lambda)}{\hat{a}(\lambda)}\left(\frac{1}{\hat{a}(\lambda)}-(C^{-1}AC+B)\right)^{-1}Cx$$
$$=\lim_{\lambda\to+\infty}\lambda\frac{\hat{k}(\lambda)}{\hat{a}(\lambda)}\left(\frac{1}{\hat{a}(\lambda)}-C^{-1}AC\right)^{-1}Cx$$
$$+\lim_{\lambda\to+\infty}\lambda\frac{\hat{k}(\lambda)}{\hat{a}(\lambda)}\left(\frac{1}{\hat{a}(\lambda)}-C^{-1}AC\right)^{-1}C$$
$$\times C^{-1}B\left(\frac{1}{\hat{a}(\lambda)}-C^{-1}AC\right)^{-1}C\left(I-C^{-1}B\left(\frac{1}{\hat{a}(\lambda)}-C^{-1}AC\right)^{-1}C\right)^{-1}x$$
$$=k(0)Cx+0=k(0)Cx,\ x\in E,$$

and the proof follows again from an application of Theorem 2.2.5. □

Remark 2.6.23. The proof of Theorem 2.2.5 implies that there exists $\omega_0>0$ such that, for every $x\in E$ and for every $\lambda\in\mathbb{C}$ with $\operatorname{Re}\lambda>\omega_0$:

$$\hat{k}(\lambda)\left(I-\hat{a}(\lambda)(C^{-1}AC+B)\right)^{-1}Cx$$
$$=\hat{k}(\lambda)\left(I-\hat{a}(\lambda)C^{-1}(C^{-1}AC+B)C\right)^{-1}Cx=\int_0^\infty e^{-\lambda t}R_B(t)x\,dt. \tag{205}$$

By Theorem 2.1.5, we obtain that (148) holds with $A+B$, $k_1(t)$ and $g_\alpha(t)$ replaced respectively by $C^{-1}(C^{-1}AC+B)C$, $k(t)$ and $a(t)$ therein; clearly, the above assertion remains true with the operator $C^{-1}(C^{-1}AC+B)C$ replaced by $C^{-1}AC+B$, provided that (iii) or (iv) holds. Taking the Laplace transform, (205) simply implies that $C^{-1}(C^{-1}AC+B)C$ is, in fact, the integral generator of $(R_B(t))_{t\geqslant 0}$.

Example 2.6.24. Let $u(t)$ be a solution of (3). Set $a_\alpha(t):=\mathcal{L}^{-1}(\frac{\lambda^\alpha}{\lambda+1})(t)$, $t\geqslant 0$, $k_\alpha(t):=e^{-t}$, $t\geqslant 0$ and $\upsilon(t)=u(t)+(1*u)(t)$, $t\geqslant 0$. Then $u(t)=\upsilon(t)-(e^{-t}*\upsilon)(t)$, $t\geqslant 0$ and $\upsilon(t)=A(a_\alpha*\upsilon)(t)+(1*f)(t)$, $t\geqslant 0$, which implies that the notion of an (a_α, k)-regularized C-resolvent family is important in the study of (3). In [392], the authors mainly use the following conditions: $k(t) = k_\alpha(t)$, $C = I$ and A is the generator of a bounded analytic C_0-semigroup. Set $\delta := \min(\frac{\pi}{2}, \frac{\pi\alpha}{2(1-\alpha)})$ and assume, more generally, that for every $\eta \in (0, (\frac{\pi}{2} + \delta)(1 - \alpha))$, there exists $\omega_\eta > 0$ such that the family

(206) $$\left\{(1 + |\lambda|)^{1-r}(\lambda - A)^{-1}C : \lambda \in \omega_\eta + \Sigma_\eta\right\}$$

is equicontinuous ($r \geqslant 0$) and the mapping

(207) $$\lambda \mapsto (\lambda - A)^{-1}\, Cx,\ \lambda \in \omega_\eta + \Sigma_\eta \text{ is continuous for every fixed } x \in E.$$

Notice that (206)-(207) hold provided that A is a subgenerator of an exponentially equicontinuous r-times integrated C-semigroup $(R_r(t))_{t\geqslant 0}$; furthermore, if

(208) $$\exists M \geqslant 1\ \exists \omega \geqslant 0 : p(R_r(t)x) \leqslant Me^{\omega t}p(x),\ x \in E,\ p \in ⊛,$$

then, for every $\eta \in (0, \frac{\pi}{2})$ and $\omega_\eta > \omega$, there exists $M_\eta > 0$ such that

$$p((\lambda - A)^{-1}\, Cx) \leqslant M_\eta(1 + |\lambda|)^{r-1}p(x),\ x \in E,\ \lambda \in \omega_\eta + \Sigma_\eta,\ p \in ⊛.$$

We refer the reader to [531, Chapter 1] for examples of differential operators generating exponentially equicontinuous, r-times integrated C-semigroups satisfying (208). Assume, further, that there exist $\omega > \max(0, \operatorname{abs}(k))$ and an analytic function $\hat{k} : \omega + \Sigma_{\frac{\pi}{2}+\delta} \to \mathbb{C}$ such that $\hat{k}(\lambda) = \tilde{k}(\lambda)$, $\operatorname{Re} \lambda > \omega$, $\hat{k}(\lambda) \neq 0$, $\lambda \in \omega + \Sigma_{\frac{\pi}{2}+\delta}$ and $|\hat{k}(\lambda)| = O(|\lambda|^{-1})$, $\lambda \in \omega + \Sigma_{\frac{\pi}{2}+\delta}$. Let $\gamma \in (0, \delta)$, and let $(\frac{\pi}{2} + \gamma)(1 - \alpha) < \eta < \frac{\pi}{2}$. Then there exists a sufficiently large $\omega'_\gamma > \omega$ such that $\frac{\lambda+1}{\lambda^\alpha} = \lambda^{1-\alpha} + \lambda^{-\alpha} \in \omega_\eta + \Sigma_\eta$ for all $\lambda \in \omega'_\gamma + \Sigma_{\frac{\pi}{2}+\gamma}$, which implies with (206)-(207) and Proposition 2.1.14(iii) that the mapping $\lambda \mapsto \frac{\hat{k}(\lambda)}{\hat{a}(\lambda)}(\frac{1}{\hat{a}(\lambda)} - A)^{-1}Cx$, $\lambda \in \omega'_\gamma + \Sigma_{\frac{\pi}{2}+\gamma}$ is analytic ($x \in E$) and that, for every $\sigma \geqslant r(1 - \alpha)$, the family $\{(\lambda - \omega'_\gamma)\frac{\hat{k}(\lambda)}{|\lambda|^\sigma}\frac{1}{\hat{a}(\lambda)}(\frac{1}{\hat{a}(\lambda)} - A)^{-1}C : \lambda \in \omega'_\gamma + \Sigma_{\frac{\pi}{2}+\gamma}\}$ is equicontinuous (if (208) holds, then there exists $N_\gamma > 0$ such that $p((\lambda - \omega'_\gamma)\frac{\hat{k}(\lambda)}{|\lambda|^\sigma}\frac{1}{\hat{a}(\lambda)}(\frac{1}{\hat{a}(\lambda)} - A)^{-1}Cx) \leqslant N_\gamma p(x)$, $x \in E$, $p \in ⊛$, $\lambda \in \Sigma_\eta$). Using Theorem 2.2.5 and because η was arbitrary, we get that A is a subgenerator of an exponentially equicontinuous, analytic $(a_\alpha, k * g_\zeta)$- regularized C-resolvent family $(R(t))_{t\geqslant 0}$ of angle δ, where

$$\zeta = \begin{cases} r(1 - \alpha), & \text{if } \overline{D(A)} = E \\ > r(1 - \alpha), & \text{if } \overline{D(A)} \neq E \end{cases}$$

(if (208) holds, then for every $\eta \in (0, \delta)$ there exist $\omega_\eta > 0$ and $L_\eta > 0$ such that $p(R(z)x) \leqslant L_\eta e^{\omega_\eta \operatorname{Re} z}p(x)$, $x \in E$, $p \in ⊛$). This is a significant improvement of [392, Theorem 3.1]. In what follows, we will provide the basic information on the C-wellposedness of (3). Given $\beta \in (0, 1)$ and $T > 0$, set

$$C^0_\beta([0, T] : E) := \left\{f \in C([0, T] : E) : f(0) = 0,\ |f|_{\beta,T,p} < \infty \text{ for all } p \in ⊛\right\},$$

where

$$|f|_{\beta,T,p} := \sup_{0\leqslant s<t\leqslant T} \frac{p(f(t)-f(s))}{(t-s)^{\beta}}.$$

Let A be densely defined, let $r = 0$ and let $\beta \in (0, 1)$ be such that $C^{-1}(1 * f_{|[0,T]}) \in C^0_\beta([0, T] : E)$ for all $T > 0$. Then $\zeta = 0$ and the proof of [463, Theorem 2.4] combined with the Cauchy integral formula implies that the function

$$\upsilon(t) = R(t)C^{-1}(1 * f)(t) + \int_0^t R'(t-s)\,(C^{-1}(1 * f)(s) - C^{-1}(1 * f)(t))\,ds,\ t \geqslant 0$$

satisfies $A(a_\alpha * \upsilon)(t) = \upsilon(t) - (1 * f)(t)$, $t \geqslant 0$ and that, for every $T > 0$, one has $\upsilon_{|[0,T]} \in C^0_\beta([0, T] : E)$; in the above formula, we assume that $(R(t))_{t\geqslant 0}$ is the exponentially equicontinuous, analytic (a_α, C)-regularized resolvent family of angle δ. It is obvious that the function $t \mapsto u(t) = \upsilon(t) - (e^{-t} * \upsilon)(t)$, $t \geqslant 0$ is a unique function satisfying (3) in integrated form

(209) $$u(t) - A\,(g_{1-\alpha} * u)(t) + (1 * u)(t) = (1 * f)(t),\ t \geqslant 0,\ u(0) = 0$$

and that $u_{|[0,T]} \in C^0_\beta([0, T] : E)$ for all $T > 0$. If $x \in E$, $x \neq 0$ and $C^{-1}(1 * (f(\cdot) - x)_{|[0,T]}) \in C^0_\beta([0, T] : E)$ for all $T > 0$, then we obtain similarly the unique solution $u(t)$ of the problem

(210) $$u(t) - x - A(g_{1-\alpha} * (u(\cdot) - x))(t) + (1 * u)(t) = (1 * f)(t),\ t \geqslant 0,\ u(0) = x;$$

furthermore, $u_{|[0,T]} \in C^0_\beta([0, T] : E)$ for all $T > 0$. Since $a_\alpha \notin BV_{loc}([0, \infty))$, the above described method does not work in the case $r > 0$.

We are turning back to the case in which A is not necessarily densely defined. Let $C^{-1}f, AC^{-1}f \in L^1_{loc}([0, \infty) : E)$, and let $(R_\alpha(t))_{t\geqslant 0}$ denote the $(a_\alpha, k_\alpha * g_\zeta)$-regularized C-resolvent family with a subgenerator A. By the proofs of [392, Theorem 3.5, Corollary 3.6], it follows that, for every $x \in R(C)$, there exists a unique solution of the problem

(211) $$u(t) - A(g_{1-\alpha} * u)(t) + (1 * u)(t) = (1 * f * g_\zeta)(t) + g_{\zeta+1}(t)x,\ t \geqslant 0,$$

given by $t \mapsto u(t) = R_\alpha(t)C^{-1}x + \int_0^t R_\alpha(t-s)C^{-1}f(s)\,ds$, $t \geqslant 0$. Only after assuming some additional conditions, one can differentiate the formulae (209)-(211), obtaining in such a way (3) or its slight modification. Now we are interested in the perturbation properties of (3). Assume $r \in [0, 1]$ and A is a subgenerator of an exponentially equicontinuous, r-times integrated C-semigroup satisfying (208). Let B be a linear operator such that $D(A) \subseteq D(B)$, $BCx = CBx$, $x \in D(A)$, and let b, $c \geqslant 0$ satisfy $p(C^{-1}Bx) \leqslant bp(Ax) + cp(x)$, $x \in D(A)$, $p \in \circledast$. By Remark 2.6.23 and the proof of Theorem 2.6.22, we have the following:

(i) If $r = \zeta = 0$, b is sufficiently small and $|C|_\circledast > 0$ satisfies $p(Cx) \leqslant |C|_\circledast p(x)$, $x \in E$, $p \in \circledast$, then $C^{-1}(A + B)C$ is the integral generator of an exponentially equicontinuous, analytic (a_α, k)-regularized C-resolvent family $(R_B(t))_{t\geqslant 0}$ of angle δ (cf. [175, Chapter III] and [450, Chapter 7] for corresponding examples).

(ii) If $b = 0$, c is sufficiently small, $r = 1$ and $\zeta = 1 - \alpha$, then $A + B$, resp. $C^{-1}(A + B)C$, is a subgenerator, resp. the integral generator, of an exponentially equicontinuous, analytic $(a_\alpha, k * g_\zeta)$-regularized C-resolvent family $(R_B(t))_{t\geqslant 0}$ of angle δ.

(iii) If $b = 0$, $0 \leqslant r < 1$ and $\zeta \geqslant r(1 - \alpha)$, then $A + B$, resp. $C^{-1}(A + B)C$, is a subgenerator, resp. the integral generator, of an exponentially equicontinuous, analytic $(a_\alpha, k * g_\zeta)$-regularized C-resolvent family $(R_B(t))_{t\geqslant 0}$ of angle δ.

We continue this example by observing that A. Karczewska and C. Lizama [251] have recently analyzed the following stochastic fractional oscillation equation

(212) $$u(t) + \int_0^t (t-s)\,[A\mathbf{D}_s^\alpha u(s) + u(s)]\,ds = W(t),\ t > 0,$$

where $1 < \alpha < 2$, A is the generator of a bounded analytic C_0-semigroup on a Hilbert space H and $W(t)$ denotes an H-valued Wiener process defined on a stochastic basis $(\Omega, \mathcal{F}, P)$. The theory of (a, k)-regularized resolvent families (cf. [251, Theorem 3.1, Theorem 3.2]) is essentially applied in the study of deterministic counterpart of the equation (212) in integrated form

(213) $$u(t) + \int_0^t g_{2-\alpha}(t-s)Au(s)\,ds + \int_0^t (t-s)u(s)\,ds = \int_0^t (t-s)f(s)\,ds,\ t > 0,$$

where $f \in L^1_{loc}([0, \infty) : E)$. The equation (213) models an oscillation process with fractional damping term and after differentiation becomes, in some sense,

(214) $$u''(t) + A\mathbf{D}_t^\alpha u(t) + u(t) = f(t),\ t \geqslant 0.$$

Without any essential changes, one can consider the C-wellposedness and perturbation properties of (213).

Example 2.6.25. ([511], [79], [453])
Let $\alpha \in (0, 1)$, $m \in \mathbb{N}$, let Ω be a bounded domain in $\mathbb{R}^n$ with boundary of class C^{4m} and let $E := C^\alpha(\overline{\Omega})$. Consider the operator $A : D(A) \subseteq C^\alpha(\overline{\Omega}) \to C^\alpha(\overline{\Omega})$ given by

$$Au(x) := \sum_{|\beta|\leqslant 2m} a_\beta(x)D^\beta u(x) \text{ for all } x \in \overline{\Omega}$$

with domain $D(A) := \{u \in C^{2m+\alpha}(\overline{\Omega}) : D^\beta u_{|\partial\Omega} = 0 \text{ for all } |\beta| \leqslant m - 1\}$. Here $\beta \in \mathbb{N}_0^n$, $|\beta| = \Sigma_{i=1}^n \beta_j$, $D^\beta = \Pi_{i=1}^n (\frac{1}{i}\frac{\partial}{\partial x_i})^{\beta_i}$, and $a_\beta : \overline{\Omega} \to \mathbb{C}$ satisfy the following conditions:

(i) $a_\beta(x) \in \mathbb{R}$ for all $x \in \overline{\Omega}$ and $|\beta| = 2m$.
(ii) $a_\beta \in C^\alpha(\overline{\Omega})$ for all $|\beta| \leqslant 2m$, and
(iii) there exists $M > 0$ such that

$$M^{-1}|\xi|^{2m} \leqslant \sum_{|\beta|=2m} a_\beta(x)\xi^\beta \leqslant M|\xi|^{2m} \text{ for all } \xi \in \mathbb{R}^n \text{ and } x \in \overline{\Omega}.$$

Then there exists a sufficiently large $\sigma > 0$ such that the operator $-A_\sigma \equiv -(A+\sigma)$ satisfies $\Sigma_\omega \cup \{0\} \subseteq \rho(-A_\sigma)$ with some $\omega \in (\frac{\pi}{2}, \pi)$ and

$$\text{(215)} \qquad ||R(\lambda : -A_\sigma)|| = O\big(|\lambda|^{\frac{\alpha}{2m}-1}\big),\ \lambda \in \Sigma_\omega.$$

Notice that A is not densely defined since $D(A) \subseteq \{u \in C^\alpha(\overline{\Omega}) : u_{|\partial\Omega} = 0\}$. Let $\varsigma \in [1, \frac{2\omega}{\pi})$ and $\tau \in (\frac{\alpha}{2m}, 1)$. By (215) and Theorem 2.2.5, we get that $-A_\sigma$ is the integral generator of an exponentially bounded, analytic $(g_\varsigma, g_{\varsigma\tau+1})$-regularized resolvent family of angle $\delta = \frac{\omega}{\varsigma} - \frac{\pi}{2} \in (0, \frac{\pi}{2})$. Assume now that $B : D(B) \subseteq C^\alpha(\overline{\Omega}) \to C^\alpha(\overline{\Omega})$ is a linear operator satisfying $D(A) \subseteq D(B)$ and $||Bu|| \leqslant c||u||$, $u \in D(A)$ for some $c > 0$. Applying Theorem 2.6.22(iv), we obtain that the operator $-(A+B)$ is the integral generator of an exponentially bounded, analytic $(g_\varsigma, g_{\varsigma\tau+1})$-regularized resolvent family of angle δ. Suppose, for example, $m = n = 1$ and $\Omega = (0, 1)$. Let $\varphi, \psi \in L^1[0, 1]$, and let the operator $B : C^\alpha[0, 1] \to C^\alpha[0, 1]$ be defined by

$$Bu(x) := \int_0^x (u(x-s) - u(0))\varphi(s)\, ds + \int_0^x (u(1-x+s) - u(1))\psi(s)\, ds,\ x \in [0, 1].$$

Then B satisfies the conditions stated above since $B \in L(C^\alpha[0, 1])$ and $||Bu|| \leqslant (||\varphi||_{L^1[0,1]} + ||\psi||_{L^1[0,1]})||u||$, $u \in C^\alpha[0, 1]$. Finally, it could be interesting to construct an example in which there does not exist $\hat{B} \in L(C^\alpha[0, 1])$ such that $\hat{B}x = Bx$ for all $x \in D(A)$.

In the remaining part of this subsection, which is mainly motivated by reading of the paper [19] by W. Arendt and C. J. K. Batty, we assume that E is a Banach space. We consider rank–1 perturbations of ultradistribution semigroups and sines whose generators possess polynomially bounded resolvent; our intention is also to prove generalizations of [19, Theorem 4.3] and [160, Theorem 1.3] for abstract time-fractional equations.

Given $a \in E$, $b^* \in E^*$ and $C \in L([D(A)] : E)$, we consider the rank–1 perturbation $B \in L([D(A)] : E)$ of A given by

$$Bx := b^*(Cx)a,\ x \in D(A).$$

We also denote this operator B by ab^*C. Denote $B_\delta(a, b^*) := \{(x, y^*) \in E \times E^* : ||x - a|| \leqslant \delta, ||x^* - b^*|| \leqslant \delta\}$ $(a \in E, b^* \in E^*, \delta > 0)$.

For the sake of convenience to the reader, we will repeat the assertion of [19, Theorem 1.3].

Lemma 2.6.26. *Let A be a closed linear operator on E, let $C \in L([D(A)] : E)$ and let $\varepsilon > 0$. Assume that $\Omega_n \subseteq \rho(A)$ and $\sup_{\lambda\in\Omega_n} ||CR(\lambda : A)x|| < \infty$ for all x in a dense subset E_n of E and all $n \in \mathbb{N}$. Let $g_n : \Omega_n \to (0, \infty)$ $(n \in \mathbb{N})$. Assume that for each $(a, b^*) \in B_\varepsilon(0, 0)$ there exists $n \in \mathbb{N}$ such that $\Omega_n \subseteq \rho(A + ab^*C)$ and $||R(\lambda : A + ab^*C)|| \leqslant g_n(\lambda)$, $\lambda \in \Omega_n$. Then there exists $m \in \mathbb{N}$ such that* $\sup_{\lambda\in\Omega_m} ||CR(\lambda : A)|| < \infty$.

Let (M_p) be a sequence of positive real numbers such that $M_0 = 1$ and the conditions (M.1)-(M.3)' are fulfilled. Recall that the associated function of (M_p) is defined by $M(t) := \sup_{p\in\mathbb{N}} \ln(t^p/M_p)$, $t > 0$ and $M(0) := 0$. The function $t \mapsto M(t)$, $t \geqslant 0$ is increasing, $\lim_{t\to\infty} M(t) = \infty$ and $\lim_{t\to\infty} (M(t)/t) = 0$.

Following [286] and [292], a closed linear operator A is said to be the generator of an ultradistribution sine of (M_p)-class iff the operator $\mathcal{A} := \begin{pmatrix} 0 & I \\ A & 0 \end{pmatrix}$ generates an ultradistribution semigroup of (M_p)-class (cf. [87], [101], [286] and [412] for the notion). The following well known lemma (cf. [101, Theorem 1.5], [286, Theorem 9] and [292, Chapter 3]) will be helpful in our further work.

Lemma 2.6.27. (i) *Let A be a closed densely defined operator on E. Then A generates an ultradistribution semigroup of (M_p)-class iff there exist $l \geqslant 1$, $\alpha > 0$ and $\beta \in \mathbb{R}$ such that*

$$\Lambda_{l,\alpha,\beta} := \{\lambda \in \mathbb{C} : \operatorname{Re}\lambda \geqslant \alpha M(l|\operatorname{Im}\lambda|) + \beta\} \subseteq \rho(A) \tag{216}$$

and

$$\|R(\lambda : A)\| = O(\exp(M(l|\lambda|))),\ \lambda \in \Lambda_{l,\alpha,\beta}.$$

(ii) *Let A be a closed densely defined operator on E. Then A generates an ultradistribution sine of (M_p)-class iff there exist $l \geqslant 1$, $\alpha > 0$ and $\beta \in \mathbb{R}$ such that*

$$\{\lambda^2 : \lambda \in \Lambda_{l,\alpha,\beta}\} \subseteq \rho(A) \tag{217}$$

and

$$\|R(\lambda^2 : A)\| = O(\exp(M(l|\lambda|))),\ \lambda \in \Lambda_{l,\alpha,\beta}.$$

Theorem 2.6.28. *Let $l \geqslant 1$, $\alpha > 0$, $\beta \in \mathbb{R}$, $k \in \mathbb{N}$ and $c > 0$. Let A be a closed densely defined operator on E.*

(i) *Assume* (217) *and*

$$\|R(\lambda^2 : A)\| \leqslant c(1 + |\lambda|)^k,\ \lambda \in \Lambda_{l,\alpha,\beta}. \tag{218}$$

Let $\varepsilon > 0$ and $z \in \mathbb{C}$ be such that for each $(a, b^*) \in B_\varepsilon(0, 0)$ the operator $A+ab^*(z - A)$ generates an ultradistribution sine of (M_p)-class. Then A must be bounded.

(ii) *Assume* (216) *and*

$$\|R(\lambda : A)\| \leqslant c(1 + |\lambda|)^k,\ \lambda \in \Lambda_{l,\alpha,\beta}. \tag{219}$$

Let $\varepsilon > 0$ and $z \in \mathbb{C}$ be such that for each $(a, b^*) \in B_\varepsilon(0, 0)$ the operator $A + ab^*(z - A)$ generates an ultradistribution semigroup of (M_p)-class. Then A generates an analytic C_0-semigroup.

Proof. We will only prove the first part of the theorem. Put $\Omega_n := \{\lambda \in \mathbb{C} : \operatorname{Re}\lambda \geqslant nM(n|\operatorname{Im}\lambda|) + n\}$. Then $\Omega_n \subseteq \Lambda_{l,\alpha,\beta}$ for all $n \geqslant \max(l, \alpha, |\beta|)$. By the generalized resolvent equation, it follows that for each $x \in Y_n \equiv D(A^{\lceil k/2 \rceil + 2})$, the set $\{||R(\lambda : A)x|| : \lambda \in \Omega_n\}$ is bounded. The prescribed assumption combined with Lemma 2.6.27(ii) implies that for each $(a, b^*) \in B_\varepsilon(0, 0)$ there exist $n \in \mathbb{N}$ and a function $g_n : \Omega_n \to (0, \infty)$ such that $\Omega_n^2 := \{\lambda^2 : \lambda \in \Omega_n\} \subseteq \rho(A + ab^*(z - A))$ and $||R(\lambda^2 : A + ab^*(z - A))|| \leqslant g_n(\lambda)$, $\lambda \in \Omega_n$. By Lemma 2.6.26, we obtain $m \in \mathbb{N}$ such that $\sup_{\lambda \in \Omega_m^2} ||\lambda R(\lambda : A)|| < \infty$. Let $\xi + i\eta = \lambda \in \partial(\Omega_m^2)$. Assume $|\eta_1| \leqslant |\eta|$ and $\mu = \xi + \eta_1$. Then $\xi = (mM(m|t|) + m)^2 - t^2$ and $\eta = 2t(mM(m|t|) + m)$ for some $t \in \mathbb{R}$. Since $\lim_{\lambda \to \infty}(M(t)/t) = 0$, we easily infer that there exist $t_0 > 0$ and $L \geqslant 1$ such that, for any $|t| \geqslant t_0$:

$$||(\lambda - \mu)R(\lambda : A)|| \leqslant 2|\eta|\,||R(\lambda : A)|| \leqslant \frac{2L\,|\,\eta\,|}{|\,\lambda\,|}$$

$$\leqslant \frac{4L\,|\,t\,|\,(mM(m\,|\,t\,|)+m)}{\left(((mM(m\,|\,t\,|)+m)^2 - t^2)^2 + 4t^2(mM(m\,|\,t\,|)+m)^2\right)^{1/2}} \leqslant \frac{1}{2},$$

which implies that $R(\mu : A)$ exists and $||R(\mu : A)|| \leqslant 2||R(\lambda : A)|| \leqslant 2L/|\lambda| \leqslant 2L$. Therefore, there exists $\omega_0 > 0$ such that $||R(\cdot : A)||$ is polynomially bounded on $\{\lambda \in \mathbb{C} : \operatorname{Re} \lambda \leqslant -\omega_0\} \setminus \{\lambda^2 : \lambda \in \Lambda_{l,\alpha,\beta}\}$. The set $\{\lambda \in \mathbb{C} : \operatorname{Re} \lambda \geqslant -\omega_0\} \setminus \{\lambda^2 : \lambda \in \Lambda_{l,\alpha,\beta}\}$ is compact, which completes the proof by [19, Lemma 2.3]. □

Remark 2.6.29. (i) It is worth noting that Theorem 2.6.28(ii) is an extension of [19, Theorem 3.1], and that Theorem 2.6.28(i) is an extension of [19, Theorem 2.2] provided $k > 0$ in the formulation of this result. Consider now the situation of [19, Theorem 2.2] with A being the generator of an exponentially bounded α-times integrated cosine function ($\alpha \geqslant 0$). Then there exists $\omega_A > 0$ such that $\sup_{\lambda > \omega_A} ||\lambda^{(2-\alpha)/2} R(\lambda : A)|| < \infty$. Let $\omega > \omega_A$. Proceeding as in [292, Section 1.4], one can construct, for every $\gamma \in \mathbb{R}$, the fractional power $A_\gamma := (\omega - A)^\gamma$. Assuming $0 \leqslant \alpha < 2$ and $0 < \gamma \leqslant (2 - \alpha)/2$, we obtain from [292, Theorem 1.4.10(iii),(x)] that $D(A) \subseteq D(A_\gamma)$ and $A_\gamma \in L([D(A)], E)$, which implies that one can define the rank–1 perturbation $B_\gamma := A + ab^*A_\gamma$ of A; notice that the case $\gamma = 1$ has been already considered in Theorem 2.6.28. Obviously, $A_\gamma R(\lambda : A)x = A_{\gamma-1} R(\lambda : A)(\omega - A)x$ for all $x \in D(A)$ and $\lambda \in \rho(A)$. By the proof of [19, Theorem 2.2], one gets that there exists $n \in \mathbb{N}$ such that $\{\lambda^2 : \operatorname{Re} \lambda \geqslant n\} \subseteq \rho(A)$ and $\sup_{\operatorname{Re} \lambda \geqslant n} ||A_\gamma R(\lambda^2 : A)|| < \infty$. Unfortunately, it is not clear whether the above conclusions together with [19, Lemma 2.4] (cf. also [71, Lemma 2.3]) imply that $\sup_{\operatorname{Re} \lambda \geqslant n} ||\lambda^{2\gamma} R(\lambda^2 : A)|| < \infty$, unless $\alpha = 0$. Notice also that the assumption $\gamma = 1$ must be imposed in the case $\alpha \geqslant 2$.

(ii) In the formulation of Theorem 2.6.28(ii), resp. Theorem 2.6.28(i), we do not assume that the operator $A + ab^*(z - A)$ has polynomially bounded resolvent on the square of $\Lambda_{l,\alpha,\beta}$, resp. on $\Lambda_{l,\alpha,\beta}$. Furthermore, we may assume that the operator $A + ab^*(z - A)$ has a slightly different spectral properties (cf. [19, Remark 2.5] and the formulation of Theorem 2.6.30 below).

(iii) Given $\varepsilon \in (0, 1)$ and $C_\varepsilon > 0$, set

$$\Omega_\varepsilon := \{\lambda \in \mathbb{C} : \operatorname{Re} \lambda \geqslant \varepsilon|\lambda| + C_\varepsilon\}.$$

The proof of Theorem 2.6.28(i), resp. Theorem 2.6.28(ii), does not work any longer if, for every $\varepsilon > 0$, the estimate (218), resp. (219), holds with $\Lambda_{l,\alpha,\beta}$ replaced by Ω_{ε}. Therefore, it is not clear whether Theorem 2.6.28 can be reformulated for certain classes of hyperfunction semigroups and sines [449], [280] and [292] for more details about hyperfunction solutions of abstract differential equations.

Recall that a local (a, k)-regularized C-resolvent family $(R(t))_{t\in[0,\tau)}$ having A as a subgenerator is of class C^L if the following holds:

(i) the mapping $t \mapsto R(t)$, $t \in (0, \tau)$ is infinitely differentiable (in the uniform operator topology), and

(ii) to each compact set $K \subseteq (0, \tau)$ there exists $h_K > 0$ such that

$$\sup_{t\in K,\, p\in\mathbb{N}_0} \left\| \frac{h_K^p \frac{d^p}{dt^p} R(t)}{M_p} \right\| < \infty;$$

$(R(t))_{t\in[0,\tau)}$ is said to be ρ-hypoanalytic, $1 \leqslant \rho < \infty$, if $(R(t))_{t\in[0,\tau)}$ is of class C^L with $M_p = p!^{\rho}$.

By [236, Theorem 5.5] and [302, Theorem 2.23], a C_0-semigroup $(T(t))_{t\geqslant 0}$ is ρ-hypoanalytic for some $\rho \geqslant 1$ if $(T(t))_{t\geqslant 0}$ is in the Crandall-Pazy class of semigroups. Recall that $(T(t))_{t\geqslant 0}$ is in the Crandall-Pazy class [236] if there exist $\gamma \in (0, 1]$, $b > 0$, $k > 0$ and $c \in \mathbb{R}$ such that

(220) $E_{\gamma,b,c} := \{\lambda \in \mathbb{C} : \operatorname{Re}\lambda \geqslant c - b|\operatorname{Im}\lambda|^{\gamma}\} \subseteq \rho(A)$ and $\|R(\lambda : A)\| \leqslant c,\ \lambda \in E_{\gamma,b,c}$.

Keeping in mind (220), the subsequent theorem can be viewed as a generalization of [19, Theorem 4.3]. Observe that the operator $(\omega - A)^{\gamma}$ $(\gamma \in \mathbb{R})$ is defined for a sufficiently large $\omega > 0$, provided that A generates an exponentially bounded (g_α, g_β)-regularized resolvent family.

Theorem 2.6.30. *Suppose* $0 < \alpha < 2$, $(\alpha - 1)/\alpha < \gamma < 1$, $z \in \mathbb{C}$, $\beta \geqslant 0$ *and a densely defined operator* A *generates an exponentially bounded* (g_α, g_β)*-regularized resolvent family* $(R(t))_{t\geqslant 0}$.

(i) *Assume that* $b = 0$ *and for each* $(a, b) \in B_\varepsilon(0, 0)$ *there exists a kernel* $k_{a,b*}(t)$ *satisfying (P1)-(P2) so that the operator* $A+ab*(\omega - A)^{\gamma}$ *generates an exponentially bounded* $(g_\alpha, k_{a,b*})$*-regularized resolvent family. Then* $(R(t))_{t\geqslant 0}$ *is* $(1/(\alpha\gamma + 1 - \alpha))$*-hypoanalytic.*

(ii) *Assume that* $b > 0$ *and for each* $(a, b) \in B_\varepsilon(0, 0)$ *there exists a kernel* $k_{a,b*}(t)$ *satisfying (P1)-(P2) so that the operator* $A+ab*(z - A)$ *generates an exponentially bounded* $(g_\alpha, k_{a,b*})$*-regularized resolvent family. Then* A *generates an exponentially bounded, analytic* $(g_\alpha, 1)$*-regularized resolvent family.*

Proof. Given $\omega \geqslant 0$, set $\Phi_{\alpha,\omega} := \{\lambda^{\alpha} : \operatorname{Re}\lambda \geqslant \omega\}$. Making use of [19, Lemma 2.4], Theorem 2.1.5 and Lemma 2.6.26, we get that there exist $n \in \mathbb{N}$ and $m > 0$ such that $\Psi_{n,\alpha} := \Phi_{\alpha,n} \cap \{z \in \mathbb{C} : |z - \lambda| \leqslant m|\lambda|^{\gamma}$ for some $\lambda \in \partial(\Phi_{\alpha,n})\} \subseteq \rho(A)$ and $\|R(\lambda : A)\| = O(|\lambda|^{-\gamma})$, $\lambda \in \Phi_{n,\alpha}$. Let $\varepsilon \in (0, 1)$ and let $K_{\varepsilon} > 0$ be such that:

$$\alpha K_{\varepsilon}(1 + K_{\varepsilon})^{\alpha-1} \leqslant m. \tag{221}$$

Put $\rho := 1/(\alpha\gamma + 1 - \alpha)$. Notice that $\rho \geqslant 1$ since $\alpha > 0$ and $(\alpha - 1)/\alpha < \gamma < 1$. With the help of (221) and the Darboux inequality, we obtain that for each $\eta \in \mathbb{R}$:

$$\begin{aligned}
&|(n - K_{\varepsilon}|\eta|^{1/\rho} + i\eta)^{\alpha} - (n + i\eta)^{\alpha}| \\
&\leqslant \alpha K_{\varepsilon}|\eta|^{1/\rho} \sup_{v \in [n+i\eta, n+i\eta-K_{\varepsilon}|\eta|^{1/\rho}]} |v|^{\alpha-1} \\
&\leqslant \alpha K_{\varepsilon}|\eta|^{1/\rho}(|n+i\eta| + K_{\varepsilon}|\eta|^{1/\rho})^{\alpha-1} \\
&\leqslant \alpha K_{\varepsilon}(1 + K_{\varepsilon})^{\alpha-1}|\eta|^{1/\rho}|n+i\eta|^{\alpha-1} \\
&\leqslant \alpha K_{\varepsilon}(1 + K_{\varepsilon})^{\alpha-1}|\eta + i\eta|^{\alpha-1+(1/\rho)} \leqslant m|n + i\eta|^{\alpha\gamma},
\end{aligned}$$

which implies that $\{\lambda \in \mathbb{C} : \operatorname{Re}\lambda \geqslant -K_{\varepsilon}|\operatorname{Im}\lambda|^{1/\rho} + n\} \subseteq \Psi_{n,\alpha}$. The proof of (i) is completed by an application of [302, Theorem 2.23]. Suppose now that the assumptions of (ii) hold. Then $\omega - A$ need not be sectorial, in general. We obtain similarly the existence of an integer $n \in \mathbb{N}$ and a number $c \in (0, 1)$ such that $\Upsilon_{n,\alpha} := \Phi_{\alpha,n} \cap \{z \in \mathbb{C} : |z - \lambda| \leqslant m|\lambda|$ for some $\lambda \in \partial(\Phi_{\alpha,n})\} \subseteq \rho(A)$ and $\|R(\lambda : A)\| \leqslant c|\lambda|^{-1}$, $\lambda \in \Upsilon_{n,\alpha}$.
Then it readily follows from Theorem 2.2.5 that A generates an exponentially bounded, analytic $(g_{\alpha}, 1)$-regularized resolvent family of angle $(\arcsin c)/\alpha$. □

Now we will extend the assertion of [160, Theorem 1.3] to abstract time-fractional equations. For $a \in E$ and $b^* \in E^*$, define A_{a,b^*} by $D(A_{a,b^*}) := \{x \in E : x + \langle b^*, x\rangle a \in D(A)\}$ and $A_{a,b^*}x := A(x + \langle b^*, x\rangle a)$, $x \in D(A_{a,b^*})$.

We need the following auxiliary lemma (cf. the proofs and formulations of [160, Lemma 2.1, Lemma 2.2]).

Lemma 2.6.31. *Let $\omega \geqslant 0$, $\alpha \in (0, 2)$, $a \in E$, $b^* \in E^*$ and $z \in \mathbb{C}_{\omega}$, where $\mathbb{C}_{\omega} := \{z \in \mathbb{C} : \operatorname{Re} z > \omega\}$.*

- (i) *Then z^{α} is an eigenvalue of both, A_{a,b^*} and $A + ab^*A$, with $A_{a,b^*}(AR(z^{\alpha} : A)a) = z^{\alpha}AR(z^{\alpha} : A)a$ and $(A + ab^*A)(R(z^{\alpha} : A)a) = z^{\alpha}R(z^{\alpha} : A)a$.*
- (ii) *Let $\langle b^*, AR(z^{\alpha} : A)a\rangle = 1$ and $\langle b^*, AR(z^{\alpha} : A)a\rangle \neq 0$. Then for each $\varepsilon > 0$ there exists $\delta > 0$ such that for all $(a_1, b_1^*) \in B_{\delta}(a, b^*)$ there exists some $z_1 \in B(z, \varepsilon)$ such that $\langle b_1^*, AR(z_1^{\alpha} : A)a_1\rangle = 1$ and $\langle b_1^*, AR(z_1^{\alpha} : A)a_1\rangle \neq 0$.*

The following fractional analogue of [160, Lemma 2.3] will be essentially utilized in the proof of Theorem 2.6.33 stated below.

Lemma 2.6.32. *Suppose $\alpha \in (0, 2)$, $n \in \mathbb{N}$, $\omega \geqslant 0$, $\varepsilon > 0$ and A is the generator of an exponentially bounded, non-analytic $(g_{\alpha}, 1)$-regularized resolvent family $(S_{\alpha}(t))_{t \geqslant 0}$ satisfying $\|S_{\alpha}(t)\| \leqslant Me^{\omega t}$, $t \geqslant 0$ for some $M \geqslant 1$. Let $r > \omega$, $k = \lceil \frac{1}{\alpha} \rceil$, $a \in D(A^k)$, b^**

$\in D((A^*)^k)$ *and* $z_j \in \mathbb{C}_\omega$ *be such that* $\langle b^*, AR(z_j^\alpha : A)a_1\rangle = 1$ *and* $\langle b^*, AR(z_j^\alpha : A)a\rangle \neq 0$ $(1 \leqslant j \leqslant n)$. *Then there exist* $(a_1, b_1^*) \in B_\varepsilon(a, b^*) \cap (D(A^k) \times D((A^*)^k))$ *and* ${}_1z, \cdot\cdot\cdot, {}_{n+1}z \in \mathbb{C}_\omega$ *such that* $\mathrm{Re}({}_{n+1}z) = r$; $|{}_jz - z_j| < \varepsilon$ $(1 \leqslant j \leqslant n)$, $\langle b_1^*, AR(({}_{n+1}z)^\alpha : A)a_1\rangle = 1$ *and* $\langle b_1^*, AR(({}_{n+1}z)^\alpha : A)a_1\rangle \neq 0$.

Proof. We will only outline the main details of the proof. First of all, notice that $\|R(z^\alpha : A^*)\| = \|R(z^\alpha : A)\| \leqslant M/((\mathrm{Re}\, z - \omega)|z|^{\alpha-1})$, $\mathrm{Re}\, z > \omega$. By [292, Lemma 1.4.10], we get that A^* is stationary dense ([332]) with $n(A^*) \leqslant 1$, which implies that $\overline{D((A^*)^k)} = \overline{D(A^*)}$ in the strong topology of E^*. Let the numbers $\delta_1 > 0, \cdot\cdot\cdot, \delta_n > 0$ be given by Lemma 2.6.31(i) and let $\delta := \min(1, \varepsilon, \delta_1, \cdot\cdot\cdot, \delta_n)$. By the generalized resolvent equation and the fact that $* : L(E) \to L(E^*)$ is an isometrically isomorphism, we obtain that for each $(a, b^*) \in D(A^k) \times D((A^*)^k)$, the following supremum

$$\sup_{\mathrm{Re}\, z=r} \max\bigl(1, \|A^*R(z^\alpha : A^*)b^*\|, \|AR(z^\alpha : A)a\|, |\langle b^*, AR(z^\alpha : A)a\rangle|\bigr) := K$$

is finite. The non-analyticity of $(S_\alpha(t))_{t\geqslant 0}$ yields that $\sup_{\mathrm{Re}\, z=r} \|AR(z^\alpha : A)\| = \infty$. By the denseness of $D(A^k)$ in E, we get the existence of an element $u \in D(A^k)$ and a complex number ${}_{n+1}z := z$ such that $\mathrm{Re}\, z = r$, $\|AR(z^\alpha : A)u\| > \frac{10K}{\delta^2}$, $AR(z^\alpha : A)^2u \neq 0$ and $\|u\| < 1$. Now one can proceed as in the proof of [160, Lemma 2.3] so as to obtain the existence of a functional $v^* \in D(A^*)$ such that $|\langle v^*, AR(z^\alpha : A)u\rangle| > \frac{10K}{\delta^2}$, $\|v^*\| < 1$ and $\langle b^* + \frac{\delta}{2}v^*, AR(z^\alpha : A)^2u\rangle \neq 0$. Since $D((A^*)^k)$ is dense in $D(A^*)$ with respect to the strong topology of E^*, we may assume that $v^* \in D((A^*)^k)$. Copying the final part of the proof of the aforementioned lemma, with $AR(z : A)$ and $AR(z : A)^2$ replaced by $AR(z^\alpha : A)$ and $AR(z^\alpha : A)^2$ there, we obtain that there exists $(a_1, b_1^*) \in B_\varepsilon(a, b^*) \cap (D(A^k) \times D((A^*)^k))$ with required properties (cf. [160, p. 474, l. 1-4]). □

If $\alpha \in (0, 2)$ and A is the generator of an exponentially bounded $(g_\alpha, 1)$-regularized resolvent family $(S_\alpha(t))_{t\geqslant 0}$ satisfying the properties stated above, then one can simply prove that for each $r > \omega$ there exist $z \in \mathbb{C}_\omega$ with $\mathrm{Re}\, z = r$ and $(a, b^*) \in D(A^k) \times D((A^*)^k)$ such that $\langle b^*, AR(z^\alpha : A)a\rangle = 1$ and $\langle b^*, AR(z^\alpha : A)^2a\rangle \neq 0$. Using induction, Lemma 2.6.32 and the proof of [160, Theorem 1.3], we can simply prove the validity of the following theorem.

Theorem 2.6.33. *Suppose* $\alpha \in (0, 2)$, $n \in \mathbb{N}$, $\omega \geqslant 0$, $\varepsilon > 0$ *and* A *is the generator of an exponentially bounded, non-analytic* $(g_\alpha, 1)$*-regularized resolvent family* $(S_\alpha(t))_{t\geqslant 0}$ *satisfying* $\|S_\alpha(t)\| \leqslant Me^{\omega t}$, $t \geqslant 0$ *for some* $M \geqslant 1$. *Let* $I_j \subseteq (\omega^\alpha, \infty)$ *be an open interval* $(1 \leqslant j < \infty)$. *Then there exist* $a \in E$ *and* $b^* \in E^*$ *such that the operators* A_{a,b^*} *and* $A + ab^*A$ *have a sequence of eigenvalues* z_j *with* $\mathrm{Re}\, z_j \in I_j$ *for all* $j = 1, 2, \ldots$.

2.6.3. Time-dependent perturbations of abstract Volterra equations. Albeit a great part of results presented here can be formulated in the setting of Fréchet spaces, we shall assume henceforth for the sake of convenience, that $(E, \|\cdot\|)$ is a non-trivial complex Banach space. We shall work only with (a, C)-regularized resolvent families because our results cannot be so easily reword if we assume that the operator family $(V(t))_{t\in[0,\tau)}$ appearing in the formulation of condition (H)' below, is a general (a, k)-regularized C-resolvent family.

Multiplicative (time-dependent) perturbations of abstract Volterra equations have been considered in [83], [302], [353]-[356], [385], [533], [536]-[537] and [541]. We start by strengthening results on multiplicative time-dependent Desch-Schappacher type perturbations of abstract Volterra equations established by T.-J Xiao, J. Liang and J. van Casteren in [536], the paper of fundamental importance in our analysis. Keeping in mind the arguments given in the above-mentioned paper, the proofs of subsequent assertions become technical and are therefore omitted.

In this subsection, the notion of Sobolev space $W^{1,1}([0, T] : E)$ will be understood in the following sense. We define $W^{1,1}([0, T] : E) := \{f \in L^1([0, T] : E) : f(s) = f(s_0) + \int_{s_0}^{s} g(\sigma)\, d\sigma$ for some $s_0 \in [0, T]$ and $g \in L^1([0, T] : E)\}$. Our standing hypothesis will be:

(H)': A is a subgenerator of an (a, C)-regularized resolvent family $(V(t))_{t\in[0,\tau)}$ such that:

$$V(t)x = Cx + A\int_0^t a(t-s)V(s)x\, ds,\ t \in [0, \tau),\ x \in E, \tag{222}$$

where $0 < \tau \leqslant \infty$. Unless stated otherwise, we assume that $T \in (0, \tau)$.

Theorem 2.6.34. *Assume (H)' holds, $a \in BV[0, T]$, $G_0 \in L(C([0; T] : E))$, $G = CG_0 + I$ and the following conditions*:

(a) $G_0(\psi) \in W^{1,1}([0, T] : E)$ *for all* $\psi \in C^1([0, T] : E)$.

(b) $\|G_0(\psi)(t)\| \leqslant M_0 \sup_{0\leqslant s\leqslant t}\|\psi(s)\|$, $\psi \in C([0, T] : E)$, $t \in [0, T]$, *for an appropriate constant* $M_0 > 0$.

(c) *For every* $\psi \in C([0, T] : E)$, $\int_0^t \tilde{V}(t-\sigma)G_0(\psi)(\sigma)\, d\sigma \in D(A)$ *and there exists* $M > 0$ *such that, for every* $t \in [0, T]$ and $\psi \in C([0, T] : E)$,

$$\left\| A\int_0^t \tilde{V}(t-\sigma)G_0(\psi)(\sigma)\, d\sigma \right\| \leqslant M\int_0^t \sup_{0\leqslant s\leqslant\sigma}\|\psi(s)\|\, d\sigma, \tag{223}$$

where $\tilde{V}(\sigma)x := a(0)V(\sigma)x + \int_0^\sigma V(\sigma-\tau)x\, da(\tau)$, $\sigma \in [0, T]$, $x \in E$.

Then the following holds:

(i) *If* $C^{-1}f \in W^{1,1}([0, T] : [D(A)])$, *then there exists a unique solution* $\mathcal{V}_f \in C([0, T] : [D(A)])$ *of the integral equation*

$$\mathcal{V}(t) = f(t) + \int_0^t a(t-s)G(A\mathcal{V})(s)\, ds,\ t \in [0, T], \tag{224}$$

which is given by $\mathcal{V}_f(t) := \sum_{m=0}^{\infty} \upsilon_m(t)$, $t \in [0, T]$, *where*

$$\upsilon_0(t) := V(t)C^{-1}f(0) + \int_0^t V(t-s)(C^{-1}f)'(s)\, ds,\ t \in [0, T]$$

and

$$v_m(t) := \int_0^t \widetilde{V}(t-s)G_0(Av_{m-1})(s)\,ds,\ m \in \mathbb{N},\ t \in [0, T]. \tag{225}$$

(ii) *If* $C^{-1}f \in W^{1,1}([0, T] : E)$, *then there exists a unique solution* $\mathcal{W}_f \in C([0, T] : E)$ *of the integral equation*

$$\mathcal{W}(t) = f(t) + A\int_0^t a(t-s)G(\mathcal{W})(s)\,ds,\ t \in [0, T], \tag{226}$$

which is given by $\mathcal{W}_f(t) := \Sigma_{m=0}^{\infty} w_m(t)$, $t \in [0, T]$, *where* $w_0(t) := v_0(t)$, $t \in [0, T]$ *and*

$$w_m(t) := A\int_0^t \widetilde{V}(t-s)G_0(w_{m-1})(s)\,ds,\ m \in \mathbb{N},\ t \in [0, T].$$

Corollary 2.6.35. *Assume (H)' holds,* $a \in BV[0, T]$, *the function* $B_0 : [0, T] \to L(E)$ *is strongly continuously differentiable and* $B(\sigma) := CB_0(\sigma) + I$, $\sigma \in [0, T]$. *Suppose* $M > 0$, $\int_0^t \widetilde{V}(t-\sigma)B_0(\sigma)\psi(\sigma)\,d\sigma \in D(A)$, $\psi \in C([0, T] : E)$ *and (223) holds with* $G_0(\psi)(\cdot)$ *replaced by* $B_0(\cdot)\psi(\cdot)$. *Then (224) has a unique solution* $\mathcal{V}_f$ *provided* $C^{-1}f \in W^{1,1}([0, T] : [D(A)])$ *and (226) holds a unique solution* $\mathcal{W}_f$ *provided* $C^{-1}f \in W^{1,1}([0, T] : E)$.

Suppose $0 < \varepsilon < T < \tau$ and $a(t) > 0$, $t \in (0, \varepsilon)$. Then the Favard class of $(V(t))_{t\in[0,\tau)}$ is defined by

$$F_V := \left\{ x \in E : \overline{\lim}_{t\to 0+} \left\| \left(\int_0^t a(s)\,ds \right)^{-1} (V(t)x - Cx) \right\| < \infty \right\}.$$

Equipped with the norm

$$||x||_{F_V} := ||x|| + \overline{\lim}_{t\to 0+} \left\| \left(\int_0^t a(s)\,ds \right)^{-1} (V(t)x - Cx) \right\|,$$

the Favard class F_V becomes a Banach space (see e.g. [302, (2.51), Theorem 2.26]).

Corollary 2.6.36. *Assume (H)' holds,* $\varepsilon \in (0, T)$ *and* $a \in BV[0, T]$.

(i) *Let* $B_0 : [0, T] \to L(E : [D(A)])$ *be strongly continuous, or*
(ii) *Let* $a(t) > 0$, $t \in (0, \varepsilon)$, *let* $B_0 : [0, T] \to L(E : F_V)$ *be strongly continuous and let* $a(t) - \alpha t^k = o(t^k)$ $(t \to 0+)$ *for certain* $k \in \mathbb{N}_0$ *and* $\alpha \neq 0$.

Then the conclusions of Corollary 2.6.35 hold.

Corollary 2.6.37. *Assume (H)' holds and* $\varepsilon \in (0, T)$.

(i) *Let* $B_0 : [0, T] \to L(E : [D(A)])$ *be strongly measurable and* $||B_0||_{E\to[D(A)]} \in L^{\infty}[0, T]$, *or*

(ii) *Let* $a(t) > 0$, $t \in (0, \varepsilon)$, *let* $B_0 : [0, T] \to L(E : F_V)$ *be strongly measurable,* $\|B_0\|_{E\to F_V} \in L^\infty[0, T]$, *and let* $a(t) - \alpha t^k = o(t^k)$ $(t \to 0+)$ *for certain* $k \in \mathbb{N}_0$ *and* $\alpha \neq 0$.

If $a \in AC[0, T]$, *then the conclusions of Corollary 2.6.35 hold.*

The following corollary is an extension of [83, Theorem 2.2], [353, Theorem 2.1], [537, Theorem 2.1], [536, Corollary 2.6] and [541, Theorem 3] (cf. also [354, Theorem 2.3]). The existence of a unique strongly continuous operator family $(V_{B,C}(t))_{t\in[0,\tau)}$ satisfying $A(I+B)\int_0^t a(t-s)V_{B,C}(s)x\,ds = V_{B,C}(t)x - Cx$, $t \in [0, \tau)$, $x \in E$ can be proved even if the condition $C_1A(I+B) \subseteq A(I+B)C_1$ is not included in the analysis; in such a way, we can prove an extension of [537, Theorem 2.3]. Notice also that the condition (227) holds provided $R(C^{-1}B) \subseteq D(A)$.

Corollary 2.6.38. *Let (H)' hold and let* $a \in BV_{loc}([0, \tau))$. *Suppose* $B \in L(E)$, $R(B) \subseteq R(C)$, *there exists an injective operator* $C_1 \in L(E)$ *satisfying* $R(C_1) \subseteq R(C)$, $C_1A(I + B) \subseteq A(I + B)C_1$ *and, for every* $T \in (0, \tau)$, *there exists* $M_T > 0$ *such that, for every* $\psi \in C([0, T] : E)$,

$$\left\| A\int_0^t \tilde{V}(t-\sigma)C^{-1}B\psi(\sigma)\,d\sigma \right\| \leqslant M_T\int_0^t \sup_{0\leqslant s\leqslant\sigma}\|\psi(s)\|\,d\sigma. \tag{227}$$

Then $A(I + B)$ *is a subgenerator of an* (a, C_1)*-regularized resolvent family* $(VB(t))_{t\in[0,\tau)}$ *satisfying*

$$V_B(t)x = V(t)C^{-1}C_1x + A\int_0^t \tilde{V}(t-s)C^{-1}BV_B(s)x\,ds,\ x \in E,\ t \in [0, \tau),$$

and (222) with A, $(V(t))_{t\in[0,\tau)}$ *and* C *replaced by* $A(I + B)$, $(V_B(t))_{t\in[0,\tau)}$ *and* C_1, *respectively. Furthermore, if* $\rho((I + B)A) \neq \emptyset$ *and* $BC_1 = C_1B$, *then* $(I + B)A$ *is a subgenerator of an* (a, C_1)*-regularized resolvent family* $(V_B^1(t))_{t\in[0,\tau)}$ *satisfying (222) with* A, $(V(t))_{t\in[0,\tau)}$ *and* C *replaced by* $(I + B)A$, $(V_B^1(t))_{t\in[0,\tau)}$ *and* C_1, *respectively.*

The following is an insignificant modification of [537, Example 2.10].

Example 2.6.39. Let Ω be a bounded domain in $\mathbb{R}^3$ with a smooth boundary, let $\alpha \geqslant 1$ and let the Dirichlet Laplacian $A := \Delta$ on $E := L^2(\Omega)$ be defined by $D(A) := H^2(\Omega) \cap H_0^1(\Omega)$ and $Af := \Delta f$, $f \in D(A)$. Assume $\gamma \in (0, \frac{\pi}{2})$, $d \in (0, 1]$ and $\frac{1}{\alpha} < \beta < \frac{\pi}{2\gamma}$. Denote by Γ_γ the boundary of $\Sigma_\gamma \cup \{z \in \mathbb{C} : |z| \leqslant d\}$ and assume that Γ_γ is oriented in such a way that $\operatorname{Im}\lambda$ decreases along Γ_γ. Define, for every $\varepsilon > 0$,

$$S_\varepsilon(t)f := \frac{1}{2\pi i}\int_{\Gamma_\gamma} E_\alpha(t^\alpha\lambda)e^{-\varepsilon\lambda^\beta}(\lambda + A)^{-1}f\,d\lambda,\ f \in E,\ t \geqslant 0.$$

Then one can simply prove that, for every $\varepsilon > 0$, $(S_\varepsilon(t))_{t\geqslant 0}$ is a global (not exponentially bounded) $(g_\alpha, S_\varepsilon(0))$-regularized resolvent family with a subgenerator $-A$ and $R(S_\varepsilon(0))$ is dense in E. Let $n \in \mathbb{N}$, $\lambda_i \in (-\infty, 0)$, g_i, $\omega_i \in E$, $Ag_i = \lambda_i g_i$ $(1 \leqslant i \leqslant n)$,

$$B_0 u := \sum_{i=1}^{n} (u, \omega_i)_{L^2(\Omega)} g_i \text{ and } B := A^{-1}B_0.$$

Then $R(S_\varepsilon(0)^{-1}B_0) \subseteq D(A)$, $A(I + B) = A + B_0$ and $R(B_0) \subseteq R(S_\varepsilon(0)^{-1})$ for all $\varepsilon > 0$. Applying Corollary 2.6.38 we get that, for every $\alpha \geqslant 1$ and $\varepsilon > 0$, there exists a unique strongly continuous operator family $(V_{B,\varepsilon}(t))_{t\geqslant 0}$ satisfying $(A + B_0) \int_0^t g_\alpha(t-s)V_{B,\varepsilon}(s)f\, ds = V_{B,\varepsilon}(t)f - S_\varepsilon(0)f$, $t \geqslant 0$, $f \in E$. Assume now $\varepsilon > 0$, $x_i \in D(A)$ and $Ax_i \in R(S_\varepsilon(0))$ $(0 \leqslant i \leqslant \lceil \alpha \rceil)$. Define

$$u_\varepsilon(t) := \sum_{i=0}^{\lceil \alpha \rceil - 1} \left[\frac{t^i}{i!} x_i + \int_0^t \frac{(t-s)^{\alpha+i-1}}{\Gamma(\alpha+i)} S_\varepsilon(s) S_\varepsilon(0)^{-1} Ax_i \right] ds, \ t \geqslant 0.$$

Keeping in mind that $S_\varepsilon(t)f = \Sigma_{n=0}^{\infty} (t^{\alpha n}/\Gamma(\alpha n+1))\, A^n S_\varepsilon(0)f$, $t \geqslant 0$, $f \in E$, it readily follows that $u_\varepsilon(t)$ is a unique solution of problem (61) with $f(t) \equiv 0$.

In a similar manner, one can prove the following results on time-dependent additive perturbations of integral Volterra equations (cf. [536, Section 3]).

Theorem 2.6.40. *Assume that the condition (H)' holds, $a \in BV[0, T]$, $G_0 \in L(C([0, T] : [D(A)]), C([0, T] : E))$ and the following conditions hold:*

(a) *$G_0(\psi) \in W^{1,1}([0, T] : E)$ for all $\psi \in C^1([0, T] : E)$.*
(b) *$\|G_0(\psi)(t)\| \leqslant M_0 \sup_{0\leqslant s\leqslant t} \|\psi(s)\|_{[D(A)]}$, $\psi \in C([0, T] : [D(A)])$, $t \in [0, T]$, for an appropriate constant $M_0 > 0$.*
(c) *For every $\psi \in C([0, T] : [D(A)])$, $\int_0^t \tilde{V}(t-\sigma)G_0(\psi)(\sigma)\, d\sigma \in D(A)$ and there exists $M > 0$ such that, for every $t \in [0, T]$ and $\psi \in C([0, T] : E)$,*

$$\left\| A \int_0^t \tilde{V}(t-\sigma)G_0(\psi)(\sigma)\, d\sigma \right\| \leqslant M \int_0^t \sup_{0\leqslant s\leqslant \sigma} \|\psi(s)\|_{[D(A)]}\, d\sigma. \tag{228}$$

If $C^{-1}f \in W^{1,1}([0, T] : E)$, then the integral equation

$$u(t) = f(t) + \int_0^t a(t-s)\big(Au(s) + CG_0(u)(s)\big)\, ds,\ t \in [0, T], \tag{229}$$

has a unique solution in $C([0, T] : [D(A)])$, which is given by $u(t) = \Sigma_{m=0}^{\infty} v_m(t)$, $t \in [0, T]$, where $v_m(t)$ $(m \in \mathbb{N}$, $t \in [0, T])$ is given by replacing $G_0(Av_{m-1})$ in (225) with $G_0(v_{m-1})$.

Example 2.6.41. Assume (H)' holds, $a, b \in BV[0, T]$, $C^{-1}B_1 : [0, T] \to L([D(A)] : E)$ is strongly continuous and $G_0(\psi)(t) = (b * C^{-1}B_1\psi)(t)$, $t \in [0, T]$, $\psi \in C([0, T] : [D(A)])$. If $C^{-1}f \in W^{1,1}([0, T] : [D(A)])$, then the integral equation

$$u(t) = f(t) + (a * (Au + b * B_1 u))(t),\ t \in [0, \mathrm{T}],$$

has a unique solution in $C([0, T] : [D(A)])$.

Corollary 2.6.42. *Assume (H)' holds, $a \in BV[0, T]$, $M > 0$ and $B_0 : [0, T] \to L([D(A)] : E)$ is strongly continuously differentiable. If $C^{-1}f \in W^{1,1}([0, T] : [D(A)])$,*

$\int_0^t \widetilde{V}(t-s)B_0(s)\,\psi(s)\,ds \in D(A)$, $t \in [0, T]$, $\psi \in C([0, T] : [D(A)])$ *and (228) holds with* $G_0(\cdot)$ *replaced by* $B_0(\cdot)$, *then the integral equation (229), with* $CG_0(\cdot)$ *replaced by* $CB_0(\cdot)$ *therein, has a unique solution in* $C([0, T] : [D(A)])$, *which is given by* $u(t) = \Sigma_{m=0}^{\infty} \upsilon_m(t)$, $t \in [0, T]$, *where* $\upsilon_m(t)$ $(m \in \mathbb{N},\ t \in [0, T])$ *is given by replacing* $G_0(A\upsilon_{m-1})(s)$ *in (225) by* $B_0(s)\upsilon_{m-1}(s)$.

Corollary 2.6.43. *Assume (H)' holds,* $a \in BV[0, T]$ *and:*

(i) $B_0 : [0, T] \to L([D(A)])$ *is strongly continuous, or*
(ii) *There exists* $\varepsilon \in (0, T)$ *such that* $a(t) > 0$, $t \in (0, \varepsilon)$, $B_0 : [0, T] \to L([D(A)] : F_V)$ *is strongly continuous, and* $a(t) - \alpha t^k = o(t^k)$ $(t \to 0+)$ *for certain* $k \in \mathbb{N}_0$ *and* $\alpha \neq 0$.

Then the conclusions of Corollary 2.6.42 hold.

Corollary 2.6.44. *Assume (H)' holds and:*

(i) $B_0 : [0, T] \to L([D(A)])$ *is strongly measurable and* $||B(\cdot)||_{L([D(A)])} \in L^{\infty}[0, T]$, *or*
(ii) *There exists* $\varepsilon \in (0, T)$ *such that* $a(t) > 0$, $t \in (0, \varepsilon)$, $B_0 : [0, T] \to L([D(A)] : F_V)$ *is strongly measurable,* $||B(\cdot)||_{L([D(A)]:F_V)} \in L^{\infty}[0, T]$, *and* $a(t) - \alpha t^k = o(t^k)$ $(t \to 0+)$ *for certain* $k \in \mathbb{N}_0$ *and* $\alpha \neq 0$.

Then the conclusions of Corollary 2.6.42 hold provided $a \in AC[0, T]$.

Assuming $s = 0$, the following corollary can be simply formulated and proved for fractional resolvent families.

Corollary 2.6.45. (i) *Assume A is a subgenerator of a C-regularized semigroup* $(V(t))_{t\in[0,\tau)}$, $C^{-1}B : [0, T] \to L([D(A)] : F_V)$ *is strongly measurable,* $||C^{-1}B(\cdot)||_{L([D(A)]:F_V)} \in L^{\infty}[0, T]$ *and* $B(\cdot)x \in C([0, T] : E)$, $x \in D(A)$. *Then, for every* $s \in [0, T]$ *and* $x \in D(A)$, *the following initial value problem*

$$\begin{cases} u'(t) = (A + B(t))u(t),\ t \in [s, T], \\ u(s) = Cx, \end{cases}$$

has a unique solution $\mathcal{U}(\cdot, s) \in C^1([s, t] : E) \cap C([s, t] : [D(A)])$, *which is given by* $\mathcal{U}(t, s) := \Sigma_{m=0}^{\infty} u_m(t, s)x$, $s \leqslant t \leqslant T$, *where* $u_0(t, s)x := V(t-s)x$, $s \leqslant t \leqslant T$ *and* $u_m(t, s) := \int_0^t V(t-\sigma)C^{-1}B(\sigma)u_{m-1}(\sigma, s)\,d\sigma$, $m \in \mathbb{N}$, $s \leqslant t \leqslant T$.

(ii) *Assume A is a subgenerator of a C-regularized cosine function* $(V(t))_{t\in[0,\tau)}$, $C^{-1}B : [0, T] \to L([D(A)] : F_V)$ *is strongly measurable,* $||C^{-1}B(\cdot)||_{L([D(A)]:F_V)} \in L^{\infty}[0, T]$ *and* $B(\cdot)x \in C([0, T] : E)$, $x \in D(A)$. *Then, for every* $s \in [0, T]$ *and* $x, y \in D(A)$, *the following initial value problem*

$$\begin{cases} u''(t) = (A + B(t))u(t),\ t \in [s, T], \\ u(s) = Cx,\ u'(s) = Cy, \end{cases}$$

has a unique solution $\mathcal{C}(\cdot, s) \in C^2([s, t] : E) \cap C([s, t] : [D(A)])$, *which is given by* $\mathcal{C}(t, s) := \Sigma_{m=0}^{\infty}(c_m(t, s)x + s_m(t, s)y)$, $s \leqslant t \leqslant T$, *where*

$$\begin{cases} s_0(t,s)x := \int_0^{t-s} V(\sigma)x\,d\sigma,\ 0 \leqslant s \leqslant t \leqslant T, \\ s_m(t,s)x := \int_s^t s_0(t,\sigma)C^{-1}B(\sigma)s_{m-1}(\sigma,s)x\,d\sigma,\ m \in \mathbb{N},\ 0 \leqslant s \leqslant t \leqslant T \end{cases}$$

and

$$\begin{cases} c_0(t,s)x := V(t-s)x,\ s \leqslant t \leqslant T, \\ c_m(t,s)x := \int_s^t s_0(t,\sigma)C^{-1}B(\sigma)c_{m-1}(\sigma,s)x\,d\sigma,\ m \in \mathbb{N},\ 0 \leqslant s \leqslant t \leqslant T. \end{cases}$$

The subsequent theorem is closely related to [49, Theorem 2.26] and can be applied to (coercive) differential operators considered in Section 2.5.

Theorem 2.6.46. *Suppose $\alpha > 1$, $M \geqslant 1$, $\omega \geqslant 0$ and A is a subgenerator of a (local) (g_α, C)-regularized resolvent family $(S_\alpha(t))_{t\in[0,\tau)}$ satisfying $\|S_\alpha(t)\| \leqslant Me^{\omega t}$, $t \in [0,\tau)$, and (222) with $V(\cdot)$ and $a(t)$ replaced by $S_\alpha(\cdot)$ and g_α, respectively.*

(i) *Let $(B(t))_{t\in[0,\tau)} \subseteq L(E)$, $R(B(t)) \subseteq R(C)$, $t \in [0,\tau)$ and $C^{-1}B(\cdot) \in C([0,\tau) : L(E))$. If $C^{-1}f \in W^{1,1}_{loc}([0,\tau) : E)$, then there exists a unique solution of the integral equation*

$$u(t,f) = f(t) + A\int_0^t g_\alpha(t-s)u(s,f)\,ds + \int_0^t g_\alpha(t-s)B(s)u(s,f)\,ds \tag{230}$$

in $C([0,\tau) : E)$. The solution $u(t,f)$ is given by $u(t,f) := \Sigma_{n=0}^{\infty} S_{\alpha,n}(t)$, $t \in [0,\tau)$, where we define $S_{\alpha,n}(t)$ $(t \in [0,\tau))$ recursively by $S_{\alpha,0}(t) := v_0(t)$ (cf. the formulation of Theorem 2.6.34) and

$$S_{\alpha,n}(t) := \int_0^t \int_0^{t-\sigma} g_{\alpha-1}(t-\sigma-s)S_\alpha(s)C^{-1}B(\sigma)S_{\alpha,n-1}(\sigma)\,ds\,d\sigma.$$

Denote, for every $T \in (0,\tau)$, $K_T := \max_{t\in[0,T]} \|C^{-1}B(t)\|$ and $F_T := \|C^{-1}f(0)\| + \int_0^T e^{-\omega s}\|(C^{-1}f)'(s)\|\,ds$. Then

$$\|u(t,f)\| \leqslant Me^{\omega t}E_\alpha(MK_T t^\alpha)F_T,\ t \in [0,T] \tag{231}$$

and

$$\|u(t,f) - v_0(t)\| \leqslant Me^{\omega t}\big(E_\alpha(MK_T t^\alpha) - 1\big)F_T,\ t \in [0,T]. \tag{232}$$

(ii) *Let $(B(t))_{t\in[0,\tau)} \subseteq L([D(A)])$ be strongly continuous and let $C^{-1}B(\cdot) \in C([0,\tau) : L([D(A)]))$. If $C^{-1}f \in W^{1,1}_{loc}([0,\tau) : [D(A)])$, then there exists a unique solution of the integral equation (230) in $C([0,\tau) : [D(A)])$. Denote, for every $T \in (0,\tau)$, $K_{T,A} := \max_{t\in[0,T]} \|C^{-1}B(t)\|_{L([D(A)])}$ and $F_{T,A} := \|C^{-1}f(0)\|_{[D(A)]} + \int_0^T e^{-\omega s}\|(C^{-1}f)'(s)\|_{[D(A)]}\,ds$. Then*

$$\|u(t,f)\|_{[D(A)]} \leqslant Me^{\omega t}E_\alpha(MK_{T,A}t^\alpha)F_{T,A},\ t \in [0,T]$$

and

$$\|u(t,f) - v_0(t)\|_{[D(A)]} \leqslant Me^{\omega t}\big(E_\alpha(MK_{T,A}\, t^\alpha) - 1\big)F_{T,A},\ t \in [0, T].$$

Proof. We will only prove the first part of theorem. Inductively, we obtain that $\|S_{\alpha,n}(t)\| \leqslant M^{n+1}K_T^n F_T e^{\omega t} g_{\alpha n+1}(t)$, $t \in [0, T]$, $n \in \mathbb{N}_0$, which implies that the series $\Sigma_{n=0}^{\infty} S_{\alpha,n}(t)$ converges uniformly on compact subsets of $[0, \tau)$ and (231)-(232) hold. Clearly, $u(t,f) = v_0(t) + \int_0^t (g_{\alpha-2} * S_\alpha)(t-s)C^{-1}B(s)u(s,f)\, ds$, $t \in [0, T]$. This implies

$$u(t,f) = f(t) + A\int_0^t g_\alpha(t-s)v_0(s)\, ds + \big[g_{\alpha-2} * S_\alpha * C^{-1}B(\cdot)u(\cdot,f)\big](t)$$

$$= f(t) + A\int_0^t g_\alpha(t-s)\big[u(s,f) - (g_{\alpha-2} * S_\alpha * C^{-1}B(\cdot)u(\cdot,f))(s)\big]\, ds$$

$$+ \big[g_{\alpha-2} * S_\alpha * C^{-1}B(\cdot)u(\cdot,f)\big](t)$$

$$= f(t) + A\int_0^t g_\alpha(t-s)u(s,f)\, ds$$

$$- A\int_0^t g_\alpha(t-s)\,(g_{\alpha-2} * S_\alpha * C^{-1}B(\cdot)u(\cdot,f))(s)\, ds$$

$$+ \big[g_{\alpha-2} * S_\alpha * C^{-1}B(\cdot)u(\cdot,f)\big](t)$$

$$= f(t) + A\int_0^t g_\alpha(t-s)u(s,f)\, ds - \big[g_{\alpha-1} * (S_\alpha(\cdot) - C) * C^{-1}Bu(\cdot)u(\cdot,f)\big](t)$$

$$+ \big[g_{\alpha-2} * S_\alpha * C^{-1}B(\cdot)u(\cdot,f)\big](t)$$

$$= f(t) + A\int_0^t g_\alpha(t-s)u(s,f)\, ds + \int_0^t g_\alpha(t-s)B(s)u(s,f)\, ds,\ t \in [0, \tau).$$

Therefore, $u(t, x)$ is a solution of (230). The uniqueness of solutions is left to the reader as an easy exercise. □

Theorem 2.6.47. ([317])

(i) *Assume* $L^1_{loc}([0, \tau)) \ni a$ *is a kernel, (H)' holds,*

$$A(t) = a(t)A + (a * B_1)(t) + B_0(t),\ t \in [0, \tau),$$

where $B_0(\cdot)$ *and* $B_1(\cdot)$ *satisfy the following conditions* $(B_0(t))_{t\in[0,\tau)} \subseteq L([D(A)]) \cap L(E : [R(C)])$, $(B_1(t))_{t\in[0,\tau)} \subseteq L([D(A)] : [R(C)])$,

(a) $C^{-1}B_0(\cdot)y \in BV_{loc}([0, \tau) : [D(A)])$ *for all* $y \in D(A)$, $C^{-1}B_0(\cdot)x \in BV_{loc}([0, \tau) : E)$ *for all* $x \in E$,

(b) $C^{-1}B_1(\cdot)y \in BV_{loc}([0, \tau) : E)$ *for all* $y \in D(A)$, *and*

(c) $CB(t)y = B(t)Cy$, $y \in D(A)$, $t \in [0, \tau)$.

Then there exists an a-regular A-regularized C-resolvent family $(R(t))_{t\in[0,\tau)}$.

(ii) *Assume A is a subgenerator of a C-regularized semigroup $(S(t))_{t\in[0,\tau)}$. If $B_0(\cdot)$ and $B_1(\cdot)$ satisfy the assumptions stated in (i), then for every $x \in D(A)$ there exists a unique solution of the problem*

$$\begin{cases} u \in C^1([0, \tau) : E) \cap C([0, \tau) : [D(A)]), \\ u'(t) = Au(t) + (dB_0 * u)(t)x + (B_1 * u)(t) + Cx, \ t \in [0, \tau), \\ u(0) = 0. \end{cases}$$

Furthermore, the mapping $t \mapsto u(t)$, $t \in [0, \tau)$ is locally Lipschitz continuous in $[D(A)]$.

(iii) *Assume A is a subgenerator of a C-regularized cosine function $(C(t))_{t\in[0,\tau)}$. If $B_0(\cdot)$ and $B_1(\cdot)$ satisfy the assumptions stated in (i), then for every $x \in D(A)$ there exists a unique solution of the problem*

$$\begin{cases} u \in C^2([0, \tau) : E) \cap C([0, \tau) : [D(A)]), \\ u''(t) = Au(t) + (dB_0 * u')(t)x + (B_1 * u)(t) + Cx, \ t \in [0, \tau), \\ u(0) = u'(0) = 0. \end{cases}$$

Furthermore, the mapping $t \mapsto u(t)$, $t \in [0, \tau)$ is continuously differentiable in $[D(A)]$ and the mapping $t \mapsto u'(t)$, $t \in [0, \tau)$ is locally Lipschitz continuous in $[D(A)]$.

Before proceeding further, we would like to note that the existing theory of time-dependent perturbations for abstract evolution equations of second order ([370], [482]) leans heavily on the notion of Kisyński's space [276]. Contrary to Corollary 2.6.45, the results obtained in the aforementioned papers cannot be proved for abstract time-fractional equations without undergoing further non-trivial analyses.

We recall the following result from [302].

Proposition 2.6.48. *Let $B \in L(E)$ and $BC = CB$.*

(i) *Assume BA is a subgenerator of an (a, k)-regularized C-resolvent family $(R(t))_{t\in[0,\tau)}$ satisfying (22) with A replaced by BA. Then AB is a subgenerator of an (a, k)-regularized C-resolvent family $(R(t))_{t\in[0,\tau)}$.*

(ii) *Assume AB is a subgenerator of an (a, k)-regularized C-resolvent family $(R(t))_{t\in[0,\tau)}$ satisfying (22) with A replaced by AB. Then BA is a subgenerator of an (a, k)-regularized C-resolvent family, provided $\rho(BA) \neq \emptyset$.*

V. Keyantuo and M. Warma analyzed in [266] the generation of fractionally integrated cosine functions in L^p-spaces by elliptic differential operators with variable coefficients. Notice that Proposition 2.6.48(i) can be applied to these operators (cf. [266, Theorem 2.2 and pp. 78-79] and [469, Example 3.1] for more details). Finally, we want to refer the reader to [158], [179], [198], [229], [252], [330], [335]-[336], [368], [408], [422], [457] and [484] for more details about perturbation properties of abstract differential equations.

2.7 Approximation and convergence of (*a*, *k*)-regularized *C*-resolvent families

The approximation theory covers a great deal of mathematical territory nowdays and we can freely say that there is no monograph which would be able to provide a fairly complete information on approximation and convergence of abstract differential equations, even those of first order. The main theme of our investigation in this section is the approximation and convergence of (a, k)-regularized C-resolvent families. We start by stating the following theorem.

Theorem 2.7.1. *Assume that, for every* $n \in \mathbb{N}_0$, $a_n(t)$ *and* $k_n(t)$ *satisfy (P1) and that* A_n *is a subgenerator of an* (a_n, k_n)*-regularized* C_n*-resolvent family* $(R_n(t))_{t\geqslant 0}$ *which satisfies (22) with* $a(t)$, $(R(t))_{t\geqslant 0}$ *and* $k(t)$ *replaced respectively by* $a_n(t)$, $(R_n(t))_{t\geqslant 0}$ *and* $k_n(t)$ $(n \in \mathbb{N}_0)$. *Assume further that there exists* $\omega \geqslant \sup_{n\in\mathbb{N}_0} \max(0, abs(a_n), abs(k_n))$ *such that, for every* $p \in \circledast$, *there exist* $c_p > 0$ *and* $r_p \in \circledast$ *with*

$$p(e^{-\omega t}R_n(t)x) \leqslant c_p r_p(x), \quad t \geqslant 0, x \in E, n \in \mathbb{N}_0. \tag{233}$$

Let $\lambda_0 \geqslant \omega$. *Put* $\mathfrak{T} := \{\lambda > \lambda_0 : \tilde{k}_n(\lambda) \neq 0$ *for all* $n \in \mathbb{N}_0\}$. *Then the following assertions are equivalent.*

(i) $\lim_{n\to\infty} \tilde{k}_n(\lambda)(I - \tilde{a}_n(\lambda)A_n)^{-1}C_n x = \tilde{k}(\lambda)(I - \tilde{a}(\lambda)A)^{-1}Cx$, $\lambda \in \mathfrak{T}$, $x \in E$ *and the sequence* $(R_n(t)x)_n$ *is equicontinuous at each point* $t \geqslant 0$ $(x \in E)$.
(ii) $\lim_{n\to\infty} R_n(t)x = R(t)x$, $t \geqslant 0$, $x \in E$, *uniformly on compacts of* $[0, \infty)$.

Proof. It is clear that, for every $\lambda \in \mathfrak{T}$, the operator $I - \tilde{a}_n(\lambda)A_n$ is injective, $R(C_n) \subseteq R(I - \tilde{a}_n(\lambda)A_n)$, and

$$\tilde{k}_n(\lambda)(I - \tilde{a}_n(\lambda)A_n)^{-1} C_n x = \int_0^\infty e^{-\lambda t} R_n(t)x\, dt, \; x \in E, \lambda \in \mathfrak{T}, n \in \mathbb{N}_0.$$

Furthermore, it is very simple to prove that the set $\mathfrak{T}$ is dense in (λ_0, ∞). By these facts, it readily follows that

$$\lim_{n\to\infty} \int_0^\infty e^{-\lambda t} R_n(t)x\, dt = \int_0^\infty e^{-\lambda t} R(t)x\, dt, \; x \in E, \lambda > \lambda_0.$$

Now the required assertion follows immediately from an application of Theorem 1.2.6. □

One of the main results of this section is given as follows.

Theorem 2.7.2. *Assume that, for every* $n \in \mathbb{N}_0$, $a_n(t)$ *and* $k_n(t)$ *satisfy (P1) and that* A_n *is a subgenerator of an* (a_n, k_n)*-regularized* C_n*-resolvent family* $(R_n(t))_{t\geqslant 0}$ *which satisfies (22) with* $a(t)$, $(R(t))_{t\geqslant 0}$ *and* $k(t)$ *replaced respectively by* $a_n(t)$, $(R_n(t))_{t\geqslant 0}$ *and* $k_n(t)$ $(n \in \mathbb{N}_0)$. *Assume further that there exists* $\omega \geqslant \sup_{n\in\mathbb{N}_0} \max(0, abs(a_n), abs(k_n))$ *such that, for every* $p \in \circledast$, *there exist* $c_p > 0$ *and* $r_p \in \circledast$ *so that (233) holds. Let* $\lambda_0 \geqslant \omega$ *and* $\mathfrak{T} = \{\lambda > \lambda_0 : \tilde{k}_n(\lambda) \neq 0$ *for all* $n \in \mathbb{N}_0\}$. *Set* $\mathfrak{T}_0 := \{\lambda > \lambda_0 : \tilde{a}_n(\lambda)\tilde{k}_n(\lambda) \neq 0$ *for all* $n \in \mathbb{N}_0\}$. *Assume that the following conditions hold*:

(i) *The sequence* $(k_n(t))_n$ *is equicontinuous at each point* $t \geqslant 0$.

(ii) *For every convergent sequence* $(x_n)_{n\in\mathbb{N}}$ *in E, we have* $\sup_{n\in\mathbb{N}} p(C_n x_n) < \infty$.

(iii) *There exists* $\lambda' \in \mathfrak{T}_0$ *such that* $R((\frac{1}{\tilde{a}(\lambda')} - A)^{-1}C)$ *is dense in E as well as that the sequences* $(\tilde{k}_n(\lambda')\,\tilde{a}_n(\lambda')^{-1})_{n\in\mathbb{N}}$ *and* $(\tilde{a}_n(\lambda')^{-1})_{n\in\mathbb{N}}$ *are bounded.*

(iv) *For every* $\varepsilon > 0$ and $t \geqslant 0$, *there exist* $\delta \in (0, 1)$ *and* $n_0 \in \mathbb{N}$ *such that*

$$\int\limits_0^{\min(t,s)} |a_n(\max(t, s) - r) - a_n(\min(t, s) - r)|\ dr + \int\limits_{\min(t,s)}^{\max(t,s)} |a_n(\max(t, s) - r)|\ dr < \varepsilon,$$

provided $|t - s| < \delta$, $s \geqslant 0$ *and* $n \geqslant n_0$.

Then, to say that

$$\lim_{n\to\infty} \tilde{k}_n(\lambda)(I - \tilde{a}_n(\lambda)A_n)^{-1}\,C_n x = \tilde{k}(\lambda)(I - \tilde{a}(\lambda)A)^{-1}Cx,\ \lambda \in \mathfrak{T}, x \in E \tag{234}$$

is equivalent to saying that $\lim_{n\to\infty} R_n(t)x = R(t)x$, $t \geqslant 0$, $x \in E$, *uniformly on compacts of* $[0, \infty)$.

Proof. Put $H_n(\lambda) := \tilde{k}_n(\lambda)(I - \tilde{a}_n(\lambda)A_n)^{-1}\,C_n$ $(n \in \mathbb{N}_0, \lambda \in \mathfrak{T})$. By Theorem 2.7.1 and the density of $R(H(\lambda'))$ in E, it suffices to show that the sequence $(R_n(t)H(\lambda')x)_{n\in\mathbb{N}}$ is equicontinuous at each point $t \geqslant 0$ $(x \in E)$. Since

$$\begin{aligned} R_n(t)H(\lambda')\,x - R_n(s)H(\lambda')x \\ &= [R_n(t)H(\lambda')x - R_n(t)H_n(\lambda')x] + [R_n(t)H_n(\lambda')x - R_n(s)H_n(\lambda')x] \\ &+ [R_n(s)H_n(\lambda')x - R_n(s)H(\lambda')x], \end{aligned}$$

the functional equation of (a, k)-regularized C-resolvent families combined with (233) implies that:

$$\begin{aligned} &p(R_n(t)H(\lambda')x - R_n(s)H(\lambda')x) \leqslant 2c_p e^{\omega(t+1)} r_p(H_n(\lambda')x - H(\lambda')x) \\ &\quad + |k_n(t) - k_n(s)| p(C_n H_n(\lambda')x) \\ &\quad + p\left(\int_0^t a_n(t - r)R_n(r)A_n H_n(\lambda')x\,dr - \int_0^s a_n(s - r)R_n(r)A_n H_n(\lambda')x\,dr\right) \\ &\quad \leqslant 2c_p e^{\omega(t+1)} r_p(H_n(\lambda')x - H(\lambda')x) + |k_n(t) - k_n(s)|\,p(C_n H_n(\lambda')x) \\ &\quad + c_p e^{\omega(t+1)} \Bigg[\int\limits_0^{\min(t,s)} \big|a_n(\max(t, s) - r) - a_n(\min(t, s) - r)\big| r_p(A_n H_n(\lambda')x)\,dr \\ &\quad + \int\limits_{\min(t,s)}^{\max(t,s)} |a_n(\max(t, s) - r)|\, r_p(A_n H_n(\lambda')x)\,dr\Bigg], \end{aligned}$$

for any $x \in E$ and $t, s \geqslant 0$ with $|t - s| \leqslant 1$. Clearly, the sequence $(H_n(\lambda')x)_{n\in\mathbb{N}}$ is convergent; since the sequences $(\tilde{k}_n(\lambda')\,\tilde{a}_n(\lambda')^{-1})_{n\in\mathbb{N}}$ and $(\tilde{a}_n(\lambda')^{-1})_{n\in\mathbb{N}}$ are bounded, the resolvent equation implies that the sequence $(r_p(A_n H_n(\lambda')x))_{n\in\mathbb{N}}$ is bounded, too. Now the assertion of theorem simply follows from the conditions (i)-(iv).

Remark 2.7.3. (i) Suppose that the assumptions (i), (ii) and (iv) quoted in the formulation of Theorem 2.7.2 hold, and that the assumption (iii) does not hold. Denote by $\mathfrak{T}_0'$ the set which consists of those numbers $\lambda' \in \mathfrak{T}_0$ such that the sequences $\overline{(\tilde{k}_n(\lambda')\tilde{a}_n(\lambda')^{-1})_{n\in\mathbb{N}} \text{ and } (\tilde{a}_n(\lambda')^{-1})_{n\in\mathbb{N}}}$ are bounded. Then (234) implies that, for every $x \in \bigcup_{\lambda\in\mathfrak{T}_0'} R((I-\tilde{a}(\lambda)A)^{-1}C)$, we have that $\lim_{n\to\infty} R_n(t)x = R(t)x$, $t \geqslant 0$, uniformly on compacts of $[0, \infty)$. Taken together with this observation, Theorem 2.7.2 provides an extension of [337, Proposition 3.4].

(ii) It is very simple to ascertain that the conditions (i), (iii) and (iv) hold provided that $R(C)$ and $D(A)$ are dense in E as well as that $k_n(t) = k(t)$ and $a_n(t) = g_{\alpha_n}(t)$, where $(\alpha_n)_{n\in\mathbb{N}}$ is a sequence of positive real numbers with $\lim_{n\to\infty} \alpha_n = \alpha > 0$. Therefore, Theorem 2.7.2 provides a proper extension of [364, Theorem 4.2].

(iii) Since

$$\int_0^{\min(t,s)} \left|a_n(\max(t,s)-r)-a_n(\min(t,s)-r)\right| dr + \int_{\min(t,s)}^{\max(t,s)} \left|a_n(\max(t,s)-r)\right| dr$$

$$\leqslant \int_0^{\min(t,s)} \left|a(\max(t,s)-r)-a(\min(t,s)-r)\right| dr + \int_{\min(t,s)}^{\max(t,s)} \left|a(\max(t,s)-r)\right| dr$$

$$+2\int_0^{\min(t,s)} \left|a_n(r)-a(r)\right| dr, \quad n\in\mathbb{N},\ t,s \geqslant 0,$$

the condition

$$\lim_{n\to\infty} \int_0^t |a_n(r)-a(r)|\, dr = 0, \quad t \geqslant 0,$$

implies the validity of (iv).

(iv) In [319], the authors have considered the well-posedness of equation (2) with $C = I$, $\alpha_1 > 0$, $\alpha = 0$, $A_j = c_j I$, $j \in \mathbb{N}_{n-1}$ $(c_j \in \mathbb{C})$ and A being densely defined. Set

$$a(t) := \mathcal{L}^{-1}\left(\frac{1}{\lambda^{\alpha_n} + \sum_{j=1}^{n-1} c_j \lambda^{\alpha_j}}\right)(t),\ t > 0,$$

and

$$k(t) := \mathcal{L}^{-1}\left(\frac{\lambda^{\alpha_n - 1}}{\lambda^{\alpha_n} + \sum_{j=1}^{n-1} c_j \lambda^{\alpha_j}}\right)(t),\ t \geqslant 0.$$

Using Theorem 1.2.5, we may conclude that $k(t)$ is an exponentially bounded continuous function on $[0, \infty)$ with $k(0) = 1$ as well as that the functions $a(t)$ and $k(t)$ can be analytically extended to the sector $\Sigma_{\pi/2}$. Moreover, it is very simple to prove with the help of [319, Theorem 3.2] and Theorem 2.1.5 that a resolvent family $(R(t))_{t\geqslant 0}$ for (2), introduced in [319, Definition 2.2], is just an (a, k)-regularized resolvent family with $a(t)$ and $k(t)$ defined above, and that Theorem 2.7.2 provides a proper extension of [319, Theorem 5.1]; avoiding detailed explanation we only remark that our interest here can also be the question whether the assertion of [319, Theorem 5.2] can be derived from Theorem 2.7.2 and Theorem 2.7.5 below. Notice finally that the proofs of [364, Theorem 4.2] and [319, Theorem 5.1] are much more complicated than that of Theorem 2.7.2.

The following extension of [364, Theorem 4.4] can be also proved with the help of Theorem 2.7.2:

Theorem 2.7.4. *Suppose $\alpha > 0$, $\beta \geqslant 1$, A is a subgenerator of an exponentially equicontinuous (g_α, g_β)-regularized C-resolvent family $(R(t))_{t\geqslant 0}$ satisfying (22) with $a(t) = g_\alpha(t)$ and $k(t) = g_\beta(t)$, and $\overline{R(C)} = \overline{D(A)} = E$. Let $(\alpha_n)_{n\in\mathbb{N}}$ be an increasing sequence of positive real numbers with $\lim_{n\to\infty} \alpha_n = \alpha$, and let $\gamma_n = \alpha_n/\alpha$ $(n \in \mathbb{N})$. Then, for every $n \in \mathbb{N}$, the operator A is a subgenerator of an exponentially equicontinuous $(g_{\alpha_n}, g_{1+\gamma_n(\beta-1)})$-regularized C-resolvent family $(R_n(t))_{t\geqslant 0}$ satisfying (22) with $a(t) = g_{\alpha_n}(t)$, $k(t) = g_{1+\gamma_n(\beta-1)}(t)$ and $(R(t))_{t\geqslant 0}$ replaced by $(R_n(t))_{t\geqslant 0}$. Furthermore, $\lim_{n\to\infty} R_n(t)x = R(t)x$, $t \geqslant 0$, $x \in E$, uniformly on compacts of $[0, \infty)$.*

Proof. Suppose that, for every $p \in ⊛$, there exist $c_p > 0$ and $r_p \in ⊛$ such that $p(e^{-\omega t}R(t)x) \leqslant c_p r_p(x)$, $t \geqslant 0$, $x \in E$. Then the use of Theorem 2.4.2 implies that, for every $n \in \mathbb{N}$, the operator A is a subgenerator of an exponentially equicontinuous $(g_{\alpha_n}, g_{1+\gamma_n(\beta-1)})$-regularized C-resolvent family $(R_n(t))_{t\geqslant 0}$. The condition (22) holds for $(R_n(t))_{t\geqslant 0}$ because it is locally equicontinuous and $D(A)$ is dense in E (cf. (H1)). By the proof of Theorem 2.4.2, we get that there exists an absolute constant $M \geqslant 1$ such that, for every $p \in ⊛$ and for every $n \in \mathbb{N}$,

$$p(e^{-\omega^{\alpha/\alpha_n}t}R_n(t)x) \leqslant c_p r_p(x), \quad t \geqslant 0, x \in E.$$

Since $\lim_{n\to\infty} g_{1+\gamma_n(\beta-1)}(t) = g_\beta(t)$, $t \geqslant 0$, uniformly on compacts of $[0, \infty)$, the result immediately follows from an application of Theorem 1.2.6. □

Theorem 2.7.5. *Assume that, for every $n \in \mathbb{N}_0$, $a_n(t)$ and $k_n(t)$ satisfy (P1) and that A is a subgenerator of an (a_n, k_n)-regularized C_n-resolvent family $(R_n(t))_{t\geqslant 0}$ which satisfies (22) with $a(t)$, $(R(t))_{t\geqslant 0}$ and $k(t)$ replaced respectively by $a_n(t)$, $(R_n(t))_{t\geqslant 0}$ and $k_n(t)$ $(n \in \mathbb{N}_0)$. Assume further that there exists $\omega \geqslant \sup_{n\in\mathbb{N}_0} \max(0, abs(a_n), abs(k_n))$ such that, for every $p \in ⊛$, there exist $c_p > 0$ and $r_p \in ⊛$ such that (233) holds. Let $\lambda_0 \geqslant \omega$, and let $\mathfrak{T} = \{\lambda > \lambda_0 : \tilde{k}_n(\lambda) \neq 0$ for all $n \in \mathbb{N}_0\}$. Suppose $l \in \mathbb{N}$ and the following holds:*

(i) $\lim_{n\to\infty} \tilde{k}_n(\lambda)(I - \tilde{a}_n(\lambda)A)^{-1} C_n x = \tilde{k}(\lambda)(I - \tilde{a}(\lambda)A)^{-1}Cx$, $\lambda \in \mathfrak{T}$, $x \in D(A^l)$.
(ii) *The sequences $(k_n(t))_n$, $((a_n * k_n)(t))_n$, $\cdots$, and $((a_n^{*,l-1} * k_n)(t))_n$ are equicontinuous at each point $t \geqslant 0$.*
(iii) *The sequence $(C_n x)_n$ is bounded for any $x \in D(A^l)$.*
(iv) *The condition (iv) of Theorem 2.7.2 holds with the function $a_n(t)$ replaced by $a_n^{*,l}(t)$.*

Then, for every $x \in \overline{D(A^l)}$, *one has* $\lim_{n\to\infty} R_n(t)x = R(t)x$, $t \geqslant 0$, *uniformly on compacts of* $[0, \infty)$.

Proof. The proof of Theorem 2.7.1 shows that

$$\lim_{n\to\infty} \int_0^\infty e^{-\lambda t} R_n(t)x\, dt = \int_0^\infty e^{-\lambda t} R(t)x\, dt, \quad x \in \overline{D(A^l)}, \lambda > \lambda_0.$$

Since (233) and (i) hold, it suffices to show that the sequence $(R_n(t)x)_n$ is equicontinuous at each point $t \geqslant 0$ ($x \in D(A^l)$). Keeping in mind the conditions (ii)-(iv), and the identity

$$\begin{aligned} R_n(t)x - R_n(s)x &= [k_n(t) - k_n(s)]\, C_n x + [(a_n * k_n)(t) - (a_n * k_n)(s)] C_n Ax \\ &+ \cdots + [(a_n^{*,l-1} * k_n)(t) - (a_n^{*,l-1} * k_n)(s)]\, C_n A^{l-1}x \\ &+ [(a_n^{*,l} * R_n(\cdot)A^l x)(t) - (a_n^{*,l} * R_n(\cdot)A^l x)(s)], \end{aligned} \tag{235}$$

the required assertion follows by repeating literally the arguments given in the proof of Theorem 2.7.2. □

In the following theorem, the existence of an (a, k)-regularized C-resolvent family subgenerated by A is not automatically satisfied.

Theorem 2.7.6. *Assume that, for every* $n \in \mathbb{N}_0$, $a_n(t)$ *and* $k_n(t)$ *satisfy (P1) and* A_n *is a closed linear operator,* $\lambda_0 > \omega \geqslant \sup_{n\in\mathbb{N}_0} \max(0, abs(a_n), abs(k_n))$. *Set* $\mathfrak{T}' := \{\lambda > 0 : \tilde{k}(\lambda)\tilde{a}(\lambda) \neq 0\}$, *and assume that* $\lim_{n\to\infty} \tilde{a}_n(\lambda) = \tilde{a}(\lambda)$, $\lambda \in \mathfrak{T}'$ *and* $\lim_{n\to\infty} \tilde{k}_n(\lambda) = \tilde{k}(\lambda)$, $\lambda \in \mathfrak{T}'$. *Suppose that* $L(E) \ni \tilde{k}(\lambda)(I - \tilde{a}(\lambda)A)^{-1}C$, $\lambda \in \mathfrak{T}'$ *and, for every* $n \in \mathbb{N}$, A_n *is a subgenerator of an* (a_n, k_n)*-regularized* C_n*-resolvent family* $(R_n(t))_{t\geqslant 0}$ *which satisfies (22) with* $a(t)$, $(R(t))_{t\geqslant 0}$ *and* $k(t)$ *replaced respectively by* $a_n(t)$, $(R_n(t))_{t\geqslant 0}$ *and* $k_n(t)$. *Let (233) hold for* $t \geqslant 0$, $x \in E$ *and* $n \in \mathbb{N}$, *and let*

$$\lim_{n\to\infty} \tilde{k}_n(\lambda)(I - \tilde{a}_n(\lambda)A_n)^{-1}\, C_n x = \tilde{k}(\lambda)(I - \tilde{a}(\lambda)A)^{-1}Cx, \quad x \in E, \lambda \in \mathfrak{T}'. \tag{236}$$

Suppose, further, that for each $\lambda \in \mathfrak{T}'$ *there exists an open ball* $\Omega_\lambda \subseteq \{z \in \mathbb{C} : \operatorname{Re} z > \lambda_0\}$, *with center at point* λ *and radius* $2\varepsilon_\lambda > 0$, *so that* $\tilde{a}_n(z)\tilde{k}_n(z) \neq 0$, $z \in \Omega_\lambda$, $n \in \mathbb{N}_0$. *Then the following holds:*

(i) *For each* $r \in (0, 1]$, A *is a subgenerator of a global* $(a, k*g_r)$*-regularized* C*-resolvent family* $(R^r(t))_{t\geqslant 0}$ *satisfying (22) with* $k(t)$ *and* $(R(t))_{t\geqslant 0}$ *replaced respectively by* $(k * g_r)(t)$ *and* $(R^r(t))_{t\geqslant 0}$. *Moreover, for every* $p \in \circledast$ *and* $B \in \mathcal{B}$, *(30) holds and the mapping* $t \mapsto p_B(R^r(t))$, $t \geqslant 0$ *is locally Hölder continuous with exponent* r.

(ii) *If* A *is densely defined, then* A *is a subgenerator of a global* (a, k)*-regularized* C*-resolvent family* $(R(t))_{t\geqslant 0}$ *satisfying (22) and that the family* $\{e^{-\omega t}R(t) : t \geqslant 0\}$ *is equicontinuous.*

Proof. It is clear that $\mathfrak{T}'$ is an open, dense subset of (λ_0, ∞). Fix, for a moment, $\lambda \in \mathfrak{T}'$. Then $\tilde{a}_n(z)\tilde{k}_n(z) \neq 0$, $z \in \Omega_\lambda$, $n \in \mathbb{N}_0$, and Theorem 2.1.5 implies that $\tilde{k}_n(z)$

$(I - \tilde{a}_n(z)A_n)^{-1} C_n x = \int_0^\infty e^{-zt} R_n(t)x\, dt$, $x \in E$, $z \in \Omega_\lambda$, $n \in \mathbb{N}$. By (233) and Lemma 1.2.4, we may conclude that for each $x \in E$ there exists an analytic mapping $z \mapsto F_x(z)$, $z \in \Omega_\lambda$ such that $\lim_{n\to\infty} \tilde{k}_n(\lambda)(I - \tilde{a}_n(\lambda)A_n)^{-1} C_n x = F_x(z)$, $z \in \Omega_\lambda$. In particular, $F_x(z) = \tilde{k}(\lambda)(I - \tilde{a}(\lambda)A)^{-1}Cx$, $x \in E$, $z \in \Omega_\lambda \cap \mathbb{R}$, the mapping $\lambda \mapsto H(\lambda) := \tilde{k}(\lambda)(I - \tilde{a}(\lambda)A)^{-1}Cx$, $\lambda \in \mathfrak{T}'$ is infinitely differentiable. Keeping in mind the condition (233), we obtain from the Cauchy integral formula and the dominated convergence theorem that, for every $\lambda \in \mathfrak{T}'$, $n \in \mathbb{N}_0$ and $x \in E$,

$$\frac{d^n}{d\lambda^n} H(\lambda) = \frac{d^n}{d\lambda^n} F_x(\lambda) = \frac{n!}{2\pi i} \oint_{|z-\lambda|=\varepsilon_\lambda} \frac{F_x(z)}{(z-\lambda)^{n+1}}\, dz$$

$$= \lim_{m\to\infty} \frac{n!}{2\pi i} \oint_{|z-\lambda|=\varepsilon_\lambda} \int_0^\infty e^{-zt}(z-\lambda)^{-n-1} R_m(t)x\, dt\, dz$$

$$= \lim_{m\to\infty} \left(\frac{d^n}{dz^n} \int_0^\infty e^{-zt} R_m(t)x\, dt \right)_{z=\lambda}$$

$$= \lim_{m\to\infty} \left(\int_0^\infty e^{-zt}(-1)^n t^n R_m(t)x\, dt \right)_{z=\lambda}$$

$$= \lim_{m\to\infty} \int_0^\infty e^{-\lambda t}(-1)^n t^n R_m(t)x\, dt.$$

Using again (233), it readily follows that the operator family $\{n!^{-1}(\lambda - \omega)^{n+1} H^{(n)}(\lambda) : \lambda \in \mathfrak{T}'\} \subseteq L(E)$ is equicontinuous. Then the proof completes an application of Theorem 2.1.6. □

Remark 2.7.7. (i) Observe that Theorem 2.7.2 and Theorem 2.7.6 can be compared to [383, Theorem 2.5, Corollary 2.7]. Notice that we do not require here that the functions $a_n(t)$ are absolutely continuous or exponentially bounded for $t \geqslant 0$ and $n \in \mathbb{N}_0$ as well as that $\tilde{a}_n(\lambda) \neq 0$ for $\lambda > \omega$ and $n \in \mathbb{N}_0$.

(ii) In the formulation of [463, Corollary 6.5, p. 170], the author has implicitly assumed that $\tilde{a}_n(\lambda) \neq 0$ for $\lambda > \omega$ and $n \in \mathbb{N}_0$ (cf. also [463, p. 43] and [292, Theorem 2.1.27(xvi)]). Furthermore, it is not difficult to prove that the assumptions of afore-mentioned corollary, which only treats the case $k_n(t) \equiv 1$, $n \in \mathbb{N}_0$, imply (iv) of Theorem 2.7.2 as well as that $\lim_{n\to\infty} \tilde{a}_n(\lambda) = \tilde{a}(\lambda)$, $\lambda > \omega$ and (236) holds. Therefore, Theorem 2.7.2 and Theorem 2.7.6 taken together provide an extension of [463, Corollary 6.5]; notice also that in the formulation of this corollary we do not have any assumption on the absolute continuity of functions $a_n(t)$ for $t \geqslant 0$.

(iii) If we go into a situation of Theorem 2.7.2 (cf. also Remark 2.7.3(i)), then, for every $r \in [0, 1]$ and for every $x \in \overline{\bigcup_{\lambda\in\mathfrak{T}'} R((I - \tilde{a}(\lambda)A)^{-1}C)}$, the following holds: $\lim_{n\to\infty}(g_r * R_n)(t)x = R^r(t)x$, $t \geqslant 0$, uniformly on compacts of $[0, \infty)$ (with $R^0(t) \equiv R(t)$, $t \geqslant 0$, in the case that A is densely defined).

Now we state the following important extension of [364, Corollary 4.4]:

Corollary 2.7.8. *Suppose* $(\alpha_n)_{n\in\mathbb{N}}$, *resp.* $(\beta_n)_{n\in\mathbb{N}}$, *is a convergent sequence in* $(0, \infty)$, *resp.* $[1, \infty)$, $\lim_{n\to\infty} \alpha_n = \alpha > 0$ *and* $\lim_{n\to\infty} \beta_n = \beta$. *Set* $\alpha' := \max_{n\in\mathbb{N}} \alpha_n$. *Let* $\lambda_0 > \omega \geqslant 0$, *and for each* $n \in \mathbb{N}_0$ *let the operator* A_n *be a subgenerator of an* $(g_{\alpha_n}, g_{\beta_n})$-*regularized C-resolvent family* $(S_n(t))_{t\geqslant 0}$ *which satisfies (22) with* $a(t)$, $(R(t))_{t\geqslant 0}$ *and* $k(t)$ *replaced respectively by* $g_{\alpha_n}(t)$, $(S_n(t))_{t\geqslant 0}$ *and* $g_{\beta_n}(t)$. *Let (233) hold for* $t \geqslant 0$, $x \in E$ *and* $n \in \mathbb{N}$, *and let*

$$(237) \qquad \lim_{n\to\infty} \lambda^{\alpha_n-\beta_n}(\lambda^{\alpha_n} - A_n)^{-1} Cx = \lambda^{\alpha-\beta} R(\lambda^{\alpha})x,\ x \in E,\ \lambda > \lambda_0,$$

where $(R(\lambda))_{\lambda>(\lambda_0)^{\alpha'}} \subseteq L(E)$. *Let* $\lambda' > (\lambda_0)^{\alpha'}$ *satisfy that the operator* $R(\lambda')$ *is injective. Then the following holds:*

(i) *Define* $D(A) := C^{-1}R(R(\lambda'))$ *and* $Ax := (\lambda - R(\lambda')^{-1}C)x$, $x \in D(A)$. *Then the operator* $A = C^{-1}AC$ *is well defined, linear, closed and satisfies* $R(\lambda) = (\lambda - A)^{-1}C$, $\lambda > (\lambda_0)^{\alpha'}$.

(ii) *For each* $r \in (0, 1]$, A *is a subgenerator of a global* $(g_{\alpha'}, g_{\beta+r})$-*regularized C-resolvent family* $(S^r(t))_{t\geqslant 0}$ *satisfying (22) with* $a(t) = g_\alpha(t)$, $k(t) = g_{\beta+r}(t)$ *and* $(R(t))_{t\geqslant 0}$ *replaced by* $(S^r(t))_{t\geqslant 0}$. *Furthermore, for every* $p \in \circledast$, *(30) holds with* $(R_r(t))_{t\geqslant 0}$ *replaced by* $(S^r(t))_{t\geqslant 0}$, *for every* $p \in \circledast$ *and* $B \in \mathcal{B}$, *the mapping* $t \mapsto p_B(S^r(t))$, $t \geqslant 0$ *is locally Hölder continuous with exponent r, and* $\lim_{n\to\infty}(g_r * S_n)(t)x = S^r(t)x$, $t \geqslant 0$, $x \in \overline{R(R(\lambda'))}$, *uniformly on compacts of* $[0, \infty)$.

(iii) *If* $C^{-1}R(R(\lambda'))$ *is dense in E, then A is a subgenerator of a global* $(g_{\alpha'}, g_\beta)$-*regularized C-resolvent family* $(S(t))_{t\geqslant 0}$ *satisfying (22) with* $a(t) = g_\alpha(t)$, $k(t) = g_\beta(t)$ *and* $(R(t))_{t\geqslant 0}$ *replaced by* $(S(t))_{t\geqslant 0}$. *Furthermore, the family* $\{e^{-\omega t}S(t) : t \geqslant 0\}$ *is equicontinuous and* $\lim_{n\to\infty} S_n(t)x = S(t)x$, $t \geqslant 0$, $x \in \overline{R(R(\lambda'))}$, *uniformly on compacts of* $[0, \infty)$.

Proof. We shall content ourselves with sketching it. Owing to (233) and (237), it is not difficult to prove that the mapping $\lambda \mapsto R(\lambda)$, $\lambda > (\lambda_0)^{\alpha'}$ is a C-pseudoresolvent in the sense of [367, Definition 3.1]. Since the operator $R(\lambda')$ is injective, we may apply [367, Lemma 3.2] in order to see that $R(\lambda)$ is injective for all $\lambda > (\lambda_0)^{\alpha'}$, and that $R(R(\lambda)) = R(R(\lambda'))$ for all $\lambda > (\lambda_0)^{\alpha'}$. The assertion (i) is an immediate consequence of [367, Theorem 3.4], whereas the assertions (ii) and (iii) follow more or less straightforwardly from Theorem 2.7.6, Theorem 2.7.2 and Remark 2.7.3(i). □

Concerning the convergence of C-resolvents, it should be noted that the assertions of [337, Proposition 3.1, Proposition 3.3] continue to hold, with appropriate technical changes, in SCLCSs:

Proposition 2.7.9. (i) *Suppose* $(A_n)_{n\in\mathbb{N}_0}$ *is a sequence of closed linear operators on* E, $\lambda \in \bigcap_{n\in\mathbb{N}_0} \rho_C(A_n)$ *and* $\lim_{n\to\infty} (\lambda - A_n)^{-1}Cx = (\lambda - A)^{-1}Cx$ *for all* $x \in E$. *Then* *for each* $x \in R((\lambda - A)^{-1}C)$ *there exists a sequence* $(x_n)_{n\in\mathbb{N}}$ *in E satisfying* $x_n \in R((\lambda - A_n)^{-1}C)$ *for all* $n \in \mathbb{N}$ *as well as* $\lim_{n\to\infty} x_n = x$ *and* $\lim_{n\to\infty} A_n x_n = Ax$.

(ii) *Suppose* $(A_n)_{n\in\mathbb{N}_0}$ *is a sequence of closed linear operators on E,* $\lambda \in \bigcap_{n\in\mathbb{N}_0} \rho_C(A_n)$ *and* D_λ *is a linear subspace of D(A) satisfying*

$$D_\lambda \subseteq R((\lambda - A)^{-1}C) \subseteq \overline{D_\lambda}^{[D(A)]}$$

and that, for every $x \in D_\lambda$, *there exists a sequence* $(x_n)_{n\in\mathbb{N}}$ *in E such that* $\lim_{n\to\infty} x_n = x$ *and* $\lim_{n\to\infty} A_n x_n = Ax$. *If for each* $p \in \circledast$ *there exist* $c_p > 0$ *and* $r_p \in \circledast$ *such that* $p((\lambda - A_n)^{-1}Cx) \leqslant c_p r_p(x)$, $x \in E$, $n \in \mathbb{N}_0$, *then* $\lim_{n\to\infty}(\lambda - A_n)^{-1}Cx = (\lambda - A)^{-1} Cx$ *for all* $x \in \overline{R(C)}$.

Proposition 2.7.10. (i) *Assume that the functions* $a(t)$ *and* $k(t)$ *satisfy (P1),* $\omega \geqslant \max(0, abs(a), abs(k))$ *and, for every* $n \in \mathbb{N}_0$, A_n *is a subgenerator of an (a, k)-regularized C-resolvent family* $(R_n(t))_{t\geqslant 0}$ *which satisfies (22) with* $(R(t))_{t\geqslant 0}$ *replaced by* $(R_n(t))_{t\geqslant 0}$. *Let (233) hold, and let* $\lambda' \in \mathbb{C}$ *satisfy* $\operatorname{Re} \lambda' > \omega$, $\tilde{a}(\lambda')\tilde{k}(\lambda') \neq 0$ *and*

$$\lim_{n\to\infty} \tilde{k}(\lambda')(I - \tilde{a}(\lambda')A_n)^{-1}Cx = \tilde{k}(\lambda')(I - \tilde{a}(\lambda')A)^{-1}Cx, \quad x \in E. \tag{238}$$

Then the following holds, for every $x \in \overline{R(C)}$ *and for every* $\lambda \in \mathbb{C}$ *with* $\operatorname{Re} \lambda > \omega$ *and* $\tilde{a}(\lambda)\tilde{k}(\lambda) \neq 0$,

$$\lim_{n\to\infty} \tilde{k}(\lambda)(I - \tilde{a}(\lambda)A_n)^{-1} Cx = \tilde{k}(\lambda)(I - \tilde{a}(\lambda)A)^{-1}Cx. \tag{239}$$

(ii) *Assume the functions* $a(t)$ *and* $k(t)$ *satisfy (P1),* $\omega \geqslant \max(0, abs(a), abs(k))$ *and, for every* $n \in \mathbb{N}$, A_n *is a subgenerator of an (a, k)-regularized resolvent family* $(R_n(t))_{t\geqslant 0}$ *which satisfies (22) with* $C = I$ *and* $(R(t))_{t\geqslant 0}$ *replaced by* $(R_n(t))_{t\geqslant 0}$. *Suppose that there exists* $c_\circledast > 0$ *such that*

$$p(e^{-\omega t}R_n(t)x) \leqslant c_\circledast p(x), x \in E, t \geqslant 0, n \in \mathbb{N}. \tag{240}$$

Let $\lambda' \in \mathbb{C}$ *satisfy* $\operatorname{Re} \lambda' > \omega$, $\tilde{a}(\lambda')\tilde{k}(\lambda') \neq 0$ *and (238). Then (239) holds for all* $x \in E$ *and for all* $\lambda \in \mathbb{C}$ *with* $\operatorname{Re} \lambda > \omega$ and $\tilde{a}(\lambda)\tilde{k}(\lambda) \neq 0$.

Proof. The proof of assertion (i) simply follows from Proposition 2.7.9 and the fact that the function $R_n : D(R_n) \to L(E)$, defined by $D(R_n) := \{\tilde{a}(\lambda)^{-1} : \operatorname{Re} \lambda > \omega, \tilde{a}(\lambda)\tilde{k}(\lambda) \neq 0\}$ and $R_n(z)x := (z - A_n)^{-1}Cx$, $z \in D(R_n)$, $n \in \mathbb{N}$, $x \in E$ is a C-pseudoresolvent. This implies by [367, Lemma 3.2] that, for every $n \in \mathbb{N}$, one has $R(R(z)) = R(R(z'))$, $z, z' \in D(R_n)$. The second assertion can be proved by a modification of arguments given in the proof of [175, Proposition 3.4.4]. Put $S := \{\lambda \in \mathbb{C} : \operatorname{Re} \lambda > \omega, \tilde{a}(\lambda)\tilde{k}(\lambda) \neq 0\}$, $W := \{\tilde{a}(\lambda)^{-1} : \operatorname{Re} \lambda > \omega, \tilde{a}(\lambda)\tilde{k}(\lambda) \neq 0\}$ and $\Omega := \{z = \tilde{a}(\lambda)^{-1} \in W : \lim_{n\to\infty}(\tilde{a}(\lambda)^{-1} - A_n)^{-1}x$ exists for all $x \in E\}$. It is clear that (240) implies

$$p((\tilde{a}(\lambda)^{-1} - A_n)^{-1}x) \leqslant cp(x)\frac{|\tilde{a}(\lambda)|}{|\tilde{k}(\lambda)|(\operatorname{Re}\lambda - \omega)}, \quad n \in \mathbb{N}, x \in E, \lambda \in S.$$

Inductively, we obtain

$$(241)\ p\left((\tilde{a}(\lambda)^{-1} - A_n)^{-s}x\right) \leqslant c^s p(x)\left[\frac{|\tilde{a}(\lambda)|}{|\tilde{k}(\lambda)|(\operatorname{Re}\lambda - \omega)}\right]^s, \quad n \in \mathbb{N}, s \in \mathbb{N}_0, x \in E, \lambda \in S.$$

By (241), it readily follows that for each $z_0 = \tilde{a}(\lambda_0)^{-1} \in \Omega$ and for every $z = \tilde{a}(\lambda_0)^{-1} \in W$ such that

$$(242) \qquad |\lambda_0 - \lambda| < \operatorname{dist}(\lambda_0, S) \text{ and } |z_0 - z| < \frac{|\tilde{k}(\lambda_0)|(\operatorname{Re}\lambda_0 - \omega)}{c|\tilde{a}(\lambda_0)|},$$

the series

$$(\tilde{a}(\lambda)^{-1} - A_n)^{-1}x = \sum_{k=0}^{\infty}(\tilde{a}(\lambda')^{-1} - z)^k (\tilde{a}(\lambda')^{-1} - A_n)^{-k-1}x$$

is convergent since the sequence of its partial sums is one of Cauchy's. Observing that $\frac{|\tilde{k}(\lambda_0)|(\operatorname{Re}\lambda_0-\omega)}{c|\tilde{a}(\lambda_0)|} \to \frac{|\tilde{k}(\lambda)|(\operatorname{Re}\lambda-\omega)}{c|\tilde{a}(\lambda)|}$ as $\lambda_0 \to \lambda$, for any $\lambda \in \mathbb{C}$ with $\operatorname{Re}\lambda > \omega$ and $\tilde{a}(\lambda)\,\tilde{k}(\lambda) \neq 0$, it can be easily seen that, for every $z = \tilde{a}(\lambda_0)^{-1} \in W$, we can find an appropriate point $z_0 = \tilde{a}(\lambda_0)^{-1} \in \Omega$ satisfying (242). Therefore, $W = \Omega$ and the proof of (ii) is completed through a routine argument. □

Remark 2.7.11. (i) Notice that Theorem 2.7.1, Theorem 2.7.6(i) and Proposition 2.7.10(i) can be reformulated for certain classes of (a, k)-regularized C_1-existence families. The situation is far less obvious for the class of (a, k)-regularized C_2-uniqueness families (cf. [308, Theorem 2.4, Proposition 2.7] and Section 2.8).

(ii) Notice that Proposition 2.7.10(ii) is an extension of [528, Lemma 3.2]. Keeping in mind the first part of Proposition 2.7.10 as well as Theorem 2.7.2, Theorem 2.7.6 and Remark 2.7.3(i), one can simply formulate and prove extensions of [528, Corollaries 3.4-3.5, 3.7-3.8 and Theorem 3.6] for abstract Volterra equations in locally convex spaces. We leave this for the reader to make explicit.

In the remaining part of this section, we shall prove some results on approximation and convergence of (A, k)-regularized C-pseudeoresolvent families; for the sake of comfortableness, we shall work only in Banach spaces. Suppose X and Y are Banach spaces, Y is continuously embedded in X and $C \in L(X)$ is injective. Let $k \in C([0, \infty))$, $k \neq 0$ and $A \in L^1_{loc}([0, \infty) : L(Y, X))$. Unless otherwise specified, the convergence of sequences is supposed to be taken with respect to the topology of X. We need the following result from [292].

Lemma 2.7.12. *Assume $A \in L^1_{loc}([0, \tau) : L(Y, X))$, the function $k(t)$ satisfy (P1), $\varepsilon_0 \geqslant 0$ and*

$$\int_0^{\infty} e^{-\varepsilon t}\,\|A(t)\|_{L(Y,X)}\,dt < \infty,\ \varepsilon > \varepsilon_0.$$

Let $(S(t))_{t\geqslant 0}$ be an (A, k)-regularized C-pseudoresolvent family such that there exists $\omega \geqslant 0$ with

$$\sup_{t>0} e^{-\omega t}\Big(\|S(t)\|_{L(X)} + \sup_{0<s<t}(t-s)^{-1}\|U(t)-U(s)\|_{L(Y)}\Big) < \infty.$$

Put $\omega_0 := \max(\omega, abs(k), \varepsilon_0)$ and $H(\lambda)x := \int_0^\infty e^{-\lambda t}S(t)x\,dt$, $x \in X$, $\operatorname{Re}\lambda > \omega_0$. Then the following holds:

(i) $C(Y) \subseteq Y$, $(\tilde{A}(\lambda))_{\operatorname{Re}\lambda>\varepsilon_0}$ *is analytic in* $L(Y,X)$, $R(C_{|Y}) \subseteq R(I-\tilde{A}(\lambda))$, $\operatorname{Re}\lambda > \omega_0$, $\tilde{k}(\lambda) \neq 0$, *and* $I-\tilde{A}(\lambda)$ *is injective*, $\operatorname{Re}\lambda > \omega_0$, $\tilde{k}(\lambda) \neq 0$.

(ii) $H(\lambda)y = \lambda\tilde{U}(\lambda)y$, $y \in Y$, $\operatorname{Re}\lambda > \omega_0$, $(I-\tilde{A}(\lambda))^{-1}C_{|Y} \in L(Y)$, $\operatorname{Re}\lambda > \omega_0$, $\tilde{k}(\lambda) \neq 0$, $(H(\lambda))_{\operatorname{Re}\lambda>\omega_0}$ *is analytic in both spaces,* $L(X)$ *and* $L(Y)$, $H(\lambda)C = CH(\lambda)$, $\operatorname{Re}\lambda > \omega_0$, *and for every* $y \in Y$ *and for every* $\lambda \in \mathbb{C}$ *with* $\operatorname{Re}\lambda > \omega_0$ *and* $\tilde{k}(\lambda) \neq 0$:

$$H(\lambda)(I-\tilde{A}(\lambda))y = (I-\tilde{A}(\lambda))H(\lambda)y = \tilde{k}(\lambda)Cy.$$

Furthermore,

$$\sup_{n\in\mathbb{N}_0}\ \sup_{\lambda>\omega_0,\tilde{k}(\lambda)\neq 0}\frac{(\lambda-\omega)^{n+1}}{n!}\left(\left\|\frac{d^n}{d\lambda^n}H(\lambda)\right\|_{L(X)} + \left\|\frac{d^n}{d\lambda^n}H(\lambda)\right\|_{L(Y)}\right) < \infty.$$

The proof of following theorem can be deduced by making use of Lemma 2.7.12 and Theorem 1.2.6.

Theorem 2.7.13. *Suppose $A_n \in L^1_{loc}([0,\infty) : L(Y_n, X))$, $C_n \in L(X)$ is injective, the function $k_n(t)$ satisfies (P1), and $(S_n(t))_{t\geqslant 0}$ is an (A_n, k_n)-regularized C_n-pseudoresolvent family ($n \in \mathbb{N}_0$). Let $\omega_0 > 0$ satisfy*

$$\int_0^\infty e^{-\omega_0 t}\|A_n(t)\|_{L(Y_n,X)}\,dt < \infty, \quad n \in \mathbb{N}_0, \tag{243}$$

let $\omega \geqslant \sup_{n\in\mathbb{N}_0}\max(\omega_0, abs(k_n))$, and let $\mathfrak{T} = \{\lambda > \lambda_0 : \tilde{k}_n(\lambda) \neq 0$ for all $n \in \mathbb{N}_0\}$. If

$$\sup_{t>0, n\in\mathbb{N}_0} e^{-\omega t}\Big(\|S_n(t)\|_{L(X)} + \sup_{0<s<t}(t-s)^{-1}\|U_n(t)-U_n(s)\|_{L(Y_n)}\Big) < \infty, \tag{244}$$

then the following assertions are equivalent:

(i) $\lim_{n\to\infty}\tilde{k}_n(\lambda)(I-\tilde{A}_n(\lambda))^{-1}C_n y = \tilde{k}(\lambda)(I-\tilde{A}(\lambda))^{-1}Cy$, $y \in \bigcap_{n\in\mathbb{N}_0} Y_n$, $\lambda \in \mathfrak{T}$, *and the sequence* $(S_n(t)y)_{n\in\mathbb{N}}$ *is equicontinuous at each point* $t \geqslant 0$ $(y \in \bigcap_{n\in\mathbb{N}_0} Y_n)$.

(ii) $\lim_{n\to\infty} S_n(t)y = S(t)y$, $t \geqslant 0$, $y \in \overline{\bigcap_{n\in\mathbb{N}_0} Y_n}^X$, *uniformly on compacts of* $[0,\infty)$.

If the assumptions of preceding theorem hold and $y \in \bigcap_{n\in\mathbb{N}_0} Y_n$, $\lambda' \in \mathfrak{T}$, then we have by (244), and (ii) of Lemma 2.7.12, that there exists $M \geqslant 1$ such that, for every $n \in \mathbb{N}_0$,

$$\|H_n(\lambda')y\|_{Y_n} \leqslant |\lambda'|\int_0^\infty e^{-t\operatorname{Re}\lambda'}\|U_n(t)y\|_{Y_n}\,dt \leqslant \frac{M}{\operatorname{Re}\lambda'-\omega}\|y\|_{Y_n}.$$

Using also the fact that $H_n(\lambda')C_n y = C_n H_n(\lambda')y$, $n \in \mathbb{N}_0$, it is very simple to prove an analog of Theorem 2.7.2 for non-scalar equations:

Theorem 2.7.14. *Suppose $A_n \in L^1_{loc}([0,\infty) : L(Y_n, X))$, $C_n \in L(X)$ is injective, the function $k_n(t)$ satisfies (P1), and $(S_n(t))_{t\geqslant 0}$ is an (A_n, k_n)-regularized C_n-pseudoresolvent family ($n \in \mathbb{N}_0$). Let $\omega_0 > 0$ satisfy (243), let $\omega \geqslant \sup_{n\in\mathbb{N}_0} \max(\omega_0, abs(k_n))$, and let $\mathfrak{T} = \{\lambda > \lambda_0 : \tilde{k}_n(\lambda) \neq 0$ for all $n \in \mathbb{N}_0\}$. Suppose (244) and the following conditions hold:*

(i) *The sequence $(k_n(t))_n$ is equicontinuous at each point $t \geqslant 0$.*

(ii) $\lim_{n\to\infty} \tilde{k}_n(\lambda)(I-\tilde{A}_n(\lambda))^{-1} C_n y = \tilde{k}(\lambda)(I-\tilde{A}(\lambda))^{-1}Cy$, $y \in \bigcap_{n\in\mathbb{N}_0} Y_n$, $\lambda \in \mathfrak{T}$.

(iii) *For every $\varepsilon > 0$ and $t \geqslant 0$, there exist $\delta \in (0, 1)$ and $n_0 \in \mathbb{N}$ such that*

$$\int_0^{\min(t,s)} \left\| A_n(\max(t,s)-r) - A_n(\min(t,s)-r) \right\|_{L(Y_n,X)} dr$$

$$+ \int_{\min(t,s)}^{\max(t,s)} \| A_n(\max(t,s)-r) \|_{L(Y_n,X)} dr < \varepsilon,$$

provided $|t-s| < \delta$, $s \geqslant 0$ and $n \geqslant n_0$.

Set

$$\mathfrak{Y} := \{ y \in \bigcap_{n\in\mathbb{N}_0} Y_n : (\|y\|_{Y_n})_n, (\|C_n y\|_{Y_n})_n \in l_\infty \}.$$

Then $\lim_{n\to\infty} S_n(t)z = S(t)z$, $t \geqslant 0$, $z \in \overline{\bigcup_{\lambda\in\mathfrak{T}'} (I-\tilde{A}(\lambda))^{-1}C(\mathfrak{Y})}^X$, uniformly on compacts of $[0,\infty)$.

To the best knowledge of the author, the only known result on approximation of (A, k)-regularized C-pseudoresolvent families is [463, Theorem 6.3, p. 167], where it has been assumed that $k_n(t) \equiv 1$, $C_n \equiv I$ and $\overline{Y}^X = X$. The proof of this theorem is workable only in the case that $\omega = 0$ (see e.g. [463, (6.41), p. 168]), so that the importance of Theorem 2.7.14, which is more similar to the classical Trotter-Kato theorem, lies in the fact that the assumption on subexponential growth of kernels $A_n(t)$ does not play any crucial role in our analysis. In connection with Theorem 2.7.14, it is also worthwhile to mention the following fact. Since $\lim_{t\downarrow 0} S(t)x = k(0)Cx$, $x \in E$, we have by [20, Proposition 4.1.3] that $\lim_{\lambda\to\infty} \lambda\int_0^\infty e^{-\lambda t}S(t)x\,dt = k(0)Cx$, $x \in E$. Hence, in the case that $k(0) \neq 0$ and $Y_n \equiv Y$ ($n \in \mathbb{N}$), the following inclusion holds $\overline{C(\mathfrak{Y})}^X \subseteq \overline{\bigcup_{\lambda\in\mathfrak{T}'} (I-\tilde{A}(\lambda))^{-1}C(\mathfrak{Y})}^X$.

2.8 (a, k)-Regularized (C_1, C_2)-existence and uniqueness families

In this section, we shall introduce and analyze the class of (mild) (a, k)-regularized (C_1, C_2)-existence and uniqueness families in the setting of sequentially complete locally convex spaces. The classes of (mild) (a, k)-regularized C_1-existence families and (mild) (a, k)-regularized C_2-uniqueness families are also defined and considered. The subordination principles as well as many other structural characterizations of (local) exponentially equicontinuous (a, k)-regularized (C_1, C_2)-existence and uniqueness families are proved.

The introduction of the following definition of a (local) (a, k)-regularized C-resolvent family is motivated by the recent researches of C. Chen, M. Li [89] and C. Lizama, F. Poblete [390].

Definition 2.8.1. Suppose $0 < \tau \leqslant \infty$, $k \in C([0, \tau))$, $k \neq 0$, $a \in L^1_{loc}([0, \tau))$ and $a \neq 0$. Then a strongly continuous operator family $(R(t))_{t\in[0,\tau)}$ is called a (local, if $\tau < \infty$) (a, k)-regularized C-resolvent family if the following conditions hold:

(i) $R(0) = k(0)C$, $R(t)C = CR(t)$, $t \in [0, \tau)$ and $R(t)R(s) = R(s)R(t)$, $t, s \in [0, \tau)$.
(ii) $R(s)(a * R)(t) - (a * R)(s)R(t) = k(s)(a * R)(t)C - k(t)(a * R)(s)C, \quad t, s \in [0, \tau)$.

The notions of integral generator and local equicontinuity of $(R(t))_{t\in[0,\tau)}$, as well as the notions of (exponential, q-exponential) equicontinuity of $(R(t))_{t\geqslant 0}$ and (exponential, q-exponential) analyticity of $(R(t))_{t\geqslant 0}$ are understood in the sense of Definition 2.1.1 and the consideration given after Proposition 2.1.3. By a subgenerator of $(R(t))_{t\in[0,\tau)}$ we mean any closed linear operator A on E satisfying $CA \subseteq AC$, $R(t)A \subseteq AR(t)$, $t \in [0, \tau)$ and the condition (iii) of Definition 2.1.1.

Now we would like to compare Definition 2.1.1 and Definition 2.8.1. Suppose that A is a subgenerator of a non-degenerate, locally equicontinuous (a, k)-regularized C-resolvent family $(R(t))_{t\in[0,\tau)}$ in the sense of Definition 2.1.1 and (22) holds. Keeping in mind the proof of [390, Theorem 3.1], we easily infer that $(R(t))_{t\in[0,\tau)}$ is an (a, k)-regularized C-resolvent family in the sense of Definition 2.8.1. Furthermore, if $a(t)$ is kernel the operator $\hat{A}$ equals $C^{-1}AC$ and is a subgenerator (the integral generator, in fact) of an (a, k)-regularized C-resolvent family $(R(t))_{t\in[0,\tau)}$ in the sense of Definition 2.8.1. Suppose, conversely, that $(R(t))_{t\in[0,\tau)}$ is a non-degenerate, locally equicontinuous (a, k)-regularized C-resolvent family in the sense of Definition 2.8.1 if $a(t)$ is kernel. Then the operator $\hat{A}$ is a subgenerator (the integral generator) of an (a, k)-regularized C-resolvent family $(R(t))_{t\in[0,\tau)}$ in the sense of Definition 2.1.1, and (22) holds with A replaced by $\hat{A}$ therein.

Definition 2.8.2. Suppose $0 < \tau \leqslant \infty$, $k \in C([0, \tau))$, $k \neq 0$, $a \in L^1_{loc}([0, \tau))$, $a \neq 0$ and A is a closed linear operator on E.

(i) Then it is said that A is a subgenerator of a (local, if $\tau < \infty$) mild (a, k)-regularized $(C_1\ C_2)$-existence and uniqueness family $(R_1(t), R_2(t))_{t\in[0,\tau)} \subseteq L(E) \times L(E)$ if the mapping $t \mapsto (R_1(t)x, R_2(t)x)$, $t \in [0, \tau)$ is continuous for every fixed $x \in E$ and if the following conditions hold:

 (a) $R_i(0) = k(0)C_i$, $i = 1, 2$,
 (b) C_2 is injective,
 (c)

(245) $$A\int_0^t a(t-s)R_1(s)x\,ds = R_1(t)x - k(t)C_1x,\ t \in [0,\tau),\ x \in E \text{ and}$$

(246) $$\int_0^t a(t-s)R_2(s)Ax\,ds = R_2(t)x - k(t)C_2x,\ t \in [0,\tau),\ x \in D(A).$$

(ii) Let $(R_1(t))_{t\in[0,\tau)} \subseteq L(E)$ be strongly continuous. Then it is said that A is a subgenerator of a (local, if $\tau < \infty$) mild (a, k)-regularized C_1-existence family $(R_1(t))_{t\in[0,\tau)}$ if $R_1(0) = k(0)C_1$ and (245) holds.

(iii) Let $(R_2(t))_{t\in[0,\tau)} \subseteq L(E)$ be strongly continuous. Then it is said that A is a subgenerator of a (local, if $\tau < \infty$) mild (a, k)-regularized C_2-uniqueness family $(R_2(t))_{t\in[0,\tau)}$ if $R_2(0) = k(0)C_2$, C_2 is injective and (246) holds.

The notions of (q-)exponential equicontinuity, analyticity and (q-)exponential analyticity of mild (a, k)-regularized C_1-existence families (C_2-uniqueness families) are understood in the sense of Definition 2.1.1. A mild (a, k)-regularized (C_1, C_2)-existence and uniqueness family $(R_1(t), R_2(t))_{t\geqslant 0}$ having A as subgenerator is said to be (q-)exponentially equicontinuous (analytic, (q-)exponentially analytic) if both $(R_1(t))_{t\geqslant 0}$ and $(R_2(t))_{t\geqslant 0}$ are.

If $a(t) = g_\alpha(t)$ for some $\alpha > 0$, then it is not difficult to prove that, for every (a, k)-regularized C_1-existence family $(R_1(t))_{t\in[0,\tau)}$ with subgenerator A, the following holds: $\bigcup_{t\in[0,\tau)} \overline{R(R_1(t))} \subseteq \overline{D(A)}$. In the case that A is a subgenerator of a mild (g_α, k)-regularized (C_1, C_2)-existence and uniqueness family $(R_1(t), R_2(t))_{t\in[0,\tau)}$, we have intuitively that $R_1(t) = E_\alpha(t^\alpha A)C_1$ and $R_2(t) = C_2E_\alpha(t^\alpha A)$ for $t \in [0, \tau)$. Further on, it is clear that the notion of a mild (a, k)-regularized C_2-uniqueness family is more general than that of an (a, k)-regularized C-resolvent family. Observe also that the notion of a mild (a, k)-regularized C_1-existence family extends the notion of a global n times integrated C-existence family ($n \in \mathbb{N}_0$), introduced by S. W. Wang [515, Definition 3.1] in the Banach space setting (for the exponential case and the initial ideas, cf. R. deLaubenfels [153]). It could be interesting to prove an extension of [515, Theorem 3.3] for mild (a, k)-regularized (C_1, C_2)-existence and uniqueness families (cf. also [316, Section 2] for some other recent results in this direction).

Notice that (245)-(246) together imply that, for every $0 \leqslant t, s < \tau$ and $x \in E$,

$$\begin{aligned}(a*R_2)(s)R_1(t)x &= (a*R_2)(s)\,[A(a*R_1)(t)x + k(t)C_1x]\\ &= k(t)(a*R_2)(s)C_1x - (a*R_2)(s)A(a*R_1)(t)x\\ &= k(t)(a*R_2)(s)C_1x + R_2(s)(a*R_1)(t)x - k(s)C_2(a*R_1)(t)x.\end{aligned}$$

This motivates the introduction of the following definition of a mild (a, k)-regularized (C_1, C_2)-existence and uniqueness family, slightly different from the previous one.

Definition 2.8.3. Suppose $0 < \tau \leqslant \infty$, $k \in C([0, \tau))$, $k \neq 0$, $a \in L^1_{loc}([0, \tau))$ and $a \neq 0$. Then it is said that a strongly continuous operator family $(R_1(t), R_2(t))_{t\in[0,\tau)} \subseteq L(E) \times L(E)$ is a (local, if $\tau < \infty$) mild (a, k)-regularized (C_1, C_2)-existence and uniqueness family if the following conditions hold:

(i) $R_i(0) = k(0)C_i$, $i = 1, 2$,
(ii) C_2 is injective,
(iii) for every $0 \leqslant t, s < \tau$ and $x \in E$, the following equality holds:

$$(a*R_2)(s)R_1(t)x - R_2(s)(a*R_1)(t)x \\ = k(t)(a*R_2)(s)C_1x - k(s)C_2(a*R_1)(t)x. \tag{247}$$

A closed linear operator A on E is said to be a subgenerator of $(R_1(t), R_2(t))_{t\in[0,\tau)}$ if (245)-(246) hold.

No matter which one of the introduced definitions of an (a, k)-regularized (C_1, C_2)-existence and uniqueness family one employs, the condition that $a(t)$ is a kernel implies that we can define the integral generator $\hat{A}$ of $(R_1(t), R_2(t))_{t\in[0,\tau)}$ by setting

$$\hat{A} := \left\{(x, y) \in E \times E : R_2(t)x - k(t)C_2x = \int_0^t a(t-s)R_2(s)y\, ds,\ t \in [0, \tau)\right\}. \tag{248}$$

Certainly, $\hat{A}$ is a linear operator and the local equicontinuity of $(R_2(t))_{t\in[0,\tau)}$ implies that $\hat{A}$ is closed. Moreover, if $R_2(t)C_2 = C_2R_2(t)$, $t \in [0, \tau)$, then $C_2^{-1}\hat{A}C_2 = \hat{A}$; the notion of integral generator of a mild (a, k)-regularized C_2-uniqueness family $(R_2(t))_{t\in[0,\tau)}$ can be also understood in the sense of (248). Suppose now that $(R_1(t), R_2(t))_{t\in[0,\tau)}$ is a mild (a, k)-regularized (C_1, C_2)-existence and uniqueness family in the sense of Definition 2.8.3. If, additionally, $(R_2(t))_{t\in[0,\tau)}$ is locally equicontinuous and the function $a(t)$ is a kernel holds, then it readily follows from (247) that the integral generator $\hat{A}$ is a maximal subgenerator of $(R_1(t), R_2(t))_{t\in[0,\tau)}$ with respect to the set inclusion.

Remark 2.8.4. Suppose $a(t) \equiv t$, $k(t) \equiv 1$, E is a Banach space and A is a subgenerator of a mild (a, k)-regularized (C_1, C_2)-existence and uniqueness family in the sense of Definition 2.8.2. Then the proof of implication (b) ⇒ (a) of [547, Theorem 1.8] implies that

$$2R_2(t)R_1(s) = C_2[R_1(t+s) + R_1(|t-s|)] \\ = [R_2(t+s) + R_2(|t-s|)]\, C_1,\ 0 \leqslant t, s, t+s < \tau. \tag{249}$$

In particular, $(R_1(t), R_2(t))_{t\in[0,\tau)}$ is a mild (C_1, C_2)-regularized cosine existence and uniqueness family in the sense of [547, Definition 1.1], provided that $\tau = \infty$. Suppose, conversely, that $(R_1(t), R_2(t))_{t\in[0,\tau)}$ is a strongly continuous operator family, $R_i(0) = C_i$, $i = 1, 2$, C_2 is injective and (249) holds. Then we may define the infinitesimal generator $\breve{A}$ of $(R_1(t), R_2(t))_{t\in[0,\tau)}$ by

$$\breve{A} := \left\{(x, y) \in E \times E : \lim_{t\to 0+} \frac{2}{t^2} (R_2(t)x - C_2x) = C_2y\right\}.$$

Proceeding as in the proof of [547, Theorem 1.6], we get that $\breve{A}$ is a subgenerator of a mild (a, k)-regularized (C_1, C_2)-existence and uniqueness family $(R_1(t), R_2(t))_{t\in[0,\tau)}$ in the sense of Definition 2.8.2 (Definition 2.8.3); moreover, $\breve{A}$ coincides with the integral generator of $(R_1(t), R_2(t))_{t\in[0,\tau)}$. The previous conclusions can be reformulated in the case that $a(t) \equiv k(t) \equiv 1$ (cf. [141, Section XVI]) or that E is a general SCLCS.

The proof of following theorem is left to the reader as an easy exercise.

Theorem 2.8.5. *Suppose A is a closed linear operator on E,* $C_1, C_2 \in L(E)$, C_2 *is injective,* $\omega_0 \geqslant 0$, $a(t)$, $k(t)$ *satisfy (P1) and* $\omega \geqslant \max(\omega_0, abs(a), abs(k))$.

(i) *Let* $(R_1(t), R_2(t))_{t\geqslant 0}$ *be strongly continuous and let the family* $\{e^{-\omega t}R_i(t) : t \geqslant 0\}$ *be equicontinuous for* $i = 1, 2$.

(a) *Suppose* $(R_1(t), R_2(t))_{t\geqslant 0}$ *is a mild (a, k)-regularized* (C_1, C_2)*-existence and uniqueness family with a subgenerator A. Then, for every* $\lambda \in \mathbb{C}$ *with* $\operatorname{Re}\lambda > \omega$ *and* $\tilde{k}(\lambda) \neq 0$, *the operator* $I - \tilde{a}(\lambda)A$ *is injective,* $R(C_1) \subseteq R(I - \tilde{a}(\lambda)A)$,

$$\tilde{k}(\lambda)(I - \tilde{a}(\lambda)A)^{-1} C_1x = \int_0^\infty e^{-\lambda t}R_1(t)x\, dt,\ x \in E, \tag{250}$$

$$\left\{\frac{1}{\tilde{a}(z)} : \operatorname{Re} z > \omega,\ \tilde{k}(z)\,\tilde{a}(z) \neq 0\right\} \subseteq \rho_{C_1}(A) \tag{251}$$

and

$$\tilde{k}(\lambda)C_2x = \int_0^\infty e^{-\lambda t}[R_2(t)x - (a * R_2)(t)Ax]\, dt,\ x \in D(A). \tag{252}$$

(b) *Let* $R_2(0) = k(0)C_2x$, $x \in E\setminus\overline{D(A)}$, *let (251) hold, and let (250) and (252) hold for any* $\lambda \in \mathbb{C}$ *with* $\operatorname{Re}\lambda > \omega$ *and* $\tilde{k}(\lambda) \neq 0$. *Then* $(R_1(t), R_2(t))_{t\geqslant 0}$ *is a mild (a, k)-regularized* (C_1, C_2)*-existence and uniqueness family with a subgenerator A.*

(ii) *Let* $(R_1(t))_{t\geqslant 0}$ *be strongly continuous, and let the family* $\{e^{-\omega t}R_i(t) : t \geqslant 0\}$ *be equicontinuous. Then* $(R_1(t))_{t\geqslant 0}$ *is a mild (a, k)-regularized* C_1*-existence family with a subgenerator A if for every* $\lambda \in \mathbb{C}$ *with* $\operatorname{Re}\lambda > \omega$ *and* $\tilde{k}(\lambda) \neq 0$, *we have* $R(C_1) \subseteq R(I - \tilde{a}(\lambda)A)$ *and*

$$\tilde{k}(\lambda)C_1x = (I - \tilde{a}(\lambda)A)\int_0^\infty e^{-\lambda t}R_1(t)x\, dt,\quad x \in E.$$

(iii) *Let* $(R_2(t))_{t\geq 0}$ *be strongly continuous, let* $R_2(0) = k(0)C_2x$, $x \in E\setminus\overline{D(A)}$, *and let the family* $\{e^{-\omega t}R_i(t) : t \geqslant 0\}$ *be equicontinuous. Then* $(R_2(t))_{t\geq 0}$ *is a mild* (a, k)*-regularized* C_2*-uniqueness family with a subgenerator* A *if for every* $\lambda \in \mathbb{C}$ *with* $\operatorname{Re}\lambda > \omega$ *and* $\tilde{k}(\lambda) \neq 0$, *the operator* $I - \tilde{a}(\lambda)A$ *is injective and (252) holds.*

The subsequent theorem can be shown following the lines of the proof of Theorem 2.4.2 (cf. also Theorem 2.8.5(b), [303, Theorem 3.9] and [49, Section 3]).

Theorem 2.8.6. *Assume* $k_\beta(t)$ *satisfies (P1),* $0 < \alpha < \beta$, $\gamma = \alpha/\beta$ *and* A *is a subgenerator of an exponentially equicontinuous* (g_β, k_β)*-regularized* C_1*-existence family* $(R_{1,\beta}(t))_{t\geqslant 0}$, *resp. a q-exponentially equicontinuous* (g_β, k_β)*-regularized* C_2*-uniqueness family* $(R_{2,\beta}(t))_{t\geqslant 0}$ *satisfying that the family* $\{e^{-\omega t}R_{1,\beta}(t) : t \geqslant 0\}$ *is equicontinuous for some* $\omega \geqslant 0$, *resp. (105) with* $R(\cdot)$ *replaced by* $R_{2,\beta}(\cdot)$ *therein. Assume that there exist a continuous function* $k_\alpha(t)$ *satisfying (P1) and a number* $\upsilon > 0$ *such that* $k_\alpha(0) = k_\beta(0)$ *and*

$$\tilde{k}_\alpha(\lambda) = \lambda^{\gamma-1}\tilde{k}_\beta(\lambda^\gamma),\ \lambda > \upsilon.$$

Then A *is a subgenerator of an exponentially equicontinuous, resp. a q-exponentially equicontinuous, mild* (g_α, k_α)*-regularized* C_1*-existence family* $(R_{1,\alpha}(t))_{t\geqslant 0}$, *resp. mild* (g_α, k_α)*-regularized* C_2*-uniqueness family* $(R_{2,\alpha}(t))_{t\geqslant 0}$, *given by* $R_{i,\alpha}(0) := k_\alpha(0)C_i$, $i = 1, 2$ *and*

$$R_{i,\alpha}(t)x := \int_0^\infty t^{-\gamma}\,\Phi_\gamma(st^{-\gamma})\,R_{i,\beta}(s)x\,ds,\ x \in E,\ t > 0,\ i = 1, 2.$$

Furthermore,

$$p(R_{2,\alpha}(t)x) \leqslant c_\gamma M_p \exp(\omega_p^{1/\gamma}t)q_p(x),\ p \in \circledast,\ t \geqslant 0,\ x \in E.$$

Let $p \in \circledast$. *Then the following estimate holds*

$$p(R_{2,\alpha}(t)x) \leqslant c_\gamma M_p \exp(\omega_p^{1/\gamma}t)q_p(x),\ t \geqslant 0,\ x \in E,$$

and the condition

$$p(R_{2,\beta}(t)x) \leqslant M_p(1 + t^{\xi_p})e^{\omega_p t}q_p(x),\ t \geqslant 0,\ x \in E\ (\xi_p \geqslant 0),$$

resp.,

$$p(R_{2,\beta}(t)x) \leqslant M_p t^{\xi_p}e^{\omega_p t}q_p(x),\ t \geqslant 0,\ x \in E,$$

implies that there exists $M_p' \geqslant 1$ *such that*

$$p(R_{2,\alpha}(t)x) \leqslant M_p'(1 + t^{\xi_p\gamma})(1 + \omega_p t^{\xi_p(1-\gamma)})\exp(\omega_p^{1/\gamma}t)q_p(x),\ t \geqslant 0,\ x \in E,$$

resp.,

$$p(R_{2,\alpha}(t)x) \leqslant M_p' t^{\xi_p\gamma}(1 + \omega_p t^{\xi_p(1-\gamma)})\exp(\omega_p^{1/\gamma}t)q_p(x),\ t \geqslant 0,\ x \in E.$$

Furthermore, in the above inequalities we may replace $R_{2,\alpha}(\cdot)$ and ω_p by $R_{1,\alpha}(\cdot)$ and ω respectively. We also have the following:

(i) *The mapping $t \mapsto R_{i,\alpha}(t)$, $t > 0$ admits an extension to $\Sigma_{\min((\frac{1}{\gamma}-1)\frac{\pi}{2},\pi)}$ and, for every $x \in E$, the mapping $z \mapsto R_{i,\alpha}(z)x$, $z \in \Sigma_{\min((\frac{1}{\gamma}-1)\frac{\pi}{2},\pi)}$ is analytic ($i = 1, 2$).*

(ii) *Let $\varepsilon \in (0, \min((\frac{1}{\gamma}-1)\frac{\pi}{2}, \pi))$. If, for every $p \in \circledast$, one has $\omega_p = 0$, then $(R_{1,\alpha}(t))_{t\geqslant 0}$ is an equicontinuous analytic (g_α, k_α)-regularized C_1-existence family of angle $\min((\frac{1}{\gamma}-1)\frac{\pi}{2}, \pi)$, resp. $(R_{2,\alpha}(t))_{t\geqslant 0}$ is an equicontinuous analytic (g_α, k_α)-regularized C_2-uniqueness family of angle $\min((\frac{1}{\gamma}-1)\frac{\pi}{2}, \pi)$.*

(iii) *If $\omega_p > 0$ for some $p \in \circledast$, then $(R_{1,\alpha}(t))_{t\geqslant 0}$ is an exponentially equicontinuous, analytic (g_α, k_α)-regularized C_1-existence family of angle $\min((\frac{1}{\gamma}-1)\frac{\pi}{2}, \frac{\pi}{2})$ and $(R_{2,\alpha}(t))_{t\geqslant 0}$ is a q-exponentially equicontinuous, analytic (g_α, k_α)-regularized C_2-uniqueness family of angle $\min((\frac{1}{\gamma}-1)\frac{\pi}{2}, \frac{\pi}{2})$.*

Let us note that it is not clear how one can rephrase Theorem 2.8.6 for a general q-exponentially equicontinuous (g_β, k_β)-regularized C_1-existence family $(R_\beta(t))_{t\geqslant 0}$.

The main objective in the subsequent theorem is to transmit the assertion of subordination principles stated in Theorem 2.1.8 to mild exponentially equicontinuous (a, k)-regularized (C_1, C_2)-existence and uniqueness families.

Theorem 2.8.7. *Suppose $C_1, C_2 \in L(E)$ and C_2 is injective.*
(i) Let $a(t)$, $b(t)$ and $c(t)$ satisfy (P1), and let $\int_0^\infty e^{-\beta t}|b(t)|\,dt < \infty$ for some $\beta \geqslant 0$. Let

$$\alpha = \tilde{c}^{-1}\left(\frac{1}{\beta}\right) \text{ if } \int_0^\infty c(t)\,dt > \frac{1}{\beta},\ \alpha = 0 \text{ otherwise,}$$

and let $\tilde{a}(\lambda) = \tilde{b}(\frac{1}{\tilde{c}(\lambda)})$, $\lambda \geqslant \alpha$. Let A be a subgenerator of a (b, k)-regularized C_1-existence family $(R_1(t))_{t\geqslant 0}$ ((b, k)-regularized C_2-uniqueness family $(R_2(t))_{t\geqslant 0}$) satisfying that the family $\{e^{-\omega_b t}R_1(t) : t \geqslant 0\}$ ($\{e^{-\omega_b t}R_2(t) : t \geqslant 0\}$) is equicontinuous for some $\omega_b \geqslant 0$. Assume, further, that $c(t)$ is completely positive and that there exists a scalar-valued continuous kernel $k_1(t)$ satisfying (P1) and

$$\tilde{k}_1(\lambda) = \frac{1}{\lambda\tilde{c}(\lambda)}\tilde{k}\left(\frac{1}{\tilde{c}(\lambda)}\right),\ \lambda > \omega_0,\ \tilde{k}\left(\frac{1}{\tilde{c}(\lambda)}\right) \neq 0, \text{ for some } \omega_0 > 0.$$

Let

$$\omega_a = \tilde{c}^{-1}\left(\frac{1}{\omega_b}\right) \text{ if } \int_0^\infty c(t)\,dt > \frac{1}{\omega_b},\ \omega_a = 0 \text{ otherwise.}$$

Then, for every $r \in (0, 1]$, A is a subgenerator of a global $(a, k_1 * g_r)$-regularized C_1-existence family $(R_{r,1}(t))_{t\geqslant 0}$ ($(a, k_1 * g_r)$-regularized C_2-uniqueness family $(R_{r,2}(t))_{t\geqslant 0}$) such that the family $\{e^{-\omega_a t}R_{r,i}(t) : t \geqslant 0\}$ is equicontinuous and that the mapping $t \mapsto R_{r,i}(t)$, $t \geqslant 0$ is locally Hölder continuous with exponent r, if $\omega_b = 0$ or $\omega_b\tilde{c}(0) \neq 1$ ($i = 1, 2$), resp., for every $\varepsilon > 0$, there exists

$M_\varepsilon \geqslant 1$ *such that the family* $\{e^{-\varepsilon t} R_{r,i}(t) : t \geqslant 0\}$ *is equicontinuous and that the mapping* $t \mapsto R_{r,i}(t)$, $t \geqslant 0$ *is locally Hölder continuous with exponent r, if* $\omega_b > 0$ *and* $\omega_b \tilde{c}(0) = 1$ *(i = 1, 2). Furthermore, if A is densely defined, then A is a subgenerator of a global (a,* k_1*)-regularized* C_1*-existence family* $(R_1(t))_{t\geqslant 0}$ *((a,* k_1*)-regularized* C_2*-uniqueness family* $(R_2(t))_{t\geqslant 0}$*) such that the family* $\{e^{-\omega_a t} R_i(t) : t \geqslant 0\}$ *is equicontinuous, resp., for every* $\varepsilon > 0$*, the family* $\{e^{-\varepsilon t} R_{r,i}(t) : t \geqslant 0\}$ *is equicontinuous (i = 1, 2).*

(ii) *Suppose* $\alpha \geqslant 0$*, A is a subgenerator of a global exponentially equicontinuous* $(1, g_{\alpha+1})$*-regularized* C_1*-existence family ((1,* $g_{\alpha+1}$*)-regularized* C_2*-uniqueness family), a(t) is completely positive and satisfies (P1), k(t) satisfies (P1) and* $\tilde{k}(\lambda) = \tilde{a}(\lambda)^\alpha$*,* λ *sufficiently large. Then, for every* $r \in (0, 1]$*, A is a subgenerator of a locally Hölder continuous (with exponent r), exponentially equicontinuous (a,* $k * g_r$*)-regularized* C_1*-existence family ((a,* $k * g_r$*)-regularized* C_2*-uniqueness family); if* $\alpha = n \in \mathbb{N}$*, resp.* $\alpha = 0$*, then A is a subgenerator of a locally Hölder continuous (with exponent r), exponentially equicontinuous (a,* $a^{*,n} * g_r$*)-regularized* C_1*-existence family ((a,* $a^{*,n} * g_r$*)-regularized* C_2*-uniqueness family) if* $\alpha = n \in \mathbb{N}$*, resp. (a,* g_{r+1}*)-regularized* C_1*-existence family ((a,* g_{r+1}*)-regularized* C_2*-uniqueness family). If, additionally, A is densely defined, then A is a subgenerator of an exponentially equicontinuous (a,* $1 * k$*)-regularized* C_1*-existence family ((a,* $1 * k$*)-regularized* C_2*-uniqueness family); if* $\alpha = n \in \mathbb{N}$*, resp.,* $\alpha = 0$*, then A is a subgenerator of an exponentially equicontinuous (a,* $1 * a^{*,n}$*)-regularized* C_1*-existence family ((a,* $1 * a^{*,n}$*)-regularized* C_2*-uniqueness family), resp. (a, 1)-regularized* C_1*-existence family ((a, 1)-regularized* C_2*-uniqueness family).*

(iii) *Suppose* $\alpha \geqslant 0$ *and A is a subgenerator of an exponentially equicontinuous (t,* $g_{\alpha+1}$*)-regularized* C_1*-existence family ((t,* $g_{\alpha+1}$*)-regularized* C_2*-uniqueness family). Let* $L^1_{loc}([0, \infty)) \ni c$ *be completely positive and let* $a(t) = (c * c)(t)$*,* $t \geqslant 0$*. (Given* $L^1_{loc}([0, \infty)) \ni a$ *in advance, such a function c(t) always exists provided a(t) is completely positive or* $a(t) \neq 0$ *is a creep function and* $a_1(t)$ *is log-convex.) Assume k(t) satisfies (P1) and* $\tilde{k}(\lambda) = \tilde{c}(\lambda)^\alpha/\lambda$*,* λ *sufficiently large. Then, for every* $r \in (0, 1]$*, A is a subgenerator of a locally Hölder continuous (with exponent r), exponentially equicontinuous (a,* $k * g_r$*)-regularized* C_1*-existence family ((a,* $k * g_r$*)-regularized* C_2*-uniqueness family); if* $\alpha = n \in \mathbb{N}$*, resp.* $\alpha = 0$*, then A is a subgenerator of a locally Hölder continuous (with exponent r), exponentially equicontinuous (a,* $c^{*,n} * g_r$*)-regularized* C_1*-existence family ((a,* $c^{*,n} * g_r$*)-regularized* C_2*-uniqueness family), resp. (a,* g_{r+1}*)-regularized* C_1*-existence family ((a,* g_{r+1}*)-regularized* C_2*-uniqueness family). If, additionally, A is densely defined, then A is a subgenerator of an exponentially equicontinuous (a,* $1 * k$*)-regularized* C_1*-existence family ((a,* $1 * k$*)-regularized* C_2*-uniqueness family); if* $\alpha = n \in \mathbb{N}$*, resp.* $\alpha = 0$*, then A is a subgenerator of an exponentially equicontinuous (a,* $1 * c^{*,n}$*)-regularized* C_1*-existence family ((a,* $1 * c^{*,n}$*)-regularized* C_2*-uniqueness family), resp. (a, 1)-regularized* C_1*-existence family ((a, 1)-regularized* C_2*-uniqueness family).*

Now we state the following proposition.

Proposition 2.8.8. *Suppose $(R_1(t), R_2(t))_{t\in[0,\tau)}$ is a mild (a, k)-regularized (C_1, C_2)-existence and uniqueness family with a subgenerator A, the family $\{R_2(t) : t \in [0, \tau)\}$ is locally equicontinuous, and the function $a(t)$ is a kernel.*
Then $C_2R_1(t) = R_2(t)C_1$, $t \in [0, \tau)$.

Proof. Let $x \in E$ be fixed. Then the mapping $t \mapsto (a * R_2)(t)x$, $t \in [0, \tau)$ is continuous. Due to the local equicontinuity of family $\{R_2(t) : t \in [0, \tau)\}$, the mappings $t \mapsto (R_2 * (a * R_1))(t)x$, $t \in [0, \tau)$ and $t \mapsto ((a * R_2) * R_1)(t)x$, $t \in [0, \tau)$ are continuous and coincide. Therefore, for every $0 \leqslant t < \tau$,

$$\begin{aligned}R(t, x) &:= -[(a * R_2) * (R_1(\cdot) - k(\cdot)C_1)](t)x \\ &\quad + [(k(0) - k(\cdot)) * C_2(a * R_1)](t)x + (R_2 * (a * R_1))(t)x \\ &= [(k(0) - k(\cdot)) * C_2\,(a * R_1)](t)x - [(a * R_2) * k(\cdot)C_1](t)x.\end{aligned}$$

On the other hand, a trivial computation involving the equalities (245)-(246) shows that

$$R(t, x) = [k(0) * C_2(a * R_1)](t)x, \quad 0 \leqslant t < \tau.$$

The above implies $(a*k*C_2R_1)(t)x = (a*k*R_2C_1)(t)x$, $t \in [0, \tau)$, which completes the proof. □

Let $(R_1(t), R_2(t))_{t\in[0,\tau)}$ be a mild (a, k)-regularized (C_1, C_2)-existence and uniqueness family with a subgenerator A. Then it is clear that the function $t \mapsto R_1(t)x$, $t \in [0, \tau)$, resp. $t \mapsto R_2(t)x$, $t \in [0, \tau)$, is a mild solution of (1) with $f(t) = k(t)C_1x$, $t \in [0, \tau)$ $(x \in E)$, resp. a strong solution of (1) with $f(t) = k(t)C_2x$, $t \in [0, \tau)$ $(x \in D(A))$, provided additionally in the last case that $R_2(t)x \in D(A)$, $t \in [0, \tau)$ and $AR_2(t)x = R_2(t)Ax$, $t \in [0, \tau)$.

Suppose now that $(R_2(t))_{t\in[0,\tau)}$ is a locally equicontinuous C_2-uniqueness family with a subgenerator A. By the proof of [382, Theorem 2.7], we easily infer that every strong solution $u(t)$ of (1) satisfies the following equality:

$$(R_2 * f)(t) = (kC_2 * u)(t), \quad 0 \leqslant t < \tau. \tag{253}$$

Since $k(t)$ is a kernel and C_2 is injective, the above equality implies that (1) has at most one strong solution. Now we will prove the uniqueness of mild solutions of the problem (1). Towards this end, suppose $u_1(t)$ and $u_2(t)$ are two such solutions. Put $u(t) := u_1(t) - u_2(t)$, $t \in [0, \tau)$. Then $A(a * u)(t) = u(t)$, $t \in [0, \tau)$ and $(a * A(a * u))(t) = (a * u)(t)$, $t \in [0, \tau)$, which implies that the function $U(t) := (a * u)(t)$, $t \in [0, \tau)$ is a strong solution of (1) with $f(t) \equiv 0$. Therefore, $u(t) = AU(t) = A0 = 0$, $t \in [0, \tau)$ and we have reached the following conclusion.

Proposition 2.8.9. *Suppose $(R_2(t))_{t\in[0,\tau)}$ is a locally equicontinuous k-regularized C_2-uniqueness family with a subgenerator A. Then every strong solution $u(t)$ of*

(1) satisfies (253). Furthermore, the problem (1) has at most one strong (mild) solution.

We continue by stating the following proposition.

Proposition 2.8.10. *Assume* $\tau \in (0, \infty]$, $L^1_{loc}([0, \tau)) \ni a_1(t)$ *is a kernel,* $L^1_{loc}([0, \tau)) \ni k(t)$ *is a kernel,* $a(t) = (a_1 * a_1)(t)$, $t \in [0, \tau)$ *and* $k_1(t) = (k * a_1)(t)$, $t \in [0, \tau)$. *Let* A *be a closed linear operator on* E, *let* $C_1, C_2 \in L(E)$, *and let* C_2 *be injective. Put* $\mathcal{A} \equiv \begin{pmatrix} 0 & I \\ A & 0 \end{pmatrix}$ *and* $\mathcal{C}_i \equiv \begin{pmatrix} C_i & 0 \\ 0 & C_i \end{pmatrix}$, $i = 1, 2$.
Then A *is a subgenerator of an* (a, k)*-regularized* C_1*-existence family* $(R_1(t))_{t\in[0,\tau)}$ *(*(a, k)*-regularized* C_2*-uniqueness family* $(R_2(t))_{t\in[0,\tau)}$*) if* $\mathcal{A}$ *is a subgenerator of an* (a_1, k_1)*-regularized* $\mathcal{C}_1$*-existence family* $(S_1(t))_{t\in[0,\tau)}$ *(*(a_1, k_1)*-regularized* $\mathcal{C}_2$*-uniqueness family* $(S_2(t))_{t\in[0,\tau)}$*).*

Observe, however, that many other assertions appearing in [292], like [292, Theorem 2.5.1] and [292, Theorem 2.5.3], can be reformulated for $(1, \Theta)$-convoluted C_1-existence (C_2-uniqueness) families.

To round off this section, we shall illustrate results obtained so far with some examples. First of all, it is worth noting that there exist examples of exponentially bounded, analytic (g_α, k)-regularized (C_1, C_2)-existence and uniqueness families whose angle of analyticity can be strictly greater than $\pi/2$ $(0 < \alpha < 1)$. To verify this, we will make use of the following adaptation of [547, Example 3.1] (see [153] for the first examples of such kind). Let $E := \{f \in C(\mathbb{R})\ ;\ \lim_{|x|\to\infty} f(x)e^{x^2} = 0\}$. Then E, furnished with the norm $\|f\| := \sup_{x\in\mathbb{R}} |f(x)e^{x^2}|$, $f \in E$, is a Banach space. Let $A := d^2/dx^2$ act on E with its maximal domain and let $(C_i f)(x) := e^{-x^2} f(x)$, $x \in \mathbb{R}$, $f \in E$, $i = 1, 2$. Put, for every $t \geqslant 0$, $f \in E$ and $x \in \mathbb{R}$:

$$(C_1(t)f)(x) := \frac{1}{2}\Big(e^{-(x+t)^2} f(x + t) + e^{-(x-t)^2} f(x - t)\Big)$$

and

$$(C_2(t)f)(x) := \frac{1}{2} e^{-x^2} (f(x + t) + f(x - t)).$$

Then $(C_1(t), C_2(t))_{t\geqslant 0}$ is a contractive mild $(C_1\ C_2)$-regularized cosine existence and uniqueness family generated by A, which implies by Theorem 2.8.6(ii) that, for every $\alpha \in (0, 2)$, A is the integral generator of an exponentially bounded analytic $(g_\alpha, 1)$-regularized (C_1, C_2)-existence and uniqueness family of angle $\min((\frac{2}{\alpha}-1)\frac{\pi}{2}, \pi)$. Suppose now that $L^1_{loc}([0, \infty)) \ni c$ is completely positive and $a(t) = (c * c)(t)$, $t \geqslant 0$. By Theorem 2.8.7(iii), A is the integral generator of an exponentially bounded $(a, 1)$-regularized (C_1, C_2)-existence and uniqueness family.

It ought to be observed that the conclusions established in [310, Example 36(iii)] and [292, Example 3.1.35(ii)] are false. In the following example, we shall correct some inconsistencies in the first of two above-mentioned examples, providing in such a way a new application of subordination principles given in Theorem 2.8.6/Theorem 2.8.7.

Example 2.8.11. We deal with the space $L^p_\varrho(\Omega, \mathbb{C})$, where Ω is an open nonempty subset of $\mathbb{R}^n$, $\varrho : \Omega \to (0, \infty)$ is a locally integrable function, m_n is the Lebesgue measure in $\mathbb{R}^n$, $1 \leqslant p < \infty$, and the norm of an element $f \in L^p_\varrho(\Omega, \mathbb{C})$ is given by $\|f\|_p := (\int_\Omega |f(\cdot)|^p \varrho(\cdot)\, dm_n)^{1/p}$. Set $|x| := (x_1^2 + \cdots + x_n^2)^{1/2}$, $x = (x_1, \cdots, x_n) \in \mathbb{R}^n$, $\varrho(x) := \exp(-|x|)$, $x \in \mathbb{R}^n$, $E := L^p_\varrho(\mathbb{R}^n, \mathbb{C})$,

$$(T_1(t)f)(x) := e^{-(|e^t x|^2+1)} f(e^t x),\ t \in \mathbb{R}, x \in \mathbb{R}^n, f \in E,$$

$$(T_2(t)f)(x) := e^{-(|x|^2+1)} f(e^t x),\ t \in \mathbb{R}, x \in \mathbb{R}^n, f \in E,$$

$C_1 = C_2 := T_1(0)$ and

$$C_i(t) := \frac{1}{2}(T_i(t) + T_i(-t)),\ t \in \mathbb{R},\ i = 1, 2.$$

Let A be the closure of the operator $f \mapsto \Sigma_{i=1}^n x_i \frac{\partial f}{\partial x_i}$, $f \in D(A) \equiv \{g \in C^1(\mathbb{R}^n : E) :$ supp(g) is a compact subset of $\mathbb{R}^n\}$. Then the operator A^2 is the integral generator of a mild (C_1, C_2)-regularized cosine existence and uniqueness family $(C_1(t), C_2(t))_{t\in\mathbb{R}}$. Furthermore, $(C_1(t))_{t\geqslant 0}$ is exponentially bounded and $(C_2(t))_{t\geqslant 0}$ is not exponentially bounded.

Notice, finally, that the matricial operators can be used for construction of some genuine examples of (g_α, g_β)-regularized C_1-existence families (C_2-uniqueness families), with $\alpha > 0$ and $\beta \geqslant 1$ (see e.g. [153, Section 7] and [547]).

2.9 Complex powers of (*a*, *b*, *C*)-nonnegative operators and fractional resolvent families generated by them

The theory of fractional powers of operators dates from a paper of E. Hille who investigated semigroups formed by the fractional powers of a bounded linear operator in 1939 (cf. also [67] for the construction of fractional powers of the negative Laplace operator). One of the most important papers for further development of the theory was written in 1960 by A. V. Balakrishnan [31]. In this paper, he introduced a method for constructing fractional powers of a wide class of closed linear operators. From the period 1960 onwards, many different approaches have been proposed for construction of fractional powers, and it would be really difficult to summarize here the whole theory and its applications. The main purpose of this section, consisting of six separate subsections, is to develop the theory of complex powers of almost C-nonnegative operators. Many other interesting subjects with regards to powers, like moment inequality (cf. [403, Lemma 3.1.7, Corollary 5.1.13], [217, Corollary 7.2], [90, Theorem 2.16]) and square root reduction problem for fractional operator families, will be reconsidered somewhere else. Concerning possible applications made to abstract fractional differential equations, we would like to mention once more our investigation of certain classes of incomplete abstract Cauchy problems with modified Liouville right-sided time-fractional derivatives.

We shall always assume in this section that $C^{-1}AC = A$ and that, for any considered operator A with $(-\infty, 0) \subseteq \rho_C(A)$, the mapping $\lambda \mapsto (\lambda + A)^{-1}Cx$, $\lambda > 0$ is continuous ($x \in E$). Similar assumptions will be used in the formulations of conditions (H) and (HQ), as well as in any case in which $\rho_C(A)$ contains a suitable region around negative real axis.

We employ the following definition of an (exponentially equicontinuous, analytic) ζ-times integrated C-semigroup ($\zeta > 0$).

Definition 2.9.1. Let $\zeta > 0$. A strongly continuous operator family $(S_\zeta(t))_{t\geqslant 0}$ is called a (global) ζ-times integrated C-semigroup if the following conditions hold:

(a) $S_\zeta(0) = 0$,
(b) $S_\zeta(t)C = CS_\zeta(t)$, $t \geqslant 0$ and
(c) $S_\zeta(t)S_\zeta(s)x = [\int_0^{t+s} - \int_0^t - \int_0^s] g_\zeta(t + s - r)S_\zeta(r)Cx\, dr$ for all $x \in E$ and $t, s \geqslant 0$.

$(S_\zeta(t))_{t\geqslant 0}$ is said to be non-degenerate if the assumption $S_\zeta(t)x = 0$ for all $t \geqslant 0$ implies $x = 0$. For a non-degenerate ζ-times integrated C-semigroup $(S_\zeta(t))_{t\geqslant 0}$ we define its (integral) generator $\hat{A}$ by

$$\hat{A} := \left\{(x, y) \in E \times E : S_\zeta(t)x - g_{\zeta+1}(t)Cx = \int_0^t S_\zeta(s)y\, ds \text{ for all } t \geqslant 0\right\}.$$

The notion of (exponentially equicontinuous) analyticity of $(S_\zeta(t))_{t\geqslant 0}$ is defined in the obvious way.

The following definition has been recently introduced in [90]; cf. also [132]-[133], [292], [446], [452], [493], [500] and [546] for further information about semigroups of growth order $r > 0$.

Definition 2.9.2. (i) An operator family $(T(t))_{t>0} \subseteq L(E)$ is said to be a C-regularized semigroup of growth order $r > 0$ if the following holds:
(a) $T(t + s)C = T(t)T(s)$ for all $t, s \geqslant 0$,
(b) for every $x \in E$, the mapping $t \mapsto T(t)x$, $t > 0$ is continuous,
(c) the family $\{t^r T(t) : t \in (0, 1]\}$ is equicontinuous, and
(d) $T(t)x = 0$ for all $t > 0$ implies $x = 0$.

(ii) Suppose $0 < \alpha \leqslant \pi/2$, $(T(t))_{t>0}$ is a C-regularized semigroup of growth order $r > 0$, and the mapping $t \mapsto T(t)x$, $t > 0$ has an analytic extension to the sector Σ_α, denoted by the same symbol. If there exists $\omega \in \mathbb{R}$ such that, for every $\delta \in (0, \alpha)$, the family $\{|z|^r e^{-\omega \operatorname{Re} z} T(z) : z \in \Sigma_\delta\}$ is equicontinuous, then $(T(z))_{z\in\Sigma_\alpha}$ is said to be an analytic C-regularized semigroup of growth order r.

The integral generator $\hat{G}$, resp. the infinitesimal generator G, of $(T(t))_{t>0}$, is defined by

$$\hat{G} := \left\{(x, y) \in E \times E : T(t)x - T(s)x = \int_s^t T(r)y\, dr \text{ for all } t, s > 0 \text{ with } t \geqslant s\right\},$$

resp.,

$$G := \left\{(x, y) \in E \times E : \lim_{t\to 0+} \frac{T(t)x - Cx}{t} = Cy\right\}.$$

The integral generator $\hat{G}$ is a closed linear operator which satisfies $C^{-1}\hat{G}C = \hat{G}$. Moreover, $G \subseteq \hat{G}$ and G is a closable linear operator. The closure of G, denoted by $\overline{G}$, is said to be the complete infinitesimal generator, in short, the c.i.g. of $(T(t))_{t>0}$. The integral generator $\hat{G}$ contains the c.i.g. $\overline{G}$ and satisfies $\hat{G} = \{(x, y) \in E \times E : (T(s)x, T(s)y) \in G \text{ for all } s > 0\}$. The set $\{x \in E : \lim_{t\to 0+} T(t)x = Cx\}$, resp. $\{x \in E : \lim_{z\to 0, z\in\Sigma_\delta} T_b(z)x = Cx \text{ for all } \delta \in (0, \alpha)\}$, is said to be the continuity set of $(T(t))_{t>0}$, resp. $(T(z))_{z\in\Sigma_\alpha}$.

We need the following useful lemma.

Lemma 2.9.3. *Let l, $n \in \mathbb{N}$ and $\rho_C(A) \neq \emptyset$. Then the operator $C^{-1}A^nC$ is closed and $C^{-1}A^nC = C^{-l}A^nC^l$.*

Proof. It is not difficult to prove that our standing hypothesis $C^{-1}AC = A$ implies $C^{-l}AC^l = A$. Furthermore, the assumption $0 \in \rho_C(A)$ implies inductively that, for every non-zero complex polynomial $p_n(\cdot)$, and for every two elements $x, y \in E$, the following holds:

$$p_n(A)C^lx = C^ly \Rightarrow p_n(A)Cx = Cy, \tag{254}$$

where the operator $p_n(A)$ is defined in the usual way. We will prove the assertion of lemma by induction. Suppose that the operator $C^{-1}A^{n-1}C$ is closed for some $n \in \mathbb{N}\setminus\{1, 2\}$. Let $\lambda \in \rho_C(A)$, and let (x_τ) be a net satisfying $x_\tau \to x$, $\tau \to \infty$ and $C^{-1}A^nCx_\tau \to y$, $\tau \to \infty$. Then $A(\lambda - A)^{-1}Cx_\tau \to A(\lambda - A)^{-1}Cx$, $\tau \to \infty$ and $C^{-1}A^{n-1}CA(\lambda - A)^{-1}Cx_\tau \to (\lambda - A)^{-1}Cy$, $\tau \to \infty$. By the closedness of $C^{-1}A^{n-1}C$, we get that $C^{-1}A^{n-1}CA(\lambda - A)^{-1}Cx = (\lambda - A)^{-1}Cy$ and $A^nC^2x = C^2y$, i.e., $\Sigma_{j=1}^n \binom{n}{j} \lambda^j(A - \lambda)^jC^2x = C^2(y - x)$. Then (254) implies $\Sigma_{j=1}^n \binom{n}{j} \lambda^j(A - \lambda)^jCx = C(y - x)$, $A^nCx = Cy$ and, finally, the closedness of the operator $C^{-1}A^nC$. The equality $C^{-1}A^nC = C^{-l}A^nC^l$ can be proved similarly, by using (254) and induction, again. □

2.9.1. Complex powers of a C-sectorial operator A satisfying $0 \in (\rho_C(A))^\circ$. We start this section with the following definition.

Definition 2.9.4. (i) (see [90, Definition 2.2]) A closed linear operator A on E is called C-nonnegative if $(-\infty, 0) \subseteq \rho_C(A)$ and the family

$$\{\lambda(\lambda + A)^{-1}\, C : \lambda > 0\}$$

is equicontinuous; moreover, a C-nonnegative operator A is called C-positive if, in addition, $0 \in \rho_C(A)$.

(ii) (see [90, Definition 2.1]) Let $0 \leqslant \omega < \pi$. Then a closed linear operator A on E is called C-sectorial of angle ω, in short $A \in Sect_C(\omega)$, if $\mathbb{C}\setminus\overline{\Sigma_\omega} \subseteq \rho_C(A)$ and the family

$$\{\lambda(\lambda - A)^{-1}\, C : \lambda \notin \Sigma_{\omega'}\}$$

is equicontinuous for every $\omega < \omega' < \pi$; if this is the case, then the C-spectral angle of A is defined by $\omega_C(A) := \inf\{\omega \in [0, \pi) : A \in Sect_C(\omega)\}$.

(iii) Let $m \in \mathbb{R}$. Then it is said that the operator A belongs to the class $\mathcal{M}_{C,m}$ if $(-\infty, 0) \subseteq \rho_C(A)$ and the family

$$\{(\lambda^{-1} + \lambda^{m})^{-1}(\lambda + A)^{-1} C : \lambda > 0\}$$

is equicontinuous. Furthermore, it is said that A is almost C-nonnegative if there exists $m \in \mathbb{R}$ such that A belongs to the class $\mathcal{M}_{C,m}$.

The important case in our analysis is $m \geqslant -1$, which covers the operators with C-polynomially bounded resolvent (see e.g. [90] and [404]) and C-nonnegative operators; the complete construction of powers is given only in this case. If $A \in \mathcal{M}_{C,m}$ for some $m < -1$ or, more generally, if A is (a, b, C)-nonnegative with $\min(a, b) < -1$ and $\max(a, b) \geqslant 1$ (cf. Subsection 2.9.6), then the construction relies upon the injectivity of A (see e.g. [20, Example 8.3.8] and [548, Example 4.1] for an example of such an operator A with $\rho(A) = \emptyset$). As another illustration, we would like to quote the non-injective operator $A := -\xi\Delta^2 + i\varrho\Delta$ acting on $E := L^2(\mathbb{R}^n)$ with its maximal distributional domain ($\xi > 0$, $\varrho \in \mathbb{R}\setminus\{0\}$). Then it is not difficult to prove that A belongs to the class $\mathcal{M}_{I,(-1)/2}$; furthermore, we have by [404, Theorem 2.1] that there does not exist $m \in [-1, (-1)/2)$ such that A belongs to the class $\mathcal{M}_{I,m}$.

The following proposition will be helpful in our further work (cf. also [90, Proposition 2.4(iii)] and [404, Remark 2.4]).

Proposition 2.9.5. *Suppose A is injective, $m \in \mathbb{R}$ and A belongs to the class $\mathcal{M}_{C,m}$. Then A^{-1} belongs to the class $\mathcal{M}_{C,-m-2}$.*

If A belongs to the class $\mathcal{M}_{C,m}$ for some $m \leqslant -1$, then it is not difficult to prove that the following equality holds:

$$\lim_{\lambda \to +\infty} [\lambda(\lambda + A)^{-1} C]^n x = C^n x, \quad x \in \overline{D(A)},\ n \in \mathbb{N}. \tag{255}$$

We continue by observing that a C-sectorial operator A has to be C-nonnegative; even in the case that E is a Fréchet space and $C = 1$ (in this section, the identity operator on E will be also denoted by 1), the converse statement is not true, in general (cf. [403, Subsection 1.4.1]). Notice also that a C-positive operator A on a Banach space E need not be C-sectorial unless $C = 1$. In order to illustrate this, consider the negative of the operator A; considered in the paragraph following Definition 2.9.4. Then $-A$ is A^{-1}-positive and $-A$ is not A^{-1}-sectorial (see [292, Example 3.5.30(ii)]). It is also worthwhile to mention here that the assumption on C-sectoriality, used in the construction of fractional powers of operators established in [90], can be slightly weakened (cf. [292], [403]-[404] and Remark 2.9.8 for further information in this direction).

Some basic properties of C-nonnegative operators are collected in the following proposition (for information about how to get proofs, see [403, Chapter 1] and [303, Remark 2.2]).

Proposition 2.9.6. (i) *If* $0 \in \rho(C)$, *then* A *is* C-*nonnegative iff* A *is nonnegative.*
(ii) *If* A *is* C-*positive, then the family* $\{(\lambda + C)(\lambda + A)^{-1} C : \lambda > 0\}$ *is equicontinuous. If the above operator family is equicontinuous, then the following holds:*
(ii.1) *The nonnegativity of* C *implies that* A *is* C-*nonnegative.*
(ii.2) *The positivity of* C *implies that* A *is* C-*positive.*
(iii) *Let* A *be* C-*nonnegative, let* $n \in \mathbb{N}$, *and let* $x \in E$ *satisfy* $Cx \in R((\lambda+A)^n)$, $\lambda > 0$. *Then the following assertions hold.*
(iii.1) *The family* $\{A(\lambda + A)^{-1} C : \lambda > 0\}$ *is equicontinuous.*
(iii.2) *If* A *is injective, then* $\lambda((\lambda + A)^{-1})^{-1}C = A(\lambda^{-1} + A)^{-1}C$ *for all* $\lambda > 0$. *Hence,* A^{-1} *is* C-*nonnegative.*
(iii.3) $\lim_{\lambda\to\infty} A^n(\lambda + A)^{-n}Cx = 0 \Leftrightarrow \lim_{\lambda\to\infty} \lambda^n(\lambda + A)^{-n} Cx = Cx$.
(iii.4) $\lim_{\lambda\to 0} \lambda^n(\lambda + A)^{-n} Cx = 0 \Leftrightarrow \lim_{\lambda\to 0} A^n(\lambda + A)^{-n}Cx = Cx$.
(iii.5) *Let* E *be barreled, and let* $D(A)$ *and* $R(C)$ *be dense in* E. *Then* A^* *is* C^*-*nonnegative in* E^*.

Let $0 \leqslant \omega < \varphi \leqslant \pi$ and $0 < d \leqslant 1$. Up to the beginning of Subsection 2.9.2, we shall always assume that a closed linear operator A satisfies the following condition.

(HC) : A is C-sectorial of angle ω, $B_{d_1} \subseteq \rho_C(-A)$, the family $\{(z - A)^{-1}C : z \in B_{d_1}\}$ is equicontinuous for all $d_1 \in (0, d)$, and the mapping $z \mapsto (z - A)^{-1} Cx$ is continuous on $\Lambda_{\omega,d} := ((\mathbb{C}\setminus\Sigma_\omega) \cup B_d)^{\circ}$ for every $x \in E$.

For a closed linear operator A satisfying (HC), one can introduce the H^∞-functional calculus $f(A)$ for appropriate holomorphic functions $f(\cdot)$. Denote by $H(\Sigma_\varphi)$ the space of all holomorphic functions on the sector Σ_φ and by $H^\infty(\Sigma_\varphi)$ the space which consists of those functions $f \in H(\Sigma_\varphi)$ such that

$$|f(z)| \leqslant M|z|^{-s} \quad (z \in \Sigma_\varphi)$$

for some constants $M, s > 0$. Notice that the mapping $z \mapsto (z - A)^{-1}Cx$ is analytic in $\Lambda_{\omega,d}$ as well as that $(z - A)^{-n}C \in L(E)$ and

$$(256) \qquad \frac{d^{n-1}}{dz^{n-1}}(z - A)^{-1}Cx = (-1)^{n-1}(n - 1)!(z - A)^{-n}Cx, \; x \in E, \; z \in \Lambda_{\omega,d}, \; n \in \mathbb{N}.$$

Now we are in a position to define the H^∞-functional calculus $f_C(A)$ for the operator A as follows

$$(257) \qquad f_C(A)x := \frac{1}{2\pi i} \int_{\Gamma_{\omega',d'}} f(z)(z - A)^{-1} Cx \, dz, \quad x \in E,$$

where $\Gamma_{\omega',d'} = \partial(\Sigma_{\omega'} \setminus B_{d'})$, the boundary of $\Sigma_{\omega'} \setminus B_{d'}$, is oriented in such a way that $\operatorname{Im} z$ increases along $\Gamma_{\omega',d'}$, with $\omega' \in (\omega, \varphi)$ and $d' \in (0, d)$ arbitrary. Then an application of Cauchy's theorem shows that the above definition does not depend on the choice of numbers ω' and d'. Furthermore, the mapping $f \mapsto f_C(A)$ is a homomorphism from $H^\infty(\Sigma_\varphi)$ into $L(E)$ in the following sense:

$$(258) \qquad f_C(A)g_C(A) = (fg)_C(A)C, \quad f, g, fg \in H^\infty(\Sigma_\varphi).$$

It immediately follows from (258) that

(259)
$$(z^{-b})_C(A)\left(\frac{1}{\lambda+z^b}\right)_C(A)=\left(\frac{z^{-b}}{\lambda+z^b}\right)_C(A)C$$
$$=\frac{1}{2\pi i}\int_{\Gamma_{\omega',d'}}\frac{z^{-b}}{\lambda+z^b}(z-A)^{-1}C^2\,dz,$$

provided $0 < b < \pi/\varphi$ and $\lambda > 0$. Given $b \in \mathbb{C}$ with $\operatorname{Re} b > 0$, set $A_C^{-b} := (z^{-b})_C(A)$ and $A_C^{-0} := C$. Obviously, $A_C^{-n} = A^{-n}C$ ($n \in \mathbb{N}$), $A_C^{-b}\,C = CA_C^{-b}\,C$ ($\operatorname{Re} b > 0$), the mapping $b \mapsto A_C^{-b}x$, $\operatorname{Re} b > 0$ is analytic for every fixed $x \in E$, and the following holds:

$$\frac{d}{db}A_C^{-b}x=\frac{(-1)}{2\pi i}\int_{\Gamma_{\omega',d'}}(\ln z)z^{-b}(z-A)^{-1}Cx\,dz,\quad x\in E,\ \operatorname{Re} b>0.$$

Applying the equality (258) once more, we get that

$$A_C^{-b_1}A_C^{-b_2}=A_C^{-(b_1+b_2)}C,\quad \operatorname{Re} b_1, \operatorname{Re} b_2>0.$$

Notice also that the mapping $z \mapsto z^{-b}$, $z \in \Sigma_\pi$ is analytic, which implies that, for every $b \in \mathbb{C}$ with $0 < \operatorname{Re} b < 1$, we can take $\omega' = \pi$ in the integration appearing in (257). In such a way, we obtain that:

$$A_C^{-b}x=\lim_{\varepsilon\to 0+}\frac{1}{2\pi i}\int_{\Gamma_{\pi,\varepsilon}}z^{-b}(z-A)^{-1}Cx\,dz$$
$$=-\frac{\sin\pi b}{\pi}\int_0^\infty\lambda^{-b}(\lambda+A)^{-1}\,Cx\,d\lambda,\quad 0<\operatorname{Re} b<1,\quad x\in E.$$

By [90, Lemma 2.5, Lemma 2.6], the family $\{A_C^{-b} : 0 < b < 1\}$ is equicontinuous. Furthermore, the operator A_C^{-b} is injective for every $b \in \mathbb{C}$ with $\operatorname{Re} b > 0$, and in the case that $D(A)$ and $R(C)$ are dense in E, we have that $(A_C^{-b})_{b>0}$ is a C-regularized semigroup on E. Define now the powers with negative imaginary part of exponent by

$$A_{-b} := C^{-1}A_C^{-b},\quad \operatorname{Re} b > 0.$$

Then A_{-b} is closed and injective ($b \in \mathbb{C}_+$), and $A_{-n} = C^{-1}A^{-n}C$ ($n \in \mathbb{N}$). Also, by Lemma [90, Lemma 2.6], we can define the powers with positive imaginary part of exponent by

$$A_b := (A_{-b})^{-1} = (A_C^{-b})^{-1}C,\quad \operatorname{Re} b > 0.$$

Clearly, $A_n = C^{-1}A^nC$ for every $n \in \mathbb{N}$, and A_b is closed (injective) due to the closedness (injectiveness) of A_{-b} ($b \in \mathbb{C}_+$). Following [403, Definition 7.1.2], we introduce the purely imaginary powers of A as follows: Let $\tau \in \mathbb{R}\setminus\{0\}$. Then the power $A_{i\tau}$ is defined by

$$A_{i\tau} := C^{-2}(A+1)_2A_{-1}A_{1+i\tau}(A+1)_{-2}C^2.$$

It is clear that $A_{i\tau}$ is a linear operator. Furthermore, $(A+1)A_{-1}A_{1+i\tau}(A+1)_{-2}C^2 \in L(E)$ and, for every $x \in E$,

$$(A+1)A_{-1}A_{1+i\tau}(A+1)_{-2}C^2x = \frac{1}{2\pi i}\int_{\Gamma_{\omega',d'}} z^{-1+i\tau}\frac{z}{z+1}(z-A)^{-1}C^2x\,dz;$$

cf. [90] for more details. Keeping in mind that $C^{-1}(A+1)_1C = (A+1)_1 = C^{-1}(A+1)C$, it readily follows that $x \in D(A_{i\tau})$ if $\frac{1}{2\pi i}\int_{\Gamma_{\omega',d'}} z^{-1+i\tau}\frac{z}{z+1}(z-A)^{-1}Cx\,dz \in D(C^{-2}(A+1)C)$. If this is the case, we have the following equality:

$$A_{i\tau}x = C^{-2}(A+1)C\frac{1}{2\pi i}\int_{\Gamma_{\omega',d'}} z^{-1+i\tau}\frac{z}{z+1}(z-A)^{-1}Cx\,dz. \tag{260}$$

The closedness of $A_{i\tau}$ now follows from (260), along with the closedness of the operator $A+1$ and the dominated convergence theorem. Notice that the operator $A_{i\tau}$ can be introduced equivalently by

$$A_{i\tau} = C^{-j}(A+\lambda)_q A_{-p} A_{p+i\tau}(A+\lambda)_{-q}C^j,$$

where $p, q, j \in \mathbb{N}$, $q > p$ and $\lambda > 0$, and that the following holds (see [90, Theorem 2.10]):

Theorem 2.9.7. *Let $\tau, \tau_1, \tau_2 \in \mathbb{R}$ and $k \in \mathbb{N}$. Then the following holds:*

(i) $C(D(A^k)) \cup \bigcup_{\lambda\in\Lambda_{\omega,d}} R((\lambda-A)^{-k}C) \subseteq D(A_{i\tau})$.
(ii) $C^{-1}A_{i\tau}C = A_{i\tau}$.
(iii) $A_{i\tau}$ *is injective and* $A_{i\tau} = (A_{-i\tau})^{-1}$.
(iv) $A_{i\tau_1}, A_{i\tau_2} \subseteq A_{i(\tau_1+\tau_2)}$, $C(D(A^k)) \cup \bigcup_{\lambda\in\Lambda_{\omega,d}} R((\lambda-A)^{-k}C) \subseteq D(A_{i\tau_1}A_{i\tau_2})$,

$$A_{i(\tau_1+\tau_2)} = ((\lambda-A)^{-k}C)^{-1}A_{i\tau_1}A_{i\tau_2}(\lambda-A)^{-k}C,$$

$A_{i\tau_1}A_{i\tau_2}$ *is closable,* $C^{-1}\overline{A_{i\tau_1}A_{i\tau_2}}C \subseteq A_{i(\tau_1+\tau_2)}$, *with the equality in the case that* $D(A)$ *and* $R(C)$ *are dense in* E.
(v) *Let* $\operatorname{Re} b < 0$ *and* $\tau \in \mathbb{R}$. *Then the following holds:*

$$A_{i\tau}A_b \subseteq A_{b+i\tau}, \tag{261}$$

$$A_{i\tau}A_b \subseteq A_{i\tau+b}, \tag{262}$$

$$A_{b+i\tau} = ((\lambda-A)^{-k}C)^{-1}A_bA_{i\tau}(\lambda-A)^{-k}C, \quad k\in\mathbb{N}, \lambda\in\Lambda_{\omega,d}, \tag{263}$$

$$A_{b+i\tau} = ((\lambda-A)^{-k}C)^{-1}A_{i\tau}A_b(\lambda-A)^{-k}C, \quad k\in\mathbb{N}, \lambda\in\Lambda_{\omega,d}, \tag{264}$$

the operators $A_{i\tau}A_b$ *and* $A_bA_{i\tau}$ *are closable,* $C^{-1}\overline{A_{i\tau}A_b}C \subseteq A_{b+i\tau}$ *and* $C^{-1}\overline{A_bA_{i\tau}}C \subseteq A_{b+i\tau}$. *If* $D(A)$ *and* $R(C)$ *are dense in* E, *then the converse inclusions hold as well.*
(vi) *Let* $\operatorname{Re} b > 0$ *and* $\tau \in \mathbb{R}$. *Then (261)-(262) hold. In the case that* $k \geqslant \lceil \operatorname{Re} b \rceil$, *we have (263)-(264). Furthermore, the operators* $A_{i\tau}A_b$ *and* $A_bA_{i\tau}$ *are closable,* $C^{-1}\overline{A_{i\tau}A_b}C \subseteq A_{b+i\tau}$ *and* $C^{-1}\overline{A_bA_{i\tau}}C \subseteq A_{b+i\tau}$. *If* $D(A)$ *and* $R(C)$ *are dense in* E, *then the converse inclusions hold as well.*

(vii) *Suppose* $\tau \in \mathbb{R}$, $x \in E$ *and* $\lambda \in \Lambda_{\omega,d}$. *Then the following equality holds:* $\lim_{b\to i\tau, \mathrm{Re}\, b>0} A_b(\lambda - A)^{-1}Cx = A_{i\tau}(\lambda - A)^{-1}Cx$ *and* $\lim_{b\to i\tau, \mathrm{Re}\, b<0} A_C^b(\lambda - A)^{-1}Cx = A_{i\tau}(\lambda - A)^{-1}C^2x.$

It would take too long us to consider some other properties and applications of purely imaginary powers of C-sectorial operators. For further information in this direction, the reader may consult [403, Sections 7-10], [465], [542, pp. 105-116] and references cited there.

Remark 2.9.8. (i) Given $\beta \geqslant -1$, $\varepsilon \in (0, 1]$ and $c \in (0, 1)$, put $P_{\beta,\varepsilon,c} := \{\xi + i\eta : \xi \geqslant \varepsilon, \eta \in \mathbb{R}, |\eta| \leqslant c(1 + \xi)^{-\beta}\}$. Assume $(E, \|\cdot\|)$ is a Banach space, $\alpha \geqslant -1$ and A is a closed linear operator on E with the following properties:

$$(0, \infty) \subseteq \rho(A) \text{ and } \sup_{\lambda>0}(1 + |\lambda|)^{-\alpha}\,\|(\lambda - A)^{-1}\| < \infty.$$

By the usual series argument, we have that there exist $d \in (0, 1]$, $c \in (0, 1)$ and $\varepsilon \in (0, 1]$ such that $(\varepsilon, c(1 + \varepsilon)^{-\alpha}) \in \partial B_d$,

$$P_{\alpha,\varepsilon,c} \cup B_d \subseteq \rho(A) \text{ and } \sup_{\lambda\in P_{\alpha,\varepsilon,c}\cup B_d}(1 + |\lambda|)^{-\alpha}\,\|(\lambda - A)^{-1}\| < \infty.$$

Put $n_\alpha := \lfloor\alpha\rfloor + 2$ if $\alpha \notin \mathbb{Z}$, and $n_\alpha := \alpha + 1$, otherwise. Denote by $(-A)^b$ ($b \in \mathbb{C}$) the complex power defined in [292, Section 1.4]. Observe now that the method we have developed in this subsection, with $C = (-A)^{-n_\alpha}$, gives the definition of power $(-A)_b$. It is not difficult to prove that $(-A)^b \subseteq (-A)_b$ for all $b \in \mathbb{C}$. Moreover, the set appearing in [300, Remark 4.1, p. 61, l. -7], resp. [300, Remark 4.1, p. 61, l. -6], coincides with $D((-A_{\omega+\sigma})_{\alpha+\varepsilon})$, resp. $D((-A_\sigma)_{\alpha+\varepsilon})$, and the equality $(-A)_b = (-A)^{k+n_\alpha}(-A)^b(-A)^{-(k+n_\alpha)}$ holds provided that $\mathrm{Re}\, b \geqslant 0$ and $k \in \mathbb{N}$.

(ii) It is also worth noting that the method described above can be employed in a more general situation. Let $\alpha \geqslant -1$, $\varepsilon \in (0, 1]$, $c \in (0, 1)$, $d \in (0, 1]$ and n_α be as in the previous part of this remark, and let $\Omega_{\alpha,\varepsilon,c,d}$ be an open neighborhood of the region $P_{\alpha,\varepsilon,c} \cup B_d$. Suppose that the following condition holds:

(HC1) : $\Omega_{\alpha,\varepsilon,c,d} \subseteq \rho_C(-A)$, the family $\{(1+|z|)^{-\alpha}(z + A)^{-1}C : z \in \Omega_{\alpha,\varepsilon,c,d}\}$ is equicontinuous, and the mapping $z \mapsto (z + A)^{-1}Cx$, $z \in \Omega_{\alpha,\varepsilon,c,d}$ is continuous for every $x \in E$.

Then there exists a sufficiently small number $\kappa > 0$ such that the operator $\mathcal{C} := (d+\kappa-A)^{-n_\alpha}C \in L(E)$ is injective and commutes with A (cf. (256)). It can be easily seen that, for every $z \in P_{\alpha,\varepsilon,c} \cup B_d$,

$$(z + A)^{-1}\,\mathcal{C}x = \frac{(z + A)^{-1}Cx}{(d + \kappa + z)^{n_\alpha}} + \sum_{i=1}^{n_\alpha}\frac{(d + \kappa - A)^{-i}\,Cx}{(d + \kappa + z)^{n_\alpha+1-i}},$$

and that the family $\{z(z + A)^{-1}\mathcal{C} : z \in P_{\alpha,\varepsilon,c} \cup B_d\}$ is equicontinuous. Therefore, we are in a position to construct the power A_b ($b \in \mathbb{C}$). Notice that such a construction does not depend on the choice of numbers α, ε, c, d, κ and n_α, and that the assertion of [90, Theorem 3.1] can be reworded in the context of this remark (with some obvious additional difficulties in the case $\alpha \notin \mathbb{Z}$).

Theorem 2.9.9. *([90])*

(i) *Let* $\alpha, \gamma \in \mathbb{C}$, *let* $-\infty < \operatorname{Re}\alpha < \operatorname{Re}\gamma < +\infty$ *and* $x \in D(A_\gamma)$. *Then* $Cx \in D(A_\alpha)$ *and* $A_\alpha Cx = A_C^{\alpha-\gamma} A_\gamma x$.

(ii) *Let* $n \in \mathbb{N}_0$, *let* $b \in \mathbb{C}$, *and let* $\operatorname{Re} b \in (0, n+1)\setminus\mathbb{N}$. *Then, for every* $x \in E$,

$$A_C^{-b}x = \frac{(-1)^n n!}{(1-b)(2-b)\cdots(n-b)}\frac{\sin\pi(n-b)}{\pi}\int_0^\infty t^{n-b}(t+A)^{-(n+1)}Cx\,dt,$$

where $(1-b)(2-b)\cdots(n-b) := 1$ *for* $n = 0$.

(iii) *Suppose* $\alpha_0, \beta_0 \in \mathbb{C}$, $\operatorname{Re}\alpha_0 > \operatorname{Re}\beta_0 > 0$, $n \in \mathbb{N}_0$ *and* $\operatorname{Re}\alpha_0 \in (n, n+1]$. *Then, for every* $p \in \circledast$, *there exist* $c_{p,\alpha_0,\beta_0} > 0$ *and* $q_p \in \circledast$ *such that:*

$$p(CA_C^{-\beta_0}x) \leqslant c_{p,\alpha_0,\beta_0} q_p(A_C^{-\alpha_0}x)^{\frac{\operatorname{Re}\beta_0}{\operatorname{Re}\alpha_0}} q_p(Cx)^{\frac{\operatorname{Re}\alpha_0-\operatorname{Re}\beta_0}{\operatorname{Re}\beta_0}}, \quad x \in E, \tag{265}$$

and

$$p(A_C^{-\beta_0}x) \leqslant c_{p,\alpha_0,\beta_0} q_p(A_{-\alpha_0}x)^{\frac{\operatorname{Re}\beta_0}{\operatorname{Re}\alpha_0}} q_p(x)^{\frac{\operatorname{Re}\alpha_0-\operatorname{Re}\beta_0}{\operatorname{Re}\beta_0}}, \quad x \in D(A_{-\alpha_0}). \tag{266}$$

(iv) *(The Moment Inequality) Suppose* $\alpha, \beta, \gamma \in \mathbb{C}$, $-\infty < \operatorname{Re}\alpha < \operatorname{Re}\beta < \operatorname{Re}\gamma < +\infty$. *Then, for every* $p \in \circledast$, *there exist* $c_{p,\alpha,\beta,\gamma} > 0$ *and* $q_p \in \circledast$ *such that:*

$$p(CA_\beta Cx) \leqslant c_{p,\alpha,\beta,\gamma}\, q_p(A_\alpha Cx)^{\frac{\operatorname{Re}\gamma-\operatorname{Re}\beta}{\operatorname{Re}\gamma-\operatorname{Re}\alpha}} q_p(A_\gamma Cx)^{\frac{\operatorname{Re}\beta-\operatorname{Re}\alpha}{\operatorname{Re}\gamma-\operatorname{Re}\alpha}}, x \in D(A_\gamma),$$

and

$$p(A_\beta Cx) \leqslant c_{p,\alpha,\beta,\gamma}\, q_p(A_\alpha x)^{\frac{\operatorname{Re}\gamma-\operatorname{Re}\beta}{\operatorname{Re}\gamma-\operatorname{Re}\alpha}} q_p(A_\gamma x)^{\frac{\operatorname{Re}\beta-\operatorname{Re}\alpha}{\operatorname{Re}\gamma-\operatorname{Re}\alpha}}, x \in D(A_{\alpha-\gamma}A_\gamma).$$

(v) *Suppose* $b \in (0, 1)$. *Then the family* $\{C^{-1}\lambda^b A_C^{-b} A(\lambda + A)^{-1}C : \lambda > 0\}$ *is equicontinuous in* $L(E)$.

The following application of moment inequality is for illustration purposes only.

Example 2.9.10. Let E be one of the spaces $L^p(\mathbb{R}^n)$ $(1 \leqslant p \leqslant \infty)$, $C_0(\mathbb{R}^n)$, $C_b(\mathbb{R}^n)$, $BUC(\mathbb{R}^n)$ and let $0 \leqslant l \leqslant n$. Let $\mathbf{T}_l$ and $\mathbf{C}_{r,l}$ possess the same meaning as before, let $m \in \mathbb{N}$, $a_\alpha \in \mathbb{C}$, $0 \leqslant |\alpha| \leqslant m$, and let $P(D)f = \Sigma_{|\alpha|\leqslant m} a_\alpha f^{(\alpha)}$ act with its maximal distributional domain. Set $P(x) := \Sigma_{|\alpha|\leqslant m} a_\alpha i^{|\alpha|}x^\alpha$, $x \in \mathbb{R}^n$, and assume that $\sup_{x\in\mathbb{R}^n} \operatorname{Re}(P(x)) < 0$. Suppose $-\infty < \varsigma < \tau < \upsilon < +\infty$. By [534, Theorem 2.2], the operator $-P(D)$ is $\mathbf{C}_{r,l}$-sectorial and, since the condition (HC) holds, we can construct the powers of $-P(D)$. Then the moment inequality and the arguments used in its proof show that, for every $\alpha \in \mathbb{N}_0^l$, there exists a constant $M_\alpha < \infty$ such that the following differential inequality holds for each $f \in D((-P(D))_{\varsigma-\upsilon}(-P(D))_\upsilon)$:

$$q_\alpha\big(((-P(D))_\tau \mathbf{C}_{r,l} f\big) \leqslant M_\alpha q_\alpha\big(((-P(D))_\varsigma f\big)^{\frac{\upsilon-\tau}{\upsilon-\varsigma}} q_\alpha\big(((-P(D))_\upsilon f\big)^{\frac{\tau-\varsigma}{\upsilon-\varsigma}}.$$

In the case that (HC) or (HC1) holds, the reader may consult [90] for more details on the powers of C-nonnegative operators and semigroups generated by them. In the following subsection we will consider the general case; the Balakrishnan operators will be used as an auxiliary tool in the construction.

2.9.2. The Balakrishnan operators. In this subsection, we shall always assume that the operator A is C-nonnegative.

Definition 2.9.11. (cf. A. V. Balakrishnan [31] and [403, Chapter 3] for the case $C = 1$) Let $\alpha \in \mathbb{C}_+$. Then:

(i) If $0 < \operatorname{Re} \alpha < 1$, $D(J_C^\alpha) := D(A)$ and

$$J_C^\alpha x := \frac{\sin \alpha\pi}{\pi} \int_0^\infty \lambda^{\alpha-1}(\lambda+A)^{-1}CAx\, d\lambda, \quad x \in D(A).$$

(ii) If $\operatorname{Re} \alpha = 1$, $D(J_C^\alpha) := D(A^2)$ and

$$J_C^\alpha x := \frac{\sin \alpha\pi}{\pi} \int_0^\infty \lambda^{\alpha-1}\left[(\lambda+A)^{-1}C - \frac{\lambda C}{\lambda^2+1}\right]Ax\, d\lambda + \sin\frac{\alpha\pi}{2}CAx,$$

for any $x \in D(A^2)$.

(iii) If $n < \operatorname{Re} \alpha < n + 1$, $n \in \mathbb{N}$, $D(J_C^\alpha) := D(A^{n+1})$ and

$$J_C^\alpha x := J_C^{\alpha-n} A^n x, \quad x \in D(A^{n+1}).$$

(iv) If $\operatorname{Re} \alpha = n + 1$, $n \in \mathbb{N}$, $D(J_C^\alpha) := D(A^{n+2})$ and

$$J_C^\alpha x := J_C^{\alpha-n} A^n x, \quad x \in D(A^{n+2}).$$

The following proposition can be proved by using the methods established in [403].

Proposition 2.9.12. *Let* $\alpha \in \mathbb{C}_+$.

(i) *Suppose* $n > \operatorname{Re} \alpha$, $n \in \mathbb{N}$. *Then* $\lim_{\beta\to\alpha} J_C^\beta x = J_C^\alpha x$, $x \in D(A^n)$.
(ii) $R(J_C^\alpha) \subseteq \overline{D(A) \cap R(A)}$.
(iii) J_C^α *commutes with* $(\lambda + A)^{-1}C$, $\lambda \in -\rho_C(A)$.
(iv) *If* $Ax = \mu x$ *for some* $x \in E$ *and* $\mu \in \mathbb{C}\setminus(-\infty, 0)$, *then* $J_C^\alpha x = \mu^\alpha Cx$.
(v) $J_C^n x = CA^n x$, $x \in D(J_C^n) = D(A^{n+1})$.

The continuity properties of Balakrishnan operators are stated in the following theorem, which can be shown by making use of the equality (255) and the proof of [403, Theorem 1.3.6].

Theorem 2.9.13. (i) *Let* S_1 *be a fixed sector about* 1 *and contained in* $\{\alpha \in \mathbb{C} : 0 < \operatorname{Re} \alpha < 1\}$. *Then*

$$\lim_{\alpha\to 1, \alpha\in S_1} J_C^\alpha x = CAx$$

for any $x \in D(A)$ *with* $Ax \in \overline{D(A)}$.

(ii) *Let S_0 be a fixed sector about* 0 *and contained in* $\{\alpha \in \mathbb{C} : 0 < \text{Re}\,\alpha < 1\}$. *Then*

$$\lim_{\alpha \to 0, \alpha \in S_0} J_C^{\alpha} x = Cx$$

for any $x \in D(A)$ *with* $\lim_{\lambda \to 0} \lambda(\lambda + A)^{-1}Cx = 0$.

The closability of operator J_C^{α} $(\alpha \in \mathbb{C}_+)$ follows from the proof of [31, Lemma 2.1] and the fact that, for every $n \in \mathbb{N}$ and $\lambda > 0$, the operator $A^n((\lambda + A)^{-1}C)^n$ is bounded. Furthermore, the analyticity of mapping $\alpha \to J_C^{\alpha}x$, $0 < \text{Re}\,\alpha < n$, for $x \in D(A^n)$, follows almost directly from Definition 2.9.11.

Proposition 2.9.14. (i) *Let* $\alpha \in \mathbb{C}_+$. *Then the operator* J_C^{α} *is closable.*
(ii) *Let* $x \in D(A^n)$. *Then the mapping* $\alpha \to J_C^{\alpha}x$, $0 < \text{Re}\,\alpha < n$ *is analytic.*

Notice that the assumption $A \in L(E)$ implies that, for every $\alpha \in \mathbb{C}_+$, we have $J_C^{\alpha} \in L(E)$. Set $A_{C,\alpha} := C^{-1}\overline{J_C^{\alpha}}$ $(\alpha \in \mathbb{C}_+)$. Then $A_{C,\alpha}$ is a closed linear operator on E, and it is not difficult to prove with the help of Lemma 2.9.3 that $A_{C,n} \subseteq C^{-1}A^nC$, $n \in \mathbb{N}$; unfortunately, the existence of an element $\phi \in E$ satisfying $C^{n+1}\phi \notin \overline{C(D(A))}$ implies that $C^{-1}A^nC \not\subseteq A_{C,n}$ (see e.g. [403, p. 64 and the proof of Corollary 1.1.4(iv), p. 5]). Therefore, in general case, we will have to find some other method for construction of powers of C-nonnegative operators.

The proof of following proposition is almost trivial and therefore omitted (cf. also [90, Proposition 2.19]).

Proposition 2.9.15. *Let* $n \in \mathbb{N}$, *let* $\alpha \in \mathbb{C}$, *and let* $n - 1 < \text{Re}\,\alpha < n$. *Then the following holds:*

$$A_{C,\alpha}x = (-1)^n \frac{\sin \pi\alpha}{\pi} C^{-1} \int_0^{\infty} \lambda^{\alpha-n} A^n(\lambda+A)^{-1}Cx \, d\lambda, \quad x \in C(D(A^n)).$$

A slight modification of the proof of [31, Lemma 2.5, p. 422] implies the semigroup property of Balakrishnan operators.

Lemma 2.9.16. *Let* $x \in D(A^2)$, *let* $\alpha, \beta \in \mathbb{C}_+$, *and let* $0 < \text{Re}(\alpha + \beta) < 1$. *Then*

$$J_C^{\alpha+\beta}Cx = J_C^{\alpha}J_C^{\beta}x.$$

The following important representation formula, obtained by H. Komatsu [288, Theorem 2.10] in the case $C = 1$, is a consequence of the proof of [403, Proposition 3.1.3] and the equality

$$\frac{d}{d\lambda}((\lambda + A)^{-1}C^2)^j x = -j((\lambda + A)^{-1}C)^{j+1}C^{j-1}x, \quad j \in \mathbb{N}, x \in E, \tag{267}$$

which can be proved by induction.

Proposition 2.9.17. *Let* $\alpha \in \mathbb{C}_+$, *let* $m, n \in \mathbb{N}$, *and let* $m \geqslant n > \text{Re}\,\alpha > 0$. *Then*

$$C^{m-1}J_C^{\alpha}x = \frac{\Gamma(m)}{\Gamma(\alpha)\Gamma(m-\alpha)} \int_0^{\infty} \lambda^{\alpha-1}[A(\lambda+A)^{-1}C]^m x \, d\lambda, \quad x \in D(A^n). \tag{268}$$

Remark 2.9.18. If $(\lambda + A)^{-j}C \in L(E)$, $\lambda > 0$, $0 \leqslant j \leqslant m$, which holds for the operators considered in [144]-[145] and [90], then the formula (268) takes the following form:

$$J_C^\alpha x = \frac{\Gamma(m)}{\Gamma(\alpha)\Gamma(m-\alpha)} \int_0^\infty \lambda^{\alpha-1} \sum_{j=0}^{m} (-1)^j (\lambda + A)^{-j} Cx \, d\lambda, \quad x \in D(A^n).$$

Notice, however, that several results obtained by H. Komatsu in a series of his papers ([287]-[289]) cannot be so easily used in a further study of powers unless $C = 1$. For example, one can find the following extension of [403, Proposition 3.1.9, Theorem 3.1.10] unsatisfactory in case $C \neq 1$.

Proposition 2.9.19. (i) *Suppose $l, n \in \mathbb{N}$, $\alpha \in \mathbb{C}_+$, $0 < \operatorname{Re} \alpha < n$ and $-\mu \in \rho_C(A)$. If $x \in E$ and $[A(\mu + A)^{-1}C]^l x \in D(\overline{J_C^\alpha})$, then $[A(\mu + A)^{-1}C]^{l-1}C^n x \in D(\overline{J_C^\alpha})$.*

(ii) *Suppose $\alpha \in \mathbb{C}_+$, $n \in \mathbb{N}$, $0 < \operatorname{Re} \alpha < n$, $x, y \in D(A)$ and*

$$\int_0^\infty \lambda^{\alpha-1} [A(\lambda+A)^{-1}C]^n x \, d\lambda = C^n y.$$

Then $C^n x \in D(\overline{J_C^\alpha})$ and $\overline{J_C^\alpha} C^n x = \frac{\Gamma(n)}{\Gamma(\alpha)\Gamma(n-\alpha)} C^{n+1} y.$

(iii) *Suppose $\alpha \in \mathbb{C}_+$, $n \in \mathbb{N}$, $0 < \operatorname{Re} \alpha < n$, $(\lambda + A)^{-j}C \in L(E)$, $\lambda > 0, j \in \mathbb{N}$, $x,y \in E$ $\lim_{\lambda \to \infty} \lambda^n(\lambda + A)^{-n}Cx = Cx$, $\lim_{\lambda \to \infty} \lambda^n(\lambda + A)^{-n}Cy = Cy$ and*

$$\int_0^\infty \lambda^{\alpha-1} \sum_{j=0}^{n} (-1)^j (\lambda + A)^{-j} Cx \, d\lambda = y.$$

Then $Cx \in D(\overline{J_C^\alpha})$ and $\overline{J_C^\alpha} Cx = \frac{\Gamma(n)}{\Gamma(\alpha)\Gamma(n-\alpha)} Cy.$

2.9.3. Complex powers of almost *C*-nonnegative operators. Throughout this section, we shall always assume that the operator A belongs to the class $\mathcal{M}_{C,m}$ for some $m \in \mathbb{R}$. Recall that $D_\infty(A) = \bigcap_{n\in\mathbb{N}} D(A^n)$ and $p_n(x) = \Sigma_{i=0}^n p(A^i x)$, $x \in D_\infty(A)$, $p \in ⊛$, $n \in \mathbb{N}_0$. Set $A_\infty := A_{|D_\infty(A)}$ and $C_\infty := C_{|D_\infty(A)}$. Then the system $(p_n)_{p\in ⊛, n\in\mathbb{N}_0}$ induces a Hausdorff sequentially complete locally convex topology on $D_\infty(A)$, $A_\infty \in L(D_\infty(A))$ and $C_\infty \in L(D_\infty(A))$ is injective.

The following proposition plays an important role in our analysis (cf. [404, Proposition 3.6] and its proof).

Proposition 2.9.20. *Let $m \geqslant -1$. Then the operator A_∞ is C_∞-nonnegative in $D_\infty(A)$. Furthermore, for every $q \in ⊛$, there exist $c_q > 0$ and $r_q \in ⊛$ such that, for every $x \in D_\infty(A)$, $\lambda > 0$ and $n \in \mathbb{N}_0$,*

$$\sum_{j=0}^{n} q\left(\lambda(\lambda + A_\infty)^{-1} C_\infty A_\infty^i x\right) \leqslant c_q \sum_{j=0}^{n+\lfloor m+2 \rfloor} r_q(A_\infty^j x). \tag{269}$$

Therefore, we can construct the power $A_{\infty,\alpha} \equiv (A_\infty)C_{\infty,\alpha}$ in the space $D_\infty(A)$ ($\alpha \in \mathbb{C}_+$); since there is no risk for confusion, the corresponding Balakrishnan operator will be denoted by $J_{\infty,\alpha}$. Then the following holds:

(i) Let $0 < \operatorname{Re} \alpha < 1$ and $q \in ⊛$. Invoking Definition 2.9.11, we obtain from (269) that there exist $c_q > 0$ and $r_q \in ⊛$ such that:

$$q(J_{\infty,\alpha}x) \leqslant \left|\frac{\sin \alpha\pi}{\pi}\right| \left[\int_0^1 \lambda^{\alpha-1} q\big(C_\infty x - \lambda(\lambda + A_\infty)^{-1} C_\infty x\big)\, d\lambda + \int_1^\infty \lambda^{\alpha-2} q\big(\lambda(\lambda + A_\infty)^{-1} C_\infty A_\infty x\big)\, d\lambda\right]$$

(270)
$$\leqslant c_q \sum_{j=0}^{\lfloor m+3 \rfloor} r_q(A_\infty^j x),\ x \in D_\infty(A).$$

(ii) Let $\operatorname{Re} \alpha = 1$ and $q \in ⊛$. Then we have

$$\left[(\lambda + A)^{-1}C - \frac{\lambda C}{\lambda^2 + 1}\right] Ax = \frac{1}{\lambda^2 + 1}[Cx - \lambda(\lambda + A)^{-1}Cx - \lambda(\lambda + A)^{-1}CA^2 x],$$

for any $x \in D(A^2)$ and $\lambda > 0$. Exploiting this equality, one can similarly prove that there exist $c_q > 0$ and $r_q \in ⊛$ such that:

(271)
$$q(J_{\infty,\alpha}x) \leqslant c_q \sum_{j=0}^{\lfloor m+4 \rfloor} r_q(A_\infty^j x),\quad x \in D_\infty(A).$$

(iii) Let $n < \operatorname{Re} \lambda < n + 1$ and $q \in ⊛$. Then the estimate (270) implies that there exist $c_q > 0$ and $r_q \in ⊛$ such that:

$$q(J_{\infty,\alpha}x) \leqslant c_q \sum_{j=0}^{\lfloor m+3+n \rfloor} r_q(A_\infty^j x),\quad x \in D_\infty(A).$$

(iv) Let $\operatorname{Re} \lambda = n + 1$ and $q \in ⊛$. Then the estimate (271) implies that there exist $c_q > 0$ and $r_q \in ⊛$ such that:

$$q(J_{\infty,\alpha}x) \leqslant c_q \sum_{j=0}^{\lfloor m+4+n \rfloor} r_q(A_\infty^j x),\quad x \in D_\infty(A).$$

Let us write down the above results as

Lemma 2.9.21. *Let $m \geqslant -1$, let $A \in \mathcal{M}_{C,m}$, and let $\alpha \in \mathbb{C}_+$ satisfy $0 < \operatorname{Re} \alpha < n$ for some $n \in \mathbb{N}$. Set $p := \lfloor m + 2 \rfloor$. Then, for every $q \in ⊛$, there exist $c_{q,\alpha} > 0$ and $r_q \in ⊛$ such that:*

$$q(J_{\infty,\alpha}x) \leqslant c_{q,\alpha} \sum_{j=0}^{p+n} r_q(A_\infty^j x),\quad x \in D_\infty(A).$$

In particular, $J_{\infty,\alpha} \in L(D_\infty(A))$.

Now we introduce the operator family $(S_\gamma(t))_{t>0} \subseteq L(E)$ for $0 < \gamma < 1/2$ (cf. [404] for more details). First of all, define

$$f_t(\lambda) := \frac{1}{\pi} e^{-t\lambda^\gamma \cos \pi\gamma} \sin(t\lambda^\gamma \sin \pi\gamma)$$

$$= \frac{1}{2\pi i}\left(e^{-t\lambda^\gamma e^{-i\pi\gamma}} - e^{-t\lambda^\gamma e^{i\pi\gamma}}\right), \quad t > 0, \lambda > 0. \tag{272}$$

This function enjoys the following properties:

1. $|f_t(\lambda)| \leqslant \pi^{-1} e^{-\lambda^\gamma \varepsilon_t}$, $\lambda > 0$, where $\varepsilon_t := t \cos \pi\gamma > 0$.
2. $|f_t(\lambda)| \leqslant \gamma t \lambda^\gamma e^{-t\lambda^\gamma \sin \varepsilon_t}$, $\lambda > 0$.
3. $\int_0^\infty \lambda^n f_t(\lambda)\, d\lambda = 0$, $n \in \mathbb{N}_0$, $t > 0$.
4. Let $m \geqslant -1$. Then the improper integral $\int_0^\infty \lambda^n f_t(\lambda)\,(\lambda + A)^{-1}C \cdot d\lambda$ is absolutely convergent and defines a bounded linear operator on E ($n \in \mathbb{N}_0$).

Put now

$$S_\gamma(t)x := \int_0^\infty f_t(\lambda)\,(\lambda + A)^{-1}Cx\, d\lambda, \quad t > 0, x \in E. \tag{273}$$

Then $S_\gamma(t) \in L(E)$, $t > 0$ and the following holds (cf. [404, Proposition 3.3, Proposition 3.5] for the proof).

Proposition 2.9.22. *Let* $m > -1$, $A \in \mathcal{M}_{C,m}$, $t > 0$, $n \in \mathbb{N}$ *and* $0 < \gamma < 1/2$.

(i) *Then* $R(S_\gamma(t)) \subseteq D_\infty(A)$, $A^n S_\gamma(t) \in L(E)$ *and*

$$A^n S_\gamma(t)x = (-1)^n \int_0^\infty \lambda^n f_t(\lambda)\,(\lambda + A)^{-1}Cx\, d\lambda, \quad x \in E. \tag{274}$$

(ii) *We have* $\lim_{t\to 0+} S_\gamma(t)x = Cx$ *for all* $x \in D(A^p)$.

Notice that the mapping $t \mapsto S_\gamma(t)x$, $t > 0$ can be analytically extended to the sector $\Sigma_{\frac{\pi}{2}-\pi\gamma}$ and that the following formula holds (cf. (272) and apply the dominated convergence theorem):

$$\frac{d^n}{dz^n} S_\gamma(z)x = \frac{1}{2\pi i}\int_0^\infty \Big[(-\lambda^\gamma e^{-i\pi\gamma})^n e^{-z\lambda^\gamma e^{-\pi\gamma i}} - (-\lambda^\gamma e^{-i\pi\gamma})^n e^{-z\lambda^\gamma e^{\pi\gamma i}}\Big]\,(\lambda + A)^{-1}Cx\, d\lambda, \tag{275}$$

for any $x \in E$, $z \in \Sigma_{\frac{\pi}{2}-\pi\gamma}$ and $n \in \mathbb{N}_0$.

Let $m \geqslant -1$ and $A \in \mathcal{M}_{C,m}$. Then the closability of the operator $A_{\infty,\alpha}$ in E follows from the proof of [404, Proposition 3.7] combined with Proposition 2.9.22(i) and the equality

$$\lim_{t\to 0+} S_\gamma(t)((1 + A)^{-1}C)^p x = C((1 + A)^{-1}C)^p x, \quad x \in E. \tag{276}$$

We shall slightly improve in the following proposition the estimate [404, (23); Proposition 3.9].

Proposition 2.9.23. *Let $m \geqslant -1$, let $A \in \mathcal{M}_{C,m}$, and let $\alpha \in \mathbb{C}_+$ satisfy $0 < \operatorname{Re} \alpha < n$ for some $n \in \mathbb{N}$. Then*

$$C^2(D(A^{2p+n})) \subseteq D(\overline{A_{\infty,\alpha}}).$$

Proof. Let $x \in D(A^{2p+n})$ and $0 < \gamma < 1/2$. Noticing that the operator $S_\gamma(t)$ commutes with A^n for any $t > 0$ and $n \in \mathbb{N}_0$, we have by (276) that $\lim_{k\to\infty} A^j S_\gamma(1/k)x = CA^j x$, provided $j \in \mathbb{N}_0$ and $j \leqslant p + n$. This simply implies that $(A_{\infty,\alpha} S_\gamma(1/k)Cx)_{k\in\mathbb{N}} = (C_\infty^{-1} J_{\infty,\alpha} S_\gamma(1/k)Cx)_{k\in\mathbb{N}} = (J_{\infty,\alpha} S_\gamma(1/k)x)_{k\in\mathbb{N}}$ is a Cauchy sequence in E and therefore convergent. Since $\lim_{k\to\infty} S_\gamma(1/k)Cx = C^2x$, it immediately follows that $C^2x \in D(\overline{A_{\infty,\alpha}})$, as required. □

Definition 2.9.24. Suppose $m \geqslant -1$, $n \in \mathbb{N}$, $p = \lfloor m+2 \rfloor$, $A \in \mathcal{M}_{C,m}$, $\alpha \in \mathbb{C}_+$ and $0 < \operatorname{Re} \alpha < n$. Then we define the power A_α as follows

$$A_\alpha := C^{-2}((1 + A)^{-1}C)^{-n(p+1)-p}\overline{A_{\infty,\alpha}}((1 + A)^{-1}C)^{n(p+1)+p}\, C^2.$$

Remark 2.9.25. Notice that the techniques of complex analysis can be hardly used for the construction of power A_α (see e.g. [403, Example 1.4.3]).

The closedness of operator A_α follows from that of $\overline{A_{\infty,\alpha}}((1+A)^{-1}C)^{n(p+1)+p}C^2$, and Definition 2.9.24 does not depend on the particular choice of numbers n and m (cf. [404, Remark 3.1]).

Lemma 2.9.26. *Let $\alpha \in \mathbb{C}_+$. Then $C^{-1}A_\alpha C = A_\alpha$.*

Proof. It is clear that $A_\alpha \subseteq C^{-1}A_\alpha C$. Suppose now $(x, y) \in C^{-1}A_\alpha C$, i.e., $\overline{A_{\infty,\alpha}}((1+A)^{-1}C)^{n(p+1)+p}C^3x = \overline{A_{\infty,\alpha}}((1+A)^{-1}C)^{n(p+1)+p}C^3y$. Notice that $\overline{A_{\infty,\alpha}}$ commutes with C, and that Proposition 2.9.23 implies $C^2((1+A)^{-1}C)^{n(p+1)+p}Cx \in D(\overline{A_{\infty,\alpha}})$. Hence, $\overline{A_{\infty,\alpha}}((1+A)^{-1}C)^{n(p+1)+p}C^2x = \overline{A_{\infty,\alpha}}((1+A)^{-1}C)^{n(p+1)+p}C^2y$, $A_\alpha x = y$ and $(x, y) \in A_\alpha$. □

Remark 2.9.27. Taken together, Proposition 2.9.23, Lemma 2.9.26 and the consideration given in [404, Remark 3.1] imply that, for any natural numbers k and l with $k \geqslant 2p + n$ and $l \geqslant 2$, the following equality holds:

$$A_\alpha = C^{-l}((1 + A)^{-1}C)^{-k}\overline{A_{\infty,\alpha}}((1 + A)^{-1}C)^k C^l.$$

Proposition 2.9.28. *Let $\alpha, \beta \in \mathbb{C}_+$. Then $A_\alpha A_\beta \subseteq A_{\alpha+\beta}$, and for every $x \in D(A_{\alpha+\beta}) \cap D(A_\beta)$, one has $A_\beta x \in D(A_\alpha)$ and $A_\alpha A_\beta x = A_{\alpha+\beta}x$.*

Proof. Suppose $n > \operatorname{Re}(\alpha+\beta) + 1$, $k \geqslant (n + 1)(p + 1) + p$, $0 < \gamma < 1/2$ and $A_\alpha A_\beta x = y$. By Remark 2.9.27, we get that

$$C^{-4}((1 + A)^{-1}C)^{-k}\overline{A_{\infty,\alpha}}\,\overline{A_{\infty,\beta}}((1 + A)^{-1}C)^k C^4 x = y.$$

Therefore, the proof of proposition will be completed if we prove the following equality:

$$\overline{A_{\infty,\alpha}}\,\overline{A_{\infty,\beta}}((1 + A)^{-1}C)^k C^4 x = \overline{A_{\infty,\alpha+\beta}}\,((1 + A)^{-1}C)^k C^4 x.$$

Towards this end, notice that Proposition 2.9.23 implies

$$((1 + A)^{-1}C)^k C^4 x \in D(\overline{A_{\infty,\alpha}}\,\overline{A_{\infty,\beta}}) \cap D(\overline{A_{\infty,\alpha+\beta}}).$$

Keeping in mind the equality (276) as well as the commutation of the operator A_ζ with the operators C, $(\lambda + A)^{-1}C$ and $S_\gamma(t)$ ($\zeta \in \mathbb{C}_+$, $\lambda > 0$, $t > 0$), it suffices to show that:

$$S_\gamma(t)C^2((1 + A)^{-1}C)^p \overline{A_{\infty,\alpha}}\,\overline{A_{\infty,\beta}}((1 + A)^{-1}C)^k C^4 x$$

$$(277) \qquad = S_\gamma(t)C^2((1 + A)^{-1}C)^p \overline{A_{\infty,\alpha+\beta}}((1 + A)^{-1}C)^k C^4 x, \quad t > 0.$$

The following equality is true, for every $t > 0$,

$$S_\gamma(t)C^2((1 + A)^{-1}C)^p \overline{A_{\infty,\alpha}}\,\overline{A_{\infty,\beta}}((1 + A)^{-1}C)^k C^4 x$$

$$(278) \qquad = ((1 + A)^{-1}C)^p J_{\infty,\alpha} J_{\infty,\beta} S_\gamma(t)((1 + A)^{-1}C)^k C^4 x.$$

In combination, Proposition 2.9.14(ii) and Lemma 2.9.16 (cf. also the proof of [403, Theorem 5.1.2]) imply that

$$J_{\infty,\alpha+\beta} Cx = J_{\infty,\alpha} J_{\infty,\beta} x, \quad \alpha, \beta \in \mathbb{C}_+, x \in D_\infty(A).$$

This simply implies (278), (277) and completes the proof of inclusion $A_\alpha A_\beta \subseteq A_{\alpha+\beta}$. The remaining part of the proof can be left to the reader. □

Proposition 2.9.29. *Let $n \in \mathbb{N}$. Then $C^{-1}A^nC = A_n$.*

Proof. Set, for every $\alpha \in \mathbb{C}_+$ with $0 < \operatorname{Re} \alpha \leqslant n$,

$$A'_\alpha := C^{-1}((1+A)^{-1}C)^{-(n+1)(p+1)-p} \overline{A_{\infty,\alpha}}((1+A)^{-1}C)^{(n+1)(p+1)+p}C.$$

We will prove that $C^{-1}A^nC = A'_n = A_n$. Suppose first that $(x, y) \in A'_n$. Then $\overline{A_{\infty,n}}((1+A)^{-1}C)^{(n+1)(p+1)+p}Cx = ((1+A)^{-1}C)^{(n+1)(p+1)+p}Cy$, and by the closedness of the operator $C^{-1}A^nC$ (cf. Lemma 2.9.3) and the fact that $A_{\infty,n} = A^n_\infty$, we obtain $C^{-1}A^nC((1 + A)^{-1}C)^{(n+1)(p+1)+p}Cx = ((1 + A)^{-1}C)^{(n+1)(p+1)+p}Cy$, i.e.,

$$[((1 + A)^{-1}C)^{-(n+1)(p+1)+p}C]C^{-1}A^nC[((1 + A)^{-1}C)^{(n+1)(p+1)+p}C]x = y.$$

Using Lemma 2.9.3 again we have that $C^{-2}A^nC^2 = C^{-1}A^nC$, which enables one to see that $[((1 + A)^{-1}C)^{-l}]C^{-1}A^nC[((1 + A)^{-1}C)^l] = C^{-1}A^nC$ ($l \in \mathbb{N}$). Therefore, $C^{-1}A^nCx = y$ and $A'_n \subseteq C^{-1}A^nC$. In order to see that the converse inclusion also holds, suppose that $C^{-1}A^nCx = y$. Then we will have to prove that

$$\overline{A_{\infty,n}}((1+A)^{-1}C)^{(n+1)(p+1)+p}Cx = ((1 + A)^{-1}C)^{(n+1)(p+1)+p}Cy.$$

Let $(t_n)_{n\in\mathbb{N}}$ be a sequence of positive real numbers tending to 0. Making use of Proposition 2.9.22(ii), we get that

$$\lim_{n\to\infty} S_\gamma(t_n)((1+A)^{-1}C)^{(n+1)(p+1)+p}x = ((1 + A)^{-1}C)^{(n+1)(p+1)+p}Cx$$

and, since $A_{\infty,n}x = A^n x$, $x \in D_\infty(A)$,

$$\begin{aligned}\lim_{n\to\infty} A_{\infty,n} S_\gamma(t_n)((1+A)^{-1}C)^{(n+1)(p+1)+p}x &= \lim_{n\to\infty} S_\gamma(t_n)((1+A)^{-1}C)^{(n+1)(p+1)+p}x \\ &= ((1+A)^{-1}C)^{(n+1)(p+1)+p}A^nCx \\ &= ((1+A)^{-1}C)^{(n+1)(p+1)+p}Cy.\end{aligned}$$

This, in turn, implies $C^{-1}A^nC \subseteq A'_n$ and $C^{-1}A^nC = A'_n$. Applying again Lemma 2.9.3, we finally conclude that $C^{-1}A^nC = C^{-2}A^nC^2 = C^{-1}C^{-1}A^nCC = C^{-1}A'_nC = A_n$. □

Lemma 2.9.30. *Suppose A is injective and $\alpha \in \mathbb{C}_+$. Then A_α is injective as well.*

Proof. Let $n \in \mathbb{N}$ satisfy $0 < \operatorname{Re} \alpha < n$, and let $A_\alpha x = 0$ for some $x \in E$. Then Proposition 2.9.28-Proposition 2.9.29 imply that $C^{-1}A^nCx = 0$, so that $x = 0$. □

In the case that the operator A is injective, we can introduce the power A_ζ for any complex number ζ; notice only that $(A + 1) \in \mathcal{M}_{C,m}$, provided $m \geqslant -1$ and $A \in \mathcal{M}_{C,m}$.

Definition 2.9.31. Suppose A is injective, $m \geqslant -1$, $A \in \mathcal{M}_{C,m}$, $\alpha \in \mathbb{C}_+$ and $\tau \in \mathbb{R}\backslash\{0\}$. Then we define

$$A_{-\alpha} := (A_\alpha)^{-1},\ A_0 := 1 \text{ and}$$

$$A_{i\tau} := C^{-3}((1 + A)^{-1}C)^{-(3p+3)} A^{-1}A_{1+i\tau}((1 + A)^{-1}C)^{3p+3}\, C^3.$$

In the following definition, we consider the case in which the injective operator A satisfies $A \in \mathcal{M}_{C,m}$ for some $m < -1$ (cf. Proposition 2.9.5).

Definition 2.9.32. Suppose A is injective, $m < -1$, $A \in \mathcal{M}_{C,m}$ and $\alpha \in \mathbb{C}$. Then we define the power A_α by

$$A_\alpha := (A^{-1})_{-\alpha}.$$

As before, Definition 2.9.31 (Definition 2.9.32) is independent on the choice of number $m \geqslant -1$ ($m < -1$). If $C = 1$ and $m = -1$, then the powers constructed in this paper coincide with the usual ones (cf. [403] and [404, Remark 3.3]).

Remark 2.9.33. (i) Suppose $0 \leqslant \omega < \varphi \leqslant \pi$, $0 < d \leqslant 1$ and A satisfies the condition (HC). Then the method developed in Subsection 2.9.1 enables one to construct the power A_b for any $b \in \mathbb{C}$. In order to avoid confusion henceforth, we shall denote the above power by ${}_bA$. Our intention is to prove that ${}_bA = A'_b = A_b$, $b \in \mathbb{C}$. It can be simply verified that the above holds if, for every $b \in \mathbb{C}_+$, one has ${}_bA = A'_b$. It is not difficult to prove with the help of Proposition 2.9.15 that $C^2x = A_C^{-\alpha} J_{\infty,\alpha}x$, $x \in C^2(D_\infty(A))$, $\alpha \in \{z \in \mathbb{C}_+ : \operatorname{Re} z \notin \mathbb{N}\}$. By making use of the analyticity of mappings $z \mapsto A_C^{-z}x$, $z \in \mathbb{C}_+$ ($x \in E$) and $z \mapsto J_{\infty,z}x$, $z \in \mathbb{C}_+$ ($x \in D_\infty(A)$), as well as the uniqueness theorem for analytic functions, we get that

(279) $$C^2x = A_C^{-\alpha} J_{\infty,\alpha} x, \quad x \in C^2(D_\infty(A)), \alpha \in \mathbb{C}_+.$$

Let $n \in \mathbb{N}$ satisfy $0 < \operatorname{Re} b < n$, and let $0 < \gamma < 1/2$. Suppose first that $(x, y) \in A'_b$, i.e., $\overline{A_{\infty,b}}(1 + A)^{-(2n+1)}C^{2n+2}x = (1 + A)^{-(2n+1)}C^{2n+2}y$.
Then it readily follows that, for every $t > 0$,

$$J_{\infty,b}(1 + A)^{-(2n+1)} C^{2n+1}S_\gamma(t)x = (1 + A)^{-(2n+1)} C^{2n+2}S_\gamma(t)y,$$

and

$$A_C^{-b}J_{\infty,b}(1 + A)^{-(2n+1)} C^{2n+1}S_\gamma(t)x = A_C^{-b}(1 + A)^{-(2n+1)} C^{2n+2}S_\gamma(t)y.$$

Owing to (279), we have that, for every $t > 0$,

$$C^2(1 + A)^{-(2n+1)} C^{2n+1}S_\gamma(t)x = A_C^{-b}(1 + A)^{-(2n+1)}C^{2n+2}S_\gamma(t)y.$$

This equality and (276) imply that $Cx = A_C^{-b}y$ and $(x, y) \in {}_bA$, hence $A'_b \subseteq {}_bA$. Let $(x, y) \in {}_bA$ and $t > 0$. Put $X_t := (1+A)^{-(2n+1)}C^{2n+2}S_\gamma(t)x$ and $Y_t := (1 + A)^{-(2n+1)}C^{2n+2}S_\gamma(t)y$. Then it is checked at once that $CX_t = A_C^{-b}Y_t$, which implies by (279) and the injectiveness of A_C^{-b} that $A_{\infty,b}X_t = Y_t$. By Proposition 2.9.23 and (276), the above implies that, for every $t > 0$,

$$S_\gamma(t)\overline{A_{\infty,b}}(1 + A)^{-(2n+1)}C^{2n+2}S_\gamma(t)x = S_\gamma(t)(1 + A)^{-(2n+1)}C^{2n+2}S_\gamma(t)y,$$

$(1+A)^{-(2n+1)}C^{2n+2}S_\gamma(t)x = (1+A)^{-(2n+1)}C^{2n+2}S_\gamma(t)y$ and $(x, y) \in A'_b$, hence $A'_b \subseteq {}_bA$ and $A_b = A'_b = {}_bA$, as claimed. Notice, finally, that it is not clear whether the equality $A_b = A'_b$, $b \in \mathbb{C}_+$ holds in general case $A \in \mathcal{M}_{C,m}$ $(m \geqslant -1)$.

(ii) Suppose $C = 1$. Then, in any of the above considered cases, the constructed powers coincide with those of [404]; we will only outline the main details of proof of this fact in the case that $A \in \mathcal{M}_{1,m}$ is injective and that the exponent of corresponding power is a purely imaginary nonzero number. Let $\tau \in \mathbb{R}\setminus\{0\}$. Keeping in mind Proposition 2.9.23, [404, Proposition 3.9] and elementary definitions, it suffices to show that:

(280) $$\overline{A_\infty^{1+i\tau}}(A + 1)_{-(3p+3)}x = \overline{A_\infty^{i\tau}}A(A + 1)_{-(3p+3)}x, \ x \in E,$$

where the operator $A_\infty^{i\tau}$ is defined by $D(A_\infty^{i\tau}) := D(A_\infty) \cap R(A_\infty)$ and

$$A_\infty^{i\tau}x := \frac{\sinh \pi\tau}{\pi\tau} \int_0^\infty \lambda^{i\tau} (\lambda + A)^{-2} Ax \, d\lambda, \ x \in D(A_\infty^{i\tau}).$$

Multiplying both sides of the equality (280) with the operator $S_\gamma(t)$ $(0 < \gamma < 1/2$ and $t > 0)$, and using the procedure described in the first part of this remark, as well as the commutation of the operator $\overline{A_\infty^{i\tau}}$ with $(A+1)_{-(3p+3)}$ and $S_\gamma(t)$, it is enough to prove that $\overline{A_\infty^{1+i\tau}}\, x = \overline{A_\infty^{i\tau}}Ax$, $x \in D(A_\infty)$. But, this equality is an immediate consequence of [404, Proposition 3.8(i)].

Here it is also worth noticing that the assertions of [404, Proposition 4.3, Remark 4.2, Corollary 4.4] remain true in the case $m \geqslant 0$. If the assumptions of

[404, Proposition 4.3] hold, then $\rho(A)$ contains 0 as an interior point and we can construct the power A^z ($z\in\mathbb{C}$) by the method developed in [292, Section 1.4]. Suppose $\operatorname{Re}\alpha > m+1$. Using [90, Remark 2.12(i)], [292, Theorem 1.4.12(iii)] as well as the statements proved by now in this remark, we get that $D(A_{-\alpha}) = R(A_{\alpha}) = R(A^{m+3}A^{\alpha}A^{-(m+3)}) = R(A^{\alpha}) = E$. Therefore, the Closed Graph Theorem implies $A_{-\alpha} \in L(E)$.

(iii) Suppose now that E is a Banach space and the operator A is injective. Then the powers constructed in [145, Section 5] coincide with those introduced in this section (cf. also the paper [144] for the construction in which $D(A)$ and $R(C)$ are dense in E, and A is not necessarily injective). More precisely, suppose that A is C-sectorial operator of angle $\theta\in[0,\pi)$ and the mapping $\lambda\to(\lambda-A)^{-1}C\in L(E)$, $\lambda\in\mathbb{C}\setminus\overline{\Sigma_\theta}$ is analytic. Let $0<\gamma<1/2$. Define, for every $\alpha\in\mathbb{C}$ and $x\in E$,

$$W(\alpha)x := \int_0^\infty \lambda^\alpha e^{-(\lambda^\gamma+\lambda^{-\gamma})\cos(\gamma\pi)} \times \sin\big(\alpha\pi + (\lambda^\gamma-\lambda^{-\gamma})\sin(\gamma\pi)\big)\,(\lambda+A)^{-1}Cx\,\frac{d\lambda}{\pi}.$$

Then $(W(\alpha))_{\alpha\in\mathbb{C}}$ is an entire $W(0)$-regularized group, the operator A satisfies $A = W(0)^{-1}W(1)$, and R. deLaubenfels-J. Pastor introduce the power A^α, for any $\alpha\in\mathbb{C}$, by $A^\alpha := W(0)^{-1}W(\alpha)$. In what follows, we shall use the regularized version of the Hirsch functional calculus in SCLCSs (cf. [403, Section 4] and [145] for the terminology and further information). Set, for any $f = (\alpha,\mu)\in\mathcal{T}$ and $x\in E$,

$$Tx := \int_{[0,1]} (1+tA)^{-1}Cx\,d\mu(t) \text{ and } Sx := \int_{(1,\infty)} A(1+tA)^{-1}Cx\,d\mu(t).$$

The closed linear operator $f(A)$ is defined by $f(A) := aC + AT + S$ with its maximal domain. Suppose now $\alpha\in\mathbb{C}_+$ and $(x,y)\in A^\alpha$, i.e., $W(\alpha)x = W(0)y$. In order to prove that $(x,y)\in A_\alpha$, we may assume without loss of generality that $x, y\in D_\infty(A)$. By Proposition 2.9.20, the operator A_∞ is C_∞-nonnegative in $D_\infty(A)$; furthermore, the mappings $z\mapsto W(z)x$, $z\in\mathbb{C}$ and $z\mapsto J^z_{C\infty}x$, $z\in\mathbb{C}_+$ are analytic for the topology of $D_\infty(A)$. It is obvious that (cf. [403, Example 4.1.1] and [145, Lemma 3.4, Lemma 3.6]):

(281) $$J^z_{C\infty} = (z^\alpha)(A_\infty),\ 0<\operatorname{Re} z<1, \text{ and}$$

(282) $$W(z)x = -(z^\alpha e^{-(z^\gamma+z^{-\gamma})})(A_\infty)x,\ z\in\mathbb{C}.$$

By the proof of [404, Theorem 4.2.3], we have $f(A_\infty)g(A_\infty)\subseteq h(A_\infty)C_\infty$, provided $f, g, h\in\mathcal{T}$ and $h = fg$. Making use of this equality, (281)-(282) and the foregoing arguments, we get that

(283) $$W(0)J^\alpha_C x = CW(\alpha)x,$$

which simply implies $(x,y)\in A_\alpha$. Keeping in mind (283) and the commutation of the operator $W(\alpha)$ with $S_\gamma(t)$ for $\alpha\in\mathbb{C}$, $t>0$ and $0<\gamma<1/2$, the converse inclusion $A_\alpha \subseteq A^\alpha$ as well as the equality $A^z = A_z$, $z \in i\mathbb{R}$ can be proved in a similar fashion.

The previous analysis shows that the powers constructed in this section can be viewed as a unification of those appearing in many other papers and monographs. Of course, it could be interesting to transfer the results obtained here to multivalued linear operators.

Suppose A is injective, $m \geqslant -1$, $A \in \mathcal{M}_{C,m}$, $\alpha \in \mathbb{C}_+$ and $\tau \in \mathbb{R}\setminus\{0\}$. Then we define the operators $J_{\infty,-\alpha}$ and $J_{\infty,i\tau}$ by $D(J_{\infty,-\alpha}) := D_\infty(A) \cap R(A_\infty^{\lfloor \operatorname{Re}\alpha\rfloor+1})$, $D(J_{\infty,i\tau}) := D_\infty(A) \cap R(A_\infty)$,

$$J_{\infty,-\alpha}x := \frac{\Gamma(\lfloor \operatorname{Re}\alpha \rfloor+1)}{\Gamma(\alpha)\Gamma(\lfloor \operatorname{Re}\alpha \rfloor+1-\alpha)} \int_0^\infty \lambda^{\lfloor \operatorname{Re}\alpha\rfloor-\alpha}[(\lambda + A)^{-1}C]^{\lfloor \operatorname{Re}\alpha\rfloor+1}x\, d\lambda,$$

for any $x \in D_\infty(A) \cap R(A_\infty^{\lfloor \operatorname{Re}\alpha\rfloor+1})$, and

$$J_{\infty,i\tau}x := \frac{\sinh \pi\tau}{\pi\tau} \int_0^\infty \lambda^{i\tau}((\lambda + A)^{-1}C)^2 Ax\, d\lambda, \quad x \in D_\infty(A) \cap R(A_\infty).$$

Integration by parts shows that, for every $x \in D_\infty(A) \cap R(A_\infty^n)$ and for every $n \in \mathbb{N}$ with $n > \operatorname{Re}\alpha$, the following equality holds:

$$J_{\infty,-\alpha}x = \frac{\Gamma(n)}{\Gamma(\alpha)\Gamma(n-\alpha)} \int_0^\infty \lambda^{n-\alpha-1}[(\lambda + A)^{-1}C]^n x\, d\lambda.$$

Furthermore, for every $q \in ⊛$, there exist $c_q > 0$ and $r_q \in ⊛$ such that

$$q(J_{\infty,-\alpha}x) \leqslant c_q \sum_{j=-(\lfloor \operatorname{Re}\alpha\rfloor+1)}^{(\lfloor \operatorname{Re}\alpha\rfloor+1)p} r_q(A^j x), \quad x \in D_\infty(A) \cap R(A_\infty^{\lfloor \operatorname{Re}\alpha\rfloor+1}), \tag{284}$$

$$q(J_{\infty,i\tau}x) \leqslant c_q \sum_{j=-1}^{2p+1} r_q(A^j x), \quad x \in D_\infty(A) \cap R(A_\infty), \tag{285}$$

and that the operators $A_{\infty,-\alpha} := C_\infty^{-(\lfloor \operatorname{Re}\alpha\rfloor+1)} J_{\infty,-\alpha}$ and $A_{\infty,i\tau} := C_\infty^{-2} J_{\infty,i\tau}$ are closable for the topology of E. By the estimates (284)-(285) and the proof of Proposition 2.9.23, we obtain that

$$C^{\lfloor \operatorname{Re}\alpha\rfloor+2}(D(A^{p(\lfloor \operatorname{Re}\alpha\rfloor+2)})) \cap R(A^n) \subseteq \overline{A_{\infty,-\alpha}},$$

and

$$C^3(D(A^{3p+1})) \cap R(A) \subseteq \overline{A_{\infty,i\tau}}.$$

Hence, the operators $\overline{A_{\infty,-\alpha}}A^{\lfloor \mathrm{Re}\,\alpha\rfloor+1}((1+A)^{-1}C)^{(\lfloor \mathrm{Re}\,\alpha\rfloor+1)(p+1)+p}C^{\lfloor \mathrm{Re}\,\alpha\rfloor+2}$ and $\overline{A_{\infty,i\tau}}A((1+A)^{-1}C)^{3p+3}C^3$ are closed and defined on the whole space E. Using the equality (267) with $j = 1$, Proposition 2.9.12(i), as well as the arguments given on pages 60 and 175 of [403], we get that

$$J_{\infty,1+i\tau}Cx = J_{\infty,i\tau}Ax, \quad x \in D_{\infty}(A).$$

Applying the procedure described in Remark 2.9.33(i), the above equality shows that

$$A_{i\tau} = C^{-3}((1+A)^{-1}C)^{-(3p+3)}A^{-1}\overline{A_{\infty,i\tau}}A((1+A)^{-1}C)^{3p+3}C^3,$$

which implies without any substantial difficulties that $C^{-1}A_{i\tau}C = A_{i\tau}$ is a closed linear operator on E.

Proposition 2.9.34. *Let* $\tau \in \mathbb{R}\setminus\{0\}$. *Then the operator* $A_{i\tau}$ *is closed,* $C^{-1}A_{i\tau}C = A_{i\tau}$ *and* $A_{i\tau}^{-1} = A_{-i\tau}$.

Notice also that the operator $A_{i\tau}$ can be equivalently introduced by

$$A_{i\tau} = C^{-l}((1+A)^{-1}C)^{-k}A^{-1}A_{1+i\tau}((1+A)^{-1}C)^{k}C^{l},$$

where $l \geqslant 3$ and $k \geqslant 3p+3$. One can similarly prove that the following proposition holds.

Proposition 2.9.35. *Let* $\alpha \in \mathbb{C}_+$, *and let* $n \in \mathbb{N}$ *satisfy* $0 < \mathrm{Re}\,\alpha < n$. *Then* $A_{-n} = C^{-1}A^{-n}C$, $C^{-1}A_{-\alpha}C = A_{-\alpha}$ *is a closed linear operator on E, and*

$$A_{-\alpha} = C^{-l}((1+A)^{-1}C)^{-k}A^{-(l-1)}\overline{A_{\infty,-\alpha}}A^{l-1}((1+A)^{-1}C)^{k}C^{l},$$

for any $l \geqslant \lfloor \mathrm{Re}\,\alpha\rfloor + 2$ *and* $k \geqslant (\lfloor \mathrm{Re}\,\alpha\rfloor + 1)(p+1) + p$.

We continue by observing that our framework will be capable of reformulation of [404, Proposition 3.8]. Making use of the foregoing arguments as well the basic definitions and obtained representation formulae, one can simply prove that the following theorem holds good.

Theorem 2.9.36. (i) *Let* $m \in \mathbb{R}$, *let* $A \in \mathcal{M}_{C,m}$ *be injective, and let* $\alpha \in \mathbb{C}$. *Then* A_α *is injective as well, and the following equality holds:*

$$A_{-\alpha} = (A_\alpha)^{-1} = (A^{-1})_\alpha.$$

(ii) *Let* $m \in \mathbb{R}$, *let* $A \in \mathcal{M}_{C,m}$ *be injective, and let* $\alpha, \beta \in \mathbb{C}$. *Then* $A_\alpha A_\beta \subseteq A_{\alpha+\beta}$, *and for every* $x \in D(A_{\alpha+\beta}) \cap D(A_\beta)$, *one has* $A_\beta x \in D(A_\alpha)$ *and* $A_\alpha A_\beta x = A_{\alpha+\beta}x$. *Furthermore, the supposition* $A_{-\alpha} \in L(E)$ *implies* $A_\alpha A_\beta = A_{\alpha+\beta}$, *which continues to hold in the case that A is not injective and* $\alpha, \beta \in \mathbb{C}_+$.

Although the inclusion $A_{\alpha+\beta} \subseteq A_\alpha A_\beta$ cannot be generally expected (cf. [404, Example 4.1]), one can simply prove the equalities like

$$A_\alpha A_\beta C^l((1+A)^{-1}C)^k = A_{\alpha+\beta}C^l((1+A)^{-1}C)^k, \tag{286}$$

where $\alpha, \beta \in \mathbb{C}_+$, $l \geqslant 2$ and $k \geqslant 2p + 1 + \lfloor \mathrm{Re}(\alpha + \beta) \rfloor$. We shall skip details for the sake of brevity.

2.9.4. The case $m = -1$. The case $m = -1$ is most important, without any doubt, and here we can refine some of results stated in the previous subsections. To the best knowledge of the author, the power A_α ($\alpha \in \mathbb{C}_+$) has not been defined elsewhere even in the case that E is a Banach space and $A \in \mathcal{M}_{C,-1}$ is a non-injective operator with $\overline{D(A)} \neq E$.

Remark 2.9.37. Suppose $m \geqslant -1$, $A \in \mathcal{M}_{C,m}$ and $\alpha \in \mathbb{C}_+$. Put $C_k := C((1 + A)^{-1} C)^k$, $k \in \mathbb{N}$. Then the use of generalized resolvent equation implies that, for every $k \in \mathbb{N}$ with $k \geqslant m + 1$, the operator A is C_k-nonnegative (under some additional assumptions, the above continues to hold with the operator C_k replaced by $C_k' \equiv (1 + A)^{-k} C)$. Therefore, we are in a position to introduce the power $A_{\alpha,k}$ by replacing respectively the operator C and the number p in Definition 2.9.24 by the operator C_k and the number 1. Keeping in mind Definition 2.9.24 as well as Proposition 2.9.17 and the procedure described in Remark 2.9.33(i), it is not so difficult to prove that $A_\alpha = A_{\alpha,k} = A_{\alpha,l}$, provided $k, l \in \mathbb{N}$ and $k, l \geqslant m+1$. Notice, finally, that it is not clear how one can significantly reduce the length of this section, without losing some valuable information, by using the above described method for construction of powers.

Remark 2.9.38. Suppose $A \in \mathcal{M}_{C,-1}$, $\alpha \in \mathbb{C}_+$, $0 < \gamma < 1/2$ and, temporarily, $A_{C,\alpha} x = y$. Then there exists a net $(x_\tau)_{\tau \in I}$ such that $x_\tau \in D(J_C^\alpha)$, $\tau \in I$ as well as that $\lim_{\tau \to \infty} x_\tau = x$ and $\lim_{\tau \to \infty} J_C^\alpha x_\tau = Cy$ (cf. also the paragraph preceding Proposition 2.9.15). Then (274) implies that, for every $t > 0$, we have $\lim_{\tau \to \infty} S_\gamma(t) x_\tau = S_\gamma(t) x$ in $D_\infty(A)$. Since $J_{C,\alpha} \in L(D_\infty(A))$, the above yields $J_{\infty,\alpha} S_\gamma(t) x = C S_\gamma(t) y$, $t > 0$ and $A_\alpha x = y$. Hence, $A_{C,\alpha} \subseteq C^{-1} A_{C,\alpha} C \subseteq C^{-1} A_\alpha C = A_\alpha$. If the following condition holds

(PB): $(\lambda + A)^{-n} C \in L(E)$, $\lambda > 0$, $n \in \mathbb{N}$ and $\lim_{\lambda \to \infty} \lambda^n (\lambda + A)^{-n} C x = Cx$, $n \in E$, $x \in E$,

then it can be easily seen that the assumption $A_\alpha x = y$ implies $\overline{J_C^\alpha} Cx = C^2 y$. Therefore, the validity of (PB) implies $A_\alpha = C^{-1} A_{C,\alpha} C$.

Consider now the following condition

(H): A is a C-sectorial operator of angle $\omega \in [0, \pi)$, and the mapping $\lambda \mapsto (\lambda - A)^{-1} Cx$, $\lambda \in \mathbb{C} \setminus \overline{\Sigma_\omega}$ is continuous for every fixed $x \in E$.

Its validity implies that the mapping $\lambda \mapsto (\lambda - A)^{-1} Cx$ is analytic in $\mathbb{C} \setminus \overline{\Sigma_\omega}$ as well as that $(\lambda - A)^{-n} C \in L(E)$; furthermore, the equicontinuity of the operator family $\{\lambda^n (\lambda + A)^{-n} C : \lambda > 0\}$ follows from the Cauchy integral formula ([90]).

In the following theorem, we shall further analyze the properties of operator family $(S_\gamma(t))_{t>0}$ (cf. Proposition 2.9.22 and the proof of [403, Theorem 5.5.1]).

Theorem 2.9.39. *Let* $0 < \gamma < 1/2$, *and let* $A \in \mathcal{M}_{C,-1}$. *Put* $S_\gamma(0) := C$, $S_{\gamma,\zeta}(t) := \int_0^t g_\zeta (t - s) S_\gamma(s) x \, ds$, $x \in E$, $t \geqslant 0$, $\zeta > 0$, *and* $S_{\gamma,0}(t) := S_\gamma(t)$, $t \geqslant 0$. *Then the family* $\{S_\gamma(t)$

: $t > 0\}$ is equicontinuous, and there exist operator families $(\mathbf{S}_\gamma(z))_{z\in\Sigma_{(\pi/2)-\gamma\pi}}$ and $(\mathbf{S}_{\gamma,\zeta}(z))_{z\in\Sigma_{(\pi/2)-\gamma\pi}}$ such that, for every $x \in E$ and $\zeta > 0$, the mappings $z \mapsto \mathbf{S}_\gamma(z)x$, $z \in \Sigma_{(\pi/2)-\gamma\pi}$ and $z \mapsto \mathbf{S}_{\gamma,\zeta}(z)x$, $z \in \Sigma_{(\pi/2)-\gamma\pi}$ are analytic as well as that $\mathbf{S}_\gamma(t) = S_\gamma(t)$, $t > 0$ and $\mathbf{S}_{\gamma,\zeta}(t) = S_{\gamma,\zeta}(t)$, $t > 0$. (This is the reason why we shall not make any difference between $S_\gamma(\cdot)$ and $\mathbf{S}_\gamma(\cdot)$ $[S_{\gamma,\zeta}(\cdot)$ and $\mathbf{S}_{\gamma,\zeta}(\cdot)]$ henceforward). Furthermore, the following holds:

(i) *$S_\gamma(z_1)S_\gamma(z_2) = S_\gamma(z_1 + z_2)\, C$ for all $z_1, z_2 \in \Sigma_{(\pi/2)-\gamma\pi}$.*

(ii) *$\lim_{z\to 0, z\in\Sigma_{(\pi/2)-\gamma\pi-\varepsilon}} S_\gamma(z)x = Cx$, $x \in \overline{D(A)}$, $\varepsilon \in (0, (\pi/2) - \gamma\pi)$; if the condition (H) holds, then the above equality remains true with the number $(\pi/2) - \gamma\pi$ replaced by $(\pi/2) - \omega\gamma$.*

(iii) *$S_\gamma(z)A_\alpha \subseteq A_\alpha S_\gamma(z)$, $z \in \Sigma_{(\pi/2)-\gamma\pi}$, $\alpha \in \mathbb{C}_+$.*

(iv) *If $D(A)$ is dense in E, then the operator $-A_\gamma$ is the integral generator of an equicontinuous analytic C-regularized semigroup $(S_\gamma(t))_{t>0}$ of angle $(\pi/2) - \gamma\pi$. If, additionally, the condition (H) holds, then $(S_\gamma(t))_{t>0}$ can be extended to an equicontinuous analytic C-regularized semigroup $(\mathbf{S}_\gamma(z))_{z\in\Sigma_{(\pi/2)-\gamma\omega}}$ of angle $(\pi/2) - \gamma\omega$; in this case, the equality stated in (i) holds for any $z_1, z_2 \in \Sigma_{(\pi/2)-\gamma\omega}$ and the equality stated in (iii) holds provided $z \in \Sigma_{(\pi/2)-\gamma\omega}$.*

(v) *For every $\zeta > 0$, the operator $-A_\gamma$ is the integral generator of an exponentially equicontinuous analytic ζ-times integrated C-regularized semigroup $(S_{\gamma,\zeta}(t))_{t>0}$ of angle $(\pi/2) - \gamma\pi$; furthermore, the family $\{|z|^{-\zeta}S_{\gamma,\zeta}(z) : z \in \Sigma_{(\pi/2)-\gamma\pi-\varepsilon}\}$ is equicontinuous for any $\varepsilon \in (0, (\pi/2) - \gamma\pi)$. If, additionally, the condition (H) holds, then $(S_{\gamma,\zeta}(t))_{t>0}$ can be extended to an exponentially equicontinuous analytic ζ-times integrated C-regularized semigroup of angle $(\pi/2) - \gamma\omega$, and the family*

$\{|z|^{-\zeta}S_{\gamma,\zeta}(z) : z \in \Sigma_{(\pi/2)-\gamma\omega-\varepsilon}\}$ is equicontinuous for any $\varepsilon \in (0, (\pi/2) - \gamma\omega)$.

(vi) *$R(S_\gamma(z)) \subseteq D_\infty(A)$, $A^nS_\gamma(z) \in L(E)$ and*

$$A^nS_\gamma(z)x = (-1)^{n+1}\int_0^\infty \lambda^n f_z(\lambda)\,(\lambda + A)^{-1}\, Cx\, d\lambda,\quad x \in E,\ z \in \Sigma_{(\pi/2)-\gamma\pi}. \tag{287}$$

(vii) *For every $x \in \overline{D(A)}$ and for every $l \in \mathbb{N}\setminus\{1, 2\}$, the incomplete abstract Cauchy problem*

$$(P_l):\begin{cases} u \in C^\infty((0,\infty) : E) \cap C((0,\infty) : D_\infty(A)),\\ u^{(l)}(t) = (-1)^l Au(t),\ t > 0,\\ \lim_{t\to 0+} u(t) = Cx,\\ \text{the set } \{u(t) : t > 0\} \text{ is bounded in } E, \end{cases}$$

has a solution $u(t) = S_{1/l}(t)x$, $t > 0$. Moreover, the mapping $t \mapsto u(t)$, $t > 0$ can be analytically extended to the sector $\Sigma_{(\pi/2)-(\pi/l)}$ and, for every $\delta \in (0, (\pi/2) - (\pi/l))$ and $j \in \mathbb{N}_0$, we have that the set $\{z^j u^{(j)}(z) : z \in \Sigma_\delta\}$ is bounded in E. If, additionally, the condition (H) holds, then the above statements continue to hold with the number $(\pi/2) - (\pi/l)$ replaced by $(\pi/2) - (\omega/l)$.

Proof. The function $G_{\gamma,t}(z) := \exp(-tz^{\gamma})$, $z \in \mathbb{C}\setminus(-\infty, 0]$ belongs to the space $\mathcal{H}_1$ introduced in [403, Chapter 3], and we have that $S_{\gamma}(t) = G_{\gamma,t}(A)$, $t > 0$, where the last operator is defined by means of the regularized version of Hirsch functional calculus (cf. Remark 2.9.33(iii)). Since we do not know whether the mapping $\lambda \mapsto (\lambda + A)^{-1}Cx$, $\lambda > 0$ is continuously differentiable for every fixed $x \in E$, one can only prove the following slightly weakened product formula: $[f(A)C][g(A)C] \subseteq h(A)C^3$, provided $f, g, h \in \mathcal{T}$ and $h = fg$. This, in turn, implies (i) for positive values of z_1 and z_2. Since the mapping $t \mapsto S_{\gamma}(t)$, $t > 0$ can be analytically extended to the sector $\Sigma_{(\pi/2)-\gamma\pi}$ by formula (275), we obtain the existence of operator family $(\mathbf{S}_{\gamma}(z))_{z\in\Sigma_{(\pi/2)-\gamma\pi}}$ with required properties, which implies (i) by the uniqueness theorem for analytic functions. Fix, momentarily, $\varepsilon \in (0, (\pi/2) - \gamma\pi)$. By the C-nonnegativity of A, we obtain that for each $q \in \circledast$ there exist $c_q > 0$ and $r_q \in \circledast$ such that, for every $z \in \Sigma_{(\pi/2)-\gamma\omega-\varepsilon}$ and $x \in E$,

$$q(S_{\gamma}(z)x) \leqslant c_q r_q(x) \int_0^{\infty} e^{-\lambda^{\gamma}|z|\cos(\arg(z))\cos(\pi\gamma)} \left|\sin(\lambda^{\gamma}|z|e^{i\arg(z)}\sin(\pi\gamma))\right| \frac{d\lambda}{\lambda}$$

$$\leqslant \frac{c_q}{\gamma} r_q(x) \int_0^{\infty} e^{-v}[2 + 2e^{-v\cos(\arg(z))}] \frac{dv}{2v} = \frac{c_q}{\gamma} r_q(x) \left[1 + \frac{1}{1 + \sin \pi\gamma}\right].$$

As a consequence of the proof of [403, Theorem 5.5.1], we have $\lim_{t\to 0+} S_{\gamma}(t)x = Cx$, $x \in \overline{D(A)}$. This implies the first part of assertion (ii) by Theorem 1.2.5(ii) and the equicontinuity of family $\{S_{\gamma}(z) : z \in \Sigma_{(\pi/2)-\gamma\pi-\varepsilon}\}$; the proofs of (iii) and (vi)–(vii) are simple and therefore omitted. Now it becomes apparent that, for every $\zeta > 0$, $(S_{\gamma,\zeta}(t))_{t\geqslant 0}$ is a ζ-times integrated C-regularized semigroup; furthermore, the density of $D(A)$ in E implies that $(S_{\gamma}(z))_{z\in\Sigma_{(\pi/2)-\gamma\pi}}$ is an equicontinuous analytic C-regularized semigroup of angle $(\pi/2) - \gamma\pi$. Now we will prove that the integral generator of $(S_{\gamma,\zeta}(t))_{t\geqslant 0}$ is the operator $-A_{\gamma}$. The proof of [403, Theorem 5.5.1] yields that, for every $x \in E$ and $\mu > 0$,

$$(288)\quad \int_0^{t} e^{-\mu t} S_{\gamma,\zeta}(t)x\,dt = \frac{\sin \gamma\pi}{\mu^{\zeta}\pi} \int_0^{\infty} \frac{\lambda^{\gamma}}{(\mu + \lambda^{\gamma}\cos \pi\gamma)^2 + \lambda^{2\gamma}\sin^2 \pi\gamma} (\lambda + A)^{-1} Cx\,d\lambda.$$

Keeping in mind that $C^{-1}A_{\gamma}C = A_{\gamma}$, (288) implies that we will only have to prove that, for every $x \in E$ and $\mu > 0$,

$$(289)\quad \frac{\sin \gamma\pi}{\pi} \int_0^{\infty} \frac{\lambda^{\gamma}}{(\mu + \lambda^{\gamma}\cos \pi\gamma)^2 + \lambda^{2\gamma}\sin^2 \pi\gamma} (\lambda + A)^{-1} Cx\,d\lambda = (\mu + A_{\gamma})^{-1} Cx.$$

Since $C^{-2}A_{\gamma}C^2 = A_{\gamma}$ commutes with $(\lambda + A)^{-1}C$, it suffices to show that, for every $x \in E$ and $\mu > 0$,

(290) $$(\mu + A_\gamma)C^2 \frac{\sin \gamma\pi}{\pi}\int_0^\infty \frac{\lambda^\gamma}{(\mu+\lambda^\gamma \cos \pi\gamma)^2 + \lambda^{2\gamma}\sin^2 \pi\gamma}(\lambda + A)^{-1} Cx\, d\lambda = C^3 x.$$

By definition of A_γ and arguments used in Remark 2.9.33, we may assume without loss of generality that $x \in C(D_\infty(A))$. Then it can be easily seen that the equality (290) is equivalent with

$$[C_\infty(\mu + z^\gamma)(A_\infty)]\,[C_\infty F_\mu(A_\infty)]\,x = C_\infty^4 x,$$

where $F_\mu(z) = (\mu + z^\alpha)^{-1} \in \mathcal{H}_0 \cap \mathcal{H}_1$. But, this a simple consequence of the weakened product formula. Now it is not difficult to prove that the operator $-A_\gamma$ is the integral generator of an exponentially equicontinuous analytic ζ-times integrated C-regularized semigroup $(\mathbf{S}_{\gamma,\zeta}(z))_{z\in\Sigma_{(\pi/2)-\gamma\pi}}$, where

(291) $$\mathbf{S}_{\gamma,\zeta}(z) = \int_0^z g_\zeta(z-s)\mathbf{S}_\gamma(s)x\, ds, \quad x \in E,\ z \in \Sigma_{(\pi/2)-\gamma\pi}.$$

This implies that the family $\{|z|^{-\zeta}S_{\gamma,\zeta}(z) : z \in \Sigma_{(\pi/2)-\gamma\pi-\varepsilon}\}$ is equicontinuous for all $\varepsilon \in (0, (\pi/2) - \gamma\pi)$. If A is densely defined, then we obtain that the operator $-A_\gamma$ is the integral generator of an equicontinuous analytic C-regularized semigroup $(\mathbf{S}_\gamma(z))_{z\in\Sigma_{(\pi/2)-\gamma\pi}}$, which implies by Theorem 2.2.4 that the mapping $\mu \mapsto (\mu + A_\gamma)^{-1} Cx$, $\mu \in \{re^{i\theta} : \theta \in (-(\pi - \pi\gamma), \pi - \pi\gamma)\}$ is analytic for all $x \in E$ and that the family $\{\mu(\mu + A_\gamma)^{-1}C : \mu \in \Sigma_{\pi-\pi\gamma-\varepsilon}\}$ is equicontinuous for all $\varepsilon \in (0, \pi - \pi\gamma)$. It is not difficult to see with the help of proof of [403, Proposition 5.3.2] that the equality (289) continues to hold for $\mu \in \{re^{i\theta} : \theta \in (-(\pi - \pi\gamma), \pi - \pi\gamma)\}$. Suppose now that the condition (H) holds. Then, for every $\theta \in (\omega - \pi, \pi - \omega)$, the operator $e^{i\theta}A$ is C-nonnegative and the mapping $\lambda \mapsto (\lambda + e^{i\theta}A)^{-1}Cx$, $\lambda > 0$ is continuous for every fixed $x \in E$. No matter whether $|\theta| \geqslant \pi/2$ or not, the Cauchy formula implies that, for every $x \in D_\infty(A)$,

(292) $$\int_0^\infty \lambda^{\gamma-1}(\lambda e^{-i\theta} + A)^{-1}CAx\, d\lambda = e^{-i\theta\gamma}\int_0^\infty \lambda^{\gamma-1}(\lambda + A)^{-1}CAx\, d\lambda.$$

By definition of Balakrishnan operators, the equality (292), and the method described in Remark 2.9.33, we obtain that $(e^{i\theta}A)_\gamma = e^{i\theta\gamma}A_\gamma$. Now it readily follows that, for every $\theta \in (\omega - \pi, \pi - \omega)$ and for every $x \in E$, the mapping $\mu \mapsto (\mu + (e^{i\theta}A)_\gamma)^{-1} Cx = (\mu + e^{i\theta\gamma}A_\gamma)^{-1}Cx = e^{-i\theta\gamma}(\mu e^{-i\theta\gamma} + A_\gamma)^{-1}Cx$, $\mu \in \{re^{i\beta} : \beta \in (-(\pi - \pi\gamma), \pi - \pi\gamma)\}$ is analytic; hence, the mapping $\mu \mapsto (\mu + A_\gamma)^{-1}Cx$, $\mu \in \Sigma_{\pi-\omega\gamma}$ is analytic for all $x \in E$; the equicontinuity of family $\{\mu(\mu + A_\gamma)^{-1}C : \mu \in \Sigma_{\pi-\omega\gamma-\varepsilon}\}$, for $\varepsilon \in (0, \pi - \omega\gamma)$, can be proved similarly. An application of Theorem 2.2.5 gives that $-A_\gamma$ is the integral generator of an equicontinuous analytic C-regularized semigroup $(\mathbf{S}_\gamma(z))_{z\in\Sigma_{\pi-\omega\gamma}}$, provided that $D(A)$ is dense in E. If $D(A)$ is not dense in E, then there exists a mapping $z \mapsto \mathbf{S}_\gamma(z) \in L(E)$, $z \in \Sigma_{\pi-\omega\gamma}$ such that $\mathbf{S}_\gamma(t) = S_\gamma(t)$, $t > 0$, the family $\{\mathbf{S}_\gamma(z) : z \in \Sigma_{\pi-\omega\gamma-\varepsilon}\}$ is equicontinuous for any $\varepsilon \in (0, \pi - \gamma\omega)$ and that, for every $x \in E$, the mapping $z \mapsto \mathbf{S}_\gamma(z)x$, $z \in \Sigma_{\pi-\omega\gamma}$ is analytic (cf. the proof of next theorem). The formula (291) remains true for $z \in \Sigma_{\pi-\omega\gamma}$, so that the proof of theorem is completed in a routine manner. □

In the following theorem, we shall reconsider and slightly improve Balakrishnan's results [403, Theorem 5.5.2] and [403, Theorem 6.3.2].

Theorem 2.9.40. *Let $A \in \mathcal{M}_{C,-1}$, and let the mapping $\lambda \mapsto (\lambda + A)^{-1}Cx$, $\lambda > 0$ be continuously differentiable for every fixed $x \in E$. Suppose*

$$\frac{d}{d\lambda}(\lambda + A)^{-1}Cx = -(\lambda + A)^{-2}Cx, \quad \lambda > 0, x \in E, \tag{293}$$

and $\{\lambda^2(\lambda + A)^{-2}C : \lambda > 0\}$ is an equicontinuous family in $L(E)$. Then the limit contained in the expression

$$S_{1/2}(t)x := \frac{1}{\pi} \lim_{N\to\infty} \int_0^N \sin(t\sqrt{\lambda})\,(\lambda + A)^{-1}Cx\,d\lambda \tag{294}$$

exists in $L(E)$ for every $x \in E$. Put $S_{1/2}(0) := C$, $S_{1/2,\zeta}(t)x := \int_0^t g_\zeta(t-s)S_{1/2}(s)x\,ds$, $x \in E$, $t \geqslant 0$, $\zeta > 0$, and $S_{1/2,0}(t) := S_{1/2}(t)$, $t \geqslant 0$. Then the family $\{S_{1/2}(t) : t > 0\}$ is equicontinuous and the following holds:

(i) $S_{1/2}(t)S_{1/2}(s) = S_{1/2}(t+s)C$ *for all $t, s > 0$.*
(ii) $\lim_{t\to 0+} S_{1/2}(t)x = Cx$, $x \in \overline{D(A)}$; *if the condition (H) holds, then*

$$\lim_{z\to 0,\, z\in\Sigma_{(\pi/2)-(\omega/2)-\varepsilon}} S_{1/2}(z)x = Cx,\ x \in \overline{D(A)},\ \varepsilon \in (0, (\pi/2) - (\omega/2)).$$

(iii) $S_{1/2}(t)A_\alpha \subseteq A_\alpha S_{1/2}(t)$, $t > 0$, $\alpha \in \mathbb{C}_+$.
(iv) *If $D(A)$ is dense in E, then $(S_{1/2}(t))_{t\geqslant 0}$ is an equicontinuous C-regularized semigroup with the integral generator $-A_{1/2}$. If, additionally, the condition (H) holds, then $(S_{1/2}(t))_{t>0}$ can be extended to an equicontinuous analytic C-regularized semigroup of angle $(\pi/2)-(\omega/2)$; in this case, the equality stated in (i) holds for any $z_1, z_2 \in \Sigma_{(\pi/2)-(\omega/2)}$, and the equality stated in (iii) holds for any $z \in \Sigma_{(\pi/2)-(\omega/2)}$.*
(v) *For every $\zeta > 0$, $(S_{1/2,\zeta}(t))_{t\geqslant 0}$ is a ζ-times integrated C-regularized semigroup with the integral generator $-A_{1/2}$; furthermore, the family $\{t^{-\zeta}S_{1/2,\zeta}(t) : t > 0\}$ is equicontinuous. If, additionally, the condition (H) holds, then $(S_{1/2,\zeta}(t))_{t>0}$ can be extended to an exponentially equicontinuous analytic ζ-times integrated C-regularized semigroup $(S_{1/2,\zeta}(z))_{z\in\Sigma_{(\pi/2)-(\omega/2)}}$ of angle $(\pi/2)-(\omega/2)$, and the family $\{|z|^{-\zeta}S_{1/2,\zeta}(z) : z \in \Sigma_{(\pi/2)-(\omega/2)-\varepsilon}\}$ is equicontinuous ($\varepsilon \in (0, (\pi/2)-(\omega/2))$).*
(vi) *Let the condition (H) hold. Then $R(S_{1/2}(t)) \subseteq D_\infty(A)$, $t > 0$ and, for every $x \in \overline{D(A)}$, the incomplete abstract Cauchy problem*

$$(P_2): \begin{cases} u \in C^\infty((0,\infty) : E) \cap C((0,\infty) : D_\infty(A)), \\ u''(t) = Au(t),\ t > 0, \\ \lim_{t\to 0+} u(t) = Cx, \\ \text{the set } \{u(t) : t > 0\} \text{ is bounded in } E, \end{cases}$$

has a unique solution $u(t) = S_{1/2}(t)x$, $t > 0$. *Moreover, the mapping* $t \mapsto u(t)$, $t > 0$ *can be analytically extended to the sector* $\Sigma_{(\pi/2)-(\omega/2)}$ *and, for every* $\delta \in (0, (\pi/2)-(\omega/2))$ *and* $j \in \mathbb{N}_0$, *we have that the set* $\{z^j u^{(j)}(z) : z \in \Sigma_\delta\}$ *is bounded in E.*

Proof. We will only outline the main details of the proof. The partial integration, along with the formula (293) and the equicontinuity of family $\{\lambda^2(\lambda + A)^{-2}C : \lambda > 0\}$, shows that the limit contained in (294) exists and equals

(295) $$S_{1/2}(t)x = \int_0^\infty f(\lambda, t)\,(\lambda + A)^{-2}\,Cx\,d\lambda,\ t > 0,\ x \in E,$$

where $f(\lambda, t) = 2\pi^{-1} t^{-2}[\sin(t\sqrt{\lambda}) - t\sqrt{\lambda}\cos(t\sqrt{\lambda})]$ for $\lambda > 0$ and $t > 0$. Using the change of variables $x = t\sqrt{\lambda}$, it readily follows that the operator family $\{S_{1/2}(t) : t > 0\}$ is equicontinuous and strongly continuous. The equalities

$$S_{1/2}(t)x - Cx = \frac{1}{\pi}\lim_{N\to\infty}\int_0^N \sin(t\sqrt{\lambda})\,((\lambda + A)^{-1}Cx - \lambda^{-1}Cx)\,d\lambda$$
$$= \frac{1}{\pi}\lim_{N\to\infty}\int_0^N \frac{\sin(t\sqrt{\lambda})}{\lambda}\,(\lambda + A)^{-1}\,CAx\,d\lambda,\quad x \in D(A),$$

imply that $\lim_{t\to 0} S_{1/2}(t)x = Cx$, $x \in D(A)$, which continues to hold, by the equicontinuity of family $\{S_{1/2}(t) : t \geqslant 0\}$, for all $x \in \overline{D(A)}$. The proof of [403, Theorem 5.5.2] implies that, for every $x \in E$ and $t \geqslant 0$, we have $\lim_{\gamma\to 1/2} S_\gamma(t)x = S_{1/2}(t)x$; hence, (i) holds. Now it is clear that, for every $\eta > 0$, $(S_{1/2,\zeta}(t))_{t\geqslant 0}$ is a ζ-times integrated C-regularized semigroup and that the family $\{t^{-\zeta} S_{1/2,\zeta}(t) : t > 0\}$ is equicontinuous; furthermore, the denseness of $D(A)$ in E implies that $(S_{1/2}(t))_{t\geqslant 0}$ is an equicontinuous C-regularized semigroup. Obviously, $C^{-1}A_{1/2}C = A_{1/2}$ and

(296) $$\int_0^t e^{-\mu t} S_{1/2,\zeta}(t)x\,dt = \frac{1}{\mu^\zeta}\frac{1}{\pi}\int_0^\infty \frac{\lambda^{1/2}}{\mu^2 + \lambda}\,(\lambda + A)^{-1}\,Cx\,d\lambda,\quad x \in E,\ \mu > 0.$$

The proof of the preceding theorem shows that, for every $\mu > 0$, the right hand side of the equality (296) equals $\mu^{-\zeta}(\mu + A_{1/2})^{-1}Cx$, which implies that the integral generator of $(S_{1/2,\zeta}(t))_{t\geqslant 0}$ is the operator $-A_{1/2}$. Suppose now that the condition (H) holds. Then the proof of Theorem 2.9.39 implies that, for every $x \in E$, the mapping $\mu \mapsto \mu(\mu + A_{1/2})^{-1}Cx$, $\mu \in \Sigma_{\pi-(\omega/2)}$ is analytic and that the family $\{\mu(\mu + A_{1/2})^{-1}C : \mu \in \Sigma_{\pi-(\omega/2)-\varepsilon}\}$ is equicontinuous for any $\varepsilon \in (0, \pi - (\omega/2))$. By Theorem 2.2.5, the operator $-A_{1/2}$ is the integral generator of an exponentially equicontinuous analytic once integrated C-regularized semigroup $(S_{1/2,1}(t))_{t\geqslant 0}$ of angle $(\pi/2) - (\omega/2)$; moreover, the denseness of $D(A)$ in E implies that the operator $-A_{1/2}$ is the integral generator of an equicontinuous analytic C-regularized semigroup of angle $(\pi/2) - (\omega/2)$. Even if $D(A)$ is not dense in E, the mapping $t \mapsto S_{1/2}(t) \in L(E)$, $t > 0$ has an extension to the sector $\Sigma_{(\pi/2)-(\omega/2)}$ which satisfies that, for every $x \in E$, the mapping $z \mapsto S_{1/2}(z)x,\ z \in \Sigma_{(\pi/2)-(\omega/2)}$ is analytic. It is not difficult to prove that, for every $\varepsilon \in (0, (\pi/2) - (\omega/2))$, the representation formula

$$S_{1/2}(z)x = \frac{1}{2\pi i} \int_{\Gamma} e^{\lambda z} (\lambda + A_{1/2})^{-1} Cx \, d\lambda, \quad x \in E,\ z \in \Sigma_{(\pi/2)-(\omega/2)-\varepsilon}$$

holds, where Γ is oriented counterclockwise and consists of $\Gamma_{\pm} := \{re^{i(\pi - \frac{\omega}{2} - \varepsilon)} : r \geqslant |z|^{-1}\}$ and $\Gamma_0 := \{|z|^{-1}e^{i\theta} : |\theta| \geqslant \pi - \frac{\omega}{2} - \varepsilon\}$. By the computation given in the proof of [20, Theorem 2.6.1] (cf. also Theorem 1.2.5), we obtain that the operator family $\{S_{1/2}(z) : z \in \Sigma_{(\pi/2)-(\omega/2)-\varepsilon}\}$ is equicontinuous, which simply implies (ii). The formula (291) continues to hold with $\zeta = 1$, so that the proof of (iv)-(v) completes a routine argument; the proof of (iii) is omitted. In order to prove (vi), fix an element $x \in \overline{D(A)}$. Using the assertion (iv), it readily follows that the mapping $t \mapsto S_{1/2}(t)x,\ t > 0$ can be analytically extended to the sector $\Sigma_{(\pi/2)-(\omega/2)}$. The identity $(-A_{1/2}) \int_0^t S_{1/2}(s)x \, ds = S_{1/2}(t)x - Cx,\ t > 0$ implies along with the closedness of the operator $A_{1/2}$ that $u'(t) = -A_{1/2}u(t),\ t > 0$. Taken together with Proposition 2.9.28, the above equality shows that $u''(t) = Au(t),\ t > 0$. Then we obtain by induction that $u^{(2n)}(t) = A^n u(t),\ t > 0$, which yields that $R(S_{1/2}(t)) \subseteq D_{\infty}(A),\ t > 0$ and that the function $t \mapsto u(t),\ t > 0$ is a solution of (P_2). To prove the uniqueness, suppose that the function $t \mapsto v(t),\ t > 0$ is a non-trivial solution of the problem (P_2) with $x = 0$. Set $C_1 := C^4((1+A)^{-1}C)^4$ and

(297) $$f(s) := -\,[C_1(1 + A_{1/2})^{-1} C_1 v'(s) + C_1 A_{1/2}(1 + A_{1/2})^{-1} C_1 v(s)], \quad s > 0.$$

Then $f(s) \in D(A_{1/2}),\ s > 0,\ f \in C^1((0, \infty) : E)$, and the equality (286) implies $(A_{1/2})^2 C_1 = AC_1$. A key point in the proof of uniqueness of solutions to (P_2) is to observe that (see the computation given in the proof of [145, Lemma 2.12]):

(298) $$f'(s) = A_{1/2}f(s), \quad s > 0.$$

Suppose now that there exists $x \in E$ such that $f(s) = x,\ s > 0$. Then $A_{1/2}x = 0$, $\lim_{s\to 0+} C_1(1 + A_{1/2})^{-1} C_1 v'(s) = -\,x$, and differentiation of (297) with respect to s yields:

$$C_1(1 + A_{1/2})^{-1} C_1 v''(s) + C_1^2 v'(s) - C_1(1 + A_{1/2})^{-1} C_1 v'(s) = 0, \quad s > 0.$$

Multiplying the above equality with $(1 + A_{1/2})^{-1}C_1$, we easily infer that $\lim_{s\to 0+} C_1((1 + A_{1/2})^{-1}C_1)^2 v''(s) = C_1 x - (1 + A_{1/2})^{-1}C_1 x = 0$. On the other hand, it can be easily seen that $C_1Av(s) + A_{1/2}C_1v'(s) = 0,\ s > 0$ and $C_1Av'(s) + A_{1/2}C_1Av(s) = 0,\ s > 0$. Therefore, the function $F(s) := ((1 + A_{1/2})^{-1}C_1)^2 C_1 Av(s) = ((1 + A_{1/2})^{-1}C_1)^2 C_1 v''(s)$, $s > 0$ is a solution of the abstract Cauchy problem:

$$\begin{cases} F \in C^1((0, \infty) : E) \cap C([0, \infty) : E), \\ F'(s) = -A_{1/2}F(s), \quad s > 0, \\ F(0) = 0. \end{cases}$$

Then the use of Ljubich uniqueness theorem implies $F(s) = 0,\ s > 0$ and $v''(s) = 0,\ s > 0$. Due to the boundedness of function $v(\cdot)$, we get that $v(s) = 0,\ s > 0$, which contradicts the non-triviality of $v(\cdot)$; hence, the function $t \mapsto f(t)$, $t > 0$ is non-constant. It is checked at once that, for every $\varepsilon > 0$ and $t > 0$,

$$\frac{d}{ds} C(1 + A_{1/2})^{-1} Cf(t - s + \varepsilon) = -A_{1/2}C(1 + A_{1/2})^{-1} Cf(t - s + \varepsilon),\ 0 < s < t.$$

Since $-A_{1/2}$ generates once integrated C-regularized semigroup $(S_{1/2,1}(t))_{t\geq 0}$, the last equality combined with [292, Proposition 2.3.4] immediately implies that, for every $\varepsilon > 0$ and $t > 0$,

$$Cf(t - s + \varepsilon) = S_{1/2}(s)f(t + \varepsilon),\ 0 < s < t. \tag{299}$$

Letting $\varepsilon \to 0+$, we get that for $0 < s < t$:

$$f(t - s) = -S_{1/2}(s)[C^{-1}C_1(1 + A_{1/2})^{-1} C_1v'(t) + C^{-1}C_1A_{1/2}(1 + A_{1/2})^{-1} C_1v(t)]. \tag{300}$$

Making use of the equality $(A_{1/2})^2C_1 = AC_1$, the properties of function $v(\cdot)$, and the formula (300) with $s = t/2 > 0$, we get that:

$$\begin{aligned} f(t) = {} & S'_{1/2}(t)A_{1/2}(1 + A_{1/2})^{-1} C^{-1}C_1^2 \int_1^{2t} v(r)\, dr \\ & - S_{1/2}(t)C^{-1}C_1(1 + A_{1/2})^{-1} C_1v'(1) \\ & - S_{1/2}(t)C^{-1}C_1A_{1/2}(1 + A_{1/2})^{-1} C_1v(2t) := f_1(t) + f_2(t) + f_3(t),\ t > 0. \end{aligned} \tag{301}$$

The Cauchy integral formula implies that the family $\{tS'_{1/2}(t) : t > 0\}$ is equicontinuous, and the last fact taken together with the boundedness of the operator $A_{1/2}(1 + A_{1/2})^{-1}C^{-1}C_1^2$ implies that $\{f_1(t) : t > 0\}$ is a bounded subset of E. This is also clear for $\{f_2(t) : t > 0\}$ and $\{f_3(t) : t > 0\}$, so that the function $t \mapsto f(t)$, $t > 0$ is bounded. It can be easily seen that the function $t \mapsto f(t)$, $t > 0$ is non-constant. Due to (299), we have that $Cf(r) = S_{1/2}(t - r)f(t)$ provided $0 < r < t$, which implies $-A_{1/2}Cf(r) = -A_{1/2}S_{1/2}(t - r)f(t) = S'_{1/2}(t - r)f(t) \to 0$ as $t \to +\infty$. Returning to (298) implies that there exists $y \in E$ such that $Cf(r) = y$ for all $r > 0$, which contradicts the fact that $f(\cdot)$ cannot be constant. □

Example 2.9.41. Suppose $-A$ generates an equicontinuous (g_β, C)-regularized resolvent family for some $\beta \in (0, 2]$. Then the operator A is C-sectorial of angle $\pi - ((\beta\pi)/2)$, so that we can construct the fractional powers of A (observe, however, that the converse statement does not hold even in the case that $\beta = 1$ and $C = I$; cf. [88, Remark 2.8(b)] for further information). Suppose now $\omega > 0$, $q \in [2, \infty)$ and the following conditions hold:

(i) $\emptyset \neq \Omega \subseteq \mathbb{R}^n$ is a bounded domain of class C^m $(m, n \in \mathbb{N})$;

(ii) For each multi-indices α, β with $0 \leqslant |\alpha| = |\beta| \leqslant m$,

$$a_{\alpha\beta}(\cdot) \in C(\overline{\Omega}),\ \overline{a_{\alpha\beta}(x)} = a_{\beta\alpha}(x),\ x \in \Omega,$$

and when $n = 2$, each of $a_{\alpha\beta}(\cdot)$ for $|\alpha| = |\beta| = m$ is real-valued;

(iii) The operator

$$A(x, D) := \sum_{k=0}^{m} \sum_{|\alpha|=|\beta|=k} D^{\alpha}(a_{\alpha\beta}(x)D^{\beta})$$

is uniformly strongly elliptic, i.e., there exists $b > 0$ such that for all $x \in \Omega$, $\xi \in \mathbb{R}^n$,

$$(-1)^m \operatorname{Re}\left\{ \sum_{|\alpha|=|\beta|=k} a_{\alpha\beta}(x)\xi^{\alpha}\xi^{\beta} \right\} \geqslant b|\xi|^{2m},$$

where $|\alpha| = \alpha_1 + \cdots + \alpha_n$, $x^{\alpha} = x_1^{\alpha_1} \cdots x_n^{\alpha_n}$ and, with a little abuse of notation, $D^{\alpha} = \partial^{\alpha}/\partial x_1^{\alpha_1} \cdots \partial x_n^{\alpha_n}$.

Then we associate with $A(x, D)$ the operator $A_q(x, D)$, or simply A_q, acting on $L^q(\Omega)$, by

$$A_q(x, D)u := A(x, D)u, \quad u \in D(A_q(x, D)) := W^{2m,q}(\Omega) \cap W_0^{m,q}(\Omega).$$

By [532, Theorem 3.2], we know that for each $r \geqslant \frac{n}{2m}(1/2- 1/q)$ the operator iA_q is the integral generator of a bounded $(\omega + A_q)^{-r}$-regularized semigroup on $L^q(\Omega)$. By Theorem 2.9.39-Theorem 2.9.40, we obtain that, for every $l \in \mathbb{N}\backslash\{1\}$ and for every $u_0 \in L^q(\Omega)$, the incomplete abstract Cauchy problem

$$(P_l): \begin{cases} u \in C^{\infty}((0, \infty): L^q(\Omega)) \cap C((0, \infty): D_{\infty}(A_q)), \\ \frac{\partial^l}{\partial t^l} u(t, x) = \pm(-1)^l iA_q u(t, x),\ t > 0,\ x \in \Omega, \\ \lim_{t\to 0+} u(t, x) = (\omega + A_q)^{-r} u_0(x), \\ \text{the set } \{u(t, \cdot) : t > 0\} \text{ is bounded in } L^q(\Omega), \end{cases}$$

has a solution $u(t, x) = (S_{1/l}(t)u_0)(x)$, $t > 0$, $x \in \Omega$. Moreover, the mapping $t \mapsto u(t, \cdot)$ can be analytically extended to the sector $\Sigma_{(\pi/2)-(\pi/2l)}$ and, for every $\delta \in (0,(\pi/2) - (\pi/2l))$ and $j \in \mathbb{N}_0$, we have that the set $\{z^j u^{(j)}(z, \cdot) : z \in \Sigma_{\delta}\}$ is bounded in $L^q(\Omega)$; clearly, the uniqueness of solution holds provided $l = 2$. It is also worth noting that the operator $-A_q$ generates an analytic C_0-semigroup of contractions on $L^q(\Omega)$ ([2]) and that, in the case $m = 1$, the operator iA_q is sectorial with the spectral angle $\omega \geqslant (\pi/2) + \arctan((q - 2)Mn/2b\sqrt{q-1})$, where $M = \max\{\|a_{ij}\|_{\infty} : 1 \leqslant i, j \leqslant n\}$ (see [403, Theorem 2.4.4, p. 48, 1.6]). In general, the difference $\pi - \omega$ can be sufficiently small so that we obtain the larger sector of analyticity of solution to (P_l) by applying the additional operator $C = (\omega + A_q)^{-r}$ to the initial value function $u_0(\cdot)$. Observe, finally, that one can similarly formulate the corresponding incomplete problem for differential operators considered in [532, Theorem 3.2, Theorem 4.2, Theorem 5.4] and [141, Example 4.18], as well as that the results established in [532] can serve one to provide possible applications of Theorem 2.9.44 and Theorem 2.9.48(ix) stated below.

Repeating literally the arguments used in the proofs of Theorem 2.9.39 and [403, Proposition 5.3.2], we can show that the following proposition holds true.

Proposition 2.9.42. *Let* $\alpha \in \mathbb{C}$ *and* $\operatorname{Re} \alpha > |\alpha|^2$. *Then the set*

$$\Omega_\alpha := \mathbb{C}^* \setminus \{\lambda^\alpha e^{i\theta\alpha} : \lambda > 0,\ \theta \in [-\pi, \pi]\}$$

is a non-empty subset of $\rho_C(A_\alpha)$ *and, for every* $x \in E$ *and* $-\mu \in \Omega_\alpha$,

$$(\mu + A_\alpha)^{-1} Cx = \frac{\sin \alpha\pi}{\pi} \int_0^\infty \frac{\lambda^\alpha}{(\mu + \lambda^\alpha \cos \pi\alpha)^2 + \lambda^{2\alpha} \sin^2 \pi\alpha} (\lambda + A)^{-1} Cx\, d\lambda.$$

If $0 < \alpha < 1$, *then* A_α *is C-sectorial, with C-spectral angle less than or equal to* $\alpha\pi$ *and, for every* $p \in \circledast$ *and* $x \in E$, *the following holds:*

$$\sup_{\lambda > 0} p(\lambda(\lambda + A_\alpha)^{-1} Cx) \leqslant \sup_{\lambda > 0} p(\lambda(\lambda + A)^{-1} Cx).$$

A similar line of reasoning shows that the following extension of [403, Proposition 5.3.3, Proposition 5.3.4] holds (cf. also [403, Theorem 5.4.1, Proposition 5.4.4] and their proofs).

Proposition 2.9.43. *Let* $\alpha \in \mathbb{C}$ *and* $\operatorname{Re} \alpha > |\alpha|^2$.

(i) *Suppose* $s \in \mathbb{C} \setminus (-\infty, 0]$. *Then* $s \in \rho_C(A)$ *iff* $s^\alpha \in \rho_C(A_\alpha)$. *If this is the case, we have:*

$$(A_\alpha - s^\alpha)^{-1} Cx = \alpha^{-1} s^{1-\alpha} (A - s)^{-1} Cx$$

$$+ \frac{\sin \alpha\pi}{\pi} \int_0^\infty \frac{\lambda^\alpha}{\lambda^{2\alpha} - 2\lambda^\alpha s^\alpha \cos \pi\alpha + s^{2\alpha}} (\lambda + A)^{-1} Cx\, d\lambda,\ \ x \in E.$$

(ii) *Suppose* $r > 0$. *Then* $r^\alpha e^{\pm i\pi\alpha} \in \rho_{C^2}(A_\alpha)$ *and, for every* $x \in E$,

$$(A_\alpha - r^\alpha e^{\pm i\pi\alpha})^{-1} C^2 x = -r^{1-\alpha} e^{\mp i\pi\alpha} (A + r)^{-1} C^2 x$$

$$+ \frac{\sin \alpha\pi}{\pi} \int_0^\infty \frac{(\lambda - r)\lambda^{\alpha-1}}{(\lambda^\alpha - r^\alpha)(\lambda^\alpha - e^{\pm 2i\pi\alpha} r^\alpha)} A(\lambda + A)^{-1} C(r + A)^{-1} Cx\, d\lambda.$$

Now we state the following generalization of [365, Theorem 3.1(a), (c)] and [90, Theorem 3.1(ii)]; notice that the assertion of [90, Theorem 3.1(i)] cannot be so easily reconsidered for operators whose C-resolvent set does not contain zero as an interior point.

Theorem 2.9.44. *Suppose* $D(A)$ *and* $R(C)$ *are dense in* E, $0 < \alpha < 2$, $0 < \gamma < 2$, $b \in (0, (2 - \gamma)/(2 - \alpha))$, $\omega \in (\pi - (\pi\alpha)/2, \min(\pi, (\pi - (\pi\gamma)/2)/b)]$, *and* $-A$ *is the integral generator of an equicontinuous* (g_α, C)*-regularized resolvent family* $(S_\alpha(t))_{t \geqslant 0}$. *Define*

$$f^b_{\gamma,\alpha}(t, s) := \frac{1}{2\pi i} \int_{\Gamma_\omega} E_\gamma(-\lambda^b t^\gamma)\, (-\lambda)^{\frac{1}{\alpha} - 1}\, e^{-(-\lambda)^{\frac{1}{\alpha}} s}\, d\lambda,$$

where the contour Γ_ω *is oriented in such a way that* $\operatorname{Im}\lambda$ *increases along* Γ_ω. *Put*

$$S^b_\gamma(t)x := \int_0^\infty f^b_{\gamma,\alpha}(t,s)S_\alpha(s)x\,ds,\ t>0,\ x\in E \text{ and } S^b_\gamma(0) := C.$$

Then $-A_b$ *is densely defined and generates the equicontinuous analytic* (g_γ, C)*-regularized resolvent family* $(S^b_\gamma(t))_{t\geq 0}$ *of angle* $\theta := \min(\pi, (\pi(1-b)/\gamma)+\pi((\alpha b/\gamma)-1)/2)$.

Proof. Let $n \in \mathbb{N}$ satisfy $n > b$. Taken together, Proposition 2.9.42, Theorem 2.2.4 and the proof of Theorem 2.9.39, show that the operator $A_{b/n}$ is C-sectorial with C-spectral angle less than or equal to $(\pi b/n)(1-(\alpha/2))$, as well as that the mapping $\lambda \mapsto (\lambda + A_{b/n})^{-1}Cx$, $\lambda \in \mathbb{C}\setminus\overline{\Sigma_{(\pi b/n)(1-(\alpha/2))}}$ is analytic for all $x \in E$. Using the procedure from Remark 2.9.33, it is not difficult to prove that the operator A_b is C-sectorial with C-spectral angle less than or equal to $(\pi b)(1-(\alpha/2))$, as well as that the mapping $\lambda \mapsto (\lambda + A_b)^{-1}Cx$, $\lambda \in \mathbb{C}\setminus\overline{\Sigma_{(\pi b)(1-(\alpha/2))}}$ is analytic for all $x \in E$; furthermore,

$$(A_b - z)^{-1}\,Cx = C^{1-n}\prod_{k=1}^{n}(A_{b/n} - z_k)^{-1}Cx = (-1)^n\sum_{k=1}^{n}\frac{(A_{b/n}-z_k)^{-1}Cx}{\prod_{1\leqslant i\leqslant n,\ i\neq k}(z_i - z_k)}$$

for all $x \in E$ and $z \in \mathbb{C}\setminus\overline{\Sigma_{(\pi b)(1-(\alpha/2))}}$, where we have denoted by $z_1, \cdots, z_n$ the n-th roots of z (cf. the proof of [403,Therorem 5.4.1]). By Theorem 2.2.5, the above immediately implies that, for every $\gamma' \in (\gamma, 2(1-b(1-(\alpha/2)))$, the operator $-A_b$ is the integral generator of an exponentially equicontinuous analytic $(g_{\gamma'}, C)$-regularized resolvent family $(S^{b,1}_{\gamma'}(t))_{t\geq 0}$ of angle $\theta_{\gamma'} = \min\left((\pi/2), \frac{\pi((1-b(1-(\alpha/2)))}{\gamma'} - (\pi/2)\right)$. Then the subordination principle clarified in Theorem 2.4.2 yields that, for every such a number γ', the operator $-A_b$ is the integral generator of an exponentially equicontinuous analytic (g_γ, C)-regularized resolvent family $(S^{b,1}_{\gamma'}(t))_{t\geq 0}$ of angle $\min(\pi, (\pi/2)(\gamma'/\gamma - 1))$. Letting $\gamma' \to 2(1-b(1-(\alpha/2)))-$, we obtain that the operator $-A_b$ is the integral generator of an exponentially equicontinuous analytic (g_γ, C)-regularized resolvent family $(S^{b,1}_{\gamma'}(t))_{t\geq 0}$ of angle θ; it is clear that A_b is densely defined. Further on, assume that $\alpha' \in (0, \alpha)$ satisfy $b \in (0, (2-\gamma)/(2-\alpha'))$ and $\omega > \pi - (\alpha'\pi)/2$. By Theorem 2.2.5, we infer that, for every $\varepsilon \geqslant 0$, the operator $-(A+\varepsilon)$ is the integral generator of an equicontinuous analytic $(g_{\alpha'}, C)$-regularized resolvent family $(S^\varepsilon_{\alpha'}(t))_{t\geq 0}$ of angle $((\alpha/\alpha') - 1)\pi/2$. Let $\delta > 0$ satisfy $\alpha'((\pi/2)+\delta) < \alpha\pi/2$. For $s > 0$ fixed, we have:

$$S^\varepsilon_{\alpha'}(s)x = \frac{1}{2\pi i}\int_\Gamma e^{\lambda s}\lambda^{\alpha'-1}(\lambda^{\alpha'} + A + \varepsilon)^{-1}Cx\,d\lambda,\ \ s>0,\ x\in E,\ \varepsilon \geqslant 0,$$

where $\Gamma = \{re^{-i(\delta+(\pi/2))} : r \geqslant 1/s\} \cup \{e^{i\varphi}/s : |\phi| \leqslant \delta + (\pi/2)\} \cup \{re^{i(\delta+(\pi/2))} : r \geqslant 1/s\}$ is oriented counterclockwise. In spite of this fact and the proof of [20, Theorem

2.6.1], it is possible to deduce that, for every $q \in ⊛$, there exist $c_q > 0$ and $r_q \in ⊛$ such that:

(302) $$q(S^{\varepsilon}_{\alpha'}(s)x) \leqslant c_q r_q(x),\ s \geqslant 0,\ x \in E,\ \varepsilon \in [0, 1].$$

Put $S^b_{\gamma,\varepsilon}(t)x := \int_0^{\infty} f^b_{\gamma,\alpha'}(t, s)S^{\varepsilon}_{\alpha'}(s)x\, ds$, $t > 0$, $x \in E$, $\varepsilon \geqslant 0$. By [90, Theorem 3.1(ii)], we have that, for every $\varepsilon > 0$, the operator $-(A + \varepsilon)_b$ is the integral generator of an equicontinuous (g_{γ}, C)-regularized resolvent family $(S^b_{\gamma,\varepsilon}(t))_{t\geqslant 0}$. The dominated convergence theorem, along with [89, Proposition 3.1] and (302), implies $\lim_{\varepsilon\to 0+} S^{\varepsilon}_{\alpha'}(s)x = S^0_{\alpha'}(s)x$, $s > 0$, $x \in E$, and $S^{b,0}_{\gamma,\alpha'}(t)x := \lim_{\varepsilon\to 0+} S^b_{\gamma,\varepsilon}(t)x = \int_0^{\infty} f^b_{\gamma,\alpha'}(t, s)$ $S_{\alpha'}(s)x\, ds$, $t > 0$, $x \in E$; the equicontinuity and strong continuity of operator family $(S^{b,0}_{\gamma,\alpha'}(t))_{t>0}$ can be proved in a similar fashion. Now we will prove that $S^{b,1}_{\gamma}(t)x = S^{b,0}_{\gamma,\alpha'}(t)x$ for all $t > 0$ and $x \in E$. Owing to Theorem 2.1.5 and the uniqueness theorem for the Laplace transform, it suffices to show that:

(303) $$\lambda^{\gamma-1}Cx = (\lambda^{\gamma} + A_b)\int_0^{\infty} e^{-\lambda t}\int_0^{\infty} f_{\gamma,\alpha'}(t, s)S_{\alpha'}(s)x\, ds\, dt.$$

Due to the procedure described in Remark 2.9.33, we have to actually prove that, for every $x \in C^2(D_{\infty}(A))$, the following equality holds:

$$\lambda^{\gamma-1}\left[C^3x - \lambda C^2\int_0^{\infty} e^{-\lambda t}\int_0^{\infty} f_{\gamma,\alpha'}(t, s)S_{\alpha'}(s)x\, ds\, dt\right]$$

(304) $$= J_{\infty,b}\int_0^{\infty} e^{-\lambda t}\int_0^{\infty} f_{\gamma,\alpha'}(t, s)S_{\alpha'}(s)Cx\, ds\, dt.$$

The proof of (304) in case $b \in \mathbb{N}$ is simple and as such will not be given. Suppose now $b \notin \mathbb{N}$. Due to [90, Theorem 3.1(ii)], the equation (304) holds with $S_{\alpha'}(\cdot)$ replaced by $S^{\varepsilon}_{\alpha'}(\cdot)$, so that Proposition 2.9.15 implies:

$$\lambda^{\gamma-1}\left[C^2x - \lambda C\int_0^{\infty} e^{-\lambda t}\int_0^{\infty} f_{\gamma,\alpha'}(t, s)S^{\varepsilon}_{\alpha'}(s)x\, ds\, dt\right]$$

$$= (-1)^{\lceil b\rceil}\frac{\sin \pi b}{\pi}\int_0^{\infty} \mu^{b-\lceil b\rceil}(A + \varepsilon)^{\lceil b\rceil}(\mu + A + \varepsilon)^{-1}C$$

$$\times \int_0^{\infty} e^{-\lambda t}\int_0^{\infty} f_{\gamma,\alpha'}(t, s)S_{\alpha'}(s)x\, ds\, dt\, d\mu.$$

Applying the dominated convergence theorem several times, we obtain by letting $\varepsilon \to 0+$ in the above equality (split the above integrals into the sum of corresponding ones along the intervals (0, 1) and (1, ∞)) that (304) holds. This, in turn, implies (303) and $S^{b,1}_{\gamma}(t)x = S^{b,0}_{\gamma,\alpha'}(t)x$ for all $t > 0$ and $x \in E$. Making use of [89, Theorem 2.10], we get that, for every $t > 0$ and $x \in E$, $\lim_{\alpha'\to\alpha-}(g_{\alpha'} * S_{\alpha'})(t)x =$

$(g_\alpha * S_\alpha)(t)x$. Together with the denseness of $D(A)$ in E and the functional equations of $(S_{\alpha'}(t))_{t\geq 0}$ and $(S_\alpha(t))_{t\geq 0}$, the above simply implies that, for every $t > 0$ and $x \in E$, we have $\lim_{\alpha' \to \alpha-} S_{\alpha'}(t)Cx = S_\alpha(t)Cx$. The dominated convergence implies $CS_\gamma^{b,1}(t)x = \lim_{\alpha' \to \alpha-} S_{\gamma,\alpha'}^{b,0}(t)Cx = CS_\gamma^\beta(t)x$, $t > 0$, $x \in E$. The proof of theorem is thereby completed. □

Remark 2.9.45. (i) The assertions of [365, Corollary 3.3, Remark 3.4, Proposition 3.5] can be simply proved for the class of regularized fractional resolvent families in SCLCSs. We leave details to the interested reader.

(ii) If $D(A)$ or $R(C)$ is not densely defined in E, and all remaining assumptions quoted in the formulation of Theorem 2.9.44 hold, then one can similarly prove that for each $\sigma > 0$ the operator $-A_b$ is the integral generator of an analytic $(g_\gamma, g_{\sigma+1})$-regularized C-resolvent family of angle θ and of subexponential growth.

Theorem 2.9.46. *Let $A \in \mathcal{M}_{C,-1}$. Then the following holds:*

(i) *Suppose $b_1, b_2 \in \mathbb{C}_+$, $k \in \mathbb{N}$ and $k > \mathrm{Re}(b_1 + b_2)$, if $b_1 + b_2 \notin \mathbb{N}$, resp. $k = b_1 + b_2$, if $b_1 + b_2 \in \mathbb{N}$. Set, for every $k \in \mathbb{N}$, $\Omega_k := \{\lambda \in \mathbb{C} \setminus (0,\infty) : (\lambda + A)^{-k}C \in L(E)\}$ and $\Omega'_k := \{\lambda \in \mathbb{C} : (\lambda + A)^{-k}C \in L(E)\}$. Then*

$$\Lambda_k := C(D(A^k)) \cup \bigcup_{\lambda \in \Omega_k} R((\lambda + A)^{-k}C) \subseteq D(A_{b_1}A_{b_2}) \cap D(A_{b_1+b_2}), \tag{305}$$

$$A_{b_1}A_{b_2}x = A_{b_1+b_2}x,\ x \in \bigcup_{\lambda \in \Omega_k} R((\lambda + A)^{-k}C), \tag{306}$$

and

$$A_{b_1+b_2} = ((\lambda + A)^{-k}C)^{-1}A_{b_1}A_{b_2}(\lambda + A)^{-k}C,\ \lambda \in \Omega_k; \tag{307}$$

if the condition (H) holds, then the equations (305)-(307) remain true with the set Ω_k replaced by Ω'_k. Furthermore, $C^{-1}\overline{A_{b_1}A_{b_2}}C \subseteq A_{b_1+b_2}$, with the equality in the case that $\lim_{\lambda \to +\infty} \lambda^n(\lambda + A)^{-n}Cx = Cx$ *for all $x \in E$ and $n \in \mathbb{N}$ (by* [90, Lemma 2.7], *the above is true provided that the condition (H) holds, as well as that $D(A)$ and $R(C)$ are dense in E).*

(ii) *Let* $\mathrm{Re}\, b > 0$. *Then* $\lim_{b' \to b} A_{b'}x = A_b x$ *for all* $x \in C(D(A^{1+\lfloor \mathrm{Re}\, b \rfloor}))$.

Proof. First suppose $\mathrm{Re}(b_1 + b_2) < 1$ and $k = 1$. Passing to the space $D_\infty(A)$, and to the operators A_∞ and C_∞, it can be easily proved that, for every $x = Cy \in C(D(A))$,

$$A_{b_1+b_2}x = \frac{\sin(b_1 + b_2)\pi}{\pi} \int_0^\infty \lambda^{b_1+b_2-1}(\lambda + A)^{-1}CAy\, d\lambda.$$

The first part of assertion (i) in this case follows from an application of Proposition 2.9.28. Similarly, if $\mathrm{Re}(b_1 + b_2) = 1$ and $k = 2$, then for every $x = Cy \in C(D(A^2))$,

$$A_{b_1}A_{b_2}x = A_{b_1+b_2}x$$

$$(308) = \frac{\sin(b_1+b_2)\pi}{\pi}\int_0^{\infty}\lambda^{b_1+b_2-1}\Big[(\lambda+A)^{-1}C - \frac{\lambda C}{\lambda^2+1}Ay\,d\lambda + \sin\frac{(b_1+b_2)\pi}{2}\Big]CAy\,d\lambda.$$

The above results can prove that $C(D(A^k)) \subseteq D(A_{b_1+b_2}) \cap D(A_{b_1}A_{b_2})$ and $A_{b_1+b_2}x = A_{b_1+b_2-\lceil \mathrm{Re}(b_1+b_2)-1\rceil}A^{\lceil \mathrm{Re}(b_1+b_2)-1\rceil}x$, $x \in C(D(A^k))$; if $b_1 + b_2 \in \mathbb{N}$, then the above conclusions trivially follow from Proposition 2.9.29. Suppose now $\mu \in \Omega_k$ and $y = (\mu + A)^{-k}Cx$ for some $x \in E$. Then $Cx \in D(A^k)$ and, by the foregoing arguments, we have:

$$\begin{aligned} A_{b_1}A_{b_2}C(\mu+A)^{-k}Cx &= A_{b_1+b_2}C(\mu+A)^{-k}Cx \\ &= A_{b_1+b_2-\lceil \mathrm{Re}(b_1+b_2)-1\rceil}A^{b_1+b_2-\lceil \mathrm{Re}(b_1+b_2)-1\rceil}(\mu+A)^{-k}C^2x \\ &= A_{b_1+b_2-\lceil \mathrm{Re}(b_1+b_2)-1\rceil} \end{aligned}$$

$$(309) \qquad \times \sum_{l=0}^{\lceil \mathrm{Re}(b_1+b_2)-1\rceil}(-1)^l\binom{\lceil \mathrm{Re}(b_1+b_2)-1\rceil}{l}\mu^l(\mu+A)^{\lceil \mathrm{Re}(b_1+b_2)-1\rceil-k-l}C^2x.$$

Put $\alpha := b_1 + b_2 - \lceil \mathrm{Re}(b_1+b_2) - 1\rceil$. Now we will prove that, for every $m \in \mathbb{N}$,

$$(310) \qquad A_\alpha(\mu+A)^{-m}C^2x = \int_0^{\infty}\lambda^{\alpha-1}(\lambda+A)^{-1}CA(\mu+A)^{-m}Cx\,d\lambda \in R(C).$$

Observe first that, for every $z \in \rho_C(A)\setminus\{\lambda\}$,

$$(z+A)^{-1}C(\lambda+A)^{-m}Cx = \frac{(-1)^m}{(z-\lambda)^k}(z+A)^{-1}C^2x + \sum_{i=1}^{m}\frac{(-1)^{m-i}(\lambda+A)^{-i}C^2x}{(z-\lambda)^{m+1-i}}.$$

If Re $\alpha \in (0, 1)$, then we have:

$$\begin{aligned} &\int_0^{\infty}\lambda^{\alpha-1}(\lambda+A)^{-1}CA(\mu+A)^{-m}Cx\,d\lambda \\ &\quad = \left(\int_0^1 + \int_1^{\infty}\right)\lambda^{\alpha-1}(\lambda+A)^{-1}CA(\mu+A)^{-m}Cx\,d\lambda \\ &\quad = \int_0^1\lambda^{\alpha-1}[C - \lambda(\lambda+A)^{-1}C](\mu+A)^{-m}Cx\,d\lambda \end{aligned}$$

$$(311) \qquad + \int_1^{\infty}\lambda^{\alpha-1}A\left[\frac{(-1)^m}{(\lambda-\mu)^k}(\lambda+A)^{-1}C^2x + \sum_{i=1}^{m}\frac{(-1)^{m-i}(\mu+A)^{-i}C^2x}{(\lambda-\mu)^{m+1-i}}\right]d\lambda.$$

The equality (310) immediately follows from (311). The consideration is quite similar in the case that Re $\alpha = 1$, so that (309) implies $A_{b_1+b_2}C(\mu+A)^{-k}Cx \in R(C)$.

Since $C^{-1}A_{b_1+b_2}C = A_{b_1+b_2}$, we get that $(\mu + A)^{-k}Cx \in D(A_{b_1+b_2})$; now we complete the proof of (305)-(307) quite easily. Suppose that the condition (H) holds. Due to the first part of the proof, we have that:

(312) $$((\lambda + e^{i\theta}A)^{-k}C)^{-1}(e^{i\theta}A)_{b_1}(e^{i\theta}A)_{b_2}(\lambda + e^{i\theta}A)^{-k}\, C = (e^{i\theta}A)_{b_1+b_2},$$

for any $\lambda \in \mathbb{C}\setminus(-\infty, 0)$ such that $(\lambda + e^{i\theta}A)^{-k}\, C \in L(E)$. If $\operatorname{Re}(b_1 + b_2) = 1$, then

(313) $$(e^{i\theta}A)_{b_1} = e^{i\theta b_1}A_{b_1} \text{ and } (e^{i\theta}A)_{b_2} = e^{i\theta b_2}A_{b_2};$$

taken together with the equality

$$((\lambda e^{-i\theta} + A)^{-k}C)^{-1}\, A_{b_1}A_{b_2}\,(\lambda e^{-i\theta} + A)^{-k}\, C = A_{b_1+b_2},$$

which holds for any $\lambda \in \mathbb{C}$ such that $\lambda e^{-i\theta} \in \Omega_k$, (312)-(313) imply $(e^{i\theta}A)_{b_1+b_2} = e^{i\theta(b_1+b_2)}A_{b_1+b_2}$. Let $t > 0$ satisfy $(t + A)^{-k}\, C \in L(E)$, and let $x \in E$. Then (312) implies that, for every $\beta \in \mathbb{C}$ with $\operatorname{Re}\beta \in (0, 1]$ and $k > \operatorname{Re}\beta$, one has $R((t + A)^{-k}\, C) = R((te^{i\theta} + e^{i\theta}A)^{-k}\, C) \subseteq D((e^{i\theta}A)_\beta) = D(e^{i\theta\beta}A_\beta) = D(A_\beta)$. Keeping in mind the equation (309), with the number μ replaced by t, we obtain that $y = (t + A)^{-k}\, Cx \in D(A_{b_1+b_2-\lceil \operatorname{Re}(b_1+b_2)-1\rceil}A^{\lceil \operatorname{Re}(b_1+b_2)-1\rceil})$. Hence,Theorem 2.9.28(ii) yields that $y \in D(A_{b_1}A_{b_2}) \cap D(A_{b_1+b_2})$, $A_{b_1}A_{b_2}y = A_{b_1+b_2}y$ and that the equation (307) holds with the number λ replaced by t. Using again Theorem 2.9.28(ii), as well as (307) and the equality $\lim_{\lambda\to+\infty}\lambda^n(\lambda + A)^{-n}Cx = Cx$ $(x \in E,\ n \in \mathbb{N})$, we get that $A_{b_1+b_2} = C^{-1}\,\overline{A_{b_1}A_{b_2}}\,C$. This completes the proof of (i). The proof of assertion (ii) in case $\operatorname{Re} b \notin \mathbb{N}$ is trivial, suppose now $\operatorname{Re} b = n + 1 \in \mathbb{N}$ and $x = Cy \in C(D(A^{n+2}))$. By the first part of proof and the analysis given on page 420 of [31] (more precisely, the equivalence of equalities [31, (2.1) and (2.2)] for $0 < \operatorname{Re}\alpha < 2$ and $x \in D(A^2)$), we obtain that, for every $b' \in \mathbb{C}$ with $\operatorname{Re} b' \in (n, n + 1)$, $A_{b'}x$ is given by the formula (308) with $b_1 + b_2$ and Ay replaced by b' and $A^{n+1}y$ therein; now one can apply the dominated convergence theorem in order to see that $\lim_{b'\to b,\operatorname{Re} b < n+1} A_{b'}x = A_bx$. Similarly we have that $\lim_{b'\to b,\operatorname{Re} b > n+1} A_{b'}x = A_bx$, which completes the proof of (ii). □

Remark 2.9.47. The assertion of [90, Theorem 2.8(iii.2)] cannot be so easily rephrased for injective C-nonnegative operators. One can prove, for example, that for every $b \in \mathbb{C}$ with $\operatorname{Re} b < 0$, the equality $\lim_{b'\to b} A_{b'}x = A_bx$ holds for all $x \in C(R(A^{1+\lfloor \operatorname{Re}(-b)\rfloor}))$ as well as that the equality $\lim_{b'\to 0} A_{b'}Cy = Cy$ holds for all $y \in R(A)$ with $\lim_{\lambda\to+\infty} A(\lambda + A)^{-1}Cy = 0$ (cf. Theorem 2.9.13(ii)). It is also worth noting that the assertion of [90, Theorem 2.8(ii)] does not continue to hold, in general, for complex powers with exponents of negative real part. Concerning the purely imaginary powers of injective C-nonnegative operators, the following facts should be stated (cf. [403, Chapter 7] and [90, Theorem 2.10] for further information in this direction). If $x = Ay$ for some $y \in C(D(A^2))$ (this, in particular, implies $x \in C(D(A)) \cap C(R(A))$), then $x \in D(A_{i\tau})$ for all $\tau \in \mathbb{R}$; if this is the case, the use of Theorem 2.1.43(ii) shows that, for every $\tau_1, \tau_2 \in \mathbb{R}$, we have $x \in D(A_{i\tau_1}A_{i\tau_2}) \cap D(A_{i(\tau_1+\tau_2)})$ and $A_{i(\tau_1+\tau_2)}x = A_{i\tau_1}A_{i\tau_2}x$.

2.9.5. Semigroups generated by fractional powers of almost C-sectorial operators. Let $0 \leqslant \omega < \pi$. Then a closed linear operator A on E is called almost C-sectorial of angle ω, if $\mathbb{C}\setminus\overline{\Sigma_\omega} \subseteq \rho_C(A)$ and there exists $m \in \mathbb{R}$ such that the family

$$\{(|\lambda|^{-1} + |\lambda|^m)^{-1} (\lambda - A)^{-1}C : \lambda \notin \Sigma_{\omega'}\} \tag{314}$$

is equicontinuous for $\omega < \omega' < \pi$. We need to introduce the following condition.

(HQ): The operator A is almost C-sectorial of angle $\omega \in [0, \pi)$, and the mapping $\lambda \mapsto (\lambda - A)^{-1}Cx$, $\lambda \in \mathbb{C}\setminus\overline{\Sigma_\omega}$ is continuous for every fixed $x \in E$.

The value of number m satisfying (314) will be clear in any use of the condition (HQ).

In the subsequent theorem, we continue the analysis raised in Theorem 2.9.39 and [90, Theorem 3.5]. The Liouville right-sided fractional derivative of order β (see [271, (2.3.4)] for the scalar-valued case) is defined for those continuous functions $u : (0, \infty) \to E$ for which $\lim_{T\to\infty} \int_t^T g_{\lceil\beta\rceil-\beta}(s-t)u(s)\,ds = \int_t^\infty g_{\lceil\beta\rceil-\beta}(s-t)u(s)\,ds$ exists and defines an $\lceil\beta\rceil$-times continuously differentiable function on $(0, \infty)$, by

$${}^{\#}\mathbf{D}_{-}^{\beta} u(t) := (-1)^{\lceil\beta\rceil} \frac{d^{\lceil\beta\rceil}}{dt^{\lceil\beta\rceil}} \int_t^\infty g_{\lceil\beta\rceil-\beta}(s-t)u(s)\,ds,\ t > 0.$$

The function $H_n : \mathbb{C} \times (\mathbb{C}\setminus(-\infty, 0]) \to \mathbb{C}$ defined by

$$H_n(\omega, z) := \frac{d^n}{dz^n} \exp(-\omega z^\gamma),\ \omega \in \mathbb{C},\ z \in \mathbb{C}\setminus(-\infty, 0],$$

is analytic in $\mathbb{C}\setminus(-\infty, 0]$ for every fixed ω, and entire in $\mathbb{C}$ for every fixed z (cf. also the proof of [404, Proposition 3.5]).

Theorem 2.9.48. *Let $0 < \gamma < 1/2$, and let $A \in \mathcal{M}_{C,m}$ where $m + \gamma > -1$. Put $S_\gamma(0) := C$, $S_{\gamma,\zeta}(t)x := \int_0^t g_\zeta(t-s)\,S_\gamma(s)x\,ds$, $x \in E$, $t \geqslant 0$, $\zeta > 0$, and $S_{\gamma,0}(t) := S_\gamma(t)$, $t \geqslant 0$. Then the family $\{(1 + t^{-(m+1)/\gamma})^{-1} S_\gamma(t) : t > 0\}$ is equicontinuous, and there exist operator families $(\mathbf{S}_\gamma(z))_{z\in\Sigma_{(\pi/2)-\gamma\pi}}$ and $(\mathbf{S}_{\gamma,\zeta}(z))_{z\in\Sigma_{(\pi/2)-\gamma\pi}}$ such that, for every $x \in E$ and $\zeta > 0$, the mappings $z \mapsto \mathbf{S}_\gamma(z)x$, $z \in \Sigma_{(\pi/2)-\gamma\pi}$ and $z \mapsto \mathbf{S}_{\gamma,\zeta}(z)x$, $z \in \Sigma_{(\pi/2)-\gamma\pi}$ are analytic as well as that $\mathbf{S}_\gamma(t) = S_\gamma(t)$, $t > 0$ and $\mathbf{S}_{\gamma,\zeta}(t) = S_{\gamma,\zeta}(t)$, $t > 0$. Furthermore, the following holds:*

(i) *$S_\gamma(z_1)S_\gamma(z_2) = S_\gamma(z_1 + z_2)C$ for all $z_1, z_2 \in \Sigma_{(\pi/2)-\gamma\pi}$.*

(ii) *If $-1 - \gamma < m < -1$, then $\lim_{z\to 0, z\in\Sigma_{(\pi/2)-\gamma\pi-\varepsilon}} S_\gamma(z)x = Cx$, $x \in \overline{D(A)}$, $\varepsilon \in (0, (\pi/2) - \gamma\pi)$; if the condition (H)' holds, then the above equality remains true with the number $(\pi/2) - \gamma\pi$ replaced by $(\pi/2) - \omega\gamma$.*

(iii) *$S_\gamma(z)A_\alpha \subseteq A_\alpha S_\gamma(z)$, $z \in \Sigma_{(\pi/2)-\gamma\pi}$, $\alpha \in \mathbb{C}_+$, where we assume that the operator A is injective provided $-1 - \gamma < m < -1$; if the condition (H)' holds, then the above inclusion remains true with the number $(\pi/2) - \gamma\pi$ replaced by $(\pi/2) - \omega\gamma$.*

[#]*Modified Liouville right-sided fractional derivative...* — see Addendum.

(iv) (iv.1) *Suppose* $m \geqslant -1$. *Then* $\lim_{z\to 0, z\in\Sigma_{(\pi/2)-\gamma\pi-\varepsilon}} S_\gamma(z)x = Cx$, $x \in D(A^p)$, $\varepsilon \in (0, (\pi/2)-\gamma\pi)$; *if the condition (H)' holds, then the above equality remains true with the number* $(\pi/2)-\gamma\pi$ *replaced by* $(\pi/2)-\omega\gamma$. *Moreover, the operator* $-A_\gamma$ *is the integral generator of an equicontinuous analytic C-regularized semigroup* $(S_\gamma(z))_{z\in\Sigma_{(\pi/2)-\gamma\pi}}$ *of growth order* $(m+1)/\gamma$, *provided that* $m > -1$. *If, additionally, the condition (H)' holds, then* $(S_\gamma(t))_{t>0}$ *can be extended to an equicontinuous analytic C-regularized semigroup* $(\mathbf{S}_\gamma(z))_{z\in\Sigma_{(\pi/2)-\gamma\omega}}$ *of growth order* $(m+1)/\gamma$; *in this case, the equality stated in (i) holds for any* $z_1, z_2 \in \Sigma_{(\pi/2)-\gamma\omega}$, *and the equality stated in (iii) holds for any* $z \in \Sigma_{(\pi/2)-\gamma\omega}$.

(iv.2) *Suppose* $-1-\gamma < m < -1$ *and* $\zeta > 0$. *Then* $(S_{\gamma,\zeta}(t))_{t\geqslant 0}$ *is an exponentially equicontinuous analytic* ζ*-times integrated C-regularized semigroup of angle* $(\pi/2)-\gamma\pi$; *furthermore, the family* $\{|z|^{-\zeta}(1+|z|^{-(m+1)/\gamma})^{-1}S_{\gamma,\zeta}(z) : z \in \Sigma_{(\pi/2)-\gamma\pi-\varepsilon}\}$ *is equicontinuous for any* $\varepsilon \in (0, (\pi/2)-\gamma\pi)$. *If, additionally, the condition (H)' holds, then* $(S_{\gamma,\zeta}(t))_{t\geqslant 0}$ *can be extended to an exponentially equicontinuous analytic* ζ*-times integrated C-regularized semigroup of angle* $(\pi/2)-\gamma\omega$, *and the family* $\{|z|^{-\zeta}(1+|z|^{-(m+1)/\gamma})^{-1}S_{\gamma,\zeta}(z) : z \in \Sigma_{(\pi/2)-\gamma\omega-\varepsilon}\}$ *is equicontinuous for any* $\varepsilon \in (0, (\pi/2)-\gamma\omega)$.

(iv.3) *Suppose* $-1-\gamma < m < -1$ *and A is densely defined. Then* $(S_\gamma(t))_{t\geqslant 0}$ *is an exponentially equicontinuous analytic C-regularized semigroup of angle* $(\pi/2)-\gamma\pi$; *furthermore, the family* $\{(1+|z|^{-(m+1)/\gamma})^{-1}S_\gamma(z) : z \in \Sigma_{(\pi/2)-\gamma\pi-\varepsilon}\}$ *is equicontinuous for any* $\varepsilon \in (0, (\pi/2)-\gamma\pi)$. *If, additionally, the condition (H)' holds, then* $(S_\gamma(t))_{t\geqslant 0}$ *can be extended to an exponentially equicontinuous analytic C-regularized semigroup of angle* $(\pi/2)-\gamma\omega$, *and the family* $\{(1+|z|^{-(m+1)/\gamma})^{-1}S_\gamma(z) : z \in \Sigma_{(\pi/2)-\gamma\omega-\varepsilon}\}$ *is equicontinuous for any* $\varepsilon \in (0, (\pi/2)-\gamma\omega)$.

(iv.4) *Suppose* $-1-\gamma < m < -1$ *and A is injective. Then the integral generator of* $(S_{\gamma,\zeta}(t))_{t\geqslant 0}$, *resp.* $(S_\gamma(t))_{t\geqslant 0}$, *is the operator* $-A_\gamma$ *(see (iv.2)-(iv.3)). Denote by* $-_\gamma A$ *the integral generator of* $(S_{\gamma,\zeta}(t))_{t\geqslant 0}$.

(v) *Let* $m \geqslant -1$ *and* $z_0 \in \mathbb{C}_+$. *Then, for every* $x \in D(A^{\lfloor m+\gamma\rfloor+2})$ *and* $\varepsilon \in (0, (\pi/2)-\gamma\pi)$,

$$\lim_{z\to 0, z\in\Sigma_{(\pi/2)-\gamma\pi-\varepsilon}} \frac{S_\gamma(z)x - Cx}{z}$$

$$= -z_0^\gamma Cx + \sum_{k=2}^{\lfloor m+\gamma\rfloor+2} \frac{(-1)^{k-1}}{(k-1)!} H_{k-1}(0, z_0)(z_0 - A)^{k-1}Cx \tag{315}$$

$$-\sin(\pi\gamma)\int_0^\infty \lambda^\gamma \frac{(\lambda+A)^{-1}C(z_0-A)^{\lfloor m+\gamma\rfloor+2}x}{(\lambda+z_0)^{\lfloor m+\gamma\rfloor+2}}\,d\lambda;$$

if the condition (H)' holds, then the formula (315) remains true for every $x \in D(A^{\lfloor m+\gamma \rfloor+2})$, *with the number* $(\pi/2) - \gamma\pi$ *replaced by* $(\pi/2) - \omega\gamma$.

(vi) *The operator* ${}_{\gamma}A$ *satisfies the condition (H)' with* $\omega = \gamma\pi$ *and* $m' = ((m+1)/\gamma) - 1$; *moreover, if the condition (H)' holds for A (with* ω *and m), then the same condition holds for* ${}_{\gamma}A$ *with* $\gamma\omega$ *and* $m' = ((m+1)/\gamma) - 1$. *Furthermore, the suppositions* $0 < \beta < 1/2$ *and* $m + \gamma\beta > -1$ *imply* ${}_{\beta}({}_{\gamma}A) = {}_{\beta\gamma}A$.

(vii) $R(S_{\gamma}(z)) \subseteq D_{\infty}(A)$, $A^{n}S_{\gamma}(z) \in L(E)$ *and (287) holds for all* $x \in E$ *and* $z \in \Sigma_{(\pi/2)-\gamma\pi}$.

(viii) *Let* $l \in \mathbb{N} \setminus \{1, 2\}$.

(viii.1) *Suppose* $A \in \mathcal{M}_{C,m}$, *where* $m > -1$. *Denote by* Ω_{l}, *resp.* Ψ_{l}, *the set of continuity of* $(S_{1/l}(t))_{t>0}$, *resp.* $(S_{1/l}(z))_{z \in \Sigma_{(\pi/2)-(\pi/l)}}$. *Then, for every* $x \in \Omega_{l}$, *the incomplete abstract Cauchy problem*

$$(P_l): \begin{cases} u \in C^{\infty}((0,\infty) : E) \cap C((0,\infty) : D_{\infty}(A)), \\ u^{(l)}(t) = (-1)^{l} A u(t),\ t > 0, \\ \lim_{t \to 0+} u(t) = Cx, \\ \text{the set } \{u(t) : t > 0\} \text{ is bounded in } E, \end{cases}$$

has a solution $u(t) = S_{1/l}(t)x$, $t > 0$, *which can be analytically extended to the sector* $\Sigma_{(\pi/2)-(\pi/l)}$. *If, additionally,* $x \in \Psi_{l}$, *then for every* $\delta \in (0, (\pi/2) - (\pi/l))$ *and* $j \in \mathbb{N}_0$, *we have that the set* $\{z^{j}u^{(j)}(z) : z \in \Sigma_{\delta}\}$ *is bounded in E. Assuming the condition (H)', the above statements continue to hold with the number* $(\pi/2) - (\pi/l)$ *replaced by* $(\pi/2) - (\omega/l)$.

(viii.2) *Let* $x \in \overline{D(A)}$, *and let* $A \in \mathcal{M}_{C,m}$, *where* $-1 - l^{-1} < m < -1$. *Then the incomplete abstract Cauchy problem*

$$(P_{l,m}): \begin{cases} u \in C^{\infty}((0,\infty) : E) \cap C((0,\infty) : D_{\infty}(A)), \\ u^{(l)}(t) = (-1)^{l} A u(t),\ t > 0, \\ \lim_{t \to 0+} u(t) = Cx, \\ \text{the set } \{(1 + t^{-l(m+1)})^{-1}u(t) : t > 0\} \text{ is bounded in } E, \end{cases}$$

has a solution $u(t) = S_{1/l}(t)x$, $t > 0$, *which can be analytically extended to the sector* $\Sigma_{(\pi/2)-(\pi/l)}$. *Moreover, for every* $\delta \in (0, (\pi/2) - (\pi/l))$ *and* $j \in \mathbb{N}_0$, *we have that the set* $\{|z|^{j}(1 + |z|^{-l(m+1)})^{-1}u^{(j)}(z) : z \in \Sigma_{\delta}\}$ *is bounded in E. If, additionally, the condition (H)' holds, then the above statements remain true with the number* $(\pi/2) - (\pi/l)$ *replaced by* $(\pi/2) - (\omega/l)$.

(ix)(ix.1) *Suppose* $\beta > 0$. *Denote by* $\Omega_{\theta,\gamma}$, *resp.* Ψ_{γ}, *the continuity set of* $(S_{\gamma}(te^{i\theta}))_{t>0}$, *resp.* $(S_{\gamma}(z))_{z \in \Sigma_{(\pi/2)-\gamma\pi}}$. *Then, for every* $x \in \Omega_{\theta,\gamma}$, *the incomplete abstract Cauchy problem*

$$(FP_\beta): \begin{cases} u \in C^\infty((0,\infty):E) \cap C((0,\infty):D_\infty(A)), \\ D_-^\beta u(t) = e^{i\theta\beta} A_{\gamma\beta} u(t), \ t > 0, \\ \lim_{t\to 0+} u(t) = Cx, \\ \textit{the set } \{u(t) : t > 0\} \textit{ is bounded in } E, \end{cases}$$

has a solution $u(t) = S_\gamma(te^{i\theta})x$, $t > 0$, *which can be analytically extended to the sector* $\Sigma_{(\pi/2)-\gamma\pi-|\theta|}$. *If, additionally,* $x \in \Psi_\gamma$, *then for every* $\delta \in (0, (\pi/2) - \gamma\pi)$ *and* $j \in \mathbb{N}_0$, *we have that the set* $\{z^j u^{(j)}(z) : z \in \Sigma_\delta\}$ *is bounded in E. Assuming the condition (H)', the above statements continue to hold with the number* $(\pi/2) - \gamma\pi$ *replaced by* $(\pi/2) - \omega\gamma$.

(ix.2) *Let* $x \in \overline{D(A)}$, *let* $\beta > (-1-m)/\gamma$, *and let* $\theta \in (-((\pi/2) - \gamma\pi), (\pi/2) - \gamma\pi)$. *Then the incomplete abstract Cauchy problem*

$$(FP_{\beta,m}): \begin{cases} u \in C^\infty((0,\infty):E) \cap C((0,\infty):D_\infty(A)), \\ D_-^\beta u(t) = e^{i\theta\beta} C^{-1} \lim_{\varepsilon\to 0+} (A+\varepsilon)_{\gamma\beta} Cu(t), \ t > 0, \\ \lim_{t\to 0+} u(t) = Cx, \\ \textit{the set } \{(1 + t^{-(m+1)/\gamma})^{-1} u(t) : t > 0\} \textit{ is bounded in } E, \end{cases}$$

has a solution $u(t) = S_\gamma(te^{i\theta})x$, $t > 0$, *which can be analytically extended to the sector* $\Sigma_{(\pi/2)-\gamma\pi-|\theta|}$. *Moreover, for every* $\delta \in (0, (\pi/2) - \gamma\pi)$ *and* $j \in \mathbb{N}_0$, *we have that the set* $\{|z|^j (1 + |z|^{-(m+1)/\gamma})^{-1} u^{(j)}(z) : z \in \Sigma_\delta\}$ *is bounded in E. If, additionally, the condition (H)' holds, then the above statements remain true with the number* $(\pi/2) - \gamma\pi$ *replaced by* $(\pi/2) - \omega\gamma$.

Proof. We shall omit the proofs of (iii), (iv.3), (vii) and (viii). Suppose $\delta \in (0, (\pi/2) - \pi\gamma)$ and $q \in \circledast$. Then there exists $c_\delta > 0$ such that $\operatorname{Re} z \cos \pi\gamma - |\operatorname{Im} z| \sin \pi\gamma \geqslant c_\delta |z|$, $z \in \Sigma_\delta$. Noticing also that $|\sin(it)| \leqslant |t|(e^t + e^{-t})$, $t \in \mathbb{R}$, the above inequality together with prescribed assumptions imply that there exist $r_q \in \circledast$, $m_q > 0$ and $c_{q,\delta} > 0$ such that, for every $x \in E$ and $z \in \Sigma_\delta$,

$$q(S_\gamma(z)x)$$

$$\leqslant m_q r_q(x) \int_0^\infty e^{-\lambda^\gamma \cos \pi\gamma \operatorname{Re} z} \Bigg| \frac{1}{2} \sin(\lambda^\gamma \sin \pi\gamma \operatorname{Re} z) | e^{\lambda^\gamma \sin \pi\gamma \operatorname{Im} z} + e^{-\lambda^\gamma \sin \pi\gamma \operatorname{Im} z} |$$

$$+ \cos(\lambda^\gamma \sin \pi\gamma \operatorname{Re} z) | \lambda^\gamma \sin \pi\gamma \operatorname{Im} z) | (e^{\lambda^\gamma \sin \pi\gamma \operatorname{Im} z} + e^{-\lambda^\gamma \sin \pi\gamma \operatorname{Im} z}) \Bigg| (\lambda^{-1} + \lambda^m) \, d\lambda$$

$$\leqslant 2m_q r_q(x) \int_0^\infty e^{-\lambda^\gamma(\cos \pi\gamma \operatorname{Re} z - |\operatorname{Im} z| \sin \pi\gamma)} \lambda^\gamma \sin \pi\gamma |z| (\lambda^{-1} + \lambda^m) \, d\lambda$$

$$\leqslant 2m_q r_q(x) \int_0^\infty e^{-c_\delta |z| \lambda^\gamma} \lambda^\gamma \sin \pi\gamma |z| (\lambda^{-1} + \lambda^m) \, d\lambda \leqslant c_{q,\delta} \left(1 + |z|^{-\frac{m+1}{\gamma}}\right).$$

It can be easily seen that, for every $x \in E$, the mapping $z \mapsto S_\gamma(z)x$, $z \in \Sigma_{(\pi/2)-\pi\gamma}$ is analytic and that the formula (275) holds; the semigroup property for $(S_\gamma(t))_{t>0}$ can be proved by using the direct computation similar to that established in [31, Section 3]. Keeping in mind the equality (276) we obtain that, in the case $m > -1$, $(S_\gamma(z))_{z\in\Sigma_{(\pi/2)-\pi\gamma}}$ is an analytic C-regularized semigroup of growth order $(m+1)/\gamma$. In the case $m \geqslant -1$, the first equality stated in (iv.1) follows from the validity of the equality [404, (13)] for any complex number $z \in \Sigma_{(\pi/2)-\gamma\pi}$, the proof of [404, Proposition 3.5] and the computation given in the estimation of the term $q(S_\gamma(z)x)$, taken together with the inequality $\lambda^\gamma|z| \exp(-c_\delta|z|\lambda^\gamma) \leqslant c_\delta^{-1}$, $\lambda > 0$, $z \in \Sigma_\delta$. Now we will prove that the operator $-A_\gamma$ is the integral generator of $(S_\gamma(t))_{t>0}$; it suffices to show that the following equivalence relation holds:

$$S_\gamma(t)x - S_\gamma(s)x = \int_s^t S_\gamma(r)y\,dr \text{ for } t > s > 0 \text{ iff } (x, y) \in -A_\gamma. \tag{316}$$

In order to see that (316) holds, we may assume without loss of generality that $x, y \in D_\infty(A)$. But, in this case, (316) follows from the fact that $(S_\gamma(t)_{|D_\infty(A)})_{t>0}$ is a global C_∞-regularized semigroup in $D_\infty(A)$ with the integral generator $-A_{\infty,\gamma}$, and Theorem 2.9.39(iv). If the condition (HQ) holds, then for each $\theta \in (\omega - \pi, \pi - \omega)$, the operator $(e^{i\theta}A)_\gamma = e^{i\theta\gamma}A_\gamma$ is the integral generator of an exponentially equicontinuous analytic C-regularized semigroup $(S_{\theta,\gamma}(z))_{z\in\Sigma_{(\pi/2)-\gamma\pi}}$ of growth order $(m+1)/\gamma$, where $S_{\theta,\gamma}(\cdot)$ is given by

$$S_{\theta,\gamma}(z)x = \frac{1}{2\pi i}\int_0^\infty f_z(\lambda)(\lambda + e^{i\theta}A)^{-1}Cx\,d\lambda,\ x \in E,\ z \in \Sigma_{(\pi/2)-\gamma\pi}.$$

Let $\varepsilon > 0$ be sufficiently small. Then one can take numbers $\theta_1 \in (0, \pi - \omega)$ and $\theta_2 \in (\omega - \pi, 0)$ such that $(\pi/2) - \gamma\pi + \gamma\theta_1 > (\pi/2) - \gamma\omega - \varepsilon$ and $\gamma\pi - (\pi/2) + \gamma\theta_2 > \omega\gamma - (\pi/2)$. It can be simply proved that the operator $-A_\gamma$ is the integral generator of an exponentially equicontinuous analytic C-regularized semigroup $(S_\gamma(z))_{z\in\Sigma_{(\pi/2)-\gamma\omega-\varepsilon}}$ of growth order $(m+1)/\gamma$, where $S_\gamma(\cdot)$ is given by

$$S_\gamma(z)x = \begin{cases} S_\gamma(z)x,\ z \in \Sigma_{(\pi/2)-\gamma\pi}, \\ S_{\theta_1,\gamma}(ze^{-\gamma i\theta_1}),\ \text{if } z \in e^{\gamma i\theta_1}\Sigma_{(\pi/2)-\gamma\pi}, \\ S_{\theta_2,\gamma}(ze^{-\gamma i\theta_2}),\ \text{if } z \in e^{\gamma i\theta_2}\Sigma_{(\pi/2)-\gamma\pi}. \end{cases}$$

This implies the second equality in (iv.1); the remaining part of proof of (iv.1) is simple. The equality (315) follows from the formula

$$\begin{aligned} \frac{S_\gamma(z)-Cx}{z} &= \frac{e^{-zz_0^\gamma}x - Cx}{z} + \sum_{k=2}^{\lfloor m+\gamma\rfloor+2} \frac{(-1)^{k-1}}{(k-1)!}H_{k-1}(z,z_0)(z_0 - A)^{k-1}Cx \\ &-\frac{1}{z}\int_0^\infty e^{-z\lambda^\gamma\cos\pi\gamma}\sin(z\lambda^\gamma\sin\pi\gamma)\frac{(\lambda+A)^{-1}C(z_0-A)^{\lfloor m+\gamma\rfloor+2}x}{(\lambda+z_0)^{\lfloor m+\gamma\rfloor+2}}\,d\lambda, \end{aligned} \tag{317}$$

and the dominated convergence theorem; in the case that the condition (HQ) holds, then the equality in (v) holds on closed subsectors of $\Sigma_{(\pi/2)-\omega\gamma}$ by the construction of $S_{\theta,\gamma}(\cdot)$. Suppose now $-1-\gamma<m<-1$ and $\zeta>0$. The equality (255) and the arguments given in the proof of [403, Theorem 5.5.1] imply $\lim_{t\to 0+}S_\gamma(t)x=Cx$, $x\in\overline{D(A)}$. Keeping in mind Theorem 1.2.5 and the growth rate of $(S_\gamma(z))_{z\in\Sigma_{(\pi/2)-\gamma\pi}}$, we obtain that, for every $\delta\in(0,(\pi/2)-\pi\gamma)$, we have $\lim_{z\to 0,z\in\Sigma_\delta}S_\gamma(z)x=Cx$, $x\in\overline{D(A)}$. Now it readily follows that $(S_{\gamma,\zeta}(t))_{t\geq 0}$ is an exponentially equicontinuous analytic (non-degenerate) ζ-times integrated C-regularized semigroup of angle $(\pi/2)-\gamma\pi$; furthermore, the family $\{|z|^{-\zeta}(1+|z|^{-(m+1)/\gamma})^{-1}S_{\gamma,\zeta}(z):z\in\Sigma_{(\pi/2)-\gamma\pi-\varepsilon}\}$ is equicontinuous for any $\varepsilon\in(0,(\pi/2)-\gamma\pi)$. This completes the proof of (iv.2). Suppose, for the time being, that the operator A is injective as well as that $x\in E$ and $\mu>0$. Using again the proof of [403, Theorem 5.5.1], we obtain that:

$$\int_0^\infty e^{-\mu t}S_{\gamma,\zeta}(t)x\,dt=\mu^{-\zeta}\frac{\sin\gamma\pi}{\pi}\int_0^\infty\frac{\lambda^\gamma}{(\mu+\lambda^\gamma\cos\pi\gamma)^2+\lambda^{2\gamma}\sin^2\pi\gamma}(\lambda+A)^{-1}Cx\,d\lambda.$$

Now we will prove that the equality (290) holds. Towards this end, observe that Proposition 2.9.5 implies $A^{-1}\in\mathcal{M}_{C,-m-2}$ and that the equality (290) is equivalent with:

$$\begin{aligned}&\overline{A^{-1}_{\infty,\gamma}}\left((1+A^{-1})^{-1}C\right)^{2\lfloor -m\rfloor+1}C^2\Bigg[C^3x-\mu C^2\left((1+A^{-1})^{-1}C\right)^{2\lfloor -m\rfloor+1}\\&\times\frac{\sin\gamma\pi}{\pi}\int_0^\infty\frac{\lambda^\gamma}{(\mu+\lambda^\gamma\cos\pi\gamma)^2+\lambda^{2\gamma}\sin^2\pi\gamma}(\lambda+A)^{-1}Cx\,d\lambda\Bigg]\\&=C^4\left((1+A^{-1})^{-1}C\right)^{2\lfloor -m\rfloor+1}\\&\times\frac{\sin\gamma\pi}{\pi}\int_0^\infty\frac{\lambda^\gamma}{(\mu+\lambda^\gamma\cos\pi\gamma)^2+\lambda^{2\gamma}\sin^2\pi\gamma}(\lambda+A)^{-1}Cx\,d\lambda.\end{aligned}$$

Multiplying both sides of the above equality with $S_\gamma^{-1}(t):=\int_0^\infty f_t(\lambda)(\lambda+A^{-1})^{-1}C\,d\lambda$ $(t>0)$, and making use of the equality (276) with $S_\gamma(\cdot)$ replaced by $S_\gamma^{-1}(\cdot)$ therein, we may assume without loss of generality that $x\in D_\infty(A^{-1})$. Therefore, it suffices to show that

$$\begin{aligned}&\overline{A^{-1}_{\infty,\gamma}}\left[Cx-\mu\frac{\sin\gamma\pi}{\pi}\int_0^\infty\frac{\lambda^\gamma}{(\mu+\lambda^\gamma\cos\pi\gamma)^2+\lambda^{2\gamma}\sin^2\pi\gamma}(\lambda+A)^{-1}Cx\,d\lambda\right]\\&=\frac{\sin\gamma\pi}{\pi}\int_0^\infty\frac{\lambda^\gamma}{(\mu+\lambda^\gamma\cos\pi\gamma)^2+\lambda^{2\gamma}\sin^2\pi\gamma}(\lambda+A)^{-1}Cx\,d\lambda.\end{aligned}$$

Since $(\lambda + A)^{-1}Cx = \lambda^{-1}(\lambda^{-1} + A^{-1})^{-1}CA^{-1}x = \lambda^{-1}(\lambda^{-1} + A^{-1})^{-1}CA^{-1}x = (\lambda + (A_{\infty}^{-1})^{-1})^{-1}Cx$, the integral appearing in the last equality is convergent in the topology of $D_{\infty}(A^{-1})$, which is continuously embedded in E. Set $B := A_{\infty}^{-1}$, $C_{\infty,-1} := C_{| D_{\infty}(A^{-1})}$ and $E_1 := D_{\infty}(A^{-1})$. Then the operator $B \in L(E_1)$ is injective and $C_{\infty,-1}$-nonnegative in E_1, so that B^{-1} is also $C_{\infty,-1}$-nonnegative in E_1. Keeping in mind these observations, we will have to prove that

$$
\begin{aligned}
(318)\quad & C_{\infty,-1}x - \mu\frac{\sin\gamma\pi}{\pi}\int_0^{\infty}\frac{\lambda^{\gamma}}{(\mu+\lambda^{\gamma}\cos\pi\gamma)^2+\lambda^{2\gamma}\sin^2\pi\gamma}(\lambda+B^{-1})^{-1}C_{\infty,-1}x\,d\lambda \\
& = B_{-\gamma}\frac{\sin\gamma\pi}{\pi}\int_0^{\infty}\frac{\lambda^{\gamma}}{(\mu+\lambda^{\gamma}\cos\pi\gamma)^2+\lambda^{2\gamma}\sin^2\pi\gamma}(\lambda+B^{-1})^{-1}C_{\infty,-1}x\,d\lambda,
\end{aligned}
$$

where the integrals are taken in the topology of E_1. Due to Proposition 2.9.42 and Theorem 2.9.36, we obtain:

$$
\begin{aligned}
& B_{-\gamma}\frac{\sin\gamma\pi}{\pi}\int_0^{\infty}\frac{\lambda^{\gamma}}{(\mu+\lambda^{\gamma}\cos\pi\gamma)^2+\lambda^{2\gamma}\sin^2\pi\gamma}(\lambda+B^{-1})^{-1}C_{\infty,-1}x\,d\lambda \\
& = B_{-\gamma}(\mu+B_{-\gamma})^{-1}C_{\infty,-1}x \\
& = C_{\infty,-1}x - \mu(\mu+B_{-\gamma})^{-1}C_{\infty,-1}x \\
& = C_{\infty,-1}x - \mu\frac{\sin\gamma\pi}{\pi}\int_0^{\infty}\frac{\lambda^{\gamma}}{(\mu+\lambda^{\gamma}\cos\pi\gamma)^2+\lambda^{2\gamma}\sin^2\pi\gamma}(\lambda+B^{-1})^{-1}C_{\infty,-1}x\,d\lambda.
\end{aligned}
$$

This implies (318), (290) and (289), then the remainder of proof of (iv.4) becomes very simple and can be omitted. Let the condition (HQ) hold. Suppose $-1-\gamma < m < -1$ and $\zeta > 0$. The assertion of [292, Corollary 2.4.3] can be simply reformulated in SCLCSs, which implies that the operator family

$$
\left\{\left(|\lambda|^{-1} + |\lambda|^{\frac{m+1}{\gamma}-1}\right)^{-1}(\lambda + {}_{\gamma}A)^{-1}C : \lambda \in \Sigma_{\pi-\pi\gamma-\varepsilon}\right\} \subseteq L(E)
$$

is equicontinuous and pointwise analytic for any $\varepsilon \in (0, \pi - \pi\gamma)$. Furthermore, for every $\theta \in (\omega - \pi, \pi - \omega)$, the operator $e^{i\theta}A$ is in the class $\mathcal{M}_{C,m}$. Since $-\operatorname{Re}(e^{-i\theta\gamma}) < 0$, the first part of proof and Proposition 2.9.42 imply at once that, for every $\mu > 0$ and $x \in E$,

$$
\begin{aligned}
& (\mu + e^{i\theta\gamma}{}_{\gamma}A)^{-1}Cx = \frac{e^{-i\theta\gamma}\sin\gamma\pi}{\pi} \\
(319)\quad & \times\int_0^{\infty}\frac{\lambda^{\gamma}}{(\mu e^{-i\theta\gamma}+\lambda^{\gamma}\cos\pi\gamma)^2+\lambda^{2\gamma}\sin^2\pi\gamma}(\lambda+A)^{-1}Cx\,d\lambda,
\end{aligned}
$$

and

(320)

$$(\mu\ +\ {}_{\gamma}(e^{i\theta}A))^{-1}Cx=\frac{\sin\gamma\pi}{\pi}\int\limits_{0}^{\infty}\frac{\lambda^{\gamma}}{(\mu+\lambda^{\gamma}\cos\pi\gamma)^{2}+\lambda^{2\gamma}\sin^{2}\pi\gamma}(\lambda+e^{i\theta}A)^{-1}Cx\,d\lambda.$$

By (319)-(320) and the Cauchy formula, we get that $C(\mu+{}_{\gamma}(e^{i\theta}A))^{-1}C=C(\mu+e^{i\theta\gamma}\,{}_{\gamma}A)^{-1}C$, $\mu>0$. This, in turn, implies ${}_{\gamma}(e^{i\theta}A)=e^{i\theta\gamma}\,{}_{\gamma}A$. Now it is not difficult to prove that the operator family

$$\left\{\left(|\lambda|^{-1}+|\lambda|^{\frac{m+1}{\gamma}-1}\right)^{-1}(\lambda\ +\ {}_{\gamma}A)^{-1}C:\lambda\in\Sigma_{\pi-\omega\gamma-\varepsilon}\right\}\subseteq L(E)$$

is equicontinuous and pointwise analytic for any $\varepsilon\in(0,\pi-\omega\gamma)$. Making use of Theorem 2.2.5, we get that the operator $-{}_{\gamma}A$ is the integral generator of an exponentially equicontinuous ζ-times integrated C-regularized semigroup $(S_{\gamma,\zeta}(t))_{t\geqslant 0}$ of angle $(\pi/2)-\omega\gamma$. Let $\omega>0$ and $\delta\in(0,(\pi/2)-\gamma\omega)$ be fixed, and let $\varepsilon\in(0,\delta)$. It remains to be proved that the family $\{|z|^{-\zeta}(1+|z|^{-(m+1)/\gamma})^{-1}S_{\gamma,\zeta}(z):z\in\Sigma_{\delta-\varepsilon}\}$ is equicontinuous. As before, there exists a mapping $z\mapsto S_{\gamma}(z)\in L(E)$, $z\in\Sigma_{(\pi/2)-\gamma\omega}$ such that, for every $\varepsilon>0$ and for every $x\in E$, the family $\{e^{-\varepsilon|z|}S_{\gamma}(z):z\in\Sigma_{(\pi/2)-\gamma\omega-\varepsilon}\}$ is equicontinuous and that the mapping $z\mapsto S_{\gamma}(z)x$, $z\in\Sigma_{(\pi/2)-\omega\gamma}$ is analytic. This implies the second part of the assertion (ii). Furthermore, the representation formula

$$S_{\gamma}(z)x=\frac{1}{2\pi i}\int\limits_{\Gamma_{\omega}}e^{\lambda z}(\lambda\ +\ {}_{\gamma}A)^{-1}\ Cx\,d\lambda,\ x\in E,\ z\in\Sigma_{\delta-\varepsilon}$$

holds, where Γ_{ω} is oriented counterclockwise and consists of $\Gamma_{\pm}:=\{\omega+re^{i((\pi/2)+\delta)}:r\geqslant|z|^{-1}\}$ and $\Gamma_{0}:=\{\omega+|z|^{-1}e^{i\theta}:|\theta|\leqslant(\pi/2)+\delta\}$. It can be simply proved that there exists an absolute constant $c>0$ that, for every $\lambda=\omega+|z|^{-1}e^{i\theta}\in\Gamma_{0}$, one has $|\lambda|\geqslant c(\omega+|z|^{-1})$. Inspecting the proof of [20, Theorem 2.6.1] again, we obtain that for every $q\in\circledast$ there exist $c_{q,\varepsilon}>0$ and $r_{q}\in\circledast$ such that, for every $x\in E$,

$$q(S_{\gamma}(z)x)\leqslant c_{q,\varepsilon}r_{q}(x)e^{\omega\operatorname{Re}z}(1+|z|^{-(m+1)/\gamma}),\ z\in\Sigma_{\delta-\varepsilon}.$$

Letting $\omega\to 0+$, we obtain that the operator family $\{(1+|z|^{-(m+1)/\gamma})^{-1}S_{\gamma}(z):z\in\Sigma_{\delta-\varepsilon}\}$ is equicontinuous. Keeping in mind the obvious estimate (291), this fact completes the proof of (viii). The first part of assertion (vi) follows immediately from (iv.2). Suppose now $0<\beta<1/2$ and $m+\beta\gamma>-1$. Then it makes sense to consider the operators ${}_{\gamma\beta}A$ and ${}_{\beta}({}_{\gamma}A)$. In order to see that ${}_{\beta}({}_{\gamma}A)={}_{\gamma\beta}A$, it suffices to prove the equality of corresponding once integrated C-regularized semigroups generated by these operators. Keeping in mind the representation of $(\mu+{}_{\gamma}A)^{-1}C$ for $\mu>0$, we will only have to prove that, for every $x\in E$,

$$\int_0^\infty e^{-t\lambda^{\beta\gamma}\cos\beta\gamma\pi}\sin(t\lambda^{\beta\gamma}\sin\beta\gamma\pi)\,(\lambda+A)^{-1}Cx\,d\lambda$$

(321)

$$=\frac{\sin\gamma\pi}{\pi}\int_0^\infty\int_0^\infty e^{-t\lambda^{\beta}\cos\beta\pi}\sin(t\lambda^{\beta}\sin\beta\pi)\mu^{\gamma}\frac{(\mu+A)^{-1}Cx}{(\lambda+\mu^{\gamma}\cos\pi\gamma)^2+\mu^{2\gamma}\sin^2\pi\gamma}\,d\mu\,d\lambda.$$

Fix, for the time being, a number $\varepsilon > 0$. We will first prove that the equality (321) holds with A replaced by $A+\varepsilon$ therein. In order to see this, we may assume without loss of generality that $x \in C^2(D_\infty(A+\varepsilon)) = C^2(D_\infty(A))$. Due to [90, Theorem 2.8(iv)] and Proposition 2.9.42, we have that the equality (321) holds with A replaced by $(A+\varepsilon)_\alpha$ therein ($\alpha \in (0, 1)$). Using Proposition 2.9.42 again, we obtain that for each $q \in ⊛$ and $\alpha \in (0, 1)$ there exist $r_q \in ⊛$ and $c_q > 0$ such that

$$\begin{aligned}&q\big((\lambda+(A+\varepsilon)_\alpha)^{-1}C^2x-(\lambda+(A+\varepsilon))^{-1}C^2x\big)\\ (322)\qquad&=q\big((\lambda+(A+\varepsilon)_\alpha)^{-1}C(\lambda+(A+\varepsilon))^{-1}C[(\lambda+A+\varepsilon)x-(\lambda+(A+\varepsilon)_\alpha)x]\big)\\ &\leqslant c_q\big((\lambda+\varepsilon)^{-1}+(\lambda+\varepsilon)^{-m}\big)^2 r_q\big((\lambda+A+\varepsilon)x-(\lambda+(A+\varepsilon)_\alpha)x\big).\end{aligned}$$

By Proposition 2.9.14(ii), the mapping $\alpha \mapsto (A+\varepsilon)_\alpha Cx$, $1/2 < \operatorname{Re}\alpha < 2$ is analytic and therefore bounded in a neighborhood of point 1. Taken together with (322), the dominated convergence theorem and the injectiveness of C, this fact implies by letting $\alpha \to 1-$ that (321) holds for $A+\varepsilon$. Applying the dominated convergence theorem again, we obtain now by letting $\varepsilon \to 0+$ that (321) holds, as claimed. This completes the proof of (vi). We will prove the assertion (ix) only in the non-trivial case $\beta\gamma \notin \mathbb{N}$, consider first the case $m \geqslant -1$. Fix, for a moment, an element $x \in D_\infty(A)$. Then, for every $\mu \geqslant 0$ and $t > 0$,

$$(z^{\mu}e^{-tz^{\gamma}})(A_\infty)x=\frac{1}{2\pi i}\int_0^\infty \lambda^{\mu}[e^{-i\mu\pi}e^{-t\lambda^{\gamma}e^{-i\pi\gamma}}-e^{i\mu\pi}e^{-t\lambda^{\gamma}e^{i\pi\gamma}}]\,(\lambda+A)^{-1}Cx\,d\lambda.$$

Taking into account the weakened product formula, we get that:

$$\begin{aligned}J_{\infty,\gamma\beta}C^2S_\gamma(t)x&=J_{\infty,\gamma\beta-\lceil\gamma\beta\rceil}J_{\infty,\lceil\gamma\beta\rceil}CS_\gamma(t)x\\&=C^2J_{\infty,\gamma\beta-\lceil\gamma\beta\rceil}(z^{\lceil\gamma\beta\rceil}e^{-tz^{\gamma}})(A_\infty)x\\&=C^3(z^{\gamma\beta}e^{-tz^{\gamma}})(A_\infty)x.\end{aligned}$$

The above computation implies that, for every $x \in E$ and $z \in \Sigma_{(\pi/2)-\gamma\pi}$,

$$A_{\infty,\gamma\beta}S_\gamma(z)x=\frac{1}{2\pi i}\int_0^\infty \lambda^{\gamma\beta}[e^{-i\gamma\beta\pi}e^{-z\lambda^{\gamma}e^{-i\pi\gamma}}-e^{i\gamma\beta\pi}e^{-z\lambda^{\gamma}e^{i\pi\gamma}}]\,(\lambda+A)^{-1}Cx\,d\lambda.$$

As before, the last equality shows that, for every $x \in E$ and $z \in \Sigma_{(\pi/2)-\gamma\pi}$, the following holds:

$$(323)\qquad A_{\gamma\beta}S_\gamma(z)x=\frac{1}{2\pi i}\int_0^\infty \lambda^{\gamma\beta}[e^{-i\gamma\beta\pi}e^{-z\lambda^{\gamma}e^{-i\pi\gamma}}-e^{i\gamma\beta\pi}e^{-z\lambda^{\gamma}e^{i\pi\gamma}}]\,(\lambda+A)^{-1}Cx\,d\lambda.$$

On the other hand, a slight modification of the proof of [90, Theorem 3.5(i)/(b)'] shows that, for every $x \in E$ and $t > 0$, the following equality holds, with $z = te^{i\theta}$,

(324) $$D_{-}^{\beta}S_{\gamma}(te^{i\theta})x = \frac{e^{i\theta\beta}}{2\pi i}\int_{0}^{\infty} \lambda^{\gamma\beta}\left[e^{-i\gamma\beta\pi}e^{-z\lambda^{\gamma}e^{-i\pi\gamma}} - e^{i\gamma\beta\pi}e^{-z\lambda^{\gamma}e^{i\pi\gamma}}\right](\lambda + A)^{-1}Cx\,d\lambda.$$

The proof of (ix.1) follows immediately from (323)-(324). We will prove the assertion (ix.2) in the case $\theta = 0$ and $\gamma\beta \notin \mathbb{N}$. Observe first that the equality (324) remains true provided $-1 - \gamma < m < -1$ and $\gamma\beta > -1 - m$, so that the only non-trivial part of the proof is to show the validity of (323) in the case that $-1 - \gamma < m < -1$ and $\gamma\beta > -1 - m$, with the operator $A_{\gamma\beta}$ replaced by $C^{-1}\lim_{\varepsilon\to 0+}(A + \varepsilon)_{\gamma\beta}C$. It is clear that, for every $\varepsilon > 0$, the operator $A + \varepsilon$ is C-nonnegative, which implies that the equality (323) holds with the operator A replaced by $A + \varepsilon$. Define $S_{\gamma,\varepsilon}(t)x := \frac{1}{2\pi i}\int_0^{\infty} f_t(\lambda)(\lambda + A + \varepsilon)^{-1}Cx\,d\lambda$, $x \in E$, $t, \varepsilon > 0$. By the dominated convergence theorem and the binomial formula, it follows that

(325) $$\lim_{\varepsilon\to 0+}(A + \varepsilon)^{n}[S_{\gamma,\varepsilon}(t)x - S_{\gamma}(t)x] = 0,\ x \in E,\ n \in \mathbb{N},\ t > 0.$$

Using the definition of Balakrishnan operators and the inequality $\gamma\beta > -1 - m$, it is not difficult to prove that, for every $\varepsilon > 0$ and $q \in ⊛$, there exist $c_{q,\gamma,\beta} > 0$ and $r_q \in ⊛$ such that

(326) $$q((A + \varepsilon)_{\beta\gamma - \lfloor\beta\gamma\rfloor}Cx) \leqslant c_{q,\gamma,\beta}(r_q(x) + r_q((A + \varepsilon)x)),\ x \in D_{\infty}(A).$$

Keeping in mind Theorem 2.9.36(ii) and the estimates (325)-(326), we obtain that, for every $\varepsilon > 0$, $x \in E$ and $q \in ⊛$,

$$\begin{aligned} &q\big((A + \varepsilon)_{\beta\gamma}C\,[S_{\gamma,\varepsilon}(t)x - S_{\gamma}(t)x]\big)\\ &= q\big((A + \varepsilon)_{\beta\gamma - \lfloor\beta\gamma\rfloor}(A + \varepsilon)_{\lfloor\beta\gamma\rfloor}C[S_{\gamma,\varepsilon}(t)x - S_{\gamma}(t)x]\big)\\ &\leqslant c_{q,\gamma,\beta}\big[r_q\big((A + \varepsilon)_{\lfloor\beta\gamma\rfloor}[S_{\gamma,\varepsilon}(t)x - S_{\gamma}(t)x]\big)\\ &\quad + r_q\big((A + \varepsilon)_{\lfloor\beta\gamma+1\rfloor}[S_{\gamma,\varepsilon}(t)x - S_{\gamma}(t)x]\big)\big] \to 0,\ \varepsilon \to 0+. \end{aligned}$$

Making use of the dominated convergence theorem and the assertion (ix.1), one gets that $\lim_{\varepsilon\to 0}(A + \varepsilon)_{\beta\gamma}S_{\gamma,\varepsilon}(t)x$ exists and equals the right hand side of (323), which remains true, by the previous computation, for $C^{-1}\lim_{\varepsilon\to 0}(A + \varepsilon)_{\beta\gamma}CS_{\gamma}(t)x$. The proof of (ix.2) now follows instantly. □

Remark 2.9.49. (i) Since $\arctan(\cos \pi\gamma) < (\pi/2) - \gamma\pi$ for $0 < \gamma < 1/2$, we have improved the angle of analyticity of C-regularized semigroups of growth order $r > 0$ appearing in the formulations of [90, Theorem 3.5(i), (ii)-(ii)'], [300, Theorem 3.1] (cf. also [292, Theorem 1.4.15-Theorem 1.4.16]), and in the formulation of [90, Theorem 3.7(i)], provided that $0 < b < 1/2$ (cf. the assertion (ii) of the above-mentioned theorem with $n \geqslant 3$). Furthermore, we have proved that the limit appearing in the formulation of [90, Theorem 3.5(i)-(d)] exists in the case $\lfloor b + \alpha\rfloor < 0$.

(ii) The proof of the preceding theorem shows that the second formula in the formulation of [90, Theorem 3.5(i)-(b)'] holds for all $x \in E$. Therefore, we have significantly improved the assertion of [90, Theorem 3.5(ii)'] (cf. also [90, Theorem 3.7(ii)'] with $0 < b < 1/2$). Notice, finally, that the assertion of [90, Theorem 3.7] does not admit a satisfactory reformulation in the case that $b > 1/2$ and that there does not exist $d \in (0, 1]$ such that the family $\{(\lambda + A)^{-1} C : |\lambda| \leqslant d\}$ is equicontinuous (cf. also Theorem 2.9.51 and Remark 2.9.52 for further information concerning the case $b = 1/2$).

(iii) Let A be injective. Then the operator $C^{-1} \lim_{\varepsilon \to 0+} (A + \varepsilon)_{\gamma\beta} C$, appearing in the formulation of the assertion (ix.2), can be replaced by the operator $A_{\gamma\beta}$. In order to see this, one can apply Proposition 2.9.35 and the method described in Remark 2.9.33; the only non-trivial part is to show that, for every $x \in C_\infty(D_\infty(A^{-1})) \cap C_\infty(D_\infty(A))$,

$$\frac{\Gamma(n)}{\Gamma(\beta\gamma)\Gamma(n-\beta\gamma)} \int_0^\infty \lambda^{n-\beta\gamma-1} [(\lambda + A^{-1})^{-1} C]^n S_\gamma(t) x \, d\lambda$$

$$= \frac{C^{n-1}}{2\pi i} \frac{e^{i\theta\beta}}{2\pi i} \int_0^\infty \lambda^{\gamma\beta} [e^{-i\gamma\beta\pi} e^{-z\lambda^\gamma e^{-i\pi\gamma}} - e^{i\gamma\beta\pi} e^{-z\lambda^\gamma e^{i\pi\gamma}}] (\lambda + A)^{-1} Cx \, d\lambda, \tag{327}$$

where $n \in \mathbb{N}$ satisfies $n > \gamma\beta$. Notice that the assertion (ix.1) implies that the equality (327) holds with the operator A replaced by $A + \varepsilon \in \mathcal{M}_{C,-1}$ $(\varepsilon > 0)$. Letting $\varepsilon \to 0+$, one obtains with the help of the dominated convergence theorem that the equality (327) holds provided $-1 - m < \gamma\beta < 1$. If $\gamma\beta \geqslant 1$, then the claimed assertion follows from the fact that the both sides of (327) are analytic in $\beta \in \{z \in \mathbb{C} : (-1 - m)/\gamma < \operatorname{Re} z < n/\gamma\}$.

(iv) Notice that, for certain values of parameters $m \in ((-3)/2, -1)$ and $\beta, \gamma \in (0, 1/2)$, the assertion (vi) removes the suppositions of injectiveness and boundedness from the formulation of second part of [404, Theorem 4.6].

Theorem 2.9.50. *(Multiplicativity, see* [403, Theorem 5.2.5, Theorem 5.4.3 and Theorem 7.1.3] *for the case* $C = 1$*)*

(i) *Suppose* $A \in \mathcal{M}_{C,-1}$ *and* $0 < \alpha < 1$. *Then* A_α *is C-sectorial with C-spectral angle less than or equal to* $\alpha\pi$ *and, for every* $\beta \in \mathbb{C}_+$, *we have that:*

$$(A_\alpha)_\beta = A_{\alpha\beta}. \tag{328}$$

(ii) *Suppose A is C-sectorial with C-spectral angle* $\omega \in [0, \pi)$, *and* $0 < \alpha < \pi/\omega$. *Then* A_α *is C-sectorial with C-spectral angle less than or equal to* $\alpha\omega$, *and for every* $\beta \in \mathbb{C}_+$, *(328) holds.*

(iii) *Suppose A is injective and C-sectorial with C-spectral angle* $\omega \in [0, \pi)$. *Then, for every* $\alpha \in (-\pi/\omega, \pi/\omega)$, *the operator* A_α *is C-sectorial and, for every* $\beta \in \mathbb{C}$, *(328) holds.*

Proof. Suppose first $A \in \mathcal{M}_{C,-1}$, $0 < \alpha < 1$ and $\beta \in \mathbb{C}_+$. Then Proposition 2.9.42 implies that A_α is C-sectorial with C-spectral angle less than or equal to $\alpha\pi$. Let

$(x, y) \in (A_\alpha)_\beta$, and let $n \in \mathbb{N}$ be sufficiently large (this number may vary throughout the proof, but the essence remains the same). It is evident that the assertions (iv.2)-(iv.4) and (vi) of Theorem 2.9.48 continue to hold for $m = -1$; in this case, ${}_\gamma A = A_\gamma$ for $\gamma \in (0, 1/2)$ and the equation (328) holds provided $\alpha, \beta \in (0, 1/2)$. Now we will prove that (328) holds in general case, i.e., that:

(329) $$\overline{A_{\infty,\alpha\beta}}\,((1 + A)^{-1}C)^n\, C^2x = ((1 + A)^{-1}C)^n\, C^2y.$$

Since the operator $S_\gamma(t)C^n$ $(t > 0)$ commutes with $\overline{A_{\infty,\alpha\beta}}$ and $(A_\alpha)_\beta$, we may assume without loss of generality that $x \in C^n(D_\infty(A))$. Then the equation (329) reads as follows

$$\overline{A_{\infty,\alpha\beta}}\,x = (A_{\infty,\alpha})_\beta x.$$

Keeping in mind the additivity of powers, Proposition 2.9.14(ii) and the uniqueness theorem for analytic functions, it suffices to show that, for every $0 < \beta < 1/2$, we have:

(330) $$J_{\alpha\beta}x = \frac{\sin \pi\beta}{\beta} \int_0^\infty \lambda^{\beta-1}\big(\lambda + (A_\infty)_\alpha\big)^{-1}Cx\, d\lambda.$$

Noticing that (330) holds for $\alpha, \beta \in (0, 1/2)$ and that both sides of this equality are analytic with respect to α, which belongs to some open connected subset of complex plane containing the interval (0, 1) (cf. Proposition 2.9.42), the claimed assertion immediately follows; hence, $(A_\alpha)_\beta \subseteq A_{\alpha\beta}$. By using a similar reasoning, we get that $A_{\alpha\beta} \subseteq (A_\alpha)_\beta$, which completes the proof of (i). The second part of theorem can be proved by using the same technique, and a key point is to show that, for every fixed $\beta \in (0, 1/2)$ and $x \in C^n(D_\infty(A))$, the mapping

$$\alpha \mapsto \frac{\sin \pi\beta}{\beta} \int_0^\infty \lambda^{\beta-1}\big(\lambda + (A_\infty)_\alpha\big)^{-1}Cx\, d\lambda$$

is analytic on some open complex neighborhood of the interval $(0, \pi/\omega)$. This follows from the dominated convergence theorem and the representation formulae

$$(\lambda + (A_\infty)_\alpha)^{-1}Cx = (-1)^n \sum_{k=1}^{n} \frac{((A_\infty)_{\alpha/n} - \lambda_k)^{-1}Cx}{\prod_{\substack{1\leqslant i\leqslant n\\ i\neq k}}(\lambda_i - \lambda_k)}$$

$$= (-1)^n \frac{\sin \alpha\pi/n}{\pi} \sum_{k=1}^{n} \int_0^\infty \frac{u^{\alpha/n}}{\mu^{2\alpha/n} - 2\mu^{\alpha/n}\cos\frac{\alpha\pi}{n}\lambda_k + \lambda_k^2} (\mu + A)^{-1}Cx\, d\mu$$

$$\times \frac{1}{\prod_{\substack{1\leqslant i\leqslant n\\ i\neq k}}(\lambda_i - \lambda_k)}, \quad \lambda > 0,$$

where $\lambda_1, \cdot\cdot\cdot, \lambda_n$ denote the n-th roots of $-\lambda$. Suppose now that A is injective and C-sectorial with C-spectral angle $\omega \in [0, \pi)$. Then it can be easily seen that the operator A^{-1} is C-sectorial with the same C-spectral angle. Due to Theorem 2.9.36(i), the assertion (iii) immediately follows if we prove that $(A_\alpha)_{i\tau} = A_{\alpha i\tau}$ for $0 < \alpha < \pi/\omega$ and $\tau \in \mathbb{R}\setminus\{0\}$. Towards this end, notice that the analysis given in Remark 2.9.33(ii) shows that

(331) $$(A_\alpha)_{i\tau}A_\alpha x = (A_\alpha)_{1+i\tau}x,\ x \in C^n(D_\infty(A)) \cap C^n(D_\infty(A^{-1})).$$

Clearly, $(A_\alpha)_{1+i\tau} = A_{\alpha+i\alpha\tau}$, which implies along with (331) and additivity of powers that $A_{1+i\alpha\tau}x = A(A_\alpha)_{i\tau}x$, $x \in C^n(D_\infty(A)) \cap C^n(D_\infty(A^{-1}))$. Using this equality, the proof of (i) and definition of purely imaginary powers, one gets that $(A_\alpha)_{i\tau} \subseteq A_{\alpha i\tau}$. The converse inclusion $A_{\alpha i\tau} \subseteq (A_\alpha)_{i\tau}$ can be proved similarly. □

Theorem 2.9.51. (i) *Let* $-1 < m < (-1)/2$, *and let the condition (HQ) hold.*

(i.1) *Then the operator* $A_{1/2}$ *satisfies the condition (HQ) with* m *and* ω *replaced respectively by* $2m + 1$ *and* $\omega/2$.

(i.2) *For every* $\zeta > 0$, *the operator* $-A_{1/2}$ *is the integral generator of an exponentially equicontinuous* $(2m+2+\zeta)$*-times integrated C-regularized semigroup* $(S_{1/2,2m+2+\zeta}(t))_{t\geqslant 0}$ *of angle* $(\pi/2) - (\omega/2)$. *Furthermore, for every* $q \in ⊛$ *and* $\delta \in (0, (\pi/2) - (\omega/2))$, *there exist* $c_{q,\delta} > 0$ *and* $r_q \in ⊛$ *(independent of* $\zeta > 0$*) such that:*

$$q(S_{1/2,2m+2+\zeta}(z)x) \leqslant c_{q,\delta}r_q(x)\,(|z|^{2m+2+\zeta} + |z|^\zeta),\ x \in E,\ z \in \Sigma_\delta.$$

Define $S_{1/2}(z)x := S'_{1/2,1}(z)x$, $x \in E$, $z \in \Sigma_{(\pi/2)-(\omega/2)}$.

(i.3) *If* $D(A)$ *is dense in* E, *then the operator* $-A_{1/2}$ *is the integral generator of an exponentially equicontinuous* $(2m + 2)$*-times integrated C-regularized semigroup* $(S_{1/2,2m+2}(t))_{t\geqslant 0}$ *of angle* $(\pi/2) - (\omega/2)$. *Furthermore, for every* $q \in ⊛$ *and* $\delta \in (0, (\pi/2) - (\omega/2))$, *there exist* $c_{q,\delta} > 0$ *and* $r_q \in ⊛$ *such that*

$$q(S_{1/2,2m+2}(z)x) \leqslant c_{q,\delta}r_q(x)\,(|z|^{2m+2} + |z|^\zeta),\ x \in E,\ z \in \Sigma_\delta.$$

(ii) *Let* $-1 > m > (-3)/2$, *and let the condition (HQ) hold.*

(ii.1) *Then, for every* $\varepsilon > 0$, *the operator* $A + \varepsilon$ *satisfies the condition (H) with the same* ω. *Denote by* $S^\varepsilon_{1/2,\zeta}(\cdot)$ *the* ζ*-times integrated C-semigroup with the integral generator* $-(A + \varepsilon)_{1/2}$ *(cf. Theorem 2.9.40);* $S^\varepsilon_{1/2}(\cdot) := S^\varepsilon_{1/2,0}(\cdot))$. *Then, for every* $\zeta \geqslant 0$ *and* $z \in \Sigma_{(\pi/2)-(\omega/2)}$, *the limit* $\lim_{\varepsilon\to 0+} S^\varepsilon_{1/2,\zeta}(z)x := S_{1/2,\zeta}(z)x$ *exists for all* $x \in E$; *here* $S_{1/2,0}(\cdot) = S_{1/2}(\cdot)$.

(ii.2) *For every* $\zeta > 0$, $(S_{1/2,\zeta}(t))_{t\geqslant 0}$ *is the exponentially equicontinuous analytic* ζ*-times integrated C-regularized semigroup of angle* $(\pi/2) - (\omega/2)$; *furthermore, if* $D(A)$ *is dense in* E, *then* $(S_{1/2}(t))_{t\geqslant 0}$ *is the exponentially equicontinuous analytic C-regularized semigroup of angle* $(\pi/2) - (\omega/2)$. *Furthermore, for every* $q \in ⊛$ *and* $\delta \in (0, (\pi/2) - (\omega/2))$, *there exist* $c_{q,\delta} > 0$ *and* $r_q \in ⊛$ *such that, for every* $\zeta \geqslant 0$,

(332) $$q(S_{1/2,\zeta}(z)x) \leqslant c_{q,\delta}r_q(x)\,(|z|^{2m+2+\zeta} + |z|^\zeta),\ x \in E,\ z \in \Sigma_\delta.$$

(ii.3) *Denote by* ${}_{1/2}A$ *the integral generator of* $(S_{1/2,1}(t))_{t\geqslant 0}$. *Then the operator* ${}_{1/2}A$ *satisfies the condition (HQ) with* m *and* ω *replaced respectively by* $2m+1$ *and* $\omega/2$.

(iii) (iii.1) *Let* $-1 < m < (-1)/2$, *and let the condition (HQ) hold. Set* $\Omega_{1/2} := \{x \in E : \lim_{z\to 0,\, z\in\Sigma_\delta} S_{1/2}(z)x = Cx$ *for every* $\delta \in (0, (\pi/2)-(\omega/2))\}$. *Then* $D(A_{1/2}) \cup D(A) \subseteq \Omega_{1/2}$ *and* $R(S_{1/2}(z)) \subseteq D_\infty(A)$, $z \in \Sigma_{(\pi/2)-(\omega/2)}$. *Furthermore, for every* $x \in \Omega_{1/2}$, *the incomplete abstract Cauchy problem*

$$(P_{2,m}): \begin{cases} u \in C^\infty(\Sigma_{(\pi/2)-(\omega/2)} : E) \cap C(\Sigma_{(\pi/2)-(\omega/2)} : D_\infty(A)), \\ u''(z) = Au(z),\ z \in \Sigma_{(\pi/2)-(\omega/2)}, \\ \lim_{z\to 0,\, z\in\Sigma_\delta} u(z) = Cx, \text{ for every } \delta \in (0, (\pi/2)-(\omega/2)), \\ \text{the set } \{(1+|z|^{-(2m+2)})^{-1} u(z) : z \in \Sigma_\delta\} \text{ is bounded in } E \\ \text{for every } \delta \in (0, (\pi/2)-(\omega/2)), \end{cases}$$

has a unique solution $u(z) = S_{1/2}(z)x$, $z \in \Sigma_{(\pi/2)-(\omega/2)}$. *Moreover, for every* $\delta \in (0, (\pi/2)-(\omega/2))$ *and* $j \in \mathbb{N}_0$, *we have that the set* $\{|z|^j(1+|z|^{-(2m+2)})^{-1} u^{(j)}(z) : z \in \Sigma_\delta\}$ *is bounded in E.*

(iii.2) *Let* $-1 > m > (-3)/2$, *and let the condition (HQ) hold. Define the set* $\Omega_{1/2}$ *as above. Then* $D(A_{1/2}) \cup D(A) \subseteq \Omega_{1/2}$, $R(S_{1/2}(z)) \subseteq D_\infty(A)$, $z \in \Sigma_{(\pi/2)-(\omega/2)}$, *and for every* $x \in \Omega_{1/2}$, *the problem* $(P_{2,m})$ *has a unique solution* $u(z) = S_{1/2}(z)x$, $z \in \Sigma_{(\pi/2)-(\omega/2)}$. *Moreover, for every* $\delta \in (0, (\pi/2)-(\omega/2))$ *and* $j \in \mathbb{N}_0$, *we have that the set* $\{|z|^j(1+|z|^{-(2m+2)})^{-1} u^{(j)}(z) : z \in \Sigma_\delta\}$ *is bounded in E.*

Proof. Suppose $-1 < m < (-1)/2$ and $A \in \mathcal{M}_{C,m}$. Then it can be easily seen that, for every $\mu \in \mathbb{C}_+$, we have $\mu \in \rho_C(-A_{1/2})$ and

$$(\mu + A_{1/2})^{-1}Cx = \frac{1}{\pi}\int_0^\infty \frac{\lambda^{1/2}}{\lambda+\mu^2}(\lambda + A)^{-1}Cx\, d\lambda,\ x \in E; \tag{333}$$

cf. the first part of proof of [404, Theorem 4.6]. Observe further that for each $\varepsilon \in (0, \pi/2)$ there exists $a_\varepsilon > 0$ such that $|\lambda+\mu^2| \geqslant a_\varepsilon(\lambda + |\mu|^2)$, $\lambda > 0$, $\mu \in \Sigma_{(\pi/2)-\varepsilon}$. Therefore, for every $q \in \circledast$ and $\varepsilon \in (0, \pi/2)$, there exist $c_q > 0$, $a_\varepsilon > 0$ and $r_q \in \circledast$ such that, for every $x \in E$,

$$\begin{aligned} &q((\mu + A_{1/2})^{-1}Cx) \\ &\leqslant c_q\pi^{-1}r_q(x)\int_0^\infty \frac{\lambda^{1/2}}{|\lambda+\mu^2|}\left[|\lambda|^{-1}+|\lambda|^m\right]d\lambda \\ &\leqslant c_q\pi^{-1}a_\varepsilon^{-1}r_q(x)\int_0^\infty \frac{\lambda^{1/2}}{|\lambda|+|\mu|^2}\left[|\lambda|^{-1}+|\lambda|^m\right]d\lambda \\ &= c_q\pi^{-1}a_\varepsilon^{-1}r_q(x)\left[|\mu|^{-1}\int_0^\infty \frac{\upsilon^{(-1)/2}}{\upsilon+1}d\upsilon + |\mu|^{2m+1}\int_0^\infty \frac{\upsilon^{m+(1/2)}}{\upsilon+1}d\upsilon\right]. \end{aligned} \tag{334}$$

The assertion (i.1) immediately follows from the estimate (334) and the fact that, for every $\theta \in (\omega - \pi, \pi - \omega)$, $e^{i\theta} A \in \mathcal{M}_{C,m}$ and $(e^{i\theta} A)_{1/2} = e^{i\theta/2} A_{1/2}$. The first part of assertion (i.2) follows from (i.1) and Theorem 2.2.8. Let $\delta \in (0, (\pi/2) - (\omega/2))$. Then it can be simply justified that, for every $\omega > 0$, $\varepsilon > 0$, $\varepsilon \in (0, \delta)$ and $x \in E$, the family $\{e^{-\varepsilon|z|}S_{1/2,1}(z) : z \in \Sigma_{\delta-\varepsilon}\}$ is equicontinuous and that the representation formula

$$S_{1/2,1}(z)x = \frac{1}{2\pi i}\int_{\Gamma_\omega} e^{\lambda z}\lambda^{-1}(\lambda + A_{1/2})^{-1}Cx\, d\lambda,\ x \in E,\ z \in \Sigma_{\delta-\varepsilon}$$

holds, where Γ_ω is oriented counterclockwise and consists of $\Gamma_\pm := \{\omega+re^{i((\pi/2)+\delta)} : r \geqslant |z|^{-1}\}$ and $\Gamma_0 := \{\omega+|z|^{-1}e^{i\theta} : |\theta| \leqslant (\pi/2) + \delta\}$. Applying the dominated convergence theorem, one gets that:

$$S_{1/2}(z)x = \frac{1}{2\pi i}\int_{\Gamma_\omega} e^{\lambda z}(\lambda + A_{1/2})^{-1}Cx\, d\lambda,\ x \in E,\ z \in \Sigma_{\delta-\varepsilon}.$$

Moreover, there exists an absolute constant $c > 0$ that, for every $\lambda = \omega+|z|^{-1}e^{i\theta} \in \Gamma_0$, one has $|\lambda| \geqslant c(\omega+|z|^{-1})$. By the proof of [20, Theorem 2.6.1], we obtain that for each $q \in ⊛$ there exist $c_{q,\varepsilon} > 0$ and $r_q \in ⊛$ such that, for every $x \in E$,

$$q(S_{1/2}(z)x) \leqslant c_{q,\varepsilon} r_q(x)e^{\omega \operatorname{Re} z}(1 + |z|^{-2(m+1)}),\ z \in \Sigma_{\delta-\varepsilon}.$$

Letting $\omega \to 0+$, we obtain that the operator family $\{(1 + |z|^{-2(m+1)})^{-1}S_{1/2}(z) : z \in \Sigma_{\delta-\varepsilon}\}$ is equicontinuous. Then the proofs of (i.2) and (i.3) can be completed without any substantial difficulty. Notice also that the Cauchy integral formula implies that, for every $j \in \mathbb{N}_0$, the operator family $\{|z|^j(1 + |z|^{-2(m+1)})^{-1}S_{1/2}(z) : z \in \Sigma_{\delta-\varepsilon}\}$ is equicontinuous. It can be simply proved that $(-A_{1/2}) \int_0^z S_{1/2,1}(s)x\, ds = S_{1/2,1}(z)x - zCx$, $x \in E$, $z \in \Sigma_{(\pi/2)-(\omega/2)}$, which implies $(-A_{1/2})S_{1/2,1}(z)x = S_{1/2}(z)x - Cx$, $S'_{1/2}(z)x = (-A_{1/2})S_{1/2}(z)x$ ($x \in E$, $z \in \Sigma_{(\pi/2)-(\omega/2)}$) and $D(A_{1/2}) \subseteq \Omega_{1/2}$. Using the proof of Theorem 2.9.40, we get that $S_{1/2}^{(2n)}(z)x = A^n S_{1/2}(z)x$ for all $x \in E$ and $z \in \Sigma_{(\pi/2)-(\omega/2)}$. Hence, $R(S_{1/2}(z)) \subseteq D_\infty(A)$, $z \in \Sigma_{(\pi/2)-(\omega/2)}$, and for every $x \in \Omega_{1/2}$, the function $u(z) = S_{1/2}(z)x$, $z \in \Sigma_{(\pi/2)-(\omega/2)}$ is a solution of the problem $(P_{2,m})$ with the required growth rate of derivatives. The uniqueness of solution of $(P_{2,m})$ can be proved as in Theorem 2.9.40. Let $x \in D(A)$. Then, for every $\lambda > 0$, we have $(\lambda + A)^{-1}Cx = \lambda^{-1}(Cx - (\lambda + A)^{-1}CAx)$. Keeping in mind this equality, (333) and the computation used in proving (334), we get that $\lim_{\lambda\to+\infty} \lambda^{-(2m+1)}(\lambda + A_{1/2})^{-1} Cx = 0$. Applying Theorem 1.2.5, we easily infer that, for every $\delta \in (0, (\pi/2) - (\omega/2))$, we have $\lim_{z\to 0, z\in\Sigma_\delta} S_{1/2,2m+2}(z)x = 0$. This simply implies $\lim_{z\to 0, z\in\Sigma_\delta} S_{1/2,2m+2}(z)y = 0$ for every $\delta \in (0, (\pi/2) - (\omega/2))$ and $y \in \overline{D(A)}$, which completes the proof of (i.3). Suppose now that $-1 > m > (-3)/2$ and A satisfies the condition (HQ). Then it can be simply checked that, for every $\varepsilon > 0$, the operator $A + \varepsilon$ satisfies the condition (H) with the same ω. Assume $0 < \gamma < (\pi/2) - (\omega/2)$, $0 < \varepsilon' < \gamma$, $\theta \in (\omega - \pi, -2\gamma)$. Let $\omega' > 0$ be arbitrarily chosen. Then it is not difficult to see that, for every $z \in \Sigma_{\gamma-\varepsilon'}$ and $\varepsilon > 0$,

$$S^{\varepsilon}_{1/2}(z)x = \frac{e^{i\theta/2}}{2\pi^2 i}\int_{\Gamma_{1,\omega'}} e^{\lambda z}\int_0^{\infty}\frac{\upsilon^{1/2}}{\lambda^2 e^{i\theta}+\upsilon}(\upsilon+e^{i\theta}(A+\varepsilon))^{-1}Cx\,d\upsilon\,d\lambda$$

(335) $$+\frac{e^{-i\theta/2}}{2\pi^2 i}\int_{\Gamma_{1,\omega'}} e^{\lambda z}\int_0^{\infty}\frac{\upsilon^{1/2}}{\lambda^2 e^{-i\theta}+\upsilon}(\upsilon+e^{-i\theta}(A+\varepsilon))^{-1}Cx\,d\upsilon\,d\lambda,$$

where $\Gamma_{1,\omega'} = \{\omega'+re^{i((\pi/2)+\gamma)} : r \geqslant |z|^{-1}\} \cup \{\omega'+|z|^{-1}e^{i\theta} : \theta \in [0, (\pi/2)+\gamma]\}$ and $\Gamma_{2,\omega'} = \{\omega'+re^{-i((\pi/2)+\gamma)} : r \geqslant |z|^{-1}\} \cup \{\omega'+|z|^{-1}e^{i\theta} : \theta \in [-(\pi/2)-\gamma, 0]\}$ are oriented counterclockwise. The dominated convergence theorem implies that, for every $x \in E$ and $z \in \Sigma_{\gamma-\varepsilon''}$,

$$\lim_{\varepsilon\to 0+} S^{\varepsilon}_{1/2}(z)x = \frac{e^{i\theta/2}}{2\pi^2 i}\int_{\Gamma_{1,\omega'}} e^{\lambda z}\int_0^{\infty}\frac{\upsilon^{1/2}}{\lambda^2 e^{i\theta}+\upsilon}(\upsilon+e^{i\theta}A)^{-1}Cx\,d\upsilon\,d\lambda$$

(336) $$+\frac{e^{-i\theta/2}}{2\pi^2 i}\int_{\Gamma_{1,\omega'}} e^{\lambda z}\int_0^{\infty}\frac{\upsilon^{1/2}}{\lambda^2 e^{-i\theta}+\upsilon}(\upsilon+e^{-i\theta}A)^{-1}Cx\,d\upsilon\,d\lambda.$$

This completes the proof of (ii.1). Furthermore, the convergence in (336) is uniform on compacts of $\Sigma_{\gamma-\varepsilon''}$ so that for each $x \in E$ the function $z \mapsto S_{1/2}(z)x$, $z \in \Sigma_{(\pi/2)-(\omega/2)}$ is analytic by Lemma 1.2.4. Similarly, as in the first part of the proof of this theorem, we have that for each $q \in ⊛$ there exist $c_{q,\varepsilon'} > 0$ and $r_q \in ⊛$ such that

$$q(S^{\varepsilon}_{1/2}(z)x) \leqslant c_{q,\varepsilon'}r_q(x)e^{\omega' \operatorname{Re} z}(1+|z|^{-2(m+1)}),\ x \in E,\ z \in \Sigma_{\gamma-\varepsilon'}.$$

Letting $\omega \to 0+$ and $\varepsilon \to 0+$, we get that for each $q \in ⊛$ there exist $c_{q,\varepsilon'} > 0$ and $r_q \in ⊛$ such that an estimate of the form (332) holds; moreover, the Cauchy integral formula implies that, for every $j \in \mathbb{N}_0$, the family $\{|z|^j(1+|z|^{-2(m+1)})^{-1}S^{(j)}_{1/2}(z) : z \in \Sigma_{\gamma-\varepsilon'}\}$ is equicontinuous. Fix, for a moment, an element $x \in \overline{D(A)}$. We will prove that $\lim_{z\to 0, z\in\Sigma_{\gamma-\varepsilon'}} S_{1/2}(z)x = Cx$, let $\varepsilon'' > 0$ be given in advance. Clearly, by Theorem 2.9.40, it suffices to show that for each $q \in ⊛$ there exists a sufficiently small $\varepsilon > 0$ such that, for every $z \in \Sigma_{\gamma-\varepsilon'}$ with $|z| \leqslant 1$, we have $q(S_{1/2}(z)x - S^{\varepsilon}_{1/2}(z)x) < \varepsilon''$. Notice that there exists $c > 0$ such that, for every $\varepsilon > 0$ and $\upsilon > 0$,

$$q\big((\upsilon e^{-i\theta}+A+\varepsilon)^{-1}Cx-(\upsilon e^{-i\theta}+A)^{-1}Cx\big) = q\left(\int_{\upsilon e^{-i\theta}}^{\upsilon e^{-i\theta}+\varepsilon}\frac{d}{d\lambda}(\lambda+A)^{-1}Cx\,d\lambda\right)$$

$$\leqslant \int_0^{\varepsilon}\frac{1}{|\upsilon e^{-i\theta}+t|}[|\upsilon e^{-i\theta}+t|^{-1}+|\upsilon e^{-i\theta}+t|^{m}]\,dt$$

$$\leqslant c\int_0^{\varepsilon}\frac{1}{\upsilon+t}[(\upsilon+t)^{-1}+(\upsilon+t)^{m}]\,dt = c\left[\frac{\varepsilon}{\upsilon(\upsilon+\varepsilon)}+(\upsilon^{m}-(\upsilon+\varepsilon)^{m})\right].$$

(337)

Then we derive from (337) that:

$$q(S_{1/2}(z)x - S^{\varepsilon}_{1/2}(z)x) \leqslant c \int_{\Gamma_{1,\omega'}\cup\Gamma_{2,\omega'}} \frac{e^{\mathrm{Re}(\lambda z)}}{|\lambda-\omega'|}$$

$$\times \int_0^{\infty} \frac{|\lambda|+\omega'}{|\lambda|^2+\upsilon} \upsilon^{1/2} [(\upsilon^{-1} - (\upsilon+\varepsilon)^{-1}) + (\upsilon^m - (\upsilon+\varepsilon)^m)]\, d\upsilon\, |d\lambda|.$$

It is obvious that there exists $c' > 0$ such that, for every $\lambda \in \Gamma_{1,\omega'} \cup \Gamma_{2,\omega'}$, we have: $(|\lambda| + \omega')/(|\lambda|^2 + \upsilon) \leqslant c'/(1 + \upsilon)$, $\upsilon \in (0, 1)$, and $|\lambda| + \omega' \leqslant 2c'|\lambda|$. Therefore,

$$q(S_{1/2}(z)x - S^{\varepsilon}_{1/2}(z)x)$$

$$\leqslant c \int_{\Gamma_{1,\omega'}\cup\Gamma_{2,\omega'}} \frac{e^{\mathrm{Re}(\lambda z)}}{|\lambda-\omega'|} \int_0^1 \frac{c'\upsilon^{1/2}}{1+\upsilon} [(\upsilon^{-1} - (\upsilon+\varepsilon)^{-1}) + (\upsilon^m - (\upsilon+\varepsilon)^m)]\, d\upsilon\, |d\lambda|$$

$$+ c \int_{\Gamma_{1,\omega'}\cup\Gamma_{2,\omega'}} \frac{e^{\mathrm{Re}(\lambda z)}}{|\lambda-\omega'|} \int_1^{\infty} \frac{|\lambda|+\omega'}{2|\lambda|\sqrt{\upsilon}} \upsilon^{1/2} [(\upsilon^{-1} - (\upsilon+\varepsilon)^{-1}) + (\upsilon^m - (\upsilon+\varepsilon)^m)]\, d\upsilon\, |d\lambda|$$

$$\leqslant cc' \int_0^1 \frac{c'\upsilon^{1/2}}{1+\upsilon} [(\upsilon^{-1} - (\upsilon+\varepsilon)^{-1}) + (\upsilon^m - (\upsilon+\varepsilon)^m)]\, d\upsilon \times \int_{\Gamma_{1,\omega'}\cup\Gamma_{2,\omega'}} \frac{e^{\mathrm{Re}(\lambda z)}|d\lambda|}{|\lambda-\omega'|}$$

$$+ cc' \int_1^{\infty} [(\upsilon^{-1} - (\upsilon+\varepsilon)^{-1}) + (\upsilon^m - (\upsilon+\varepsilon)^m)]\, d\upsilon \times \int_{\Gamma_{1,\omega'}\cup\Gamma_{2,\omega'}} e^{\mathrm{Re}(\lambda z)} \frac{|d\lambda|}{|\lambda-\omega'|}$$

By the dominated convergence theorem, we reveal that:

$$\lim_{\varepsilon\to 0+} \int_0^1 \frac{c'\upsilon^{1/2}}{1+\upsilon} [(\upsilon^{-1} - (\upsilon+\varepsilon)^{-1}) + (\upsilon^m - (\upsilon+\varepsilon)^m)]\, d\upsilon = 0.$$

On the other hand, direct computation shows that:

$$\int_1^{\infty} [(\upsilon^{-1} - (\upsilon+\varepsilon)^{-1}) + (\upsilon^m - (\upsilon+\varepsilon)^m)]\, d\upsilon = \varepsilon + |(m+1)^{-1}|(1 - (1+\varepsilon)^{m+1}).$$

Since $\int_{\Gamma_{1,\omega'}\cup\Gamma_{2,\omega'}} e^{\mathrm{Re}(\lambda z)}|d\lambda|/|\lambda - \omega'| < \infty$, we finally get that $\lim_{z\to 0, z\in\Sigma_{\gamma-\varepsilon'}} S_{1/2}(z)x = Cx$, as claimed. The semigroup property of $(S_{1/2}(t))_{t\geqslant 0}$ follows from (336), the remaining part of proofs of (ii.2) and (ii.3) are simple and therefore omitted. Clearly, the equality $(d^2/dz^2)S^{\varepsilon}_{1/2}(z)x = (A+\varepsilon)S^{\varepsilon}_{1/2}(z)x$, $z \in \Sigma_{\gamma-\varepsilon''}$, $x \in E$, implies along with the closedness of the operator A that $\lim_{\varepsilon\to 0}(d^2/dz^2)S^{\varepsilon}_{1/2}(z)x = S''_{1/2}(z)x = AS_{1/2}(z)x$, $z \in \Sigma_{\gamma-\varepsilon''}$, $x \in E$. Recommencing this procedure, we obtain by induction that $(d^{2n}/dz^{2n}) S_{1/2}(z)x = A^n S_{1/2}(z)x$, $z \in \Sigma_{\gamma-\varepsilon''}$, $x \in E$. The proof of (iii.2) now becomes standard and here we will only outline the main details of proof of uniqueness of solutions to

$(P_{2,m})$. It can be simply proved that there exists $l \in \mathbb{N}$ such that the equality ${}_{1/2}A\,{}_{1/2}AC_2 = AC_2 \in L(E)$ holds with $C_2 = C^l((1 + {}_{1/2}A)^{-1}C)^l\,((1 + A)^{-1}C)^l$. Arguing as in the proof of Theorem 2.9.40, with the operators C_1 and $A_{1/2}$ replaced respectively by C_2 and ${}_{1/2}A$, we obtain that the function $f(s) = -[C_2(1 + {}_{1/2}A_{1/2})^{-1}C_2\upsilon'(s) + C_2\,{}_{1/2}A\,(1 + {}_{1/2}A)^{-1}C_2\upsilon(s)]$, $s > 0$ satisfies $Cf(r) = S_{1/2}(t - r)\,f(t)$ for $0 < r < t$. This implies $AC\,f(r) = S''_{1/2}(t - r)\,f(t) \to 0$ as $t \to \infty$. Since $f''(r) = ({}_{1/2}A)^2 f(r)$, $r > 0$, the above implies $f''(r) = 0$, $r > 0$ and the existence of elements $x, y \in E$ such that $f(r) = rx + y$, $r > 0$. The obvious equality $f'(r) = {}_{1/2}Af(r)$, $r > 0$ implies $x = {}_{1/2}A(rx + y)$, $r > 0$ as well as ${}_{1/2}Ay = x$ and ${}_{1/2}Ax = 0$. Then we obtain from the equality $(-{}_{1/2}A)\int_0^r S_{1/2}(s)x\,ds = S_{1/2}(r)x - Cx$, $r > 0$ that $S_{1/2}(r)x = Cx$, $r > 0$. Since $C\,f(r) = S_{1/2}(r)\,f(2r)$, $r > 0$, we further get that $C(rx + y) = S_{1/2}(r)(2rx + y) = 2rCx + S_{1/2}(r)y$, $r > 0$, i.e., $rCx = S_{1/2}(r)y - Cy$, $r > 0$. Using the growth rate of $(S_{1/2}(r))_{r>0}$, it follows that $Cx = x = 0$ and that the function $f(\cdot)$ is constant, which is a contradiction. It remains to be proved that, in the case $(-1)/2 > m > -1$, we have $D(A) \subseteq \Omega_{1/2}$. Let $x \in D(A)$ and $\omega' > 0$ be fixed. Since $(S_{1/2}(z) \equiv C)_{z\in\Sigma_{\pi/2}}$ is the analytic C-regularized semigroup with the integral generator $0_{1/2} = 0 \in L(E)$, the arguments used in the proof of (ii) show that:

$$S_{1/2}(z)x - Cx$$

$$= \frac{e^{i\theta/2}}{2\pi^2 i}\int_{\Gamma_{1,\omega'}} \frac{e^{\lambda z}}{(\lambda-\omega')^2}\int_0^\infty \frac{(\lambda-\omega')^2\upsilon^{1/2}}{\lambda^2 e^{i\theta}+\upsilon}\left[(\upsilon+e^{i\theta}A)^{-1}Cx - \frac{Cx}{\upsilon}\right]d\upsilon\,d\lambda \tag{338}$$

$$+\frac{e^{-i\theta/2}}{2\pi^2 i}\int_{\Gamma_{1,\omega'}} \frac{e^{\lambda z}}{(\lambda-\omega')^2}\int_0^\infty \frac{(\lambda-\omega')^2\upsilon^{1/2}}{\lambda^2 e^{-i\theta}+\upsilon}\left[(\upsilon+e^{-i\theta}A)^{-1}Cx - \frac{Cx}{\upsilon}\right]d\upsilon\,d\lambda.$$

Denote by $\Delta_{\omega'}$ the region on the right of $\Gamma_{\omega'} := \Gamma_{1,\omega'} \cup \Gamma_{2,\omega'}$. Notice that there exists $d > 0$ such that, for every $q \in ⊛$, there exist $c'_q > 0$ and $r'_q \in ⊛$ such that, for every $\lambda \in \Gamma_{1,\omega''}$

$$\int_0^\infty \left|\frac{(\lambda-\omega')^2\upsilon^{1/2}}{\lambda^2 e^{i\theta}+\upsilon}\right| q\left((\upsilon+e^{i\theta}A)^{-1}Cx - \frac{Cx}{\upsilon}\right)d\upsilon$$

$$\leqslant d_1c'_qr'_q(x)\int_0^1 \frac{|\lambda-\omega'|^2\upsilon^{-(1/2)}}{|\lambda|^2+\upsilon}d\upsilon + d_1\int_1^\infty \frac{|\lambda-\omega'|^2\upsilon^{1/2}}{|\lambda|^2+\upsilon}q((\upsilon+e^{i\theta}A)^{-1}CAx)\,d\upsilon$$

$$\leqslant 2d_1c'_qr'_q(x)\,|\lambda-\omega'|^2|\lambda|^{-2} + d_1c'_qr'_q(Ax)\,|\lambda-\omega'|^2|\lambda|^{-2}\int_1^\infty \upsilon^{m-(3/2)}\,d\upsilon.$$

Applying the same trick to the integral appearing in the second addend of term on the right hand side of (338), we obtain that there exists $d_2 > 0$ such that, for every $q \in ⊛$, there exist $c''_q > 0$ and $r''_q \in ⊛$ such that, for every $\lambda \in \Gamma_{2,\omega''}$

$$\int_0^\infty \left|\frac{(\lambda-\omega')^2 \upsilon^{1/2}}{\lambda^2 e^{-i\theta}+\upsilon}\right| q\left((\upsilon+e^{-i\theta}A)^{-1}Cx-\frac{Cx}{\upsilon}\right) d\upsilon \leqslant d_2 c_q''[r_q''(x)+r_q''(Ax)].$$

Define $F : \Gamma_{1,\omega'} \cup \Delta_{\omega'} \to E$ by

$$F(\lambda) := e^{i\theta/2} \int_0^\infty \frac{(\lambda-\omega')^2 \upsilon^{1/2}}{\lambda^2 e^{i\theta}+\upsilon}\left[(\upsilon+e^{i\theta}A)^{-1}Cx-\frac{Cx}{\upsilon}\right] d\upsilon$$

for $\lambda \in D(F) \cap \{z \in \mathbb{C} : \operatorname{Im} z \geqslant 0\}$, and

$$F(\lambda) := e^{-i\theta/2} \int_0^\infty \frac{(\lambda-\omega')^2 \upsilon^{1/2}}{\lambda^2 e^{-i\theta}+\upsilon}\left[(\upsilon+e^{-i\theta}A)^{-1}Cx-\frac{Cx}{\upsilon}\right] d\upsilon$$

for $\lambda \in D(F) \cap \{z \in \mathbb{C} : \operatorname{Im} z \leqslant 0\}$. We have proved that, for every $q \in ⊛$, there exist $c_q > 0$ and $r_q \in ⊛$ such that:

(339) $$q(F(\lambda)) \leqslant c_q[r_q(x)+r_q(Ax)],\ \lambda \in \Gamma_{1,\omega'} \cup \Delta_{\omega'}.$$

Obviously, the function $\lambda \mapsto F(\lambda)$, $\lambda \in \Gamma_{1,\omega'} \cup \Delta_{\omega'}$ can be analytically extended to an open connected subset containing its domain. Combining the dominated convergence theorem with (339), it follows that:

$$\lim_{z\to 0, z\in\Sigma_{\gamma-\varepsilon'}} [S_{1/2}(z)x - Cx] = \frac{1}{2\pi^2 i}\int_{\Gamma_{\omega'}} \frac{F(\lambda)}{(\lambda-\omega')^2}\, d\lambda.$$

The last integral equals zero by the residue theorem. □

Remark 2.9.52. If the assumptions quoted in the formulation of (i) hold, then the operator $-A_{1/2}$ is the integral generator of the analytic C-regularized semigroup $(S_{1/2}(z))_{z\in\Sigma_{(\pi/2)-(\omega/2)}}$ of growth order $2m + 2$; observe also that we have slightly improved the angle of analyticity of semigroup $(T_{1/2}(t))_{t>0}$ (i.e. $(S_{1/2}(t))_{t>0}$ in our notation) appearing in the formulation of [90, Theorem 3.7]. The assertion (iii) continues to hold, with appropriate modifications, if one considers the problem $(P_{2,m})$ for functions defined on $(0, \infty)$ (cf. Theorem 2.9.40, then the set $\overline{D(A)} \cup \overline{D(A_{1/2})}$ is contained in the angular set of continuity of $(S_{1/2}(z))_{z\in\Sigma_{(\pi/2)-(\omega/2)}}$).

Example 2.9.53. ([92]) Suppose $n \in \mathbb{N}\setminus\{1\}$, $c_1, \cdots, c_{n-1} \in \mathbb{C}$, $0 \leqslant \alpha_1 < \cdots < \alpha_n \leqslant 2$, $0 < \beta < 1$, $1 < \gamma \leqslant 2$, $k_\beta, k_\gamma > 0$ and $L > 0$. The following scalar multi-term time-space Caputo-Riesz fractional advection diffusion equation, (MT-TSCR-FADE) for short,

(340)

$$\mathbf{D}_t^{\alpha_n} u(t,x) + c_{n-1}\mathbf{D}_t^{\alpha_{n-1}} u(t,x) + \cdots + c_1 \mathbf{D}_t^{\alpha_1} u(t,x) = k_\beta \frac{\partial^\beta u(t,x)}{\partial |x|^\beta} + k_\gamma \frac{\partial^\gamma u(t,x)}{\partial |x|^\gamma}$$

$$\left(\frac{\partial^k}{\partial t^k} u(t,x)\right)_{t=0} = u_k(x),\ k = 0, \cdots, \lceil \alpha_n \rceil - 1,\ 0 \leqslant x \leqslant L,$$

where $\frac{\partial^{\beta}u(t,x)}{\partial|x|^{\beta}}$ denotes the Riesz fractional operator of order β, has been recently analyzed by H. Jiang, F. Liu, I. Turner and K. Burrage in [238]. In the case of Dirichlet boundary conditons, [238, Lemma 1] yields that the equality $\frac{\partial^{\beta}u(t,x)}{\partial|x|^{\beta}} = -(-\Delta)^{\beta/2}$ holds in a certain sense; furthermore, it should be noted that the existing literature is somewhat controversial about the validity of the above formula (cf. for example [27, p. 190, l. 23]). In what follows, we would like to present two different evolution models of problem (340). In the first one, we rewrite (340) in the form of the following multi-term fractional differential equation:

$$\text{(341)}\qquad \begin{aligned} &\mathbf{D}_t^{\alpha_n}u(t) + c_{n-1}\,\mathbf{D}_t^{\alpha_{n-1}}u(t) + \cdots + c_1\,\mathbf{D}_t^{\alpha_1}u(t) = -k_\beta A_{\beta/2} - k_\gamma A_{\gamma/2}\\ &u^{(k)}(0) = u_k,\ k = 0, \cdots, \lceil \alpha_n \rceil - 1, \end{aligned}$$

where the operator A belongs to the class $\mathcal{M}_{C,m}$ for some $m \in \mathbb{R}$, and acts on an appropriately chosen space of functions defined on $[0, L]$. It is clear that the analysis of problem (341) is very difficult in general case, and here we would like to inscribe, without giving full details, some very special results on the existence and uniqueness of analytical solutions to (341); cf. also [27] and [215]. We employ the standard hypotheses from the theory of sectorial operators: $(E, \|\cdot\|)$ denotes a Banach space and A denotes a sectorial operator of angle $\omega \in [0, \pi)$, with $D(A)$ and $R(A)$ not necessarily dense in E. Let the so-called parabolicity condition $2\pi > (\beta + \gamma)\omega$ hold, and let $\alpha_n^{-1}(\pi - (\gamma\omega/2)) - (\pi/2) > 0$. Then it is not difficult to prove with the help of Da Prato-Grisvard theorem (see e.g. [215, Theorem 9.3.1, Corollary 9.3.2]) that the operator $k_\beta A_{\beta/2} + k_\gamma A_{\gamma/2}$ is sectorial of angle $\gamma\pi/2$. By [318, Theorem 2.12], we easily infer that for each $\sigma > 0$ the operator $-k_\beta A_{\beta/2} - k_\gamma A_{\gamma/2}$ is the integral generator of an analytic $g_{\sigma+1}$-regularized resolvent propagation family $((R_0(t))_{t\geq 0}, \cdots, ((R_{\lceil \alpha_n \rceil -1}(t))_{t\geq 0})$ for (341) of angle $\delta \equiv \min(\pi/2, \alpha_n^{-1}(\pi - (\gamma\omega/2)) - (\pi/2))$ and of subexponential growth (cf. [318, Definition 2.2(iii)] and the next section for the notion used henceforth); furthermore, A is the integral generator of a bounded analytic resolvent propagation family of angle δ, provided that A is densely defined. These facts simply imply the existence and uniqueness of mild analytical solutions of the corresponding integral equations associated with (341). In the second approach, we consider a sequence $(M_p)_{p\in\mathbb{N}_0}$ of positive real numbers satisfying $M_0 = 1$, (M.1), (M.2) and (M.3)'. Put

$$E := \left\{ f \in C^{\infty}[0,1]\ ;\ \|f\| := \sup_{p\in\mathbb{N}_0} \frac{\left\|f^{(p)}\right\|_{\infty}}{M_p} < \infty \right\},$$

$A := -d/ds$, $D(A) := \{f \in E : f' \in E, f(0) = 0\}$ and $E^{(M_p)}(A) := \{f \in D_{\infty}(A) : \sup_{p\in\mathbb{N}_0}(h^p\|f^{(p)}\|_{\infty})/M_p < \infty$ for all $h > 0\}$. As mentioned before, A generates a non-dense ultradistribution semigroup of (M_p)-class on E, $\rho(A) = \mathbb{C}$ and there exists an injective operator $C \in L(E)$ such that $E^{(M_p)}(A) \subseteq C(D_{\infty}(A))$ and A generates a bounded C-regularized semigroup $(S(t))_{t\geq 0}$ on E (cf. [90, Example 3.3] and [292]). Albeit the final conclusion on the existence and uniqueness of analytical solutions of problem [90, (53)], considered in the above-mentioned example, is completely true, it ought to be observed that we have made several serious mistakes in the corresponding analysis. First of all, we would like to stress that the formula [90, (52)] holds, with the exception of integer values of parameter b, only for those functions $f \in E$ for which $f^{(i)}(0) = 0$, $i \in \mathbb{N}_0$; if this is not the case, then the function $g_b * f(\cdot)$ cannot be an element of E, provided $b \in (0, \infty)\setminus\mathbb{N}$. Furthermore, it is not clear whether the corresponding powers of $-A$ possess the following property: $(f, g) \in (-A)_b$ iff $f(t) = \int_0^t g_b(t-s)g(s)\,ds$, $t \in [0, 1]$ $(b > 0)$. Speaking matter-of-factly, it is not difficult to prove that $(f, g) \in (-A)_b$ iff $C^2 f(t) = \int_0^t g_b(t-s)C^2 g(s)\,ds$, $t \in [0, 1]$ $(b > 0)$, and that a pair $(f, g) \in D_\infty(A) \times D_\infty(A)$ belongs to $(-A)_b$ iff $f(t) = \int_0^t g_b(t-s)g(s)\,ds$, $t \in [0, 1]$ $(b > 0)$. In what follows, we analyze the C-wellposedness of the backwards equation (MT-TSCR-FADE):

$$\mathbf{D}_t^{\alpha_n} u(t) + c_{n-1}\,\mathbf{D}_t^{\alpha_{n-1}} u(t) + \cdots + c_1\,\mathbf{D}_t^{\alpha_1} u(t) = -k_\beta A_\beta - k_\gamma A_\gamma;\ u(0) = u_0, \tag{342}$$

where $0 < \beta < 1$, $1 < \gamma < 2$, $k_\beta, k_\gamma > 0$, $0 \leqslant \alpha_1 < \cdots < \alpha_n < 1$ and $\alpha_n^{-1}(\pi - (\gamma\pi/2)) > \pi/2$. The parabolicity condition $(\beta\pi)/2 + (\gamma\pi)/2 < \pi$, i.e., $\beta + \gamma < 2$ will be used again. Clearly, the topology induced by the following family $(\|\cdot\|_n \equiv \|\cdot\| + \cdots + \|A^n \cdot\|)_{n\in\mathbb{N}}$ of seminorms turns $D_\infty(A)$ into a Fréchet space. Our intention is to prove that the operator $\mathcal{A} \equiv k_\beta A_{\infty,\beta} + k_\gamma A_{\infty,\gamma}$ is C_∞-sectorial of angle $\gamma\pi/2$ (in $D_\infty(A)$). Towards this end, fix an element $f \in D_\infty(A)$ and observe that, for every $\lambda \in \mathbb{C}$, the equation $(\lambda + \mathcal{A})^{-1} C_\infty f = g$ is equivalent with

$$\lambda(g_\gamma * g)(t) + k_\beta(g_{\gamma-\beta} * g)(t) + k_\gamma g(t) = (Cf * g_\gamma)(t),\ t \in [0, 1],$$

and because of $g(0) = g'(0) = 0$, with equation

$$\mathbf{D}_t^\gamma g(t) + \frac{k_\beta}{k_\gamma}\mathbf{D}_t^\beta g(t) + \frac{\lambda}{k_\gamma} g(t) = \frac{Cf(t)}{k_\gamma},\ t \in [0, 1]. \tag{343}$$

By [395, Theorem 4.1], it follows that the unique solution of (343) is given by the following formula:

$$g(t) = \int_0^t x^{\gamma-1} E_{(\cdot),\gamma}(x) C f(t-x)\,dx,\ t \in [0, 1], \tag{344}$$

where $E_{(\cdot),\gamma}(x) := E_{(\gamma-\beta,\gamma),\gamma}(-(k_\beta/k_\gamma)x^{\gamma-\beta}, -(\lambda/k_\gamma)x^\gamma)$ and the two dimensional Mittag-Leffler function $E_{(\gamma-\beta,\gamma),\gamma}(z_1, z_2)$ is defined by

$$E_{(\gamma-\beta,\gamma),\gamma}(z_1, z_2) := \sum_{k=0} \sum_{l_1+l_2=k;\, l_1, l_2 \geqslant 0} \frac{k!}{l_1!\, l_2!} \frac{z_1^{l_1}}{\Gamma(l_1(\gamma-\beta)+\gamma)} \frac{z_2^{l_2}}{\Gamma(l_2\gamma+\gamma)}, \tag{345}$$

for any $z_1, z_2 \in \mathbb{C}$ ([395]). Then we obtain from (344) that

$$g^{(p)}(t) = \int_0^t x^{\gamma-1} E_{(\cdot),\gamma}(x)(Cf)^{(p)}(t-x)\,dx,\ t \in [0, 1],\ p \in \mathbb{N}_0. \tag{346}$$

By (345) and an elementary argumentation, we get that there exists a locally bounded function $F : \mathbb{C} \rightarrow [0, \infty)$ such that $\rho_{C_\infty}(\mathcal{A}) = \mathbb{C}$ and that, for every $n \in \mathbb{N}$,

$$\|(\lambda + \mathcal{A})^{-1}Ch\|_n \leqslant F(|\lambda|)\|h\|_n, \; \lambda \in \mathbb{C}, \, h \in D_\infty(A).$$

Our goal is to prove that, for every $\zeta \in (0, \pi - (\gamma\pi)/2)$, there exist $c_\zeta > 0$ and $R_\zeta > 0$ such that, for every $\lambda \in \Sigma_\zeta$ with $|\lambda| \geqslant R_\zeta$, the following holds:

$$(347) \qquad \|(\lambda + \mathcal{A})^{-1}Ch\|_n \leqslant c_\zeta|\lambda|^{-1}\|h\|_n, n \in \mathbb{N}, \, h \in D_\infty(A).$$

By (346), the integral representation [395, (47)] of function $E_{(\cdot),\gamma}(x)$ and the Fubini theorem, we obtain that there exists $\tau > 0$ such that, for every $\lambda \in \mathbb{C}$ with $|\lambda| \geqslant 1$, one has:

$$g^{(p)}(t) = \frac{1}{2\pi i}\int_{Ha(\varsigma+)}\int_0^t \frac{e^{sx}}{k_\gamma s^\gamma + k_\beta s^\beta + \lambda}(Cf)^{(p)}(t-x)\,dx\,ds, \quad t \in [0, 1], p \in \mathbb{N}_0,$$

where $\varsigma = \tau|\lambda|^{1/(\gamma-\beta)}$ and the Hankel path $Ha(\varsigma+)$ starts from $-\infty$ along the lower side of the negative real axis, encircles the circular disc $|\eta| = \varsigma$, and ends at $-\infty$ at the upper side of the negative real axis. The condition $\beta + \gamma < 2$ implies the existence of a sufficiently small number $\varepsilon > 0$ such that $k_\gamma s^\gamma + k_\beta s^\gamma + \lambda \neq 0$ for all $\lambda \in \Sigma_{\pi-(\gamma\pi)/2}$ and $s \in \Sigma_{(\pi/2)+\varepsilon}$. Taken together with the Cauchy theorem, this fact implies that one can replace the path of integration $Ha(\varsigma+)$ with the path Γ_+, obtained from $Ha(\varsigma+)$ by replacing the contour $\{\varsigma e^{i\theta} : -((\pi/2)+\varepsilon) \leqslant \theta \leqslant ((\pi/2)+\varepsilon)\}$ with union $\Gamma_{+,1} \cup \Gamma_{+,2} \cup \Gamma_{+,3}$, where $\Gamma_{+,1} := \{ve^{-i((\pi/2)+\varepsilon)} : 1 \leqslant v \leqslant \varsigma\}$, $\Gamma_{+,2} := \{e^{i\theta} : -((\pi/2)+\varepsilon) \leqslant \theta \leqslant ((\pi/2) + \varepsilon)\}$ and $\Gamma_{+,3} := \{ve^{i((\pi/2)+\varepsilon)} : 1 \leqslant v \leqslant \varsigma\}$. Carrying out a straightforward integral computation, the above shows that the estimate (347) holds. By the foregoing arguments, we also have that $A_{\infty,b} \in L(D_\infty(A))$ and $\mathcal{A} \in L(D_\infty(A))$ $(b > 0)$. Applying again [318, Theorem 2.12], we get that the operator $\mathcal{A}$ is the integral generator of an analytic equicontinuous regularized C_∞-resolvent propagation family $(R_0(t))_{t\geqslant 0}$ for (342) of angle $\delta \equiv \min(\pi/2, \alpha_n^{-1}(\pi - (\gamma\pi/2)) - (\pi/2))$. Hence, for every $u_0 \in E^{(M_p)}(A)$, there exists a unique solution $u(t)$ of the associated integral equation

$$[u(\cdot) - u_0] + \sum_{j=1}^{n-1} c_j g_{\alpha_n - \alpha_j} * [u(\cdot) - u_0] = g_{\alpha_n - \alpha} * \mathcal{A}\,[u(\cdot) - u_0],$$

given by $u(t) = R_0(t)C^{-1}u_0$, $t \geqslant 0$; furthermore, this solution can be analytically extended to the sector Σ_δ.

2.9.6. Fractional powers of (*a*, *b*, *C*)-nonnegative operators and semigroups generated by them. We introduce the notion of an (a, b, C)-nonnegative operator as follows.

Definition 2.9.54. (i) Let $a, b \in \mathbb{R}$. Then it is said that a closed linear operator A on E is (a, b, C)-nonnegative (or, equivalently, that the operator A belongs to the class $\mathcal{M}_C(a, b)$) if $(-\infty, 0) \subseteq \rho_C(A)$ and the family

$$\{(\lambda^a + \lambda^b)^{-1}(\lambda + A)^{-1}C : \lambda > 0\}$$

is equicontinuous. Set $\mathcal{M}_C := \bigcup_{a,b\in\mathbb{R}} \mathcal{M}_C(a, b)$.

(ii) Let $0 \leqslant \omega < \pi$. Then it is said that a closed linear operator A on E is (a, b, C)-sectorial of angle ω (or, equivalently, that A belongs to the class $\mathcal{S}_{C,\omega}(a, b)$) if $\mathbb{C}\setminus\overline{\Sigma_\omega} \subseteq \rho_C(A)$ and the family

$$\{(|\lambda|^a + |\lambda|^b)^{-1}(\lambda - A)^{-1}C : \lambda \notin \Sigma_{\omega'}\}$$

is equicontinuous for every $\omega < \omega' < \pi$. Set $\mathcal{S}_{C,\omega} := \bigcup_{a,b\in\mathbb{R}} \mathcal{S}_{C,\omega}(a, b)$.

It is clear that a closed linear operator A belongs to the class $\mathcal{M}_{C,m}$ iff it is (a, b, C)-nonnegative with $a = -1$ and $b = m$ $(m \in \mathbb{R})$, and that the notions introduced in Definition 2.9.54 can be understood even in the case that the operator C is not injective.

In the following illustrative example, we shall present how one can construct (a, b, C)-sectorial operator on product spaces.

Example 2.9.55. Suppose E_i is an SCLCS and the abbreviation $\circledast_i$ denotes the fundamental system of seminorms which defines the topology of E_i, $i = 1, 2$. Let A_i be an (a_i, b_i, C_i)-nonnegative operator on E_i, resp. an (a_i, b_i, C_i)-sectorial operator of angle $\omega_i \in [0, \pi)$ on E_i; $a_i, b_i \in \mathbb{R}$, $i = 1, 2$. Suppose, further, that $B : D(A_2) \to R(C_1)$ is a linear mapping satisfying that for each $q \in \circledast_1$ there exist $c > 0$ and $r \in \circledast_2$ such that

$$q(C_1^{-1}By) \leqslant c[r(y) + r(A_2y)],\ y \in D(A_2).$$

Define $E := E_1 \times E_2$, $A(x, y) := (A_1x + By, A_2y)$, $(x, y) \in D(A) \equiv D(A_1) \times D(A_2)$, $C(x, y) := (C_1x, C_2y)$, $(x, y) \in D(C) \equiv E$, $S := \{a_1, b_1, a_2, b_2, a_1 + a_2, b_1 + b_2, a_1 + b_2, b_1 + a_2\}$ and $\omega := \max(\omega_1, \omega_2)$. Then (see the proof of [542, Theorem 2.16]) the operator A is $(\min(S), \max(S), C)$-nonnegative in E, resp. $(\min(S), \max(S), C)$-sectorial of angle ω in E.

It is very simple to prove the following assertions:

(i) If the operator A is injective and belongs to the class $\mathcal{M}_C(a, b)$ for some $a, b \in \mathbb{R}$, then the operator A^{-1} belongs to the class $\mathcal{M}_C(\min(-1, -\max(a, b) - 2), \max(-1, -\min(a, b) - 2))$.

(ii) If $-\infty < a_1 \leqslant a_2 \leqslant b \leqslant a_3 \leqslant a_4 < \infty$, then $\mathcal{M}_C(a_2, b) \subseteq \mathcal{M}_C(a_1, b)$ and $\mathcal{M}_C(a_3, b) \subseteq \mathcal{M}_C(a_4, b)$.

The following proposition can be proved by using the van Neumann expansion in SCLCSs.

Proposition 2.9.56. *Suppose a, $b > -1$ and there exists $c_A > 0$ such that the inequality*

$$p((\lambda + A)^{-1}Cx) \leqslant c_A(\lambda^a + \lambda^b)p(x),\ x \in E,\ p \in \circledast,\ \lambda > 0 \tag{348}$$

holds with $C = 1$. Then $\rho(A)$ contains an open neighborhood of 0. *Moreover, if (348) holds with a, $b > 0$, then $D(A) = \{0\}$.*

Before proceeding further, we would like to observe that the assertion of [404, Theorem 2.1] can be reconsidered for $(a, b, 1)$-nonnegative operators; this is not a question of crucial importance in our analysis and we shall skip details for the sake of brevity. Suppose now that the operator A is (a, b, C)-nonnegative for some a, $b \in \mathbb{R}$, our intention is to construct the complex powers of A. In order to do that, we first observe that there exist four possible cases:

(i) $a \geqslant -1$ and $b \geqslant -1$. Then we have constructed in Subsection 2.9.3 the power A_α for any exponent $\alpha \in \mathbb{C}_+$ as well as the power A_α for any exponent $\alpha \in \mathbb{C}$ provided, in addition, that the operator A is injective. Although the analysis given in Theorem 2.9.58 and Theorem 2.9.60 below can be done for (a, b, C)-nonnegative operators with indices a and b greater than or equal to -1, we shall omit the corresponding discussion from the practical point of view.

(ii) $a \geqslant -1$ and $b < -1$.

(iii) $a < -1$ and $b \geqslant -1$. Thanks to the symmetry of indices a and b in Definition 2.9.54, this case can be reduced to the previous one.

(iv) $a < -1$ and $b < -1$. This case is rather non-fascinating because the above inequalities simply imply that E must be trivial. Indeed,

$$\lambda(\lambda + A)^{-1}C(1 + A)^{-1}Cx = \frac{\lambda}{1-\lambda}[(\lambda + A)^{-1}C^2x - (1 + A)^{-1}C^2x],\ \lambda > 1,\ x \in E. \tag{349}$$

Letting $\lambda \to +\infty$ in (349), we get $(1+ A)^{-1}C^2x = 0$, $x \in E$, and therefore, $E = \{0\}$.

Keeping in mind that each complex power of an injective almost C-nonnegative operator A has been constructed in Subsection 2.9.3, the above consideration shows that we have essentially only two unsolved problems to consider in this subsection:

(1) Let $a \geqslant -1$, let $b < -1$, and let A be a non-injective (a, b, C)-nonnegative operator. Construct the power A_α for any $\alpha \in \mathbb{C}_+$.

(2) Let $a > -1$, let $b < -1$, and let A be an injective (a, b, C)-nonnegative operator. Construct the power A_α for any $\alpha \in \mathbb{C}$.

Now we shall explain how one can simply solve the problem (2), recall that $C_k = C((1 + A)^{-1}C)^k$, $k \in \mathbb{N}$. Because $C^{-1}AC = A$, we have that $C_k^{-1}AC_k = A$, $k \in \mathbb{N}$. Moreover, if $a \geqslant -1$ and $b < -1$, then for every $p \in \circledast$, there exist $c_p > 0$ and $q_p \in \circledast$ such that the following holds:

$$p\left((\lambda+A)^{-1}C\left((1+A)^{-1}C\right)^{k}x\right)$$

$$=p\left(\frac{(-1)^{k}}{(\lambda-1)^{k}}(\lambda+A)^{-1}C^{k+1}x+\sum_{i=1}^{k}\frac{(-1)^{k-i}\left((1+A)^{-1}C\right)^{i}C^{k+1-i}x}{(\lambda-1)^{k-1-i}}\right)$$

$$\leqslant c_{p}q_{p}(x)\left[\frac{\lambda^{a}+\lambda^{b}}{(\lambda-1)^{k}}+\frac{1}{\lambda-1}\right],\quad \lambda\geqslant 2,\ x\in E,$$

and

$$p((\lambda+A)^{-1}C((1+A)^{-1}C)^{k}x)\leqslant c_{p}q_{p}(x)(\lambda^{a}+\lambda^{b}),\ \lambda\in(0,2],\ x\in E.$$

This, in turn, implies that, for every $k \in \mathbb{N}$ with $k \geqslant \lceil a \rceil + 1$, the operator A belongs to the class $\mathcal{M}_{C_k,b}$. Therefore, the problem (2) has an almost trivial solution: the definition of power A_α ($\alpha \in \mathbb{C}$) can be simply obtained by using the method described in Subsection 2.9.3, with the operator C replaced by C_k for any $k \geqslant \lceil a \rceil + 1$. It is not difficult to prove that this definition does not depend on the particular choice of number k. Algebraic properties of powers clarified so far continue to hold and, among many other equalities stated in Subsection 2.9.3, one can prove that $C^{-1}A^kC = A_k$ ($k \in \mathbb{Z}$) and $C^{-1}A_\alpha C = A_\alpha$ ($\alpha \in \mathbb{C}$).

In what follows, we shall briefly describe a few unsuccessful attempts to solve the problem (1). First of all, notice that the previous computation shows that the operator A belongs to the class $\mathcal{M}_{C_k,b}$ provided that $a \geqslant -1$ and $b < -1$. Therefore, the problem (1) can be always reduced to the problem of construction of power A_α ($\alpha \in \mathbb{C}_+$) of a non-injective operator A belonging the class $\mathcal{M}_{C,m}$ for some $m < -1$.

Remark 2.9.57. Let $A \in \mathcal{M}_{C,m}$ be non-injective for some $m < -1$.

(i) Then the method described in Theorem 2.9.39 is applicable only in the case that $-1 > m > (-3)/2$, and gives the definition of power A_α for some very special values of exponent α.

(ii) It is not difficult to prove that the operator A is C'_k-nonnegative with C'_k being defined by $C'_k := C(A(1+A)^{-1}C)^k$ ($k \in \mathbb{N}$, $k \geqslant -1 - m$); albeit not used in the sequel, the above shows that, for any injective operator $A \in \mathcal{M}_C(a, b)$, one can find another regularizing injective operator C' such that $A \in \mathcal{M}_{C',-1}$. The operator C'_k is not injective, unfortunately, and the power A_α cannot be constructed as before. Consider, for example, the case $0 < \alpha < 1/2$. If so, we can define the degenerate C'_k-regularized semigroup $(S_\alpha(t))_{t\geqslant 0}$ by replacing the operator C with C'_k. Then the integral generator $\mathcal{A}$ of $(S_\alpha(t))_{t\geqslant 0}$, given by

$$\mathcal{A} := \left\{(x, y) \in E \times E : S_\alpha(t)x - C'_k x = \int_0^t S_\gamma(s)y\, ds \text{ for all } t > 0\right\},$$

is a multivalued linear operator in the sense of [412, Definition 1.6.1]. In the case that C'_k is injective, then $\mathcal{A} = -A_a$, so that the natural choice for $-A_\alpha$, in our concrete situation, is some of single-valued branches of $\mathcal{A}$. In the existing literature on degenerate C-regularized semigroups, we could not find corresponding results which would enable us to define the power A_α in a satisfactory way; cf. also [412, Proposition 1.6.4].

In the remainder of this subsection, we shall analyze the generation of various types of fractional resolvent families by the negatives of fractional powers of (a, b, C)-sectorial operators. We shall apply the obtained results in the study of certain classes of incomplete abstract Cauchy problems, in general with modified Liouville right-sided time-fractional derivatives. Recall that the function $H_n : \mathbb{C} \times (\mathbb{C}\backslash(-\infty, 0]) \rightarrow \mathbb{C}$, defined by $H_n(\omega, z) = (d^n/dz^n) \exp(-\omega z^\gamma)$, $\omega \in \mathbb{C}$, $z \in \mathbb{C}\backslash(-\infty, 0]$, is analytic in $\mathbb{C}\backslash(-\infty, 0]$ for every fixed ω, and entire in $\mathbb{C}$ for every fixed z.

Theorem 2.9.58. *Let $a > -1$, $b < -1$, $\omega \in [0, \pi)$ and $k \geqslant \lceil a \rceil + 1$. Suppose $0 < \gamma < 1/2$, the operator A is (a, b, C)-nonnegative and $b + \gamma > -1$. Put $P := \lfloor a + 2 \rfloor$, $P' := \lfloor a+\gamma+2 \rfloor$, $S_\gamma(0) := C$ and define, for every $t > 0$, the bounded linear operator $S_\gamma(t)$ by (273). Then there exists an operator family $(\mathbf{S}_\gamma(z))_{z\in\Sigma_{(\pi/2)-\gamma\pi}} \subseteq L(E)$ (denoted also by $S_\gamma(\cdot)$ in the sequel) such that, for every $x \in E$, the mapping $z \mapsto \mathbf{S}_\gamma(z)x$, $z \in \Sigma_{(\pi/2)-\gamma\pi}$ is analytic as well as that the family $\{(|z|^{-(a+1)/\gamma}+|z|^{-(b+1)/\gamma})^{-1}\mathbf{S}_\gamma(z) : z \in \Sigma_\delta\}$ is equicontinuous for all $\delta \in (0, (\pi/2) - \gamma\pi)$; if the operator A is (a,b,C)-sectorial of angle ω, then there exists an operator family $(\mathbf{S}_\gamma(z))_{z\in\Sigma_{(\pi/2)-\gamma\omega}} \subseteq L(E)$ such that the above holds with the number $(\pi/2) - \gamma\pi$ replaced by $(\pi/2) - \gamma\omega$. Furthermore, we have the following:*

(i) *$S_\gamma(z_1)S_\gamma(z_2) = S_\gamma(z_1 + z_2)C$ for all $z_1, z_2 \in \Sigma_{(\pi/2)-\gamma\pi}$.*
(ii) *One has $\lim_{z\rightarrow 0, z\in\Sigma_{(\pi/2)-\gamma\pi-\varepsilon}} S_\gamma(z)x = Cx$, $x \in D(A^P)$, $\varepsilon \in (0, (\pi/2) - \gamma\pi)$.*
(iii) *Suppose A is injective. Then $S_\gamma(z)A_\alpha \subseteq A_\alpha S_\gamma(z)$, $z \in \Sigma_{(\pi/2)-\gamma\pi}$, $\alpha \in \mathbb{C}_+$; if the operator A is (a, b, C)-sectorial of angle ω, then (i)-(iii) hold with the number $(\pi/2) - \gamma\pi$ replaced by $(\pi/2) - \omega\gamma$.*
(iv) (iv.1) *The operator family $(S_\gamma(z))_{z\in\Sigma_{(\pi/2)-\gamma\pi}}$ is an exponentially equicontinuous analytic C-regularized semigroup of growth order $(a + 1)/\gamma$. If, additionally, the operator A is (a, b, C)-sectorial of angle ω, then $(S_\gamma(t))_{t>0}$ can be extended to an exponentially equicontinuous analytic C-regularized semigroup $(S_\gamma(z))_{z\in\Sigma_{(\pi/2)-\gamma\omega}}$ of growth order $(a + 1)/\gamma$.*
(iv.2) *Suppose A is injective. Then the integral generator of $(S_\gamma(t))_{t\geqslant 0}$ is the operator $-A_\gamma$.*
If $0 < \zeta < 1/2$ and $b + \zeta > -1$, denote by $-A^\zeta$ the integral generator of $(S_\zeta(t))_{t\geqslant 0}$.
(v) *Let $z_0 \in \mathbb{C}_+$. Then, for every $x \in D(A^{P'})$ and $\varepsilon \in (0, (\pi/2) - \gamma\pi)$,*

(350)
$$\lim_{z\to 0, z\in\Sigma_{(\pi/2)-\gamma\pi-\varepsilon}} \frac{S_\gamma(z)x - Cx}{z} = -z_0^\gamma Cx + \sum_{k=2}^{P'} \frac{(-1)^{k-!}}{(k-1)!} H_{k-1}(0,z_0)(z_0 - A)^{k-1} Cx - \sin(\pi\gamma)\int_0^\infty \lambda^\gamma \frac{(\lambda + A)^{-1} C(z_0 - A)^{P'} x}{(\lambda + z_0)^{P'}} d\lambda.$$

If the operator A is (a, b, C)-sectorial of angle ω, then the formula (350) remains true for every $x \in D(A^{P'})$, *with the number* $(\pi/2) - \gamma\pi$ *replaced by* $(\pi/2) - \omega\gamma$.

(vi) *The operator A belongs to the class* $\mathcal{M}_{C_k,b}$; *the suppositions* $0 < \beta < 1/2$ *and* $\gamma\beta + b > -1$ *imply that* ${}_\beta(A^\gamma) = A^{\beta\gamma}$, *where the operator* ${}_\beta(A^\gamma) = {}_\beta({}_\gamma A)$ *is defined in the sense of Theorem 2.9.48(vi), with the operator C replaced by* C_k.

(vii) $R(S_\gamma(z)) \subseteq D_\infty(A)$, $A^n S_\gamma(z) \in L(E)$ *and (287) holds for all* $x \in E$ *and* $z \in \Sigma_{(\pi/2)-\gamma\pi}$.

(viii) *Let* $l \in \mathbb{N}\setminus\{1, 2\}$ *and* $b + (1/l) > -1$. *Denote by* Ω_l, *resp.* Ψ_l, *the set of continuity of* $(S_{1/l}(t))_{t>0}$, *resp.* $(S_{1/l}(z))_{z\in\Sigma(\pi/2)-(\pi/2l)}$. *Then, for every* $x \in \Omega_l$, *the incomplete abstract Cauchy problem*

$$(P_{l,b}) : \begin{cases} u \in C^\infty((0,\infty) : E) \cap C((0,\infty) : D_\infty(A)), \\ u^{(l)}(t) = (-1)^l Au(t),\ t > 0, \\ \lim_{t\to 0+} u(t) = Cx, \\ \text{the set } \{(1 + t^{-l(b+1)})^{-1}u(t) : t > 0\} \text{ is bounded in } E, \end{cases}$$

has a solution $u(t) = S_{1/l}(t)x$, $t > 0$, *which can be analytically extended to the sector* $\Sigma_{(\pi/2)-(\pi/2l)}$. *If, additionally,* $x \in \Psi_l$, *then for every* $\delta \in (0, (\pi/2) - (\pi/2l))$ *and* $j \in \mathbb{N}_0$, *we have that the set* $\{|z|^j(1+|z|^{-l(b+1)})^{-1}u^{(j)}(z) : z \in \Sigma_\delta\}$ *is bounded in E. Assuming that the operator A is (a, b, C)-sectorial of angle ω, then the above statements continue to hold with the number* $(\pi/2) - (\pi/2l)$ *replaced by* $(\pi/2) - (\omega/2l)$.

(ix) *Let* $\beta > 0$ *be such that* $\beta\gamma + b > -1$, *and let* $\theta \in (-((\pi/2)-\gamma\pi), (\pi/2)-\gamma\pi)$. *Set* $S_{\gamma,k}(z)x := \int_0^\infty f_{\gamma,z}(\lambda)(\lambda + A)^{-1}C_k x\, d\lambda$, $z \in \Sigma_{(\pi/2)-\gamma\pi}$, $x \in E$. *Then, for every* $x \in \overline{D(A)}$, *the incomplete abstract Cauchy problem*

$$(FP_{\beta,b}) : \begin{cases} u \in C^\infty((0,\infty) : E) \cap C((0,\infty) : D_\infty(A)), \\ \mathbf{D}_-^\beta u(t) = e^{i\theta\beta} C_k^{-1} \lim_{\varepsilon\to 0+} (A + \varepsilon)_{\gamma\beta} C_k u(t),\ t > 0, \\ \lim_{t\to 0+} u(t) = C_k x, \\ \{(1 + t^{-l(b+1)})^{-1}u(t) : t > 0\} \text{ is bounded in } E, \end{cases}$$

has a solution $u(t) = S_{\gamma,k}(te^{i\theta})x$, $t > 0$, *which can be analytically extended to the sector* $\Sigma_{(\pi/2)-\gamma\pi-|\theta|}$; *furthermore, the injectiveness of operator A implies that the operator* $C_k^{-1}\lim_{\varepsilon\to 0+}(A + \varepsilon)_{\gamma\beta}C_k$ *can be replaced by the operator* $A_{\gamma\beta}$, *and for every* $\delta \in (0, (\pi/2) - \gamma\pi - |\theta|)$ *and* $j \in \mathbb{N}_0$, *the set* $\{|z|^j(1+|z|^{-(b+1)/\gamma})^{-1}u^{(j)}(z) : z \in \Sigma_\delta\}$ *is bounded in E. If, additionally, the operator A is (a, b, C)-sectorial of angle ω, then the above statements remain true with the number* $(\pi/2) - \gamma\pi$ *replaced by* $(\pi/2) - \omega\gamma$.

Proof. We shall only outline a few relevant points. The assertions (i), (iii), (vii), (viii) and (ix), as well as the corresponding argumentation in the case that the operator A is (a, b, C)-sectorial of angle ω, are simple and left to the reader as an easy exercise. Furthermore, the proof of Theorem 2.9.48 implies that, for every $\delta \in (-((\pi/2) - \gamma\pi), (\pi/2) - \gamma\pi)$, there exist $r_q \in \circledast$, $m_q > 0$ and $c_{q,\delta} > 0$ such that, for every $x \in E$ and $z \in \Sigma_\delta$,

$$q(S_\gamma(z)x) \leqslant c_{q,\delta} r_q(x)\left(|z|^{-\frac{a+1}{\gamma}} + |z|^{-\frac{b+1}{\gamma}} \right).$$

The assertion (v) as well as the analyticity of mapping $z \mapsto S_\gamma(z)x$, $z \in \Sigma_\delta$ $(x \in E)$ are also very simple consequences of the proof of the afore-mentioned theorem. For the proof of assertion (vi), observe only that the assumptions $0 < \zeta < 1/2$ and $b + \zeta > -1$ imply that the integral generator of $(S_\zeta(t))_{t\geqslant 0}$ coincides with that of $(S_{\zeta,k}(t) \equiv S_\zeta(t)((1 + A)^{-1}C)^k)_{t\geqslant 0}$ and apply after that Theorem 2.9.48(vi). □

Remark 2.9.59. (i) Concerning the assertion (vi), we want to make the following remark. Suppose, in addition, that $a < \gamma - 1$. Then the proof of Theorem 2.9.51 implies that the operator A^γ belongs to the class $\mathcal{M}_C(((a + 1)/\gamma) - 1, ((b+1)/\gamma) - 1)$; furthermore, if $A \in \mathcal{S}_{C,\omega}(a, b)$ for some $\omega \in [0, \pi)$, then $A^\gamma \in \mathcal{S}_{C,\omega\gamma}(((a + 1)/\gamma) - 1, ((b + 1)/\gamma) - 1)$. In any of these cases, the operator $(A^\gamma)^\beta$ can be defined and we have that $(A^\gamma)^\beta = {}_\beta(A^\gamma) = {}_\beta({}_\gamma A) = {}_{\beta\gamma}A = A^{\beta\gamma}$, with the meaning clear.

(ii) The operator $A + \varepsilon$, appearing in the assertion (ix), belongs to the intersection of classes $\mathcal{M}_{C,a}$ and $\mathcal{M}_{C_k,-1}$. Notice that, for every $\zeta \in \mathbb{C}_+$, the power $(A + \varepsilon)_\zeta$ does not depend on the choice of class chosen for its construction.

In the following theorem, we shall summarize the most important facts concerning the incomplete second order equation

$$(P_{2,a,b}): \begin{cases} u \in C^\infty(\Sigma_{(\pi/2)-(\omega/2)} : E) \cap C(\Sigma_{(\pi/2)-(\omega/2)} : D_\infty(A)), \\ u''(z) = Au(z),\ z \in \Sigma_{(\pi/2)-(\omega/2)}, \\ \lim_{z\to 0, z\in\Sigma_\delta} u(z) = Cx, \text{ for every } \delta \in (0, (\pi/2) - (\omega/2)), \\ \text{the set } \{(|z|^{-(2a+2)} + |z|^{-(2b+2)})^{-1} u(z) : z \in \Sigma_\delta\} \text{ is bounded in } E \\ \text{for every } \delta \in (0, (\pi/2) - (\omega/2)). \end{cases}$$

Theorem 2.9.60. *Suppose* $-1 < a < (-1)/2$, $(-3)/2 < b < -1$ *and* A *is an* (a, b, C)-*sectorial operator of angle* $\omega \in [0, \pi)$.

(i) *Then, for every* $\varepsilon > 0$ *and* $\zeta > 0$, *the operator* $A + \varepsilon$ *belongs to the class* $\mathcal{M}_{C,a}$ *and the operator* $-(A + \varepsilon)_{1/2}$ *is the integral generator of an exponentially equicontinuous* $(2a + 2 + \zeta)$-*times integrated* C-*regularized semigroup* $(S^\varepsilon_{1/2,2a+2+\zeta}(t))_{t\geqslant 0}$ *of angle* $(\pi/2) - (\omega/2)$. *Define* $S^\varepsilon_{1/2}(z)x := (d/dz)S^\varepsilon_{1/2,1}(z)x$, $x \in E$, $z \in \Sigma_{(\pi/2)-(\omega/2)}$.

(ii) *The limit* $\lim_{\varepsilon\to 0+} S^\varepsilon_{1/2}(z)x := S_{1/2}(z)x$ *exists for all* $x \in E$ *and* $z \in \Sigma_{(\pi/2)-(\omega/2)}$. *Furthermore,* $(S_{1/2}(z))_{z\in\Sigma(\pi/2)-(\omega/2)} \subseteq L(E)$ *is an exponentially equicontinuous analytic C-regularized semigroup of growth order* $(2a + 2)$ *and, for every* $q \in \circledast$ *and* $\delta \in (0, (\pi/2) - (\omega/2))$*, there exist* $c_{q,\delta} > 0$ *and* $r_q \in \circledast$ *such that*

(351) $$q(S_{1/2}(z)x) \leqslant c_{q,\delta} r_q(x)(|z|^{-(2a+2)} + |z|^{-(2b+2)}),\ x \in E,\ z \in \Sigma_\delta.$$

Denote by $-A^{1/2}$ *the integral generator of* $(S_{1/2}(t))_{t \geqslant 0}$.

(iii) *Set* $\Omega_{1/2} := \{x \in E : \lim_{z \to 0, z \in \Sigma_\delta} S_{1/2}(z)x = Cx$ *for every* $\delta \in (0, (\pi/2) - (\omega/2))\}$. *Then* $D(A^{1/2}) \cup D(A) \subseteq \Omega_{1/2}$ *and* $R(S_{1/2}(z)) \subseteq D_\infty(A)$, $z \in \Sigma_{(\pi/2)-(\omega/2)}$. *Moreover, for every* $x \in \Omega_{1/2}$*, the incomplete abstract Cauchy problem* $(P_{2,a,b})$ *has a unique solution* $u(z) = S_{1/2}(z)x$, $z \in \Sigma_{(\pi/2)-(\omega/2)}$ *and, for every* $\delta \in (0, (\pi/2) - (\omega/2))$ *and* $j \in \mathbb{N}_0$*, the set* $\{|z|^j(|z|^{-(2a+2)} + |z|^{-(2b+2)})^{-1} u^{(j)}(z) : z \in \Sigma_\delta\}$ *is bounded in E.*

Proof. A simple computation shows that, for every $\varepsilon > 0$, the operator $A + \varepsilon$ belongs to the class $\mathcal{M}_{C,a}$ as well as that, for every $\mu \in \mathbb{C}_+$, we have $\mu \in \rho_C(-(A + \varepsilon)_{1/2})$ and

$$(\mu+(A+\varepsilon)_{1/2})^{-1}Cx = \frac{1}{\pi}\int_0^\infty \frac{\lambda^{1/2}}{\lambda+\mu^2}(\lambda+A+\varepsilon)^{-1}Cx\,d\lambda,\quad x \in E.$$

The operator $(A + \varepsilon)_{1/2}$ satisfies the condition (HQ) stated in Subsection 2.9.5 with the numbers $(2a + 1)$ and $\omega/2$ and, therefore, the assertion (i) follows directly from Theorem 2.9.51(i.2). Let $\gamma \in (0, (\pi/2) - (\omega/2))$, $\omega' > 0$, $\varepsilon' \in (0, \gamma)$, $\theta \in (\omega - \pi, -2\gamma)$, and let the curves $\Gamma_{1,\omega'} = \{\omega' + re^{i((\pi/2)+\gamma)} : r \geqslant |z|^{-1}\} \cup \{\omega + |z|^{-1}e^{i\theta} : \theta \in [0, (\pi/2) + \gamma]\}$ and $\Gamma_{2,\omega'} = \{\omega + re^{-i((\pi/2)+\gamma)} : r \geqslant |z|^{-1}\} \cup \{\omega + |z|^{-1}e^{i\theta} : \theta \in [-(\pi/2)-\gamma, 0]\}$ be oriented counterclockwise. Using the proof of the second part of afore-mentioned theorem, we obtain that, for every $x \in E$ and $z \in \Sigma_{\gamma-\varepsilon'}$, (335) holds. Using the dominated convergence theorem, we get that, for every $x \in E$ and $z \in \Sigma_{\gamma-\varepsilon'}$, $S_{1/2}(z)x = \lim_{\varepsilon \to 0+} S^\varepsilon_{1/2}(z)x$ is given by the formula (336). Owing to the computation given in the proof of assertion Theorem 2.9.51(i.1) and the fact that A is (a, b, C)-sectorial of angle ω, we easily infer that, for every $q \in \circledast$, there exist $c > 0$ and $r_q \in \circledast$ such that

$$q\left(\int_0^\infty \frac{\upsilon^{1/2}}{\lambda^2 e^{i\theta}+\upsilon}\left(\upsilon+e^{i\theta}A\right)^{-1} Cx\,d\upsilon\right) \leqslant cr_q(x)\left(|\lambda|^{2a+1}+|\lambda|^{2b+1}\right),\ x \in E,\ \lambda \in \Gamma_{1,\omega'}$$

(352)

and

(353)

$$q\left(\int_0^\infty \frac{\upsilon^{1/2}}{\lambda^2 e^{-i\theta}+\upsilon}\left(\upsilon+e^{-i\theta}A\right)^{-1} Cx\,d\upsilon\right) \leqslant cr_q(x)\left(|\lambda|^{2a+1}+|\lambda|^{2b+1}\right),\ x \in E,\ \lambda \in \Gamma_{2,\omega'}.$$

The estimate (351) follows from (352)-(353) and the computation given in the proof of [20, Theorem 2.6.1]. Moreover, the convergence in (336) is uniform on compact subsets of $\Sigma_{\gamma-\varepsilon'}$, which implies that, for every $x \in E$, the mapping $z \mapsto S_{1/2}(z)x$, $z \in \Sigma_{(\pi/2)-(\omega/2)}$ is analytic. Now the proof of (ii) can be completed quite easily. In order to prove (iii), notice that the closedness of A taken together with the obvious equality $(d^2/dz^2)S^{\varepsilon}_{1/2}(z)x = AS^{\varepsilon}_{1/2}(z)x + \varepsilon S^{\varepsilon}_{1/2}(z)x$ implies by letting $\varepsilon \to 0+$ that $(d^2/dz^2)S_{1/2}(z)x = AS_{1/2}(z)x$ $(x \in E,\ z \in \Sigma_{(\pi/2)-(\omega/2)})$. Inductively, we get that $(d^{2n}/dz^{2n})S_{1/2}(z)x = A^nS_{1/2}(z)x$ for all $x \in E$ and $z \in \Sigma_{(\pi/2)-(\omega/2)}$, so that $\bigcup_{z\in\Sigma_{(\pi/2)-(\omega/2)}} R(S_{1/2}(z)) \subseteq D_{\infty}(A)$. If $x \in \Omega_{1/2}$, then it is very simple to prove that the function $u(z) = S_{1/2}(z)x$, $z \in \Sigma_{(\pi/2)-(\omega/2)}$ is a solution of problem $(P_{2,a,b})$. The Cauchy integral formula yields that, for every $\delta \in (0, (\pi/2)-(\omega/2))$ and $j \in \mathbb{N}_0$, the set $\{(|z|^j(|z|^{-(2a+2)} + |z|^{-(2b+2)})^{-1}u^{(j)}(z) : z \in \Sigma_{\delta}\}$ is bounded in E. Define now $S^1_{1/2}(\cdot)\,(\Omega^1_{1/2})$ in the same way as $S_{1/2}(\cdot)\,(\Omega_{1/2})$, with the operator C replaced by $C_1 = C(1+A)^{-1}C$. Then $\Omega_{1/2} \subseteq \Omega^1_{1/2}$ and the operator A satisfies the condition (HQ) with numbers b and ω (and the operator C replaced by C_1). If $u(\cdot)$ is a solution of problem $(P_{2,a,b})$ with $x \in \Omega_{1/2}$, then an application of Theorem 2.9.51(iii.2) gives that $(1+A)^{-1}Cu(z) = S^1_{1/2}(z)x$, $z \in \Sigma_{(\pi/2)-(\omega/2)}$. This implies that the problem $(P_{2,a,b})$ has at most one solution. It remains to be proved that the inclusion $D(A^{1/2}) \cup D(A) \subseteq \Omega_{1/2}$ holds. In order to do that, observe first that the operator $-A^{1/2}$ is the integral generator of an exponentially equicontinuous analytic once integrated C-semigroup $(\int_0^t S_{1/2}(s)\cdot ds)_{t\geqslant 0}$ of angle $(\pi/2)-\omega/2$, so that the assumption $x \in D(A^{1/2})$ implies $S_{1/2}(z)x = Cx - \int_0^z S_{1/2}(t)A^{1/2}x\, dt$, $z \in \Sigma_{(\pi/2)-\omega/2}$. Hence, $D(A^{1/2}) \subseteq \Omega_{1/2}$. The inclusion $D(A) \subseteq \Omega_{1/2}$ follows from the facts that, for every $\varepsilon \in (0, 1)$, $D(A)$ is contained in the angular set of continuity of semigroup $S^{\varepsilon}_{1/2}(z))_{z\in\Sigma(\pi/2)-\omega/2}$ (cf. Theorem 2.9.51(iii.1)) and that, for every $\varepsilon'' > 0$ and $q \in \circledast$, there exists a sufficiently small $\varepsilon > 0$ such that, for every $z \in \Sigma_{\gamma-\varepsilon'}$ with $|z| \leqslant 1$, one has $q(S_{1/2}(z)x - S^{\varepsilon}_{1/2}(z)x) < \varepsilon''$ (this can be proved following the lines of the corresponding part of proof of Theorem 2.9.51(iii.2)). □

Remark 2.9.61. (i) The formula (295) continues to hold in the situation of preceding theorem; if $z \in \Sigma_{(\pi/2)-(\omega/2)} \setminus (0, \infty)$, then the bounded linear operator $S_{1/2}(z)$ cannot be represented in the similar way.

(ii) If $(-3)/2 < b < -1$, $a \geqslant (-1)/2$, $k \geqslant \lceil a \rceil + 1$ and A is (a, b, C)-sectorial of angle $\omega \in [0, \pi)$, then Theorem 2.9.51(iii.2) can be applied with the operator C replaced by C_k (or some of its slight modifications). Observe, finally, that the method previously described does not produce optimal results in the case $a \in \bigcup_{k\in\mathbb{N}_0}[(-1)/2+k, k)$, and that it would be very difficult to prove a satisfactory analogue of Theorem 2.9.60 in the case that $(-1)/2 \leqslant a < 0$ and $(-3)/2 < b < -1$.

Example 2.9.62. (i) Suppose $\alpha \in (0, 2]$, $\zeta \geqslant 1$, $\mu \geqslant 0$, $v \geqslant 0$ and $-A$ is the integral generator of an exponentially equicontinuous (g_{α}, g_{ζ})-regularized C-resolvent family $(R(t))_{t\geqslant 0}$ satisfying that $(-A)\int_0^t g_{\alpha}(t-s)R(s)x\, ds = R(t)x - g_{\zeta}(t)Cx$, $t \geqslant 0$, $x \in E$ and that the family $\{(t^{-\mu}+t^{-v})^{-1}R(t) : t > 0\}$ is equicontinuous. By Theorem 2.1.5, we have that $\Sigma_{\alpha\pi/2} \subseteq \rho_C(-A)$ and

$$\lambda^{-\zeta}(1+\lambda^{-\alpha}A)^{-1}Cx = \int_0^{\infty} e^{-\lambda t}R(t)x\, dt,\ x \in E,\ \mathrm{Re}\,\lambda > 0.$$

This simply implies that, for every sufficiently small $\varepsilon > 0$, the family

$$\left\{\left(|\lambda|^{\frac{\alpha+\mu+1-\zeta}{\alpha}}+|\lambda|^{\frac{\alpha+\nu+1-\zeta}{\alpha}}\right)(\lambda+A)^{-1}C:\lambda\in\Sigma_{\frac{\pi\alpha}{2}-\varepsilon}\right\}$$

is equicontinuous. Suppose that

$$(\mu+1-\zeta)(\nu+1-\zeta)\geqslant 0. \tag{354}$$

Then $A \in \mathcal{M}_{C,m}$ with:

1. $m = -\max(\frac{\alpha+\mu+1-\zeta}{\alpha},\frac{\alpha+\nu+1-\zeta}{\alpha})$, provided $\min(\frac{\alpha+\mu+1-\zeta}{\alpha},\frac{\alpha+\nu+1-\zeta}{\alpha}) \geqslant 1$ or $\max(\frac{\alpha+\mu+1-\zeta}{\alpha},\frac{\alpha+\nu+1-\zeta}{\alpha}) < 1$.
2. $m = -\frac{\alpha+\mu+1-\zeta}{\alpha}$, provided $\frac{\alpha+\nu+1-\zeta}{\alpha} = 1$, and
3. $m = -\frac{\alpha+\nu+1-\zeta}{\alpha}$, provided $\frac{\alpha+\mu+1-\zeta}{\alpha} = 1$.

If $(\mu+1-\zeta)(\nu+1-\zeta)<0$, then there does not exist $m\in\mathbb{R}$ such that $A\in\mathcal{M}_{C,m}$; in this case, we have the existence of numbers $m_1 > 1$ and $m_2 < 1$ such that the operator A belongs to the class $\mathcal{M}_C(-m_1, -m_2)$ (cf. (ii) for the corresponding example). On the other hand, the inequality (354) is always true provided $\zeta = 1$ and, in the case of abstract differential operators considered in [223, Theorem 4.2], [255, Theorem 4.1] and [417, Theorem 4], (354) holds with some $\zeta > 1$. The results obtained in [90] and this paper (see, e.g., Theorem 2.9.48 and Theorem 2.9.51) can be also applied to (pseudo-)differential operators considered in [227], [531, Theorem 1.5.10], [548, Theorem 3.1-Theorem 3.3] (with $\omega = 0$, $1 < p < \infty$ and $m = -r_p - 1$) and [321, Theorem 2.1-Theorem 2.2] (with $\omega = 0$, $1 < p < \infty$ and an appropriately chosen number $m \in ((-3)/2, -1)$)-needless to say that p must be sufficiently close to 2.

(ii) Suppose $1 < p < \infty$, $m, n \in \mathbb{N}$, $E = L^p(\mathbb{R}^n)$ and $a_\alpha \in \mathbb{C}$ for each $\alpha \in \mathbb{N}_0^n$ with $|\alpha| \leqslant m$. Put $r_p := n|1/p - 1/2|$ and assume that the elliptic polynomial $p(x) = \Sigma_{|\alpha|\leqslant m}\, i^{|\alpha|} a_\alpha x^\alpha$ $(x\in\mathbb{R}^n)$ satisfies $\sup_{x\in\mathbb{R}^n} \mathrm{Re}(P(x)) = 0$. Then [531, Theorem 1.5.9] implies that the associated partial differential operator $A : D(A) \to E_l$, given by

$$Af := \sum_{|\alpha|\leqslant m} a_\alpha D^\alpha f \text{ and } D(A) := \{f\in E_l : Af \in E_l \text{ distributionally}\},$$

generates an exponentially equicontinuous r_p-times integrated semigroup $(S_{r_p}(t))_{t\geqslant 0}$ on E_l $(0 \leqslant l \leqslant n)$. Furthermore, there exists $M \geqslant 1$ such that, for every $\alpha \in \mathbb{N}_0^l$, we have $q_\alpha(S_{r_p}(t)x) \leqslant M(1 + t^{2r_p})q_\alpha(x)$, $t \geqslant 0$, $x \in E$. Hence, the operator $-A$ is $(r_p-1, -1-r_p, 1)$-sectorial of angle $\pi/2$, and Theorem 2.9.60(iii) can be applied provided that $0 \leqslant r_p < 1/2$.

2.9.7. The existence and growth of mild solutions of operators generating fractionally integrated *C*-semigroups and cosine functions in locally convex spaces. Our intention in this subsection is to incorporate some of the results on powers of *C*-sectorial operators in the study of existence and growth of mild solutions of abstract Cauchy problems involving generators of integrated

C-semigroups and cosine functions. In order to do that, we shall follow the method proposed by J.M.A.M. van Neerven and B. Straub in [431] (cf. also the paper [300] in which the assertions of [431, Theorem 1.1-Theorem 1.2] have been generalized to generators with not necessarily dense domain).

We shall repeat the following assertion for the sake of convenience of the reader (cf. [292, Theorem 2.1.11]).

Lemma 2.9.63. *Suppose $\alpha > 0$ and A is a closed linear operator on E. Then the following assertions are equivalent:*

(i) *A is a subgenerator of an α-times integrated C-cosine function $(C_\alpha(t))_{t\geqslant 0}$ in E.*

(ii) *The operator $\mathcal{A} := \begin{pmatrix} 0 & I \\ A & 0 \end{pmatrix}$ is a subgenerator of an $(\alpha + 1)$-times integrated $\mathcal{C}$-semigroup $(S_{\alpha+1}(t))_{t\geqslant 0}$ in $E \times E$, where $\mathcal{C} := \begin{pmatrix} C & 0 \\ 0 & C \end{pmatrix}$.*

In this case:

$$S_{\alpha+1}(t) = \begin{pmatrix} \int_0^t C_\alpha(s)\,ds & \int_0^t (t-s) C_\alpha(s)\,ds \\ C_\alpha(t) - g_{\alpha+1}(t) C & \int_0^t C_\alpha(s)\,ds \end{pmatrix}, \quad t \geqslant 0,$$

and the integral generators of $(C_\alpha(t))_{t\geqslant 0}$ and $(S_{\alpha+1}(t))_{t\geqslant 0}$, denoted respectively by B and $\mathcal{B}$, satisfy $\mathcal{B} = \begin{pmatrix} 0 & I \\ B & 0 \end{pmatrix}$. Furthermore, the integral generator of $(C_\alpha(t))_{t\geqslant 0}$, resp. $(S_{\alpha+1}(t))_{t\geqslant 0}$, is $C^{-1}AC$, resp. $\mathcal{C}^{-1}\mathcal{A}\mathcal{C} \equiv \begin{pmatrix} 0 & I \\ C^{-1}AC & 0 \end{pmatrix}$.

Recall that the function $u(\cdot, x_0)$ is a mild solution of the abstract Cauchy problem

$$(ACP_1) : u'(t, x_0) = Au(t, x_0),\ t \geqslant 0,\ u(0, x_0) = x_0, \text{ resp.,}$$

$$(ACP_2) : u''(t, x_0, y_0) = Au(t, x_0, y_0),\ t \geqslant 0,\ u(0, x_0, y_0) = x_0,\ u'(0, x_0, y_0) = y_0,$$

iff the mapping $t \mapsto u(t, x_0)$, $t \geqslant 0$ is continuous, $\int_0^t u(s, x_0)\,ds \in D(A)$ and $A\int_0^t u(s, x_0)\,ds = u(t, x_0) - x_0$, $t \geqslant 0$, resp., the mapping $t \mapsto u(t, x_0, y_0)$, $t \geqslant 0$ is continuous, $\int_0^t (t-s)u(s, x_0, y_0)\,ds \in D(A)$ and $A\int_0^t (t-s)u(s, x_0, y_0)\,ds = u(t, x_0, y_0) - x_0 - ty_0$, $t \geqslant 0$.

Suppose $\alpha \geqslant 0$ and A is the integral generator of a global α-times integrated C-semigroup $(S^\alpha(t))_{t\geqslant 0}$ satisfying that there exists $\omega \geqslant 0$ such that the family $\{e^{-\omega t}S^\alpha(t) : t \geqslant 0\}$ is equicontinuous. Let $\sigma \in (0, 1]$ be fixed. Then $C^{-1}AC = A$ and, for every $\gamma \in (0, \frac{\pi}{2})$, there exists $d \in (0, 1]$ such that $\Sigma(\gamma, d) \subseteq \rho(A - \omega - \sigma)$ and the family $\{(1 + |\lambda|)^{1-\alpha}(\lambda - (A - \omega - \sigma))^{-1}C : \lambda \in \Sigma(\gamma, d)\}$ is equicontinuous. Set $A_{\omega+\sigma} := -(\omega + \sigma - A)$ and, after that, $C_\alpha := (-A_{\omega+\sigma})^{-1-\lfloor\alpha\rfloor}C^2$. Then $C_\alpha^{-1}A_{\omega+\sigma}C_\alpha = A_{\omega+\sigma}$ and it is not difficult to prove that the operator $-A_{\omega+\sigma}$ is C_α-sectorial of angle $\pi/2$ and the condition (HC) holds with $d = \sigma/2$. Therefore, for every $z \in \mathbb{C}$, we can construct the power $(-A_{\omega+\sigma})_z$ following the method proposed in Subsection 2.9.1, with the operator C replaced by C_α. Then, for every $z \in \mathbb{C}$, the power $(-A_{\omega+\sigma})_z$ coincides with that constructed in Subsection 2.9.3. The following properties of powers will be used henceforth:

(P0)' For every $k \in \mathbb{Z}$, we have $(-A_{\omega+\sigma})_k = C_\alpha^{-1}(-A_{\omega+\sigma})^k C_\alpha$, where $(-A_{\omega+\sigma})^k$

denotes the usual power of the operator $-A_{\omega+\sigma}$ and $(-A_{\omega+\sigma})^0 := 1$.

(P1)' For every $z \in \mathbb{C}$, the operator $(-A_{\omega+\sigma})_z$ is injective and the following equality holds:

$$(-A_{\omega+\sigma})_{-z} = ((-A_{\omega+\sigma})_z)^{-1} = ((-A_{\omega+\sigma})^{-1})_z.$$

(P2)' Let $z_1, z_2 \in \mathbb{C}$. Then $(-A_{\omega+\sigma})_{z_1}(-A_{\omega+\sigma})_{z_2} \subseteq (-A_{\omega+\sigma})_{z_1+z_2}$, and for every $x \in D((-A_{\omega+\sigma})_{z_1+z_2}) \cap D((-A_{\omega+\sigma})_{z_2})$, one has $(-A_{\omega+\sigma})_{z_2}x \in D((-A_{\omega+\sigma})_{z_1})$ and $(-A_{\omega+\sigma})_{z_1}(-A_{\omega+\sigma})_{z_2}x = (-A_{\omega+\sigma})_{z_1+z_2}x$. If, in addition, $(-A_{\omega+\sigma})_{z_1} \in L(E)$, then we have that $(-A_{\omega+\sigma})_{z_1}(-A_{\omega+\sigma})_{z_2} = (-A_{\omega+\sigma})_{z_1+z_2}$.

(P3)' If $0 < \operatorname{Re} z < 1$, then

$$(-A_{\omega+\sigma})_{-z} C_\alpha x = \frac{\sin z\pi}{\pi} \int_0^\infty \lambda^{-z} \left(\lambda - A_{\omega+\sigma}\right)^{-1} C_\alpha x \, d\lambda, \quad x \in E.$$

(P4)' If $C = 1$, then $(-A_{\omega+\sigma})_z \in L(E)$ for any $z \in \mathbb{C}$ with $\operatorname{Re} z < -\alpha$.

Theorem 2.9.64. *Let $\alpha \in (0, \infty)\setminus\mathbb{N}$, let $\omega \geqslant 0$, and let A be the integral generator of an α-times integrated C-semigroup $(S^\alpha(t))_{t\geqslant 0}$ satisfying that the family $\{e^{-\omega t}S^\alpha(t) : t \geqslant 0\}$ is equicontinuous. Suppose $\varepsilon > 0$, $\lfloor\alpha\rfloor = \lfloor\alpha + \varepsilon\rfloor$, $x_0' \in D((-A_{\omega+\sigma})_{\alpha+\varepsilon}) \cap D((-A_{\omega+\sigma})_{\alpha+\varepsilon-\lfloor\alpha+\varepsilon\rfloor})$ and $x_0 = Cx_0'$. Then the abstract Cauchy problem (ACP_1) has a unique mild solution, denoted by $u(t, x_0)$, and for every $\varepsilon > 0$, the set $\{e^{-(\omega+\sigma+\varepsilon)t}u(t, x_0) : t \geqslant 0\}$ is bounded. If, in addition, $A_{\omega+\sigma}x_0' \in D((-A_{\omega+\sigma})_{\alpha+\varepsilon}) \cap D((-A_{\omega+\sigma})_{\alpha+\varepsilon-\lfloor\alpha+\varepsilon\rfloor})$, then the solution is classical.*

Proof. Set $x_0'' := (-A_{\omega+\sigma})_{\alpha+\varepsilon-\lfloor\alpha+\varepsilon\rfloor}x_0'$. Denote by $(S^\alpha_{\omega+\sigma}(t))_{t\geqslant 0}$ the α-times integrated C-semigroup generated by $A_{\omega+\sigma}$ (cf. Theorem 2.1.44(i)-(b)). Then, for every $\beta > \alpha$, $(S^\beta_{\omega+\sigma}(t) \equiv (g_{\beta-\alpha} * S^\alpha_{\omega+\sigma})(t))_{t\geqslant 0}$ is the β-times integrated C-semigroup generated by $A_{\omega+\sigma}$; by $(S^\beta(t))_{t\geqslant 0}$ we denote the β-times integrated C-semigroup generated by A. Then the following representation formula holds:

$$S^\beta_{\omega+\sigma}(t)x = \int_0^\infty e^{-(\omega+\sigma)(t-s)} S^\beta(t-s)x \, dg_{\omega+\sigma,\beta}, \quad x \in E, \quad t \geqslant 0,$$

with function $g_{\omega+\sigma,\beta}(\cdot)$ being defined by $g_{\omega+\sigma,\beta}(s) := \chi_{(0,\infty)}(s) + \Sigma_{k=1}^\infty \beta(\beta-1)\cdots(\beta-k+1)(\omega+\sigma)^k s^k / k!^2$, $s \geqslant 0$ (cf. [431, Proposition 3.3]). Since, by (P1)', $x_0' \in D((-A_{\omega+\sigma})_{\alpha+\varepsilon}) = R((-A_{\omega+\sigma})_{-\alpha-\varepsilon})$, we have the existence of an element $z_0 \in E$ such that $x_0'' = (-A_{\omega+\sigma})_{\alpha+\varepsilon-\lfloor\alpha+\varepsilon\rfloor}(-A_{\omega+\sigma})_{-\alpha-\varepsilon}z_0$. Keeping in mind (P0)' and (P2)', as well as Lemma 2.9.3, the above implies $x_0'' = C_\alpha^{-1}(-A_{\omega+\sigma})^{-\lfloor\alpha+\varepsilon\rfloor}C_\alpha z_0$ and $(-A_{\omega+\sigma})^{\lfloor\alpha+\varepsilon\rfloor}Cx_0'' = Cz_0$. Define now, for every $t \geqslant 0$,

$$S^{\alpha+\varepsilon-\lfloor\alpha+\varepsilon\rfloor}_{\omega+\sigma}(t)x_0'' := (-1)^{\lfloor\alpha+\varepsilon\rfloor} S^{\alpha+\varepsilon}_{\omega+\sigma}(t)z_0 + \sum_{i=0}^{\lfloor\alpha+\varepsilon\rfloor-1} g_{\alpha+\varepsilon-i}(t) A_{\omega+\sigma}^{\lfloor\alpha+\varepsilon\rfloor-1-i} Cx_0''.$$

Then [292, Proposition 2.3.3(i)] implies that, for every $t \geqslant 0$,

$$S_{\omega+\sigma}^{\alpha+\varepsilon-\lfloor\alpha+\varepsilon\rfloor}(t)x_0'' = C_\alpha^{-1}\frac{d^{\lfloor\alpha+\varepsilon\rfloor}}{dt^{\lfloor\alpha+\varepsilon\rfloor}}S_{\omega+\sigma}^{\alpha+\varepsilon}(t)C_\alpha x_0''.$$

We will prove that the mild solution in (i)-(ii) is given by the formula

$$u(t, x_0) := e^{(\omega+\sigma)t}\upsilon_{\omega+\sigma}(t, x_0''),\ t \geqslant 0,$$

where

$$\upsilon_{\omega+\sigma}(t,x_0'') := \Gamma_{\alpha,\varepsilon}\int_0^\infty \frac{ds}{s-1}\left(s^{\lfloor\alpha+\varepsilon\rfloor-\alpha-\varepsilon}S_{\omega+\sigma}^{\alpha+\varepsilon-\lfloor\alpha+\varepsilon\rfloor}(t)-\frac{1}{s}S_{\omega+\sigma}^{\alpha+\varepsilon-\lfloor\alpha+\varepsilon\rfloor}\left(\frac{t}{s}\right)\right)x_0'', \tag{355}$$

and $\Gamma_{\alpha,\varepsilon} := \frac{\sin(\alpha+\varepsilon-\lfloor\alpha+\varepsilon\rfloor)\pi}{\pi}$, see [431, Sections 3-4] and [300, Theorem 4.1]. First of all, notice that the convergence of singular integral appearing in (355), written as the sum of corresponding integrals along the intervals (0, 1/2), (1/2, 2) and $(2, \infty)$, follows from the following:

Suppose that the operator family $\{(1+t^\gamma)^{-1}e^{-\omega t}S^\alpha(t) : t \geqslant 0\}$ is equicontinuous. Put $\delta := 2^{-1}\min(\varepsilon, \alpha+\varepsilon-\lfloor\alpha+\varepsilon\rfloor)$. Then the computation given in the proofs of [431, Lemma 4.1-Lemma 4.2] shows that there exists $c_{\alpha,\varepsilon,\gamma,\omega} > 0$ such that, for every $p \in ⊛$, there exist $r_p \in ⊛$, $c_p > 0$ and $c_{p,\omega,\gamma,\varepsilon,\sigma} > 0$ such that:

$$p\left(S_{\omega+\sigma}^{\alpha+\varepsilon-\lfloor\alpha+\varepsilon\rfloor}(t)x_0''\right) \leqslant c_p\frac{\sigma^{\min(-\lfloor\alpha+\varepsilon\rfloor,\alpha+\varepsilon-\lfloor\alpha+\varepsilon\rfloor-\gamma-1)}}{\ln\left(1+\frac{\sigma}{4\omega+2\sigma}\right)}$$

$$\times\, t^{\alpha+\varepsilon-\lfloor\alpha+\varepsilon\rfloor-1}\left[r_p(z_0)+\sum_{i=0}^{\lfloor\alpha+\varepsilon\rfloor-1}p(A^iCx_0'')\right],\quad t \geqslant 0$$

and

$$p\left(S_{\omega+\sigma}^{\alpha+\varepsilon-\lfloor\alpha+\varepsilon\rfloor}(t)x_0'' - S_{\omega+\sigma}^{\alpha+\varepsilon-\lfloor\alpha+\varepsilon\rfloor}(\tau)x_0''\right)$$

$$\leqslant c_p c_{\alpha,\varepsilon,\gamma,\omega}(t-s)^\delta\left[r_p(z_0)+\sum_{i=0}^{\lfloor\alpha+\varepsilon\rfloor-1}p(A^iCx_0'')\right]$$

$$\times\,\frac{\sigma^{\min(-\lfloor\alpha+\varepsilon\rfloor,\alpha+\varepsilon-\lfloor\alpha+\varepsilon\rfloor-\gamma-\varepsilon-1)}}{\ln\left(1+\frac{\sigma}{4\omega+2\sigma}\right)},\quad 0 \leqslant \tau \leqslant t < \infty.$$

We obtain similarly as in the proofs of [431, Lemma 4.3-Lemma 4.4] that the mapping $t \mapsto \upsilon_{\omega+\sigma}(t, x_0'')$, $t \geqslant 0$ is continuous, and that the equicontinuity of operator family $\{(1+t^\gamma)^{-1}S^\alpha(t) : t \geqslant 0\}$ implies that there exists $c_{\alpha,\varepsilon,\gamma} > 0$ such that, for every $p \in ⊛$, there exist $r_p \in ⊛$ and $c_p > 0$ satisfying:

$$p\big(v_\sigma(t,x_0'')\big)\leqslant c_p c_{\alpha,\varepsilon,\gamma}\sigma^{\min(-\lfloor\alpha+\varepsilon\rfloor,\alpha-\lfloor\alpha+\varepsilon\rfloor-\gamma-1)}$$
$$\times\left[r_p(z_0)+\sum_{i=0}^{\lfloor\alpha+\varepsilon\rfloor-1}p(A^iCx_0'')\right]t^{\alpha+\varepsilon-\lfloor\alpha+\varepsilon\rfloor-1},\quad t\geqslant 2.$$

Since $\int_0^\infty e^{-\lambda t}S_{\omega+\sigma}^{\alpha+\varepsilon}(t)x\,dt=\lambda^{-\alpha-\varepsilon}(\lambda-A_{\omega+\sigma})^{-1}Cx$ for all $x\in E$ and $\lambda>0$, it is not difficult to prove, with the help of proof of [431, Lemma 6.1] and the property (P3)' of powers, that

$$\int_0^\infty e^{-\lambda t}CC_\alpha v_{\omega+\sigma}(t,x_0'')\,dt=(\lambda-A_{\omega+\sigma})^{-1}C^2(-A_{\omega+\sigma})_{\lfloor\alpha+\varepsilon\rfloor-\alpha-\varepsilon}C_\alpha x_0''.$$

The resolvent equation and the previous equality imply that:

$$A_{\omega+\sigma}\int_0^\infty e^{-\lambda t}\int_0^t CC_\alpha v_{\omega+\sigma}(s,x_0'')\,ds\,dt$$
$$=CC_\alpha\left[\int_0^\infty e^{-\lambda t}v_{\omega+\sigma}(t,x_0'')\,dt-\frac{x_0}{\lambda}\right],\quad \lambda>0.$$

Taking into account the Laplace transformability of function $t\mapsto v_{\omega+\sigma}(t,x_0'')$, $t\geqslant 0$ and the equality $(CC_\alpha)^{-1}A_{\omega+\sigma}CC_\alpha=A_{\omega+\sigma}$, we get that

$$A_{\omega+\sigma}\int_0^\infty e^{-\lambda t}\int_0^t v_{\omega+\sigma}(s,x_0'')\,ds\,dt$$
$$=\int_0^\infty e^{-\lambda t}v_{\omega+\sigma}(t,x_0'')dt-\frac{x_0}{\lambda},\quad \lambda>0.$$

Taken together with Theorem 1.2.1(vii), the previous equality implies that

$$A_{\omega+\sigma}\int_0^t v_{\omega+\sigma}(s,x_0'')\,ds=v_{\omega+\sigma}(t,x_0'')-x_0,\quad t\geqslant 0.$$

Hence, the mapping $t\mapsto u(t,x_0)$, $t\geqslant 0$ is a mild solution in (i)-(ii); the uniqueness is a simple consequence of a Ljubich type theorem. If, in addition, $A_{\omega+\sigma}x_0'\in D((-A_{\omega+\sigma})_{\alpha+\varepsilon})\cap D((-A_{\omega+\sigma})_{\alpha+\varepsilon-\lfloor\alpha+\varepsilon\rfloor})$, then (P2)' implies that the terms $A_{\omega+\sigma}(-A_{\omega+\sigma})_{\alpha+\varepsilon-\lfloor\alpha+\varepsilon\rfloor}x_0''$ and $(-A_{\omega+\sigma})_{\alpha+\varepsilon-\lfloor\alpha+\varepsilon\rfloor}A_{\omega+\sigma}x_0''$ are well defined and equal. As a simple consequence, we have that

$$A_{\omega+\sigma}v_{\omega+\sigma}(t,x_0'')=v_{\omega+\sigma}(t,A_{\omega+\sigma}x_0''),\quad t\geqslant 0$$

and that the constructed mild solution, for such an initial value x_0, is, in fact, classical. □

Before proceeding further, we would like to recommend to the reader the paper [519], and Section 7 of [431], for more details concerning the exponential type of constructed classical solutions.

Remark 2.9.65. Suppose $C = 1$.

(i) Then the assumption $x_0' \in D((-A_{\omega+\sigma})_{\alpha+\varepsilon})$ implies by (P1)'-(P2)' that $x_0' \in D((-A_{\omega+\sigma})_{\alpha+\varepsilon-\lfloor\alpha+\varepsilon\rfloor})$ and $(-A_{\omega+\sigma})_{\alpha+\varepsilon-\lfloor\alpha+\varepsilon\rfloor}x_0' \in D(A^{\lfloor\alpha+\varepsilon\rfloor})$. Using the same properties of powers, it is checked at once that the assumption $x_0' \in D((-A_{\omega+\sigma})_{1+\alpha+\varepsilon})$ implies $x_0' \in D(-A_{\omega+\sigma}) \cap D((-A_{\omega+\sigma})_{\alpha+\varepsilon}) \cap D((-A_{\omega+\sigma})_{\alpha+\varepsilon-\lfloor\alpha+\varepsilon\rfloor})$ as well as $-A_{\omega+\sigma}x_0' \in D((-A_{\omega+\sigma})_{\alpha+\varepsilon}) \cap D((-A_{\omega+\sigma})_{\alpha+\varepsilon-\lfloor\alpha+\varepsilon\rfloor})$.

(ii) It is worth noting that, for every $z \in \mathbb{C}$ with $|\mathrm{Re}\, z| > \alpha$, the domain of power $(-A_{\omega+\sigma})_z$ does not depend on the particular choice of number $\sigma \in (0,1]$. In order to better explain this, suppose that $0 < \sigma_1 < \sigma_2 \leqslant 1$. Then the operator $-A_{\sigma_1,\sigma_2} \equiv -A_{\sigma_2}(-A_{\sigma_1})^{-1}$ belongs to $L(E)$ and the computation given in the proof of [431, Lemma 5.2] shows that the operator $-A_{\sigma_1,\sigma_2}$ is positive, so that the power $(-A_{\sigma_1,\sigma_2})_z$ can be constructed in the usual way (see, e.g., [403]). Keeping in mind that $(-A_{\omega+\sigma})_z \in L(E)$, provided $\mathrm{Re}\, z < -\alpha$, it is quite easy to show that the following equalities hold, for every $z \in \mathbb{C}$ with $\mathrm{Re}\, z < -\alpha$,

$$(356) \qquad (-A_{\sigma_2})_z x = (-A_{\sigma_1,\sigma_2})_z(-A_{\sigma_1})_z x = (-A_{\sigma_1})_z(-A_{\sigma_1,\sigma_2})_z x,\ x \in E.$$

If $\mathrm{Re}\, z > \alpha$, then one can use the equality $D((-A_{\sigma_2})_z) = R((-A_{\sigma_2})_{-z})$ and (356) in order to see that $D((-A_{\sigma_2})_z) \subseteq D((-A_{\sigma_1})_z)$. The converse inclusion can be proved similarly, so that $D((-A_{\sigma_2})_z) = D((-A_{\sigma_1})_z)$ for $\mathrm{Re}\, z > \alpha$. Therefore, the supposition $x_0' \in D((-A_\sigma)_{\alpha+\varepsilon})$ implies $x_0' \in D((-A_1)_{\alpha+\varepsilon})$ and, in this case, (356) holds with $\sigma_2 = \sigma$, $\sigma_1 = \sigma$, $z = \alpha+\varepsilon$ and $x_0' = x$. Hence,

$$(357) \qquad A^jC(-A_\sigma)_{\alpha+\varepsilon-\lfloor\alpha+\varepsilon\rfloor}x_0' = (-A_{1,\sigma})_{\lfloor\alpha+\varepsilon\rfloor-(\alpha+\varepsilon)}A^jC(-A_1)_{\alpha+\varepsilon-\lfloor\alpha+\varepsilon\rfloor}x_0',$$

for $0 \leqslant j \leqslant \lfloor\alpha + \varepsilon\rfloor$.

(iii) Consider the situation of Theorem 2.9.64 with $\omega = 0$. Using again the computation given in the proof of [431, Lemma 5.2], we get that the family $\{\sigma^{-\min(0,\alpha-\gamma)}(-A_\sigma)_{\lfloor\alpha+\varepsilon\rfloor-\alpha-\varepsilon} : 0 < \sigma \leqslant 1/2\} \subseteq L(E)$ is equicontinuous. Combining this with the proof of [431, Theorem 1.2] and (357), we have that the set $\{(1 + t)^{-\max(\alpha-1+\varepsilon,\gamma+\varepsilon,2\gamma-\alpha+\varepsilon)}u(t, x_0) : t \geqslant 0\}$ is bounded; this is, certainly, one of the facts that cannot be so easily formulated in the general case $C \neq 1$.

(iv) The assertion of [300, Theorem 4.2] continues to hold, with appropriate technical modifications, in the setting of sequentially complete locally convex spaces.

(v) In a joint follow-up research with L. Abadias, C. Lizama and P. J. Miana, the author has recently considered Laguerre and Hermite expansions of solutions to abstract Volterra equations and abstract multi-term fractional differential equations. Especially, we have proved a simple and elegant representation of the mild solution appearing in (355) and the formula preceding it.

Notice that Theorem 2.9.64 and Remark 2.9.65 taken together provide a proper extension of [300, Theorem 4.1]. This enables one to simply state results concerning the growth of mild solutions of abstract Cauchy problems for elliptic differential operators acting on E_l-type spaces (apply the result stated in Remark 2.9.65(iii)); we can also prove an extension of [362, Theorem 3.7] for such operators.

Suppose now that the operator A is the integral generator of an α-times integrated C-cosine function $(C_\alpha(t))_{t\geqslant 0}$ satisfying that the family $\{e^{-\omega t}C_\alpha(t) : t \geqslant 0\}$ is equicontinuous for some $\omega \geqslant 0$. Then we know from Lemma 2.9.63 that the operator $\mathcal{A} = \binom{0\ I}{A\ 0}$ is the integral generator of an $(\alpha + 1)$-times integrated $\mathcal{C}$-semigroup $(S_{\alpha+1}(t))_{t\geqslant 0}$ in $E \times E$, where $\mathcal{C} = \binom{C\ 0}{0\ C}$. Therefore, for any $\sigma \in (0, 1]$ given in advance, the operator $-\mathcal{A}_{\omega+\sigma} \equiv \mathcal{A} - \omega - \sigma$ is $\mathcal{C}_\alpha$-sectorial of angle $\pi/2$, with $\mathcal{C}_\alpha$ being defined by $\mathcal{C}_\alpha := (-\mathcal{A}_{\omega+\sigma})^{-1-\lfloor\alpha\rfloor}\mathcal{C}^2$. Then we can construct the power $(-\mathcal{A}_{\omega+\sigma})_z$ for any $z \in \mathbb{C}$. Keeping in mind the representation formula for $(S_{\alpha+1}(t))_{t\geqslant 0}$, given in the formulation of the above-mentioned lemma, it is not difficult to prove that the following theorem holds.

Theorem 2.9.66. *Let* $\alpha \in (0, \infty)\setminus\mathbb{N}$, *let* $\varepsilon > 0$ *such that* $\lfloor\alpha\rfloor = \lfloor\alpha + \varepsilon\rfloor$, *and let* $\sigma \in (0, 1]$. *Suppose that A is the integral generator of an α-times integrated C-cosine function* $(C_\alpha(t))_{t\geqslant 0}$ *satisfying that the family* $\{e^{-\omega t}C_\alpha(t) : t \geqslant 0\}$ *is equicontinuous for some* $\omega \geqslant 0$. *Then, for every* $(x_0, y_0) \in D((-\mathcal{A}_{\omega+\sigma})_{\alpha+\varepsilon+1}) \cap D((-\mathcal{A}_{\omega+\sigma})_{\alpha+\varepsilon-\lfloor\alpha+\varepsilon\rfloor})$, *the abstract Cauchy problem* (ACP_2) *has a unique mild solution, denoted by* $u(t, x_0, y_0)$, *and for every* $\varepsilon > 0$, *the set* $\{e^{-(\omega+\sigma+\varepsilon)t}u(t, x_0, y_0) : t \geqslant 0\}$ *is bounded. If, in addition,*
$\mathcal{A}_{\omega+\sigma}x_0' \in D((-\mathcal{A}_{\omega+\sigma})_{\alpha+\varepsilon+1}) \cap D((-\mathcal{A}_{\omega+\sigma})_{\alpha+\varepsilon-\lfloor\alpha+\varepsilon\rfloor})$, *then the solution is classical.*

Remark 2.9.67. Suppose that $C = 1$ and the family $\{(1 + t^\gamma)^{-1}C_\alpha(t) : t \geqslant 0\}$ is equicontinuous for some $\gamma \geqslant 0$. By the foregoing, we have that, for every $(x_0, y_0) \in D((-\mathcal{A}_{\omega+\sigma})_{\alpha+\varepsilon+1}) \cap D((-\mathcal{A}_{\omega+\sigma})_{\alpha+\varepsilon-\lfloor\alpha+\varepsilon\rfloor})$, the mild solution $u(t, x_0, y_0)$ has the property that the set
$\{(1 + t)^{-\max(\alpha+\varepsilon, \max(\alpha,\gamma+2)+\varepsilon, 2\max(\alpha,\gamma+2)-(\alpha+1)+\varepsilon)}u(t, x_0, y_0) : t \geqslant 0\}$ is bounded.

2.9.8. Representation of powers. The standard theory of fractional powers of generators of bounded strongly continuous semigroups on Banach spaces originated with the early works of A. V. Balakrishnan [32], T. Kato [253], E. Nelson [432], M. A. Krasnoselskii-P. E. Sobolevskii [328], H. Komatsu [288] and J. L. Lions-J. Peetre [372] (cf. also [403, Subsection 3.2] and [346]). For example, A. V. Balakrishnan defined in [32] the power A_γ, $0 < \gamma < 1$, by the formula

$$A_\gamma x = \frac{\gamma}{\Gamma(1-\gamma)} \int_0^\infty \frac{x - T(t)s}{t^\gamma} \frac{dt}{t},$$

with $D(A_\gamma)$ taken to be the set of those elements $x \in E$ for which the above integral converges and $(T(t))_{t\geqslant 0}$ being the bounded strongly continuous semigroup generated by $-A$.

Unless otherwise specified, we shall always assume in the sequel of this subsection that the operator $-A$ generates an equicontinuous (g_α, C)-regularized resolvent family for some $\alpha \in (0, 2]$.

Proposition 2.9.68. *Let $\alpha \in (0, 2]$, and let $-A$ generate the equicontinuous (g_α, C)-regularized resolvent family $(S_\alpha(t))_{t\geq 0}$. Then the following statements hold.*

(i) *If $\gamma \in \mathbb{C}_+$ and $\operatorname{Re} \gamma < 1$, then $C(D(A)) \subseteq D(A_\gamma)$ and*

$$A_\gamma Cx = c_{\alpha,\gamma} \int_0^\infty \frac{S_\alpha(t)x - Cx}{t^{\gamma\alpha}} \frac{dt}{t}, \quad x \in D(A), \tag{358}$$

where $c_{\alpha,\gamma} = \frac{\alpha}{\pi}\Gamma(\gamma\alpha + 1) \sin \gamma\pi$.

(ii) *if $\gamma \in \mathbb{C}_+ \setminus \{1\}$ and $\operatorname{Re} \gamma = 1$, then $C(D(A^2)) \subseteq D(A_\gamma)$ and*

$$A_\gamma Cx = c_{\alpha,\gamma} \left\{ \int_0^a t^{-\gamma\alpha-1} \left[S_\alpha(t)x - Cx + g_{\alpha+1}(t)CAx \right] dt + \int_a^\infty t^{-\gamma\alpha-1} S_\alpha(t)x \, dt \right. \tag{359}$$
$$\left. - \frac{1}{\Gamma(\alpha+1)} \frac{a^{\alpha-\gamma\alpha}}{\alpha - \gamma\alpha} CAx - \frac{a^{-\gamma\alpha}}{\gamma\alpha} Cx \right\}, \quad x \in D(A^2), \quad a > 0.$$

Proof. Let $\gamma \in \mathbb{C}_+$ satisfy $\operatorname{Re} \gamma < 1$, and let $x \in D(A)$. Observe that the integral $\int_0^\infty t^{-\gamma\alpha} (S_\alpha(t)x - Cx) \frac{dt}{t}$ is convergent since, for every $q \in ⊛$, there exist $c_q > 0$ and $r_q \in ⊛$ such that:

$$\int_1^\infty t^{-\gamma\alpha} q(S_\alpha(t)x - Cx) \frac{dt}{t} < +\infty$$

and

$$\int_0^1 t^{-\gamma\alpha} q(S_\alpha(t)x - Cx) \frac{dt}{t} = \int_0^1 t^{-\gamma\alpha} q((g_\alpha * S_\alpha)(t)Ax) \frac{dt}{t}$$
$$\leqslant c_q r_p(Ax) \int_0^1 t^{\alpha-\gamma\alpha-1} \, dt < +\infty.$$

The following simple computation

$$\begin{aligned}
A_\gamma Cx &= \frac{\sin \gamma\pi}{\pi} \int_0^\infty \lambda^{\gamma-1} (\lambda + A)^{-1} CAx \; d\lambda \\
&= \frac{\sin \gamma\pi}{\pi} \int_0^\infty (\lambda^{1/\alpha})^{\alpha(\gamma-1)} \lambda^{1/\alpha} (\lambda^{1/\alpha})^{-1} \left[(\lambda^{1/\alpha})^\alpha - (-A) \right]^{-1} CAx \; d\lambda \\
&= \frac{\sin \gamma\pi}{\pi} \int_0^\infty (\lambda^{1/\alpha})^{\alpha(\gamma-1)} \lambda^{1/\alpha} \int_0^\infty e^{-t\lambda^{1/\alpha}} (g_\alpha * S_\alpha)(t)Ax \, dt \, d\lambda \\
&= \frac{\sin \gamma\pi}{\pi} \int_0^\infty (S_\alpha(t)x - Cx) \int_0^\infty e^{-t\lambda^{1/\alpha}} (\lambda^{1/\alpha})^{\alpha(\gamma-1)} \lambda^{1/\alpha} \, d\lambda \, dt \\
&= \frac{\alpha}{\pi} \Gamma(\gamma\alpha + 1) \sin \gamma\pi \int_0^\infty t^{-\gamma\alpha} (S_\alpha(t)x - Cx) \frac{dt}{t}
\end{aligned}$$

shows that (i) holds. In order to prove (ii), observe first that the functional equation of $(S_\alpha(t))_{t\geq 0}$ implies that, for every $k \in \mathbb{N}_0$ and $t \geqslant 0$,

$$S_\alpha(t)x - \sum_{l=0}^{k}(-1)^l g_{l\alpha+1}(t)CA^l x = (g_{\alpha(k+1)} * S_\alpha)(t)A^k x, \quad x \in D(A^k).$$

As an immediate consequence of this equality, we get that, for every $q \in \circledast$ and $k \in \mathbb{N}_0$, there exist $c_{q,k} > 0$ and $r_{q,k} \in \circledast$ such that

(360) $$q\left(S_\alpha(t)x - \sum_{l=0}^{k}(-1)^l g_{l\alpha+1}(t)CA^l x\right) \leqslant c_{q,k} r_{q,k}(A^k x) g_{\alpha(k+1)+1}(t),\ t \geqslant 0,\ x \in D(A^k),$$

which clearly implies that the first integral in (359) converges for Re $\gamma = 1, \gamma \neq 1$. For fixed $x \in D(A^2)$ and $a > 0$, denote the right hand side of (359) by $G(\gamma)$; then the mapping $\gamma \to G(\gamma)$ is analytic for $0 < \text{Re}\,\gamma < 2$, $\gamma \neq 1$, and an almost trivial computation shows that the equality $A_\gamma Cx = G(\gamma)$ holds for every $\gamma \in \mathbb{C}_+$ with Re $\gamma < 1$. Also, we have that the mapping $\gamma \mapsto A_\gamma Cx$ $(0 < \text{Re}\,\gamma < 2)$ is analytic, so that the formula (359) follows from an application of the uniqueness theorem for analytic functions. □

Remark 2.9.69. (i) Suppose $\gamma \in \mathbb{C}_+$, $k \in \mathbb{N}$ and $k > \text{Re}\,\gamma$. By the foregoing, we have that

(361) $$\Lambda_k = C(D(A^k)) \cup \bigcup_{\lambda \in \Sigma_{\alpha\pi/2}} R((\lambda + A)^{-k}C) \subseteq D(A_\gamma)$$

and

(362) $$A_\gamma x = A_{\gamma - \lceil \text{Re}\,\gamma - 1 \rceil} A^{\lceil \text{Re}\,\gamma - 1 \rceil} x,\ x \in C(D(A^k)),$$

which clearly implies that for each $x \in C(D(A^k))$ we can simply represent the element $A_\gamma x$ in terms of $(S_\alpha(t))_{t \geqslant 0}$. Suppose now $y = (\mu + A)^{-k}Cx$ for some $x \in E$ and $\mu \in \Sigma_{\alpha\pi/2}$. If Re $\gamma \notin \mathbb{N}$, then we obtain from (361)-(362) and (358):

$$CA_\gamma y = A_\gamma C_y = A_{\gamma - \lceil \text{Re}\,\gamma - 1 \rceil} CA^{\lceil \text{Re}\,\gamma - 1 \rceil} y$$
$$= c_{\alpha,\gamma'} \int_0^\infty \frac{S_\alpha(t)A^{\lceil \text{Re}\,\gamma - 1 \rceil}y - CA^{\lceil \text{Re}\,\gamma - 1 \rceil}y}{t^{\gamma'\alpha}} \frac{dt}{t},$$

where $\gamma' = \gamma - \lceil \text{Re}\,\gamma - 1 \rceil$. Hence, $\int_0^\infty \frac{S_\alpha(t)A^{\lceil \text{Re}\,\gamma - 1 \rceil}y - CA^{\lceil \text{Re}\,\gamma - 1 \rceil}y}{t^{\gamma'\alpha}} \frac{dt}{t} \in R(C)$

and

$$A_\gamma y = c_{\alpha,\gamma'} C^{-1} \int_0^\infty \frac{S_\alpha(t)A^{\lceil \text{Re}\,\gamma - 1 \rceil}y - CA^{\lceil \text{Re}\,\gamma - 1 \rceil}y}{t^{\gamma'\alpha}} \frac{dt}{t}.$$

A similar representation formula can be obtained if Re $\gamma \in \mathbb{N}$.

(ii) Suppose, for the time being, $n \in \mathbb{N}$, $n - 1 < \text{Re}\,\gamma < n$, $x \in D(A^n)$, and consider the formula (358) with γ and x replaced respectively by $\gamma - (n - 1)$ and $A^{n-1}x$. In the case of semigroups ($\alpha = 1$), the estimate (360) shows that we can apply integration by parts $(n - 1)$-times in this formula. In such a way, we obtain that

$$A_\gamma Cx = \frac{1}{\Gamma(-\gamma)}\int_0^\infty t^{-\gamma-1}\left[S_1(t)x - \sum_{k=0}^{n-1}(-1)^n \frac{t^n}{n!}A^n Cx\right] dt,$$

which is a generalization of the formula [403, (3.15), p. 66]. Suppose now that $0 < \operatorname{Re}\gamma < n$, $\gamma \notin \mathbb{N}$, $a > 0$, $x \in D(A^n)$ and $\alpha = 1$. Since the mapping

$$\gamma \mapsto F(\gamma) \equiv \frac{1}{\Gamma(-\gamma)}\left\{\int_0^\alpha t^{-\gamma-1}\left[S_1(t)x - \sum_{k=0}^{n-1}(-1)^n \frac{t^n}{n!}A^n Cx\right] dt\right.$$

$$\left. + \int_\alpha^\infty t^{-\gamma-1}S_1(t)x\, dt + \sum_{k=0}^{n-1}(-1)^k \frac{a^{k-\gamma}}{k!(k-\gamma)}A^k Cx\right\}$$

is analytic for $0 < \operatorname{Re}\gamma < n$, $\gamma \notin \mathbb{N}$, and equals $A_\gamma Cx$ for $n-1 < \operatorname{Re}\gamma < n$, the uniqueness theorem for analytic functions implies that $A_\gamma Cx$ is given by $F(\gamma)$ for those values of γ; this is an extension of formula [403, (3.14)]. Observe also that the case $\alpha \neq 1$ can be treated similarly; for example, one can use the estimate (360) and the same trick as above so as conclude that the following formula holds:

$$A_\gamma Cx = c_{\alpha,\gamma}\left\{\int_0^a t^{-\gamma\alpha-1}\left[S_\alpha(t)x - \sum_{l=0}^{k}(-1)^l g_{l\alpha+1}(t)CA^l x\right] dt\right.$$

$$\left. + \int_0^\infty \frac{S_\alpha(t)x - Cx}{t^{\gamma\alpha+1}}\, dt + \sum_{l=0}^{k}(-1)^l \frac{a^{\alpha(l-\gamma)}}{\alpha(l-\gamma)\Gamma(\alpha l+1)}CA^l x\right\},$$

provided $k, n \in \mathbb{N}$, $k \geqslant n-1$, $0 < \operatorname{Re}\gamma < n$, $\gamma \notin \mathbb{N}$ and $x \in D(A^n)$. Applying integration by parts successively we get that the following equality holds provided that $x \in D(A)$, $0 < \operatorname{Re}\gamma < 1$ and $m \in \mathbb{N}_0$,

$$\text{(363)} \qquad A_\gamma Cx = c_{\alpha,\gamma}(1+\gamma\alpha)\cdots(m+\alpha\gamma)\int_0^\infty t^{-(m+1)-\gamma\alpha}(g_{\alpha+m} * S_\alpha)(t)Ax\, dt.$$

Replacing γ and x in (363) respectively with $\gamma-(n-1)$ and $A^{n-1}x$, we obtain that

$$A_\gamma Cx = (-1)^{n+1}\frac{\alpha}{\pi}\sin\gamma\pi\Gamma(m+1+\alpha(\gamma-(n-1)))$$

$$\times \int_0^\infty t^{-(\gamma-(n-1))\alpha-(m+1)}(g_{\alpha+m} * S_\alpha)(t)A^n x\, dt,$$

provided $n-1 < \operatorname{Re}\gamma < n$, $x \in D(A^n)$ and $m \in \mathbb{N}_0$. Since the repeated partial integration can be applied in the first integral of (359), we obtain similarly that, for every $n \in \mathbb{N}$ and for every $\gamma \in \mathbb{C}_+$ with $\operatorname{Re}\gamma = n$, the following formula holds:

$$A_\gamma Cx = c_{\alpha,\gamma-(n-1)} \left\{ \prod_{l=1}^{k} (l+(\gamma-(n-1))\alpha) \right.$$

$$\times \int_0^a t^{-(\gamma-(n-1))\alpha-(k+1)} (g_{2\alpha+k} * S_\alpha)(t) A^{n+1} x \, dt$$

$$+ \int_a^\infty t^{-(\gamma-(n-1))\alpha-1} S_\alpha(t) A^{n-1} x \, dt - \frac{1}{\Gamma(\alpha+1)} \frac{a^{\alpha-(\gamma-(n-1))\alpha}}{\alpha-(\gamma-(n-1))\alpha} CA^n x$$

$$\left. - \frac{a^{-(\gamma-(n-1))\alpha}}{(\gamma-(n-1))\alpha} CA^{n-1} x \right\}, \quad x \in D(A^{n+1}), \ a > 0, \ k \in \mathbb{N}.$$

(iii) Let $\gamma \in \mathbb{C}_+$, let $m, n \in \mathbb{N}$, and let $m \geqslant n > \operatorname{Re} \gamma > 0$. Then it can be easily verified that

$$A_\gamma Cx = \frac{\Gamma(m)}{\Gamma(\gamma)\Gamma(m-\gamma)} \int_0^\infty \lambda^{\gamma-1} \left[Cx - (\lambda + A)^{-1} Cx \right. \tag{364}$$

$$\left. - \sum_{j=2}^{m} (j-1)!^{-1} \frac{d^{j-1}}{d\lambda^{j-1}} (\lambda + A)^{-1} Cx \right] d\lambda, \quad x \in D(A^n).$$

On the other hand, we have that

$$(\lambda + A)^{-1} Cx = \lambda^{(1-\alpha)/\alpha} \int_0^\infty e^{-t\lambda^{1/\alpha}} S_\alpha(t) x \, dt, \quad \lambda > 0, x \in E, \tag{365}$$

and

$$\frac{d^{j-1}}{d\lambda^{j-1}} (\lambda + A)^{-1} Cx = \sum_{k=0}^{j-1} \binom{j-1}{k} \lambda^{\frac{1-\alpha}{\alpha}-(j-1-k)} \prod_{l=0}^{j-2-k} \left(\frac{1-\alpha}{\alpha} - l \right) \tag{366}$$

$$\times \sum_{k_0=1}^{k} c_{k_0,k,\alpha} \int_0^\infty e^{-t\lambda^{1/\alpha}} t^{k_0} S_\alpha(t) x \, dt, \quad \lambda > 0, \ x \in E, \ j \geqslant 2,$$

with $c_{k_0,k,\alpha}$ being the numbers defined directly before Lemma 2.6.1. Plugging (365)-(366) into (364), we obtain that

$$A_\gamma Cx = \frac{\Gamma(m)}{\Gamma(\gamma)\Gamma(m-\gamma)} \int_0^\infty \lambda^{\gamma-1} \left[Cx - \lambda^{(1-\alpha)/\alpha} \int_0^\infty e^{-t\lambda^{1/\alpha}} S_\alpha(t) x \, dt \right.$$

$$- \sum_{j=2}^{m} (j-1)!^{-1} \sum_{k=0}^{j-1} \binom{j-1}{k} \lambda^{\frac{1-\alpha}{\alpha}-(j-1-k)} \prod_{l=0}^{j-2-k} \left(\frac{1-\alpha}{\alpha} - l \right)$$

$$\left. \times \sum_{k_0=1}^{k} c_{k_0,k,\alpha} \int_0^\infty e^{-t\lambda^{1/\alpha}} t^{k_0} S_\alpha(t) x \, dt \right], \quad x \in D(A^n).$$

In the following theorem we shall prove a slight generalization of formulae [403, (3.17)-(3.18)], known also as the Balakrishnan-Komatsu-Lions-Peetre algorithm. Notice only that a corresponding Balakrishnan operator will be denoted by J_C^{γ}.

Theorem 2.9.70. *Suppose $-A$ generates an equicontinuous C-regularized semigroup. Let $\gamma \in \mathbb{C}_+$, let m, $n \in \mathbb{N}$, and let $m \geqslant n > \operatorname{Re}\gamma > 0$. Then the following holds: For every $x \in D(A^n)$, we have*

$$A_\gamma Cx = K_{\gamma,m}\int_0^\infty t^{-\gamma-1}\left[Cx + \sum_{k=1}^{m}\binom{m}{k}(-1)^k S_1(kt)x\right]dt, \tag{367}$$

where $K_{\gamma,m} := \int_0^\infty t^{-\gamma-1}(1-e^{-t})^m\,dt$. Furthermore, the preassumptions $x, y \in E$, $\lim_{\lambda\to\infty}\lambda^n(\lambda + A)^{-n}Cx = Cx$, $\lim_{\lambda\to\infty}\lambda^n(\lambda + A)^{-n}Cy = Cy$, $K_{\gamma,m} \neq 0$ and

$$\lim_{\kappa\to 0}\int_\kappa^\infty t^{-\gamma-1}\left[Cx + \sum_{k=1}^{m}\binom{m}{k}(-1)^k S_1(kt)x\right]dt = y$$

imply that $Cx \in J_C^{\gamma}$ and $J_C^{\gamma}Cx = (K_{\gamma,m})^{-1}Cy$; in particular, $Cx \in D(A_\gamma)$ and $A_\gamma Cx = (K_{\gamma,m})^{-1}y$.

Proof. Notice that for each $t \geqslant 0$ we have:

$$Cx + \sum_{k=1}^{m}\binom{m}{k}(-1)^k S_1(kt)x = [C^{-(m-n)}(C - S_1(t))^{m-n}]C^{-(n-1)}(C - S_1(t))^n$$

$$= (-1)^n[C^{-(m-n)}(C - S_1(t))^{m-n}]\int_0^t\int_t^{x_1+t}\cdots\int_t^{x_{n-1}+t} S_1(x_{n-1})A^n x\,dx_{n-1}\cdots d_{x_1}\,ds.$$

This implies that, for every $q \in \circledast$,

$$q\left(x + \sum_{k=1}^{m}\binom{m}{k}(-1)^k S_1(kt)x\right) = O(t^n) \text{ as } t \to 0+.$$

Therefore, the integral appearing in (367) is convergent. For the rest of the proof of this equality, it suffices to consider the case in which $x \in C^{m+2}(D_\infty(A))$; then (367) is equivalent with

$$A_\gamma C^m x = K_{\gamma,m}\int_0^\infty t^{-\gamma-1}(C - S_1(t))^m x\,dt. \tag{368}$$

Observing that the functions

$$t \mapsto t^{-\alpha+q}\frac{d^q}{dt^q}(C - S_1(t))^m x,\ t > 0\ (0 \leqslant q \leqslant n-1)$$

vanish when t goes to zero or to infinity, (368) simply follows from the arguments given in the proof of [403, Theorem 3.2.2]. The remaining part of the proof can also be given by using an insignificant modification of the corresponding parts of the proof of afore-mentioned theorem, and Remark 2.9.38. □

Theorem 2.9.70 can be reformulated in the case that the operator $-A$ generates an equicontinuous C-regularized cosine function (cf. the paper [231] by R. H. W. Hoppe). Unfortunately, we do not know in the present situation whether the Balakrishnan-Komatsu-Lions-Peetre algorithm admits a satisfactory reformulation in the case that $-A$ generates an equicontinuous (g_α, C)-regularized resolvent family for some $\alpha \in (0, 2)\setminus\{1\}$.

The following lemma serves as a useful tool in the characterization of fractional powers of the operator A.

Lemma 2.9.71. *Let $n \in \mathbb{N}_0$, $\alpha \in (0, 2]$, let $b \in \mathbb{C}_+$ such that $n < \operatorname{Re} b < n + 1$, and let $-A$ generate an equicontinuous (g_α, C)-regularized resolvent family $(S_\alpha(t))_{t\geqslant 0}$ on E. Then*

$$A_b Cx = c_{\alpha,b-n}\int_0^\infty \frac{S_\alpha(t)-C}{t^{(b-n)\alpha}}A^n x\frac{dt}{t},\quad x\in D(A^{n+1}) \tag{369}$$

$$= \frac{c_{\alpha,b-n}}{(b-n)\alpha}\lim_{N\to+\infty}\int_0^N \frac{S_\alpha'(t)A^n x}{t^{(b-n)\alpha-1}}\frac{dt}{t},\quad x\in D(A^{m+n+1}), \tag{370}$$

where $c_{\alpha,b} = (\alpha/\pi)\sin(b\pi)\Gamma(b\alpha + 1)$ and $m \in \mathbb{N}_0$ satisfies that $\alpha \geqslant 1/(m + 1)$. In particular, the second equality holds for each $x \in D(A^{n+1})$ if $1 \leqslant \alpha \leqslant 2$.

Proof. The equality (369) has been already proved. Furthermore,

$$\begin{aligned} S_\alpha(t)x &= Cx + (g_\alpha * S_\alpha)(t)(-A)x \quad (x \in D(A))\\ &= Cx - g_{\alpha+1}(t)ACx + (g_{2\alpha} * S_\alpha)(t)A^2x \quad (x \in D(A^2))\\ &= Cx - g_{\alpha+1}(t)ACx + \cdots + (-1)^m g_{m\alpha+1}(t)A^m Cx\\ &\quad + (g_{(m+1)\alpha} * S_\alpha)(t)A^{m+1}x \quad (x \in D(A^{m+1})), \end{aligned}$$

for any $t \geqslant 0$. Therefore, $S_\alpha(\cdot)A^n x \in C^1((0, \infty) : E)$ for each $x \in D(A^{m+n+1})$ and the equality (370) follows immediately from (369) by applying the partial integration. □

Before we state some special consequences of Lemma 2.9.71, observe that the analyticity of $(S_\alpha(t))_{t\geqslant 0}$ implies that the integral $\int_0^\infty \frac{S_\alpha'(t)A^n x}{t^{(b-n)\alpha-1}}\frac{dt}{t}\left(\int_1^\infty \frac{S_\alpha'(t)A^n x}{t^{(b-n)\alpha-1}}\frac{dt}{t}\right)$, appearing in (370), exists for all $x \in D(A^{m+n+1})$ $(x \in D(A^n))$; cf. also the formulation of Corollary 2.9.76(ii).

Remark 2.9.72. Suppose $C = 1$, $0 < \operatorname{Re} b < 1$ and E is a Banach space.

(i) By letting $\alpha = 1$ in Lemma 2.9.71 we get

$$A_b x = \frac{1}{\pi}\sin(b\pi)\Gamma(b+1)\int_0^\infty t^{-b}(T(t)x - x)\frac{dt}{t},\quad x\in D(A),$$

where $(T(t))_{t\geqslant 0}$ is the bounded C_0-semigroup generated by $-A$. This is the well known result in the theory of C_0-semigroups of operators.

(ii) Plugging $\alpha = 2$ in Lemma 2.9.71 one obtains

$$A_b x = \frac{2}{\pi}\sin(b\pi)\Gamma(2b+1)\int_0^\infty t^{-2b}(C(t)x - x)\frac{dt}{t}$$
$$= \frac{2}{\pi}\sin(b\pi)\Gamma(2b)\lim_{N\to+\infty}\int_0^N t^{-2b+1}C'(t)x\frac{dt}{t}, \quad x \in D(A),$$

where $(C(t))_{t\geqslant 0}$ is the bounded cosine operator function generated by $-A$. The first of these two equalities has been proved by R. H. W. Hoppe in [230]. Further on, by taking $b = 1/2$ in the second of these equalities, we get an old result of H. O. Fattorini [180]-[181]:

$$A_{1/2}x = \frac{2}{\pi}\lim_{N\to+\infty}\int_0^N C'(t)x\frac{dt}{t}, \quad x \in D(A).$$

In the following extension of [403, Lemma 6.1.5], we shall explicitly represent the powers with exponents of negative real part.

Lemma 2.9.73. *Let $\alpha \in (0, 2]$, let $b \in \mathbb{C}_+$ such that $0 < \operatorname{Re} b < 1$, and let $-A$ generate an equicontinuous (g_α, C)-regularized resolvent family $(S_\alpha(t))_{t\geqslant 0}$ on E. Suppose that A is injective.*

(i) *For every $x \in R(A)$, the following holds:*

$$A_{-b}Cx = c_{\alpha,1-b}\int_0^\infty \frac{(g_\alpha * S_\alpha(\cdot)x)(t)}{t^{(1-b)\alpha}}\frac{dt}{t}. \tag{371}$$

(ii) *Suppose additionally $1 \leqslant \alpha \leqslant 2$. Then, for every $x \in R(A)$, the following holds:*

$$A_{-b}Cx = c_{\alpha,1-b}((1-b)\alpha)^{-1}\lim_{N\to+\infty}\int_0^N \frac{(g_{\alpha-1} * S_\alpha(\cdot)x)(t)}{t^{(1-b)\alpha}}\,dt. \tag{372}$$

(iii) *If $\alpha = 1$ and the family $\{e^{\eta t}S_\alpha(t) : t \geqslant 0\}$ is equicontinuous for some $\eta > 0$, then (372) holds for any $b \in \mathbb{C}_+$ and $x \in E$.*

Proof. Suppose that $x = Ay$ for some $y \in E$. By the foregoing, we have that

$$A_{-b}Cx = (A^{-1})_b Cx = (A^{-1})_{b-1}A^{-1}Cx = A_{1-b}A^{-1}Cx = A_{1-b}Cy.$$

Therefore, the assertion (i) is an immediate consequence of Proposition 2.9.68. If $1 \leqslant \alpha \leqslant 2$, then the mapping $t \mapsto S_\alpha(t)y$, $t \geqslant 0$ is continuously differentiable with $S_\alpha'(t)y = (g_{\alpha-1} * S_\alpha(\cdot)Ay)(t) = (g_{\alpha-1} * S_\alpha(\cdot)x)(t)$, $t \geqslant 0$, so that we can apply integration by parts in (371). This implies (372) and completes the proof of (ii), the proof of (iii) is simple and therefore omitted. □

The assertion of [403, Theorem 6.1.6] can be reformulated for 'exponentially decaying' C-regularized semigroups in SCLCSs. Since domains of powers of C-sectorial operators are not stable under small translations, it is not clear from H. Komatsu's proof of the above-mentioned result (cf. [288, Theorem 2.10, Theorem 4.4]) whether the same holds for equicontinuous C-regularized semigroups. In connection with this problem, we would like to mention the papers [53] by H. Berens, P. L. Butzer, U. Westphal and [477] by S. Samko. Making use of the method proposed in [53], one can prove the following: Suppose $0 < \gamma < r$, where $r \in \mathbb{N}$, and $-A$ is the integral generator of an equicontinuous C-regularized semigroup $(T(t))_{t\geqslant 0}$. Then, for every $x \in D(A_\gamma)$, we have that the limit

$$(373)\qquad y = \lim_{\varepsilon\to 0+} \int_{\varepsilon}^{\infty} t^{-\gamma-1}\left[Cx + \sum_{k=1}^{r}\binom{r}{k}(-1)^k S_1(kt)x\right]dt,$$

exists in E. Furthermore,

$$A_\gamma Cx = K_{\gamma,r}\int_0^{\infty} t^{-\gamma-1}\left[Cx + \sum_{k=1}^{r}\binom{r}{k}(-1)^k S_1(kt)x\right]dt.$$

Albeit it seems plausible, it is not clear whether a modification of the above result continues to hold if $\gamma \in \mathbb{C}_+ \setminus (0, \infty)$ and $\operatorname{Re} \gamma < r$, since it is not clear whether in this case we have $\int_0^\infty |q_{\gamma,r}(u)|\, du < \infty$, with $q_{\gamma,r}(u)$ being the function defined in the fundamental lemma of [53].

Theorem 2.9.74. *Let $n \in \mathbb{N}_0$, $\alpha \in (0, 2]$, let $b \in \mathbb{C}_+$ such that $n < \operatorname{Re} b < n + 1$, and let $-A$ generate an equicontinuous (g_α, C)-regularized resolvent family $(S_\alpha(t))_{t\geqslant 0}$ on E. Put*

$$\mathfrak{K} := \left\{x \in D(A^n)\ ;\ \int_0^1 \frac{(g_\alpha * S_\alpha)(t)}{t^{(b-n)\alpha}} A^n x \frac{dt}{t} \in D(A)\right\}.$$

Then the following holds:

(i) *If $x \in \mathfrak{K}$, then $Cx \in D(A_b)$ and*

$$(374)\qquad A_b Cx = c_{\alpha,b-n}\left\{\int_1^\infty \frac{S_\alpha(t)-C}{t^{(b-n)\alpha}} A^n x\frac{dt}{t} + A\int_0^1 \frac{(g_\alpha * S_\alpha)(t)}{t^{(b-n)\alpha}} A^n x\frac{dt}{t}\right\}.$$

(ii) *If $x \in D(A_b) \cap D(A^n)$, then $x \in \mathfrak{K}$.*

Proof. Set $P_\mu := \mu(\mu + A)^{-1}C$ for $\mu > 0$. Then it is not difficult to prove (see, e.g., [20, Proposition 4.1.3] or [303, Theorem 3.6, Theorem 3.9]) that $\lim_{\mu\to+\infty} P_\mu x = Cx$ for all $x \in E$. Let $x \in \mathfrak{K}$. Then $P_\mu x \in D(A^{n+1})$ and it follows from (369) that:

$$\begin{aligned}
A_b CP_\mu x &= c_{\alpha,b-n}\left(\int_1^\infty + \int_0^1\right)\frac{S_\alpha(t)-C}{t^{(b-n)\alpha}} P_\mu A^n x\frac{dt}{t}\\
&= c_{\alpha,b-n}\left(P_\mu\int_1^\infty \frac{S_\alpha(t)-C}{t^{(b-n)\alpha}} A^n x\frac{dt}{t} + AP_\mu\int_0^1 \frac{(g_\alpha * S_\alpha)(t)}{t^{(b-n)\alpha}} A^n x\frac{dt}{t}\right)\\
&= c_{\alpha,b-n}\left(P_\mu\int_1^\infty \frac{S_\alpha(t)-C}{t^{(b-n)\alpha}} A^n x\frac{dt}{t} + P_\mu A\int_0^1 \frac{(g_\alpha * S_\alpha)(t)}{t^{(b-n)\alpha}} A^n x\frac{dt}{t}\right)\\
&\to c_{\alpha,b-n}\left(C\int_1^\infty \frac{S_\alpha(t)-C}{t^{(b-n)\alpha}} A^n x\frac{dt}{t} + CA\int_0^1 \frac{(g_\alpha * S_\alpha)(t)}{t^{(b-n)\alpha}} A^n x\frac{dt}{t}\right)
\end{aligned}$$

as $\mu \to +\infty$. Therefore, the closedness of A_b taken together with the equality $C^{-1}A_bC = C$ implies $Cx \in D(A_b)$ and (374), finishing the proof of (i). Suppose now $x \in D(A_b) \cap D(A^n)$. Then it can easily seen that $x \in \mathfrak{K}$, and that (374) holds iff

$$C\int_0^1 \frac{(g_\alpha * S_\alpha)(t)}{t^{(b-n)\alpha}} A^n x \frac{dt}{t} - (1+A)^{-1} C\int_0^1 \frac{(g_\alpha * S_\alpha)(t)}{t^{(b-n)\alpha}} A^n x \frac{dt}{t}$$

$$(375) \quad = (1+A)^{-1} C\left[(c_{\alpha,b-n})^{-1} CA_b x - \int_1^\infty \frac{S_\alpha(t) - C}{t^{(b-n)\alpha}} A^n x \frac{dt}{t} \right].$$

Let $0 < \gamma < 1/2$. Multiplying both sides of (375) with the operator $C^l((1+A)^{-1}C)^l \int_0^\infty f_\gamma(\lambda, t)(\lambda + A)^{-1}C \cdot d\lambda$ for $t > 0$ and $l \geqslant 2$ sufficiently large, and using the procedure described in Remark 2.9.33(i), it suffices to prove the validity of (375), or equivalently (374), for $x \in C^2(D_\infty(A))$. Put $C_\infty := C_{|D_\infty(A)}$. Since $(S_\alpha(t)_{|D_\infty(A)})_{t\geqslant 0}$ is the equicontinuous (g_α, C_∞)-regularized resolvent family in $D_\infty(A)$ generated by the operator $-A_\infty \in L(D_\infty(A))$, we simply obtain that the equality (374) holds for $x \in C^2(D_\infty(A))$, with integrals taken for the topology of $D_\infty(A)$. □

If $-A$ generates an equicontinuous $(g_\alpha, 1)$-regularized resolvent family $(S_\alpha(t))_{t\geqslant 0}$, then $D(A)$ must be dense in E (see the proof of [89, Proposition 3.3(d)]). If $C = 1$, then the inclusion $D(A_b) \subseteq D(A^n)$ also holds and we can state therefore the following corollary of Theorem 2.9.74:

Corollary 2.9.75. *Let $n \in \mathbb{N}_0$, $\alpha \in (0, 2]$, let $b \in \mathbb{C}_+$ such that $n < \operatorname{Re} b < n + 1$, and let $-A$ generate an equicontinuous $(g_\alpha, 1)$-regularized resolvent family $(S_\alpha(t))_{t\geqslant 0}$ on E. Then*

$$D(A_b) = \left\{ x \in D(A^n)\ ;\ \int_0^1 \frac{(g_\alpha * S_\alpha)(t)}{t^{(b-n)\alpha}} A^n x \frac{dt}{t} \in D(A) \right\},$$

and

$$A_b x = c_{\alpha,b-n} \left\{ \int_1^\infty \frac{S_\alpha(t) - 1}{t^{(b-n)\alpha}} A^n x \frac{dt}{t} + A\int_0^1 \frac{(g_\alpha * S_\alpha)(t)}{t^{(b-n)\alpha}} A^n x \frac{dt}{t} \right\}$$

for $x \in D(A_b)$.

Corollary 2.9.76. *Let $n \in \mathbb{N}_0$, $\alpha \in (0, 2]$, let $b \in \mathbb{C}_+$ such that $n < \operatorname{Re} b < n + 1$, and let $-A$ generate an equicontinuous (g_α, C)-regularized resolvent family $(S_\alpha(t))_{t\geqslant 0}$ on E. Then the following statements hold.*

(i) *If $x \in D(A^n)$ such that the limit $\lim_{\varepsilon\to 0+} \int_\varepsilon^1 \frac{S_\alpha(t)-C}{t^{(b-n)\alpha}} A^n x \frac{dt}{t}$ exists in E, then $Cx \in D(A_b)$ and*

$$(376) \quad A_b Cx = c_{\alpha,b-n} \left[\lim_{\varepsilon\to 0+} \int_\varepsilon^1 \frac{S_\alpha(t) - C}{t^{(b-n)\alpha}} A^n x \frac{dt}{t} + \int_1^\infty \frac{S_\alpha(t) - C}{t^{(b-n)\alpha}} A^n x \frac{dt}{t} \right].$$

(ii) *Let $x \in D(A^n)$ satisfy that the mapping $t \mapsto S_\alpha(t)A^n x$, $t \in (0, 1]$ is continuously differentiable and the limit $\lim_{\varepsilon\to 0+} \int_\varepsilon^1 t^{(n-b)\alpha} S_\alpha'(t)A^n x\ dt$ exists in E. Then $Cx \in D(A_b)$ and*

$$A_b Cx = c_{\alpha,b-n}\left\{\int_1^{\infty}\frac{S_\alpha(t)-C}{t^{(b-n)\alpha}}A^n x\frac{dt}{t}+\frac{S_\alpha(1)-C}{(n-b)\alpha}A^n x\right.$$
$$\left.-\frac{1}{(n-b)\alpha}\lim_{\varepsilon\to 0+}\int_\varepsilon^1 t^{(n-b)\alpha}S'_\alpha(t)A^n x\,dt\right\}.$$

Proof. Take any $x \in D(A^n)$ such that the limit $\lim_{\varepsilon\to 0+}\int_\varepsilon^1 \frac{S_\alpha(t)-C}{t^{(b-n)\alpha}}A^n x\frac{dt}{t}$ exists in E. Since

$$\lim_{\varepsilon\to 0+}\int_\varepsilon^1 \frac{(g_\alpha * S_\alpha)(t)}{t^{(b-n)\alpha}}A^n x\frac{dt}{t}=\int_0^1 \frac{(g_\alpha * S_\alpha)(t)}{t^{(b-n)\alpha}}A^n x\frac{dt}{t}$$

and

$$A\int_\varepsilon^1 \frac{(g_\alpha * S_\alpha)(t)}{t^{(b-n)\alpha}}A^n x\frac{dt}{t}=\int_\varepsilon^1 \frac{S_\alpha(t)-C}{t^{(b-n)\alpha}}A^n x\frac{dt}{t},$$

Theorem 2.9.74(i) and the closedness of A imply $Cx \in D(A_b)$ and (376). To prove (ii), consider again the operators $P_\mu = \mu(\mu + A)^{-1}C$ for $\mu > 0$. Since the integral $\int_0^1 \frac{S_\alpha(t)-C}{t^{(b-n)\alpha}}y\frac{dt}{t}$ exists for all $y \in D(A)$, we have that:

$$AP_\mu\int_0^1 \frac{(g_\alpha * S_\alpha)(t)}{t^{(b-n)\alpha}}A^n x\frac{dt}{t}=\int_0^1 \frac{S_\alpha(t)-C}{t^{(b-n)\alpha}}A^n P_\mu x\frac{dt}{t}$$
$$=\lim_{\varepsilon\to 0+}\int_0^1 \frac{S_\alpha(t)-C}{t^{(b-n)\alpha}}A^n P_\mu x\frac{dt}{t},\quad \mu>0.$$

Applying integration by parts in the last integral, we obtain that:

$$AP_\mu\int_0^1 \frac{(g_\alpha * S_\alpha)(t)}{t^{(b-n)\alpha}}A^n x\frac{dt}{t}$$
$$=P_\mu\frac{S_\alpha(1)-C}{(n-b)\alpha}A^n x-\frac{1}{(n-b)\alpha}P_\mu\lim_{\varepsilon\to 0+}\int_\varepsilon^1 t^{(n-b)\alpha}S'_\alpha(t)A^n x\,dt. \tag{377}$$

Letting $\mu \to +\infty$ in (377) yields $\int_0^1 \frac{(g_\alpha * S_\alpha)}{t^{(b-n)\alpha}}A^n x\frac{dt}{t} \in D(A)$ and:

$$A\int_0^1 \frac{(g_\alpha * S_\alpha)(t)}{t^{(b-n)\alpha}}A^n x\frac{dt}{t}=\frac{S_\alpha(1)-C}{(n-b)\alpha}A^n x-\frac{1}{(n-b)\alpha}\lim_{\varepsilon\to 0+}\int_\varepsilon^1 t^{(n-b)\alpha}S'_\alpha(t)A^n x\,dt.$$

The remaining part of the proof follows from Theorem 2.9.74(i). □

Remark 2.9.77. (i) Let $x \in D(A^n)$ satisfy that the mapping $t \mapsto S_\alpha(t)A^n x$, $t > 0$ is continuously differentiable and the limit

$$\lim_{N\to+\infty}\int_{1/N}^{N} t^{(n-b)\alpha}S'_\alpha(t)A^n x\,dt$$

exists in E. Due to Corollary 2.9.76(ii), we have that $Cx \in D(A_b)$; if this is the case, then one can prove that the following equality holds:

(378) $$A_b Cx = \frac{c_{\alpha,b-n}}{(b-n)\alpha} \lim_{N\to+\infty} \int_{1/N}^{N} t^{(n-b)\alpha} S_\alpha'(t) A^n x \, dt.$$

In connection with (378), we also want to mention in passing that the following result holds (cf. [109, Proposition 2.10]): Let E be a UMD Banach space, and let $-A$ be the generator of a bounded cosine operator function $(C(t))_{t\geqslant 0}$ on E. Then the integral

$$\lim_{N\to+\infty} \int_{1/N}^{N} C'(t)x \frac{dt}{t}$$

exists for all $x \in D(A^{1/2})$, and (378) holds with $C = 1$ and $b = 1/2$.

(ii) Let $\alpha \in (1, 2]$, and let $-A$ generate an equicontinuous (g_α, C)-regularized resolvent family $(S_\alpha(t))_{t\geqslant 0}$ on E. Set

$$\mathfrak{F} := \left\{ x \in E\,;\, S_\alpha(\cdot)x \in C^1((0,\infty):E) \text{ and } \lim_{\varepsilon\to 0+} \int_\varepsilon^1 S_\alpha'(t)\, x \frac{dt}{t} \text{ exists in } E \right\}.$$

Then Corollary 2.9.76(ii) implies that, for every $x \in \mathfrak{F}$, we have $Cx \in D(A_{1/\alpha})$. Consider now the most important case $C = 1$. Even then, the space $\{x \in E : S_\alpha(\cdot)x \in C^1((0,\infty):E)\}$ does not necessarily belong to $D(A_{1/\alpha})$ and we can illustrate this fact by giving a simple counterexample in the case $\alpha = 2$; assuming additionally that E is a UMD Banach space, then we have the following inclusion $\{x \in E : S_2(\cdot)x \in C^1((0,\infty):E)\} \subseteq D(A_{1/2})$.

In the following theorem, we shall further analyze domains of powers whose exponents have positive integers as real parts.

Theorem 2.9.78. *Let $n \in \mathbb{N}$, $\alpha \in (0, 2]$, let $b \in \mathbb{C}_+ \setminus \mathbb{N}$ such that* Re $b = n$, *and let $-A$ generate an equicontinuous (g_α, C)-regularized resolvent family $(S_\alpha(t))_{t\geqslant 0}$ on E. Set $b' := b - (n-1)$ and*

$$\mathfrak{L} := \left\{ x \in D(A^n)\,;\, \int_0^1 t^{-b'\alpha-1}(g_{2\alpha} * S_\alpha)(t) A^n x \, dt \in D(A) \right\}.$$

Then the following statements hold.

(i) *If $x \in \mathfrak{L}$, then $Cx \in D(A_b)$ and*

$$A_b Cx = c_{\alpha,b'} \left\{ A \int_0^1 t^{-b'\alpha-1}(g_{2\alpha} * S_\alpha)(t) A^n x \, dt \right.$$

(379) $$\left. - \int_1^\infty t^{-b'\alpha-1} S_\alpha(t) A^n x \, dt - \frac{CA^n x}{(\alpha - b'\alpha)\Gamma(\alpha+1)} - \frac{CA^{n-1}x}{b'\alpha} \right\}.$$

(ii) *If $x \in D(A_b) \cap D(A^n)$, then $x \in \mathfrak{L}$.*

(iii) *If* $x \in D(A^n)$ *such that the limit*

$$\lim_{\varepsilon\to 0+}\int_\varepsilon^1 t^{-(b-(n-1))\alpha-1}[S_\alpha(t)A^{n-1}x - CA^{n-1}x + g_{\alpha+1}(t)CA^n x]\,dt$$

exists in E, then $Cx \in D(A_b)$ *and*

$$A_b Cx = c_{\alpha,b-(n-1)}\Bigg\{-\frac{CA^n x}{\Gamma(\alpha+1)(\alpha-(b-(n-1))\alpha)} - \frac{A^{n-1}Cx}{(b-(n-1))\alpha}$$

$$+\lim_{\varepsilon\to 0+}\int_\varepsilon^1 t^{-(b-(n-1))\alpha-1}\Big[S_\alpha(t)A^{n-1}x - CA^{n-1}x + g_{\alpha+1}(t)CA^n x\Big]\,dt$$

$$+\int_1^\infty t^{-(b-(n-1))\alpha-1}S_\alpha(t)A^{n-1}x\,dt\Bigg\}.$$

Proof. We will only prove the assertions (i) and (iii) provided that $n = 1$; the proof of (ii) is similar to that of Theorem 2.9.74(ii) and therefore omitted. By the foregoing, we have that $C(D(A^2)) \subseteq D(A_b)$ and

$$A_b Cx = c_{\alpha,b}\Bigg\{\int_0^1 t^{-b\alpha-1}[S_\alpha(t)x - Cx + g_{\alpha+1}(t)CAx]\,dt + \int_1^\infty t^{-b\alpha-1}S_\alpha(t)x\,dt$$

$$(380)\qquad -\frac{1}{\Gamma(\alpha+1)}\frac{CAx}{\alpha-b\alpha} - \frac{Cx}{b\alpha}\Bigg\},\ x \in D(A^2).$$

Suppose, for the time being, $x \in \mathfrak{L}$. Since $AP_\mu \in L(E)$ for $\mu > 0$, $\int_0^1 t^{-b'\alpha-1}(g_{2\alpha} * S_\alpha)(t)Ax\,dt \in D(A)$,

$$P_\mu A\int_0^1 t^{-b\alpha-1}(g_{2\alpha} * S_\alpha)(t)Ax\,dt = AP_\mu\int_0^1 t^{-b\alpha-1}(g_{2\alpha} * S_\alpha)(t)Ax\,dt$$

$$=\int_0^1 t^{-b\alpha-1}\big[S_\alpha(t)P_\mu x - CP_\mu x + g_{\alpha+1}(t)CP_\mu Ax\big]\,dt,\quad \mu > 0,$$

and (380) holds with x replaced by $P_\mu x$, we get that:

$$A_b CP_\mu x = c_{\alpha,b}\Bigg\{P_\mu A\int_0^1 t^{-b\alpha-1}(g_{2\alpha} * S_\alpha)(t)Ax\,dt$$

$$(381)\qquad +P_\mu\int_1^\infty t^{-b\alpha-1}S_\alpha(t)x\,dt - \frac{1}{\Gamma(\alpha+1)}\frac{CAP_\mu x}{\alpha-b\alpha} - \frac{CP_\mu x}{b\alpha}\Bigg\}.$$

In (381) we can take the limit as μ approaches $+\infty$. Using after that the equality $C^{-1}A_b C = A_b$, we get that $Cx \in D(A_b)$ and (379) holds. If the limit $\lim_{\varepsilon\to 0+}\int_\varepsilon^1 t^{-b\alpha-1}[S_\alpha(t)x - Cx + g_{\alpha+1}(t)CAx]\,dt$ exists in E, then we have the following equality:

$$A_b CP_\mu x = c_{\alpha,b} \left\{ P_\mu \lim_{\varepsilon\to 0+} \int_\varepsilon^1 t^{-b\alpha-1}[S_\alpha(t)x - Cx + g_{\alpha+1}(t)CAx]\, dt \right.$$

$$\left. + P_\mu \int_1^\infty t^{-b\alpha-1} S_\alpha(t)x\, dt - \frac{1}{\Gamma(\alpha+1)} \frac{CAP_\mu x}{\alpha - b\alpha} - \frac{CP_\mu x}{b\alpha} \right\}, \quad x \in D(A),$$

so that the assertion (iii) can be proved similarly. □

Example 2.9.79. J. E. Galé, P. J. Miana and R. P. Stinga have recently extended in [191] results of Caffarelli-Silvestre and Stinga-Torrea regarding a representation of fractional powers of differential operators via the so called extension problem. Although the main results of the above-mentioned paper can be reconsidered and proved for certain classes of fractionally integrated C-semigroups and cosine functions with corresponding growth order, we shall briefly explain in the following example how one can express the element $A_\sigma Cx$ of E ($x \in D(A)$) in terms of solutions of the following incomplete abstract Cauchy problem:

$$(P): \begin{cases} u''(z) + \dfrac{1-2\sigma}{z} u'(z) = Au(z), \quad z \in \Sigma_{\pi/4}, \\ \lim_{z\to 0, z\in\Sigma_{(\pi/4)-\varepsilon}} u(z) = Cx \quad (x \in E,\ 0 < \varepsilon < \pi/4), \end{cases}$$

Where $0 < \mu < \sigma < 1$ and $-A$ generates a global C-regularized semigroup $(T(t))_{t\geqslant 0}$ satisfying that the family $\{(1 + t^\mu)^{-1}T(t) : t \geqslant 0\}$ is equicontinuous. First of all, notice that we have defined the power A_σ only if $\mu = 0$, or if $\mu > 0$ and A is injective. In these two cases, the following equality holds:

$$(382) \qquad A_\sigma Cx = \frac{\sin \sigma\pi}{\pi} \int_0^\infty \lambda^{\sigma-1}(\lambda + A)^{-1}Cx\, d\lambda, \quad x \in D(A);$$

if $\mu > 0$ and A is not injective, then the integral appearing in (382) converges absolutely so that the value of $A_\sigma Cx$ can be understood in the same way. Then, for every $x \in D(A)$,

$$(383) \qquad A_\sigma Cx = c_{1,\sigma} \int_0^\infty \frac{T(t)x - Cx}{t^{1+\sigma}}\, dt.$$

The arguments given in [191] show that an analytic solution of the problem (P) can be obtained through the formula

$$u(z) = \frac{z^{2\sigma}}{4^\sigma \Gamma(\sigma)} \int_0^\infty \frac{e^{-z^2/(4t)}}{t^{1+\sigma}} T(t)x\, dt.$$

Furthermore, this solution is bounded on closed subsectors of $\Sigma_{\pi/4}$ and the following equalities hold:

$$(384) \qquad \frac{u(z) - Cx}{z^{2\sigma}} = \frac{1}{4^\sigma \Gamma(\sigma)} \int_0^\infty \frac{e^{-z^2/(4t)}}{t^{1+\sigma}} [T(t)x - Cx]\, dt, \quad z \in \Sigma_{\pi/4}$$

and

$$(385) \qquad z^{1-2\sigma}u'(z) = \frac{1}{4^{\sigma}\Gamma(\sigma)}\int_0^{\infty}\left(2\sigma - \frac{z^2}{2t}\right)\frac{e^{-z^2/(4t)}}{t^{1+\sigma}}[T(t)x - Cx]\,dt, \quad z \in \Sigma_{\pi/4}.$$

Making use of the formulae (384)-(385), (369) as well as the dominated convergence theorem, it readily follows that, for every $x \in D(A)$,

$$4^{-\sigma}\Gamma(-\sigma)^{-1}A_{\sigma}Cx = \lim_{z\to 0}\frac{u(z)-Cx}{z^{2\sigma}} = \lim_{z\to 0}(2\sigma)^{-1}z^{1-2\sigma}u'(z),$$

on closed subsectors of $\Sigma_{\pi/4}$. It could be interesting to know whether the assumption $x \in D(A_{\sigma})$ implies the existence of limit $\lim_{z\to 0+} z^{-2\sigma}(u(z) - Cx)$ appearing in (384); observe here that the proofs of Theorem 2.9.74 and Corollary 2.9.75(i) show that the existence of above limit implies $Cx \in D(A_{\sigma})$ and
$A_{\sigma}Cx = c_{1,\sigma}4^{\sigma}\Gamma(\sigma)\Gamma(-\sigma)^{-1}\lim_{z\to 0+} z^{-2\sigma}(u(z) - Cx)$.
It is also clear that the operator $A + \varepsilon$ is C-sectorial of angle $\pi/2$ and (cf. Lemma 2.9.73):

$$(386) \qquad (A+\varepsilon)_{-\sigma}Cx = \int_0^{\infty} g_{\sigma}(s)e^{-\varepsilon s}T(s)x\,ds, \quad \varepsilon > 0, x \in E.$$

Suppose additionally $\mu + \sigma < 1$. Then, for every $x \in D(A)$, we can define the element $A_{1-\sigma}Cx$ by (382) and prove that its value is given by (383), with obvious replacement of number σ with $1 - \sigma$. Inserting the representation formula for $(A + \varepsilon)_{\sigma}Cx$, given in Lemma 2.9.71, in the equation (386), with A replaced by $A + \varepsilon$ therein, we get that

$$(387) \qquad C^2x = \frac{\sin\sigma\pi}{\pi}\int_0^{\infty}\int_0^{\infty} g_{\sigma}(t)e^{-\varepsilon t}T(t)\frac{e^{-\varepsilon\upsilon}T(\upsilon)x - Cx}{\upsilon^{1+\sigma}}\,d\upsilon\,dt, \quad x \in D(A).$$

Fix, for a moment, $x \in D(A)$. Define $\Omega_1 := \{(t, v) \in (0, \infty) \times (0, \infty) : v \leqslant 1\}$, $\Omega_2 := \{(t, v) \in (0, \infty) \times (0, \infty) : t \leqslant 1, v \geqslant 1\}$, $\Omega_3 := \{(t, v) \in (0, \infty) \times (0, \infty) : t \geqslant 1, v \geqslant 1\}$ and

$$I_{i,\varepsilon} := \int_{\Omega_i} g_{\sigma}(t)e^{-\varepsilon t}T(t)\frac{(e^{-\varepsilon\upsilon}-1)T(\upsilon)x}{\upsilon^{1+\sigma}}\,d\upsilon\,dt, \quad i = 1,2,3, \ \varepsilon > 0.$$

Using the estimates $1 - e^{-\varepsilon\upsilon} \leqslant \varepsilon\upsilon$, $\varepsilon > 0$, $\upsilon \geqslant 0$, $\sup_{\varepsilon\in(0,1]}\int_1^{\infty}\frac{1-e^{-\varepsilon\upsilon}}{\upsilon^{1+\sigma-\mu}}\,d\upsilon < \infty$ and $\lim_{\varepsilon\to 0+}\int_0^1 g_{\sigma}(t)e^{-\varepsilon t}T(t)x\,dt = 0$, it follows that $\lim_{\varepsilon\to 0+}(I_{1,\varepsilon} + I_{2,\varepsilon}) = 0$. Since the operator $B \equiv 0$ generates the C_0-semigroup $(S(t) \equiv 1)_{t\geqslant 0}$ on $\mathbb{C}$, and $\lim_{\varepsilon\to 0+}(\varepsilon + B)_{-(\sigma+\mu)}B_{\sigma+\mu}z = z$, $z \in \mathbb{C}$, we obtain similarly that $\lim_{\varepsilon\to 0+} I_{3,\varepsilon} = 0$. Hence,

$$\lim_{\varepsilon\to 0+}\int_0^{\infty}\int_0^{\infty} g_{\sigma}(t)e^{-\varepsilon t}T(t)\frac{(e^{-\varepsilon\upsilon}-1)T(\upsilon)x}{\upsilon^{1+\sigma}}\,dt\,d\upsilon = 0,$$

which further implies that

$$(388) \qquad \lim_{\varepsilon\to 0+}(\varepsilon + A)_{-\sigma}CA_{\sigma}Cx = C^2x, \quad x \in D(A).$$

Keeping in mind the following obvious equality

$$\frac{z^{2\sigma}}{4^{\sigma}\Gamma(\sigma)}\int_0^{\infty}\frac{e^{-z^2/(4t)}}{t^{1+\sigma}}e^{-\varepsilon t}T(t)\int_0^{\infty}g_{\sigma}(s)e^{-\varepsilon s}T(s)x\,ds\,dt=\int_0^{\infty}\frac{e^{-z^2/(4t)}}{\Gamma(\sigma)t^{1-\sigma}}e^{-\varepsilon t}T(t)Cx\,dt,$$

(388) shows that

$$Cu(z)=\lim_{\varepsilon\to 0+}\int_0^{\infty}\frac{e^{-z^2/(4t)}e^{-\varepsilon t}}{\Gamma(\sigma)t^{1-\sigma}}T(t)A_{\sigma}Cx\,dt,\quad x\in D(A). \tag{389}$$

On the other hand, it is not difficult to prove that

$$\frac{1}{\Gamma(\sigma)}\lim_{N\to+\infty}\int_0^{N}t^{\sigma-1}T(t)A_{\sigma}CA(1+A)^{-1}\,Cx\,dt=A(1+A)^{-1}C^3x,\quad x\in D(A). \tag{390}$$

Taking into account (389)-(390), we get that

$$Cu(z)=C^2x+\frac{1}{\Gamma(\sigma)}\int_0^{\infty}\frac{e^{-z^2/(4t)}-1}{t^{1-\sigma}}T(t)A_{\sigma}Cx\,dt,\quad x\in D(A),\ z\in\overline{\Sigma_{\pi/4}}.$$

In this way, we have proved an extension of [191, Theorem 2.1].

We continue by observing that V. Keyantuo investigated in [259] the generation of fractionally integrated semigroups and cosine functions by using some results from the theory of interpolation of semigroups; cf. [12]-[13] and [108] for some pioneering results in this direction. In the remaining part of this section, we shall continue the research contained in [259]. Our basic concepts are given as follows: By X we denote another SCLCS that is continuously embedded in E, the symbol $\circledast_X$ stands for the fundamental system of seminorms which defines the topology of X. Suppose that $-A$ is the integral generator of an equicontinuous C-regularized semigroup $(T(t))_{t\geqslant 0}$ on E, and $C(X)\subseteq X$. Define the operator A_X by

$$D(A_X):=\{x\in D(A)\cap X: Ax\in X\}\text{ and }A_Xx:=Ax,\ x\in D(A_X).$$

Then A_X is a closed linear operator on X. Since we have assumed that $C(X)\subseteq X$, the operator C_X, defined in a similar manner, is a bounded linear operator on X.

The proof of subsequent lemma follows, more or less, immediately from the arguments given in that of [259, Lemma 2.4], along with Corollary 2.9.76(i) and the equality $C^{-1}A_{\alpha}C=A_{\alpha}$; notice only that the assumption $0<\alpha<\beta\leqslant 1$ implies that, for every $p\in\circledast$, $s>0$ and $x\in E$, we have the following estimate:

$$p\left(\int_0^{s}g_{\beta}(s-\tau)T(t+\tau)x\,d\tau-\int_0^{s}g_{\beta}(s-\tau)T(\tau)x\,d\tau\right)=O(t^{\beta}),\quad t\in[0,1]. \tag{391}$$

Lemma 2.9.80. *Let* $0<\alpha<\beta\leqslant 1$. *Then, for every* $s\geqslant 0$ *and* $x\in E$, *one has* $(g_{\beta}*T)(s)x\in D(A_{\alpha})$ *and*

$$A_{\alpha}(g_{\beta}*T)(s)x=c_{1,\alpha}\int_0^{\infty}\int_0^{s}\frac{g_{\beta}(s-\tau)[T(t+\tau)x-T(\tau)x]}{t^{\alpha+1}}\,d\tau\,dt. \tag{392}$$

Now we are in a position to state the following theorem.

Theorem 2.9.81. *Suppose that $-A$ is the integral generator of an equicontinuous C-regularized semigroup $(T(t))_{t\geqslant 0}$ on E, and $C(X)\subseteq X$. Let $0<\alpha<\beta\leqslant 1$, let $D(A_\alpha)\subseteq X$, and let for each $p_X\in ⊛_X$ there exist $c_{pX}>0$ and $q_{pX}\in ⊛$ such that*

$$(393)\qquad p_X(x)\leqslant c_{pX}\,[q_{pX}(x)+q_{pX}(A_\alpha x)],\ x\in D(A_\alpha).$$

Define $U_\beta(t)x:=(g_\beta * T)(t)x$, $t\geqslant 0$, $x\in X$. Then $(U_\beta(t))_{t\geqslant 0}$ is a β-times integrated C_X-semigroup on X, the family $\{(1+t^\beta)^{-1}U_\beta(t) : t\geqslant 0\}\subseteq L(X)$ is equicontinuous, and the operator A_X is the integral generator of $(U_\beta(t))_{t\geqslant 0}$.

Proof. The proof is very similar to that of [259, Theorem 2.5], and we will only outline the main details. By Lemma 2.9.80 and (393), we have that the operator $U_\beta(t)$ belongs to the space $L(X)$ for all $t\geqslant 0$. The equicontinuity of operator family $\{(1+t^\beta)^{-1}U_\beta(t) : t\geqslant 0\}\subseteq L(X)$ is a consequence of (393) and a simple computation involving (392). Furthermore, it is not difficult to prove, with the help of (391)-(393) and the dominated convergence theorem, that $(U_\beta(t))_{t\geqslant 0}$ is a strongly continuous operator family in $L(X)$, so that the verification of fact that $(U_\beta(t))_{t\geqslant 0}$ is a non-degenerate β-times integrated C_X-semigroup with the integral generator A_X becomes almost trivial. □

The situation is a little complicated in the case that $D(A_{n+\alpha})\subseteq X$ for some $n\in\mathbb{N}$ and $\alpha\in(0,1)$. In order to better explain this, assume also that $\beta>\alpha$ and the condition (H) holds with $\omega=\pi/2$ and some $d\in(0,1]$. We adapt (392) by assuming that for each $p_X\in ⊛_X$ there exist $c_{pX}>0$ and $q_{pX}\in ⊛$ such that

$$(394)\qquad p_X(x)\leqslant c_{pX}\,[q_{pX}(x)+q_{pX}(A_{n+\alpha}x)],\ x\in D(A_{n+\alpha}).$$

Let $(\lambda+A)^{-k}C(X)\subseteq X$ for all $\lambda\in\mathbb{C}_+\cup B_d$ and $k\in\mathbb{N}$. Then the operator $((\lambda+A)^{-k}C)_X$ belongs to $L(X)$ for such values of parameters λ and k. Define $S_{n+\beta}(t)x:=(g_{n+\beta} * T)(t)x$, $t\geqslant 0$, $x\in X$. The proof of [259, Corollary 2.6] combined with the equality $A_{-n}=C^{-1}A^{-n}C$ $(n\in\mathbb{N})$ and the additivity of powers, shows that

$$(395)\qquad S_{n+\beta}(t)Cx=A^{-n}C(g_\beta * T)(t)x-\sum_{k=0}^{n-1}g_{\beta+k+1}(t)A^{k-n}Cx$$

$$(396)\qquad =A_{-(n+\alpha)}A_\alpha(g_\beta * T)(t)x-\sum_{k=0}^{n-1}g_{\beta+k+1}(t)A^{k-n}Cx,\quad t\geqslant 0,\ x\in E.$$

Keeping in mind (395)-(396), one can simply prove that the following holds:

(i) if $0\in\rho(A)$, or $A_{-(n+\alpha)}\in L(E)$ and $A_{-(n+\alpha)}(X)\subseteq X$, then $(S_{n+\beta}(t))_{t\geqslant 0}$ is an $(n+\beta)$-times integrated C_X-semigroup on X generated by A_X.

(ii) $(S_{n+\beta}(t)C_X)_{t\geqslant 0}$ is an $(n+\beta)$-times integrated C_X^2-semigroup on X generated by A_X.

We may state the following analog of Theorem 2.9.81 for equations of second order as follows.

Theorem 2.9.82. *Suppose $n \in \mathbb{N}_0$, $d \in (0, 1]$, $-A$ is the integral generator of an equicontinuous C-regularized cosine function $(C(t))_{t\geqslant 0}$ on E, and $C(X) \subseteq X$. Let $0 < \alpha < \beta \leqslant 1$, let $D(A_{n+\alpha}) \subseteq X$, and let (394) hold. Define $V_{2n+2\beta}(t)x := (g_{2n+2\beta} * C)(t)x$, $t \geqslant 0$, $x \in X$. Then the family $\{(1 + t^{2n+2\beta})^{-1} V_{2n+2\beta}(t) : t \geqslant 0\} \subseteq L(X)$ is equicontinuous and the following holds:*

(i) *$(V_{2n+2\beta}(t))_{t\geqslant 0}$ is a $(2n + 2\beta)$-times integrated C_X-semigroup on X generated by the operator A_X, provided that one of the following conditions holds:*
(a) $n = 0$,
(b) *$n \geqslant 1$, (HC) holds with $\omega = \pi/2$ and d, and $0 \in \rho(A)$,*
(c) *$n \geqslant 1$, (HC) holds with $\omega = \pi/2$ and d, $A_{-(n+\alpha)} \in L(E)$ and $A_{-(n+\alpha)}(X) \subseteq X$.*

(ii) *If $n \geqslant 1$ and (HC) holds with $\omega = \pi/2$ and $d \in (0,1]$, then $(V_{2n+2\beta}(t)C_X)_{t\geqslant 0}$ is a $(2n + 2\beta)$-times integrated C_X^2-semigroup on X generated by the operator A_X.*

It is an open problem to extend the assertions of Theorem 2.9.81-Theorem 2.9.82 to the case in which $-A$ generates an equicontinuous (g_α, C)-regularized resolvent family for some $\alpha \in (0, 2)\setminus\{1\}$.

Before proceeding to the next section, we would like to recommend for the reader the references [233], [299], [409], [452], [497], [521] and [543] for more details about fractional powers of operators and incomplete abstract Cauchy problems.

2.10 Abstract multi-term fractional differential equations

A great number of abstract fractional differential equations appearing in engineering, mathematical physics, chemistry and biology can be modeled through the abstract Cauchy problem (2). The aim of this section is to develop some operator-theoretical methods for solving abstract time-fractional equations of the form (2).
We assume that $n \in \mathbb{N}\setminus\{1\}$, A and $A_1, \cdots, A_{n-1}$ are closed linear operators on a sequentially complete locally convex space E, $0 \leqslant \alpha_1 < \cdots < \alpha_n$, $0 \leqslant \alpha < \alpha_n$ and $f(t)$ is an E-valued function. As before, $\mathbf{D}_t^\alpha$ denotes the Caputo fractional derivative of order α. We introduce and systematically analyze various classes of k-regularized (C_1, C_2)-existence and uniqueness (propagation) families for (2), thus continuing the researches raised in [141], [153], [308], [319] and [529].

Recall that a global (a, k)-regularized (C_1, C_2)-existence and uniqueness family $(R_1(t), R_2(t))_{t\geqslant 0}$ having A as a subgenerator, is locally equicontinuous (exponentially equicontinuous, (q-)exponentially equicontinuous, analytic, (q-)exponentially analytic, · · ·) if both $(R_1(t))_{t\geqslant 0}$ and $(R_2(t))_{t\geqslant 0}$ are. Henceforth we assume that $k, k_1, k_2, \cdots$ are scalar-valued kernels and $a \neq 0$ in $L^1_{loc}([0, \tau))$. All considered operator families will be non-degenerate. Set, finally, $m_j := \lceil \alpha_j \rceil$, $1 \leqslant j \leqslant n$, $m := m_0 := \lceil \alpha \rceil$, $A_0 := A$ and $\alpha_0 := \alpha$.

2.10.1. k-Regularized (C_1, C_2)-existence and uniqueness propagation families for (2). In this subsection, we shall clarify the main structural properties of

k-regularized (C_1, C_2)-existence and uniqueness propagation families. First of all, we need to introduce the following definition.

Definition 2.10.1. A function $u \in C^{m_n-1}([0, \infty) : E)$ is called a (strong) solution of (2) if $A_i \mathbf{D}_t^{\alpha_i} u \in C([0, \infty) : E)$ for $0 \leqslant i \leqslant n-1$, $g_{m_n-\alpha_n} * (u - \Sigma_{k=0}^{m_n-1} u_k g_{k+1}) \in C^{m_n}([0, \infty) : E)$ and (2) holds. The abstract Cauchy problem (2) is said to be (strongly) C-wellposed if:

(i) For every $u_0, \cdots, u_{m_n-1} \in \bigcap_{0 \leqslant j \leqslant n-1} C(D(A_j))$, there exists a unique solution $u(t; u_0, \cdots, u_{m_n-1})$ of (2).
(ii) For every $T > 0$ and $q \in \circledast$, there exist $c > 0$ and $r \in \circledast$ such that, for every $u_0, \cdots, u_{m_n-1} \in \bigcap_{0 \leqslant j \leqslant n-1} C(D(A_j))$, the following holds:

$$q(u(t; u_0, \cdots, u_{m_n-1})) \leqslant c \sum_{k=0}^{m_{n-1}} r(C^{-1}u_k), \quad t \in [0, T].$$

In the case of abstract Cauchy problem (ACP_n), the definition of C-wellposedness introduced above is slightly different from the corresponding definition introduced by T.-J. Xiao and J. Liang [531, Definition 5.2, p. 116] in the Banach space setting (cf. also [531, Definition 1.2, p. 46] for the case $C = I$). Recall that the notion of a strong C-propagation family is important in the study of existence and uniqueness of strong solutions of the abstract Cauchy problem (ACP_n); cf. [531, Section 3.5, pp. 115–130] for further information in this direction. Suppose now that $u(t) \equiv u(t; u_0, \cdots, u_{m_n-1})$, $t \geqslant 0$ is a strong solution of (2), with $f(t) \equiv 0$ and initial values $u_0, \cdots, u_{m_n-1} \in R(C)$. Convoluting both sides of (2) with $g_{\alpha_n}(t)$, and making use of the equality [49, (1.21)], it readily follows that $u(t)$, $t \geqslant 0$ satisfies the following:

$$u(\cdot) - \sum_{k=0}^{m_{n-1}} u_k g_{k+1}(\cdot) + \sum_{j=1}^{n-1} g_{\alpha_n - \alpha_j} * A_j \left[u(\cdot) - \sum_{k=0}^{m_j - 1} u_k g_{k+1}(\cdot) \right]$$

$$= g_{\alpha_n - \alpha} * A \left[u(\cdot) - \sum_{k=0}^{m-1} u_k g_{k+1}(\cdot) \right]. \tag{397}$$

In the sequel of this section, we shall primarily consider various types of solutions of the integral equation (397).

Given $i \in \mathbb{N}^0_{m_n-1}$ in advance, set $D_i := \{j \in \mathbb{N}_{n-1} : m_j - 1 \geqslant i\}$. Then it is clear that $D_{m_n-1} \subseteq \cdots \subseteq D_0$. Plugging $u_j = 0$, $0 \leqslant j \leqslant m_n - 1$, $j \neq i$, in (397), one gets:

$$\begin{aligned} &[u(\cdot; 0, \cdots, u_i, \cdots, 0) - u_i g_{i+1}(\cdot)] \\ &+ \sum_{j \in D_i} g_{\alpha_n - \alpha_j} * A_j [u(\cdot; 0, \cdots, u_i, \cdots, 0) - u_i g_{i+1}(\cdot)] \\ &+ \sum_{j \in \mathbb{N}_{n-1} \setminus D_i} [g_{\alpha_n - \alpha_j} * A_j u(\cdot; 0, \cdots, u_i, \cdots, 0)] \\ &= \begin{cases} g_{\alpha_n - \alpha} * Au(\cdot; 0, \cdots, u_i, \cdots, 0), & m - 1 < i, \\ g_{\alpha_n - \alpha} * A[u(\cdot; 0, \cdots, u_i, \cdots, 0) - u_i g_{i+1}(\cdot)], & m - 1 \geqslant i, \end{cases} \end{aligned} \tag{398}$$

where u_i appears in the i-th place $(0 \leqslant i \leqslant m_n - 1)$ starting from 0. Suppose now $0 < \tau \leqslant \infty$, $0 \neq K \in L^1_{loc}([0, \tau))$ and $k(t) = \int_0^t K(s)\,ds$, $t \in [0, \tau)$. Denote $R_i(t)C^{-1}u_i = (K * u(\cdot; 0, \cdots, u_i, \cdots, 0))(t)$, $t \in [0, \tau)$, $0 \leqslant i \leqslant m_n - 1$. Convoluting formally both sides of (398) with $K(t)$, $t \in [0, \tau)$, one obtains that, for $0 \leqslant i \leqslant m_n - 1$:

$$[R_i(\cdot)C^{-1}u_i - (k * g_i)(\cdot)u_i] + \sum_{j \in D_i} g_{\alpha_n - \alpha_j} * A_j [R_i(\cdot)C^{-1}u_i - (k * g_i)(\cdot)u_i]$$

$$+ \sum_{j \in \mathbb{N}_{n-1} \setminus D_i} [g_{\alpha_n - \alpha_j} * A_j R_i(\cdot)C^{-1}u_i]$$

$$= \begin{cases} (g_{\alpha_n - \alpha} * AR_i)(\cdot)C^{-1}u_i, & m - 1 < i, \\ g_{\alpha_n - \alpha} * A[R_i(\cdot)C^{-1}u_i - (k * g_i)(\cdot)u_i], & m - 1 \geqslant i. \end{cases}$$

Motivated by the above analysis, we introduce the following definition.

Definition 2.10.2. Suppose $0 < \tau \leqslant \infty$, $k \in C([0, \tau))$, $C, C_1, C_2 \in L(E)$, C and C_2 are injective. A sequence $((R_0(t))_{t \in [0,\tau)}, \cdots, (R_{m_n - 1}(t))_{t \in [0,\tau)})$ of strongly continuous operator families in $L(E)$ is called a (local, if $\tau < \infty$):

(i) k-regularized C_1-existence propagation family for (2) if $R_i(0) = (k * g_i)(0)C_1$ and the following holds:

$$[R_i(\cdot)x - (k * g_i)(\cdot)C_1 x] + \sum_{j \in D_i} A_j [g_{\alpha_n - \alpha_j} * (R_i(\cdot)x - (k * g_i)(\cdot)C_1 x)]$$

$$+ \sum_{j \in \mathbb{N}_{n-1} \setminus D_i} A_j (g_{\alpha_n - \alpha_j} * R_i)(\cdot)x$$

$$= \begin{cases} A(g_{\alpha_n - \alpha} * R_i)(\cdot)x, & m - 1 < i, x \in E, \\ A[g_{\alpha_n - \alpha} * (R_i(\cdot)x - (k * g_i)(\cdot)C_1 x)](\cdot), & m - 1 \geqslant i, x \in E, \end{cases} \tag{399}$$

for any $i = 0, \cdots, m_n - 1$.

(ii) k-regularized C_2-uniqueness propagation family for (2) if $R_i(0) = (k * g_i)(0) C_2$ and

$$[R_i(\cdot)x - (k * g_i)(\cdot)C_2 x] + \sum_{j \in D_i} g_{\alpha_n - \alpha_j} * [R_i(\cdot)A_j x - (k * g_i)(\cdot)C_2 A_j x]$$

$$+ \sum_{j \in \mathbb{N}_{n-1} \setminus D_i} (g_{\alpha_n - \alpha_j} * R_i(\cdot)A_j x)(\cdot)$$

$$= \begin{cases} (g_{\alpha_n - \alpha} * R_i(\cdot)Ax)(\cdot), & m - 1 < i, \\ g_{\alpha_n - \alpha} * [R_i(\cdot)Ax - (k * g_i)(\cdot)C_2 Ax](\cdot), & m - 1 \geqslant i, \end{cases} \tag{400}$$

for any $x \in \bigcap_{0 \leqslant j \leqslant n-1} D(A_j)$ and $i \in \mathbb{N}^0_{m_n - 1}$.

(iii) k-regularized C-resolvent propagation family for (2), in short k-regularized C-propagation family for (2), if $((R_0(t))_{t \in [0,\tau)}, \cdots, (R_{m_n - 1}(t))_{t \in [0,\tau)})$ is a k-regularized C-uniqueness propagation family for (2), and if for every $t \in [0, \tau)$, $i \in \mathbb{N}^0_{m_n - 1}$ and $j \in \mathbb{N}^0_{n-1}$, one has $R_i(t)A_j \subseteq A_j R_i(t)$, $R_i(t)C = CR_i(t)$ and $CA_j \subseteq A_j C$.

The above classes of propagation families can be defined by purely algebraic equations (cf. [89], [308] and [319]). We shall not go into further details about this topic here.

As announced before, we shall consider only non-degenerate k-regularized C-resolvent propagation families for (2). In case $k(t) = g_{\zeta+1}(t)$, where $\zeta \geqslant 0$, it is also said that $((R_0(t))_{t\in[0,\tau)}, \cdots, (R_{m_n-1}(t))_{t\in[0,\tau)})$ is a ζ-times integrated C-resolvent propagation family for (2); 0-times integrated C-resolvent propagation family for (2) is simply called C-resolvent propagation family for (2). For a k-regularized (C_1, C_2)-existence and uniqueness family $((R_0(t))_{t\in[0,\tau)}, \cdots, (R_{m_n-1}(t))_{t\in[0,\tau)})$, it is said that is locally equicontinuous (exponentially equicontinuous, (q-)exponentially equicontinuous, analytic, (q-)exponentially analytic, ···) if each single operator family $(R_0(t))_{t\in[0,\tau)}, \cdots, (R_{m_n-1}(t))_{t\in[0,\tau)}$ is. The above terminological agreements and abbreviations can be simply understood for the classes of k-regularized C_1-existence propagation families and k-regularized C_2-uniqueness propagation families. The class of k-regularized (C_1, C_2)-existence and uniqueness propagation families for (2) can also be introduced.

If $A_j = c_j I$, where $c_j \in \mathbb{C}$ for $1 \leqslant j \leqslant n-1$, then it is also said that the operator A is a subgenerator of $((R_0(t))_{t\in[0,\tau)}, \cdots, (R_{m_n-1}(t))_{t\in[0,\tau)})$. Assuming that A is a subgenerator of a k-regularized C-resolvent propagation family $((R_0(t))_{t\in[0,\tau)}, \cdots, (R_{m_n-1}(t))_{t\in[0,\tau)})$ for (2), then, in general, there do not exist $a_i \in L^1_{loc}([0, \tau))$, $i \in \mathbb{N}^0_{m_n-1}$ and $k_i \in C([0, \tau))$ such that $(R_i(t))_{t\in[0,\tau)}$ is an (a_i, k_i)-regularized C-resolvent family with subgenerator A; the same holds for the classes of k-regularized C_1-existence propagation families and k-regularized C_2-uniqueness propagation families. Despite this fact, the structural results for k-regularized C-resolvent propagation families can be derived by using appropriate modifications of the proofs of corresponding results for (a, k)-regularized C-resolvent families. Furthermore, these results can be clarified for any single operator family $(R_i(t))_{t\in[0,\tau)}$ of the tuple $((R_0(t))_{t\in[0,\tau)}, \cdots, (R_{m_n-1}(t))_{t\in[0,\tau)})$.

Let $((R_0(t))_{t\in[0,\tau)}, \cdots, (R_{m_n-1}(t))_{t\in[0,\tau)})$ be a k-regularized C-resolvent propagation family with subgenerator A. Then one can simply prove that the validity of condition (H5) implies the following functional equation:

$$[R_i(\cdot)x - (k * g_i)(\cdot)Cx] + \sum_{j=1}^{n-1} c_j g_{\alpha_n-\alpha_j} * [R_i(\cdot)x - (k * g_i)(\cdot)Cx]$$

$$+ \sum_{j\in\mathbb{N}_{n-1}\setminus D_i} c_j [g_{\alpha_n-\alpha_j+i} * k](\cdot)Cx$$

$$= \begin{cases} A[g_{\alpha_n-\alpha} * R_i](\cdot)x, & m-1 < i,\ x \in E, \\ A[g_{\alpha_n-\alpha} * (R_i(\cdot)x - (k * g_i)(\cdot)Cx)], & m-1 \geqslant i,\ x \in E, \end{cases} \tag{401}$$

for any $i = 0, \cdots, m_n - 1$. The set consisting of all subgenerators of $((R_0(t))_{t\in[0,\tau)}, \cdots, (R_{m_n-1}(t))_{t\in[0,\tau)})$, denoted by $\chi(R)$, need not be finite. If $A \in \chi(R)$, then $C^{-1}AC \in \chi(R)$ as well. The integral generator $\hat{A}$ of $((R_0(t))_{t\in[0,\tau)}, \cdots, (R_{m_n-1}(t))_{t\in[0,\tau)})$ is defined as the set of those pairs $(x, y) \in E \times E$ such that, for every $i = 0, \cdots, m_n - 1$ and for every $t \in [0, \tau)$, the following holds:

$$[R_i(\cdot)x - (k * g_i)(\cdot)Cx] + \sum_{j=1}^{n-1} c_j g_{\alpha_n-\alpha_j} * [R_i(\cdot)x - (k * g_i)(\cdot)Cx]$$

$$+ \sum_{j\in \mathbb{N}_{n-1}\setminus D_i} c_j[g_{\alpha_n-\alpha_j+i} * k](\cdot)Cx$$

$$= \begin{cases} [g_{\alpha_n-\alpha} * R_i](\cdot)y, & m-1 < i, \\ g_{\alpha_n-\alpha} * [R_i(\cdot)y - (k * g_i)(\cdot)Cy)], & m-1 \geqslant i. \end{cases}$$

It is a linear operator on E which extends any subgenerator $A \in \chi(R)$ and satisfies $\hat{A} = C^{-1}\hat{A}C$. We have the following:

(i) $R_i(t)(\lambda - A)^{-1}C = (\lambda - A)^{-1}CR_i(t)$, $t \in [0, \tau)$, provided $A \in \chi(R)$, $\lambda \in \rho_C(A)$ and $0 \leqslant i \leqslant m_n - 1$.

(ii) Let $((R_0(t))_{t\in[0,\tau)}, \cdots, (R_{m_n-1}(t))_{t\in[0,\tau)})$ be locally equicontinuous. Then:
 (a) $\hat{A}$ is a closed linear operator.
 (b) $\hat{A} \in \chi(R)$, if $R_i(t)R_i(s) = R_i(s)R_i(t)$, $0 \leqslant t, s < \tau$, $i \in \mathbb{N}^0_{m_n-1}$.
 (c) $\hat{A} = C^{-1}AC$, if $A \in \chi(R)$ and (H5) holds. Furthermore, the condition (H5) can be replaced by (401).

(iv) Let $\{A, B\} \subseteq \chi(R)$. Then $Ax = Bx$, $x \in D(A) \cap D(B)$, and $A \subseteq B \Leftrightarrow D(A) \subseteq D(B)$. Assume that (401) holds for A, and that (401) holds with A replaced by B. Then the following holds:
 (a) $C^{-1}AC = C^{-1}BC$ and $C(D(A)) \subseteq D(B)$.
 (b) A and B have the same eigenvalues.
 (c) $A \subseteq B \Rightarrow \rho_C(A) \subseteq \rho_C(B)$.

Albeit similar assertions can be considered in a general case, we shall omit the corresponding discussion even in the case that $A_j \in L(E)$ for $1 \leqslant j \leqslant n - 1$.

Proposition 2.10.3. *Let* $i \in \mathbb{N}^0_{m_n-1}$, *and let* $((R_0(t))_{t\in[0,\tau)}, \cdots, (R_{m_n-1}(t))_{t\in[0,\tau)})$ *be a locally equicontinuous k-regularized C-resolvent propagation family for (2). If (399) holds with* $C_1 = C$, *then the following holds:*

(i) *The equality*

$$R_i(t)R_i(s) = R_i(s)R_i(t),\ 0 \leqslant t, s < \tau \tag{402}$$

holds, provided that $m - 1 < i$ *and the condition*
 (◊) *Any of the assumptions* $f(t)+\Sigma_{j\in D_i} A_j(g_{\alpha_n-\alpha_j} * f)(t) = 0$, $t \in [0, \tau)$, *or* A $(g_{\alpha_n-\alpha} * f)(t) = 0$, *for some* $f \in C([0, \tau) : E)$, *implies* $f(t) = 0$, $t \in [0, \tau)$, *holds.*

(ii) *The equality (402) holds provided* $m - 1 \geqslant i$, $\mathbb{N}_{n-1}\setminus D_i \neq \emptyset$, *and the condition*
 (◊◊) *If* $\Sigma_{j\in\mathbb{N}_{n-1}\setminus D_i} A_j(g_{\alpha_n-\alpha_j} * f)(t) = 0$, $t \in [0, \tau)$, *for some* $f \in C([0, \tau) : E)$, *then* $f(t) = 0$, $t \in [0, \tau)$, *holds.*

Proof. Let $x \in E$ and $s \in [0, \tau)$ be fixed. Define $u_i(t) := R_i(t)R_i(s)x - R_i(s)R_i(t)x$, $t \in [0, \tau)$. Keeping in mind (399), it can be simply shown that:

$$(403)\quad A\int_0^t g_{\alpha_n-\alpha}(t-r)u(r)\,dr = u(t)+\sum_{j=1}^{n-1}\int_0^t A_j(g_{\alpha_n-\alpha_j} * u)(r)\,dr = 0,\quad t\in[0,\tau).$$

Let $m-1<i$. Convoluting both sides of (403) with $R_i(\cdot)$, we easily infer that $u(t)+\Sigma_{j=1}^{n-1}A_j(g_{\alpha_n-\alpha_j} * u)(t)=0$, $t\in[0,\tau)$ and $A(g_{\alpha_n-\alpha} * u)(t)=0$, $t\in[0,\tau)$. Now the equality (402) follows from (◊). The proof is quite similar in the case $m-1\geqslant i$. □

Proposition 2.10.4. *Suppose* $((R_{j,0}(t))_{t\in[0,\tau)}, \cdot\cdot\cdot, (R_{j,m_n-1}(t))_{t\in[0,\tau)})$ *is a locally equicontinuous* k_j*-regularized C-resolvent propagation family for (2),* $j=1,2$, *and* $0\leqslant i\leqslant m_n-1$. *Then we have the following.*
(i) *If* $m-1<i$ *and (◊) holds, then*

$$(404)\qquad (k_1 * R_{2,i})(t)x=(k_2 * R_{1,i})(t)x,\quad x\in\bigcap_{j=0}^{n-1}D(A_j),\ t\in[0,\tau).$$

If, additionally,

$$(405)\qquad \bigcap_{j=0}^{n-1}D(A_j)\ \text{is dense in } E,$$

then (404) holds for all $x\in E$.
(ii) *The equality (404) holds provided* $m-1\geqslant i$, $\mathbb{N}_{n-1}\backslash D_i\neq\emptyset$ *and (◊◊); assuming additionally (405), we have the validity of (404) for all* $x\in E$.

Proof. We will only prove the second part of proposition. Let $x\in\bigcap_{j=0}^{n-1}D(A_j)$. Then the functional equation of $(R_{j,i}(t))_{t\in[0,\tau)}$ $(j=1,2)$ implies:

$$\begin{aligned}
&[(k_2 * g_i) * (R_{1,i}(\cdot)x-(k_1 * g_i)(\cdot)Cx)](\cdot)\\
&=\Big\{R_{2,i}(\cdot)+\sum_{j\in D_i} g_{\alpha_n-\alpha_j} * [R_{2,i}(\cdot)A_j-(k * g_i)(\cdot)CA_j]\\
&\quad+\sum_{j\notin D_i} g_{\alpha_n-\alpha_j} * R_{2,i}(\cdot)A_j - g_{\alpha_n-\alpha} * [R_{2,i}(\cdot)A-(k * g_i)(\cdot)CA]\Big\}\\
&\quad * [R_{1,i}(\cdot)x-(k * g_i)(\cdot)Cx](\cdot)\\
&=\Big\{R_{2,i}(\cdot)+\sum_{j\in D_i} g_{\alpha_n-\alpha_j} * [R_{2,i}(\cdot)A_j-(k * g_i)(\cdot)CA_j]\\
&\quad+\sum_{j\notin D_i} g_{\alpha_n-\alpha_j} * R_{2,i}(\cdot)A_j\Big\} * [R_{1,i}(\cdot)-(k_1 * g_i)(\cdot)Cx](\cdot)\\
&\quad-[R_{2,i}(\cdot)x-(k_2 * g_i)(\cdot)C] * A(g_{\alpha_n-\alpha} * [R_{1,i}(\cdot)x-(k_1 * g_i)(\cdot)Cx])(\cdot),
\end{aligned}$$

which yields after a tedious computation:

$$\sum_{j\notin D_i} g_{\alpha_n-\alpha_j} * A_j[(k_2 * R_{1,i})(\cdot)-(k_1 * R_{2,i})(\cdot)]\equiv 0.$$

In view of (◊◊), the above equality shows that $(k_2 * R_{1,i})(t)x=(k_1 * R_{2,i})(t)x$, $t\in[0,\tau)$. It can be verified that the condition (405) implies the validity of (402) for all $x\in E$. □

Proposition 2.10.5. *Let $((R_0(t))_{t\in[0,\tau)}, \cdots, (R_{m_n-1}(t))_{t\in[0,\tau)})$ be a locally equicontinuous k-regularized C_1-existence propagation family (k-regularized C_2-uniqueness propagation family, k-regularized C-resolvent propagation family) for (2), and let $b \in L^1_{loc}([0, \tau))$ be a kernal. Then the tuple $(((b * R_0)(t))_{t\in[0,\tau)}, \cdots, ((b * R_{m_n-1})(t))_{t\in[0,\tau)})$ is a locally equicontinuous $(k * b)$-regularized C_1-existence propagation family ($(k * b)$-regularized C_2-uniqueness propagation family, $(k * b)$-regularized C-resolvent propagation family) for (2).*

Suppose now E is complete, (2) is C-wellposed, $\bigcap_{j=0}^{n-1} D(A_j)$ is dense in E and $0 \leqslant i \leqslant m_n - 1$. Set $R_i(t)x := u(t; 0, \cdots, Cx, \cdots, 0)(t)$, $t \geqslant 0$, $x \in \bigcap_{j=0}^{n-1} D(A_j)$, where $0 \leqslant i \leqslant m_n - 1$ and Cx appears in the i-th place in the preceding expression. Since we have assumed that E is complete, the operator $R_i(t)$ $(t \geqslant 0)$ can be uniquely extended to a bounded linear operator on E. It can be easily proved that $((R_0(t))_{t\in[0,\tau)}, \cdots, (R_{m_n-1}(t))_{t\in[0,\tau)})$ is a locally equicontinuous C-uniqueness propagation family for (2), and that the assumption $CA_j \subseteq A_jC$, $j \in \mathbb{N}^0_{n-1}$ implies $R_i(t)C = CR_i(t)$, $t \geqslant 0$. In the case that $A_j = c_jI$, where $c_j \in \mathbb{C}$ for $1 \leqslant j \leqslant n - 1$, one can apply the arguments given in the proof of [463, Proposition 1.1, p. 32] in order to see that $((R_0(t))_{t\in[0,\tau)}, \cdots, (R_{m_n-1}(t))_{t\in[0,\tau)})$ is a locally equicontinuous C-resolvent propagation family for (2). Regrettably, it is not clear how one can prove in general case that $R_i(t)A_j \subseteq A_jR_i(t)$, $j \in \mathbb{N}^0_{n-1}$, $t \geqslant 0$.

Definition 2.10.6. Let $T > 0$ and $f \in C([0, T] : E)$. Consider the following inhomogeneous equation:

$$u(t) + \sum_{j=1}^{n-1} (g_{\alpha_n-\alpha_j} * A_j u)(t) = f(t) + (g_{\alpha_n-\alpha} * Au)(t), \quad t \in [0, T]. \tag{406}$$

A function $u \in C([0, T] : E)$ is said to be:

(i) a strong solution of (406) if $A_j u \in C([0, T] : E)$, $j \in \mathbb{N}^0_{n-1}$ and (406) holds for every $t \in [0, T]$.

(ii) a mild solution of (406) if $(g_{\alpha_n-\alpha_j} * u)(t) \in D(A_j)$, $t \in [0, T]$, $j \in \mathbb{N}^0_{n-1}$ and

$$u(t) + \sum_{j=1}^{n-1} A_j(g_{\alpha_n-\alpha_j} * u)(t) = f(t) + A(g_{\alpha_n-\alpha} * u)(t), \quad t \in [0, T].$$

It is clear that every strong solution of (406) is also a mild solution of the same problem. The converse statement is not true, in general. One can similarly define the notion of a strong (mild) solution of the problem (397).

Let $0 < \tau \leqslant \infty$, and let $T \in (0, \tau)$. Then the following holds:

(a) If $((R_0(t))_{t\in[0,\tau)}, \cdots, (R_{m_n-1}(t))_{t\in[0,\tau)})$ is a C_1-existence propagation family for (2), then the function $u(t) = \sum_{i=0}^{m_n-1} R_i(t)x_i$, $t \in [0, T]$ is a mild solution of (397) with $u_i = C_1x_i$ for $0 \leqslant i \leqslant m_n - 1$.

(b) If $((R_0(t))_{t\in[0,\tau)}, \cdots, (R_{m_n-1}(t))_{t\in[0,\tau)})$ is a C_2-uniqueness propagation family for (2), and $A_jR_i(t)x = R_i(t)A_jx$, $t \in [0, T]$, $x \in \cap_{j=0}^{n-1} D(A_j)$, $i \in \mathbb{N}^0_{m_n-1}$, $j \in \mathbb{N}^0_{n-1}$, then the function $u(t) = \Sigma_{i=0}^{m_n-1} R_i(t)C_2^{-1}u_i$, $t \in [0, T]$ is a strong solution of (397), provided $u_i \in C_2(\cap_{j=0}^{n-1} D(A_j))$ for $0 \leqslant i \leqslant m_n - 1$.

Theorem 2.10.7. *Suppose $((R_0(t))_{t\in[0,\tau)}, \cdots, (R_{m_n-1}(t))_{t\in[0,\tau)})$ is a locally equicontinuous k-regularized C_2-uniqueness propagation family for (2), (399) holds, $T \in (0, \tau)$ and $f \in C([0, T] : E)$. Then the following holds:*

(i) *If $m - 1 < i$, then any strong solution $u(t)$ of (406) satisfies the equality:*

$$(R_i * f)(t) = (k * g_i * C_2u)(t) + \sum_{j\in D_i} (g_{\alpha_n-\alpha_j+i} * k * C_2A_ju)(t),$$

for any $t \in [0, T]$. Therefore, there is at most one strong (mild) solution for (406), provided that (◊) holds.

(ii) *If $m - 1 \geqslant i$, then any strong mild solution $u(t)$ of (406) satisfies the equality:*

$$(R_i * f)(t) = -\sum_{j\in\mathbb{N}_{n-1}\setminus D_i} (g_{\alpha_n-\alpha_j+i} * k * C_2A_ju)(t),\ t \in [0, T].$$

Therefore, there is at most one strong (mild) solution for (406), provided that $\mathbb{N}_{n-1}\setminus D_i \neq \emptyset$ and (◊◊) holds.

Proof. We will only prove the second part of theorem. Let $m - 1 \geqslant i$. Taking into account (400), we get:

$$\begin{aligned}
&[R_i - (k * g_iC)] * f \\
&= [R_i - (k * g_iC)] * \left\{u + \sum_{j=1}^{n-1}(g_{\alpha_n-\alpha_j} * A_ju) - (g_{\alpha_n-\alpha} * Au)\right\} \\
&= [R_i - (k * g_iC)] * \left(u + \sum_{j=1}^{n-1}(g_{\alpha_n-\alpha_j} * A_ju)\right) \\
&= \Bigg\{[R_i - (k * g_iC)] + \sum_{j\in D_i}[g_{\alpha_n-\alpha_j} * (R_i(\cdot)A_jx - (k * g_i)(\cdot)C_2A_jx)] \\
&+ \sum_{j\notin D_i}(g_{\alpha_n-\alpha_j} * R_i(\cdot)A_jx)\Bigg\} * u \\
&= -\sum_{\mathbb{N}_{n-1}\setminus D_i}(g_{\alpha_n-\alpha_j+i} * k * C_2A_ju), \quad t \in [0, T].
\end{aligned}$$

This implies the uniqueness of strong solutions to (406), provided that $\mathbb{N}_{n-1}\setminus D_i \neq \emptyset$ and (◊◊) holds. The uniqueness of mild solutions in the above case follows from the fact that, for every such a solution $u(t)$, there exists a sufficiently large $\zeta > 0$ such that the function $(g_\zeta * u)(\cdot)$ is a strong solution of (406), with $f(\cdot)$ replaced by $(g_\zeta * f)(\cdot)$ therein. □

The subsequent theorems can be shown by modifying the arguments given in the proof of [292, Theorem 2.2.1].

Theorem 2.10.8. *Suppose $k(t)$ satisfies (P1), $\omega \geqslant \max(0, abs(k))$, $(R_i(t))_{t\geqslant 0}$ is strongly continuous, and the family $\{e^{-\omega t}R_i(t) : t \geqslant 0\}$ is equicontinuous, provided $0 \leqslant i \leqslant m_n - 1$. Let A be a closed linear operator on E, let $C_1, C_2 \in L(E)$, and let C_2 be injective. Set*

$$P_\lambda := \lambda^{\alpha_n-\alpha} + \sum_{j=1}^{n-1} \lambda^{\alpha_j-\alpha} A_j - A, \lambda \in \mathbb{C}\setminus\{0\}.$$

(i) *Suppose $A_j \in L(E)$, $j \in \mathbb{N}_{n-1}$. Then $((R_0(t))_{t\geqslant 0}, \cdots, (R_{m_n-1}(t))_{t\geqslant 0})$ is a global k-regularized C_1-existence propagation family for (2) if the following conditions hold.*

(a) *The equality*

$$P_\lambda \int_0^\infty e^{-\lambda t} R_i(t)x\,dt = \lambda^{\alpha_n-\alpha-i}\,\tilde{k}(\lambda)C_1x + \sum_{j\in D_i} \lambda^{\alpha_j-\alpha-i}\,\tilde{k}(\lambda)A_jC_1x, \tag{407}$$

holds provided $x \in E$, $i \in \mathbb{N}^0_{m_n-1}$, $m - 1 < i$ and $\operatorname{Re}\lambda > \omega$.

(b) *The equality*

$$\begin{aligned} &P_\lambda \int_0^\infty e^{-\lambda t}[R_i(t)x - (k * g_i)(t)C_1x]\,dt \\ &= -\sum_{j\in\mathbb{N}_{n-1}\setminus D_i} \lambda^{\alpha_j-\alpha-i}\,\tilde{k}(\lambda)A_jC_1x, \end{aligned} \tag{408}$$

holds provided $x \in E$, $i \in \mathbb{N}^0_{m_n-1}$, $m - 1 \geqslant i$ and $\operatorname{Re}\lambda > \omega$.

(ii) *Suppose $R_i(0) = (k * g_i)(0)C_2x$, $x \in E\setminus\overline{\bigcap_{0\leqslant j\leqslant n-1} D(A_j)}$, $i \in \mathbb{N}^0_{m_n-1}$. Then $((R_0(t))_{t\geqslant 0}, \cdots, (R_{m_n-1}(t))_{t\geqslant 0})$ is a global k-regularized C_2-uniqueness propagation family for (2) if, for every $\lambda \in \mathbb{C}$ with $\operatorname{Re}\lambda > \omega$, and for every $x \in \bigcap_{0\leqslant j\leqslant n-1} D(A_j)$, the following equality holds:*

$$\begin{aligned} &\int_0^\infty e^{-\lambda t}[R_i(t)x - (k * g_i)(t)C_2x]\,dt \\ &\quad + \sum_{j\in D_i} \lambda^{\alpha_j-\alpha_n} \int_0^\infty e^{-\lambda t}[R_i(t)x - (k * g_i)(t)C_2A_jx]\,dt \\ &\quad + \sum_{j\in\mathbb{N}_{n-1}\setminus D_i} \lambda^{\alpha_j-\alpha_n} \int_0^\infty e^{-\lambda t} R_i(t)A_jx\,dt \\ &\quad = \begin{cases} \lambda^{\alpha-\alpha_n} \int_0^\infty e^{-\lambda t} R_i(t)Ax\,dt, \\ \lambda^{\alpha-\alpha_n} \int_0^\infty e^{-\lambda t}[R_i(t)Ax - (k * g_i)(t)C_2Ax]\,dt, \; m-1 \geqslant i. \end{cases} \end{aligned}$$

Theorem 2.10.9. *Suppose $k(t)$ satisfies (P1), $\omega \geqslant \max(0, abs(k))$, $(R_i(t))_{t\geqslant 0}$ is strongly continuous, and the family $\{e^{-\omega t}R_i(t) : t \geqslant 0\}$ is equicontinuous, provided $0 \leqslant i \leqslant m_n - 1$. Let $CA_j \subseteq A_jC$, $j \in \mathbb{N}^0_{n-1}$, $A_j \in L(E)$, $j \in \mathbb{N}_{n-1}$, $A_iA_j = A_jA_i$, $i, j \in \mathbb{N}_{n-1}$*

and $A_jA \subseteq AA_j$, $j \in \mathbb{N}_{n-1}$. *Assume, additionally, that the operator* $\lambda^{\alpha_n - i} + \Sigma_{j \in D_i} \lambda^{\alpha_j - i} A_j$ *is injective for every* $i \in \mathbb{N}^0_{m_n - 1}$ *with* $m - 1 < i$ *and for every* $\lambda \in \mathbb{C}$ *with* $\operatorname{Re} \lambda > \omega$ *and* $\tilde{k}(\lambda) \neq 0$, *and that the operator* $\Sigma_{j \in \mathbb{N}_{n-1} \setminus D_i} \lambda^{\alpha_j - i} A_j$ *is injective for every* $i \in \mathbb{N}^0_{m_n - 1}$ *with* $m - 1 \geqslant i$ *and for every* $\lambda \in \mathbb{C}$ *with* $\operatorname{Re} \lambda > \omega$ *and* $\tilde{k}(\lambda) \neq 0$. *Then* $((R_0(t))_{t \geqslant 0}, \cdots, (R_{m_n - 1}(t))_{t \geqslant 0})$ *is a global k-regularized C-resolvent propagation family for (2), and (399) holds, iff the equalities (407)-(408) are fulfilled.*

Keeping in mind Theorem 2.10.9, one can simply clarify the most important Hille-Yosida type theorems for exponentially equicontinuous *k*-regularized *C*-resolvent propagation families. Notice also that similar assertions can be proved for *k*-regularized (C_1, C_2)-existence and uniqueness resolvent propagation families.

The analytical properties of *k*-regularized *C*-resolvent propagation families are stated in the following two theorems whose proofs are omitted.

Theorem 2.10.10. *Suppose* $\beta \in (0, \frac{\pi}{2}]$, $((R_0(t))_{t \geqslant 0}, \cdots, (R_{m_n - 1}(t))_{t \geqslant 0})$ *is an analytic k-regularized C-resolvent propagation family for (2), k(t) satisfies (P1), (399) holds, and* $\tilde{k}(\lambda)$ *can be analytically continued to a function* $\hat{k} : \omega + \Sigma_{\frac{\pi}{2} + \beta} \to \mathbb{C}$, *where* $\omega \geqslant \max(0, abs(k))$. *Suppose* $CA_j \subseteq A_jC$, $j \in \mathbb{N}^0_{n-1}$, $A_j \in L(E)$, $j \in \mathbb{N}_{n-1}$, $A_iA_j = A_jA_i$, $i, j \in \mathbb{N}_{n-1}$ *and* $A_jA \subseteq AA_j$, $j \in \mathbb{N}_{n-1}$. *Let the family* $\{e^{-\omega z} R_i(z) : z \in \Sigma_\gamma\}$ *be equicontinuous, provided* $i \in \mathbb{N}^0_{m_n - 1}$ *and* $\gamma \in (0, \beta)$, *and let the set*

$$\left\{(\lambda - \omega)\hat{k}(\lambda)\lambda^{-i} : \lambda \in \omega + \Sigma_{\frac{\pi}{2} + \gamma}\right\}$$

be bounded provided $\gamma \in (0, \beta)$ *and* $m - 1 \geqslant i$. *Set*

$$N_i := \left\{\lambda \in \omega + \Sigma_{\frac{\pi}{2} + \beta} : \hat{k}(\lambda)\left(\lambda^{\alpha_n} + \sum_{j \in D_i} \lambda^{\alpha_j} A_j\right) \text{ is injective}\right\},$$

provided $m - 1 < i$, *and*

$$N_i := \left\{\lambda \in \omega + \Sigma_{\frac{\pi}{2} + \beta} : \hat{k}(\lambda)\left(\lambda^{\alpha_n} + \sum_{j \in \mathbb{N}_{n-1} \setminus D_i} \lambda^{\alpha_j} A_j\right) \text{ is injective}\right\},$$

provided $m - 1 \geqslant i$. *Suppose* N_i *is an open connected subset of* $\mathbb{C}$, *and the set* $N_i \cap \{\lambda \in \mathbb{C} : \operatorname{Re} \lambda > \omega\}$ *has a limit point in* $\{\lambda \in \mathbb{C} : \operatorname{Re} \lambda > \omega\}$, *for any* $i \in \mathbb{N}^0_{m_n - 1}$. *Then the operator* P_λ *is injective for every* $\lambda \in N_i$ *and* $i \in \mathbb{N}^0_{m_n - 1}$,

$$\lim_{\lambda \to +\infty, \lambda \in N_i} \lambda \tilde{k}(\lambda) P_\lambda^{-1} \left(\lambda^{\alpha_n - \alpha - i} + \sum_{j \in D_i} \lambda^{\alpha_j - \alpha - i} A_j\right) Cx = (k * g_i)(0)Cx,$$

provided $m - 1 < i$ *and* $x \in E$, *and*

$$\lim_{\lambda \to +\infty, \lambda \in N_i} \lambda \tilde{k}(\lambda) P_\lambda^{-1} \sum_{j \in \mathbb{N}_{n-1} \setminus D_i} \lambda^{\alpha_j - \alpha - i} A_j Cx = 0,$$

provided $m - 1 \geqslant i$ *and* $x \in E$. *Suppose, additionally, that there exists* $\mu \in \mathbb{C}$ *such that* $P_\mu^{-1} C \in L(E)$. *Then the family*

$$\left\{(\lambda-\omega)\,\hat{k}(\lambda)\left(\lambda^{\alpha_n-\alpha}+\sum_{j=1}^{n-1}\lambda^{\alpha_j-\alpha}A_j-C^{-1}AC\right)^{-1}\right.$$

$$\left.\times\left(\lambda^{\alpha_n-\alpha-i}C+\sum_{j\in D_i}\lambda^{\alpha_j-\alpha-i}A_jC\right):\lambda\in N_i\cap\left(\omega+\sum_{\frac{\pi}{2}+\gamma}\right)\right\} \text{ is equicontinuous,}$$

provided $m-1<i$ *and* $\gamma\in(0,\beta)$, *resp., the family*

$$\left\{(\lambda-\omega)\,\hat{k}(\lambda)\left(\lambda^{\alpha_n-\alpha}+\sum_{j=1}^{n-1}\lambda^{\alpha_j-\alpha}A_j-C^{-1}AC\right)^{-1}\sum_{j\in\mathbb{N}_{n-1}\setminus D_i}\lambda^{\alpha_j-\alpha-i}A_jC\right.$$

$$\left.:\lambda\in N_i\cap\left(\omega+\sum_{\frac{\pi}{2}+\gamma}\right)\right\} \text{ is equicontinuous,}$$

provided $m-1\geqslant i$ *and* $\gamma\in(0,\beta)$, *the mapping*

$$\lambda\mapsto\left(\lambda^{\alpha_n-\alpha}+\sum_{j=1}^{n-1}\lambda^{\alpha_j-\alpha}A_j-C^{-1}AC\right)^{-1}\left(\lambda^{\alpha_n-\alpha-i}C+\sum_{j\in D_i}\lambda^{\alpha_j-\alpha-i}A_jC\right)x,$$

defined for $\lambda\in N_i$, *is analytic, provided* $m-1<i$ *and* $x\in E$, *and the mapping*

$$\lambda\mapsto\left(\lambda^{\alpha_n-\alpha}+\sum_{j=1}^{n-1}\lambda^{\alpha_j-\alpha}A_j-C^{-1}AC\right)^{-1}\sum_{j\in\mathbb{N}_{n-1}\setminus D_i}\lambda^{\alpha_j-\alpha-i}A_jCx,\ \lambda\in N_i,$$

is analytic, provided $m-1\geqslant i$ *and* $x\in E$.

Theorem 2.10.11. *Assume* $k(t)$ *satisfies (P1),* $\omega\geqslant\max(0,abs(k))$, $\beta\in(0,\frac{\pi}{2}]$ *and, for every* $i\in\mathbb{N}^0_{m_n-1}$ *with* $m-1\geqslant i$, *the function* $(k*g_i)(t)$ *can be analytically extended to a function* $k_i:\Sigma_\beta\to\mathbb{C}$ *satisfying that, for every* $\gamma\in(0,\beta)$, *the set* $\{e^{-\omega z}k_i(z):z\in\Sigma_\gamma\}$ *is bounded. Let* $CA_j\subseteq A_jC$, $j\in\mathbb{N}^0_{n-1}$, $A_j\in L(E)$, $j\in\mathbb{N}_{n-1}$, $A_iA_j=A_jA_i$, $i,j\in\mathbb{N}_{n-1}$ *and* $A_jA\subseteq AA_j$, $j\in\mathbb{N}_{n-1}$. *Assume, additionally, that for each* $i\in\mathbb{N}^0_{m_n-1}$ *the set* $V_i:=N_i\cap\{\lambda\in\mathbb{C}:\operatorname{Re}\lambda>\omega\}$ *contains the set* $\{\lambda\in\mathbb{C}:\operatorname{Re}\lambda>\omega,\tilde{k}(\lambda)\neq0\}$, *and that* $R(\lambda^{\alpha_n}C+\Sigma_{j\in D_i}\lambda^{\alpha_j}A_jC)\subseteq R(P_\lambda)$, *provided* $m-1<i$ *and* $\lambda\in V_i$, *resp.* $R(\lambda^{\alpha_n}C+\Sigma_{j\in\mathbb{N}_{n-1}\setminus D_i}\lambda^{\alpha_j}A_jC)\subseteq R(P_\lambda)$, *provided* $m-1\geqslant i$ *and* $\lambda\in V_i$ *(cf. the formulation of preceding theorem). Suppose also that the operator* $\lambda^{\alpha_n}I+\Sigma_{j\in D_i}\lambda^{\alpha_j}A_j$ *is injective, provided* $m-1<i$ *and* $\lambda\in V_i$, *and that the operator* $\lambda^{\alpha_n}I+\Sigma_{j\in\mathbb{N}_{n-1}\setminus D_i}\lambda^{\alpha_j}A_j$ *is injective, provided* $m-1\geqslant i$ *and* $\lambda\in V_i$. *Let* $q_i:\omega+\Sigma_{\frac{\pi}{2}+\beta}\to L(E)$ $(0\leqslant i\leqslant m_n-1)$ *satisfy that, for every* $x\in E$, *the mapping* $\lambda\mapsto q_i(\lambda)x$, $\lambda\in\omega+\Sigma_{\frac{\pi}{2}+\beta}$ *is analytic as well as that:*

$$q_i(\lambda)x=\tilde{k}(\lambda)P_\lambda^{-1}\left(\lambda^{\alpha_n-\alpha-i}C+\sum_{j\in D_i}\lambda^{\alpha_j-\alpha-i}A_jC\right)x,\quad x\in E,\ \lambda\in V_i,$$

provided $m-1<i$,

$$q_i(\lambda)x=-\tilde{k}(\lambda)P_\lambda^{-1}\sum_{j\in\mathbb{N}_{n-1}\setminus D_i}\lambda^{\alpha_j-\alpha-i}A_jCx,\quad x\in E,\ \lambda\in V_i,$$

provided $m-1\geqslant i$,

the family $\left\{(\lambda-\omega)q_i(\lambda):\lambda\in\omega+\sum_{\frac{\pi}{2}+\gamma}\right\}$ *is equicontinuous for all* $\gamma\in(0,\beta)$,

and, in the case $\overline{D(A)}\neq E$,

$$\lim_{\lambda\to+\infty} \lambda q_i(\lambda)x = \begin{cases} (k * g_i)(0)Cx, & x \notin \overline{D(A)},\ m-1 < i, \\ 0, & x \notin \overline{D(A)},\ m-1 \geqslant i. \end{cases}$$

Then there exists an exponentially equicontinuous, analytic k-regularized C-resolvent propagation family $((R_0(t))_{t\geqslant 0}, \cdots, (R_{m_n-1}(t))_{t\geqslant 0})$ *for (2). Furthermore, the family* $\{e^{-\omega z}R_i(z) : z \in \Sigma_\gamma\}$ *is equicontinuous for all* $i \in \mathbb{N}^0_{m_n-1}$ *and* $\gamma \in (0, \beta)$*, (399) holds, and* $R_i(z)A_j \subseteq A_j R_i(z)$, $z \in \Sigma_\beta$, $j \in \mathbb{N}^0_{n-1}$.

The entire solutions of abstract Cauchy problem (ACP_n) have been analyzed by T.-J. Xiao and J. Liang in [540, Theorem 2.1]. Notice that their result cannot be so easily interpreted for the general equation of the kind (2) and that it would be very interesting to clarify some results on the existence and uniqueness of solutions of (2) which can be analytically extended to the region $\mathbb{C}\setminus(-\infty, 0]$. Also, the interested reader may try to state some results about differential properties of k-regularized C-resolvent (propagation) families (cf. [292] and Section 2.2 for more details).

In the following theorem, which has several obvious consequences, we consider q-exponentially equicontinuous k-regularized I-resolvent propagation families in complete locally convex spaces (cf. Section 2.4 for more details).

Theorem 2.10.12. (i) *Suppose* $k(0) \neq 0$, $((R_0(t))_{t\geqslant 0}, \cdots, (R_{m_n-1}(t))_{t\geqslant 0})$ *is a q-exponentially equicontinuous k-regularized I-resolvent propagation family for (2),* $A_j \in L_{\circledast}(E)$, $j \in \mathbb{N}_{n-1}$*, and for every* $p \in \circledast$*, there exist* $M_p \geqslant 1$ *and* $\omega_p \geqslant 0$ *such that:*

$$p(R_i(t)x) \leqslant M_p e^{\omega_p t} p(x),\ t \geqslant 0,\ x \in E,\ 0 \leqslant i \leqslant m_n - 1. \tag{409}$$

Then A is a compartmentalized operator and, for every seminorm $p \in \circledast$, $((\overline{R_{0,p}(t)})_{t\geqslant 0}, \cdots, (\overline{R_{m_n-1,p}(t)})_{t\geqslant 0})$ *is an exponentially bounded k-regularized* $\overline{I_p}$*-resolvent propagation family for (2), in* $\overline{E_p}$*, with* A_j *replaced by* $\overline{A_{j,p}}$ $(0 \leqslant j \leqslant n-1)$*. Furthermore,*

$$\|\overline{R_{i,p}(t)}\| \leqslant M_p e^{\omega_p t},\ t \geqslant 0,\ 0 \leqslant i \leqslant m_n - 1, \tag{410}$$

and $((\overline{R_{0,p}(t)})_{t\geqslant 0}, \cdots, (\overline{R_{m_n-1,p}(t)})_{t\geqslant 0})$ *is a q-exponentially equicontinuous, analytic k-regularized* $\overline{I_p}$*-resolvent propagation family of angle* $\beta \in (0, \pi]$*, provided that* $((R_0(t))_{t\geqslant 0}, \cdots, (R_{m_n-1}(t))_{t\geqslant 0})$ *is. Assume additionally that (399) holds. Then, for every* $p \in \circledast$*, (399) holds with* A_j *and* $((R_0(t))_{t\geqslant 0}, \cdots, (R_{m_n-1}(t))_{t\geqslant 0})$ *replaced by* $\overline{A_{j,p}}$ *and* $((\overline{R_{0,p}(t)})_{t\geqslant 0}, \cdots, (\overline{R_{m_n-1,p}(t)})_{t\geqslant 0})$*, respectively.*

(ii) *Suppose* $k(t)$ *satisfies (P1), E is complete, A is a compartmentalized operator in E,* $A_j = c_j I$ *for some* $c_j \in \mathbb{C}$ $(1 \leqslant j \leqslant n-1)$ *and, for every* $p \in \circledast$, $\overline{A_p}$ *is a subgenerator (the integral generator, in fact) of an exponentially bounded k-regularized* $\overline{I_p}$*-resolvent propagation family* $((\overline{R_{0,p}(t)})_{t\geqslant 0}, \cdots, (\overline{R_{m_n-1,p}(t)})_{t\geqslant 0})$ *in* $\overline{E_p}$ *satisfying (410), and (399) with A and* $((R_0(t))_{t\geqslant 0}, \cdots, (R_{m_n-1}(t))_{t\geqslant 0})$ *replaced respectively by* $\overline{A_p}$ *and* $((\overline{R_{0,p}(t)})_{t\geqslant 0}, \cdots, (\overline{R_{m_n-1,p}(t)})_{t\geqslant 0})$*. Suppose, additionally,* *that* $\mathbb{N}_{n-1} \setminus D_i \neq \emptyset$ *and* $\Sigma_{j \in \mathbb{N}_{n-1} \setminus D_i} |c_j|^2 > 0$, *provided* $m - 1 \geqslant i$. *Then, for every* $p \in \circledast$, *(409) holds* $(0 \leqslant i \leqslant m_n - 1)$ *and* A *is a subgenerator (the integral generator, in fact) of a q-exponentially equicontinuous k-regularized I-resolvent propagation family* $((R_0(t))_{t \geqslant 0}, \cdots, (R_{m_n-1}(t))_{t \geqslant 0})$ *satisfying (399). Furthermore,* $((R_0(t))_{t \geqslant 0}, \cdots, (R_{m_n-1}(t))_{t \geqslant 0})$ *is a q-exponentially equicontinuous, analytic k-regularized I-resolvent propagation family of angle* $\beta \in (0, \pi]$ *provided that, for every* $p \in \circledast$, $((\overline{R_{0,p}(t)})_{t \geqslant 0}, \cdots, (\overline{R_{m_n-1,p}(t)})_{t \geqslant 0})$ *is a q-exponentially bounded, analytic k-regularized* $\overline{I_p}$*-resolvent propagation family of angle* β.

Proof. The proof is very similar to that of Theorem 2.4.3, and we will only outline a few relevant facts needed for the proof of (i). Suppose $x, y \in D(A)$ and $p(x) = p(y)$ for some $p \in \circledast$. Then (400) in combination with (409) implies that $\Psi_p(R_i(t) A(x - y)) = 0$, $t \geqslant 0$, provided $m - 1 < i$, and $\Psi_p(R_i(t)A(x - y) - (k * g_i)(t)(x - y)) = 0$, $t \geqslant 0$, provided $m - 1 \geqslant i$. At any rate, $\Psi_p(R_i(t)A(x - y)) = 0$, $t \geqslant 0$, which implies $p(R_i(t)A(x - y)) = 0$, $t \geqslant 0$, and in particular $p(k(0)A(x - y)) = 0$. Since $k(0) \neq 0$, we obtain $p(Ax - Ay) = 0$ and $p(Ax) = p(Ay)$. Therefore, A is a compartmentalized operator. It is clear that (410) holds and that the mapping $t \mapsto \overline{R_{i,p}(t)} x_p$, $t \geqslant 0$ is continuous for any $x_p \in E_p$. This implies by the standard limit procedure that the mapping $t \mapsto \overline{R_{i,p}(t)} \overline{x_p}$, $t \geqslant 0$ is continuous for any $\overline{x_p} \in \overline{E_p}$. Now we will prove that, for every $p \in \circledast$, the operator A_p is closable for the topology of $\overline{E_p}$. For that, suppose (x_n) is a sequence in $D(A)$ with $\lim_{n\to\infty} \Psi_p(x_n) = 0$ and $\lim_{n\to\infty} \Psi_p(Ax_n) = y$, in $\overline{E_p}$. Using the dominated convergence theorem, (400) and (409), we get that $\int_0^t g_{\alpha_n - \alpha}(t - s)\overline{R_{i,p}(s)} y \, ds = \lim_{n\to\infty} \int_0^t g_{\alpha_n - \alpha}(t - s) \overline{R_{i,p}(s)} \Psi_p(Ax_n) \, ds = 0$, for any $t \geqslant 0$. Taking the Laplace transform, one obtains $\overline{R_{i,p}(t)} y = 0$, $t \geqslant 0$. Since $\overline{R_{i,p}(0)} = k(0)\overline{I_p}$, we get that $y = 0$ and that A_p is closable, as claimed. Suppose $0 \leqslant i \leqslant m_n - 1$. It is checked at once that $\overline{R_{i,p}(t)}\ \overline{A_{j,p}} \subseteq \overline{A_{j,p}}\ \overline{R_{i,p}(t)}$, $t \geqslant 0$, $i \in \mathbb{N}^0_{m_n-1}$, $j \in \mathbb{N}_{n-1}$. The functional equation (400) for the operators $\overline{A_{j,p}}$, $0 \leqslant j \leqslant n - 1$ and $((\overline{R_{0,p}(t)})_{t \geqslant 0}, \cdots, (\overline{R_{m_n-1,p}(t)})_{t \geqslant 0})$ can be trivially verified, which also holds for the functional equation (400) in case of its validity for the operators A_j, $0 \leqslant j \leqslant n-1$, and $((R_0(t))_{t \geqslant 0}, \cdots, (R_{m_n-1}(t))_{t \geqslant 0})$. The remaining part of the proof can be obtained by duplicating the final part of the proof of Theorem 2.4.3(i). □

Remark 2.10.13. In the second part of the Theorem 2.10.12, we must restrict ourselves to the case in which $A_j = c_j I$ for some $c_j \in \mathbb{C}$ $(1 \leqslant j \leqslant n - 1)$. As a matter of fact, it is not clear how one can prove that the operator $\lambda^{\alpha_n} \overline{I_p} + \Sigma_{j \in D_i} \lambda^{\alpha_j} \overline{A_{j,p}}$ is injective, provided $m - 1 < i$, $\operatorname{Re} \lambda > \omega$, and $\tilde{k}(\lambda) \neq 0$, as well as that the operator $\Sigma_{j \in \mathbb{N}_{n-1} \setminus D_i} \lambda^{\alpha_j} \overline{A_{j,p}}$ is injective, provided $m - 1 \geqslant i$, $\operatorname{Re} \lambda > \omega$ and $\tilde{k}(\lambda) \neq 0$. Then Theorem 2.10.9 is inapplicable, which implies that the argumentation used in the proof of [304, Theorem 3.1(ii)] does not work for the proof of fact that, for every $i \in \mathbb{N}^0_{m_n-1}$ and $t > 0$, $\{\overline{R_{i,p}(t)} : p \in \circledast\}$ is a projective family of operators.

2.10.2. *k*-Regularized (C_1, C_2)-existence and uniqueness families for (2). We continue our research of abstract multi-term problems by enquiring into the basic structural properties of k-regularized (C_1, C_2)-existence and uniqueness families for (2). We shall always assume that X and Y are sequentially complete locally convex spaces. By $L(Y, X)$ we denote the space which consists of all bounded linear operators from Y into X. The fundamental system of seminorms which defines the topology on X, resp. Y, is denoted by $\circledast_X$, resp. $\circledast_Y$. The symbol I designates the identity operator on X.

Let $0 < \tau \leqslant \infty$. A strongly continuous operator family $(W(t))_{t\in[0,\tau)} \subseteq L(Y, X)$ is said to be locally equicontinuous if, for every $T \in (0, \tau)$ and for every $p \in \circledast_X$, there exist $q_p \in \circledast_Y$ and $c_p > 0$ such that $p(W(t)y) \leqslant c_p q_p(y)$, $y \in Y$, $t \in [0, T]$; the notion of equicontinuity of $(W(t))_{t\in[0,\tau)}$ is defined similarly. Notice that $(W(t))_{t\in[0,\tau)}$ is automatically locally equicontinuous in case that the space Y is barreled.

Following T.-J. Xiao and J. Liang [529], we introduce the following definition.

Definition 2.10.14. Suppose $0 < \tau \leqslant \infty$, $k \in C([0, \tau))$, $C_1 \in L(Y, X)$ and $C_2 \in L(X)$ is injective.

(i) A strongly continuous operator family $(E(t))_{t\in[0,\tau)} \subseteq L(Y, X)$ is said to be a (local, if $\tau < \infty$) k-regularized C_1-existence family for (2) if, for every $y \in Y$, the following holds: $E(\cdot)y \in C^{m_n-1}([0, \tau) : X)$, $E^{(i)}(0)y = 0$ for every $i \in \mathbb{N}_0$ with $i < m_n - 1$, $A_j(g_{\alpha_n-\alpha_j} * E^{(m_n-1)})(\cdot)y \in C([0, \tau): X)$ for $0 \leqslant j \leqslant n-1$, and

$$E^{(m_n-1)}(t)y + \sum_{j=1}^{n-1} A_j(g_{\alpha_n-\alpha_j} * E^{(m_n-1)})(t)y$$

$$- A(g_{\alpha_n-\alpha} * E^{(m_n-1)})(t)y = k(t)C_1y, \tag{411}$$

for any $t \in [0, \tau)$.

(ii) A strongly continuous operator family $(U(t))_{t\in[0,\tau)} \subseteq L(X)$ is said to be a (local, if $\tau < \infty$) k-regularized C_2-uniqueness family for (2) if, for every $\tau \in [0, \tau)$ and $x \in \bigcap_{0\leqslant j\leqslant n-1} D(A_j)$, the following holds:

$$U(t)x + \sum_{j=1}^{n-1} (g_{\alpha_n-\alpha_j} * U(\cdot)A_jx)(t)$$

$$- (g_{\alpha_n-\alpha} * U(\cdot)Ax)(t)y = (k * g_{m_n-1})(t)C_2x.$$

(iii) A strongly continuous family $((E(t))_{t\in[0,\tau)}, (U(t))_{t\in[0,\tau)}) \subseteq L(Y, X) \times L(X)$ is said to be a (local, if $\tau < \infty$) k-regularized (C_1, C_2)-existence and uniqueness family for (2) if $(E(t))_{t\in[0,\tau)}$ is a k-regularized C_1-existence family for (2), and $(U(t))_{t\in[0,\tau)}$ is a k-regularized C_2-uniqueness family for (2).

(iv) Suppose $Y = X$ and $C = C_1 = C_2$. Then a strongly continuous operator family $(R(t))_{t\in[0,\tau)} \subseteq L(X)$ is said to be a (local, if $\tau < \infty$) k-regularized C-resolvent family for (2) if $(R(t))_{t\in[0,\tau)}$ is a k-regularized C-uniqueness family for (2), $R(t)A_j \subseteq A_jR(t)$, for $0 \leqslant j \leqslant n-1$ and $t \in [0, \tau)$, as well as $R(t)C = CR(t)$, $t \in [0, \tau)$, and $CA_j \subseteq A_jC$, for $0 \leqslant j \leqslant n-1$.

In case $k(t) = g_{\zeta+1}(t)$, where $\zeta \geqslant 0$, it is also said that $(E(t))_{t\in[0,\tau)}$ is a ζ-times integrated C_1-existence family for (2); 0-times integrated C_1-existence family for (2) is also said to be a C_1-existence families for (2). The definition of (exponential) analyticity of C_1-existence families for (2) is taken in the obvious way; the above terminological agreements can be simply understood for all other classes of uniqueness and resolvent families introduced in Definition 2.10.14.

Integrating both sides of (411) sufficiently many times, we easily infer that (cf. [529, Definition 2.1, p. 151; and (2.8), p. 153]):

$$(412) \qquad E^{(l)}(t)y + \sum_{j=1}^{n-1} A_j(g_{\alpha_n-\alpha_j} * E^{(l)})(t)y - A(g_{\alpha_n-\alpha} * E^{(l)})(t)y = (k * g_{m_n-1-l})(t)C_1 y,$$

for any $t \in [0, \tau)$, $y \in Y$ and $l \in \mathbb{N}^0_{m_n-1}$. In this place, it is worth noting that the identity (412), with $k(t) = 1$, $l = 0$, $\tau = \infty$ and $\alpha_j = j$ $(0 \leqslant j \leqslant n-1)$, has been used in [529] for the definition of a C_1-existence family for (ACP_n). It can be simply proved that this definition is equivalent with the corresponding one given by Definition 2.10.14.

Proposition 2.10.15. *Let* $((E(t))_{t\in[0,\tau)}, (U(t))_{t\in[0,\tau)})$ *be a k-regularized* (C_1, C_2)-*existence and uniqueness family for (2), and let* $(U(t))_{t\in[0,\tau)}$ *be locally equicontinuous. If* $A_j \in L(X)$, $j \in \mathbb{N}_{n-1}$ *or* $\alpha \leqslant \min(\alpha_1, \cdots, \alpha_{n-1})$, *then* $C_2E(t)y = U(t)C_1 y$, $t \in [0, \tau)$, $y \in Y$.

Proof. Let $y \in Y$ be fixed. By the local equicontinuity of $(U(t))_{t\in[0,\tau)}$, we easily infer that the mappings $t \mapsto ((g_{\alpha_n-\alpha} * U) * E(\cdot)y)(t)$, $t \in [0, \tau)$ and $t \mapsto (U * (g_{\alpha_n-\alpha} * E(\cdot)y))(t)$, $t \in [0, \tau)$ are continuous and coincide. The prescribed assumptions also imply that, for every $j \in \mathbb{N}_{n-1}$, $t \in [0, \tau)$ and $y \in Y$,

$$\left(g_{\alpha_n-\alpha} * U * A_j(g_{\alpha_n-\alpha_j} * E(\cdot)y)\right)(t)y = \left(g_{\alpha_n-\alpha} * UA_j * g_{\alpha_n-\alpha} * E(\cdot)y\right)(t)y.$$

Keeping in mind (412) and the foregoing arguments, we get that:

$$\begin{aligned} g_{\alpha_n-\alpha} * U * \left[E(\cdot)y + +\sum_{j=1}^{n-1} A_j(g_{\alpha_n-\alpha_j} * E)(\cdot)y - k(\cdot)C_1 y\right] \\ = g_{\alpha_n-\alpha} * UA * [g_{\alpha_n-\alpha} * E](\cdot)y \\ = \left[U(\cdot) + \sum_{j=1}^{n-1}(g_{\alpha_n-\alpha_j} * U(\cdot)A_j) - k(\cdot)C_2\right] * g_{\alpha_n-\alpha} * E(\cdot)y. \end{aligned}$$

This implies the required equality $C_2E(t)y = U(t)C_1 y$, $t \in [0, \tau)$. □

Definition 2.10.16. Suppose $0 \leqslant i \leqslant m_n - 1$. Then we define $D_i' := \{j \in \mathbb{N}^0_{n-1} : m_j - 1 \geqslant i\}$, $D_i'' := \mathbb{N}^0_{n-1} \setminus D_i'$ and

$$\mathbf{D}_i := \left\{x \in \bigcap_{j\in D_i''} D(A_j) : A_j u_i \in R(C_1), j \in D_i''\right\}.$$

In the first part of the subsequent theorem (cf. also [529, Remark 2.2, Example 2.5, Remark 2.6]), we shall consider the most important case $k(t) = 1$. The analysis is similar if $k(t) = g_{n+1}(t)$ for some $n \in \mathbb{N}$.

Theorem 2.10.17. (i) *Suppose* $(E(t))_{t\in[0,\tau)}$ *is a* C_1*-existence family for (2),* $T\in(0,\tau)$*, and* $u_i\in \mathbf{D}_i$ *for* $0\leqslant i\leqslant m_n-1$. *Then the function*

$$u(t)=\sum_{i=0}^{m_n-1}u_i g_{i+1}(t)-\sum_{i=0}^{m_n-1}\sum_{j\in\mathbb{N}_{n-1}\setminus D_i}(g_{\alpha_n-\alpha_j}*E^{(m_n-1-i)})(t)v_{i,j}$$

$$+\sum_{i=m}^{m_n-1}(g_{\alpha_n-\alpha}*E^{(m_n-1-i)})(t)v_{i,0},\quad 0\leqslant t\leqslant T, \tag{413}$$

is a strong solution of the problem (397) on $[0, T]$, *where* $v_{i,j}\in Y$ *satisfy* $A_j u_i = C_1 v_{i,j}$ *for* $0\leqslant j\leqslant n-1$.

(ii) *Suppose* $(U(t))_{t\in[0,\tau)}$ *is a locally equicontinuous k-regularized* C_2*-uniqueness family for (2), and* $T\in(0,\tau)$. *Then there exists at most one strong (mild) solution of (397) on* $[0, T]$, *with* $u_i=0$, $i\in\mathbb{N}^0_{m_n-1}$.

Proof. Making use of (412), it can be easily verified that:

$$u(\cdot)-\sum_{i=0}^{m_n-1}u_i g_{i+1}(\cdot)+\sum_{j=1}^{n-1}Aj\left(g_{\alpha_n-\alpha_j}*\left[u(\cdot)-\sum_{i=0}^{m_j-1}u_i g_{i+1}(\cdot)\right]\right)$$

$$=-\sum_{i=0}^{m_n-1}\sum_{j\in\mathbb{N}_{n-1}\setminus D_i}(g_{\alpha_n-\alpha_j}*R^{(m_n-1-i)})(\cdot)\,v_{i,j}+\sum_{i=m}^{m_n-1}(g_{\alpha_n-\alpha}*R^{(m_n-1-i)})(\cdot)\,v_{i,0}$$

$$+\sum_{j=1}^{n-1}A_j\left(g_{\alpha_n-\alpha_j}*\left\{\sum_{i=m_j}^{m_n-1}g_{i+1}(\cdot)u_i-\sum_{i=0}^{m_n-1}\sum_{l\in\mathbb{N}_{n-1}\setminus D_i}(g_{\alpha_n-\alpha_l}*R^{(m_n-1-i)})(\cdot)v_{i,l}\right.\right.$$

$$\left.\left.+\sum_{i=m}^{m_n-1}(g_{\alpha_n-\alpha}*R^{(m_n-1-i)})(\cdot)v_{i,0}\right\}\right)$$

$$=-\sum_{i=0}^{m_n-1}\sum_{j\in\mathbb{N}_{n-1}\setminus D_i}(g_{\alpha_n-\alpha_j}*R^{(m_n-1-i)})(\cdot)v_{i,j}+\sum_{i=m}^{m_n-1}(g_{\alpha_n-\alpha}*R^{(m_n-1-i)})(\cdot)v_{i,0}$$

$$+\sum_{j=1}^{n-1}\sum_{i=m_j}^{m_n-1}C_1v_{i,j}g_{\alpha_n-\alpha_j+i+1}(\cdot)-\sum_{i=0}^{m_n-1}\sum_{l\in\mathbb{N}_{n-1}\setminus D_i}g_{\alpha_n-\alpha_l}*\Big[-R^{(m_n-1-i)}(\cdot)v_{i,l}$$

$$\Big.+A(g_{\alpha_n-\alpha}*R^{(m_n-1-i)})(\cdot)v_{i,l}+g_{i+1}(\cdot)C_1v_{i,l}\Big]$$

$$+\sum_{i=m}^{m_n-1}g_{\alpha_n-\alpha}*[-R^{(m_n-1-i)}(\cdot)v_{i,0}+A(g_{\alpha_n-\alpha}*R^{(m_n-1-i)})(\cdot)v_{i,0}+g_{i+1}(\cdot)C_1v_{i,0}]$$

$$=g_{\alpha_n-\alpha}*A\left[u(\cdot)-\sum_{i=0}^{m-1}u_i g_{i+1}(\cdot)\right],$$

since

$$\sum_{j=1}^{n-1}\sum_{i=m_j}^{m_n-1}C_1v_{i,j}g_{\alpha_n-\alpha_j+i+1}(\cdot)=\sum_{i=0}^{m_n-1}\sum_{j\in\mathbb{N}_{n-1}\setminus D_i}C_1v_{i,j}g_{\alpha_n-\alpha_j+i+1}(\cdot).$$

This implies that $u(t)$ is a mild solution of (397) on $[0, T]$. In order to complete the proof of (i), it suffices to show that $\mathbf{D}_t^{\alpha_n} u(t) \in C([0, T] : X)$ and $A_i\mathbf{D}_t^{\alpha_i} u \in C([0, T] : X)$ for all $i \in \mathbb{N}_{n-1}^0$. Towards this end, notice that the partial integration implies that, for every $t \in [0, T]$,

$$g_{m_n-\alpha_n} * \left[u(\cdot) - \sum_{i=0}^{m_n-1} u_i g_{i+1}(\cdot)\right](t)$$

$$= \sum_{i=m}^{m_n-1} (g_{m_n-\alpha+i} * E^{(m_n-1)})(t)v_{i,0} - \sum_{i=0}^{m_n-1} \sum_{j\in\mathbb{N}_{n-1}\setminus D_i} (g_{m_n-\alpha_j+i} * E^{(m_n-1)})(t)v_{i,j}.$$

Therefore, $\mathbf{D}_t^{\alpha_n} u \in C([0, T] : X)$ and, for every $t \in [0, T]$,

$$\mathbf{D}_t^{\alpha_n} u(t) = \frac{d^{m_n}}{dt^{m_n}}\left\{g_{m_n-\alpha_n} * \left[u(\cdot) - \sum_{i=0}^{m_n-1} u_i g_{i+1}(\cdot)\right](t)\right\}$$

$$= \sum_{i=m}^{m_n-1} (g_{i-\alpha} * E^{(m_n-1)})(t)v_{i,0} - \sum_{i=0}^{m_n-1} \sum_{j\in\mathbb{N}_{n-1}\setminus D_i} (g_{i-\alpha_j} * E^{(m_n-1)})(t)v_{i,j}.$$

Suppose provisionally $i \in \mathbb{N}_{n-1}^0$. Then $A_iu_j \in R(C_1)$ for $j \geqslant m_i$. Moreover, the inequality $l \geqslant \alpha_j$ holds provided $0 \leqslant l \leqslant m_n - 1$ and $j \in \mathbb{N}_{n-1}\setminus D_l$, and $A_j(g_{\alpha_n-\alpha_j} * E^{(m_n-1)})(\cdot)y \in C([0, T] : X)$ for $0 \leqslant j \leqslant n-1$ and $y \in Y$. Now it is not difficult to prove that:

$$A_i\mathbf{D}_t^{\alpha_i} u(\cdot)$$

$$= \sum_{j=m_i}^{m_n-1} g_{j+1-\alpha_i}(\cdot)A_iu_j - \sum_{l=0}^{m_n-1} \sum_{j\in\mathbb{N}_{n-1}\setminus D_l} [g_{l-\alpha_j} * A_i(g_{\alpha_n-\alpha_i} * E^{(m_n-1)})](\cdot)v_{l,j}$$

$$+ \sum_{l=m}^{m_n-1} [g_{l-\alpha} * A_i(g_{\alpha_n-\alpha_i} * E^{(m_n-1)})](\cdot)v_{l,0} \in C([0, T] : X),$$

finishing the proof of (i). The second part of theorem can be proved as follows. Suppose $u(t)$ is a strong solution of (397) on $[0, T]$, with $u_i = 0$, $i \in \mathbb{N}_{m_n-1}^0$. This fact and the equality

$$\int_0^t \int_0^{t-s} g_{\alpha_n-\alpha_j}(r)U(t-s-r)A_ju(s)\,dr\,ds = \int_0^t \int_0^s g_{\alpha_n-\alpha_j}(r)U(t-s)A_ju(s-r)\,dr\,ds,$$

holding for any $t \in [0, T]$ and $j \in \mathbb{N}_{n-1}^0$, imply that (for more general results, see [317, Proposition 2.4(i) and [463, p. 155]):

$$(U * u)(t) = (k * g_{m_n-1}C_2 * u)(t)$$

$$+ \int_0^t \int_0^{t-s} [g_{\alpha_n-\alpha_j}(r)U(t-s-r)A_ju(s) - g_{\alpha_n-\alpha}(r)U(t-s-r)Au(s)]\,dr\,ds$$

$$= (k * g_{m_n-1}C_2 * u)(t) + (U * u)(t), \quad t \in [0, T].$$

Therefore, $(k * g_{m_{n-1}}C_2 * u)(t) = 0$, $t \in [0, T]$ and $u(t) = 0$, $t \in [0, T]$. □

Before proceeding further, we would like to notice that the solution $u(t)$, given by (413), need not be of class $C^1([0, T] : X)$, in general. Integration by parts yields that (413) is an extension of the formula [529, (2.5); Theorem 2.4, p. 152]. Notice, finally, that the proof of Theorem 2.10.17(ii) is much simpler than that of [529, Theorem 2.4(ii)].

The standard proofs of subsequent theorems are omitted (see, e.g., [529, Theorem 2.7, Remark 2.8, Theorem 2.9], Section 1.2 and Section 2.1).

Theorem 2.10.18. *Suppose $k(t)$ satisfies (P1), $(E(t))_{t\geqslant 0} \subseteq L(Y, X)$, $(U(t))_{t\geqslant 0} \subseteq L(X)$, $\omega \geqslant \max(0, abs(k))$, $C_1 \in L(Y, X)$ and $C_2 \in L(X)$ is injective. Set $\mathbf{P}_\lambda := I + \Sigma_{j=1}^{n-1}\lambda^{\alpha_j-\alpha_n}A_j - \lambda^{\alpha-\alpha_n}A$, $\operatorname{Re}\lambda > 0$.*

(i) (a) *Let $(E(t))_{t\geqslant 0}$ be a k-regularized C_1-existence family for (2), let the family $\{e^{-\omega t}E(t) : t \geqslant 0\}$ be equicontinuous, and let the family $\{e^{-\omega t}A_j(g_{\alpha_n-\alpha_j} * E)(t) : t \geqslant 0\}$ be equicontinuous $(0 \leqslant j \leqslant n-1)$. Then the following holds:*

$$\mathbf{P}_\lambda \int_0^\infty e^{-\lambda t}E(t)y\,dt = \tilde{k}(\lambda)\lambda^{1-m_n}C_1y, \quad y \in Y,\ \operatorname{Re}\lambda > \omega.$$

(b) *Let the operator $\mathbf{P}_\lambda$ be injective for every $\lambda > \omega$ with $\tilde{k}(\lambda) \neq 0$. Suppose, additionally, that there exist strongly continuous operator families $(W(t))_{t\geqslant 0} \subseteq L(Y, X)$ and $(W_j(t))_{t\geqslant 0} \subseteq L(Y, X)$ such that $\{e^{-\omega t}W(t) : t \geqslant 0\}$ and $\{e^{-\omega t}W_j(t) : t \geqslant 0\}$ are equicontinuous $(0 \leqslant j \leqslant n-1)$ as well as that:*

$$\int_0^\infty e^{-\lambda t}W(t)y\,dt = \tilde{k}(\lambda)\mathbf{P}_\lambda^{-1}C_1y$$

and

$$\int_0^\infty e^{-\lambda t}W_j(t)y\,dt = \tilde{k}(\lambda)\lambda^{\alpha_j-\alpha_n}A_j\mathbf{P}_\lambda^{-1}C_1y,$$

for every $\lambda \in \mathbb{C}$ with $\operatorname{Re}\lambda > \omega$ and $\tilde{k}(\lambda) \neq 0$, $y \in Y$ and $j \in \mathbb{N}_{n-1}^0$. Then there exists a k-regularized C_1-existence family for (2), denoted by $(E(t))_{t\geqslant 0}$. Furthermore, $E^{(m_n-1)}(t)y = W(t)y$, $t \geqslant 0$, $y \in Y$ and $A_j(g_{\alpha_n-\alpha_j} * E^{(m_n-1)})(t)y = W_j(t)y$, $t \geqslant 0$, $y \in Y$, $j \in \mathbb{N}_{n-1}^0$.

(ii) *Let the assumptions of (i) hold with $k(t) = 1$. If $m_n > 1$, then we suppose additionally that, for every $j \in \mathbb{N}_{n-1}^0$, there exists a strongly continuous operator family $(V_j(t))_{t\geqslant 0} \subseteq L(Y, X)$ such that $\{e^{-\omega t}V_j(t) : t \geqslant 0\}$ is equicontinuous as well as that*

$$\int_0^\infty e^{-\lambda t}V_j(t)y\,dt = \lambda^{\alpha_j-\alpha_n-1}\mathbf{P}_\lambda^{-1}A_jC_1y,$$

for every $\lambda \in \mathbb{C}$ *with* $\operatorname{Re}\lambda > \omega$, *and* $y \in D(A_jC_1)$. *Let* $u_i \in \mathbf{D}_i$, *and let* $C_1 v_i = u_i$ *for some* $v_i \in Y$ $(0 \leqslant i \leqslant m_n - 1)$. *Then for every* $p \in \circledast_X$, *there exist* $c_p > 0$ *and* $q_p \in \circledast_Y$ *such that the corresponding solution* $u(t)$ *satisfies the following estimate:*

$$\begin{cases} p(u(t)) \leqslant c_p e^{\omega t} \sum_{i=0}^{m_n-1} q_p(v_i), & t \geqslant 0, \text{ if } \omega > 0, \text{ and} \\ p(u(t)) \leqslant c_p g_{m_n}(t) \sum_{i=0}^{m_n-1} q_p(v_i), & t \geqslant 0, \text{ if } \omega = 0. \end{cases}$$

(iii) *Suppose* $(U(t))_{t\geqslant 0}$ *is strongly continuous and the operator family* $\{e^{-\omega t}U(t) : t \geqslant 0\}$ *is equicontinuous. Then* $(U(t))_{t\geqslant 0}$ *is a k-regularized* C_2*-uniqueness family for (2) if, for evey* $x \in \bigcap_{j=0}^{n-1} D(A_j)$, *the following holds:*

$$\int_0^\infty e^{-\lambda t} U(t)\mathbf{P}_\lambda x\, dt = \tilde{k}(\lambda)\lambda^{1-m_n} C_2 x, \quad \operatorname{Re}\lambda > \omega.$$

Theorem 2.10.19. (i) *Suppose X is barreled,* $\zeta > 0$, $(R(t))_{t\in[0,\tau)}$ *is a k-regularized C-resolvent family for (2), and* $\overline{\bigcap_{j=0}^{n-1} D(A_j)} = \overline{R(C)} = X$.
Then $((g_\zeta * R(\cdot)^*)(t))_{t\in[0,\tau)}$ *is a k-regularized* C^**-resolvent family for (2), with* A_j *replace by* A_j^* $(0 \leqslant j \leqslant n-1)$.

(ii) *Suppose X is barreled,* $(R(t))_{t\in[0,\tau)}$ *is a (local, global exponentially equicontinuous) k-regularized C-resolvent family for (2), and* $\overline{\bigcap_{j=0}^{n-1} D(A_j)} = \overline{R(C)} = X$. *Put* $Z := \overline{\bigcap_{j=0}^{n-1} D(A_j^*)}$. *Then* $(R(t)^*_{|Z})_{t\in[0,\tau)}$, *is a (local, global exponentially equicontinuous) k-regularized* $C^*_{|Z}$*-resolvent family for (2), in Z.*

(iii) *Suppose* $(R(t))_{t\in[0,\tau)}$ *is a locally equicontinuous k-regularized C-resolvent family for (2), and* $\overline{\bigcap_{j=0}^{n-1} D(A_j)} = \overline{R(C)} = X$. *Then* $(R(t)^*)_{t\in[0,\tau)}$ *is a locally equicontinuous k-regularized* C^**-resolvent family for (2), in* $(X^*, C(X^*, X))$, *with* A_j *replaced by* A_j^* $(0 \leqslant j \leqslant n-1)$. *Furthermore, if* $(R(t))_{t\geqslant 0}$ *is exponentially equicontinuous, then* $(R(t)^*)_{t\geqslant 0}$ *is also exponentially equicontinuous.*

Let $f \in C([0, T] : X)$. Convoluting both sides of (2) with $g_{\alpha_n}(t)$, we get that:

$$u(\cdot) - \sum_{k=0}^{m_n-1} u_k g_{k+1}(\cdot) + \sum_{j=1}^{n-1} g_{\alpha_n-\alpha_j} * A_j\left[u(\cdot) - \sum_{k=0}^{m_j-1} u_k g_{k+1}(\cdot)\right]$$

$$= g_{\alpha_n-\alpha} * A\left[u(\cdot) - \sum_{k=0}^{m-1} u_k g_{k+1}(\cdot)\right] + (g_{\alpha_n} * f)(\cdot), \quad t \in [0, T]. \tag{414}$$

In the following theorem, we shall analyze the inhomogeneous Cauchy problem (414) in more detail.

Theorem 2.10.20. *(cf.* [529, Theorem 3.1(i)] *and* [318]) *Suppose* $(E(t))_{t\in[0,\tau)}$ *is a locally equicontinuous* C_1*-existence family for (2),* $T \in (0, \tau)$, *and* $u_i \in \mathbf{D}_i$ *for* $0 \leqslant i \leqslant m_n - 1$. *Let* $f \in C([0, T] : X)$, *let* $g \in C([0, T] : Y)$ *satisfy* $C_1 g(t) = f(t)$, $t \in [0, T]$, *and let* $G \in C([0, T] : Y)$ *satisfy* $(g_{\alpha_n - m_n + 1} * g)(t) = (g_1 * G)(t)$, $t \in [0, T]$. *Then the function*

$$u(t) = \sum_{i=0}^{m_n - 1} u_i g_{i+1}(t) - \sum_{i=0}^{m_n - 1} \sum_{j \in \mathbb{N}_{n-1} \setminus D_i} (g_{\alpha_n - \alpha_j} * E^{(m_n - 1 - i)})(t) v_{i,j}$$

$$+ \sum_{i=m}^{m_n - 1} (g_{\alpha_n - \alpha} * E^{(m_n - 1 - i)})(t) v_{i,0} + \int_0^t E(t-s) G(s)\, ds, \quad 0 \leqslant t \leqslant T, \tag{415}$$

is a mild solution of the problem (414) on $[0, T]$, *where* $v_{i,j} \in Y$ *satisfy* $A_j u_i = C_1 v_{i,j}$ *for* $0 \leqslant j \leqslant n - 1$. *If, additionally,* $g \in C^1([0, T] : Y)$ *and* $(E^{(m_n - 1)}(t))_{t \in [0,\tau)} \subseteq L(Y, X)$ *is locally equicontinuous, then the solution* $u(t)$, *given by (415), is a strong solution of (2) on* $[0, T]$.

Remark 2.10.21. Suppose that all conditions quoted in the first part of the above theorem hold, and the family $(E^{(m_n - 1)}(t))_{t \in [0,\tau)} \subseteq L(Y, X)$ is locally equicontinuous. We assume, instead of condition $g \in C^1([0, T] : Y)$, that there exists a locally equicontinuous C_2-uniqueness family for (2) on $[0, \tau)$, as well as that there exist functions $h_j \in L^1([0, T] : Y)$ such that $A_j f(t) = C_1 h_j(t)$, $t \in [0, T]$, $0 \leqslant j \leqslant n - 1$ (cf. the formulation of [529, Theorem 3.1(ii)]). Using the functional equation of $(E(t))_{t \in [0,\tau)}$, one can simply prove that, for every $\sigma \in [0, T]$, the function

$$r_\sigma(\cdot) = E(\cdot) g(\sigma) - g_{m_n}(\cdot) f(\sigma)$$

$$+ \sum_{j=1}^{n-1} (g_{\alpha_n - \alpha_j} * E(\cdot) h_j(\sigma))(\cdot) - (g_{\alpha_n - \alpha} * E(\cdot) h_0(\sigma))(\cdot),$$

is a mild solution of the problem

$$u(t) + \sum_{j=1}^{n-1} A_j (g_{\alpha_n - \alpha_j} * u)(t) - A(g_{\alpha_n - \alpha} * u)(t) = 0, \quad t \in [0, T].$$

By the uniqueness of the solutions, we have that the following holds:

$$E(t - \sigma) g(\sigma) - g_{m_n}(t - \sigma) f(\sigma) + \sum_{l=1}^{n-1} (g_{\alpha_n - \alpha_l} * E(\cdot) h_l(\sigma))(t - \sigma)$$

$$- (g_{\alpha_n - \alpha} * E(\cdot) h_0(\sigma))\, (t - \sigma) = 0, \tag{416}$$

provided $0 \leqslant t, \sigma \leqslant T$ and $\sigma \leqslant t$. Fix an integer $i \in \mathbb{N}^0_{n-1}$. Then (416) implies that, for every $j \in \mathbb{N}^0_{m_n - 1}$ with $j \leqslant \min(\lfloor \alpha_i + m_n - \alpha_{n-1} - 1 \rfloor, \lfloor \alpha_i + m_n - \alpha - 1 \rfloor)$, we have:

$$A_i E^{(j)}(t - \sigma) g(\sigma) - g_{m_n - j}(t - \sigma) C_1 h_i(\sigma) + \sum_{l=1}^{n-1} A_i (g_{\alpha_n - \alpha_l} * E^{(j)}(\cdot) h_l(\sigma))(t - \sigma)$$

$$- A_i (g_{\alpha_n - \alpha} * E^{(j)}(\cdot) h_0(\sigma))\, (t - \sigma) = 0, \tag{417}$$

provided $0 \leqslant t, \sigma \leqslant T$ and $\sigma \leqslant t$. For such an index j, we may conclude from (417) that the mapping $t \mapsto \int_0^t A_i E^{(j)}(t - \sigma) g(\sigma)\, d\sigma$, $t \in [0, T]$ is continuous. Observe now that the condition

(418) $\alpha_n - \alpha_i - m_n + \min(\lfloor \alpha_i + m_n - \alpha_{n-1} - 1 \rfloor, \lfloor \alpha_i + m_n - \alpha - 1 \rfloor) \geqslant 0,\ i \in \mathbb{N}^0_{n-1}$,

which holds in the case of abstract Cauchy problem (ACP_n), shows that the mapping $t \mapsto A_i[g_{\alpha_n-\alpha_i-m_n+j} * E^{(j)} * g](t)$, $t \in [0, T]$ is continuous as well as that the mapping $t \mapsto \frac{d}{dt}[E^{(m_n-1)} * g](t)$, $t \in [0, T]$ is continuous. Hence, the validity of condition (418) implies that the function $u(t)$, given by (415), is a strong solution of (2) on $[0, T]$.

The proof of following subordination principle can be derived by using Theorem 2.10.18 and the arguments from [49, Section 3].

Theorem 2.10.22. *Suppose $C_1 \in L(Y, X)$, $C_2 \in L(X)$ is injective and $\gamma \in (0, 1)$.*

(i) *Let $\omega > \max(0, abs(k))$, and let the assumptions of Theorem 2.10.18(i)-(b) hold. Put*

(419) $$W_\gamma(t) := \int_0^\infty t^{-\gamma}\Phi_\gamma(t^{-\gamma}s)W(s)y\,ds,\ t > 0,\ y \in Y \text{ and } W_\gamma(0) := W(0).$$

Define, for every $j \in \mathbb{N}^0_{n-1}$ and $t \geqslant 0$, $W_{j,\gamma}(t)$ by replacing $W(t)$ in (419) with $W_j(t)$. Suppose that there exist a number $v > 0$ and a continuous kernel $k_\gamma(t)$ satisfying (P1) and $\widetilde{k_\gamma}(\lambda) = \lambda^{\gamma-1}\tilde{k}(\lambda^\gamma)$, $\lambda > v$. Then there exists an exponentially bounded k_γ-regularized C_1-existence family $(E_\gamma(t))_{t\geqslant 0}$ for (2), with α_j replaced by $\alpha_j\gamma$ therein $(0 \leqslant j \leqslant n-1)$. Furthermore, the family $\{(1 + t^{\lceil \alpha_n\gamma \rceil - 2})^{-1}e^{-\omega^{1/\gamma}t}E_\gamma(t) : t \geqslant 0\}$ is equicontinuous.

(ii) *Let $\omega \geqslant 0$, let the assumptions of Theorem 2.10.18(ii) hold, and let $k(t) = k_\gamma(t) = 1$. Define, for every $j \in \mathbb{N}^0_{n-1}$ and $t \geqslant 0$, $V_{j,\gamma}(t)$ by replacing $W(t)$ in (419) with $V_j(t)$. Then, for every $j \in \mathbb{N}^0_{n-1}$, the family $\{e^{-\omega^{1/\gamma}t}V_{j,\gamma}(t) : t \geqslant 0\}$ is equicontinuous,*

$$\int_0^\infty e^{-\lambda t}V_{j,\gamma}(t)y\,dt = \lambda^{\alpha_j\gamma-\alpha_n\gamma-1}\mathbf{P}^{-1}_{\lambda^\gamma}A_jC_1y,$$

for every $\lambda \in \mathbb{C}$ with $\mathrm{Re}(\lambda^\gamma) > \omega$, and $y \in D(A_jC_1)$. Let $u_i \in \mathbf{D}_{i,\gamma}$ (defined in the obvious way), and let $C_1v_i = u_i$ for some $v_i \in Y$ $(0 \leqslant i \leqslant \lceil \alpha_n\gamma \rceil - 1)$. Then, for every $p \in \circledast_X$, there exist $c_p > 0$ and $q_p \in \circledast_Y$ such that the corresponding solution $u(t)$ satisfies the following estimate:

$$\begin{cases} p(u(t)) \leqslant c_p e^{\omega^{1/\gamma}t} \sum_{i=0}^{\lceil \alpha_n\gamma \rceil - 1} q_p(v_i), & t \geqslant 0, \text{ if } \omega > 0, \text{ and} \\ p(u(t)) \leqslant c_p g_{\lceil \alpha_n\gamma \rceil}(t) \sum_{i=0}^{\lceil \alpha_n\gamma \rceil - 1} q_p(v_i), & t \geqslant 0, \text{ if } \omega = 0. \end{cases}$$

(iii) *Suppose $(U(t))_{t\geqslant 0}$ is a k-regularized C_2-uniqueness family for (2), and the family $\{e^{-\omega t}U(t) : t \geqslant 0\}$ is equicontinuous. Define, for every $t \geqslant 0$, $U_\gamma(t)$ by replacing $W(t)$ in (419) with $U(t)$. Suppose that there exist a number $v > 0$ and a continuous kernel $k_\gamma(t)$ satisfying (P1) and $\widetilde{k_\gamma}(\lambda) = \lambda^{\gamma(2-m_n)-2+\lceil \alpha_n\gamma \rceil}\tilde{k}(\lambda^\gamma)$,*

$\lambda > v$. Then there exists a k_γ-regularized C_2-uniqueness family for (2), with α_j replaced by $\alpha_j\gamma$ therein $(0 \leqslant j \leqslant n-1)$. Furthermore, the family $\{e^{-\omega 1/\gamma t} U_\gamma(t) : t \geqslant 0\}$ is equicontinuous.

Remark 2.10.23. (i) In the situation of Theorem 2.10.22(ii), we have the following obvious equality $(k * g_{m_n-1})(0) = (k_\gamma * g_{\lceil \alpha_n \gamma \rceil -1})(0)$. If $\sigma \geqslant 1$, $k(t) = g_\sigma(t)$ and $(\sigma - 1 + m_n - 1)\gamma + 1 - \lceil \alpha_n \gamma \rceil \geqslant 0$ (this inequality holds provided $\sigma \geqslant 2$), then $k_\gamma(t) = g_{(\sigma-1+m_n-1)\gamma+2-\lceil \alpha_n \gamma \rceil}(t)$.

(ii) Let $b \in L^1_{loc}([0, \tau))$ be a kernel, and let $(U(t))_{t\in[0,\tau)}$ be a (local) k-regularized C_2-uniqueness family for (2). Then $((b * U)(t))_{t\in[0,\tau)}$ is a $(b * k)$-regularized C_2-uniqueness family for (2).

(iii) Concerning the analytical properties of k_γ-regularized C_1-existence families in Theorem 2.10.22(i), the following facts should be stated:

(a) The mapping $t \mapsto E_\gamma(t)$, $t > 0$ admits an extension to $\Sigma_{\min((\frac{1}{\gamma}-1)\frac{\pi}{2},\pi)}$ and, for every $y \in Y$, the mapping $z \mapsto E_\gamma(z)y$, $z \in \Sigma_{\min((\frac{1}{\gamma}-1)\frac{\pi}{2},\pi)}$ is analytic.

(b) Let $\varepsilon \in (0, \min((\frac{1}{\gamma}-1)\frac{\pi}{2}, \pi))$, and let $(W(t))_{t\geqslant 0}$ be equicontinuous. Then $(E_\gamma(t))_{t\geqslant 0}$ is an exponentially equicontinuous, analytic k_γ-regularized C_1-existence family of angle $\min((\frac{1}{\gamma}-1)\frac{\pi}{2}, \pi)$, and for every $p \in \circledast_X$, there exist $M_{p,\varepsilon} > 0$ and $q_{p,\varepsilon} \in \circledast_Y$ such that

$$p(E_\gamma(z)y) \leqslant M_{p,\varepsilon} q_{p,\varepsilon}(y)\,(1 + |z|^{\lceil \alpha_n \gamma \rceil - 1}), \quad z \in \Sigma_{\min((\frac{1}{\gamma}-1)\frac{\pi}{2},\pi)-\varepsilon},\ y \in Y.$$

(c) $(E_\gamma(t))_{t\geqslant 0}$ is an exponentially equicontinuous, analytic k_γ-regularized C_1-existence family of angle $\min((\frac{1}{\gamma}-1)\frac{\pi}{2}, \frac{\pi}{2})$.

Similar statements can be clarified for the k_γ-regularized C_2-uniqueness family $(U_\gamma(t))_{t\geqslant 0}$ in Theorem 2.10.22(iii).

The results on k-regularized (C_1, C_2)-existence and uniqueness families can be applied in the study of the following abstract Volterra equation:

$$u(t) = f(t) + \sum_{j=0}^{n-1} (a_j * A_j u)(t),\ t \in [0, \tau), \tag{420}$$

where $0 < \tau \leqslant \infty$, $f \in C([0, \tau) : X)$, $a_0, \cdots, a_{n-1} \in L^1_{loc}([0, \tau))$, and $A_0, \cdots, A_{n-1}$ are closed linear operators on X. As in Definition 2.10.6, by a mild solution, resp. strong solution, of (420), we mean any function $u \in C([0, \tau) : X)$ such that $A_j(a_j * u)(t) \in C([0, \tau) : X)$, $j \in \mathbb{N}^0_{n-1}$ and

$$u(t) = f(t) + \sum_{j=0}^{n-1} A_j(a_j * u)(t), \quad t \in [0, \tau),$$

resp. any function $u \in C([0, \tau) : X)$ such that $u(t) \in \bigcap_{j=0}^{n-1} D(A_j)$, $t \in [0, \tau)$ and (420) holds.

We need the following definition.

Definition 2.10.24. Suppose $0 < \tau \leqslant \infty$, $k \in C([0, \tau))$, $C_1 \in L(Y, X)$, and $C_2 \in L(X)$ is injective.

(i) A strongly continuous operator family $(E(t))_{t\in[0,\tau)} \subseteq L(Y, X)$ is said to be a (local, if $\tau < \infty$) k-regularized C_1-existence family for (420) if

$$E(t)y = k(t)C_1 y + \sum_{j=0}^{n-1} A_j(a_j * E)(t)y, \quad t \in [0, \tau), y \in Y.$$

(ii) A strongly continuous operator family $(U(t))_{t\in[0,\tau)} \subseteq L(X)$ is said to be a (local, if $\tau < \infty$) k-regularized C_2-uniqueness family for (420) if

$$U(t)x = k(t)C_2 x + \sum_{j=0}^{n-1} (a_j * A_j U)(t)x, \; t \in [0, \tau), x \in \bigcap_{j=0}^{n-1} D(A_j).$$

The classes of k-regularized (C_1, C_2)-existence and uniqueness families and k-regularized C-resolvent families for (420) can also be introduced; cf. Definition 2.10.14. The following facts are clear:

(i) Suppose $(E(t))_{t\in[0,\tau)}$ is a k-regularized C_1-existence family for (420).Then, for every $y \in Y$, the function $u(t) = E(t)y$, $t \in [0, \tau)$ is a mild solution of (420) with $f(t) = k(t)C_1 y$, $t \in [0, \tau)$.

(ii) Let $(U(t))_{t\in[0,\tau)}$ be a locally equicontinuous k-regularized C_2-uniqueness family for (420). Then there exists at most one mild (strong) solution of (420).

The proof of following subordination principle is non-trivial but can be carried through with relatively elementary machinery (cf. the proofs of [463, Theorem 4.1, p. 101] and [529, Theorem 2.7]).

Theorem 2.10.25. (i) *Suppose there is an exponentially equicontinuous k-regularized C_1-existence family for (2). Let $c(t)$ be completely positive, let $c(t)$, $k(t)$ and $k_1(t)$ satisfy (P1), and let $\omega_0 > 0$ be such that, for every $\lambda > \omega_0$ with $\tilde{c}(\lambda) \neq 0$ and $\tilde{k}(1/\tilde{c}(\lambda)) \neq 0$, the following holds:*

$$\tilde{a}_j(\lambda) = -\tilde{k}_1(\lambda)\tilde{c}(\lambda)^{1+\alpha_n-\alpha_j} \frac{\lambda}{\tilde{k}(1/\tilde{c}(\lambda))}, \; j \in \mathbb{N}_{n-1},$$

and

$$\tilde{a}_0(\lambda) = -\tilde{k}_1(\lambda)\tilde{c}(\lambda)^{1+\alpha_n-\alpha} \frac{\lambda}{\tilde{k}(1/\tilde{c}(\lambda))}.$$

Assume, additionally, that there exist a number $z \in \mathbb{C}$ and a function $k_2(t)$ satisfying (P1) so that, for every $\lambda > \omega_0$ with $\tilde{c}(\lambda) \neq 0$ and $\tilde{k}(1/\tilde{c}(\lambda)) \neq 0$, we have:

$$\frac{\widetilde{k_1}(\lambda)}{\tilde{k}(1/\tilde{c}(\lambda))} = z + \widetilde{k_2}(\lambda).$$

Then there exists an exponentially equicontinuous k_1-regularized C_1-existence family for (420).

(ii) *Suppose there is an exponentially equicontinuous k-regularized C_2-uniqueness family for (2). Let c(t) be completely positive, let c(t), k(t) and $k_1(t)$ satisfy (P1), and let $\omega_0 > 0$ be such that, for every $\lambda > \omega_0$ with $\tilde{c}(\lambda) \neq 0$ and $\tilde{k}(1/\tilde{c}(\lambda)) \neq 0$, the following holds:*

$$\widetilde{a_j}(\lambda) = \tilde{c}(\lambda)^{\alpha_n - \alpha_j},\ j \in \mathbb{N}^0_{n-1} \text{ and } \widetilde{k_1}(\lambda) = \lambda^{-1}\tilde{c}(\lambda)^{m_n-2}\tilde{k}(1/\tilde{c}(\lambda)).$$

Then there exists an exponentially equicontinuous k_1-regularized C_2-uniqueness family for (420).

It is not difficult to rephrase Theorem 2.10.25 for the class of strong C-propagation families (see Example 2.10.33 below). Notice, however, that the question whether the concept of strong C-propagation families can or cannot be set up for the equations of form (2) is very difficult to answer in general, albeit it might not be too difficult to answer in particular cases or for some particular classes of abstract multi-term fractional differential equations.

2.10.3. Approximation and convergence of *k*-regularized *C*-resolvent propagation families; further results, examples and applications. This subsection discusses various problems about k-regularized C-resolvent (propagation) families. Our first task will be to clarify the most useful approximation and convergence type theorems for exponentially equicontinuous k-regularized C-resolvent propagation families. We will shorten the proofs of these results to a large extent since they can be derived by using appropriate modifications of the proofs already given in Section 2.7. We also consider additive perturbation theorems for C_1-existence families and $(1,C_2)$-uniqueness families (being introduced in Definition 2.10.35 for the first time), as well as the Ljubich uniqueness theorem for abstract multi-term problems and certain results linking the classses of (k, C_1, C_2)-existence and uniqueness families and k-regularized (C_1, C_2)-existence and uniqueness propagation families. In this subsection, and in the following one, we shall illustrate our theoretical results with numerous interesting examples.

Put $G_i(\lambda) := \lambda^{\alpha_n - i} + \Sigma_{j\in D_i} \lambda^{\alpha_j - i}A_j$ for every $i \in \mathbb{N}^0_{m_n-1}$ with $m - 1 < i$, and $G_i(\lambda) := -\Sigma_{j\in \mathbb{N}_{n-1}\setminus D_i} \lambda^{\alpha_j - i}A_j$ for every $i \in \mathbb{N}^0_{m_n-1}$ with $m - 1 \geqslant i$ $(\lambda \in \mathbb{C}\setminus\{0\})$. Then the assumptions of Theorem 2.10.9 imply that the operator P_λ is injective for any $\lambda \in \mathbb{C}$ with Re $\lambda > \omega$ and $\tilde{k}(\lambda) \neq 0$. Furthermore, for such a value of complex parameter λ, we have that $P_\lambda^{-1}G_i(\lambda)C \in L(E)$. We shall employ the condition

(HCC): $CA_j \subseteq A_jC$, $j \in \mathbb{N}^0_{n-1}$, $A_j \in L(E)$, $j \in \mathbb{N}_{n-1}$, $A_iA_j = A_jA_i$, $i, j \in \mathbb{N}_{n-1}$, $A_jA \subseteq AA_j$, $j \in \mathbb{N}_{n-1}$.

Suppose now $n \in \mathbb{N}\setminus\{1\}$, $\alpha_{n,l} > \alpha_{n-1,l} > \cdots > \alpha_{1,l} \geqslant 0$, $\alpha_{n,l} > \alpha_l$ $(l \in \mathbb{N}_0)$ and, for every $j \in \mathbb{N}^0_{n-1}$, $(A_{j,l})_{l\in\mathbb{N}_0}$ is a sequence of closed linear operators on E; set $\alpha_{0,l} := \alpha_l$ $(l \in \mathbb{N}_0)$. We consider the abstract Cauchy problem (2) with tuples $(\alpha_n, \cdots, \alpha_0)$ and $(A_{n-1}, \cdots, A_0)$ replaced respectively by $(\alpha_{n,l}, \cdots, \alpha_{0,l})$ and $(A_{n-1,l}, \cdots, A_{0,l})$. The meanings of numbers $m_{j,l}$ for $0 \leqslant j \leqslant n$, the sets $D_{i,l}$ for $0 \leqslant i \leqslant m_{n,l} - 1$, and the operators $P_{\lambda,l}$, $G_{i,l}(\lambda)$ for $0 \leqslant i \leqslant m_{n,l} - 1$, in the case that there exists a corresponding k_l-regularized C_l-resolvent propagation family $((R_{0,l}(t))_{t\geqslant 0}, \cdots, (R_{m_n,l-1}(t))_{t\geqslant 0})$ satisfying the properties stated in the formulation of Theorem 2.10.9, are clear $(l \in \mathbb{N}_0)$.

Theorem 2.10.26. *Suppose that the function $k_l(t)$ satisfies (P1), $\lambda_0 > \omega \geqslant \sup_{l\in\mathbb{N}_0}$ $\max(0, abs(k_l))$, $\sup_{l\in\mathbb{N}_0, t\geqslant 0} e^{-\omega t}|k_l(t)| < \infty$, and the condition (HCC) holds with $(A_{n-1}, \cdots, A_0)$ and C replaced respectively by $(A_{n-1,l}, \cdots, A_{0,l})$ and C_l $(l \in \mathbb{N}_0)$. Let $0 \leqslant i \leqslant \min_{l\in\mathbb{N}_0} m_{n,l} - 1$, and for each $p \in ⊛$ let there exist $c_p > 0$ and $r_p \in ⊛$ such that*

$$p\big(R_{i,l}(t)x\big) \leqslant c_p e^{\omega t} r_p(x), \quad t \geqslant 0, x \in E, l \in \mathbb{N}_0. \tag{421}$$

Assume, additionally, that the operator $G_{i,l}(\lambda)$ is injective for every $i \in \mathbb{N}^0_{m_{n,l}-1}$ and for every $\lambda \in \mathbb{C}$ with $\operatorname{Re} \lambda > \omega$ and $\tilde{k}_l(\lambda) \neq 0$ $(l \in \mathbb{N}_0)$. Let $((R_{0,l}(t))_{t\geqslant 0}, \cdots, (R_{m_{n,l}-1}(t))_{t\geqslant 0})$ be the corresponding k_l-regularized C_l-resolvent propagation family, and let $\mathfrak{T} = \{\lambda > \lambda_0 : \tilde{k}_l(\lambda) \neq 0$ for all $l \in \mathbb{N}_0\}$. Set, for every $\lambda \in \mathbb{C}$ with $\operatorname{Re} \lambda > \omega$ and $\tilde{k}_l(\lambda) \neq 0$ for all $l \in \mathbb{N}_0$, $K_{i,l}(\lambda)x := \lambda^{-\alpha_l} \tilde{k}_l(\lambda) P^{-1}_{\lambda,l} G_{i,l}(\lambda) C_l x$, $l \in \mathbb{N}_0$, $x \in E$. Suppose that the following conditions hold:

(i) $\lim_{l\to\infty} k_l(t) = k(t)$, $t \geqslant 0$, *uniformly on compacts of* $[0, \infty)$.
(ii) $\lim_{l\to\infty} \alpha_{j,l} = \alpha_j$ *for any* $j \in \mathbb{N}^0_{n-1}$.
(iii) *For every bounded sequence $(x_l)_l$ in E and for every $j \in \mathbb{N}_{n-1}$, the sequences $(C_l x_l)_l$ and $(A_{j,l} x_l)_l$ are bounded in E, too.*
(iv) $\lim_{l\to\infty} P^{-1}_{\lambda,l} G_{i,l}(\lambda) C_l x = P_\lambda^{-1} G_i(\lambda) C x$, *provided* $\lambda \in \mathfrak{T}$ *and* $x \in R(K_i(\lambda))$.

Then $\lim_{l\to\infty} R_{i,l}(t)x = R_i(t)x$, $t \geqslant 0$, $x \in \overline{\bigcup_{\lambda\in\mathfrak{T}} R(K_i(\lambda))}$, uniformly on compacts of $[0, \infty)$.

Proof. If $l \in \mathbb{N}_0$, then we have by Theorem 2.10.9 that, for every $\lambda \in \mathbb{C}$ with $\operatorname{Re} \lambda > \omega$ and $\tilde{k}_l(\lambda) \neq 0$,

$$\int_0^\infty e^{-\lambda t} R_{i,l}(t)x\,dt = \lambda^{-\alpha_l} \tilde{k}_l(\lambda) P^{-1}_{\lambda,l} G_{i,l}(\lambda) C_l x, \quad x \in E, \tag{422}$$

provided that $m_l - 1 < i$, and

$$\int_0^\infty e^{-\lambda t} R_{i,l}(t)x\,dt = \lambda^{-i} \tilde{k}_l(\lambda) + \lambda^{-\alpha_l} \tilde{k}_l(\lambda) P^{-1}_{\lambda,l} G_{i,l}(\lambda) C_l x, \quad x \in E, \tag{423}$$

provided that $m_l - 1 \geqslant i$. By (i)-(ii) and Theorem 1.2.6, we get that $\lim_{l\to\infty} \lambda^{-\alpha_l}\tilde{k}_l(\lambda) = \lambda^{-\alpha}\tilde{k}(\lambda)$, $\lambda \in \mathfrak{T}$. These facts enable one to see that, for every $x \in R(K_i(\lambda))$ and for every $\lambda \in \mathbb{C}$ with $\operatorname{Re}\lambda > \omega$, $\lim_{l\to\infty}\int_0^\infty e^{-\lambda t}R_{i,l}(t)x\,dt = \int_0^\infty e^{-\lambda t}R_i(t)x\,dt$. By (421), it suffices to show that the sequence $(R_{i,l}(t)x)_l$ is equicontinuous at each point $t \geqslant 0$ $(x \in \bigcup_{\lambda\in\mathfrak{T}} R(K_i(\lambda)))$. Towards this end, notice first that (i) implies that, for every $\zeta > 0$, the sequence $((k_l * g_\zeta)(t))_l$ is equicontinuous at each point $t \geqslant 0$. Let $\lambda \in \mathfrak{T}$, $x \in R(R(K_i(\lambda)))$ and $t \geqslant 0$ be fixed. Then, for every $s \in [0, t+1]$,

$$p\big(R_{i,l}(t)K_i(\lambda)x - R_{i,l}(s)K_i(\lambda)x\big) = p\big(\big[R_{i,l}(t)K_i(\lambda)x - R_{i,l}(t)K_{i,l}(\lambda)x\big]$$
$$+\big[R_{i,l}(t)K_{i,l}(\lambda)x) - R_{i,l}(s)K_{i,l}(\lambda)x\big] + \big[R_{i,l}(s)K_{i,l}(\lambda)x - R_{i,l}(s)K_i(\lambda)x\big]\big)$$
$$\leqslant 2c_p r_p\big(K_i(\lambda)x - K_{i,l}(\lambda)x\big)e^{\omega(t+1)} + p\big(R_{i,l}(t)K_{i,l}(\lambda)x - R_{i,l}(s)K_{i,l}(\lambda)x\big).$$

Moreover,

$$R_{i,l}(t)K_{i,l}(\lambda)x - R_{i,l}(s)K_{i,l}(\lambda)x = (k_l * g_i)(t)C_lK_{i,l}(\lambda)x - (k_l * g_i)(s)C_lK_{i,l}(\lambda)x$$
$$+\sum_{j\in D_{i,l}}\big[g_{\alpha_{n,l}-\alpha_{j,l}} * \big(R_{i,l}(\cdot)A_{j,l}K_{i,l}(\lambda)x - (k_l * g_i)(\cdot)\,C_lA_{j,l}K_{i,l}(\lambda)x\big)\big](s)$$
$$-\sum_{j\in D_{i,l}}\big[g_{\alpha_{n,l}-\alpha_{j,l}} * \big(R_{i,l}(\cdot)A_{j,l}K_{i,l}(\lambda)x - (k_l * g_i)(\cdot)\,C_lA_{j,l}K_{i,l}(\lambda)x\big)\big](t)$$
$$+\sum_{j\in\mathbb{N}_{n-1}\setminus D_{i,l}}\big[(g_{\alpha_{n,l}-\alpha_{j,l}} * R_{i,l})(s)A_{j,l}K_{i,l}(\lambda)x - (g_{\alpha_{n,l}-\alpha_{j,l}} * R_{i,l})(t)\,A_{j,l}K_{i,l}(\lambda)x\big]$$
$$= \begin{cases} (g_{\alpha_{n,l}-\alpha_l} * R_{i,l})(t)\,A_lK_{i,l}(\lambda)x - (g_{\alpha_{n,l}-\alpha_l} * R_{i,l})(s)\,A_lK_{i,l}(\lambda)x, & m_l - 1 < i,\\ \big[g_{\alpha_{n,l}-\alpha_l} * \big(R_{i,l}(\cdot)\,A_lK_{i,l}(\lambda)x - (k_l * g_i)(\cdot)\,C_lA_lK_{i,l}(\lambda)x\big)\big](t) & \\ -\big[g_{\alpha_{n,l}-\alpha_l} * (R_{i,l}(\cdot)\,A_lK_{i,l}(\lambda)x - (k_l * g_i)(\cdot)\,C_lA_lK_{i,l}(\lambda)x)\big](s), & m_l - 1 \geqslant i. \end{cases}$$

Due to (iii), the sequences $(A_{j,l}K_{i,l}(\lambda)x)_l$ and $(A_{j,l}C_lK_{i,l}(\lambda)x)_l$ are bounded in E for all $j \in \mathbb{N}^0_{n-1}$. Now the proof can be completed by literally repeating the arguments given in the proof of Theorem 2.7.2 (cf. also Remark 2.7.3(ii)). □

Theorem 2.10.27. *Suppose $q \in \mathbb{N}$, $k_l(t)$ satisfies (P1), $\lambda_0 > \omega \geqslant \sup_{l\in\mathbb{N}_0, t\geqslant 0}\max(0, abs(k_l))$, $\sup_{l\in\mathbb{N}_0} e^{-\omega t}|k_l(t)| < \infty$, and the condition (HCC) holds with $(A_{n-1}, \cdot\cdot\cdot, A_0)$ and C replaced respectively by $(A_{n-1,l}, \cdot\cdot\cdot, A_{0,l})$* and C_l *($l \in \mathbb{N}_0$). Let $A_l \equiv A$, $0 \leqslant i \leqslant \min_{l\in\mathbb{N}_0} m_{n,l} - 1$, and let for each $p \in ⊛$ there exist $c_p > 0$ and $r_p \in ⊛$ such that (421) hold. Assume, additionally, that the operator $G_{i,l}(\lambda)$ is injective for every $i \in \mathbb{N}^0_{m_{n,l}-1}$ and for every $\lambda \in \mathbb{C}$ with* Re *$\lambda > \omega$ and $\tilde{k}_l(\lambda) \neq 0$ ($l \in \mathbb{N}_0$). Let $((R_{0,l}(t))_{t\geqslant 0}, \cdot\cdot\cdot, (R_{m_{n,l}-1}(t))_{t\geqslant 0})$ be the corresponding k_l-regularized C_l-resolvent propagation family, and let $\mathfrak{T}$ and $K_{i,l}(\lambda)$ possess the same meanings as in the preceding theorem. Suppose that the conditions (i) and (ii) of Theorem 2.10.26 hold, as well as that:*

(iii) $\lim_{l\to\infty} P^{-1}_{\lambda,l}\,G_{i,l}(\lambda)C_lx = P^{-1}_\lambda G_i(\lambda)Cx$, *provided $\lambda \in \mathfrak{T}$ and $x \in D(A^q)$.*

(iv) *If* $x \in D(A^q)$, $j \in \mathbb{N}_{n-1}$ *and* $0 \leqslant r \leqslant q-1$, *then the sequences* $(C_l A^r x)_l$ *and* $(A_{j,l}A^r x)_l$ *are bounded in E.*

(v) *If there exists* $l_0 \in \mathbb{N}$ *such that* $m_l - 1 \geqslant i$, $l \geqslant l_0$, *then we also assume that the sequence* $(C_l A^q x)_l$ *is bounded in E for each* $x \in D(A^q)$.

Then $\lim_{l\to\infty} R_{i,l}(t)x = R_i(t)x$, $t \geqslant 0$, $x \in \overline{D(A^q)}$, *uniformly on compacts of* $[0, \infty)$.

Proof. The proof of this theorem is very similar to those of Theorem 2.7.5 and Theorem 2.10.26 and, because of that, we will only sketch it. As in the proof of Theorem 2.10.26, we obtain that, for every $l \in \mathbb{N}_0$ and for every $\lambda \in \mathbb{C}$ with Re $\lambda > \omega$ and $\tilde{k}_l(\lambda) \neq 0$, (422)-(423) hold. By (421), it suffices to show that the sequence $(R_{i,l}(t)x)_l$ is equicontinuous at each point $t \geqslant 0$ ($x \in D(A^q)$). This follows as in the proof of Theorem 2.7.5, by applying the functional equation (399) successively q times (cf. also (235)). Here we also use the condition (HCC) as well as the fact that, for every $\zeta > 0$, the sequence $((k_l * g_\zeta)(t))_l$ is equicontinuous at each point $t \geqslant 0$. □

The following theorem on convergence of k-regularized C-resolvent propagation families naturally corresponds to Theorem 2.7.6.

Theorem 2.10.28. *Suppose that the function* $k_l(t)$ *satisfies (P1),* $\lambda_0 > \omega \geqslant \sup_{l\in\mathbb{N}_0} \max(0, abs(k_l))$, $\sup_{l\in\mathbb{N}_0, t\geqslant 0} e^{-\omega t}|k_l(t)| < \infty$, *and the condition (HCC) holds with* $(A_{n-1}, \cdot\cdot\cdot, A_0)$ *and C replaced respectively by* $(A_{n-1,l}, \cdot\cdot\cdot, A_{0,l})$ *and* C_l $(l \in \mathbb{N}_0)$. *Let the condition (i) and (ii) of Theorem 2.10.27 hold, and let for each* $p \in ⊛$ *there exist* $c_p > 0$ *and* $r_p \in ⊛$ *such that (421) holds for* $t \geqslant 0$, $x \in E$ *and* $n \in \mathbb{N}$. *Let* $((R_{0,l}(t))_{t\geqslant 0}, \cdot\cdot\cdot, (R_{m_{n,l}-1}(t))_{t\geqslant 0})$ *be the corresponding* k_l*-regularized* C_l*-resolvent propagation family, satisfying the property (399) for any* $l \in \mathbb{N}$. *Suppose that* $m_{n,l} \geqslant m_n$, $l \in \mathbb{N}$ *and, for every* $i \in \mathbb{N}^0_{m_n-1}$ *and* $l \in \mathbb{N}_0$, *the operator* $G_{i,l}(\lambda)$ *is injective for every* $\lambda \in \mathbb{C}$ *with* $\lambda > \lambda_0$ *and* $\tilde{k}_l(\lambda) \neq 0$. *Let* $\mathfrak{T}' = \{\lambda > \lambda_0 : \tilde{k}_l(\lambda) \neq 0\}$, *and let* $K_{i,l}(\lambda)$ *possess the same meaning as in Theorem 2.10.27. Suppose that for each* $\lambda \in \mathfrak{T}'$ *there exists an open ball* $\Omega_\lambda \subseteq \{z \in \mathbb{C} : \text{Re } z > \lambda_0\}$, *with center at point* λ *and radius* $2\varepsilon_\lambda > 0$, *so that* $\tilde{k}_l(z) \neq 0$, $z \in \Omega_\lambda$, $l \in \mathbb{N}_0$. *If there exists* $i \in \mathbb{N}^0_{m_n-1}$ *such that* $m-1 \geqslant i$, *then we also assume that* $\lim_{l\to\infty} C_l x = Cx$ *for all* $x \in E$. *Then the following holds:*

(i) *For each* $r \in (0,1]$, *there exists an exponentially equicontinuous* $(k * g_r)$*-regularized C-resolvent propagation family* $((R^r_0(t))_{t\geqslant 0}, \cdot\cdot\cdot, (R^r_{m_n-1}(t))_{t\geqslant 0})$ *satisfying (399) for any* $i \in \mathbb{N}^0_{m_n-1}$ *and that, for every* $p \in ⊛$, *(30) holds with* $(R_r(t))_{t\geqslant 0}$ *replaced by* $(W^r_i(t) \equiv R^r_i(t))_{t\geqslant 0}$, *if* $m-1 < i$, *and* $(W^r_i(t) \equiv R^r_i(t) - (k * g_i)(t)C)_{t\geqslant 0}$, *if* $m-1 \geqslant i$. *Furthermore, for every* $p \in ⊛$, $i \in \mathbb{N}^0_{m_n-1}$ *and* $B \in \mathcal{B}$, *the mapping* $t \mapsto p_B(W^r_i(t))$, $t \geqslant 0$ *is locally Hölder continuous with exponent r.*

(ii) *If A is densely defined and* $\alpha_n - \alpha_j \geqslant 1$, $j \in \mathbb{N}^0_{n-1}$, *then there exists an exponentially equicontinuous k-regularized C-resolvent propagation family* $((R_0(t))_{t\geqslant 0}, \cdot\cdot\cdot, (R_{m_n-1}(t))_{t\geqslant 0})$ *satisfying (399) for any* $i \in \mathbb{N}^0_{m_n-1}$ *and that the family* $\{e^{-\omega t}R_i(t) : t \geqslant 0\}$ *is equicontinuous for all* $i \in \mathbb{N}^0_{m_n-1}$.

Proof. Although technically complicated, the proof of assertion (i) follows in almost the same way as in the proof of Theorem 2.7.6. Because of that, we shall only outline the main details needed for the proof of (ii). If $\alpha_n - \alpha_j \geqslant 1$, $j \in \mathbb{N}^0_{n-1}$, then we obtain similarly as in the proof of [20, Proposition 1.3.6] that, for every $x \in E$, $i \in \mathbb{N}^0_{m_n-1}$ and $j \in \mathbb{N}^0_{n-1}$, the mapping $t \mapsto F_{j,i}(t)x \equiv (g_{\alpha_n-\alpha_j} * W_i^1)(t)x$ is continuously differentiable for $t \geqslant 0$ and that $(d/dt)F_{j,i}(t)x = (g_{\alpha_n-\alpha_j-1} * W_i^1)(t)x$, $t \geqslant 0$; here we assume that $W_i^1(t)$ is defined as in (i) with $r = 1$. If, in addition, A is densely defined, then the proof of [531, Theorem 3.4, pp. 14–15] and the functional equation (399) imply that, for every $x \in E$, $i \in \mathbb{N}^0_{m_n-1}$, and $j \in \mathbb{N}^0_{n-1}$, the mapping $t \mapsto W_i^1(t)x$, $t \geqslant 0$ is continuously differentiable. Set $W_i(t)x := (d/dt)W_i^1(t)x$ for $t \geqslant 0$, $x \in E$ and $i \in \mathbb{N}^0_{m_n-1}$, as well as $R_i(t) := W_i(t)$, $t \geqslant 0$, if $m - 1 < i$, and $R_i(t) := W_i(t) + (k * g_i)(t)C$, $t \geqslant 0$, if $m - 1 \geqslant i$. Repeating literally the proof of the above-mentioned theorem, we obtain finally that (ii) holds. □

The approximation and convergence of k-regularized C_1-existence families for (2) and k-regularized C-resolvent families for (2) can be considered in a similar way (the best option for work occurs in the case that the assumptions of Theorem 2.10.18(i)-(b) hold).

Example 2.10.29. (i) Let $1 < p < \infty$, let $E = L^p(\mathbb{R}^n)$ and let $0 \leqslant l \leqslant n$. Suppose $0 < \alpha < 2$, $\omega \geqslant 0$, $N \in \mathbb{N}$, $r \in (0, N]$, $P(x)$ is an r-coercive complex polynomial of degree N (cf. Section 2.5 for the notion), $a \in \mathbb{C} \setminus P(\mathbb{R}^n)$, $\gamma \geqslant n|\frac{1}{p} - \frac{1}{2}|\frac{N}{r\min(1,\alpha)}$, and

$$\sup_{x \in \mathbb{R}^n} \operatorname{Re}(P(x)^{1/\alpha}) \leqslant \omega.$$

Let $P(D)f = \Sigma_{|\eta| \leqslant N} a_\eta D^\eta f$ act on E_l with its maximal distributional domain, where $D = (-i\partial/\partial x_1, \cdots, -i\partial/\partial x_n)$. As announced before, $(R_\alpha(t) \equiv \mathbf{T}_l\langle E_\alpha(t^\alpha P(x))(a - P(x))^{-\gamma}\rangle)_{t \geqslant 0}$ is an exponentially equicontinuous $(g_\alpha, R_\alpha(0))$-regularized resolvent family with the integral generator $P(D)$, and we have the following estimate

$$q_\eta(R_\alpha(t)f) \leqslant M\left(1 + t^{\max(1,\alpha)n|\frac{1}{p}-\frac{1}{2}|}\right)e^{\omega t}q_\eta(f),\ t \geqslant 0, f \in E_l,\ \eta \in \mathbb{N}_0^l,$$

with M being independent of $f \in E_l$ and $\eta \in \mathbb{N}_0^l$. Suppose $(\alpha_l)_l$ is a strictly increasing sequence in $(0, \infty)$ with $\lim_{l \to \infty} \alpha_l = \alpha$, and $(R_{\alpha_l}(t))_{t \geqslant 0}$ denotes the subordinated $(g_{\alpha_l}, R_\alpha(0))$-regularized resolvent family ($l \in \mathbb{N}$). Then Theorem 2.7.4 enables one to simply prove that $\lim_{l \to \infty} R_{\alpha_l}(t)f = R_\alpha(t)f$, $t \geqslant 0$, $f \in E_l$, uniformly on compacts of $[0, \infty)$.

(ii) Suppose $\omega \geqslant 0$, $c_{j,l} \in \mathbb{C}$ ($1 \leqslant j \leqslant n - 1$, $l \in \mathbb{N}_0$) and, for every $i \in \mathbb{N}^0_{m_{n,l}-1}$ with $m_l - 1 \geqslant i$, we have $\mathbb{N}_{n-1} \setminus D_{i,l} \neq \emptyset$ and $\Sigma_{j \in \mathbb{N}_{n-1} \setminus D_{i,l}} |c_{j,l}|^2 > 0$. Let $A_{j,l} = c_{j,l} I$ for $1 \leqslant j \leqslant n - 1$ and $l \in \mathbb{N}_0$, and let $A_l \equiv A$, $C_l \equiv C$. Suppose $0 \leqslant i \leqslant \min_{l \in \mathbb{N}_0} m_{n,l} - 1$, $0 < \delta_l \leqslant 2$, $a_l > 0$, $b_l \in (0,1)$, $\pi\delta_l/(2(\alpha_{n,l} - \alpha_l)) > \pi/2$ and $K_l(t) = \mathcal{L}^{-1}(\exp(-a_l\lambda^{b_l}))(t)$, $t \geqslant 0$ ($l \in \mathbb{N}_0$). Let for each $l \in \mathbb{N}_0$ the operator A be a subgenerator of an exponentially equicontinuous (g_{δ_l}, K_l)-regularized C-resolvent family

$(S_l(t))_{t\geqslant 0}$ satisfying that for each $p \in ⊛$ there exist $c_p > 0$ and $r_p \in ⊛$ such that (421) holds with $(R_l(t))_{t\geqslant 0}$ replaced by $(S_l(t))_{t\geqslant 0}$, and that (22) holds with $a(t) = g_{\delta_l}(t)$, $k(t) = K_l(t)$ and $(R(t))_{t\geqslant 0}$ replaced by $(S_l(t))_{t\geqslant 0}$. Suppose $\lim_{l\to\infty} \alpha_{j,l} = \alpha_j$ for any $j \in \mathbb{N}^0_{n-1}$, $\lim_{l\to\infty} a_l = a$, $\lim_{l\to\infty} b_l = b$ and $\lim_{l\to\infty} \delta_l = \delta$. Put $\theta_l := \min((\pi/2), \pi\delta_l/(2(\alpha_{n,l} - \alpha_l)) - (\pi/2))$ and $k_l(t) := \mathcal{L}^{-1}(\exp(-a_{l,1}\lambda^{b_{l,1}}))(t)$, $t \geqslant 0$, with $b_{l,1} := (\alpha_{n,l} - \alpha_l)b_l\delta_l^{-1}$ and $a_{l,1} > a_l(\cos((\alpha_{n,l} - \alpha_l)b_l\delta_l^{-1}(\frac{\pi}{2} + \theta_l)))^{-1}$ $(l \in \mathbb{N}_0)$. It is very simple to prove that $\lim_{l\to\infty} k_l(t) = k(t)$, $t \geqslant 0$, uniformly on compacts of $[0, \infty)$. Arguing as in Example 2.10.30(i)-(b) below, we get that the operator A is a subgenerator of an exponentially equicontinuous k_l-regularized C-resolvent propagation family $((R_{0,l}(t))_{t\geqslant 0}, \cdots, (R_{m_{n,l}-1}(t))_{t\geqslant 0})$. Furthermore, Theorem 2.10.26 implies that there exists $\lambda_0 \geqslant \omega$ such that (421) holds with the number ω replaced by λ_0 as well as that $\lim_{l\to\infty} R_{i,l}(t)x = R_i(t)x$, $t \geqslant 0$, $x \in \overline{\bigcup_{\lambda>\lambda_0} R(K_i(\lambda))}$, uniformly on compacts of $[0, \infty)$.

Now we shall present some other examples and applications of results obtained throughout this section.

Example 2.10.30. Suppose $c_j \in \mathbb{C}$ $(1 \leqslant j \leqslant n-1)$ and, for every $i \in \mathbb{N}^0_{m_n-1}$ with $m-1 \geqslant i$, we have $\mathbb{N}_{n-1}\setminus D_i \neq \emptyset$ and $\Sigma_{j\in\mathbb{N}_{n-1}\setminus D_i} |c_j|^2 > 0$. Let $A_j = c_jI$ for $1 \leqslant j \leqslant n-1$.

(i) (a) Suppose $0 < \delta \leqslant 2$, $\sigma \geqslant 1$, $\frac{\pi\delta}{2(\alpha_n-\alpha)} - \frac{\pi}{2} > 0$, and A is a subgenerator of an exponentially equicontinuous (g_δ, g_σ)-regularized C-resolvent family $(R_\delta(t))_{t\geqslant 0}$ which satisfies the following equality:

$$A\int_0^t g_\delta(t-s)R_\delta(s)x\,ds = R_\delta(t)x - g_\sigma(t)Cx, \quad x \in E,\ t \geqslant 0. \tag{424}$$

Put $\sigma' := \max(1, 1+(\alpha_n - \alpha)(\sigma-1)\delta^{-1})$ and $\theta := \min(\frac{\pi}{2}, \frac{\pi\delta}{2(\alpha_n-\alpha)} - \frac{\pi}{2})$. Then, for every sufficiently small $\varepsilon > 0$, there exists $\omega_\varepsilon > 0$ such that $\omega_\varepsilon + \Sigma_{\frac{\pi}{2}\delta-\varepsilon} \subseteq \rho_C(A)$ and the family $\{|\lambda|^{\frac{\delta-\sigma}{\delta}}(1+|\lambda|^{\frac{1}{\delta}})(\lambda - A)^{-1}C : \lambda \in \omega_\varepsilon + \Sigma_{\frac{\pi}{2}\alpha-\varepsilon}\}$ is equicontinuous. Notice also that

$$\begin{aligned}
&\arg\left(\lambda^{\alpha_n-\alpha} + \sum_{j=1}^{n-1} c_j\lambda^{\alpha_j-\alpha}\right)\\
&= \arg\left(\lambda^{\alpha_n-\alpha}|\lambda|^{\alpha-\frac{\alpha_{n-1}+\alpha_n}{2}} + \sum_{j=1}^{n-1} c_j\lambda^{\alpha_j-\alpha}|\lambda|^{\alpha-\frac{\alpha_{n-1}+\alpha_n}{2}}\right)\\
&\approx \arg\left(\lambda^{\alpha_n-\alpha}|\lambda|^{\alpha-\frac{\alpha_{n-1}+\alpha_n}{2}}\right)\\
&= (\alpha_n - \alpha)\arg(\lambda), \quad \lambda \to \infty,\ \arg(\lambda) < \frac{\pi}{\alpha_n - \alpha}.
\end{aligned}$$

Due to the choice of θ, we have that, for every sufficiently small $\varepsilon > 0$, there exists $\omega_\varepsilon > 0$ such that, for every $\lambda \in \omega_\varepsilon + \Sigma_{\frac{\pi}{2}+\theta-\varepsilon}$,

$$\arg\left(\lambda^{\alpha_n-\alpha}+\sum_{j=1}^{n-1} c_j\lambda^{\alpha_j-\alpha}\right)<\frac{\pi}{2}\delta-\varepsilon.$$

Therefore, we have the following: If the operator A is densely defined, then the above inequality in combination with Theorem 2.10.11 indicates that A is a subgenerator of an exponentially equicontinuous, analytic $(\sigma'-1)$-times integrated C-resolvent propagation family $((R_0(t))_{t\geqslant 0},\cdots,(R_{m_n-1}(t))_{t\geqslant 0})$ for (2), with θ being the angle of analyticity; if the operator A is not densely defined, then the above conclusion continues to hold with σ' replaced by any number $\sigma''>\sigma'$. Observe finally that the procedure described above supplies a great number of concrete examples of analytic resolvent propagation families.

(a') Suppose $0<\delta\leqslant 2$, $\sigma\geqslant 1$, $\frac{\delta(\frac{\pi}{2}+\gamma)}{(\alpha_n-\alpha)}-\frac{\pi}{2}>0$, A is a subgenerator of an exponentially equicontinuous, analytic (g_δ, g_σ)-regularized C-resolvent family $(R_\delta(t))_{t\geqslant 0}$ of angle $\gamma\in(0,\frac{\pi}{2}]$, and (424) holds. Put $\sigma_1:=\sigma'$ and $\theta_1:=\min(\frac{\pi}{2},\frac{\delta(\frac{\pi}{2}+\gamma)}{(\alpha_n-\alpha)}-\frac{\pi}{2})$. If the operator A is densely defined, then it follows from Theorem 2.2.4 and the above analysis that the operator $C^{-1}AC$ is the integral generator of an exponentially equicontinuous, analytic (σ_1-1)-times integrated C-resolvent propagation family $((R_0(t))_{t\geqslant 0},\cdots,(R_{m_n-1}(t))_{t\geqslant 0})$ for (2), with θ_1 being the angle of analyticity; if the operator A is not densely defined, then the above remains true with σ_1 replaced by any number $\sigma_2>\sigma_1$. Now we shall apply the obtained results to the following fractional analogue of the telegraph equation:

$$\mathbf{D}_t^{\alpha_2}u(t,x)+c_1\mathbf{D}_t^{\alpha_1}u(t,x)=D\Delta_x u(t,x),\quad t>0,\ x\in\mathbb{R}^n,$$

where $c_1>0$, $D>0$ and $0<\alpha_1\leqslant\alpha_2<2$. Let E be one of the spaces $L^p(\mathbb{R}^n)$ $(1\leqslant p\leqslant\infty)$, $C_0(\mathbb{R}^n)$, $C_b(\mathbb{R}^n)$, $BUC(\mathbb{R}^n)$, let $0\leqslant l\leqslant n$, and let $A:=D\Delta$ act with its maximal distributional domain. Suppose first $E\neq L^\infty(\mathbb{R}^n)$ and $E\neq C_b(\mathbb{R}^n)$. Then the operator A is the integral generator of an exponentially equicontinuous, analytic C_0-semigroup of angle $\pi/2$, which implies that A is the integral generator of an exponentially equicontinuous, analytic I-regularized resolvent propagation family $(R_0(t))_{t\geqslant 0}$, if $\alpha_2\leqslant 1$, resp. $((R_0(t))_{t\geqslant 0},(R_1(t))_{t\geqslant 0})$ if $\alpha_2>1$, of angle $\zeta=\min(\frac{\pi}{2},\frac{\pi}{\alpha_2}-\frac{\pi}{2})$; this conclusion continues to hold if we choose the Fréchet nuclear space Ξ as the underlying SCLCS ([303]). Observe, however, that it is not clear whether the angle of analyticity of constructed I-regularized resolvent propagation families, in the case that $\alpha_1<\alpha_2<1$, can be improved by allowing that ζ takes the value $\min(\pi,\frac{\pi}{\alpha_2}-\frac{\pi}{2})$. Suppose now $E=L^\infty(\mathbb{R}^n)$ or $E=C_b(\mathbb{R}^n)$. Then, for every $\sigma'>1$, the operator A is the integral generator of an exponentially equicontinuous, analytic $(\sigma'-1)$-times integrated I-regularized resolvent propagation family $(R_0(t))_{t\geqslant 0}$, if $\alpha_2\leqslant 1$, resp. $((R_0(t))_{t\geqslant 0},(R_1(t))_{t\geqslant 0})$ if $\alpha_2>1$, of angle $\min(\frac{\pi}{2},\frac{\pi}{\alpha_2}-\frac{\pi}{2})$.

(b) Suppose $0 < \delta \leqslant 2$, $\sigma \geqslant 1$, $\frac{\pi\delta}{2(\alpha_n - \alpha)} - \frac{\pi}{2} > 0$, $a > 0$, $b \in (0, 1)$, $k_{a,b}(t) := \mathcal{L}^{-1}(\exp(-a\lambda^b))(t)$, $t \geqslant 0$ and A is a subgenerator of an exponentially equicontinuous $(g_\delta, k_{a,b})$-regularized C-resolvent family $(R_{a,b}(t))_{t\geqslant 0}$ which satisfies the following equality:

$$A\int_0^t g_\delta(t-s)R_{a,b}(s)x\,ds = R_{a,b}(t)x - k_{a,b}(t)Cx, \quad x \in E,\ t \geqslant 0. \tag{425}$$

Let θ be defined as in (a). Then it can be easily seen that $(\alpha_n - \alpha)b\delta^{-1} < 1$ and $(\alpha_n - \alpha)b\delta^{-1}(\frac{\pi}{2} + \theta) < \frac{\pi}{2}$. Put $k_1(t) := k_{a_1,b_1}(t)$, $t \geqslant 0$, where $b_1 := (\alpha_n - \alpha)b\delta^{-1}$ and $a_1 > a(\cos((\alpha_n - \alpha)b\delta^{-1}(\frac{\pi}{2} + \theta)))^{-1}$. Then it is clear that, for every $\theta' \in (0, \theta)$, there exists a sufficiently large $\omega_{\theta'} > 0$ such that, for every $\lambda \in \omega_{\theta'} + \Sigma_{\frac{\pi}{2}+\theta'}$,

$$\frac{|\widetilde{k_1}(\lambda)|}{\left|\tilde{k}\left(\left(\lambda^{\alpha_n - \alpha} + \sum_{j=1}^{n-1} c_j\lambda^{\alpha_j - \alpha}\right)^{1/\delta}\right)\right|}$$

$$\leqslant |\widetilde{k_1}(\lambda)|\exp\left(a|\lambda|^{b_1} + \sum_{j=1}^{n-1}|c_j||\lambda|^{(\alpha_j - \alpha)b/\delta}\right).$$

Arguing as in (a), we reveal that A is a subgenerator of an exponentially equicontinuous analytic k_1-regularized C-resolvent propagation family $((R_0(t))_{t\geqslant 0}, \cdots, (R_{m_n - 1}(t))_{t\geqslant 0})$ for (2), with θ being the angle of analyticity.

(b') Suppose $0 < \delta \leqslant 2$, $\sigma \geqslant 1$, $\frac{\delta(\frac{\pi}{2} + \gamma)}{\alpha_n - \alpha} - \frac{\pi}{2} > 0$, $A_j = c_j I$ $(1 \leqslant j \leqslant n-1)$, $a > 0$, $b \in (0, 1)$, A is a subgenerator of an exponentially equicontinuous, analytic $(g_\delta, k_{a,b})$-regularized C-resolvent family $(R_{a,b}(t))_{t\geqslant 0}$ of angle $\gamma \in (0, \frac{\pi}{2}]$, and (425) holds. Assume, additionally, that $b(1 + \frac{2\gamma}{\pi}) \leqslant 1$. Define θ_1 as in (a'), and $k_2(t) := k_{a_2,b_2}(t)$, $t \geqslant 0$, where $b_2 := (\alpha_n - \alpha)b\delta^{-1}$ and $a_2 > a(\cos((\alpha_n - \alpha)b\delta^{-1}(\frac{\pi}{2} + \theta_1)))^{-1}$. The reader will have no difficulty in checking that $(\alpha_n - \alpha)b < \delta$ and $(\alpha_n - \alpha)b\delta^{-1}(\frac{\pi}{2} + \gamma) \leqslant \frac{\pi}{2}$. Making use of Theorem 2.2.4 and the foregoing arguments, we obtain that the operator $C^{-1}AC$ is the integral generator of an exponentially equicontinuous, analytic k_2-regularized C-resolvent propagation family $((R_0(t))_{t\geqslant 0}, \cdots, (R_{m_n-1}(t))_{t\geqslant 0})$ for (2), with θ being the angle of analyticity.

(ii) Suppose E is complete, $0 < \delta \leqslant 2$, $\frac{\pi\delta}{2(\alpha_n - \alpha)} - \frac{\pi}{2} > 0$, A is densely defined and generates a q-exponentially equicontinuous (g_δ, g_1)-regularized resolvent family $(R_\delta(t))_{t\geqslant 0}$ which satisfies that, for every $p \in \circledast$, there exist $M_p \geqslant 1$ and $\omega_p \geqslant 0$ such that $p(R_\delta(t)x) \leqslant M_p e^{\omega_p t}p(x)$, $t \geqslant 0$, $x \in E$. By Theorem 2.4.3, we know that A is a compartmentalized operator and that, for every $p \in \circledast$, the operator $\overline{A_p}$ is the integral generator of an exponentially bounded (g_δ, g_1)-regularized $\overline{I_p}$-resolvent family in $\overline{E_p}$. Then the first part of this example shows that $\overline{A_p}$ is the integral generator of an exponentially bounded, analytic $\overline{I_p}$-resolvent propagation family, with $\min(\frac{\pi}{2}, \frac{\pi\delta}{2(\alpha_n - \alpha)} - \frac{\pi}{2})$ being the angle of analyticity. By Theorem 2.10.12(ii), we obtain that A is the integral generator of a q-exponentially equicontinuous, analytic I-resolvent propagation family $((R_0(t))_{t>0}, \cdots, (R_{m_n-1}(t))_{t>0})$ for (2), and that the corresponding angle of analyticity is $\min(\frac{\pi}{2}, \frac{\pi\delta}{2(\alpha_n-\alpha)} - \frac{\pi}{2})$. It can be simply shown that, for every $p \in \circledast$ and $i \in \mathbb{N}^0_{m_n-1}$, there exist $M_{p,i} \geqslant 1$ and $\omega_{p,i} \geqslant 0$ such that $p(R_i(t)x) \leqslant M_{p,i}e^{\omega_{p,i}t}p(x)$, $t \geqslant 0, x \in E$.

Example 2.10.31. Suppose $1 \leqslant p \leqslant \infty$, $E := L^p(\mathbb{R})$, $m : \mathbb{R} \to \mathbb{C}$ is measurable, $a_j \in L^\infty(\mathbb{R})$, $(A_jf)(x) := a_j(x)f(x)$, $x \in \mathbb{R}, f \in E$ $(1 \leqslant j \leqslant n-1)$ and $(Af)(x) := m(x)f(x)$, $x \in \mathbb{R}$, with maximal domain. Assume $s \in (1,2)$, $\delta = 1/s$, $M_p = p!^s$ and $k_\delta(t) = \mathcal{L}^{-1}(\exp(-\lambda^\delta))(t)$, $t \geqslant 0$. Denote by $M(t)$ the associated function of sequence (M_p) and put $\Lambda_{\alpha',\beta',\gamma'} := \{\lambda \in \mathbb{C} : \mathrm{Re}\lambda \geqslant \gamma'^{-1}M(\alpha'\lambda) + \beta'\}$, $\alpha', \beta', \gamma' > 0$. Clearly, there exists a constant $C_s > 0$ such that $M(\lambda) \leqslant C_s|\lambda|^{1/s}$, $\lambda \in \mathbb{C}$. In continuation of this example, we assume that the following condition holds:

(CH): For every $\tau > 0$, there exist $\alpha' > 0, \beta' > 0$ and $d > 0$ such that $\tau \leqslant \frac{\cos(\frac{\delta\pi}{2})}{C_s\alpha'^{1/s}}$ and

$$\left|\lambda^{\alpha_n-\alpha} + \sum_{j=1}^{n-1}\lambda^{\alpha_j-\alpha}a_j(x) - m(x)\right| \geqslant d, \quad x \in \mathbb{R}, \lambda \in \Lambda_{\alpha',\beta',1}.$$

Notice that the above condition holds provided $n = 2$, $\alpha_2 - \alpha = 2$, $\alpha_2 - \alpha_1 = 1$ and $m(x) = \frac{1}{4}a_1^2(x) - \frac{1}{16}a_1^4(x) - 1$, $x \in \mathbb{R}$ (cf. [317]), and that the validity of condition (CH) does not imply, in general, the essential boundedness of function $m(\cdot)$. We will prove that A is the integral generator of a global (not exponentially bounded, in general) k_δ-regularized I-resolvent propagation family $((R_0(t))_{t>0}, \cdots, (R_{m_n-1}(t))_{t>0})$ for (2). Clearly, it suffices to show that, for every $\tau > 0$, A is the integral generator of a local k_δ-regularized I-resolvent propagation family for (2) on $[0, \tau)$. Suppose $\tau > 0$ is given in advance, and $\alpha' > 0$, $\beta' > 0$ and $d > 0$ satisfy (CH), with this τ. Let Γ denote the upwards oriented boundary of ultra-logarithmic region $\Lambda_{\alpha',\beta',1}$. Put, for every $t \in [0, \tau)$, $f \in E$ and $x \in \mathbb{R}$,

$$(R_i(t)f)(x) := \frac{1}{2\pi i}\int_\Gamma e^{\lambda t-\lambda^\delta}\frac{\left[\lambda^{\alpha_n-\alpha-i} + \sum_{j\in D_i}\lambda^{\alpha_j-\alpha-i}a_j(x)\right]f(x)}{\lambda^{\alpha_n-\alpha} + \sum_{j=1}^{n-1}\lambda^{\alpha_j-\alpha}a_j(x) - m(x)}\,d\lambda,$$

if $m - 1 < i$, and

$$(R_i(t)f)(x) := \frac{(-1)}{2\pi i}\int_\Gamma e^{\lambda t-\lambda^\delta}\frac{\lambda^{\alpha_j-\alpha-i}a_j(x)\,d\lambda}{\lambda^{\alpha_n-\alpha} + \sum_{j=1}^{n-1}\lambda^{\alpha_j-\alpha}a_j(x) - m(x)} + (k_\delta * g_i)(t)f(x),$$

if $m - 1 \geqslant i$. It is clear that, for every $i \in \mathbb{N}^0_{m_n-1}$, $R_i(t)A_j \subseteq A_jR_i(t)$, $t \in [0, \tau)$, $j \in \mathbb{N}^0_{n-1}$ and that $(R_i(t))_{t\in[0,\tau)} \subseteq L(E)$ is strongly continuous. Moreover, the Cauchy theorem implies that $R_i(0) = 0 = k_\delta(0)$, $i \in \mathbb{N}^0_{m_n-1}$. Now we will prove that the identity (400) holds provided $m - 1 < i$ and $C_2 = I$. Let $f \in D(A)$. Then a simple computation involving Cauchy theorem shows that (400) holds, with x replaced by $f(\cdot)$ therein, iff:

$$\frac{1}{2\pi i}\int_{\Gamma} e^{\lambda t-\lambda^{\delta}} \frac{\left[\lambda^{\alpha_n-\alpha-i}+\sum_{j\in D_i}\lambda^{\alpha_j-\alpha-i}a_j(x)\right]f(x)}{\lambda^{\alpha_n-\alpha}+\sum_{j=1}^{n-1}\lambda^{\alpha_j-\alpha}a_j(x)-m(x)}\,d\lambda$$

$$+\sum_{j=1}^{n-1}\frac{1}{2\pi i}\int_{\Gamma}\left(\int_0^t g_{\alpha_n-\alpha_j}(t-s)e^{\lambda s}\,ds\right)e^{-\lambda^{\delta}}\frac{\left[\lambda^{\alpha_n-\alpha-i}+\sum_{l\in D_i}\lambda^{\alpha_l-\alpha-i}g_l(x)\right]f(x)}{\lambda^{\alpha_n-\alpha}+\sum_{l=1}^{n-1}\lambda^{\alpha_l-\alpha}a_l(x)-m(x)}\,d\lambda$$

$$-\frac{1}{2\pi i}\int_{\Gamma}\left(\int_0^t g_{\alpha_n-\alpha}(t-s)e^{\lambda s}\,ds\right)e^{-\lambda^{\delta}}\frac{\left[\lambda^{\alpha_n-\alpha-i}+\sum_{j\in D_i}\lambda^{\alpha_j-\alpha-i}a_j(x)\right]m(x)\,f(x)}{\lambda^{\alpha_n-\alpha}+\sum_{j=1}^{n-1}\lambda^{\alpha_j-\alpha}a_j(x)-m(x)}\,d\lambda$$

$$=\frac{1}{2\pi i}\int_{\Gamma} e^{\lambda t-\lambda^{\delta}}\left[\lambda^{-i}f(x)+\sum_{j\in D_i}\lambda^{\alpha_j-\alpha_n-i}a_j(x)f(x)\right]d\lambda.$$

By [531, Lemma 5.5, p. 23] and the Cauchy theorem, the above equality is equivalent with:

$$\frac{1}{2\pi i}\int_{\Gamma} e^{\lambda t-\lambda^{\delta}} \frac{\left[\lambda^{\alpha_n-\alpha-i}+\sum_{j\in D_i}\lambda^{\alpha_j-\alpha-i}a_j(x)\right]f(x)}{\lambda^{\alpha_n-\alpha}+\sum_{j=1}^{n-1}\lambda^{\alpha_j-\alpha}a_j(x)-m(x)}\,d\lambda$$

$$+\sum_{j=1}^{n-1}\frac{1}{2\pi i}\int_{\Gamma}\frac{e^{\lambda t-\lambda^{\delta}}}{\lambda^{\alpha_n-\alpha_j}}\frac{\left[\lambda^{\alpha_n-\alpha-i}+\sum_{l\in D_i}\lambda^{\alpha_l-\alpha-i}g_l(x)\right]f(x)}{\lambda^{\alpha_n-\alpha}+\sum_{l=1}^{n-1}\lambda^{\alpha_l-\alpha}a_l(x)-m(x)}\,d\lambda$$

$$-\frac{1}{2\pi i}\int_{\Gamma}\frac{e^{\lambda t-\lambda^{\delta}}}{\lambda^{\alpha_n-\alpha}}\frac{\left[\lambda^{\alpha_n-\alpha-i}+\sum_{j\in D_i}\lambda^{\alpha_j-\alpha-i}a_j(x)\right]m(x)\,f(x)}{\lambda^{\alpha_n-\alpha}+\sum_{j=1}^{n-1}\lambda^{\alpha_j-\alpha}a_j(x)-m(x)}\,d\lambda$$

$$=\frac{1}{2\pi i}\int_{\Gamma} e^{\lambda t-\lambda^{\delta}}\left[\lambda^{-i}f(x)+\sum_{j\in D_i}\lambda^{\alpha_j-\alpha_n-i}a_j(x)f(x)\right]d\lambda,$$

which is true because the integrands appearing on both sides of this equality are equal identically. One can similarly prove that the identity (400) holds provided $m-1\geqslant i$ and $C_2=I$, so that $((R_0(t))_{t\geqslant 0},\cdots,(R_{m_n-1}(t))_{t\geqslant 0})$, defined in a very obvious way, is a k_δ-regularized I-resolvent propagation family for (2), with subgenerator A. Notice that the condition (CH) implies $m(\cdot)/(\lambda^{\alpha_n-\alpha}+\Sigma_{j=1}^{n-1}\lambda^{\alpha_j-\alpha}a_j(\cdot)-m(\cdot))\in L^\infty(\mathbb{R})$ for all $\lambda\in\Lambda_{\alpha',\beta',1}$, which has as a further consequence that $R(R_i(t))\subseteq D(A)$, provided $t\geqslant 0$ and $m-1<i$, and $R(R_i(t)-(k_\delta * g_i)(t))\subseteq D(A)$, provided $t\geqslant 0$ and $m-1\geqslant i$. The equality (399) holds for $((R_0(t))_{t\geqslant 0},\cdots,(R_{m_n-1}(t))_{t\geqslant 0})$, the integral generator of $((R_0(t))_{t\geqslant 0},\cdots,(R_{m_n-1}(t))_{t\geqslant 0})$ coincides with the operator A, which is the unique subgenerator of $((R_0(t))_{t\geqslant 0},\cdots,(R_{m_n-1}(t))_{t\geqslant 0})$. Notice that, for every compact set $K\subseteq[0,\infty)$, there exists $h_K>0$ such that

$$\sup_{t\in K, p\in \mathbb{N}_0, i\in \mathbb{N}^0_{m_n-1}} \frac{\left\| h_k^p \frac{d^p}{dt^p} R_i(t) \right\|}{p!^s} < \infty;$$

one can similarly consider the generation of local $k_{1/2}$-regularized I-resolvent propagation families which obey a modification of the property stated above with $s = 2$. Now we would like to give an example of a k_δ-regularized I-resolvent propagation family for (2) in which $A_j \notin L(E)$ for some $j \in \mathbb{N}_{n-1}$. Assume $n = 2$, $\alpha_2 - \alpha = 2$, $\alpha_2 - \alpha_1 = 1$, $a_1(x) = -2x$, $x \in \mathbb{R}$ and $m(x) = x^2 - x^4 - 1$, $x \in \mathbb{R}$. Define A_1, A and $R_i(\cdot)$ as above ($i = 0, 1$). Then the established conclusions continue to hold since, for every $\tau > 0$, there exist $\alpha' > 0$, $\beta' > 0$ and $d > 0$ such that (CH) holds as well as that:

$$\frac{x^2 + (x^4 - x^2 + 1)|\lambda|^{-2}}{|\lambda^2 - 2x\lambda + (x^4 - x^2 + 1)|} \leqslant d, \quad x \in \mathbb{R},\ \lambda \in \Lambda_{\alpha',\beta',1}.$$

The interested reader may try to construct some examples of local k-regularized C-resolvent propagation families which cannot be extended beyond its maximal interval of existence ([11], [295]).

Example 2.10.32. (see [319, Example 8.3]) Let $X := L^2[0, \pi]$ and let $A := \Delta$ with the Dirichlet boundary conditions. Suppose $0 < \gamma < 1$ and $X \ni f(x) = \Sigma_{n=1}^{\infty} c_n \sin nx$, $x \in [0, \pi]$. Then the unique solution of problem

$$\mathbf{D}_t^{2\gamma} u(t, x) + 2\mathbf{D}_t^{\gamma} u(t, x) + u(t, x) = 0, \quad t > 0, \quad x \in [0, \pi],$$
$$u(0, x) = f(x),\ u_t(0, x) = 0,\ x \in [0, \pi],$$

is given by

$$u(t, x) = \frac{1}{2} \sum_{n=1}^{\infty} c_n \sin nx \left[E_\gamma((in^2 - 1)t^\gamma) + E_\gamma(-(in^2 + 1)t^\gamma)\right]$$

$$-\frac{i}{2} \sum_{n=1}^{\infty} \frac{c_n}{n^2} \sin nx \left[E_\gamma((in^2 - 1)t^\gamma) - E_\gamma(-(in^2 + 1)t^\gamma)\right],\ t \geqslant 0,\ x \in [0, \pi].$$

Example 2.10.33. Suppose $1 \leqslant p \leqslant \infty$, $X := L^p(\mathbb{R})$, $a \in \mathbb{R}$, $r > 0$, $1/2 < \gamma \leqslant 1$, $T > 0$, $f \in C([0,T]: X)$, and $\frac{d}{dt}(g_{2\gamma-1} * \frac{d}{dx} f(t, \cdot)) \in C([0,T] : X)$. Put $A_1 := ad/dx$ and $Au := r\Delta u - \vartheta(\cdot)u$ with maximal distributional domain. Now we will focus our attention to the following fractional analogue of damped Klein-Gordon equation:

$$\text{(426)} \quad \mathbf{D}_t^{2\gamma} u(t, x) + a\frac{\partial}{\partial x} \mathbf{D}_t^{\gamma} u(t, x) - r\Delta_x u(t, x) + \vartheta(x)u(t, x) = f(t, x), \quad t > 0, x \in \mathbb{R},$$
$$u(0, x) = \phi(x),\ u_t(0, x) = \psi(x), \quad x \in \mathbb{R}.$$

Our intention in the first place is to describe results on the well-posedness of equation (426), obtained in [318] by the method proposed in [529]; in these papers, the condition $\vartheta(\cdot) \in W^{1,\infty}(\mathbb{R})$ has been required for the existence and uniquenss of solutions to (426). The case $\gamma = 1$ has been analyzed in [529, Example 4.1], showing that there exists an exponentially bounded I-uniqueness family for (426) and that, for every $\mu_0 \in \rho(A_1)$, there exists an exponentially bounded $(\mu_0 - A_1)^{-1}$-existence family for (426) with $Y = X$ (at first glance, this result seems to be non-optimal because the operator $C_1 = (\mu_0 - A_1)^{-1}$ does not depend on the choice of number p; cf. also (427)). The estimates obtained in the cited example enable one to simply verify that the conditions of Theorem 2.10.22(i)-(ii) hold with $k(t) = 1$ and $C_1 = (\mu_0 - A_1)^{-1}$, as well as that the conditions of Theorem 2.10.22(iii) hold with $k(t) = t$ and $C_2 = I$. This implies that there exists an exponentially bounded $g_{2\gamma}$-regularized I-uniqueness family $(U_\gamma(t))_{t\geqslant 0}$ for (426) with $\alpha_j = j\gamma$, $j = 0, 1, 2$, and that there exists an exponentially bounded $(\mu_0 - A_1)^{-1}$-existence family $(E_\gamma(t))_{t\geqslant 0}$ for (426) with $\alpha_j = j\gamma$, $j = 0, 1, 2$. Applying Theorem 2.10.20, we obtain that, for every $\phi \in W^{3,p}(\mathbb{R})$ and $\psi \in W^{3,p}(\mathbb{R})$, there exists a unique mild solution $u(t, x)$ of the corresponding problem (414) as well as that there exist $M \geqslant 1$ and $\omega \geqslant 0$ such that the following estimate holds for each $t \geqslant 0$:

$$\|u(t, x)\|_{L^p(\mathbb{R})} \leqslant Me^{\omega t}\Bigg[\|\phi\|_{W^{1,p}(\mathbb{R})} + \|\psi\|_{W^{1,p}(\mathbb{R})}$$

$$+ \int_0^t (t-s)^{2\gamma-2}\|f(s,\cdot)\|_{L^p(\mathbb{R})}\, ds + \int_0^t \Big\|\frac{d}{ds}\Big(g_{2\gamma-1} * \frac{d}{dx}f(s,\cdot)\Big)\Big\|_{L^p(\mathbb{R})}\Bigg]\, ds.$$

The solution $u(\cdot, x)$ is analytically extensible to the sector $\Sigma_{(\frac{1}{\gamma}-1)\frac{\pi}{2}}$, provided $f(t, x) \equiv 0$.

In the second approach, we shall suppose first that the function $\vartheta(\cdot)$ is a positive constant: $\vartheta(x) \equiv \vartheta > 0$. Set $\kappa := |1/2 - 1/p|$, provided $1 < p < \infty$, resp. $\kappa > 1/2$, provided $p \in \{1, \infty\}$, $C := (1 - \Delta) - \frac{1}{2}\kappa$ and $f(t, x) \equiv 0$. Then there exists a strong C-propagation family $\{(S_0(t))_{t\geqslant 0}, (S_1(t))_{t\geqslant 0}\}$ for the problem (426) with $\gamma = 1$ (cf. [531, Example 5.8, p. 130]), which implies, however, that for every $\phi \in W^{2,p}(\mathbb{R})$ and $\psi \in W^{2,p}(\mathbb{R})$, the function $u_\gamma(t, \cdot)$, $t > 0$, given by

$$u_\gamma(t, \cdot) := \int_0^\infty t^{-\gamma}\Phi(st^{-\gamma})\big[S_1(s)\phi + S_1'(s)\phi\big]\, ds$$

$$+ \int_0^t g_{1-\gamma}(t-s)\int_0^\infty s^{-\gamma}\Phi(rs^{-\gamma})S_1(r)\psi\, dr\, ds,$$

is a unique strong solution of the corresponding integral equation (414) with $u_0 = C\phi$ and $u_1 = C\psi$; certainly, this solution is analytically extensible to the sector $\Sigma_{(\frac{1}{\gamma}-1)\frac{\pi}{2}}$ (see, e.g., [49 (1.23), p. 12; Theorem 3.1-Theorem 3.3, pp. 40–42] and [531, Proposition 5.3(iii), p. 116]). In general the existence of a strong $(1 - \Delta)^{-\frac{1}{2}\kappa}$-

propagation family $\{(S_0(t))_{t>0}, (S_1(t))_{t>0}\}$ for the problem (426) with $\gamma = 1$ (and $\vartheta(x) \equiv \vartheta > 0$), taken together with [530, Proposition 1.6(iii)], [529, Theorem 2.7], Theorem 2.10.40 below, and Theorem 2.10.22, implies that there is an exponentially bounded $(1-\Delta)^{-\frac{1}{2}\kappa}$-existence family for (426) with $\alpha_j = j\gamma$, $j = 0, 1, 2$, provided that the mapping

(427) $$f \mapsto (1-\Delta)^{\frac{1}{2}\kappa}[(\vartheta(\cdot) - \vartheta)f(\cdot)],$$

from $L^p(\mathbb{R})$ into $L^p(\mathbb{R})$, is well-defined and continuous; in particular case $p = 2$, the essential boundedness of function $\vartheta(\cdot)$ (recall that the much stronger condition $\vartheta(\cdot) \in W^{1,\infty}(\mathbb{R})$ has been used so far, giving the much weaker results on the well-posedness of inhomogenous problem (426)) implies that there exists an exponentially bounded I-existence family for (426).

Similarly we can consider (cf. [529, Example 4.2] for more details) the results concerning the existence and uniqueness of mild solutions of the following time-fractional equation:

$$\mathbf{D}_t^{2\gamma}u(t,x) + \Big(\rho_1\frac{\partial^3}{\partial x^3} - \rho_2\frac{\partial^2}{\partial x^2}\Big)\mathbf{D}_t^{\gamma}u(t,x) + \Big(c\frac{\partial^2}{\partial x^2} + a(x)\Big)u(t,x) = f(t,x),$$
$$u(0,x) = \phi(x),\ u_t(0,x) = \psi(x).$$

Theorem 2.10.25 can be also applied in the analysis of the following integral equation:

$$u(t,x) = a\int_0^t a_1(t-s)\frac{\partial}{\partial x}u(s,x)\,ds + \int_0^t a_2(t-s)[r\Delta_x u(s,x) - \vartheta(x)u(s,x)]\,ds + f(t,x),$$

for certain classes of kernels $a_1(t)$ and $a_2(t)$, we leave details to the reader.

We wish now to take a closer look at the following slight modification of (426):

(428)
$$\mathbf{D}_t^{2\gamma}u(t,x) + a\frac{\partial}{\partial x}\mathbf{D}_t^{\gamma}u(t,x) - re^{i(2-2\gamma)\frac{\pi}{2}}\Delta_x u(t,x) + \vartheta(x)u(t,x) = f(t,x),\ t>0,\ x\in\mathbb{R},$$
$$u(0,x) = \phi(x),\ (\mathbf{D}_t^{\gamma}u(t,x))_{|t=0} = \psi(x),\ x\in\mathbb{R}.$$

Let $a \neq 0$ (for further information concerning the case $a = 0$, the references [314] and [321] may be of some importance). Although the equality $\mathbf{D}_t^{2\gamma}u(t,x) = \mathbf{D}_t^{\gamma}u(t,x)\,\mathbf{D}_t^{\gamma}u(t,x)$ does not hold in general, we would like to point out that the existence and uniqueness of mild solutions to the homogeneous counterpart of (428) cannot be so easily proved for the initial values which belong to the Sobolev space $W^{k,p}(\mathbb{R})$, for some $k \in \mathbb{N}$. Strictly speaking, we can introduce a new function $v(\cdot,\cdot)$ by $v(t,x) := \mathbf{D}_t^{\gamma}u(t,x)$. Then the equation (428) can be rewritten in the following equivalent matricial form:

$$\mathbf{D}_t^{\gamma}\Big[u(t,x)\quad v(t,x)\Big]^T = \begin{bmatrix} 0 & 1 \\ -re^{i(2-2\gamma)\frac{\pi}{2}} & -aix \end{bmatrix}(A)\Big[u(t,x)\quad v(t,x)\Big]^T,\quad t \geqslant 0,$$

where $A = -id/dx$ (cf. Section 2.3 and Section 2.5). The characteristic values of associated polynomial matrix $P(x) := \begin{bmatrix} 0 & 1 \\ -re^{i(2-2\gamma)\frac{\pi}{2}} & -aix \end{bmatrix}$ are

$$\lambda_{1,2}(x) = \frac{1}{2}\left(-aix \pm \sqrt{a^2 + 4re^{i(2-2\gamma)\frac{\pi}{2}}}\right), \; x \in \mathbb{R},$$

which implies that the condition of Petrovskii for systems of abstract time-fractional equations, i.e., $\sup_{x\in\mathbb{R}} \operatorname{Re}((\lambda_{1,2}(x))^{1/\lambda}) < \infty$, is not satisfied.

Example 2.10.34. Let $s' > 1$,

$$E := \left\{ f \in C^\infty[0, 1]; \; \|f\| := \sup_{p\geqslant 0} \frac{\| f^{(p)} \|_\infty}{p!^{s'}} < \infty \right\}$$

and

$$A := -d/ds, \; D(A) := \{f \in E; \, f' \in E, \, f(0) = 0\}.$$

Then $\rho(A) = \mathbb{C}$, and for every $\eta > 1$, $\|R(\lambda : A)\| = O(e^{\eta|\lambda|})$, $\lambda \in \mathbb{C}$. Consider now the complex non-zero polynomials $P_j(z) = \Sigma_{l=0}^{nj} a_{j,l} z^l$, $z \in \mathbb{C}$, $a_{j,n_j} \neq 0$ $(0 \leqslant j \leqslant n-1)$, and define, for every $\lambda \in \mathbb{C}$ and $j \in \mathbb{N}^0_{n-1}$, the operator $P_j(A)$ by $D(P_j(A)) := D(A^{n_j})$ and $P_j(A)f := \Sigma_{l=0}^{nj} a_{j,l} A^l f$, $f \in D(P_j(A))$. Our intention is to analyze the smoothing properties of solutions of the equation (414) with $A_j := p_j(A)$, $j \in \mathbb{N}^0_{n-1}$, $u_k = 0$, $k \in \mathbb{N}^0_{m_n-1}$, and a suitable chosen function $f(t)$. In order to do that, set $N := \max(\mathrm{dg}(P_0), \cdots, \mathrm{dg}(P_{n-1}))$, $\mathcal{P}_\lambda(z) := 1 + \Sigma_{j=1}^{n-1} \lambda^{\alpha_j - \alpha_n} P_j(z) - \lambda^{\alpha-\alpha_n} P_0(z)$ $(\lambda \in \mathbb{C}\setminus\{0\}, z \in \mathbb{C})$, and after that, $\Phi := \{\lambda \in \mathbb{C}\setminus\{0\} : \mathrm{dg}(\mathcal{P}_\lambda(\cdot)) = N, \, \mathcal{P}_\lambda(0) \neq 0\}$. Then it is not difficult to prove (cf. Example 2.6.10) that, for every $\lambda \in \mathbb{C}\setminus\{0\}$, $\mathbf{P}_\lambda^{-1} = (I + \Sigma_{j=1}^{n-1} \lambda^{\alpha_j - \alpha_n} A_j - \lambda^{\alpha-\alpha_n} A)^{-1} \in L(E)$ and

$$\text{(429)} \qquad \mathbf{P}_\lambda^{-1} = (-1)^N g(\lambda)^{-1} R(z_{1,\lambda} : A) \cdots R(z_{N,\lambda} : A), \quad \lambda \in \Phi,$$

where $z_{1,\lambda}, \cdots, z_{N,\lambda}$ denote the zeroes of $\mathcal{P}_\lambda(z)$ and $g(\lambda) := N!^{-1} \mathcal{P}_\lambda^{(N)}(0)$, $\lambda \in \Phi$. Suppose now that the following condition holds

(CCH): There exist $\sigma \in (0, 1)$, $\omega > 0$ and $m > 0$ such that, for every $j \in \mathbb{N}^0_{n-1}$, one has: $|z_{j,\lambda}| \leqslant m|\lambda|^\sigma$, $\lambda \in \Phi$, $\operatorname{Re} \lambda > \omega$.

It is well known from the elementary courses of numerical analysis ([448]) that the condition

(CCH$_1$): There exist $\sigma \in (0, 1)$, $\omega > 0$ and $m > 0$ such that, for every $j \in \mathbb{N}^0_{n-1}$, one has:

$$\left| \frac{N! \mathcal{P}_\lambda^{(j)}(0)}{j! \mathcal{P}_\lambda^{(N)}(0)} \right|^{1/(N-j)} \leqslant \frac{1}{2} m|\lambda|^\sigma, \; \lambda \in \Phi, \; \operatorname{Re} \lambda > \omega,$$

implies (CCH). The validity of (CCH) can be simply verified in a great number of concrete situations, and it seems that slightly better estimates, compared with those clarified below, can be obtained only for some very special equations of the form (2). Observe also that the condition (CCH) need not be satisfied, in general. Using (429), the inequality $\|A^l R(\mu_l : A) \cdots R(\mu_1 : A)\| \leqslant (1 + |\mu_1|\|R(\mu_1 : A)\|) \cdots (1 + |\mu_l|\|R(\mu_l : A)\|)$ ($l \in \mathbb{N}, \mu_1, \cdots, \mu_l \in \mathbb{C}$), as well as the continuity of mappings $\lambda \mapsto \mathbf{P}_\lambda^{-1}$, $\operatorname{Re}\lambda > \omega$ and $\lambda \mapsto A_j \mathbf{P}_\lambda^{-1}$, $\operatorname{Re}\lambda > \omega$, for $0 \leqslant j \leqslant n-1$, we obtain the existence of a positive polynomial $p(\cdot)$ such that:

(430)
$$\left\|\mathbf{P}_\lambda^{-1}\right\| + \sum_{j=0}^{n-1} \left\|A_j \mathbf{P}_\lambda^{-1}\right\| \leqslant p(|\lambda|) e^{mN|\lambda|^\sigma}, \operatorname{Re}\lambda > \omega.$$

In what follows, we shall use the following family of kernels. Define, for every $l > 0$, the entire function $\omega_l(\cdot)$ by $\omega_l(\lambda) := \prod_{p=1}^{\infty}(1+\frac{l\lambda}{p^s})$, $\lambda \in \mathbb{C}$, where $s := \sigma^{-1}$. Then it is clear that: $|\omega_l(\lambda)| \geqslant \sup_{k\in\mathbb{N}} \prod_{p=1}^{k} |1 + \frac{l\lambda}{p^s}| \geqslant \sup_{k\in\mathbb{N}} \prod_{p=1}^{k} \frac{l|\lambda|}{p^s} \geqslant \sup_{k\in\mathbb{N}} \frac{(l|\lambda|)^k}{p!^s}$, $\lambda \in \mathbb{C}$, $\operatorname{Re}\lambda \geqslant 0$. Consequently, $|\omega_l(\lambda)| \geqslant e^{M(l|\lambda|)}$, $\lambda \in \mathbb{C}$, $\operatorname{Re}\lambda \geqslant 0$, where $M(\lambda)$ denotes the associated function of sequence $(p!^s)$. It is well known that, for every $\zeta \in (0, \frac{\pi}{2})$, $p \in \mathbb{N}_0$ and $\lambda \in \Sigma_{\frac{\pi}{2}+\zeta}$, we have $|1 + \frac{l\lambda}{p^s}| \geqslant \frac{l|\operatorname{Im}\lambda|}{p^s} \geqslant \frac{l(1+\tan\zeta)^{-1}|\lambda|}{p^s}$ and

$$|\omega_l(\lambda)| \geqslant e^{M\left(l(1+\tan\zeta)^{-1}|\lambda|\right)}, \zeta \in (0, \pi/2), l > 0, \lambda \in \Sigma_{\frac{\pi}{2}+\zeta}.$$

Now put $K_l(t) := \mathcal{L}^{-1}(1/\omega_l(\lambda))(t)$, $t \geqslant 0$, $l > 0$. Then, for every $l > 0$, $0 \in \operatorname{supp}(K_l)$, $K_l(0) = 0$ and $K_l(t)$ is infinitely differentiable for $t \geqslant 0$. By Theorem 2.10.18(i)-(b), (iii) and (430), we easily infer that there exists $k > 0$ such that, for every $l > k$, there exists an exponentially bounded K_l-regularized I-resolvent family $(E_l(t))_{t\geqslant 0}$ for (2), with $Y = X = E$. Furthermore, the mapping $t \mapsto E_l(t)$, $t \geqslant 0$ is infinitely differentiable in the uniform operator topology of $L(E)$ and, for every compact set $K \subseteq [0, \infty)$ and for every $l > k$, there exists $h_{K,l} > 0$ such that

$$\sup_{p\geqslant 0, t\in K} \frac{h_{K,l}^p \left\|E_l^{(p)}(t)\right\|}{p!^s} < \infty.$$

One can similarly construct examples of exponentially bounded, analytic K_l-regularized I-resolvent families.

In the sequel to this section the state space will also be denoted by X; by Y we denote another SCLCS over the field of complex numbers.

Definition 2.10.35. (compare with [319, Definition 2.2]) Suppose $0 < \tau \leqslant \infty$, $k \in C([0, \tau))$, $C_1 \in L(Y, X)$, and $C_2 \in L(X)$ is injective.

(i) A strongly continuous operator family $(E(t))_{t\in[0,\tau)} \subseteq L(Y, X)$ is said to be a (local, if $\tau < \infty$) (k, C_1)-existence family for (2) if, for every $y \in Y$ and $t \in [0, \tau)$, the following holds: $A_j(g_{\alpha_n-\alpha_j} * E)(\cdot)y \in C([0, \tau) : X)$ for $0 \leqslant j \leqslant n-1$, and

(431)
$$E(t)y + \sum_{j=1}^{n-1} A_j(g_{\alpha_n-\alpha_j} * E)(t)y - A(g_{\alpha_n-\alpha} * E)(t)y = k(t)C_1 y.$$

(ii) A strongly continuous operator family $(U(t))_{t\in[0,\tau)} \subseteq L(X)$ is said to be a (local, if $\tau < \infty$) (k, C_2)-uniqueness family for (2) if, for every $t \in [0, \tau)$ and $x \in \bigcap_{0\leqslant j\leqslant n-1} D(A_j)$, the following holds:

$$U(t)x+\sum_{j=1}^{n-1}(g_{\alpha_n-\alpha_j}*U(\cdot)A_jx)(t)-(g_{\alpha_n-\alpha}*U(\cdot)Ax)(t)x=k(t)C_2x.$$

(iii) A strongly continuous family $((E(t))_{t\in[0,\tau)},(U(t))_{t\in[0,\tau)})\subseteq L(X)\times L(X)$ is said to be a (local, if $\tau<\infty$) (k, C_1, C_2)-existence and uniqueness family for (2) if $(E(t))_{t\in[0,\tau)}$ is a (k, C_1)-existence family for (2), and $(U(t))_{t\in[0,\tau)}$ is a (k, C_2)-uniqueness family for (2).

(iv) Suppose $C=C_1=C_2$. Then a strongly continuous operator family $(R(t))_{t\in[0,\tau)}\subseteq L(X)$ is said to be a (local, if $\tau<\infty$) (k, C)-resolvent family for (2) if $(R(t))_{t\in[0,\tau)}$ is a (k, C)-uniqueness family for (2), $R(t)A_j\subseteq A_jR(t)$, for $0\leqslant j\leqslant n-1$ and $t\in[0,\tau)$, as well as $R(t)C=CR(t)$, $t\in[0,\tau)$, and $CA_j\subseteq A_jC$, for $0\leqslant j\leqslant n-1$.

The (exponential) analyticity of various types of (C_1, C_2)-existence and uniqueness families for (2) is defined in the obvious way. If $k(t)=1$, $t\in[0,\tau)$, then any k-regularized (C_1, C_2)-existence and uniqueness family (k-regularized C-regularized resolvent family) will be simply called a (C_1, C_2)-existence and uniqueness family for (2) (C-resolvent family for (2)). Notice that the following assertions hold:

(i) Suppose $(E(t))_{t\in[0,\tau)}\subseteq L(Y, X)$ is a strongly continuous operator family. Then $(E(t))_{t\in[0,\tau)}$ is a (local) (k, C_1)-existence family for (2) if $((g_{m_n-1}*E)(t))_{t\in[0,\tau)}$ is a k-regularized C_1-existence family for (2).

(ii) Suppose $(U(t))_{t\in[0,\tau)}\subseteq L(X)$ is a strongly continuous operator family. Then $(U(t))_{t\in[0,\tau)}$ is a (local) $(k*g_{m_n-1}, C_2)$-uniqueness family for (2) if $(U(t))_{t\in[0,\tau)}$ is a (local) k-regularized C_2-uniqueness family for (2).

Recall that any (local) k-regularized C_1-existence propagation family $((R_0(t))_{t\in[0,\tau)},\cdots,(R_{m_n-1}(t))_{t\in[0,\tau)})$ for (2) satisfies the functional equation (399) for $0\leqslant i\leqslant m_n-1$. In the subsequent theorems, we shall prove certain relations between (k, C_1, C_2)-existence and uniqueness families ((k, C)-resolvent families) and k-regularized (C_1, C_2)-existence and uniqueness propagation families (k-regularized C-resolvent propagation families).

Theorem 2.10.36. (i) *Let $C, C_1\in L(X)$, and let C be injective. Suppose that $A_j\in L(X)$ and $A_jA_l=A_lA_j$ for $1\leqslant j, l\leqslant n-1$ and $A_jA\subseteq AA_j$ for $1\leqslant j\leqslant n-1$. Suppose $(E(t))_{t\in[0,\tau)}$ is a (k, C_1)-existence family for (2), and $(R(t))_{t\in[0,\tau)}$ is a (k, C)-resolvent family for (2). Put, for every $x\in X$, $t\in[0,\tau)$, and $i=m,\cdots,m_n-1$,*

$$E_i(t)x:=(g_i*E)(t)x+\sum_{j\in D_i}A_j(g_{\alpha_n-\alpha_j+i}*E)(t)x,$$

and, for every $x\in X$, $t\in[0,\tau)$, and $i=0,\cdots,m-1$,

$$E_i(t)x:=(k*g_i)(t)C_1x-\sum_{j\in\mathbb{N}_{n-1}\setminus D_i}A_j(g_{\alpha_n-\alpha_j+i}*E)(t)x.$$

Define also $R_i(t)x$ by replacing respectively $E(t)$ and C_1 with $R(t)$ and C in the above formulae. Then $((E_0(t))_{t\in[0,\tau)}, \cdots, (E_{m_n-1}(t))_{t\in[0,\tau)})$ is a k-regularized C_1-existence propagation family for (2) and $((R_0(t))_{t\in[0,\tau)}, \cdots, (R_{m_n-1}(t))_{t\in[0,\tau)})$ is a k-regularized C-resolvent propagation family for (2). Furthermore, (399) holds for $((R_0(t))_{t\in[0,\tau)}, \cdots, (R_{m_n-1}(t))_{t\in[0,\tau)})$, provided that (431) holds for $(R(t))_{t\in[0,\tau)}$.

(ii) *Let the following condition hold:*

(432) $$c_j \in \mathbb{C} \text{ and } A_j = c_j I \text{ for all } j \in \mathbb{N}_{n-1}.$$

Suppose $((R_0(t))_{t\in[0,\tau)}, \cdots, (R_{m_n-1}(t))_{t\in[0,\tau)})$ is a k-regularized C_1-existence propagation family for (2), resp. k-regularized C_2-uniqueness propagation family for (2) (k-regularized C-resolvent propagation family for (2)), and m = 0. Define, for every $t > 0$,

$$b(t) := \mathcal{L}^{-1}\Big(\Big(1 + \sum_{j\in D_i} c_j\lambda^{\alpha_j-\alpha_n}\Big)^{-1} - 1\Big)(t),$$

and

$$R(t)x := R_0(t)x + (b * R_0)(t)x,\ t \in [0, \tau),\ x \in X.$$

Then $(R(t))_{t\in[0,\tau)}$ is a (k, C_1)-existence family for (2), resp. (k, C_2)-uniqueness family for (2) ((k, C)-resolvent family for (2)). Furthermore, if (399) holds for $((R_0(t))_{t\in[0,\tau)}, \cdots, (R_{m_n-1}(t))_{t\in[0,\tau)})$, then (431) holds with $(R(t))_{t\in[0,\tau)}$ in place of $(E(t))_{t\in[0,\tau)}$.

Proof. We will only prove the assertion (i) in the case of existence families. Suppose first $i \in \mathbb{N}^0_{m_n-1}$ and $m - 1 < i$. Then it is clear that $(E_{1,i}(t))_{t\in[0,\tau)}$ is a strongly continuous operator family. By definitions of $(E_{1,i}(t))_{t\in[0,\tau)}$ and $(E(t))_{t\in[0,\tau)}$, as well as the equalities $A_jA_l = A_lA_j$ for $1 \leqslant j, l \leqslant n-1$ and $A_jA \subseteq AA_j$ for $1 \leqslant j \leqslant n-1$, we get that, for every $t \in [0, \tau)$ and $x \in X$,

$$\begin{aligned}
A(g_{\alpha_n-\alpha} * E_{1,i})(t)x &= g_i * \Big[E_1(\cdot)\,x - k(\cdot)C_1x + \sum_{j=1}^{n-1} A_j(g_{\alpha_n-\alpha_j} * E_1)(\cdot)x\Big](t) \\
&+ A\Big(\sum_{j\in D_i} A_j(g_{\alpha_n-\alpha_j+i+\alpha_n-\alpha} * E_1)(t)x\Big) \\
&= g_i * \Big[E_1(\cdot)x - k(\cdot)C_1x + \sum_{j=1}^{n-1} A_j(g_{\alpha_n-\alpha_j} * E_1)(\cdot)x\Big](t) \\
&+ \sum_{j\in D_i} A_j\Big(g_{\alpha_n-\alpha_j+i} * \Big[R_1(\cdot) - k(\cdot)C_1 + \sum_{l=1}^{n-1} A_l(g_{\alpha_n-\alpha_l} * R_1)(\cdot)\Big]\Big)(t)x \\
&= [R_{1,i}(\cdot)x - (k * g_i)(\cdot)C_1x](t) + \sum_{j\in\mathbb{N}\setminus D_i} A_j(g_{\alpha_n-\alpha_j} * [g_i * R_1])(t)x \\
&+ \sum_{j\in D_i} A_j\Big(g_{\alpha_n-\alpha_j+i} * \Big[R_1(\cdot) - k(\cdot)C_1 + \sum_{l=1}^{n-1} A_l(g_{\alpha_n-\alpha_l} * R_1)(\cdot)\Big]\Big)(t)x \\
&= [R_{1,i}(\cdot)x - (k * g_i)(\cdot)C_1x](t)
\end{aligned}$$

$$+\sum_{j\in\mathbb{N}\setminus D_i} A_j\Big(g_{\alpha_n-\alpha_j} * \Big[R_{1,i}(\cdot)-\sum_{l\in D_i} A_l(g_{\alpha_n-\alpha_j+i} * R_1)(\cdot)\Big]\Big)(t)x$$

$$+\sum_{j\in D_i}\Big(g_{\alpha_n-\alpha_j} * \Big[R_{1,i}(\cdot)x - \sum_{l\in D_i} A_l(g_{\alpha_n-\alpha_j+i} * R_1)(\cdot)x-(k * g_i)(\cdot)C_1x\Big]$$

$$+ g_{\alpha_n-\alpha_j+i} * \sum_{l=1}^{n-1} A_l(g_{\alpha_n-\alpha_l} * R_1)(\cdot)x\Big)(t).$$

Since, for every $t\in[0,\tau)$ and $x\in X$,

$$-\sum_{j\in\mathbb{N}_{n-1}\setminus D_i}\sum_{l\in D_i} A_jA_l(g_{\alpha_n-\alpha_l+i+\alpha_n-\alpha_j} * R_1)(t)x$$

$$-\sum_{j\in D_i}\sum_{l\in D_i} A_jA_l(g_{\alpha_n-\alpha_l+i+\alpha_n-\alpha_j} * R_1)(t)x$$

$$=\sum_{l=1}^{n-1}\sum_{j\in D_i} A_jA_l(g_{\alpha_n-\alpha_l+i+\alpha_n-\alpha_j} * R_1)(t)x,$$

(apply the substitution $(j, l) \hookrightarrow (l, j)$), the above computation implies (399), with $(R_i(t))_{t\in[0,\tau)}$ replaced by $(E_{1,i}(t))_{t\in[0,\tau)}$. The proof in case $m-1 \geqslant i$ is similar and therefore omitted. □

Let (432) hold, and let $m > 0$. Then the relations between k-regularized C-resolvent propagation families and (k, C)-resolvent families are far from clear. Here we recognize the following subcases:

(a) There exists $i\in\mathbb{N}^0_{m_n-1}$ such that $i > m-1$. Then the consideration is trivial provided that, for such an index i, one has $D_i=\emptyset$ (cf. Theorem 2.10.36); because of that, we shall assume in the further analysis of this subcase that $D_i\neq\emptyset$.

(b) $m=m_n$ and, for every $i\in\mathbb{N}^0_{m_n-1}$, one has $\mathbb{N}_{n-1}\setminus D_i=\emptyset$. That is the worst case possible and here we have that the function $u(t)=\Sigma_{k=0}^{m_n-1}u_kg_{k+1}(t)$, $t\geqslant 0$ is a strong solution of (2) since $\mathbf{D}_t^\beta g_\gamma(t)\equiv 0$, provided $\beta>\gamma-1>0$.

(c) $m=m_n$, and there exists $i\in\mathbb{N}^0_{m_n-1}$ such that $\mathbb{N}_{n-1}\setminus D_i\neq\emptyset$.

The proof of the following theorem, which considers the cases (a) and (c) in more detail, is omitted ([325]).

Theorem 2.10.37. (i) *Let $m > 0$, and let $i\in\mathbb{N}^0_{m_n-1}$ satisfy $i > m-1$ and $D_i\neq\emptyset$. Suppose $l\in\mathbb{N}$ and there exists $j\in\mathbb{N}_{n-1}\setminus D_i$ such that $\alpha_j-\alpha_n+l>i$. Let $((R_0(t))_{t\in[0,\tau)}$, $\cdots$, $(R_{m_n-1}(t))_{t\in[0,\tau)})$ be a k-regularized C_1-existence propagation family for (2), resp. k-regularized C_2-uniqueness propagation family for (2) (k-regularized C-resolvent propagation family for (2)). Suppose, additionally, that for every $x\in X$ and $p\in\circledast$, the mapping $t\mapsto R_i^{(j)}(t)x$, $t\in[0,\tau)$ is continuous with $R_i^{(j)}(0)x=0$ for all $j\in\mathbb{N}^0_{l-1}$, and that, for every $p\in\circledast$ and $t\in[0,\tau)$, there exist*

$c_{p,t} > 0$ and $q_{p,t} \in \circledast$ such that $p(R_i^{(l)}(t)x) \leqslant c_{p,t}q_{p,t}(x)$, $x \in X$. Put, for every $t \in [0, \tau)$ and $x \in X$,

$$R(t)x := \left[\mathcal{L}^{-1}\left(\frac{1}{\lambda^{l-i} + \sum_{j \in D_i} c_j \lambda^{\alpha_j - \alpha_n - i + l}}\right) * R_i^{(l)}(\cdot)\right](t)x.$$

Then $(R(t))_{t\in[0,\tau)}$ is a (k, C_1)-existence family for (2), resp. (k, C_2)-uniqueness family for (2) ((k, C)-resolvent propagation family for (2)). Furthermore, if (399) holds for $((R_0(t))_{t\in[0,\tau)}, \cdots, (R_{m_n-1}(t))_{t\in[0,\tau)})$, then (431) holds with $(R(t))_{t\in[0,\tau)}$ in place of $(E(t))_{t\in[0,\tau)}$.

(ii) *Let $m = m_n$, and let $i \in \mathbb{N}^0_{m_n-1}$ be such that $\mathbb{N}_{n-1} \setminus D_i \neq \emptyset$ and $\Sigma_{j\in\mathbb{N}_{n-1}\setminus D_i} |c_j|^2 > 0$. Suppose $((R_0(t))_{t\in[0,\tau)}, \cdots, (R_{m_n-1}(t))_{t\in[0,\tau)})$ is a k-regularized C_1-existence propagation family for (2), resp. k-regularized C_2-uniqueness propagation family for (2) (k-regularized C-resolvent propagation family for (2)). Let $l \in \mathbb{N}$, let $k_0 \in C([0, \infty))$ satisfy (P1) and let $k_0(t) = k(t)$, $t \in [0, \tau)$. Define the function $c \in C([0, \infty))$ by*

$$\tilde{c}(\lambda) = \tilde{k}(\lambda) \sum_{j\in\mathbb{N}_{n-1}\setminus D_i} c_j \lambda^{\alpha_j - \alpha_n}, \text{ for } \lambda \text{ sufficiently large,}$$

and suppose that there exists $j \in \mathbb{N}_{n-1} \setminus D_i$ such that $\alpha_j - \alpha_n + l > i$. Suppose, additionally, that for every $x \in X$ and $p \in \circledast$, the mapping $t \mapsto R_i^{(l)}(t)x$, $t \in [0, \tau)$ is continuous with $R_i^{(j)}(0)x = 0$ for all $j \in \mathbb{N}^0_{l-1}$ and that, for every $p \in \circledast$ and $t \in [0, \tau)$, there exist $c_{p,t} > 0$ and $q_{p,t} \in \circledast$ such that $p(R_i^{(l)}(t)x) \leqslant c_{p,t}q_{p,t}(x)$, $x \in X$. Put, for every $t \in [0, \tau)$ and $x \in X$,

$$R(t)x := c(t)C_1 - \left[\mathcal{L}^{-1}\left(\frac{1}{\sum_{j\in\mathbb{N}_{n-1}\setminus D_i} c_j \lambda^{\alpha_j - \alpha_n - i + l}}\right) * R_i^{(l)}(\cdot)\right](t)x.$$

Then $(R(t))_{t\in[0,\tau)}$ is a (k, C_1)-existence family for (2); if we replace the operator C_1 in the above formula with the operator C_2 (C), then $(R(t))_{[0,\tau)}$ is a (k, C_2)-uniqueness family for (2) ((k, C)-resolvent family for (2)). Furthermore, if (399) holds for $((R_0(t))_{t\in[0,\tau)}, \cdots, (R_{m_n-1}(t))_{t\in[0,\tau)})$, then (431) holds with $(R(t))_{t\in[0,\tau)}$ in place of $(E(t))_{t\in[0,\tau)}$.

Remark 2.10.38. Assume that (432) holds.

(i) Let $m = 0$. Then there exist $M \geqslant 1$ and $\omega \geqslant 0$ such that $\int_0^t |b(s)|\, ds \leqslant Me^{\omega t}$, $t \geqslant 0$, which implies that the properties of (q)-exponential equicontinuity and local equicontinuity are stable under passing from k-regularized C_1-existence propagation families (C_2-uniqueness propagation families, C-resolvent propagation families) to (k, C_1)-existence families ((k, C_2)-uniqueness families, (k, C)-resolvent families). Suppose now that $((R_0(t))_{t\geqslant 0}, \cdots, (R_{m_n-1}(t))_{t\geqslant 0})$ is an exponentially equicontinuous, analytic k-regularized C_1-existence family (C_2-uniqueness family, C-resolvent propagation family) of angle $\delta \in (0, \pi/2]$.

Owing to Theorem 1.2.5(i), we have that $(R(t))_{t\geqslant 0}$ is an exponentially equicontinuous, analytic (k, C_1)-existence family ((k, C_2)-uniqueness family, (k, C)-resolvent family) of angle δ.

(ii) Let $((R_0(t))_{t\in[0,\tau)}, \cdot\cdot\cdot, (R_{m_n-1}(t))_{t\in[0,\tau)})$ be a locally equicontinuous k-regularized C-resolvent propagation family for (2), and let $m = 0$. Then it is not so difficult to see that the following equality holds, for every $x \in D(A)$, $i \in \mathbb{N}^0_{m_n-1}$ and $t \in [0, \tau)$,

$$R_i(t)x = g_i * [R_0(\cdot)x + (b * R_0)(\cdot)x](t)$$

$$+ \sum_{j\in D_i} c_j\big[g_{\alpha_n-\alpha_j+i} * (R_0(\cdot)x + (b * R_0(\cdot)x))\big](t)x; \tag{433}$$

furthermore, (433) holds for every $x \in X$, $i \in \mathbb{N}^0_{m_n-1}$ and $t \in [0, \tau)$, provided that (399) holds. In general case, it is very complicated to clarify precise relations between single operator families $(R_0(t))_{t\in[0,\tau)}, \cdot\cdot\cdot$, and $(R_{m_n-1}(t))_{t\in[0,\tau)}$ (cf. Theorem 2.10.7. and [531, p. 116]).

(iii) Let us put ourselves in the situation of Example 2.10.30(b). Then there exists an exponentially equicontinuous, analytic k_1-regularized I-resolvent propagation family $((R_0(t))_{t\geqslant 0}, \cdot\cdot\cdot, (R_{m_n-1}(t))_{t\geqslant 0})$ for the corresponding problem (2), with $k_1(t)$ being defined by $k_1(t) = \mathcal{L}^{-1}(\exp(-a_1\lambda^{b_1}))(t)$, $t \geqslant 0$ for certain positive real numbers $a_1 > 0$ and $b_1 \in (0, 1)$. Under some extra assumptions (very natural in the theory of convoluted operator families), we may apply Theorem 2.10.37 in the construction of an exponentially equicontinuous, analytic (k_1, I)-resolvent family for (2).

Proposition 2.10.39. *Suppose (432) holds, $((R_{1,0}(t))_{t\in[0,\tau)}, \cdot\cdot\cdot, (R_{1,m_n-1}(t))_{t\in[0,\tau)})$ is a k-regularized C_1-existence propagation family for (2), $\mathbb{N}_{n-1}\backslash D_i \neq \emptyset$ provided $m-1 \geqslant i$, and $((R_{2,0}(t))_{t\in[0,\tau)}, \cdot\cdot\cdot, (R_{2,m_n-1}(t))_{t\in[0,\tau)})$ is a locally equicontinuous k-regularized C_2-uniqueness propagation family for (2). Then, for every $i \in \mathbb{N}^0_{m_n-1}$, one has $C_2R_{1,i}(t) = R_{2,i}(t)C_1$, $t \in [0, \tau)$.*

Before discussing some perturbation properties of (2), we would like to observe that Proposition 2.10.39 can be proved under slightly weakened assumptions (see, e.g., the formulation of Proposition 2.10.3). Consider now the abstract Cauchy problem:

$$\begin{aligned} &\mathbf{D}_t^{\alpha_n} u(t) + \sum_{i=1}^{n-1} (A_i + B_i)\mathbf{D}_t^{\alpha_i} u(t) = (A + B)\mathbf{D}_t^{\alpha} u(t) + f(t),\ t > 0, \\ &u^{(k)}(0) = u_k,\ k = 0, \cdot\cdot\cdot, \lceil \alpha_n \rceil - 1, \end{aligned} \tag{434}$$

where B and $B_1, \cdot\cdot\cdot, B_{n-1}$ are closed linear operators on E. Put $B_0 := B$.

The following theorem is an important extension of [153, Theorem 6.2-Theorem 6.3].

Theorem 2.10.40. (i) *Suppose $Y = X$, $(E(t))_{t\in[0,\tau)} \subseteq L(X)$ is a (local) C_1-existence family for (2), $D_j \in L(X)$ and $B_j = C_1D_j$ $(j \in \mathbb{N}^0_{n-1})$.*

Suppose that the following conditions hold:

(a) *For every* $p \in \circledast_X$ *and for every* $T \in (0, \tau)$, *there exists* $c_{p,T} > 0$ *such that*

$$p(E^{(m_n-1)}(t)x) \leqslant c_{p,T}p(x), \quad x \in X, \ t \in [0, T].$$

(b) *For every* $p \in \circledast_X$, *there exists* $c_p > 0$ *such that*

$$p(D_j x) \leqslant c_p p(x), \ j \in \mathbb{N}^0_{n-1}, \ x \in X.$$

(c) $\alpha_n - \alpha_{n-1} \geqslant 1$ and $\alpha_n - \alpha \geqslant 1$.

Then there exists a (local) C_1*-existence propagation family* $(R(t))_{t\in[0,\tau)}$ *for (434). If* $\tau = \infty$ *and if, for every* $p \in \circledast_X$, *there exist* $M \geqslant 1$ *and* $\omega \geqslant 0$ *such that*

(435) $$p(E^{(m_n-1)}(t)x) \leqslant Me^{\omega t}p(x), \ t \geqslant 0, \ x \in X,$$

then $(R(t))_{t\geqslant 0}$ *is exponentially equicontinuous, and moreover,* $(R(t))_{t\geqslant 0}$ *also satisfies the condition (435), with possibly different numbers* $M \geqslant 1$ *and* $\omega > 0$.

(ii) *Suppose* $Y = X$, $(U(t))_{t\in[0,\tau)} \subseteq L(X)$ *is a (local)* $(1, C_2)$*-uniqueness family for (2),* $D_j \in L(E)$ *and* $B_j = D_j C_2$ $(j \in \mathbb{N}^0_{n-1})$. *Suppose that (b)-(c) hold, and that (a) holds with* $(E^{(m_n-1)}(t))_{t\in[0,\tau)}$ *replaced by* $(U(t))_{t\in[0,\tau)}$ *therein. Then there exists a (local)* $(1, C_2)$*-uniqueness family* $(W(t))_{t\in[0,\tau)}$ *for (434). If* $\tau = \infty$ *and if, for every* $p \in \circledast_X$, *there exist* $M \geqslant 1$ *and* $\omega \geqslant 0$ *such that (435) holds, then* $(W(t))_{t\geqslant 0}$ *is exponentially equicontinuous, and moreover,* $(W(t))_{t\geqslant 0}$ *also satisfies the condition (435), with possibly different numbers* $M \geqslant 1$ *and* $\omega > 0$.

Proof. We shall content ourselves with sketching it. Put

$$K_0(t)x := -\left\{E^{(m_n-1)} * \left[\sum_{j=1}^{n-1} g_{\alpha_n-\alpha_j} D_j + g_{\alpha_n-\alpha} D\right]\right\}(t)x, \ t \in [0, \tau), \ x \in X.$$

Then the assumption (c) implies that, for every fixed $x \in X$, we have $K_0(\cdot)x \in C^1([0, \tau) : X)$ and

$$K_0'(t)x := -\left\{E^{(m_n-1)} * \left[\sum_{j=1}^{n-1} g_{\alpha_n-\alpha_j-1} D_j + g_{\alpha_n-\alpha-1} D\right]\right\}(t)x, \ t \in [0, \tau).$$

By assumptions (a)-(b), we have that, for every $t \in [0, \tau)$ and $x \in X$, the series

$$L(t)x := -[K_0(t)x + (K_0 * K_0')(t)x + \cdots + (K_0 * (K_0')^{*,k})(t)x + \cdots]$$

converges, uniformly on compacts of $[0, \tau)$. Furthermore, the operator family $(L(t))_{t\in[0,\tau)} \subseteq L(X)$ is strongly continuous and, for every $x \in X$, the unique solution of the following integral equation:

(436) $$R_{m_n-1}(t)x = E^{(m_n-1)}(t)x + \int_0^t K_0'(t-s)\, R_{m_n-1}(s)x \, ds, \ t \in [0, \tau),$$

is given by

(437) $$R_{m_n-1}(t)x = E^{(m_n-1)}(t)x + \int_0^t L'(t-s)\, E^{(m_n-1)}(s)x\, ds,\ t \in [0,\tau);$$

cf. also [463, Theorem 0.5, Corollary 0.3]. It is not difficult to prove that $(R_{m_n-1}(t))_{t\in[0,\tau)}$ is a strongly continuous operator family in $L(X)$. Define now $R(t)x := (g_{m_n-1} * R_{m_n-1})(t)x$, $t \in [0, \tau)$, $x \in X$. Applying the functional equation for $(E(t))_{t\in[0,\tau)}$ twice, we can prove that $(R(t))_{t\in[0,\tau)}$ satisfies

$$R^{(m_n-1)}(t)x + \sum_{j=1}^{n-1}(A_j + B_j)\,(g_{\alpha_n-\alpha_j} * R^{(m_n-1)})(t)x$$
$$-(A+B)\,(g_{\alpha_n-\alpha} * R^{(m_n-1)})(t)x = C_1x,\ t\in[0,\tau),\ x\in X,$$

iff, for every $t \in [0, \tau)$ and $x \in X$, the following holds:

$$(dK_0 * R_{m_n-1})(t)x + \Big(E^{(m_n-1)} * \sum_{j=0}^{n-1} g_{\alpha_n-\alpha_j-1}(\cdot)D_j * R_{m_n-1}\Big)(t)x$$
$$+\Big(C_1 * \sum_{j=0}^{n-1} g_{\alpha_n-\alpha_j-1}(\cdot)D_j * R_{m_n-1}\Big)(t)x$$
$$+\sum_{j=1}^{n-1} B_j(g_{\alpha_n-\alpha_j} * R_{m_n-1})(t)x - B(g_{\alpha_n-\alpha} * R_{m_n-1})(t)x = 0.$$

But, the last equality is an immediate consequence of (436). If $(E^{(m_n-1)}(t))_{t\geqslant 0}$ satisfies (435), then we can prove with the help of (436)-(437) that $(R(t))_{t\in[0,\tau)}$ also satisfies the same condition, with possibly different numbers $M \geqslant 1$ and $\omega > 0$. The proof of (ii) is quite similar. Define, for every $t \in [0, \tau)$ and $x \in X$,

$$Q_0(t)x := -\Big\{\Big[\sum_{j=1}^{n-1} g_{\alpha_n-\alpha_j} D_j + g_{\alpha_n-\alpha} D\Big] * U\Big\}(t)x,$$

and

$$Z(t)x := -[Q_0(t)x + (Q_0 * Q_0')(t)x + \cdots + (Q_0 * (Q_0')^{*,k})(t)x + \cdots].$$

The resulting $(1, C_2)$-uniqueness family $(W(t))_{t\in[0,\tau)}$ is a unique solution of the following integral equation

$$W(t)x = U(t)x + \int_0^t W(t-s)Q'(s)x\, ds,\ t\in[0,\tau),\ x\in X.$$

This solution is given by

$$W(t)x = U(t)x + \int_0^t U(t-s)Z'(s)x\, ds,\ t\in[0,\tau),\ x\in X,$$

which makes the remaining part of the proof routine. □

Remark 2.10.41. (i) It is worth noting that the proof of the preceding theorem is a slight modification of the corresponding proofs of [463, Theorem 6.1] and [317, Theorem 2.12], established for abstract Volterra equations of non-scalar type. Using the method given in the proofs of afore-mentioned theorems, one can similarly clarify some results on time-dependent perturbations of (2).

(ii) It is not clear how one can prove an analogue of Theorem 2.10.40(ii) in the case of a (local) C_2-uniqueness family for (2).

Theorem 2.10.42. *Suppose $A, A_1, \cdots, A_{n-1}$ are closed linear operators on X, $\omega > 0$, $L(X) \ni C$ is injective and $u_0, \cdots, u_{m_n-1} \in X$. Let the following conditions hold:*

(i) *The operator P_λ is injective for $\lambda > \omega$ and $D(P_\lambda^{-1}C) = X$, $\lambda > \omega$.*

(ii) *If $0 \leqslant j \leqslant n-1$, $0 \leqslant k \leqslant m_n - 1$, $m - 1 < k$, $1 \leqslant l \leqslant n-1$, $m_l - 1 \geqslant k$ and $\lambda > \omega$, then $Cu_k \in D(P_\lambda^{-1}A_l)$,*

$$A_j\Big\{\lambda^{\alpha_j}\Big[\lambda^{\alpha_n-\alpha-k-1}P_\lambda^{-1}Cu_k + \sum_{l\in D_k}\lambda^{\alpha_l-\alpha-k-1}P_\lambda^{-1}A_lCu_k\Big] - \sum_{l=0}^{m_j-1}\delta_{kl}\lambda^{\alpha_j-1-l}Cu_k\Big\} \in LT-X \tag{438}$$

and

$$\lambda^{\alpha_n}\Big[\lambda^{\alpha_n-\alpha-k-1}P_\lambda^{-1}Cu_k + \sum_{l\in D_k}\lambda^{\alpha_l-\alpha-k-1}P_\lambda^{-1}A_lCu_k\Big] - \lambda^{\alpha_n-1-k}Cu_k \in LT-X. \tag{439}$$

(iii) *If $0 \leqslant j \leqslant n-1$, $0 \leqslant k \leqslant m_n - 1$, $m - 1 \geqslant k$, $\mathbb{N}_{n-1}\backslash D_k \neq \emptyset$, $s \in \mathbb{N}_{n-1}\backslash D_k$ and $\lambda > \omega$, then $Cu_k \in D(A_s)$, $\Sigma_{l\in\mathbb{N}_{n-1}\backslash D_k}\lambda^{\alpha_l-\alpha-k-1}A_lCu_k \in D(P_\lambda^{-1})$,*

$$A_j\Big\{\lambda^{\alpha_j}\Big[\lambda^{-k-1}Cu_k - P_\lambda^{-1}\sum_{l\in\mathbb{N}_{n-1}\backslash D_k}\lambda^{\alpha_l-\alpha-k-1}A_lCu_k\Big] - \sum_{l=0}^{m_j-1}\delta_{kl}\lambda^{\alpha_j-1-l}Cu_k\Big\} \in LT-X \tag{440}$$

and

$$\lambda^{\alpha_n}\Big[\lambda^{-k-1}Cu_k - P_\lambda^{-1}\sum_{l\in\mathbb{N}_{n-1}\backslash D_k}\lambda^{\alpha_l-\alpha-k-1}A_lCu_k\Big] - \lambda^{\alpha_n-1-k}Cu_k \in LT-X. \tag{441}$$

Then the abstract Cauchy problem (2) has a strong solution, with u_k replaced by Cu_k $(0 \leqslant k \leqslant m_n - 1)$.

Proof. Suppose temporarily $0 \leqslant k \leqslant m_n - 1$ and $m - 1 < k$. Let $F_{k,n} \in C([0,\infty) : X)$ such that, for every $p \in ⊛$, there exists $M_p > 0$ satisfying $p(F_{k,n}(t)) \leqslant M_pe^{\omega t}$, $t \geqslant 0$ and

$$\int_0^\infty e^{-\lambda t}F_{k,n}(t)\,dt = \lambda^{\alpha_n}\Big[\lambda^{\alpha_n-\alpha-k-1}P_\lambda^{-1}Cu_k + \sum_{l\in D_k}\lambda^{\alpha_l-\alpha-k-1}P_\lambda^{-1}A_lCu_k\Big] - \lambda^{\alpha_n-1-k}Cu_k. \tag{442}$$

Then there exists a continuous function $u_k : [0, \infty) \to X$ such that, for every $p \in \circledast$, there exists $M'_p > 0$ satisfying $p(u_k(t)) \leqslant M'_p e^{\omega t}$, $t \geqslant 0$ and

$$\int_0^\infty e^{-\lambda t} u_k(t)\, dt = \lambda^{\alpha_n - \alpha - k - 1} P_\lambda^{-1} C u_k + \sum_{l \in D_k} \lambda^{\alpha_l - \alpha - k - 1} P_\lambda^{-1} A_l C u_k, \quad \lambda > \omega.$$

The Laplace transform can be used to prove that:

(443) $$(g_{m_n} * F_{k,n})(t) = \left[g_{m_n - \alpha_n} * (u_k(\cdot) - g_{k+1}(\cdot) C u_k)\right](t),\ t \geqslant 0.$$

This implies

$$(g_{\alpha_n} * F_{k,n})(t) = u_k(t) - g_{k+1}(t) C u_k,\ t \geqslant 0,$$

$u \in C^{m_n - 1}([0, \infty) : X)$ and $u_k^{(j)}(0) = \delta_{kj} C u_k$ for $0 \leqslant j \leqslant m_n - 1$. Due to (443), we have $\mathbf{D}_t^{\alpha_n} u_k(t) = F_{k,n}(t)$, $t \geqslant 0$. Keeping in mind (16), it can be simply proved that, for every $j \in \mathbb{N}_{n-1}^0$, $\mathbf{D}_t^{\alpha_j} u_k$ is well defined as well as that

$$\int_0^\infty e^{-\lambda t} \mathbf{D}_t^{\alpha_j} u_k(t)\, dt = \lambda^{\alpha_j} \int_0^\infty e^{-\lambda t} u_k(t)\, dt - \sum_{l=0}^{m_j - 1} u_k^{(l)}(0) \lambda^{\alpha_j - 1 - l}, \quad \lambda > \omega.$$

Observe now that the condition (438), along with Theorem 1.2.1(vii), implies that the mapping $t \mapsto A_j \mathbf{D}_t^{\alpha_j} u_k(t)$, $t \geqslant 0$ is well defined, continuous, as well as that

(444) $$\int_0^\infty e^{-\lambda t} A_j \mathbf{D}_t^{\alpha_j} u_k(t)\, dt = A_j \left[\lambda^{\alpha_j} \int_0^\infty e^{-\lambda t} u_k(t)\, dt - \sum_{l=0}^{m_j - 1} u_k^{(l)}(0) \lambda^{\alpha_j - 1 - l}\right], \quad \lambda > \omega.$$

A simple calculation involving (442), (444), and the definition of P_λ, yields that:

$$\int_0^\infty e^{-\lambda t} \left[\mathbf{D}_t^{\alpha_n} u_k(t) + A_{n-1} \mathbf{D}_t^{\alpha_{n-1}} u_k(t) + \cdots + A_1 \mathbf{D}_t^{\alpha_1} u_k(t) - A \mathbf{D}_t^{\alpha} u_k(t)\right] dt = 0,$$

which implies by the uniqueness theorem for the Laplace transform that $u_k(\cdot)$ is a strong solution of the problem (2) with $u_k^{(j)}(0) = \delta_{kj} C u_k$. Suppose now $0 \leqslant k \leqslant m_n - 1$ and $m - 1 \geqslant k$. Then one can similarly prove, with the help of conditions (440)-(441), that the function

$$t \mapsto u_k(t) := \mathcal{L}^{-1}\Big(\lambda^{-k-1} C u_k - P_\lambda^{-1} \sum_{l \in \mathbb{N}_{n-1} \setminus D_k} \lambda^{\alpha_l - \alpha - l - 1} A_l C u_k\Big)(t),\ t \geqslant 0,$$

is a strong solution of the problem (2) with $u_k^{(j)}(0) = \delta_{kj} C u_k$. Define $u(t) := \Sigma_{k=0}^{m_n - 1} u_k(t)$, $t \geqslant 0$. Then it is clear that $u(\cdot)$ is a strong solution of the abstract Cauchy problem (2). □

Remark 2.10.43. (i) Let $0 \leqslant k \leqslant m_n - 1$ and $m - 1 < k$. Then Theorem 2.10.42 continues to hold if we replace the term

$$\lambda^{\alpha_n - \alpha - k - 1} P_\lambda^{-1} C u_k + \sum_{l \in D_k} \lambda^{\alpha_l - \alpha - k - 1} P_\lambda^{-1} A_l C u_k$$

i.e., the Laplace transform of $u_k(t)$, in (438)-(439) by

$$\lambda^{-k-1}Cu_k - \sum_{l\in\mathbb{N}_{n-1}\setminus D_k} \lambda^{\alpha_l-\alpha-k-1} P_\lambda^{-1}A_lCu_k + \lambda^{-k-1}P_\lambda^{-1}ACu_k;$$

in this case, we have to assume that $Cu_k \in D(P_\lambda^{-1}A_l)$, provided $0 \leqslant l \leqslant n-1$, $k > m_l - 1$ and $\lambda > \omega$. Notice also that a similar modification can be made in the case $0 \leqslant k \leqslant m_n - 1$ and $m - 1 \geqslant k$. As a matter of fact, one can replace the term

$$\lambda^{-k-1}Cu_k - P_\lambda^{-1} \sum_{l\in\mathbb{N}_{n-1}\setminus D_k} \lambda^{\alpha_l-\alpha-k-1} A_lCu_k$$

i.e., the Laplace transform of $u_k(t)$, in (440)-(441) by

$$\lambda^{\alpha_n-\alpha-k-1}P_\lambda^{-1}Cu_k + \sum_{l\in D_k} \lambda^{\alpha_l-\alpha-k-1}P_\lambda^{-1}A_lCu_k - \lambda^{-k-1}P_\lambda^{-1}ACu_k;$$

in this case, we have to assume that $Cu_k \in D(P_\lambda^{-1}A_l)$, provided $0 \leqslant l \leqslant n-1$, $m_l - 1 \geqslant k$ and $\lambda > \omega$.

(ii) Consider now the situation of the abstract Cauchy problem (ACP_n), i.e., suppose that $\alpha_j = j$, $j \in \mathbb{N}_n^0$. Keeping in mind the proof of [531, Lemma 2.2.1, pp. 54-55], it readily follows that the condition:

$$\lambda^j A_j P_\lambda^{-1} Cu_{n-1},\ \lambda^j A_j \sum_{i=0}^{k} \lambda^{i-k-1} P_\lambda^{-1} A_i Cu_k \in LT - X,$$

for any $k \in \mathbb{N}_{n-2}^0$ and $j \in \mathbb{N}_{n-1}^0$, implies (438)-(439). Therefore, Theorem 2.10.42 can be viewed as a generalization of the above-mentioned result.

Now we shall state and prove the Ljubich uniqueness theorem for abstract multi-term fractional differential equation (2).

Theorem 2.10.44. *Let $\lambda > 0$, let $L(X) \ni C$ be injective, and let $D(P_{n\lambda}^{-1}C) = X$, $n \in \mathbb{N}$. Suppose, additionally, that $CA_j \subseteq A_jC$, $j \in \mathbb{N}_{n-1}^0$ and that for every positive real number $\sigma > 0$ and for every null sequence $(x_n)_{n\in\mathbb{N}}$ in X, one has:*

(445) $$\lim_{n\to\infty} e^{-n\lambda\sigma}P_{n\lambda}^{-1}Cx_n = 0.$$

Then, for every $u_0, \cdots, u_{m_n-1} \in X$, the abstract Cauchy problem (2) has at most one strong (integral) solution.

Proof. Clearly, it suffices to show the uniqueness of integral solutions of the abstract Cauchy problem (2) with $u^{(k)}(0) = u_k = 0$, $k \in \mathbb{N}_{m_n-1}^0$. Let $u(t)$ be such a solution. Then, for every $n \in \mathbb{N}$ and $t \geqslant 0$,

$$P_{n\lambda} \int_0^t e^{n\lambda(t-s)}Cu(s)\,ds$$
$$= (n\lambda)^{\alpha_n-\alpha} \int_0^t e^{n\lambda(t-s)}\Big[(g_{\alpha_n-\alpha} * CAu)(s) - \sum_{j=1}^{n-1}(g_{\alpha_n-\alpha_j} * CA_ju)(s)\Big]\,ds$$

$$+\sum_{j=1}^{n-1}(n\lambda)^{\alpha_j-\alpha}CA_j\int_0^t e^{n\lambda(t-s)}u(s)\,ds - CA\int_0^t e^{n\lambda(t-s)}u(s)\,ds$$

$$=\Big[(n\lambda)^{\alpha_n-\alpha}\int_0^t e^{n\lambda(t-s)}(g_{\alpha_n-\alpha}*CAu)(s)\,ds - CA\int_0^t e^{n\lambda(t-s)}u(s)\,ds\Big]$$

$$+\sum_{j=1}^{n-1}(n\lambda)^{\alpha_j-\alpha}CA_j\int_0^t e^{n\lambda(t-s)}u(s)\,ds$$

$$-(n\lambda)^{\alpha_n-\alpha}\int_0^t e^{n\lambda(t-s)}(g_{\alpha_n-\alpha_j}*CA_ju)(s)\,ds. \tag{446}$$

Repeating literally the arguments from the proof of [531, Lemma 1.5.5, p. 23], we obtain that there exist numbers $M_0,\cdots,M_{n-1}\geqslant 1$ and $k_0,\cdots,k_{n-1}\in\mathbb{N}$ such that, for every $p\in\circledast$, $t\geqslant 0$, $n\in\mathbb{N}$ and $j\in\mathbb{N}^0_{n-1}$,

$$p\left((n\lambda)^{\alpha_n-\alpha}\int_0^t e^{n\lambda(t-s)}(g_{\alpha_n-\alpha}*Au)(s)\,ds - A\int_0^t e^{n\lambda(t-s)}u(s)\,ds\right)$$

$$=p\left((n\lambda)^{\alpha_n-\alpha}\int_0^t\left(\int_0^\infty\frac{(sn\lambda+\varsigma)^{\alpha_n-\alpha-1}}{\Gamma(\alpha_n-\alpha)}e^{-\varsigma}\,d\varsigma\right)Au(t-s)\,ds\right) \tag{447}$$

$$\leqslant M_0(1+n+|\lambda|)^{k_0}\int_0^t p(Au(s))\,ds, \tag{448}$$

and

$$p\left((n\lambda)^{\alpha_j-\alpha}A_j\int_0^t e^{n\lambda(t-s)}u(s)\,ds-(n\lambda)^{\alpha_n-\alpha}\int_0^t e^{n\lambda(t-s)}(g_{\alpha_n-\alpha_j}*A_ju)(s)\,ds\right)$$

$$=p\left((n\lambda)^{\alpha_n-\alpha}\int_0^t\Big[\int_0^s e^{n\lambda(s-r)}g_{\alpha_n-\alpha_j}(r)\,dr-e^{n\lambda s}(n\lambda)^{\alpha_j-\alpha_n}\Big]A_ju(t-s)\,ds\right) \tag{449}$$

$$\leqslant M_j(1+n+|\lambda|)^{k_j}\int_0^t p(A_ju(s))\,ds. \tag{450}$$

Making use of (446)-(450), it readily follows that, for every $\sigma>0$ and $t\geqslant 0$, we have $\lim_{n\to\infty}e^{-n\lambda\sigma}\int_0^t e^{n\lambda(t-s)}Cu(s)\,ds=0$. Since $\lim_{n\to\infty}\int_{t-\sigma}^t e^{n\lambda(t-s-\sigma)}Cu(s)\,ds=0$ for $0\leqslant\sigma\leqslant t$, we obtain that $\lim_{n\to\infty}\int_0^{t-\sigma}e^{n\lambda(t-s-\sigma)}Cu(s)\,ds=0$. By Lemma 2.1.33(iii), one gets $Cu(t)=0$, $t\geqslant 0$, which completes the proof by the injectiveness of C. □

If $\alpha_n-\alpha_j\in\mathbb{N}$, $j\in\mathbb{N}^0_{n-1}$, then the formulae (447) and (449) imply that it suffices to suppose (instead of a slightly stronger condition (445)) that, for every $\sigma>0$ and $x\in X$, one has $\lim_{n\to\infty}e^{-n\lambda\sigma}P^{-1}_{n\lambda}Cx=0$; with this in mind, we can easily check that Theorem 2.10.44 provides a generalization of [292, Theorem 2.3.23] and [531, Lemma 2.3.1, pp. 67-68].

2.10.4. (*a*, *k*)-Regularized *C*-resolvent families and abstract multi-term fractional differential equations. In this subsection, we shall always assume that $A = 0$, $n \geqslant 3$ and $0 = \alpha_1 < \cdots < \alpha_n$; notice that these assumptions are not such restrictive since we can always get the term $A\mathbf{D}_t^{\alpha}u(t)$ on the left-hand side of (2) and add, after an obvious regrouping of terms, $0u(t)$ on the same side of (2), if necessary. Hence, in the formulations of our results we shall always have that $\mathbb{N}_{n-1} \setminus D_k \neq \emptyset$, $k \in \mathbb{N}^0_{m_n-1}$.

It is well-known from the theory of higher order abstract differential equations that the operator $-A_{n-1}$ plays a crucial role for the solvability of equation (2), and that the operators $-A_{n-2}, \cdots, -A_1$ are subordinated to $-A_{n-1}$ in some sense. F. Neubrander [433] was the first who investigated the well-posedness of problem (2) in the case that $\alpha_n = n - 1$, $n \in \mathbb{N}\setminus\{1, 2\}$, X is a Banach space and $-A_{n-1}$ is the integral generator of a strongly continuous semigroup on X. Concerning equations with integer order derivatives, further contributions have been obtained by R. deLaubenfels [148] and T.-J. Xiao-J. Liang [531] (cf. also [436], [479] and [538] for more details on the subject), where the authors have analyzed the well-posedness of problem (2) in the case that there exists a number $r \in \mathbb{N}_0$ such that the operator $-A_{n-1}$ is the integral generator of an exponentially bounded r-times integrated semigroup on X. Our analysis is inspired by the fact that the methods developed in [531] cannot be simply modified to cover the case where r is a non-integer number. We overcome the problem mentioned above by using an additional assumption that the operator $-A_{n-1}$ is the integral generator of an exponentially bounded C-regularized semigroup on X, for a suitable chosen injective operator $C \in L(X)$ that is practically always different from the operator $(\mu_0 - A_{n-1})^{-r}$, for some $\mu_0 \in \rho(A_{n-1})$; furthermore, we consider the case in which there exist numbers $\sigma \in [1, 2]$, $r \geqslant 0$ and $\omega \geqslant 0$ such that $(\omega^{\sigma}, \infty) \subseteq \rho(-A_{n-1})$ and the operator $-A_{n-1}$ generates an exponentially equicontinuous $(g_{\sigma}, g_{\sigma r+1})$-regularized resolvent family or an exponentially equicontuinuous (g_{σ}, C)-regularized resolvent family (cf. Theorem 2.10.45). Concerning the abstract Cauchy problem (ACP_n), such reasoning produces, on the concrete level, significant improvements of regularity properties of the initial data which guarantee the existence and uniqueness of solutions; cf. Example 2.10.47(i). On the other hand, we feel duty bound to say that Theorem 2.10.45 has some disadvantages in the case that the equation under its consideration contains more than two dominating terms, the possibility of application to real problems is then hindered because the fractional derivatives of order > 2 are allowed. This is, certainly, not the case with the assertion of Theorem 2.10.48, whose main purpose is to extend the assertions of [538, Theorem (*)] and [531, Theorem 3.4.2] to abstract multi-term fractional differential equations.

Let $\sigma > 0$ and $l \in \mathbb{N}$. Recall that, for any X-valued function $f(t)$ satisfying (P1), there exist uniquely determined real numbers $(c_{l_0,l,\sigma})_{1 \leqslant l_0 \leqslant l}$, independent of X and $f(t)$, such that (141) holds. We have $c_{l,l,\sigma} = (-1)^l \sigma^{-l}$, $l \geqslant 1$ and the existence of a number $\zeta \geqslant 1$ such that (143) holds.

Theorem 2.10.45. *Suppose* $n \in \mathbb{N}\setminus\{1, 2\}$, $\sigma \in [1, 2]$, $r \geqslant 0$, $\alpha_n - \alpha_{n-1} = \sigma$, $\alpha_{n-1} - \alpha_{n-2} \geqslant \sigma$, $\omega \geqslant 0$, $D(A_{n-1}) \subseteq \bigcap_{i=0}^{n-2} D(A_i)$ *and* $(\omega^\sigma, \infty) \subseteq \rho(-A_{n-1})$. *Put* $\breve{A}_i(\lambda) := \lambda^{\alpha_i-\alpha_{n-1}} A_i(\lambda^\sigma + A_{n-1})^{-1}$, $\lambda > \omega$, $i \in \mathbb{N}_{n-2}$ *and suppose that the following conditions hold:*

(i) $A_iA_jx = A_jA_ix$, $1 \leqslant i, j \leqslant n-1$, $x \in D(A_{n-1}^2)$ *and* $CA_j \subseteq A_jC$, $j \in \mathbb{N}_{n-2}$.
(ii) *There exists* $\omega_0 \geqslant \omega$ *such that, for every* $p \in \circledast$, *there exists* $c_p \in (0, 1/(n-2))$ *satisfying*

$$p(\breve{A}_i(\lambda)x) \leqslant c_p p(x),\ \lambda > \omega_0,\ x \in X,\ i \in \mathbb{N}_{n-2}. \tag{451}$$

If

(a) *The operator* $-A_{n-1}$ *is the integral generator of a* $(g_\sigma, g_{\sigma r+1})$*-regularized resolvent family* $(S_{\sigma,r}(t))_{t\geqslant 0}$ *on* X, *the family* $\{e^{-\omega t}S_{\sigma,r}(t) : t \geqslant 0\}$ *is equicontinuous, and* $A_lu_k \in D(A_{n-1}^{\max(\lceil \frac{1}{\sigma}(\sigma r+\alpha_l-k)\rceil,0)})$, *provided* $0 \leqslant k \leqslant m_n - 1$ *and* $l \in \mathbb{N}_{n-1}\setminus D_k$,

or

(b) *The operator* $-A_{n-1}$ *is the integral generator of a* (g_σ, C)*-regularized resolvent family* $(T_\sigma(t))_{t\geqslant 0}$ *on* X, *the family* $\{e^{-\omega t}T_\sigma(t) : t \geqslant 0\}$ *is equicontinuous, and* $u_k \in C(\bigcap_{l\in\mathbb{N}_{n-1}\setminus D_k} D(A_l))$ *for* $0 \leqslant k \leqslant m_n - 1$,

then the abstract Cauchy problem (2) has a unique strong solution.

Proof. Let $\mu_0 < -\omega_0^\sigma$. By (ii), it follows that, for every $p \in \circledast$ and $l \in \mathbb{N}$, we have $p(\breve{A}_i(\lambda)^l x) \leqslant c_p^l p(x)$, $\lambda > \omega_0$, $x \in X$, $i \in \mathbb{N}_{n-2}$. This inequality and the polynomial formula imply that, for every $p \in \circledast$,

$$p\left(\left[\sum_{i=1}^{n-2} \breve{A}_i(\lambda)\right]^k x\right) \leqslant c_p^k (n-2)^k p(x),\ \lambda > \omega_0,\ k \in \mathbb{N}_0,\ x \in X.$$

Since $c_p(n-2) < 1$, $p \in \circledast$, the above implies that, for every $x \in X$ and $\lambda > \omega_0$, the series

$$B_\lambda x \equiv \sum_{k=0}^{\infty} (\lambda^\sigma + A_{n-1})^{-1}\left[-\sum_{i=1}^{n-2} \breve{A}_i(\lambda)\right]^k x \tag{452}$$

is convergent. Put $\tilde{A}_i := A_i(\mu_0 - A_{n-1})^{-1}$, $i \in \mathbb{N}_{n-2}$. Then (451) implies $\tilde{A}_i \in L(X)$, $i \in \mathbb{N}_{n-2}$. Using the polynomial formula again, we get that, for every $\lambda > \omega$, $k \in \mathbb{N}_0$ and $j \in \mathbb{N}_0$,

$$\left[\sum_{i=1}^{n-2} \lambda^{\alpha_i-\alpha_{n-1}+\sigma} \tilde{A}_i\right]^k \left(1 + \frac{\mu_0}{\lambda^\sigma}\right)^j = \sum_{\substack{(l_1,\cdots,l_{n-2})\in\mathbb{N}_0^{n-2}\\ l_1+\cdots+l_{n-2}=k}} \sum_{s=0}^{j} \frac{k!}{l_1!\cdots l_{n-2}!}$$

$$\times (\lambda^{\alpha_1-\alpha_{n-1}+\sigma}\tilde{A}_1)^{l_1} \cdots (\lambda^{\alpha_{n-2}-\alpha_{n-1}+\sigma}\tilde{A}_{n-2})^{l_{n-2}} \binom{j}{s} \frac{\mu_0^s}{\lambda^{\sigma s}}. \tag{453}$$

Since $\alpha_{n-1} - \alpha_{n-2} \geqslant \sigma$, (453) yields that, for every $k \in \mathbb{N}_0$ and $j \in \mathbb{N}_0$ with $0 \leqslant j \leqslant k$, there exist numbers $l_{kj} \in \mathbb{N}$, $\beta_0, \cdots, \beta_{l_{kj}} \in (-\infty, 0]$ and operators $A_{kjm} \in L(X)$ $(0 \leqslant m \leqslant l_{kj})$ such that $\beta_0 > \cdots > \beta_{l_{kj}}$ and

$$\left[\sum_{i=1}^{n-2}\lambda^{\alpha_i-\alpha_{n-1}+\sigma}\tilde{A}_i\right]^k\left(1+\frac{\mu_0}{\lambda^\sigma}\right)^j=\sum_{m=0}^{l_{kj}}\lambda^{\beta_m}A_{kjm},\ \lambda>\omega_0.$$

Repeating literally the arguments given in the proof of [531, Theorem 3.2.1, p. 95], we obtain that:

$$B_\lambda x=\sum_{k=0}^{\infty}\sum_{j=0}^{k}\binom{k}{j}\sum_{m=0}^{l_{kj}}\lambda^{\beta_m}A_{kjm}\left[\frac{(-1)^j}{\lambda^{(k-j)\sigma}}(\lambda^\sigma+A_{n-1})^{-j-1}x\right],\ x\in X,\ \lambda>\omega_0.$$

Keeping in mind (452) and the equality $\alpha_n-\alpha_{n-1}=\sigma$, it can be easily seen that

(454) $$P_\lambda B_\lambda x=\lambda^{\alpha_{n-1}}x,\ x\in X,\ \lambda>\omega_0,$$

as well as that (cf. (i)): $A_i(\lambda+A_{n-1})^{-2}=(\lambda+A_{n-1})^{-1}A_i(\lambda+A_{n-1})^{-1}$, $\lambda>\omega$, $1\leqslant i\leqslant n-2$ and $A_i(\lambda+A_{n-1})^{-1}A_j(\lambda+A_{n-1})^{-1}=A_j(\lambda+A_{n-1})^{-1}A_i(\lambda+A_{n-1})^{-1}$, $\lambda>\omega$, $1\leqslant i,j\leqslant n-2$. Taken together with (454), this implies that P_λ is injective for $\lambda>\omega_0$ and $B_\lambda x=\lambda^{\alpha_{n-1}}P_\lambda^{-1}x$, $x\in X$, $\lambda>\omega_0$. Then the existence of strong solutions simply follows from Theorem 2.10.42-Remark 2.10.43 if we prove that, for every $k\in\mathbb{N}^0_{m_n-1}$ and $j\in\mathbb{N}_{n-1}$,

(455) $$\lambda^{\alpha_j}A_j\sum_{l\in\mathbb{N}_{n-1}\setminus D_k}\lambda^{\alpha_l-k-1}P_\lambda^{-1}A_lu_k\in LT-X$$

and

(456) $$\lambda^{\alpha_n}\sum_{l\in\mathbb{N}_{n-1}\setminus D_k}\lambda^{\alpha_l-k-1}P_\lambda^{-1}A_lu_k\in LT-X.$$

Clearly, the relation (455) with $j=n-1$ is equivalent to say that

(457) $$\lambda^{\alpha_{n-1}}(\mu_0-A_{n-1})\sum_{l\in\mathbb{N}_{n-1}\setminus D_k}\lambda^{\alpha_l-k-1}P_\lambda^{-1}A_lu_k\in LT-X.$$

Suppose first that (b) holds. We will prove that $\lambda^{\sigma-1}B_\lambda Cx\in LT-X$ for every fixed element $x\in X$. Since

$$(z+A_{n-1})^{-1}Cx=z^{(1-\sigma)/\sigma}\int_0^\infty e^{-z^{1/\sigma}t}T_\sigma(t)x\,dt,\ x\in X,\ z>\omega^\sigma,$$

the equality (141) taken together with

$$(-1)^j(z+A_{n-1})^{-j-1}Cx=j!^{-1}\frac{d^j}{dz^j}(z+A_{n-1})^{-1}Cx,\ z>\omega^\sigma,\ j\in\mathbb{N}_0,\ x\in X,$$

implies

$$\lambda^{\sigma-1}(\lambda^\sigma+A_{n-1})^{-j-1}Cx=\frac{(-1)^j}{j!}\sum_{l=0}^{j}\binom{j}{l}\frac{1-\sigma}{\sigma}\cdots\left(\frac{1-\sigma}{\sigma}-(j-l-1)\right)$$

(458) $$\times\sum_{l_0=1}^{l}c_{l_0,l,\sigma}\lambda^{l_0-j\sigma}\int_0^\infty e^{-\lambda t}t^{l_0}T_\sigma(t)x\,dt,\ x\in X,\ \lambda>\omega,$$

where we have put, by common consent,

$$\tfrac{1-\sigma}{\sigma}\cdots\left(\tfrac{1-\sigma}{\sigma}-(j-l-1)\right)\equiv\begin{cases}1, & \text{if } \sigma=1 \text{ and } j=l,\\ 0, & \text{if } \sigma=1 \text{ and } j>l,\end{cases} \text{ and}$$

$\Sigma_{l_0=1}^{l} c_{l_0,l,\sigma}\lambda^{l_0-j\sigma}\int_0^{\infty}e^{-\lambda t}t^{l_0}T_\sigma(t)x\,dt \equiv \lambda^{-j\sigma}\int_0^{\infty}e^{-\lambda t}T_\sigma(t)x\,dt$, $t \geqslant 0$, $x \in X$, for $l = 0$. If $k \in \mathbb{N}_0$, $0 \leqslant j \leqslant k$, $t \geqslant 0$ and $x \in X$, then we define

$$H^{\sigma}_{C,kj0}(t;0,0)x := j!^{-1}\sum_{l=0}^{j}\sum_{l_0=1}^{l}\binom{j}{l}\frac{1-\sigma}{\sigma}\cdots\left(\frac{1-\sigma}{\sigma}-(j-l-1)\right)$$
$$\times\, c_{l_0,l,\sigma}[g_{k\sigma-l_0} * \cdot^{l_0}T_\sigma(\cdot)x](t).$$

Due to the estimate (143), we obtain that, for every $x \in X$, the series

$$H^{\sigma}_{C}(t;0,0)x := \sum_{k=0}^{\infty}\sum_{j=0}^{k}\binom{k}{j}\sum_{m=0}^{l_{kj}}A_{kjm}\left(g_{-\beta_m} * H^{\sigma}_{C,kj0}(\cdot;0,0)x\right)(t)$$

converges uniformly on compacts of $[0, \infty)$. By definition of $H^{\sigma}_{C,kj0}(\cdot;0,0)$ and (16), it readily follows that:

$$\int_0^{\infty} e^{-\lambda t}H^{\sigma}_{C,kj0}(t;0,0)x\,dt = (-1)^j\lambda^{\sigma-1-(k-j)\sigma}(\lambda^{\sigma}+A_{n-1})^{-j-1}Cx, \tag{459}$$

provided $k \in \mathbb{N}_0$, $0 \leqslant j \leqslant k$, $x \in X$ and $\lambda > \omega_0$. Furthermore, there exists $c_\sigma > 0$ such that

$$\left|\frac{1-\sigma}{\sigma}\cdots\left(\frac{1-\sigma}{\sigma}-(j-l-1)\right)\right| \leqslant c_\sigma(j-l)!,$$

provided $j \in \mathbb{N}_0$ and $0 \leqslant l \leqslant j$. Taken together with the inequality (143), the last estimate yields the existence of a number $\eta \geqslant 1$ such that, for every $p \in ⊛$, there exist $c_p > 0$ and $q_p \in ⊛$ such that

$$p(H^{\sigma}_{C,kj0}(t;0,0)x) \leqslant c_p e^{\omega t}\eta^k g_{k\sigma+1}(t)q_p(x),\ x \in X,\ t \geqslant 0,\ k \in \mathbb{N}_0,\ 0 \leqslant j \leqslant k.$$

Since $E_\sigma(a(bt)^\sigma) = O(a^{1/\sigma}e^{bt})$, $t \geqslant 0$ $(a, b > 0)$, it is not difficult to show that the series appearing in the definition of $H^{\sigma}_{C}(t;0,0)$ converges uniformly in compacts of $[0, \infty)$ and that there exists $\omega' > \omega$ such that, for every $p \in ⊛$, there exist $c_p > 0$ and $q_p \in ⊛$ satisfying $p(H^{\sigma}_{C}(t;0,0)x) \leqslant c_p e^{\omega' t}q_p(x)$, $x \in X$, $t \geqslant 0$. Clearly,

$$\int_0^{\infty} e^{-\lambda t}H^{\sigma}_{C}(t;0,0)x\,dt = \lambda^{\sigma-1}B_\lambda Cx,\ x \in X,\ \lambda > \omega_0.$$

If $k \in \mathbb{N}_0$, $0 \leqslant j \leqslant k$, $l \in \mathbb{N}_{n-1} \setminus D_k$, $x \in X$ and $t \geqslant 0$, then we set

$$F^{\sigma,l}_{C,kj0}(t)x := \begin{cases}-g_{k\sigma+k+1-\alpha_l}(t)Cx, & \text{if } j=0,\\ (g_{k-\alpha_l} * H^{\sigma}_{C,k(j-1)0}(\cdot;0,0)x)(t), & \text{if } j>0\end{cases}$$
$$+\left([g_{k-\alpha_l}(\cdot)+\mu_0 g_{k+\sigma-\alpha_l}(\cdot)] * H^{\sigma}_{C,kj0}(\cdot;0,0)x\right)(t).$$

Using the resolvent equation and (459), we get that

$$\mathcal{L}^{-1}\Big((\mu_0 - A_{n-1})\lambda^{\alpha_l - k-1}B_\lambda Cx\Big)(t)$$
$$= \sum_{k=0}^{\infty}\sum_{j=0}^{k}\binom{k}{j}\sum_{m=0}^{l_{kj}} A_{kjm}(g_{-\beta_m} * F^{\sigma,l}_{C,kj0}(\cdot; 0, 0)x)(t),$$

provided $k \in \mathbb{N}^0_{m_n-1}$, $l \in \mathbb{N}_{n-1} \setminus D_k$, $x \in X$ and $t \geqslant 0$. Since $A_j(\mu_0 - A_{n-1})^{-1}(\mu_0 - A_{n-1})x = A_j x$, $1 \leqslant j \leqslant n-2$, $x \in D(A_{n-1})$, the above ensures that

$$\lambda^{\alpha_j}A_j \sum_{l \in \mathbb{N}_{n-1}\setminus D_k} \lambda^{\alpha_l - k-1}P_\lambda^{-1}CA_lC^{-1}u_k \in LT - X.$$

Suppose now that (a) holds; fix an element $x \in X$. Then we define:

$$H^\sigma_{kj0}(t; 0, \sigma r + 1 - \sigma) := j!^{-1}\sum_{l=0}^{j}\sum_{l_0=1}^{l}\binom{j}{l}$$
$$\times \frac{\sigma r + 1 - \sigma}{\sigma}\cdots\left(\frac{\sigma r+1-\sigma}{\sigma} - (j-l-1)\right) c_{l_0,l,\sigma}[g_{k\sigma - l_0} * {}^{\cdot l_0}S_{\sigma,r}(\cdot)x](t),$$

and

$$H^\sigma(t; 0, \sigma r + 1 - \sigma)$$
$$:= \sum_{k=0}^{\infty}\sum_{j=0}^{k}\binom{k}{j}\sum_{m=0}^{l_{kj}} A_{kjm}\Big(g_{-\beta_m} * H^\sigma_{kj0}(\cdot; 0, \sigma r + 1 - \sigma)x\Big)(t),$$

for any $t \geqslant 0$. Clearly,

$$(z + A_{n-1})^{-1}x = z^{(\sigma r+1-\sigma)/\sigma}\int_0^\infty e^{-z^{1/\sigma}t}S_{\sigma,r}(t)x\,dt,\ z > \omega^\sigma. \tag{460}$$

By definition of $H^\sigma_{kj0}(\cdot; 0, \sigma r + 1 - \sigma)$ and (460), it readily follows that (cf. also the equation (458)):

$$\int_0^\infty e^{-\lambda t}H^\sigma_{kj0}(t; 0, \sigma r + 1 - \sigma)x\,dt$$
$$= (-1)^j\lambda^{\sigma-1-(k-j)\sigma}\lambda^{\sigma-\sigma r-1}(\lambda^\sigma + A_{n-1})^{-j-1}x,$$

provided $k \in \mathbb{N}_0$, $0 \leqslant j \leqslant k$, $\lambda > \omega_0$, and

$$\int_0^\infty e^{-\lambda t}H^\sigma(t; 0, \sigma r + 1 - \sigma)x\,dt = \lambda^{\sigma-\sigma r-1}B_\lambda x,\ \lambda > \omega_0.$$

Assume $r_0 \in \mathbb{N}_0 \cup \{-1\}$, $r_1 \in \mathbb{R}$ and $r_1 + r_0\sigma \geqslant \sigma r + 1 - \sigma$; notice that in the previous analysis we have considered the case $r_0 = 0$. If $r_0 = -1$, then it is very simple to construct, with the help of the resolvent equation and the arguments given in the case $r_0 = 0$, the continuous function $t \mapsto H^\sigma(t; -1, \sigma r + 1)x$, $t \geqslant 0$ such that $H^\sigma(t; -1, \sigma r + 1) \in L(X)$ for $t \geqslant 0$ and

$$B_\lambda x = \lambda^{r\sigma+1}(\mu_0 - A_{n-1}) \int_0^\infty e^{-\lambda t} H^\sigma(t; -1, \sigma r + 1)x\, dt, \lambda > \omega_0.$$

Suppose now $r_0 > 0$. Then the identities

$$S_{\sigma,r}(t)y = \sum_{l=0}^{r_0-1} (-1)^l g_{\sigma r+1+l\sigma}(t)A_{n-1}^l y$$
$$+ (-1)^{r_0}\big(g_{\sigma r_0} * S_{\sigma,r}(\cdot)A_{n-1}^{r_0} y\big)(t), t \geqslant 0, y \in D(A^{r_0}),$$

and

$$\int_0^\infty e^{-\lambda t} t^{l_0}(g_{r_0\sigma} * S_{\sigma,r_0}(\cdot)x)(t)\, dt$$
$$= \lambda^{\sigma+r_1-\sigma r-1} \int_0^\infty e^{-\lambda t} \Bigg\{ \sum_{l_1=0}^{l_0} \binom{l_0}{l_1} (\sigma r + 1 - \sigma - r_1) \cdots (\sigma r - \sigma - r_1 + l_0 - l_1)$$
$$\times \big[g_{l_1-l_0}(\cdot) * (-1)^{r_0} \cdot^{l_1} (g_{r_0\sigma+\sigma+r_1-\sigma r-1} * S_{\sigma,r}(\cdot)A_{n-1}^{r_0}(\mu_0 - A_{n-1})^{-r_0} x)\big](t)\Bigg\}\, dt,$$

which hold for any $l_0 \in \mathbb{N}$ and $\lambda > \omega$ suff. large, imply that

$$\lambda^{(j-k)\sigma-r_1}(\mu_0 - A_{n-1})^{-r_0}(\lambda^\sigma + A_{n-1})^{-j-1}x$$
$$= (-1)^j \lambda^{(j-k)\sigma-r_1} j!^{-1} \sum_{l=0}^{j} \sum_{l_0=1}^{l} \binom{j}{l} \frac{\sigma r+1-\sigma}{\sigma} \cdots \left(\frac{\sigma r+1-\sigma}{\sigma} - (j-l-1)\right)$$
$$\times c_{l_0,l,\sigma} \lambda^{\sigma r+1-\sigma-(j-l)\sigma} \lambda^{l_0-l\sigma} \int_0^\infty e^{-\lambda t} t^{l_0} S_{\sigma,r}(t)\, (\mu_0 - A_{n-1})^{-r_0} x\, dt$$
$$= (-1)^j j!^{-1} \sum_{l=0}^{j} \sum_{l_0=1}^{l} \binom{j}{l} \frac{\sigma r+1-\sigma}{\sigma} \cdots \left(\frac{\sigma r+1-\sigma}{\sigma} - (j-l-1)\right) l_0! c_{l_0,l,\sigma}$$
$$\times \lambda^{-k\sigma-r_1+\sigma r+1-\sigma+l_0} \sum_{m=0}^{r_0-1} (-1)^m \widetilde{g_{\sigma r+1+l_0+\sigma m}}(\lambda) A_{n-1}^m (\mu_0 - A_{n-1})^{-r_0} x$$
$$+ (-1)^j j!^{-1} \sum_{l=0}^{j} \sum_{l_0=1}^{l} \binom{j}{l} \frac{\sigma r+1-\sigma}{\sigma} \cdots \left(\frac{\sigma r+1-\sigma}{\sigma} - (j-l-1)\right) c_{l_0,l,\sigma}$$
$$\times \lambda^{l_0-k\sigma} \int_0^\infty e^{-\lambda t} \Bigg\{ \sum_{l_1=0}^{l_0} \binom{l_0}{l_1} (\sigma r + 1 - \sigma - r_1) \cdots (\sigma r - \sigma - r_1 + l_0 - l_1)$$
$$\big[g_{l_1-l_0}(\cdot) * (-1)^{r_0} \cdot^{l_1} (g_{r_0\sigma+\sigma+r_1-\sigma r-1} * S_{\sigma,r}(\cdot)A_{n-1}^{r_0}(\mu_0 - A_{n-1})^{-r_0} x)\big](t)\Bigg\}\, dt,$$

so that $\lambda^{(j-k)\sigma-r_1}(\mu_0 - A_{n-1})^{-r_0}(\lambda^\sigma + A_{n-1})^{-j-1}x \in LT - X$. Put, for every $t \geqslant 0$,

$$H_{kj0}^\sigma(t; r_0, r_1)x := \mathcal{L}^{-1}\big(\lambda^{(j-k)\sigma-r_1}(\mu_0 - A_{n-1})^{-r_0}(\lambda^\sigma + A_{n-1})^{-j-1}x\big)(t)$$

and

$$H^{\sigma}(t; r_0, r_1)x := \sum_{k=0}^{\infty}\sum_{j=0}^{k}\binom{k}{j}\sum_{m=0}^{l_{kj}} A_{kjm}(g_{-\beta_m} * H^{\sigma}_{kj0}(\cdot; r_0, r_1)x)(t).$$

Since $r_1 + r_0\sigma \geqslant \sigma r + 1 - \sigma$, we obtain by the foregoing arguments that the mapping $t \mapsto H^{\sigma}(t; r_0, r_1)x$, $t \geqslant 0$ is continuous as well as that $H^{\sigma}(t; r_0, r_1) \in L(X)$, $t \geqslant 0$ and

(461) $$B_{\lambda}x = \lambda^{r_1}(\mu_0 - A_{n-1})^{r_0}\int_0^{\infty} e^{-\lambda t}H^{\sigma}(t; r_0, r_1)x\, dt,\ \lambda > \omega \text{ suff. large.}$$

Put $s_{l,k,\sigma} := \max(\lceil \frac{1}{\sigma}(\sigma r + \alpha_l - k)\rceil, 0)$. Using the first part of the proof, it is not difficult to see that there exists $\omega' \geqslant 0$ such that, for every $p \in \circledast$, there exist $c_p > 0$ and $q_p \in \circledast$ such that $p(H^{\sigma}(t; r_0, r_1)x) \leqslant c_p e^{\omega' t}q_p(x)$, $x \in X$, $t \geqslant 0$. Now fix an index $k \in \mathbb{N}^0_{m_n-1}$ and an integer $l \in \mathbb{N}_{n-1} \setminus D_k$. Then (457) follows on account of (461), the inequality $(\sigma + \alpha_l - k - 1) + (\sigma r + 1 - s_{l,k,\sigma}\sigma) \leqslant 0$ and the following relation:

$$\lambda^{\alpha_{n-1}+\alpha_l-k-1}(\mu_0 - A_{n-1})P_{\lambda}^{-1}A_l u_k = \lambda^{\sigma+\alpha_l-k-1}\lambda^{\sigma r+1-s_{l,k,\sigma}\sigma}$$
$$\times \left[\lambda^{-(\sigma r+1-s_{l,k,\sigma}\sigma)}(\mu_0 - A_{n-1})^{1-s_{k,l,\sigma}}B_{\lambda}(\mu_0 - A_{n-1})^{s_{k,l,\sigma}}A_l u_k\right] \in LT - X;$$

one can simply prove (455) by using (457) and decomposition $A_j x = A_j(\mu_0 - A_{n-1})^{-1}(\mu_0 - A_{n-1})x$, $1 \leqslant j \leqslant n-2$, $x \in D(A_{n-1})$. Similarly, we have by (461) and the inequality $(\sigma + \alpha_l - k - 1) + (\sigma r + 1 - \sigma - s_{l,k,\sigma}\sigma) \leqslant 0$ that

$$\lambda^{\alpha_n+\alpha_l-k-1}P_{\lambda}^{-1}A_l u_k = \lambda^{\sigma+\alpha_l-k-1}\lambda^{\sigma r+1-\sigma-s_{l,k,\sigma}\sigma}$$
$$\times \left[\lambda^{-(\sigma r+1-\sigma-s_{l,k,\sigma}\sigma)}(\mu_0 - A_{n-1})^{-s_{k,l,\sigma}}B_{\lambda}(\mu_0 - A_{n-1})^{s_{k,l,\sigma}}A_l u_k\right] \in LT - X.$$

Hence, (456) holds and the proof of the theorem is thereby completed. □

Remark 2.10.46. (i) For every $i \in \mathbb{N}_{n-2}$, the operator $\tilde{A}_i$ is closed, linear and defined on the whole space X. If we assume that $\alpha_{n-2} - \alpha_{n-1} + \sigma r < 0$, as well as that X is a webbed bornological space and that there exists $M \geqslant 1$ such that

(462) $$p(S_{\sigma,r}(t)x) \leqslant Me^{\omega t}p(x),\ p \in \circledast,\ t \geqslant 0,\ x \in X,$$

then (451) holds.

(ii) Suppose that (a) holds with some $r > 0$. Then Corollary 2.1.20 implies that the operator $-A_{n-1}$ is the integral generator of an exponentially equicontinuous $(g_{\sigma}, (\mu_0 - A_{n-1})^{-\lceil r\rceil})$-regularized resolvent family. By Theorem 2.10.45(b), we obtain that there exists a unique strong solution of (2) provided that the initial values satisfy the condition $u_k \in (\mu_0 - A_{n-1})^{-\lceil r\rceil}(\bigcap_{l\in\mathbb{N}_{n-1}\setminus D_k} D(A_l))$ for $0 \leqslant k \leqslant m_n - 1$. Since $s_{l,k,\sigma} \leqslant \lceil r\rceil$ (in many concrete situations, the above inequality is strict), the use of integrated operator solution families produces better results here, so that the choice $C \neq (\mu_0 - A_{n-1})^{-\lceil r\rceil}$ is inevitable for obtaining larger initial data sets $\mathfrak{T}_k$ such that the equation (2) has a unique strong solution provided $u_k \in \mathfrak{T}_k$ $(0 \leqslant k \leqslant m_n - 1)$.

(iii) Set, for every $k \in \mathbb{N}^0_{m_{n-1}}$ and $l \in \mathbb{N}_{n-1} \setminus D_k$, $Q_{k,l} := \max(\lceil \frac{1}{\sigma}(\sigma r + \alpha_l - k - \alpha_n) \rceil, 0)$. Suppose that (462) holds with $(S_{\sigma,r}(t))_{t \geqslant 0}$, and with $(S_{\sigma,r}(t))_{t \geqslant 0}$ replaced by $(T_\sigma(t))_{t \geqslant 0}$ therein. Then it is not difficult to see that the assumptions $0 \leqslant k \leqslant m_n - 1$ and $D_k = \emptyset$ imply

$$x - \sum_{l \in \mathbb{N}_{n-1} \setminus D_k} \lambda^{\alpha_l} P_\lambda^{-1} A_l x = \lambda^{\alpha_n} P_\lambda^{-1} x, \; x \in X.$$

In this case, the Laplace transform of strong solution $u_k(t)$ of (2) with $u_k^{(j)}(0) = \delta_{jk} u_k$ can also be computed by

$$\int_0^\infty e^{-\lambda t} u_k(t)\, dt = \lambda^{\sigma-k-1} B_\lambda u_k$$
$$= \lambda^{\sigma-k-1} \Big[\lambda^{k+1-\sigma} \int_0^\infty e^{-\lambda t} H^\sigma(t; \max(\lceil \sigma^{-1} (\sigma r - k) \rceil, 0), k + 1 - \sigma)\, u_k\, dt \Big],$$

for $\lambda > \omega$ suff. large; cf. Theorem 2.10.42-Remark 2.10.43. Then the proof of Theorem 2.10.45, taken together with the inequality

$$g_{r_0} * H^\sigma_{k'j0}(\cdot; r_0, \sigma r + 1 - \sigma - r_0)\, (\mu_0 - A_{n-1})^{r_0} x$$
$$= g_{r_0'} * H^\sigma_{k'j0}(\cdot; r_0', \sigma r + 1 - \sigma - r_0')\, (\mu_0 - A_{n-1})^{r_0'} x,$$

which holds provided $x \in D(A_{n-1}^{\max(r_0, r_0')})$ and $r_0, r_0' \in \mathbb{N}_0$, implies that the strong solution $u(t)$ of (2) has the following form:

$$u(t) = \sum_{k=0}^{m_{n-1}-1} \Bigg[g_{k+1}(t) u_k - \sum_{l \in \mathbb{N}_{n-1} \setminus D_k} \Bigg(g_{\alpha_n + k - r + \sigma Q_{k,l} - \alpha_l - \sigma r} * \sum_{k'=0}^{\infty} \sum_{j=0}^{k'} \binom{k'}{j}$$
$$\times \sum_{m=0}^{l_{k'j}} A_{k'jm} (g_{-\beta_m} * H^\sigma_{k'j0}(\cdot; Q_{k,l}, \sigma r + 1 - \sigma - Q_{k,l})$$

(463) $$(\mu_0 - A_{n-1})^{Q_{k,l}} A_l u_k) \Bigg)(t) \Bigg] + \sum_{k=m_{n-1}}^{m_n - 1} u_k(t), \; t \geqslant 0,$$

where $H^\sigma_{k'j0}(\cdot; Q_{k,l}, r - Q_{k,l})$ can be further expressed in terms of $(S_{\sigma,r}(t))_{t \geqslant 0}$. Then we get the existence of numbers $M' \geqslant 1$ and $\omega' > \omega$ such that, for every $p \in$ ⊛ and $t \geqslant 0$,

$$p(u(t)) \leqslant M' e^{\omega' t} \Bigg\{ \sum_{k=m_{n-1}}^{m_n - 1} \sum_{l=0}^{\max(\lceil \frac{1}{\sigma}(\sigma r - k) \rceil, 0)} p(A^l_{n-1} u_k)$$
$$+ \sum_{k=0}^{m_{n-1}-1} \Big[p(u_k) + \sum_{l \in \mathbb{N}_{n-1} \setminus D_k} \sum_{s=0}^{Q_{k,l}} p(A^s_{n-1} A_l u_k) \Big] \Bigg\}.$$

Similarly, if (b) holds, then $\int_0^\infty e^{-\lambda t} u_k(t)\, dt = \lambda^{-k} \lambda^{\sigma-1} B_\lambda C C^{-1} u_k$ provided that $\lambda > \omega$ is suff. large and $m_{n-1} \leqslant k \leqslant m_n - 1$. The strong solution $u(t)$ of (2) has the following form:

$$u(t) = \sum_{k=m_n-1}^{m_n-1} u_k(t)$$

(464) $$+ \sum_{k=0}^{m_{n-1}-1} \left[g_{k+1}(t)u_k - \sum_{l\in \mathbb{N}_{n-1}\setminus D_k} (g_{k-\alpha_l} * H_C^{\sigma}(\cdot;0,0)A_l C^{-1}u_k)(t) \right],$$

for any $t \geqslant 0$, and the following estimate holds:

$$p(u(t)) \leqslant M'e^{\omega' t}\left\{ \sum_{k=m_{n-1}}^{m_n-1} p(C^{-1}u_k) + \sum_{k=0}^{m_{n-1}-1} \sum_{l\in \mathbb{N}_{n-1}\setminus D_k} \left[p(u_k) + p(A_l C^{-1}u_k) \right] \right\},$$

for any $p \in \circledast$ and $t \geqslant 0$.

(iv) Suppose that (a) holds with $(S_{\sigma,r}(t))_{t\geqslant 0}$ being an exponentially equicontinuous analytic $(g_\sigma, g_{\sigma r+1})$-regularized resolvent family of angle $\theta \in (0, \pi/2]$. Then the formula appearing in the brackets of the second addend on the right side of (463) represents the solution $u_k(t)$ for each $k \in \mathbb{N}^0_{m_n-1}$. Using this fact and Lemma 1.2.4-Theorem 1.2.5, it is not difficult to prove that the mapping $t \mapsto u_k(t)$, $t > 0$ can be analytically extended to the sector Σ_θ. Similarly, if (b) holds with $(T_\sigma(t))_{t\geqslant 0}$ being an exponentially equicontinuous analytic (g_σ, C)-regularized resolvent family of angle θ, then the solution $u_k(t)$ of (2) can be analytically extended to the sector Σ_θ.

(v) It is worth noting that we do not assume in the formulation of Theorem 2.10.45(a) that $r \in \mathbb{N}_0$. In the case of abstract Cauchy problem (ACP_n), we cannot use this fact for obtaining better results on the wellposedness of (2); the situation is quite different in the case of a general multi-term fractional differential equation (2), and we shall illustrate this by the following example. Consider the equation

(465) $$u'''(t) + A_3u''(t) + A_2\mathbf{D}_t^{1/2}u(t) + A_1u(t) = 0,\ t > 0,$$
$$u(0) = 0,\ u'(0) = u_1,\ u''(0) = 0.$$

Assuming that the operator $-A_3$ generates an exponentially equicontinuous r-times integrated semigroup $(S_{1,r}(t))_{t\geqslant 0}$ for some $r \in (0, 1/2]$, the abstract Cauchy problem (465) has a unique solution for any $u_1 \in D(A_1) \cap D(A_2)$. If $r = 1$, then we obtain a weaker result on the wellposedness of (465) since we must impose the condition that $u_1 \in D(A_1) \cap D(A_3A_2)$.

(vi) In what follows, we shall consider the well-posedness results for the inhomogeneous Cauchy problem:

(466) $$\mathbf{D}_t^{\alpha_n}u(t) + \sum_{i=1}^{n-1} A_i\mathbf{D}_t^{\alpha_i}u(t) = f(t),\ t > 0,$$
$$u^{(k)}(0) = u_k,\ k = 0, \cdot\cdot\cdot, m_n - 1,$$

where $f \in C([0, \infty) : X)$. Let the estimate (462) hold with $(S_{\sigma,r}(t))_{t\geqslant 0}$, and with $(S_{\sigma,r}(t))_{t\geqslant 0}$ replaced by $(T_{\sigma}(t))_{t\geqslant 0}$ therein. Suppose first that the assumptions of Theorem 2.10.45(a) hold as well as that the mapping $t \mapsto (\mu_0 - A_{n-1})^{\lceil \sigma^{-1}(\sigma r+1)\rceil} f(t)$, $t \geqslant 0$ is continuous and satisfies that, for every $p \in \circledast$, there exists $c_p > 0$ such that

$$p\Big((\mu_0 - A_{n-1})^{\lceil \sigma^{-1}(\sigma r+1)\rceil} f(t)\Big) \leqslant c_p e^{\omega t},\ t \geqslant 0. \tag{467}$$

Then (461) implies that

$$B_\lambda \tilde{f}(\lambda) = \lambda^{\sigma}(\mu_0 - A_{n-1})^{\lceil \sigma^{-1}(\sigma r+1)\rceil} \times \int_0^\infty e^{-\lambda t} H^{\sigma}(t; \lceil \sigma^{-1}(\sigma r + 1)\rceil, -\sigma)\tilde{f}(\lambda)\, dt \in LT-X.$$

Therefore, there exists a function $v_f \in C([0, \infty) : X)$ such that

$$\lambda^{-\sigma} B_\lambda \tilde{f}(\lambda) = \int_0^\infty e^{-\lambda t} v_f(t)\, dt,\ \lambda > \omega \text{ suff. large.}$$

Set $u_f(t) := (g_{\alpha_{n-1}-\sigma} * v_f)(t)$, $t \geqslant 0$. Then $\widetilde{u_f}(\lambda) = P_\lambda^{-1}\tilde{f}(\lambda)$ for $\lambda > \omega$ suff. large, and (461) yields that, for every $j \in \mathbb{N}_{n-1}$,

$$\begin{aligned}
&\lambda^{\alpha_j}(\mu_0 - A_{n-1})\, P_\lambda^{-1}\tilde{f}(\lambda)\\
&= \lambda^{\alpha_j+\sigma-\alpha_n}(\mu_0 - A_{n-1})^{1-\lceil \sigma^{-1}(\sigma r+1)\rceil} B_\lambda (\mu_0 - A_{n-1})^{\lceil \sigma^{-1}(\sigma r+1)\rceil}\tilde{f}(\lambda)\\
&= \lambda^{\alpha_j+\sigma-\alpha_n}\lambda^{\sigma r+1-\sigma-(\lceil \sigma^{-1}(\sigma r+1)\rceil-1)\sigma}\\
&\times \int_0^\infty e^{-\lambda t} H^{\sigma}(t; \lceil \sigma^{-1}(\sigma r + 1)\rceil - 1, \sigma r + 1 - \sigma - (\lceil \sigma^{-1}(\sigma r + 1)\rceil - 1)\,\sigma)\\
&(\mu_0 - A_{n-1})^{\lceil \sigma^{-1}(\sigma r+1)\rceil}\tilde{f}(\lambda)\, dt \in LT-X,
\end{aligned}$$

because $\sigma r + 1 - \sigma - \lceil \sigma^{-1}(\sigma r + 1)\rceil\sigma + 2\sigma + \alpha_j - \alpha_n \leqslant 0$. The above implies that, for every $j \in \mathbb{N}_{n-1}$, we have $\lambda^{\alpha_j} A_j P_\lambda^{-1}\tilde{f}(\lambda) \in LT-X$. By Theorem 1.2.1(vii), we may conclude that, for every $j \in \mathbb{N}_{n-1}$, the mapping $t \mapsto A_j \mathbf{D}_t^{\alpha_j} u_f(t)$, $t \geqslant 0$ is well defined, continuous and

$$\int_0^\infty e^{-\lambda t} A_j \mathbf{D}_t^{\alpha_j} u_f(t)\, dt = \lambda^{\alpha_j} A_j P_\lambda^{-1}\tilde{f}(\lambda), \tag{468}$$

for $\lambda > \omega$ suff. large. By performing the Laplace transform, we get that (466) holds with $u_k = 0$ for $0 \leqslant k \leqslant m_n - 1$. Hence, the function $u(t) := u_f(t) + \Sigma_{k=0}^{m_n-1} u_k(t)$, $t \geqslant 0$, is a strong solution of (466), with the meaning clear. Furthermore,

$$\begin{aligned}
u_f(\cdot) = {} & H^{\sigma}(\cdot; \max(\lceil \sigma^{-1}(\sigma r + 1 - \alpha_n)\rceil, -1), \alpha_{n-1})\\
& * (\mu_0 - A_{n-1})^{\max(\lceil \sigma^{-1}(\sigma r+1-\alpha_n)\rceil, -1)} f(\cdot),
\end{aligned}$$

which implies that in the estimate of growth rate of $p(u(t))$, given after the equation (463), we need to add the term

(469) $$Me^{\omega' t}\sup_{0\leqslant s\leqslant t} p\Big((\mu_0 - A_{n-1})^{\max(\lceil \sigma^{-1}(\sigma r+1-\alpha_n)\rceil,-1)} f(s)\Big),\ t \geqslant 0.$$

In such a way, we have proved an extension of [539, Theorem 2.2]. Suppose now that (b) holds as well as that the mapping $t \mapsto (\mu_0 - A_{n-1})^{\lceil \sigma^{-1}\rceil}C^{-1}f(t)$, $t \geqslant 0$ is well defined, continuous and satisfies that, for every $p \in \circledast$, there exists $c_p > 0$ such that

$$p\Big((\mu_0 - A_{n-1})^{\lceil \sigma^{-1}\rceil}C^{-1}f(t)\Big) \leqslant c_p e^{\omega t},\ t \geqslant 0.$$

Arguing in a similar manner, we obtain that there exists a unique strong solution of (466) and that in the estimate of growth rate of $p(u(t))$ we need to add the term

$$Me^{\omega' t}\sup_{0\leqslant s\leqslant t} p\Big((\mu_0 - A_{n-1})^{-1}C^{-1}f(s)\Big),\ t \geqslant 0.$$

Example 2.10.47. (i) The conditions of [148, Theorem 3.3] (cf. also [538, Theorem (*)]) are not fulfilled in the situation of [531, Example 6.2.5, Example 6.2.6]; Theorem 2.10.45 produces much better results here compared with [531, Theorem 6.3.1]. In order to illustrate this, we shall first consider the equation

(470) $$\frac{\partial^2 u(t,x)}{\partial t^2} + \left(\rho_1\frac{\partial^3}{\partial x^3} - \rho_2\frac{\partial^2}{\partial x^2}\right)\frac{\partial u(t,x)}{\partial t} + c\frac{\partial^2 u(t,x)}{\partial x^2} = 0,\ t \geqslant 0,\ x \in \mathbb{R},$$

$$u(0,x) = \varphi(x),\ u_t(0,x) = \psi(x),\ x \in \mathbb{R},$$

where $\rho_1 \in \mathbb{R}$, $\rho_2 > 0$ and $c \in \mathbb{C}$. Let $X = L^p(\mathbb{R})$ for some $p \in (1, \infty)$, and let the fractional Sobolev space $S^{\alpha,p}(\mathbb{R}^n)$ be defined in the sense of [403, Definition 12.3.1, p. 297] ($n \in \mathbb{N}$, $\alpha \in \mathbb{C}_+$); cf. [164] for an elementary introduction to fractional Sobolev spaces. By [531, Theorem 1.5.10], the operator $-(\rho_1\frac{\partial^3}{\partial x^3} - \rho_2\frac{\partial^2}{\partial x^2})$, considered with its maximal distributional domain, generates an exponentially bounded $(I-\Delta)^{-(3/2)|1/p-1/2|}$-regularized semigroup $(T_1(t))_{t\geqslant 0}$ on X. Applying Theorem 2.10.45, we obtain that there exists a unique solution of problem (470) provided that $\varphi \in S^{2+3|1/p-1/2|,p}(\mathbb{R})$ and $\psi \in S^{3+3|1/p-1/2|,p}(\mathbb{R})$; observe, however, that the existence and uniqueness of solutions of (470) have been proved in [531, Example 6.2.5] under the assumptions $\varphi \in S^{5,p}(\mathbb{R})$, $\psi \in S^{6,p}(\mathbb{R})$. Furthermore, [301, Theorem 2.18] and the analysis given in the example preceding [301, Remark 3.9] imply that the mapping $t \mapsto T_1(t) \in L(X)$, $t > 0$ is infinitely differentiable and that, for every compact set $K \subseteq (0, \infty)$, there exists $h_K > 0$ such that $\sup_{p'\in\mathbb{N}_0, t\in K}(h_K^{p'}\|\frac{d^{p'}}{dt^{p'}}T_1(t)\|/p'!^{3/2}) < \infty$, i.e., $(T_1(t))_{t\geqslant 0}$ is $\frac{3}{2}$-hypoanalytic in the sense of [301, Definition 2.14]. Now we will prove that, for every $\varphi \in S^{2+3|1/p-1/2|,p}(\mathbb{R})$ and $\psi \in S^{3+3|1/p-1/2|,p}(\mathbb{R})$, the corresponding solutions $u_0(t)$ and $u_1(t)$ of problem (2) are also $\frac{3}{2}$-hypoanalytic (with the clear meaning). Let $K \subseteq (0, \infty)$ be a compact set. By the proofs of [301, Lemma 2.15, Theorem 2.10] and the representation formula (464), it suffices to prove that, for every $x \in X$, the mapping $t \mapsto H_C^1(t; 0, 0)x$, $t > 0$ is

$\frac{3}{2}$-hypoanalytic. With the notation used so far, we have that the mapping $t \mapsto H^1_C(t; 0, 0)x$, $t > 0$ is infinitely differentiable with

$$(471)\qquad \frac{d^{p'}}{dt^{p'}} H^1_C(\cdot; 0, 0)x = \sum_{k=0}^{\infty}\sum_{j=0}^{k}\sum_{s=0}^{j} \frac{(-1)^j}{j!}\binom{k}{j}\binom{j}{s}\mu_0^s \tilde{A}_1^{\,k}\, \frac{d^{p'}-(s+k-j)}{dt^{p'}-(s+k-j)}[\cdot^j T_1(\cdot)x],$$

for any $p' \in \mathbb{N}_0$, where we have put $\frac{d^{\upsilon}}{dt^{\upsilon}}[\cdot^j T_1(\cdot)x] \equiv g_{-\upsilon} * [\cdot^j T_1(\cdot)x]$ if $-\upsilon \in \mathbb{N}$. The $\frac{3}{2}$-hypoanalyticity of the above mapping now follows from the equality (471), the estimate

$$\sup_{t\in K}\left\| \frac{d^{p'-(s+k-j)}}{dt^{p'-(s+k-j)}}\left[\cdot^j T_1(\cdot)x\right](t)\right\|$$
$$\leqslant (1 + c_K)^{k+p'}\left[p'!^{3/2}(s+k-j)!^{(-3/2)} + (k-j+s-p'+1)!^{-1}\right],$$

which holds for any $p' \in \mathbb{N}_0$ and appropriately chosen constant $c_K > 0$, and a simple computation including the $\frac{3}{2}$-hypoanalyticity of $(T_1(t))_{t>0}$. We want also to note, without carrying out a deeper and detailed analysis, that our results can be applied to the equation

$$(472)\qquad \frac{\partial^2 u(t,x)}{\partial t^2} + \left(\rho_1 \frac{\partial^3}{\partial x^3} - \rho_2 \frac{\partial^2}{\partial x^2}\right)\frac{\partial u(t,x)}{\partial t} + \left(c\frac{\partial^2}{\partial x^2} + a(x)\right)u(t,x) = 0,\ t \geqslant 0,$$
$$u(0, x) = \varphi(x),\ u_t(0, x) = \psi(x),\ x \in \mathbb{R},$$

where $a \in L^{\infty}(\mathbb{R})$; cf. [529, Example 4.2] and Example 2.10.33 for more details. Speaking-matter-of-factly, the estimates obtained in the proof of Theorem 2.10.45(b), taken together with [529, Theorem 2.7(a)] (cf. also Theorem 2.10.18(b)), show that there exists an exponentially bounded $(I - \Delta)^{-(3/2)|1/p-1/2|}$-existence family $(E(t))_{t\geqslant 0}$ for (470), in the sense of [529, Definition 2.1], and that there exist $M \geqslant 1$ and $\omega \geqslant 0$ such that $\|E(t)\| + \|E'(t)\| \leqslant Me^{\omega t}$, $t \geqslant 0$. Designate $S^{0,2}(\mathbb{R}) := L^{\infty}(\mathbb{R})$. Then Theorem 2.10.40(i) implies that there exists an exponentially bounded $(I - \Delta)^{-(3/2)|1/p-1/2|}$-existence family $(E_0(t))_{t\geqslant 0}$ for (472), provided that $a \in L^{\infty}(\mathbb{R}) \cap S^{3|1/p-1/2|,p}(\mathbb{R})$. If the function $a(x)$ satisfies the above condition, then there exists a unique solution of (472) provided $\varphi \in S^{2+3|1/p-1/2|,p}(\mathbb{R})$, $\psi \in S^{3+3|1/p-1/2|,p}(\mathbb{R})$, $a\varphi \in S^{3|1/p-1/2|,p}(\mathbb{R})$ and $a\psi \in S^{3|1/p-1/2|,p}(\mathbb{R})$. Notice that T.-J. Xiao and J. Liang have imposed in [529, Example 4.2] much stronger conditions $a \in W^{3,\infty}(\mathbb{R})$ and $\varphi \in S^{5,p}(\mathbb{R})$, $\psi \in S^{6,p}(\mathbb{R})$. Consider now the problem

$$(473)\qquad \begin{aligned} &u_{ttt}(t, x) + i\rho\Delta u_{tt}(t, x) + \sum_{|\alpha|\leqslant 2} a_\alpha D^\alpha u_t(t,x) + \sum_{|\beta|\leqslant 2} b_\beta D^\beta u(t,x) = 0,\ t \geqslant 0,\\ &u(0, x) = \varphi(x),\ u_t(0, x) = \psi(x),\ u_{tt}(0, x) = \phi(x),\ x \in \mathbb{R}^n, \end{aligned}$$

where $\rho \in \mathbb{R}\setminus\{0\}$ and $a_\alpha, b_\beta \in \mathbb{C}$ ($|\alpha|, |\beta| \leqslant 2$). Let $X = L^p(\mathbb{R}^n)$ for some $p \in (1, \infty)$. Then the operator $-i\rho\Delta$ generates an exponentially bounded $(I-\Delta)^{-n|1/p-1/2|}$-regularized semigroup on X, and

$$\lim_{\lambda\to+\infty} \lambda^{1-\kappa}\|(\lambda + ip\Delta)^{-1}\| = 0, \quad \kappa > 0,$$

because the operator Δ generates a bounded analytic semigroup of angle $\pi/2$ on X. By Theorem 2.10.45, we know that there exists a unique solution of (473) provided $\varphi, \psi, \phi \in S^{2+2n|1/p-1/2|,p}(\mathbb{R}^n)$. In [531, Example 6.2.6], the authors have considered the case $n = 3$ and $p \in (6/5, 6)$, where the assumptions $\varphi, \psi, \phi \in S^{4,p}(\mathbb{R}^3)$ have been required for the existence and uniqueness of solutions of (473); notice that our result is better since $2 + 6|1/p-1/2| < 4$ for any $p \in (6/5, 6)$.

(ii) Let X be one of the spaces $L^p(\mathbb{R}^n)$ $(1 \leqslant p \leqslant \infty)$, $C_0(\mathbb{R}^n)$, $C_b(\mathbb{R}^n)$, $BUC(\mathbb{R}^n)$, and let $0 \leqslant l \leqslant n$. Suppose $1 \leqslant \sigma < 2$, $n = 3$, $A_2 := -e^{i(2-\sigma)\frac{\pi}{2}}\Delta$, $A_1 := \Sigma_{|\beta|\leqslant 1} a_\beta D^\beta$ $(a_\beta \in \mathbb{C}, |\beta| \leqslant 1)$, $\gamma > n/2$, resp. $\gamma = n|1/p-1/2|$ if $1 < p < \infty$ and $X = L^p(\mathbb{R}^n)$. Set $C := \mathbf{T}_0\langle(1 + |x|^2)^{-\gamma}\rangle$ and consider the equation (2) with $\alpha_3 = \alpha_2 + \sigma$ and $\alpha_2 \in [\sigma, 2)$. Then we know that the operator $-A_2$ is the integral generator of a global (g_σ, C)-regularized resolvent family $(R_\sigma(t))_{t\geqslant 0}$ satisfying that there exists $M \geqslant 1$ such that

$$q_\eta(R_\sigma(t)f) \leqslant M(1 + t^{n/2})\, q_\eta(f),\ t \geqslant 0, f \in X_l,\ \eta \in \mathbb{N}_0^l, \text{ resp.,}$$

$$q_\eta(R_\sigma(t)f) \leqslant M(1 + t^{n\left|\frac{1}{p}-\frac{1}{2}\right|})\, q_\eta(f),\ t \geqslant 0, f \in X_l,\ \eta \in \mathbb{N}_0^l. \tag{474}$$

The estimate (451) is also valid since, for every $\zeta > 0$, the operator Δ generates an exponentially bounded analytic ζ-times integrated semigroup of angle $\pi/2$ on X, satisfying additionally an estimate like (474). If $1 < \sigma < 2$, resp. $\sigma = 1$, then Theorem 2.10.45(b) shows that the equation (2) has a unique strong solution provided that $u_0, u_1 \in C(D(A_1))$ and $u_2 \in C(D(A_2))$, resp. $u_0 \in C(D(A_1))$ and $u_1 \in C(D(A_2))$; if $X = L^p(\mathbb{R}^n)$ for some $p \in (1, \infty)$, and $l = 0$, this simply means that $u_0, u_1 \in S^{2n|1/p-1/2|+1,p}(\mathbb{R}^n)$ and $u_2 \in S^{2n|1/p-1/2|+2,p}(\mathbb{R}^n)$, resp., $u_0 \in S^{2n|1/p-1/2|+1,p}(\mathbb{R}^n)$ and $u_1 \in S^{2n|1/p-1/2|+2,p}(\mathbb{R}^n)$. It can be easily seen that the use of integrated operator solution families produces weaker results here; however, it should be noted that the non-existence of an appropriate reference which systematically treats the generation of $(g_\sigma, g_{\sigma r+1})$-regularized resolvent families by coercive differential operators additionally hinders possibility of proper applications of Theorem 2.10.45(a). As an illustrative example, we would like to quote the operator $e^{i(2-\sigma)\frac{\pi}{2}}\Delta$ acting on $L^1(\mathbb{R})$ with its maximal distributional domain $(1 < \sigma < 2)$; then it is not clear whether there exists a number $\zeta \in (0, 1)$ such that $e^{i(2-\sigma)\frac{\pi}{2}}\Delta$ generates an exponentially bounded $(g_\sigma, g_{1+\zeta})$-regularized resolvent family.

Before stating the following theorem, we would like to recall that the number $s_{l,k,\sigma} = \max(\lceil \frac{1}{\sigma}(\alpha_l - k + \sigma r)\rceil, 0)$ has already been defined in the proof of Theorem 2.10.45, for any $k \in \mathbb{N}^0_{m_n-1}$ and $l \in \mathbb{N}_{n-1} \setminus D_k$.

Theorem 2.10.48. *Suppose* $n \in \mathbb{N}\setminus\{1, 2\}$, $\sigma \in (0, 2]$, $r \geqslant 0$, $\alpha_n - \alpha_{n-1} = \sigma$, $M \geqslant 1$, $\omega \geqslant 0$, $D(A_{n-1}) \subseteq \bigcap_{i=0}^{n-2} D(A_i)$ *and* $(\omega^\sigma, \infty) \subseteq \rho(-A_{n-1})$. *Put* $\check{A}_i(\lambda)x := \lambda^{\alpha_i-\alpha_{n-1}}(\lambda^\sigma + A_{n-1})^{-1} A_i x$, $b_i := \max(\lceil \sigma^{-1}(\alpha_i - \alpha_{n-1} + \sigma r + 1)\rceil, 0)$ *and* $\upsilon_i := \max(\lceil \sigma^{-1}(\alpha_i - \alpha_{n-1} + 1)\rceil, 0)$ *for* $x \in D(A_{n-1})$, $\lambda > \omega$ *and* $i \in \mathbb{N}_{n-2}$. *Let* $\mu_0 < -\omega^\sigma$. *If*

(a) *The operator* $-A_{n-1}$ *is the integral generator of a* $(g_\sigma, g_{\sigma r+1})$*-regularized resolvent family* $(S_{\sigma,r}(t))_{t\geqslant 0}$ *satisfying (462) as well as*

$$p((\mu_0 - A_{n-1})^{b_i} A_i x) \leqslant M[p(x) + p(A_{n-1}x)], \tag{475}$$

for any $x \in D(A_{n-1})$, $p \in \circledast$, $i \in \mathbb{N}_{n-2}$, *and* $A_l u_k \in D(A_{n-1}^{s_{l,k,\sigma}})$, *provided* $0 \leqslant k \leqslant m_n - 1$ *and* $l \in \mathbb{N}_{n-1}\setminus D_k$,

or

(b) *The operator* $-A_{n-1}$ *is the integral generator of a* (g_σ, C)*-regularized resolvent family* $(T_\sigma(t))_{t\geqslant 0}$ *satisfying (462) with* $(S_{\sigma,r}(t))_{t\geqslant 0}$ *replaced by* $(T_\sigma(t))_{t\geqslant 0}$ *therein, as well as*

$$p((\mu_0 - A_{n-1})^{\upsilon_i} C^{-1} A_i x) \leqslant M[p(x) + p(A_{n-1}x)], \tag{476}$$

for any $x \in D(A_{n-1})$, $p \in \circledast$, $i \in \mathbb{N}_{n-2}$, *and* $A_l u_k \in R(C)$, *provided* $0 \leqslant k \leqslant m_n - 1$ *and* $l \in \mathbb{N}_{n-1}\setminus D_k$,

or

(c) *The operator* $-A_{n-1}$ *is the integral generator of a* (g_σ, C)*-regularized resolvent family* $(T_\sigma(t))_{t\geqslant 0}$ *satisfying (462) with* $(S_{\sigma,r}(t))_{t\geqslant 0}$ *replaced by* $(T_\sigma(t))_{t\geqslant 0}$ *therein, as well as (a) holds and* $A_l u_k \in R(C)$, *provided* $0 \leqslant k \leqslant m_n - 1$ *and* $l \in \mathbb{N}_{n-1}\setminus D_k$,

then the abstract Cauchy problem (2) has a unique strong solution.

Proof. We shall only consider the case in which X is a Banach space; although technically complicated, the proof of theorem in general case is quite similar and follows from the proofs of [531, Theorem 1.1.11] and Theorem 2.10.45, along with the dominated convergence theorem and the sequential completeness of X. In any of the cases (a), (b) or (c) set out above, the uniqueness of strong solutions is a simple consequence of the Ljubich theorem; because of that, we shall only prove the existence of such solutions. Suppose first that (a) holds. By the generalized resolvent equation, we easily infer that, for every $m \in \{0, 1\}$, $i \in \mathbb{N}_{n-2}$ and $x \in X$, we have $\lambda^{\alpha_i-\alpha_{n-1}} A_{n-1}^m (\lambda^\sigma + A_{n-1})^{-1} (\mu_0 - A_{n-1})^{-b_i} x \in LT - X$. Keeping in mind (475), it readily follows that there exist $M' \geqslant M$ and $\omega' \geqslant \omega$ (universal constants in the remaining part of the proof, possibly different from line to line) such that, for every $i \in \mathbb{N}_{n-2}$, $m \in \{0, 1\}$ and $x \in D(A_{n-1})$, there exists a continuous function $t \mapsto F_{m,i}(t; x)$, $t \geqslant 0$ so that $F_{1,i}(t; x) = A_{n-1} F_{0,i}(t; x)$, $t \geqslant 0$, $x \in D(A_{n-1})$,

$$\|F_{m,i}(t; x)\| \leqslant M'e^{\omega' t}\|(\mu_0 - A_{n-1})^{b_i} A_i x\| \leqslant M'e^{\omega' t} [\|x\| + \| A_{n-1}x\|],$$

provided $t \geqslant 0$, $x \in D(A_{n-1})$, and

$$\lambda^{\alpha_i-\alpha_{n-1}} A_{n-1}^m (\lambda^\sigma + A_{n-1})^{-1} A_i x = \int_0^\infty e^{-\lambda t} F_{m,i}(t; x)\, dt,\ x \in D(A_{n-1}),\ \lambda > \omega'.$$

Setting $F_{0,i}(t)\,x := F_{0,i}(t; x)$, $t \geqslant 0$, $x \in D(A_{n-1})$, it is not difficult to prove that $(F_{0,i}(t))_{t\geqslant 0} \subseteq L([D(A_{n-1})])$ is exponentially bounded, strongly continuous and

$$\sum_{i=1}^{n-2} \breve{A}_i(\lambda)x = \int_0^\infty e^{-\lambda t} \sum_{i=1}^{n-2} F_{0,i}(t)x\, dt,\ x \in D(A_{n-1}),\ \lambda > \omega'.$$

In particular, there exists $c \in (0, 1/(n-2))$ such that, for every $x \in D(A_{n-1})$, $\lambda > \omega'$ and $i \in \mathbb{N}_{n-2}$,

$$\|\breve{A}_i(\lambda)x\| + \|A_{n-1} \breve{A}_i(\lambda)x\| \leqslant c[\|x\| + \| A_{n-1} x\|]. \tag{477}$$

Combined with (477), the proof of Theorem 2.10.45 shows that, for every $x \in D(A_{n-1})$ and $\lambda > \omega'$, the series

$$\mathcal{B}_\lambda x := \sum_{k=0}^{\infty} \left[-\sum_{i=1}^{n-2} \breve{A}_i(\lambda) \right]^k x$$

is convergent in the topology of $[D(A_{n-1})]$. Taking into account the equality $\alpha_n - \alpha_{n-1} = \sigma$, it can be easily seen that the operator P_λ is injective for $\lambda > \omega'$ as well as that

$$\mathcal{B}_\lambda(\lambda^\sigma + A_{n-1})^{-1}x = \lambda^{\alpha_{n-1}} P_\lambda^{-1}x,\ x \in X,\ \lambda > \omega'. \tag{478}$$

Define now $F_0(t) := -\Sigma_{i=1}^{n-2} F_{0,i}(t)$, $t \geqslant 0$. The foregoing arguments and the proof of [531, Theorem 1.1.11] imply that

$$\mathcal{B}_\lambda x - x = \int_0^\infty e^{-\lambda t} \sum_{k=1}^{\infty} F_0^{*,k}(t)x\, dt,\quad x \in D(A_{n-1}),\ \lambda > \omega'. \tag{479}$$

Since $A_l u_k \in D(A_{n-1}^{s_{l,k,\sigma}})$ for $0 \leqslant k \leqslant m_n - 1$ and $l \in \mathbb{N}_{n-1} \setminus D_k$, it is very simple to prove with the help of (6) that there exists a continuous function $t \mapsto G(t) \in [D(A_{n-1})]$, $t \geqslant 0$ such that $\|G(t)\| + \|A_{n-1} G(t)\| \leqslant M'e^{\omega' t}$, $t \geqslant 0$ and

$$\sum_{k=0}^{m_n-1} \sum_{l\in \mathbb{N}_{n-1}\setminus D_k} \lambda^{\alpha_l-k-1} (\lambda^\sigma + A_{n-1})^{-1} A_l u_k = \int_0^\infty e^{-\lambda t} G(t)\, dt,\ \lambda > \omega'. \tag{480}$$

Define $v(t) := G(t) + (\Sigma_{k=1}^{\infty} F_0^{*,k} * G)(t)$, $t \geqslant 0$. By (478)-(480), we get that the mapping $t \mapsto v(t) \in [D(A_{n-1})]$, $t \geqslant 0$ is continuous, exponentially bounded and that

$$\tilde{v}(\lambda) = -\lambda^{\alpha_{n-1}} P_\lambda^{-1} \sum_{k=0}^{m_n-1} \sum_{l\in \mathbb{N}_{n-1}\setminus D_k} \lambda^{\alpha_l-k-1} A_l u_k,\ \lambda > \omega'. \tag{481}$$

Taken together with Theorem 1.2.1(vii), the equalities

$$A_i\int_0^\infty e^{-\lambda t}(g_{\alpha_{n-1}-\alpha_i} * v)(t)\,dt$$
$$= A_i(\mu_0 - A_{n-1})^{-1}\lambda^{\alpha_i-\alpha_{n-1}}(\mu_0 - A_{n-1})\tilde{v}(\lambda)$$
$$= A_i(\mu_0 - A_{n-1})^{-1}\mathcal{L}\big(g_{\alpha_{n-1}-\alpha_i} * [\mu_0 v(\cdot) - A_{n-1}v(\cdot)]\big)(\lambda),\ \lambda > \omega',$$

show that the mapping $t \mapsto A_i(g_{\alpha_{n-1}-\alpha_i} * v)(t)$, $t \geqslant 0$ is well defined and continuous ($i \in \mathbb{N}_{n-1}$). Keeping in mind that $\mathbf{D}_t^{\alpha_j} g_{k+1}(t)$ identically equals 0, if $m_j - 1 \geqslant k$ and $t \geqslant 0$, resp. $g_{k+1-\alpha_j}(t)$ if $m_j - 1 < k$ and $t \geqslant 0$ ($j \in \mathbb{N}_n$, $k \in \mathbb{N}^0_{m_n-1}$), it is very simple to conclude with the help of (481) that

$$\lambda^\sigma \tilde{v}(\lambda) + \mathcal{L}\left(\sum_{i=1}^{n-1} A_i(g_{\alpha_{n-1}-\alpha_i} * v)(t)\right)(\lambda) + \tilde{I}_0(\lambda) = 0,$$

where

$$I_0(t) := \sum_{l=1}^{n-1}\sum_{k=m_l}^{m_n-1} A_l\, g_{k+1-\alpha_l}(t)u_k,\ t \geqslant 0.$$

The above simply implies that there exists a continuous, exponentially bounded function $t \mapsto V(t)$, $t \geqslant 0$ such that $v \in C^{\lceil\sigma\rceil-1}([0,\infty):X)$, $v^{(k)}(0) = 0$ for $0 \leqslant k \leqslant \lceil\sigma\rceil - 1$ and $\mathbf{D}_t^\sigma v(t) = V(t)$, $t \geqslant 0$. Then the uniqueness theorem for Laplace transform, along with the equality (20), shows that

$$\mathbf{D}_t^\sigma v(t) + \sum_{i=1}^{n-1} A_i(g_{\alpha_{n-1}-\alpha_i} * v)(t) + I_0(t) = 0,\ t \geqslant 0,$$
$$v^{(k)}(0) = 0,\ k = 0, \cdots, \lceil\sigma\rceil - 1.$$

Now it is quite easy to show that the function

$$u(t) := \sum_{k=0}^{m_n-1} g_{k+1}(t)u_k + (g_{\alpha_{n-1}} * v)(t),\ t \geqslant 0,$$

is a strong solution of (2). Observing that for each $x \in D(A_{n-1})$ there exists $y \in X$ such that (cf. (476)), for every $\lambda > \omega'$ and $m \in \{0, 1\}$,

$$\lambda^{\alpha_i-\alpha_{n-1}} A^m_{n-1}(\lambda^\sigma + A_{n-1})^{-1} CC^{-1} A_i x$$
$$= \lambda^{\alpha_i-\alpha_{n-1}} A^m_{n-1}(\lambda^\sigma + A_{n-1})^{-1} C(\mu_0 - A_{n-1})^{-v_i} y,$$

and

$$\lambda^{\alpha_l-k-1} A^m_{n-1}(\lambda^\sigma + A_{n-1})^{-1} CC^{-1} A_l u_k \in LT - X,\ m \in \{0, 1\},$$

provided $0 \leqslant k \leqslant m_n - 1$ and $l \in \mathbb{N}_{n-1} \setminus D_k$, the proof of theorem in the case that (b) holds can be deduced similarly. The proof of (c) may be omitted. □

Remark 2.10.49. (i) A careful examination of the proof of Theorem 2.10.48 shows the following. In the case that (a) holds, we have the following estimate on the growth rate of constructed solution $u(t)$:

$$p\left(u(t)-\sum_{k=0}^{m_n-1} g_{k+1}(t)u_k\right)+p\left(A_{n-1}\left[u(t)-\sum_{k=0}^{m_n-1} g_{k+1}(t)u_k\right]\right)$$

$$\leqslant M'e^{\omega' t}\sum_{k=0}^{m_n-1}\sum_{l\in \mathbb{N}_{n-1}\setminus D_k}\sum_{q=0}^{s_{l,k,\sigma}} p(A_{n-1}^{q}A_l u_k),\ t\geqslant 0,\ p\in \circledast.$$

Similarly, in the case that (b) or (c) holds, we have

$$p\left(u(t)-\sum_{k=0}^{m_n-1} g_{k+1}(t)u_k\right)+p\left(A_{n-1}\left[u(t)-\sum_{k=0}^{m_n-1} g_{k+1}(t)u_k\right]\right)$$

$$\leqslant M'e^{\omega' t}\sum_{k=0}^{m_n-1}\sum_{l\in \mathbb{N}_{n-1}\setminus D_k} p(C^{-1}A_l u_k),\ t\geqslant 0,\ p\in \circledast.$$

(ii) Keeping in mind the first part of this remark as well as the estimate (475), it can be easily seen that Theorem 2.10.48(a) provides a generalization of [538, Theorem (∗)] and [531, Theorem 3.4.2], where the cases $\sigma = 1$ and $\sigma = 2$ have been considered. Although formulated with an arbitrary number $r \geqslant 0$, the choice $\sigma r \notin \mathbb{N}$ does not produce here any refinement of already known results on the wellposedness of abstract Cauchy problems [538, (1.1)] and [531, (4.1), p. 111] (cf. also Remark 2.10.46(v)). It is also worth noting that [538, Theorem (∗)] has been generalized in [538, Proposition 3.4, Theorem 3.5]; the proofs of these results rely upon a similar analysis on the Banach space $(D(A^p), \|\cdot\|_p)$, where $p \geqslant 2$ and $\|x\|_p \equiv \|x\| + \cdots + \|A^p x\|$, $x \in D(A^p)$. Without giving full details, we wish to observe that Theorem 2.10.48(b), compared with [538, Theorem 3.5], can produce a larger set of initial data for which a strong solution of problem [538, (1.1)] exists.

(iii) There exists a larger number of concrete examples where the condition (i) stated in the formulation of Theorem 2.10.45 is not fulfilled, in many of them Theorem 2.10.48(c) is applicable and produces better results than Theorem 2.10.48(a). Notice also that Theorem 2.10.45 can be applied only in the case that $\sigma \in [1, 2]$ and $\alpha_{n-1} - \alpha_{n-2} \geqslant \sigma$. As mentioned earlier, our results from Section 2.5 provide several genuine applications of Theorem 2.10.48 with $\sigma \in (0, 1)$.

Remark 2.10.50. Concerning inhomogeneous abstract multi-term Cauchy problems, Theorem 2.10.45 produces similar results as Theorem 2.10.48 and we shall explain this fact only in the case that $\sigma \in (0, 2)$ and the assumptions of Theorem 2.10.48(a) hold. Suppose that $u(t)$ is the solution of a homogeneous counterpart of (466) with the initial values u_k $(0 \leqslant k \leqslant m_n - 1)$. Let the mapping $t \mapsto (\mu_0 - A_{n-1})^{\lceil \sigma^{-1}(\sigma r+1) \rceil} f(t)$, $t \geqslant 0$ be continuous, and let the estimate (467) hold, for any $p \in ⊛$ and a corresponding $c_p > 0$. Then the generalized resolvent equation implies, along with the formulae [49, (1.26)-(1.27)] and (467), that

(482) $$\lambda^{\sigma}(\lambda^{\sigma} + A_{n-1})^{-1} \tilde{f}(\lambda) \in LT - X.$$

Designate $x_f(t) := \mathcal{L}^{-1}(\lambda^{\sigma}(\lambda^{\sigma} + A_{n-1})^{-1} \tilde{f}(\lambda))(t)$, $t \geqslant 0$ and $y_f(t) := \mathcal{L}^{-1}((\lambda^{\sigma} + A_{n-1})^{-1} \tilde{f}(\lambda))(t)$, $t \geqslant 0$. Taking into account (482), it is very simple to prove that $(\mu_0 - A_{n-1}) y_f(t) = \mu_0 y_f(t) - f(t) + x_f(t)$, $t \geqslant 0$. In the sequel, we shall employ the same notation as in the proof of Theorem 2.10.48; recall that the operator family $(Q(t) \equiv \Sigma_{k=0}^{\infty} F_0^{*,k}(t))_{t \geqslant 0} \subseteq L([D(A_{n-1})])$ is exponentially bounded. Then the mapping $t \mapsto \int_0^t Q(t - s)y_f(s)\, ds$, $t \geqslant 0$ and $t \mapsto \int_0^t (\mu_0 - A_{n-1})Q(t - s)y_f(s)\, ds$, $t \geqslant 0$ are well defined and exponentially bounded, as well as

$$(\mu_0 - A_{n-1}) \int_0^{\infty} e^{-\lambda t} Q(t)(\lambda^{\sigma} + A_{n-1})^{-1} \tilde{f}(\lambda)\, dt$$
$$= \int_0^{\infty} e^{-\lambda t}((\mu_0 - A_{n-1})Q * y_f)(t)\, dt,$$

for $\lambda > \omega$ suff. large. For $j \in \mathbb{N}_{n-1}$ fixed, we similarly obtain that

$$\lambda^{\alpha_j}(\mu_0 - A_{n-1})P_{\lambda}^{-1}\tilde{f}(\lambda) = \lambda^{\alpha_j - \alpha_{n-1}} \Big[(\lambda^{\sigma} + A_{n-1})^{-1}(\mu_0 - A_{n-1})\tilde{f}(\lambda)$$
$$+ \int_0^{\infty} e^{-\lambda t}(\mu_0 - A_{n-1})Q(t)(\lambda^{\sigma} + A_{n-1})^{-1}\tilde{f}(\lambda)\, dt\Big] \in LT - X.$$

Using the resolvent equation, (482) and the foregoing arguments, we get that $\lambda^{\alpha_j} A_j P_{\lambda}^{-1} \tilde{f}(\lambda) \in LT - X$, $j \in \mathbb{N}_{n-1}$ and that (468) holds. Since

(483) $$\lambda^{\alpha_{n-1}} A_{n-1} P_{\lambda}^{-1} \tilde{f}(\lambda) = \tilde{f}(\lambda) - \lambda^{\alpha_n} P_{\lambda}^{-1} \tilde{f}(\lambda) - \sum_{j=1}^{n-2} \lambda^{\alpha_j} A_j P_{\lambda}^{-1} P_{\lambda}^{-1} \tilde{f}(\lambda),.$$

the above yields that $\lambda^{\alpha_n} P_{\lambda}^{-1} \tilde{f}(\lambda) \in LT - X$. Hence, there exists a unique continuous, exponentially bounded function $t \mapsto w_f(t)$, $t \geqslant 0$ such that $\mathcal{L}(w_f(t))(\lambda) = \lambda^{\alpha_n} P_{\lambda}^{-1} \tilde{f}(\lambda)$ for $\lambda > \omega$ suff. large. Set $U_f(t) := (g_{\alpha_n} * w_f)(t)$, $t \geqslant 0$. Then $U_f \in C^{m_n - 1}([0, \infty) : X)$, $(U_f)^{(k)}(0) = 0$ for $0 \leqslant k \leqslant m_n - 1$ and the Caputo derivative $\mathbf{D}_t^{\zeta} U_f(t)$ is defined for any $\zeta \in [0, \alpha_n]$. Furthermore, a simple computation involving the Laplace transform shows that the function $t \mapsto u(t) + U_f(t)$, $t \geqslant 0$ is a unique solution of the problem (466). By (483), we have that

$$\tilde{U}_f(\lambda) = P_{\lambda}^{-1} \tilde{f}(\lambda)$$
$$= \lambda^{-\alpha_n} \Big[\tilde{f}(\lambda) - \sum_{j=1}^{n-2} \lambda^{\alpha_j} A_j (\mu_0 - A_{n-1})^{-1}(\mu_0 - A_{n-1})P_{\lambda}^{-1}\tilde{f}(\lambda)$$
$$+ \lambda_{\alpha_{n-1}}(\mu_0 - A_{n-1})P_{\lambda}^{-1}\tilde{f}(\lambda) - \mu_0 \lambda^{\alpha_{n-1}} P_{\lambda}^{-1} \tilde{f}(\lambda)\Big],$$

for $\lambda > \omega$ suff. large. It can be simply checked with the help of (460) and the generalized resolvent equation (6) that

$$\lambda^{-\alpha_{n-1}}(\lambda^{\sigma} + A_{n-1})^{-1} (\mu_0 - A_{n-1})^{-\max(\lceil \sigma^{-1}(\sigma r+1-\alpha_n)\rceil, -1)}$$
$$\times \mathcal{L}\left((\mu_0 - A_{n-1})^{\max(\lceil \sigma^{-1}(\sigma r+1-\alpha_n)\rceil, -1)} f\cdot\right)(\lambda) \in LT - X$$

and that the inverse Laplace transform of this function, denoted by $z(\cdot)$, satisfies that, for every $t \geqslant 0$,

$$\|z(t)\| \leqslant Me^{\omega' t} \sup_{0 \leqslant s \leqslant t} \left\|(\mu_0 - A_{n-1})^{\max(\lceil \sigma^{-1}(\sigma r+1-\alpha_n)\rceil, -1)} f(s)\right\|, \quad t \geqslant 0. \tag{484}$$

If $\lceil \sigma^{-1}(\sigma r + 1 - \alpha_n)\rceil \geqslant 0$, then we can use (484), (475) and the equality

$$\lambda^{-\sigma} (\mu_0 - A_{n-1}) P_{\lambda}^{-1} \tilde{f}(\lambda)$$
$$= \mu_0 \lambda^{-\alpha_n} (\lambda^{\sigma} + A_{n-1})^{-1} \tilde{f}(\lambda) + \lambda^{-\alpha_{n-1}}(\lambda^{\sigma} + A_{n-1})^{-1} \tilde{f}(\lambda) - \lambda^{-\alpha_n} \tilde{f}(\lambda)$$
$$+ \lambda^{-\alpha_n} \int_0^{\infty} e^{-\lambda t}(\mu_0 - A_{n-1}) Q(t) (\lambda^{\sigma} + A_{n-1})^{-1} \tilde{f}(\lambda)\, dt,$$

so as to conclude that, in the final estimate of growth rate of $p(u(t))$, we need to add the term appearing in (469). If $\lceil \sigma^{-1}(\sigma r + 1 - \alpha_n)\rceil \leqslant -1$, then the best we can do is show (a slightly weaker estimate than (469)) that, in the final estimate of growth rate of $p(u(t))$, one can add the term

$$Me^{\omega' t} \sup_{0 \leqslant s \leqslant t} p(f(s)), \; t \geqslant 0.$$

Before proceeding to the next chapter, we would like to recommend for the reader the references [240], [250], [262], [373], [379]-[380], [389], [396], [459]-[460], [464], [501], [504] and [517]-[518] for some other questions regarding various classes of abstract Volterra integro-differential equations.

3

HYPERCYCLIC AND TOPOLOGICALLY MIXING PROPERTIES OF CERTAIN CLASSES OF VOLTERRA INTEGRO-DIFFERENTIAL EQUATIONS

3.1 Hypercyclic and topologically mixing properties of abstract first order equations

Throughout this section, we assume that X is a separable infinite-dimensional Fréchet space and that E is a sequentially complete locally convex space, both over the field $\mathbb{K} \in \{\mathbb{R}, \mathbb{C}\}$. The Montel property of the space X plays a crucial role in some statements concerning S-hypercyclicity of operator semigroups and every employment of this property will be explicitly quoted. Let us recall that a normed space is Montel iff it is finite dimensional. As mentioned in the Introduction, we assume that the topology of X is induced by the fundamental system $(p_n)_{n\in\mathbb{N}}$ of increasing seminorms. Then the translation invariant metric $d: X\times X \rightarrow [0, \infty)$, defined by:

$$d(x, y) := \sum_{n=1}^{\infty} \frac{1}{2^n} \frac{p_n(x-y)}{1+p_n(x-y)},$$

for all $x, y \in E$, satisfies the following properties: $d(x+u, y+v) \leqslant d(x, y) + d(u, v)$ and $d(cx, cy) \leqslant (|c|+1)d(x, y)$, $c \in \mathbb{K}$, $x, y, u, v \in X$. We designate by S a non-empty closed subset of $\mathbb{K}$ satisfying $S \setminus \{0\} \neq \emptyset$. In order to simplify the notation, set $\inf S := \inf\{|s| : s \in S\}$ and $\sup S := \sup\{|s| : s \in S\}$. If $\alpha \in (0, \pi]$, then we define $\Delta(\alpha) := \{re^{i\theta} : r \geqslant 0, \theta \in [-\alpha, \alpha]\}$. Suppose $\Delta \in \{[0, \infty), \mathbb{R}, \mathbb{C}\}$ or $\Delta = \Delta(\alpha)$ for an appropriate $\alpha \in (0, \frac{\pi}{2}]$. Set, for every $\delta > 0$, $\Delta_\delta := \{z \in \Delta : |z| \leqslant \delta\}$. For a closed, linear operator A

acting on E, we denote by $\sigma_p(A)$, $\sigma_c(A)$ and $\sigma_r(A)$ the point, continuous and residual spectrum of A, respectively.

One of the most general theoretical concepts in the analysis of hypercyclic properties of the first order equations appears in [294] (see, e.g., [292, Subsection 3.1.4]), where the author has considered various types of S-hypercyclic C-distribution semigroups; in the next section, we shall use a similar approach in the analysis of hypercyclic properties of the second order equations. Increasingly more facts can be stated about hypercyclic properties of strongly continuous semigroups and this will be at the center of the remaining part of Section 3.1.

As is well known, an operator family $(T(t))_{t\in\Delta}$ ($T(t) \in L(X)$, $t \in \Delta$) is a strongly continuous semigroup if:

(i) $T(0) = I$,
(ii) $T(t+s) = T(t)T(s)$, $t, s \in \Delta$ and
(iii) the mapping $t \mapsto T(t)x$, $t \in \Delta$ is continuous for every fixed $x \in X$.

For the basic theory of semigroups of operators in locally convex spaces one may refer, e.g., to [25], [100], [154], [290]-[291], [333], [423], [508]-[509] and [531].

In what follows, we will single out some very special classes of strongly continuous semigroups for special attention. If $(T(t))_{t\in\Delta}$ is a strongly continuous semigroup, then it will be said that $(T(t))_{t\in\Delta}$ is:

(i) hypercyclic, if there exists $x \in X$ whose orbit $\mathrm{Orb}(x, T) := \{T(t)x : t \in \Delta\}$ is dense in X. Such an element x is called a hypercyclic vector for $(T(t))_{t\in\Delta}$; HC(T) denotes the set of all hypercyclic vectors for $(T(t))_{t\in\Delta}$
(ii) chaotic, if $(T(t))_{t\in\Delta}$ is hypercyclic and the set of periodic points of $(T(t))_{t\in\Delta}$, defined by $\{x \in X : T(t_0)x = x \text{ for some } t_0 \in \Delta\setminus\{0\}\}$, is dense in X
(iii) topologically transitive, if for every pair of open non-empty sets U, V of X, there exists $t \in \Delta$ such that $T(t)U \cap V \neq \emptyset$
(iv) topologically mixing, if for every pair of open non-empty sets U, V of X, there exists $t_0 \in \Delta$ such that $T(t)U \cap V \neq \emptyset$ for every $t \in \Delta$ with $|t| \geqslant |t_0|$
(v) weakly mixing, if the semigroup $(T \oplus T(t))_{t\in\Delta}$ is topologically transitive in $X \oplus X$, where $T \oplus T(t)(x, y) := (T(t)x, T(t)y)$, $x, y \in X$, $t \in \Delta$
(vi) supercyclic, if there exists $x \in X$ such that the projective orbit $\{cT(t)x : c \in \mathbb{K}, t \in \Delta\}$ is dense in X; SHC(T) denotes the set of all $x \in X$ whose projective orbit is dense in X
(vii) positively supercyclic, if there exists $x \in X$ such that its positive projective orbit $\{cT(t)x : c \geqslant 0, t \in \Delta\}$ is dense in X; $\mathrm{SHC}_{\mathrm{pos}}(T)$ denotes the set of all $x \in X$ whose positive projective orbit is dense in X
(viii) S-hypercyclic, if there exists $x \in X$ such that its S-projective orbit $\{cT(t)x : c \in S, t \in \Delta\}$ is dense in X; $\mathrm{HC}_S(T)$ denotes the set of all $x \in X$ whose S-projective orbit is dense in X
(ix) S-topologically transitive, if for every pair of open non-empty sets U, V of X, there exist $c \in S$ and $t \in \Delta$ such that $cT(t)U \cap V \neq \emptyset$.

Recall also that $(T(t))_{t\in\Delta}$ is:

(x) norm continuous, if the mapping $t \mapsto T(t) \in L(X)$, $t \in \Delta$ is continuous, where we assume, as in our earlier work, that $L(X)$ is endowed with the strong operator topology

(xi) analytic semigroup of angle $\alpha \in (0, \frac{\pi}{2}]$, if the mapping $t \mapsto T(t)$, $t \in (\Delta(\alpha))^{\circ}$ is analytic ($\Delta = \Delta(\alpha)$)

(xii) locally equicontinuous, if for any $r > 0$, the family $\{T(t) : t \in \Delta_r\}$ is equicontinuous.

By the (integral) generator of a strongly continuous semigroup $(T(t))_{t\in\Delta}$ in X we mean the integral generator of $(T(t))_{t\geqslant 0}$.

There exist a great number of abstract first order differential equations which do have certain hypercyclic behaviour. For example, the hypercyclic properties of quasi-linear Lasota equation, linear transport equation and Black-Sholes equation have been analyzed in [86], [173] and [174], respectively. Chaotic strongly continuous semigroups induced by semiflows in Lebesgue and Sobolev spaces have been recently considered by J. Aroza, T. Kalmes and E. Mangino in [21], while the hypercyclic properties of (multi-dimensional) Ornstein-Uhlenbeck operators have been analyzed by E. Mangino and J. A. Conejero in [117].

The principal purpose of this section is to characterize basic structural properties of S-hypercyclic semigroups whose index set is an appropriate sector of the complex plane. The concept of S-hypercyclicity of strongly continuous semigroups is meaningful and does not coincide with hypercyclicity, resp. positive supercyclicity, if sup S $< \infty$, resp. sup S $= \infty$. The important relationship between S-topological transitivity and S-hypercyclicity of a strongly continuous semigroup $(T(t))_{t\in\Delta}$ is stated in Theorem 3.1.1-Theorem 3.1.2. In Theorem 3.1.4 and Theorem 3.1.19, we shall extend a great number of the assertions proved by W. Desch, W. Schappacher and G. F. Webb in their systematic exposition [161] to S-hypercyclic semigroups in Fréchet spaces. The spectral mapping theorem for strongly continuous semigroups in locally convex spaces (cf. Theorem 3.1.5) is multi-functionally used in this section and its proof can be given by making use of the arguments given in the monographs of K. J. Engel, R. Nagel [175] and A. Pazy [450]. In order to better explain the importance of Theorem 3.1.5 in our research, we begin with recalling a profound result of J. A. Conejero, V. Müler and A. Peris [121] which states that a strongly continuous semigroup $(T(t))_{t\geqslant 0}$ in a separable Fréchet space is hypercyclic iff every single operator $T(t)$, $t > 0$ is hypercyclic (supercyclicity of single valued operators in a supercyclic strongly semigroup $(T(t))_{t\geqslant 0}$ has been analyzed by S. Shkarin in [487]). As opposed to hypercyclic semigroups, chaotic semigroups may have some other peculiar features: F. Bayart and T. Bermúdez have proved in [48] that every single operator $T(t)$, $t > 0$ of a chaotic strongly continuous semigroup $(T(t))_{t\geqslant 0}$ need not be chaotic itself. In the case that all suppositions quoted in the formulation of [161, Theorem 3.1] hold, T. Kalmes proved that $T(t)$ must be chaotic for all $t > 0$ (cf. [244, Theorem 4.9, Corollary 4.10]). We extend the above assertions to chaotic semigroups in complex separable Fréchet spaces by means of Theorem 3.1.5. Furthermore, Theorem 3.1.5 is essentially applied in proving Proposition 3.1.9 (cf. also Proposition 3.1.10) which states that the integral generator A as well as every single operator of a hypercyclic semigroup $(T(t))_{t\in\Delta}$ (Δ is either $[0, \infty)$ or $\mathbb{R}$) has the empty residual spectrum. Further on, we systematically analyze the class of weekly mixing semigroups. It turns out that K.-G. Grosse Erdmann's collapse/blow-up version of the Hypercyclicity Criterion for single operators and operator semigroups (cf. [61, Definition 2.1] and [60]) is a natural framework for investigation of weakly mixing semigroups whose index set is an appropriate sector of the complex plane. In Theorem 3.1.13, we recollect several results obtained by J. A. Conejero, A. Peris [119], L. Bernal-Gonzáles, K.-G. Grosse Erdmann [60] and T. Kalmes [244] concerning weakly mixing properties of strongly continuous semigroups whose index set $\Delta = [0, \infty)$. One of our main results is Theorem 3.1.14 which almost completely describes weakly mixing semigroups $(T(t))_{t\in\Delta}$ in the case $\Delta \neq [0, \infty)$. Concerning hypercyclicity of products of strongly continuous semigroups whose index set is $[0, \infty)$, it is worth mentioning that W. Desch and W. Schappacher [159] have introduced a strengthened version of the Hypercyclicity Criterion (cf. [159, Definition 2.1, Proposition 2.2]) called by authors the Recurrent Hypercyclicity Criterion. Although the analysis of strongly continuous semigroups $(T(t))_{t\in\Delta}$, $\Delta \neq [0, \infty)$ which satisfy the Recurrent Hypercyclicity Criterion falls out from the framework of our study, we shall present a slight modification of [161, Example 4.11] which shows that the Recurrent Hypercyclicity Criterion is strictly stronger than the Hypercyclicity Criterion. The inheritance law for the Hypercyclicity Criterion (cf. [62] and [159, Proposition 3.1]) is clarified in Proposition 3.1.17. We profile S-hypercyclic translation semigroups as well as S-hypercyclic strongly continuous semigroups induced by semiflows on various kinds of weighted function spaces. We continue the researches of M. Matsui, M. Yamada, F. Takeo [405], [407], F. Takeo [498]-[499], T. Kalmes [245, Section 4] and carry over the assertion of the positive Supercyclicity Theorem (cf. [350, Theorem 1]) proved by F. León-Saavedra, V. Müler to operator semigroups in complex Fréchet spaces. Chaoticity and mixing properties of the translation semigroup $(T(t))_{t\in\Delta}$ on the Fréchet space $C^m(\Delta, \mathbb{K})$, $m \in \mathbb{N}_0 \cup \{\infty\}$ are proved in Example 3.1.29 with the help of extension type theorems for continuously differentiable functions and the Whitney extension theorem.

We start with the following assertion whose proof is omitted.

Theorem 3.1.1. *Let $(T(t))_{t\in\Delta}$ be a strongly continuous semigroup in X. Then the following assertions are equivalent:*

(i) *$(T(t))_{t\in\Delta}$ is S-topologically transitive.*
(ii) *$(T(t))_{t\in\Delta}$ is S-hypercyclic and $HC_S(T)$ is a dense subset of X.*
(iii) *For every $y, z \in X$ and $\varepsilon > 0$, there exist $c \in S$, $t \in \Delta$ and $\upsilon \in X$ so that $d(y, \upsilon) < \varepsilon$ and $d(z, cT(t)\upsilon) < \varepsilon$.*
(iv) *For every $\varepsilon > 0$, there exists a dense subset D of X such that, for every $z \in D$, there exists a dense subset D' of X such that, for every $y \in D'$, there exist $c \in S$, $t \in \Delta$ and $\upsilon \in X$ so that $d(y, \upsilon) < \varepsilon$ and $d(z, cT(t)\upsilon) < \varepsilon$.*

If any of the conditions (i)-(iv) holds, then $HC_S(T)$ is a dense G_δ-subset of X.

The next theorem is inspired by [61, Theorem 5.1], [118, Remark 1], [405, Lemma 1, Theorem 1] and the analysis given on [119, p. 771].

Theorem 3.1.2. (i) *Suppose S is bounded, $\alpha \in (0, \frac{\pi}{2}]$, $\Delta \in \{[0, \infty), \Delta(\alpha)\}$, $(T(t))_{t\in\Delta}$ is an S-hypercyclic strongly continuous semigroup in X and $x \in HC_S(T)$. Then the set $\{cT(t)x : c \in S, t \in \Delta\setminus\Delta_s\}$ is dense in X for all $s > 0$. If $\Delta = \Delta(\alpha)$ and $z \in (\partial\Delta)\setminus\{0\}$, then $\overline{\{cT(t)x : c \in S, t \in z + \Delta\}} = X$ or $\overline{\{cT(t)x : c \in S, t \in \overline{z} + \Delta\}} = X$.*

(ii) *Suppose $\Delta \in \{\mathbb{R}, \mathbb{C}\}$. Then the S-hypercyclicity of a strongly continuous semigroup $(T(t))_{t\in\Delta}$ is equivalent to its S-topological transitivity. The previous statement remains true if $\Delta = [0, \infty)$ and S is bounded.*

(iii) *Suppose S is bounded, $\Delta = \Delta(\alpha)$ for some $\alpha \in (0, \frac{\pi}{2}]$, $(T(t))_{t\in\Delta}$ is a strongly continuous semigroup in X and the set*

$$\{cT(t)x : c \in S, t \in \Delta(\beta)\}$$

is dense in X for an appropriate $\beta \in (0, \alpha)$ and an $x \in X$. Then the semigroup $(T(t))_{t\in\Delta}$ is S-topologically transitive.

(iv) *Suppose $\Delta = [0, \infty)$ and X is not a Montel space. Then the S-hypercyclicity of a strongly continuous semigroup $(T(t))_{t\in\Delta}$ is equivalent to its S-topological transitivity.*

Proof. Let $s > 0$ be fixed. We will prove that $\overline{\{cT(t)x : c \in \mathrm{S}, t \in \Delta\setminus\Delta_s\}} = X$ only in the case $\Delta = \Delta(\alpha)$, the consideration is quite similar if $\Delta = [0, \infty)$. Put $\Delta_{s,1} := \{t \in \Delta : \operatorname{Re} t \geqslant s\}$. Then the strong continuity of $(T(t))_{t\in\Delta}$ implies:

$$\overline{\{T(t)X : t \in \Delta_{s,1}\}} \subseteq \overline{\{cT(t)x : c \in \mathrm{S}, t \in \Delta_{s,1}\}}. \tag{485}$$

Using again the strong continuity of $(T(t))_{t\in\Delta}$, we obtain that the set $\{cT(t)x : c \in \mathrm{S}, t \in (\Delta_{s,1})^c\}$ is bounded. Let p be a continuous seminorm on X, and let M be a positive real number such that $p(x) \neq 0$ and $p(cT(t)x) < M$, $c \in \mathrm{S}$, $t \in (\Delta_{s,1})^c$. Certainly, there exist a number $k > 0$ and a sequence (c_n) in S such that $p(kx) > M$, $\lim_{n\to\infty} c_nT(t_n)x = kx$ and $\lim_{n\to\infty} p(T(c_nt_n)x) = p(kx)$. Hence, there exists a subsequence (t_{n_m}) of (t_n) and a subsequence (c_{n_m}) of (c_n) satisfying $t_{n_m} \in \Delta_{s,1}$, $m \in \mathbb{N}$, $\lim_{m\to\infty} c_{n_m}T(t_{n_m})x = kx$ and $\lim_{m\to\infty} T(t_{n_m})(c_{n_m}\frac{x}{k}) = x$. In view of (485), we have that $x \in \overline{\{cT(t)x : c \in \mathrm{S}, t \in \Delta_{s,1}\}}$. Since $x \in \mathrm{HC_S}(T)$, the previous inclusion shows that any open ball contains an element of the set $\{T(t)X : t \in \Delta_{s,1}\}$. Applying (485) again, we obtain that $\overline{\{cT(t)x : c \in \mathrm{S}, t \in \Delta_{s,1}\}} = X$. Fix now a number $z \in (\partial\Delta)\setminus\{0\}$. Suppose that

$$x_1 \notin \overline{\{cT(t)x : c \in \mathrm{S}, t \in z + \Delta\}} = \overline{\{T(t)X : t \in z + \Delta\}}$$

and

$$x_2 \notin \overline{\{cT(t)x : c \in \mathrm{S}, t \in \overline{z} + \Delta\}} = \overline{\{T(t)X : t \in \overline{z} + \Delta\}}.$$

Then the first part of the proof shows that there exist a sequence (t_n) in $\Delta_{2|z|,1}$ and a sequence (c_n) in S such that $\lim_{n\to\infty} c_n T(t_n)x = x_1 + x_2$. On the other hand, there exist a sequence (t'_k) in $z + \Delta$ and a sequence (t''_k) in $\bar{z} + \Delta$ as well as two sequences (c'_k) and (c''_k) in S such that $\lim_{k\to\infty} c'_k T(t'_k)x = x_2$ and $\lim_{k\to\infty} c''_k T(t''_k)x = x_1$. Without loss of generality, we may assume that there exists a sequence (t_{n_k}) of (t_n) satisfying $t_{n_k} \in z + \Delta$, $k \in \mathbb{N}$. Hence, $x_1 = \lim_{k\to\infty} (c_{n_k} T(t_{n_k})x - c''_k T(t''_k)x) \in \overline{\{T(t)X : t \in z + \Delta\}}$, which is a contradiction. To prove (ii), suppose $(T(t))_{t\in\Delta}$ is S-hypercyclic, where either $\Delta \in \{\mathbb{R}, \mathbb{C}\}$ or $\Delta = [0, \infty)$ and S is bounded. Taking into account (i), one obtains that the range of $T(t)$, $t \in \Delta$ is dense in X. Then an application of [210, Theorem 1, Proposition 1], with $I' = \{(c, t) : c \in \mathrm{S}, t \in \Delta\}$, gives that $(T(t))_{t\in\Delta}$ is S-topologically transitive, as required. Let us prove (iii). The prescribed assumptions imply that $(T(t))_{t\in\Delta(\beta)}$ is an S-hypercyclic strongly continuous semigroup in X, which implies by (i) that the set $\{cT(t)x : c \in \mathrm{S}\setminus\{0\}, t \in \Delta(\beta), |t| \geqslant r\}$ is dense in X for every $r > 0$. Let $t \in \Delta$ and $c \in \mathrm{S}\setminus\{0\}$ be fixed. Then there exists $R > 0$ such that $\{z \in \Delta(\beta) : \operatorname{Re} z \geqslant R\} \subseteq t + \Delta$. Therefore, $\{c'T(s)x : c' \in \mathrm{S}, s \in t + \Delta\}$ is dense in X, which implies that $R(T(t))$ and $R(cT(t))$ are also dense in X. Denote $I = \{(c, t) : c \in \mathrm{S}\setminus\{0\}, t \in \Delta\}$ and put, for every $\tau = (c, t) \in I$, $T_\tau := cT(t)$. Apply [210, Theorem 1, Proposition 1] again to end the proof of (iii). In order to prove (iv), suppose that $(T(t))_{t\geqslant 0}$ is S-hypercyclic and the set $\{cT(t)x : c \in \mathrm{S}, t \geqslant 0\}$ is dense in X for some $x \neq 0$. We will slightly alter the arguments given in the proof [405, Lemma 1] in order to see that $T(t)x \neq 0$, $t \geqslant 0$ and that the set $\{cT(t)x : c \in \mathrm{S}, t \geqslant s\}$ is dense in X for every $s \geqslant 0$. Suppose $t_0 = \min\{t \geqslant 0 : T(t)x = 0\}$; obviously, $t_0 > 0$. We will prove that, for every $y \in X$, there exist $c \in \mathrm{S}$ and $t \in [0, t_0]$ such that $y = cT(t)x$. We consider only the non-trivial case $y \neq 0$. It is evident that there exist a sequence (t_n) in $[0, t_0]$ converging to some $t \in [0, t_0]$, and a sequence (c_n) in S so that $\lim_{n\to\infty} c_n T(t_n)x = y$. Assume first that $t = t_0$. Then

$$
\begin{aligned}
d(0, T(t_0 - t_n)y) &= d(0 + 0, T(t_0 - t_n)(y - c_n T(t_n)x) + T(t_0 - t_n)(c_n T(t_n)x)) \\
&\leqslant d(0, T(t_0 - t_n)(y - c_n T(t_n)x)) + d(0, T(t_0 - t_n)(c_n T(t_n)x)) \\
&= d(0, T(t_0 - t_n)(y - c_n T(t_n)x)), \qquad (486)
\end{aligned}
$$

and moreover, the strong continuity of $(T(t))_{t\geqslant 0}$ implies $\lim_{n\to\infty} d(0, T(t_0 - t_n)y) = d(0, y)$. Let us prove that $\lim_{n\to\infty} T(t_0 - t_n)(y - c_n T(t_n)x) = 0$. So, let p be an arbitrary continuous seminorm on X. Using the equicontinuity of family $\{T(t) : t \in [0, t_0]\}$, one obtains the existence of a seminorm q on X such that:

$$
p(T(t)x) \leqslant q(x),\ t \in [0, t_0],\ x \in X. \qquad (487)
$$

Due to (487), one gets $0 \leqslant p(T(t_0 - t_n)(y - c_n T(t_n)x)) \leqslant q(y - c_n T(t_n)x) \to 0$ as $n \to \infty$. This implies $\lim_{n\to\infty} T(t_0 - t_n)(y - c_n T(t_n)x) = 0$ and one can employ (486) to conclude that $d(0, y) = 0$, i.e., $y = 0$, which is a contradiction. Suppose now $t < t_0$. Then $T(t)x \neq 0$ and there exists a continuous seminorm p on X so that $p(T(t)$

$x) \neq 0$. Since $p(T(t_n)x) \to p(T(t)x) \neq 0$ and $|c_n|p(T(t_n)x) \to p(T(t)y)$ as $n \to \infty$, we have the existence of an integer $n_0 \in \mathbb{N}$ and a positive real number m satisfying $p(T(t_n)x) \geqslant m$ and $|c_n|p(T(t_n)x) \leqslant p(T(t)y) + 1$, $n \geqslant n_0$. Therefore, $|c_n| \leqslant \frac{p(T(t)y)+1}{m}$, $n \geqslant n_0$ and the closedness of S yields that there exist a subsequence (c_{n_k}) of (c_n) and a number $c \in \mathrm{S}$ satisfying $\lim_{k\to\infty} c_{n_k} = c$ and $y = cT(t)x$. Proceeding as in the proof of [405, Lemma 1], we get that $T(t)x \neq 0$, $t \geqslant 0$, as required. Assume now that the set $\{cT(t)x : c \in \mathrm{S}, t \geqslant s\}$ is not dense in X for some $s \geqslant 0$. Then there exists an open, bounded subset $U \subseteq X$ which fulfills:

$$U \cap \overline{\{cT(t)x : c \in \mathrm{S}, t \geqslant s\}} = \emptyset \text{ and } U \subseteq \overline{\{cT(t)x : c \in \mathrm{S}, t \in [0, s]\}}.$$

Let $t \in [0, s]$ be fixed. Since $T(t)x \neq 0$, there exists a continuous seminorm p_t on X such that $p_t(T(t)x) \neq 0$. Then the strong continuity of $(T(t))_{t\geqslant 0}$ implies the existence of a number $\varepsilon_t > 0$ satisfying:

$$p_t(T(t')x) \neq 0,\ t' \in (t - \varepsilon_t, t + \varepsilon_t) \cap [0, s].$$

Therefore, there exists a finite subset $\{t_1, \cdots, t_k\}$ of $[0, s]$ such that $[0, s] \subseteq \bigcup_{i=1}^{k} (t_i - \varepsilon_{t_i}, t_i + \varepsilon_{t_i})$. Choose a continuous seminorm p on X so that $p \geqslant \max(p_{t_1}, \cdots, p_{t_k})$. This implies $p(T(t)x) > 0$, $t \in [0, s]$ and the continuity of $t \mapsto p(T(t)x)$, $t \in [0, s]$ shows that there are positive real numbers m_1 and m_2 with:

$$m_1 < p(T(t)x) < m_2,\ t \in [0, s].$$

The boundedness of U gives the existence of a positive real number M so that $p(u) < M$, $u \in U$. Let $u \in U$ be an arbitrary vector; clearly, there exist a sequence (t_n) in $[0, s]$ and a sequence (c_n) in S with $|c_n|p(T(t_n)x) \to p(u)$ as $n \to \infty$. Hence, there is an $n_0 \in \mathbb{N}$ with $|c_n| \leqslant \frac{M}{m_1}$, $n \geqslant n_0$. This enables one to see that

$$U \subseteq \overline{\{cT(t)x : c \in \mathrm{S}, |c| \leqslant Mm_1^{-1}, 0 \leqslant t \leqslant s\}}.$$

So $\overline{U}$ is a compact subset of X. Since X is not a Montel space, there exists a bounded set W such that $\overline{W}$ is not compact. In the meantime, there exist $u \in U$ and $\alpha > 0$ so that $\overline{W} \subseteq \alpha(-u + \overline{U})$, which is a compact set. Hence, $\overline{W}$ is compact and this is a contradiction. We have proved that $R(cT(s))$, $c \in \mathrm{S}\backslash\{0\}$, $s \geqslant 0$ is dense in X and the proof of (iv) follows from an application [210, Theorem 1, Proposition 1]. □

Herein the following questions arise immediately:

1. Suppose S is bounded and $\inf \mathrm{S} > 0$. Does the S-hypercyclicity of a strongly continuous semigroup $(T(t))_{t\in\Delta}$ reduce to its hypercyclicity?
2. Suppose S is bounded, $\Delta = \Delta(\alpha)$ for some $\alpha \in (0, \frac{\pi}{2}]$ and $(T(t))_{t\in\Delta}$ is an S-hypercyclic strongly continuous semigroup in X. Is $(T(t))_{t\in\Delta}$ S-topologically transitive?

As we will see later, the answers are affirmative in the case of translation semigroups and strongly continuous semigroups induced by semiflows.

We continue by introducing the following subsets of X which play an important role in the analysis of hypercyclic and chaotic semigroups in Banach spaces [161]:

X_0 : is the set of all $x \in X$ so that $\lim_{t\to\infty, t\in\Delta} T(t)x = 0$ and

X_∞ : is the set of all $x \in X$ such that, for every $\varepsilon > 0$, there exist $\omega \in X$ and $t \in \Delta\setminus\{0\}$ satisfying $d(\omega, 0) < \varepsilon$ and $d(T(t)\omega, x) < \varepsilon$.

Lemma 3.1.3. *Suppose* $(T(t))_{t\in\Delta}$ *is locally equicontinuous and* $x \in X_\infty$. *Then, for every* $s > 0$ *and* $\varepsilon > 0$, *there exist* $\omega \in X$ *and* $t \in \Delta\setminus\Delta_s$ *so that* $d(\omega, 0) < \varepsilon$ *and* $d(T(t)\omega, x) < \varepsilon$.

Proof. The assertion is trivial if $x = 0$. Suppose $x \in X_\infty\setminus\{0\}$; then one gets the existence of a sequence (t_n) in $\Delta\setminus\{0\}$ and a sequence (ω_n) in X such that $d(\omega_n, 0) < \frac{1}{n}$ and $d(T(t_n)\omega_n, x) < \frac{1}{n}$. Hence, $\lim_{n\to\infty} \omega_n = 0$ and $\lim_{n\to\infty} T(t_n)\omega_n = x$. Let p be an arbitrary continuous seminorm on X. The assumption $|t_n| \leqslant s$, $n \in \mathbb{N}$ for some $s > 0$ and the local equicontinuity of $(T(t))_{t\in\Delta}$ together imply the existence of a continuous seminorm q on X which fulfills $p(T(t)x) \leqslant q(x)$, $t \in \Delta_s$, $x \in X$. In particular, $0 \leqslant p(T(t_n)\omega_n) \leqslant q(\omega_n)$, $n \in \mathbb{N}$. Letting $n \to \infty$, we infer that $p(x) = 0$ and, because p was arbitrary, $x = 0$. Therefore, given $\varepsilon > 0$ in advance, there exists a subsequence (t_{n_k}) of (t_n) satisfying $d(\omega_{n_k}, 0) < \frac{1}{n_k}$, $d(T(t_{n_k})\omega_{n_k}, x) < \frac{1}{n_k}$, $|t_{n_k}| > s$ and $n_k > \frac{1}{\varepsilon}$, $k \in \mathbb{N}$. This completes the proof. □

With the help of Theorem 3.1.1, Theorem 3.1.2, Lemma 3.1.3 and the proofs of [161, Theorems 2.2, 2.3, 2.5; Remark 2.4], we can prove the validity of the following theorem appearing in [309].

Theorem 3.1.4. (i) *Suppose that* $(T(t))_{t\in\Delta}$ *is locally equicontinuous. If* X_0 *and* X_∞ *are dense subsets of* X, *then* $(T(t))_{t\in\Delta}$ *is topologically transitive and* $X_\infty = X$.

(ii) *Suppose* $(T(t))_{t\in\mathbb{R}}$ *is a strongly continuous group in* X *and* $\{\frac{1}{y} : y \in S\setminus\{0\}\} \subseteq S$. *Then* $(T(t))_{t\geqslant 0}$ *is S-topologically transitive if* $(T(-t))_{t\geqslant 0}$ *is S-topologically transitive. If* X *is not a Montel space or* S *is bounded, then the preceding assertions are also equivalent to the existence of an element* $x \in X$ *such that both S-projective orbits* $\{cT(t)x : c \in S, t \geqslant 0\}$ *and* $\{cT(-t)x : c \in S, t \geqslant 0\}$ *are dense in* X.

We need the following useful extensions of [450, Theorems 2.4-2.6. pp. 46-48].

Theorem 3.1.5. *Suppose that a closed linear operator* A *generates a locally equicontinuous strongly continuous semigroup* $(T(t))_{t\geqslant 0}$ *in a complex sequentially complete locally convex space* E. *Then:*

(i)

$$e^{t\sigma_p(A)} \subseteq \sigma_p(T(t)) \subseteq e^{t\sigma_p(A)} \cup \{0\},\ t \geqslant 0.$$

(ii) *If* $\lambda \in \sigma_r(A)$, $t > 0$ *and* $(\lambda + \frac{2\pi i\mathbb{Z}}{t}) \cap \sigma_p(A) = \emptyset$, *then* $e^{\lambda t} \in \sigma_r(T(t))$.

(iii) *If* $\lambda \in \mathbb{C}$, $t > 0$ *and* $e^{\lambda t} \in \sigma_r(T(t))$, *then* $(\lambda + \frac{2\pi i \mathbb{Z}}{t}) \cap \sigma_p(A) = \emptyset$ *and there exists* $k \in \mathbb{Z}$ *such that* $\lambda_k := \lambda + \frac{2\pi i k}{t} \in \sigma_r(A)$.

(iv) *If* $\lambda \in \sigma_c(A)$, $t > 0$ *and* $(\lambda + \frac{2\pi i \mathbb{Z}}{t}) \cap (\sigma_p(A) \cup \sigma_r(A)) = \emptyset$, *then* $e^{\lambda t} \in \sigma_c(A)$.

Proof. We will only prove the part (i). By the foregoing, we have that A is a closed, densely defined operator which satisfies $T(t)A \subseteq AT(t)$, $t \geqslant 0$. Furthermore, the following equality can be simply justified:

$$(A - \lambda) \int_0^t e^{-\lambda s} T(s)x \, ds = e^{-\lambda t} T(t)x - x, \; x \in E, \; \lambda \in \mathbb{C}, \; t \geqslant 0. \tag{488}$$

Therefore, the assumption $Ax = \lambda x$, for some $\lambda \in \mathbb{C}$ and $x \in E \setminus \{0\}$, implies $T(t)x = e^{\lambda t} x$, $t \geqslant 0$; in other words, $e^{t\sigma_p(A)} \subseteq \sigma_p(T(t))$, $t \geqslant 0$. In order to prove the second spectral inclusion, we must adapt the arguments given in the proof of [450, Theorem 2.4, p. 46] since $\rho(A)$ can be equal to the empty set (cf. [25, p. 164] and [291]). We consider only the non-trivial case $t > 0$. Suppose $T(t)x = e^{\lambda t} x$ for some $x \in E \setminus \{0\}$ and $\lambda \in \mathbb{C}$. It is clear that there exists $x^* \in E^*$ such that $x^*(x) \neq 0$. Further on, the function $f : [0, \infty) \to \mathbb{C}$ defined by $f(s) := x^*(e^{-\lambda s} T(s)x)$, $s \geqslant 0$ is continuous and periodic with period t. Since the function $f(\cdot)$ does not vanish identically on $[0, \infty)$, we have the existence of an integer $k \in \mathbb{Z}$ such that $\frac{1}{t} \int_0^t e^{-\frac{2\pi i k s}{t}} x^*(e^{-\lambda s} T(s)x) \, ds \neq 0$. This clearly implies that

$$x_k := \frac{1}{t} \int_0^t e^{-\frac{2\pi i k s}{t}} (e^{-\lambda s} T(s)x) \, ds \neq 0.$$

Define $\Omega := \mathbb{C} \setminus \{\lambda + \frac{2\pi n i}{t} : n \in \mathbb{Z}\}$ and the function $g : \Omega \to E$ by

$$g(\eta) := (1 - e^{(\lambda - \eta)t})^{-1} \int_0^t e^{-\eta s} T(s)x \, ds, \; \eta \in \Omega.$$

As a matter of routine, one obtains $\int_0^t e^{-\eta s} T(s)x \, ds \in D(A)$ and

$$A \int_0^t e^{-\eta s} T(s)x \, ds = e^{-\eta t} T(t)x - x + \eta \int_0^t e^{-\eta s} T(s)x \, ds, \; \eta \in \mathbb{C}.$$

Therefore, $g(\eta) \in D(A)$, $\eta \in \Omega$ and

$$\begin{aligned}(\eta - A)g(\eta) &= (1 - e^{(\lambda - \eta)t})^{-1} \left[\eta \int_0^t e^{-\eta s} T(s)x \, ds - e^{-\eta t} T(t)x + x - \eta \int_0^t e^{-\eta s} T(s)x \, ds \right] \\ &= (1 - e^{(\lambda - \eta)t})^{-1} (1 - e^{(\lambda - \eta)t})x = x, \; \eta \in \Omega.\end{aligned}$$

By definition of $g(\cdot)$, we get that

$$\lim_{\eta \to \lambda + \frac{2\pi k i}{t}} \left(\eta - \lambda - \frac{2\pi k i}{t} \right) g(\eta) = x_k \tag{489}$$

and

(490) $$\lim_{\eta \to \lambda + \frac{2\pi ki}{t}} \left(\eta - \lambda - \frac{2\pi ki}{t} \right)^2 g(\eta) = 0.$$

On the other hand, $(\lambda + \frac{2\pi ki}{t} - A)(\eta - \lambda - \frac{2\pi ki}{t})g(\eta) = (\eta - \lambda - \frac{2\pi ki}{t})[(\lambda + \frac{2\pi ki}{t})g(\eta) - Ag(\eta)] = (\eta - \lambda - \frac{2\pi ki}{t})[(\lambda + \frac{2\pi ki}{t})g(\eta) - \eta g(\eta) + x] = (\eta - \lambda - \frac{2\pi ki}{t})x - (\eta - \lambda - \frac{2\pi ki}{t})^2 g(\eta)$, $\eta \in \Omega$. Hence, (490) implies:

(491) $$\lim_{\eta \to \lambda + \frac{2\pi ki}{t}} \left(\lambda + \frac{2\pi ki}{t} - A \right) \left[\left(\eta - \lambda - \frac{2\pi ki}{t} \right) g(\eta) \right] = 0.$$

The closedness of A, (489) and (491) imply $x_k \in D(A)$, $(\lambda + \frac{2\pi ki}{t} - A)x_k = 0$, $Ax_k = (\lambda + \frac{2\pi ki}{t})x_k$ and $\lambda + \frac{2\pi ki}{t} \in \sigma_p(A)$. □

Remark 3.1.6. It could be of some interest to prove generalizations of the inclusion $\sigma_p(T(t)) \subseteq e^{t\sigma_p(A)} \cup \{0\}$, $t \geqslant 0$, as well as the statements (ii)-(iv), for (a, k)-regularized C-resolvent families in SCLCSs; cf. [364, Section 3], [388, Section 5] and [302, Theorem 2.28] for further information in this direction.

With Theorem 3.1.5 in view, one can simply prove the following generalizations of [161, Theorem 3.1] and [244, Theorem 4.9, Corollary 4.10].

Theorem 3.1.7. *Suppose X is a complex space, $(T(t))_{t \geqslant 0}$ is a strongly continuous semigroup in X generated by A and U is an open, non-empty, connected subset of $\mathbb{C}$ which intersects the imaginary axis. Suppose, further, that there exists a family $\{x_\lambda : \lambda \in U\}$ which satisfies the following properties:*

(i) *$x_\lambda \in D(A) \setminus \{0\}$ and $Ax_\lambda = \lambda x_\lambda$, $\lambda \in U$.*
(ii) *For every $\phi \in X^*$, the function $F_\phi : U \to \mathbb{C}$, defined by $F_\phi(\lambda) := \phi(x_\lambda)$, $\lambda \in U$ is analytic.*
(iii) *The function $F_\phi(\cdot)$ does not vanish identically on U unless $\phi = 0$.*

Then $(T(t))_{t \geqslant 0}$ is chaotic and, for every $t_0 > 0$, the operator $T(t_0)$ is chaotic.

Proposition 3.1.8. *Suppose X is a complex space and $(T(t))_{t \geqslant 0}$ is a strongly continuous semigroup in X generated by A. If $\sigma_p(A)$ is an open, non-empty, connected subset of $\mathbb{C}$ such that there exists a family $\{x_\lambda : \lambda \in \sigma_p(A)\}$ which satisfies the properties (i)-(iii) given in the formulation of Theorem 3.1.7, then the following assertions are equivalent:*

(i) *$T(t)$ is chaotic for every $t > 0$.*
(ii) *$(T(t))_{t \geqslant 0}$ is chaotic.*
(iii) *$(T(t))_{t \geqslant 0}$ has a non-trivial periodic point.*
(iv) *$\sigma_p(A)$ intersects the imaginary axis.*

The following proposition is an extension of [121, Corollary 2.2], [161, Theorem 3.3] and [244, Corollary 4.11].

Proposition 3.1.9. *Suppose X is a complex space, $\Delta \in \{[0, \infty), \mathbb{R}\}$ and $(T(t))_{t\in\Delta}$ is a hypercyclic strongly continuous semigroup in X generated by A. Then:*

(i) $\sigma_r(A) = \emptyset$ *and* $\sigma_r(T(t)) = \emptyset$, $t \in \Delta \setminus \{0\}$.

(ii) *Suppose* $\Delta = \mathbb{R}$. *Then:*

(ii.1) $R(T(t) - \lambda I)$ *is dense in X for every* $t \in \mathbb{R}\setminus\{0\}$ *and* $\lambda \in \mathbb{C}$ *with* $|\lambda| = 1$.

(ii.2) $R(T(t) - e^{\lambda t}I)$ *is dense in X for every* $t \in \mathbb{R}\setminus\{0\}$ *and* $\lambda \in \mathbb{C}$ *with*

$$\left(\lambda + \frac{2\pi i\mathbb{Z}}{t}\right) \cap \sigma_p(A) = \emptyset.$$

(iii) (iii.1) *If* $\Delta = [0, \infty)$, $(\alpha_1, \alpha_2) \in \mathbb{C}^2\setminus\{(0, 0)\}$, $t_1 \geqslant 0$, $t_2 \geqslant 0$ *and* $0 \leqslant t_1 < t_2$, *then* $R(\alpha_1 T(t_1) + \alpha_2 T(t_2))$ *is dense in X.*

(iii.2) *If* $\Delta = \mathbb{R}$, $(\alpha_1, \alpha_2) \in \mathbb{C}^2$, $|\alpha_1| = |\alpha_2| > 0$, $t_1 \in \mathbb{R}$, $t_2 \in \mathbb{R}$ *and* $t_1 \neq t_2$, *then* $R(\alpha_1 T(t_1) + \alpha_2 T(t_2))$ *is dense in X.*

(iv) (iv.1) *Let* Δ, (α_1, α_2) *and* (t_1, t_2) *be as in the formulation of (iii.1), and let* $x \in HC(T)$. *Then* $\alpha_1 T(t_1)x + \alpha_2 T(t_2)x \in HC(T)$.

(iv.2) *Let* Δ, (α_1, α_2) *and* (t_1, t_2) *be as in the formulation of (iii.2), and let* $x \in HC(T)$. *Then* $\alpha_1 T(t_1)x + \alpha_2 T(t_2)x \in HC(T)$.

Proof. We will prove (i) by making use of two different ideas. Suppose first $\Delta = [0, \infty)$ and $\lambda \in \sigma_r(A)$. Since $R(\lambda I - A)$ is not dense in X, there exists a functional $x^* \in X^*\setminus\{0\}$ such that $x^*(Ax - \lambda x) = 0$ for all $x \in D(A)$. Due to (488), $x^*(e^{-\lambda t}T(t)x - x) = 0$, $t \geqslant 0$, $x \in D(A)$ and the denseness of A implies:

$$x^*(T(t)x - e^{\lambda t}x) = 0,\ t \geqslant 0,\ x \in X.$$

Hence, $R(T(t) - e^{\lambda t}I)$ is not dense in X and this contradicts [121, Lemma 2.1]. Using this lemma, again, we obtain $\sigma_r(A) = \emptyset$ and (i) in the case $\Delta = [0, \infty)$. Suppose now $\Delta = \mathbb{R}$ and $\lambda \in \sigma_r(A)$. Then $\pm A$ generate strongly continuous semigroups $(T(\pm t))_{t\geqslant 0}$ and, by the foregoing, we have the existence of a functional $x^* \in X^*\setminus\{0\}$ such that:

$$x^*(T(t)x - e^{\lambda t}x) = 0,\ t \in \mathbb{R},\ x \in X.$$

Suppose now $x \in \mathrm{HC}(T)$. Since $x^* \neq 0$, it must be surjective, and therefore, $\mathbb{C} = x^*(X) = \overline{\{x^*(T(t)x) : t \in \mathbb{R}\}} = \overline{\{e^{\lambda t}x^*(x) : t \in \mathbb{R}\}}$. If $x^*(x) = 0$, the contradiction is obvious; otherwise, $\overline{\{e^{\lambda t} : t \in \mathbb{R}\}} = \mathbb{C}$. This is again a contradiction since the mapping $t \mapsto e^{\lambda t}$, $t \in \mathbb{R}$ is continuous and, for every $R > 1$, $\mathrm{card}(\{e^{\lambda t} : t \in \mathbb{R}\} \cap \{z \in \mathbb{C} : |z| = R\}) \leqslant 1$. Hence, $\sigma_r(A) = \emptyset$, and moreover, $0 \notin \sigma_r(T(t))$ since $T(t)$ is bijective for all $t \in \mathbb{R}$. Now the proof of (i) finishes an application of Theorem 3.1.5(iii). To prove (ii.1), one can repeat literally the argumentation used in the proof of [121, Lemma 2.1] while (ii.2) is a simple consequence of (i) as well as Theorem 3.1.5(i). The proofs of (iii.1) and (iii.2) follow by using the afore-mentioned lemma, (ii) and decompositions

$$\alpha_1 T(t_1) + \alpha_2 T(t_2) = \alpha_2 T(t_1)\left[\frac{\alpha_1}{\alpha_2} I + T(t_2 - t_1)\right] \text{ if } \alpha_2 \neq 0 \text{ and:}$$

$$\alpha_1 T(t_1) + \alpha_2 T(t_2) = \alpha_1 T(t_1)\left[I + \frac{\alpha_2}{\alpha_1} T(t_2 - t_1)\right], \text{ if } \alpha_1 \neq 0,$$

while (iv.1) and (iv.2) follow automatically from (iii). □

Before proceeding further, let us notice that the assertions (ii.1), (iii.1)-(iii.2) and (iv.1)-(iv.2) hold in real spaces.

Proposition 3.1.10. *Suppose $\Delta = \Delta(\alpha)$ for some $\alpha \in (0, \frac{\pi}{2})$, X is a complex space and $(T(t))_{t\in\Delta}$ is a hypercyclic strongly continuous semigroup in X. Denote by A_β the generator of $(T(te^{i\beta}))_{t\geqslant 0}$, $\beta \in (-\alpha, \alpha)$ and suppose that $A_\beta = e^{i\beta}A$, $\beta \in (-\alpha, \alpha)$. Then $\sigma_r(A) \cap (-\Delta(\frac{\pi}{2} - \alpha) \cup \Delta(\frac{\pi}{2} - \alpha)) = \emptyset$.*

Proof. Suppose $\lambda \in \sigma_r(A) \cap (-\Delta(\frac{\pi}{2} - \alpha) \cup \Delta(\frac{\pi}{2} - \alpha))$. Since $A_\beta = e^{i\beta}A$, $\beta \in (-\alpha, \alpha)$, one obtains the existence of a functional $x^* \in X^*\setminus\{0\}$ such that $x^*(A_\beta x - \lambda e^{i\beta}x) = 0$ for all $x \in D(A)$ and $\beta \in (-\alpha, \alpha)$. Arguing as in the proof of Proposition 3.1.9, we have that $x^*(T(te^{i\beta})x - e^{\lambda e^{i\beta}t}x) = 0$, $t \geqslant 0$, $x \in X$, $\beta \in (-\alpha, \alpha)$. The strong continuity of $(T(t))_{t\in\Delta}$ implies:

(492) $$x^*(T(z)x - e^{\lambda z}x) = 0,\ z \in \Delta,\ x \in X.$$

Suppose $x \in HC(T)$. Taken together, the surjectivity of x^*, (492) and the proof of Proposition 3.1.9, give $\overline{\{e^{\lambda z}x^*(x) : z \in \Delta\}} = \mathbb{C}$. Exclusion of the trivial case $x^*(x) = 0$ yields $\overline{\{e^{\lambda z} : z \in \Delta\}} = \mathbb{C}$. This equality and the assumption $\lambda \in -\Delta(\frac{\pi}{2} - \alpha) \cup \Delta(\frac{\pi}{2} - \alpha)$ imply $\overline{\{e^z : z \in \mathbb{C}, \operatorname{Re} z \geqslant 0\}} = \mathbb{C}$ or $\overline{\{e^z : z \in \mathbb{C}, \operatorname{Re} z \leqslant 0\}} = \mathbb{C}$. This is a contradiction. □

Notice that the equality $A_\beta = e^{i\beta}A$, $\beta \in (-\alpha, \alpha)$ appearing in the formulation of the preceding proposition holds if $(T(t))_{t\in\Delta}$ is an analytic semigroup of angle α.

Now we focus our attention towards the analysis of weakly mixing semigroups. The most common tool for proving hypercyclicity of single operators is the well-known Hypercyclicity Criterion which was discovered independently by C. Kitai [277, Theorem 1.4] and R. M. Gethner-J. H. Shapiro [192, Theorem 2.2]. It turned out that this criterion is equivalent to the corresponding ones given by J. Bès, A. Peris (cf. [62] and [454, Theorem 1.1]), L. Bernal-Gonzáles, K.-G. Grosse Erdmann [60], [210] and F. León-Saavedra [349]. We also refer the reader to [55], [96], [115], [124], [197] and [207]. Such criteria possess natural reformulations in the theory of operator semigroups (cf. [47], [54], [61], [115], [119], [150], [171] and [244]). Motivated by K.-G. Grosse Erdmann's collapse/blow-up definition of hypercyclicity for single operators and operator semigroups, we introduce the Hypercyclicity Criterion for strongly continuous semigroups whose index set is, in general, an appropriate sector of the complex plane.

Definition 3.1.11. Suppose $(T(t))_{t\in\Delta}$ is a strongly continuous semigroup in X. It is said that $(T(t))_{t\in\Delta}$ satisfies the Hypercyclicity Criterion if there exist dense subsets Y, Z of X and a sequence (t_n) in Δ such that:

(○) $\lim_{n\to\infty} T(t_n)y = 0$, $y \in Y$ and

(○○) for every $z \in Z$, there exists a sequence (u_n) in X such that $\lim_{n\to\infty} u_n = 0$ and $\lim_{n\to\infty} T(t_n)u_n = z$.

The proof of following auxiliary lemma is omitted.

Lemma 3.1.12. *Suppose that a strongly continuous semigroup* $(T(t))_{t\in\Delta}$ *satisfies the Hypercyclicity Criterion. Then:*

(i) $(\underbrace{T \oplus \cdots \oplus T}_{k})(t))_{t\in\Delta}$ *satisfies the Hypercyclicity Criterion for all* $k \in \mathbb{N}$.

(ii) $(\underbrace{T \oplus \cdots \oplus T}_{k})(t))_{t\in\Delta}$ *is topologically transitive for all* $k \in \mathbb{N}$.

The next theorem is a recollection of results obtained by J. A. Conejero, A. Peris [119], L. Bernal-Gonzáles, K.-G. Grosse Erdmann [60] and T. Kalmes [244]. Let us recall [119] that a backwards orbit of x under $(T(t))_{t\geqslant 0}$ is a family $\{x_t : t \geqslant 0\}$ of elements of X satisfying $x_0 = x$ and $T(t)x_s = x_{s-t}$ for all $s \geqslant t \geqslant 0$, and that a sequence (T_n) in $L(X)$ is called hypercyclic if there exists $x \in X$ so that its orbit under (T_n), defined by $\{T_n x : n \in \mathbb{N}_0\}$, is dense in X; (T_n) is said to be hereditarily hypercyclic [62] if every subsequence of (T_n) is hypercyclic.

Theorem 3.1.13. ([60], [119], [244]) *Suppose* $(T(t))_{t\geqslant 0}$ *is a strongly continuous semigroup in X. Then the following assertions are equivalent:*

(i) *There exist dense subspaces Y, Z of X, a strictly increasing sequence* (t_n) *in* $(0, \infty)$ *with* $\lim_{n\to\infty} t_n = \infty$ *and a family* $\{S(t) : Z \to X \mid t \geqslant 0\}$ *of linear (not necessarily continuous) mappings satisfying:*
 (i.1) $\lim_{n\to\infty} T(t_n)y = 0$, $y \in Y$ *and*
 (i.2) $\lim_{n\to\infty} S(t_n)z = 0$, $z \in Z$ *and* $T(t)S(t)z = z$, $t \geqslant 0$, $z \in Z$.

(ii) *There exist dense subspaces Y, Z of X, a strictly increasing sequence* (t_n) *in* $(0, \infty)$ *with* $\lim_{n\to\infty} t_n = \infty$ *and a family* $\{S(t) : Z \to X \mid t \geqslant 0\}$ *of linear (not necessarily continuous) mappings satisfying:*
 (ii.1) $\lim_{n\to\infty} T(t_n)y = 0$, $y \in Y$ *and*
 (ii.2) *every* $z \in Z$ *admits a backwards orbit* $\{z_t : t \geqslant 0\}$ *such that* $\lim_{n\to\infty} z_{t_n} = 0$.

(iii) $(T(t))_{t\geqslant 0}$ *satisfies the Hypercyclicity Criterion with a strictly increasing sequence* (t_n) *in* $(0, \infty)$ *which fulfills* $\lim_{n\to\infty} t_n = \infty$.

(iv) *For every pair of open non-empty subsets U, V of X and for every zero neighborhood W in X there exists* $t > 0$ *so that* $T(t)U \cap W \neq \emptyset$ *and* $T(t)W \cap V \neq \emptyset$.

(v) *For every pair of open non-empty subsets U, V of X, there exists* $t > 0$ *such that* $T(t)U \cap V \neq \emptyset$ *and* $T(t+1)U \cap V \neq \emptyset$.

(vi) *There exists* $\alpha > 0$ *such that, for every pair of open non-empty subsets U, V of X, there exists* $t > 0$ *such that* $T(t)U \cap V \neq \emptyset$ *and* $T(t+\alpha)U \cap V \neq \emptyset$.

(vii) *If* $I \subseteq [0, \infty)$ *is syndetic, i.e., there exists* $K > 0$ *such that* $[t, t+K] \cap I \neq \emptyset$ *for all* $t \geqslant 0$, *then the family* $\{T \oplus T(t) : t \in I\}$ *is topologically transitive.*

(viii) *There exists $K > 0$ such that, for every $I \subseteq [0, \infty)$ satisfying $[t, t+K] \cap I \neq \emptyset$ for all $t \geqslant 0$, the family $\{T \oplus T(t) : t \in I\}$ is topologically transitive.*
(ix) *$(T(t_n))$ is a hypercyclic sequence of operators for any strictly increasing sequence (t_n) in $(0, \infty)$ satisfying $\lim_{n\to\infty} t_n = \infty$ and $\sup_{n\in\mathbb{N}}(t_{n+1} - t_n) < \infty$.*
(x) *For every open, non-empty subsets U, V_1, V_2 of X, there exists $t \in \Delta$ such that $T(t)U \cap V_1 \neq \emptyset$ and $T(t)U \cap V_2 \neq \emptyset$.*
(xi) *$(T(t))_{t\geqslant 0}$ has a hereditarily hypercyclic subsequence $(T(t_n))$.*
(xii) *$(T(t))_{t\geqslant 0}$ is weakly mixing.*
(xiii) *$(T(t))_{t\geqslant 0}$ satisfies the Hypercyclicity Criterion.*

Proof. By [119, Theorem 2.1], (i) $\Leftrightarrow$ (ii) $\Leftrightarrow$ (iii) $\Leftrightarrow$ (xii), and by [61, Theorem 2.3-Theorem 2.5], (iii) $\Leftrightarrow$ (iv) $\Leftrightarrow$ (ix) $\Leftrightarrow$ (xi) $\Leftrightarrow$ (xii). The equivalence (x) $\Leftrightarrow$ (xii) follows as in the proof of [244, Theorem 2.13]. Hence, (i) $\Leftrightarrow$ (ii) $\Leftrightarrow$ (iii) $\Leftrightarrow$ (iv) $\Leftrightarrow$ (ix) $\Leftrightarrow$ (x) $\Leftrightarrow$ (xi) $\Leftrightarrow$ (xii). Arguing as in the proof of [244, Theorem 2.5], one obtains: (iv) $\Rightarrow$ (v) $\Rightarrow$ (vi) and (xii) $\Rightarrow$ (vii) $\Rightarrow$ (viii) $\Rightarrow$ (vi). The proof of implication (vi) $\Rightarrow$ (xii) follows by making use of the argumentation given in the proofs of [207, Theorem 3.2] and [244, Theorem 2.5]; we will prove this implication for the sake of reader's convenience. It is evident that (vi) implies that $(T(t))_{t\geqslant 0}$ is topologically transitive. Hence, Theorem 3.1.1 shows that the set HC(T) is a dense G_δ-subset of X. Suppose now U_i, V_i, $i = 1, 2$ are open non-empty subsets of X and $\upsilon_1 \in \mathrm{HC}(T) \cap V_1$. So, there exists $r_1 > 0$ such that $u_1 := T(r_1)\upsilon_1 \in U_1$. Since $R(T(t_1))$ is dense in X (see Theorem 3.1.2(i); Proposition 3.1.9 and [121, Lemma 2.1]), there is an $\omega_2 \in E$ such that $u_2 := T(r_1)\omega_2 \in U_2$. Further on, let us assume $L(u_2, \delta) \subseteq U_2$, $L(\upsilon_2, \delta) \subseteq V_2$ and

$$T(r_1)(L(0, \delta') + L(0, \delta')) \subseteq L(0, \delta) \text{ for a suitable } \delta' \in (0, \delta). \tag{493}$$

An application of Proposition 3.1.9 gives that $(T(\alpha) - I)\upsilon_1 \in \mathrm{HC}(T)$, and consequently, one can select two positive real numbers p_1 and q_1 satisfying:

$$\begin{aligned} &d(T(q_1)(T(\alpha) - I)\upsilon_1 - (\omega_2 - \upsilon_2), 0) < \delta' \text{ and} \\ &d(T(p_1)\upsilon_1 - (\upsilon_2 - T(q_1)\upsilon_1), 0) < \delta'. \end{aligned} \tag{494}$$

Set $y_2 := T(p_1)\upsilon_1 + T(q_1)\upsilon_1$ and $z_2 := T(p_1)u_1 + T(q_1 + \alpha)u_1$. Clearly, $z_2 = T(p_1 + r_1)\upsilon_1 + T(q_1 + \alpha + r_1)\upsilon_1$ and (494) implies that $y_2 \in V_2$. To prove that $z_2 \in U_2$, observe that

$$z_2 - u_2 = T(r_1)[(T(q_1)(T(\alpha) - I)\upsilon_1 - (\omega_2 - \upsilon_2)) + (T(p_1)\upsilon_1 - (\upsilon_2 - T(q_1)\upsilon_1))]$$

and that (493) implies

$$z_2 - u_2 \in T(r_1)(L(0, \delta') + L(0, \delta')) \subseteq L(0, \delta).$$

One can employ (vi) with $\tilde{U}_k = L(u_1, 2^{-k})$ and $\tilde{V}_k = L(\upsilon_1, 2^{-k})$ in order to obtain the existence of sequences (u_k) and $(\tilde{u}_k)$ with the limit u_1 and a sequence (t_k) in $(0, \infty)$ such that $(T(t_k)u_k)$ and $(T(t_k + \alpha)\tilde{u}_k)$ converge to υ_1. Since

$$\lim_{k\to\infty} T(t_k)\,(T(p_1)u_k + T(q_1 + \alpha)\tilde{u}_k) = T(p_1)\upsilon_1 + T(q_1)\upsilon_1 = y_2 \in V_2$$

and

$$\lim_{k\to\infty} (T(p_1)u_k + T(q_1 + \alpha)\tilde{u}_k) = T(p_1)u_1 + T(q_1 + \alpha)u_1 \in U_2,$$

one concludes that $T(t_k)U_i \cap V_i \neq \emptyset$, $i = 1, 2$ and that (xii) holds. The implication (iii) $\Rightarrow$ (xiii) is trivial and the implication (xiii) $\Rightarrow$ (xii) follows from an application of Lemma 3.1.12. This ends the proof of Theorem 3.1.13. □

The situation is more complicated if $\Delta \neq [0, \infty)$ and $(T(t))_{t\in\Delta}$ is a strongly continuous semigroup in X. An insignificant modification of the notion is made to cover a newly arisen situation: it is said that a subset I of Δ is syndetic if there exist a number $K > 0$ and a ray $R \subseteq \Delta$ starting at 0 so that:

(495) for every $t \in \Delta$ and $z \in R$ with $|z| \geqslant K : [t, t + z] \cap I \neq \emptyset$.

Theorem 3.1.14. *Suppose $(T(t))_{t\in\Delta}$ is a strongly continuous semigroup in X. Consider the following assertions:*

(i) *$(T(t))_{t\in\Delta}$ satisfies the Hypercyclicity Criterion.*
(ii) *For every pair of open non-empty sets U, $V \subseteq X$ and for every zero neighborhood W in X, there exists $t \in \Delta\backslash\{0\}$ so that $T(t)U \cap W \neq \emptyset$ and $T(t)W \cap V \neq \emptyset$.*
(iii) *For every $s \in \Delta\backslash\{0\}$ and for every pair of open non-empty sets U, $V \subseteq X$, there exists $t \in \Delta\backslash\{0\}$ such that $T(t)U \cap V \neq \emptyset$ and $T(t + s)U \cap V \neq \emptyset$.*
(iv) *There exist $s \in \Delta\backslash\{0\}$ such that for every pair of open non-empty sets U, $V \subseteq X$, there exists $t \in \Delta\backslash\{0\}$ such that $T(t)U \cap V \neq \emptyset$ and $T(t + s)U \cap V \neq \emptyset$.*
(v) *The family $\{T \oplus T(t) : t \in I\}$ is topologically transitive for every syndetic subset I of Δ.*
(vi) *There exist a number $K > 0$ and a ray $R \subseteq \Delta$ starting at 0 so that for every $I \subseteq \Delta$ satisfying (495), the family $\{T \oplus T(t) : t \in I\}$ is topologically transitive.*
(vii) *For every open, non-empty sets U, V_1, $V_2 \subseteq X$, there exists $t \in \Delta\backslash\{0\}$ such that $T(t)U \cap V_1 \neq \emptyset$ and $T(t)U \cap V_2 \neq \emptyset$.*
(viii) *$(T(t))_{t\in\Delta}$ has a hereditarily hypercyclic subsequence $(T(t_n))$.*
(ix) *$(T(t))_{t\in\Delta}$ is weakly mixing.*

Then we have:

(a) *In the case $\Delta = \mathbb{R}$, (i) $\Leftrightarrow$ (ii) $\Leftrightarrow$ (iii) $\Leftrightarrow$ (iv) $\Leftrightarrow$ (v) $\Leftrightarrow$ (vi) $\Leftrightarrow$ (vii) $\Leftrightarrow$ (viii) $\Leftrightarrow$ (ix).*
(b) *Suppose $\Delta = \mathbb{C}$. Then the following holds: (i) $\Leftrightarrow$ (ii) $\Leftrightarrow$ (v) $\Leftrightarrow$ (vi) $\Leftrightarrow$ (vii) $\Leftrightarrow$ (viii) $\Leftrightarrow$ (ix) and (ii) $\Rightarrow$ (iii) $\Rightarrow$ (iv). Suppose, in addition, that there exists $\alpha \in \Delta\backslash\{0\}$ such that $R(T(\alpha) - I)$ is dense in X. Then all assertions (i)-(ix) are mutually equivalent.*
(c) *Suppose $\Delta = \Delta(\alpha)$, for some $\alpha \in (0, \frac{\pi}{2}]$. Then (i) $\Leftrightarrow$ (ii) $\Leftrightarrow$ (v) $\Leftrightarrow$ (vi) $\Leftrightarrow$ (vii) $\Leftrightarrow$ (viii) $\Leftrightarrow$ (ix).*
   (c.1) *If, additionally, $R(T(t))$ is dense in X for every $t \in \Delta$, then (ii) $\Rightarrow$ (iii).*
   (c.2) *We have (ii) $\Rightarrow$ (iv).*

(c.3) *(iv) ⇒ (ix) under the additional assumption* $\overline{R(T(s)-I)} = X$.

(c.4) *Suppose* $\overline{R(T(t))} = X$, $t \in \Delta$ *and* $\overline{R(T(s)-I)} = X$ *for some* $s \in \Delta$. *Then we have the equivalence of all assertions (i)-(ix).*

(d) *Suppose that* $(T(t))_{t\in\Delta}$ *is topologically transitive and there exists a dense set of points* $x \in X$ *with bounded orbits* $\{T(t)x : t \in \Delta\}$. *Then (ii) holds so that* $(T(t))_{t\in\Delta}$ *is weakly mixing.*

(e) *Suppose* $\Delta \in \{[0,\infty), \mathbb{R}\}$ *and* $(T(t))_{t\in\Delta}$ *is chaotic. Then* $(T(t))_{t\in\Delta}$ *is weakly mixing.*

Proof. When $\Delta = \mathbb{R}$, we may take advantage of the assertions (ii.1) and (iv.2) of Proposition 3.1.9, Lemma 3.1.12, the local equicontinuity of $(T(t))_{t\in\mathbb{R}}$, as well as of the proofs of [244, Theorem 2.5] and Theorem 3.1.13 in order to see that (i) ⇒ (ii) ⇔ (iii) ⇔ (iv) ⇔ (v) ⇔ (vii) ⇔ (ix). By Lemma 3.1.12, we have that

$$\Big((\underbrace{T \oplus \cdots \oplus T}_{k})(t)\Big)_{t\in\mathbb{R}}$$

is topologically transitive for all $k \in \mathbb{N}$; by the proof of [244, Theorem 2.5], we have that (i) ⇒ (v) ⇒ (vi) ⇒ (iv). Note that once we prove (viii) ⇒ (ix) ⇒ (i) ⇒ (viii), the assertion (a) follows immediately. In order to see that (viii) implies (ix), we will slightly modify the proof of [62, Theorem 2.3]. Suppose that U_i, V_i, $i = 1, 2$ are open non-empty subsets of X and that $\overline{\{T(t_n)x : n \in \mathbb{N}\}} = X$; this implies that, for every $s \in \mathbb{R}$, $\overline{\{T(t_n)T(s)x : n \in \mathbb{N}\}} = X$. Hence, there exist $u_1 \in U_1$ and $n_1 \in \mathbb{N}$ such that $\overline{\{T(t_n)u_1 : n \in \mathbb{N}\}} = X$ and $T(t_{n_1})U_1 \cap V_1 \neq \emptyset$. Using the fact that $(T(t_n))$ is hereditarily hypercyclic, one obtains inductively the existence of a strictly increasing sequence (n_k) in $\mathbb{N}$ satisfying $T(t_{n_k})U_1 \cap V_1 \neq \emptyset$, $k \in \mathbb{N}$. Now the hypercyclicity of $(T(t_{n_k}))$ gives the existence of a number $k_0 \in \mathbb{N}$ such that $T(t_{n_{k_0}})U_2 \cap V_2 \neq \emptyset$ and $T(t_{n_{k_0}})U_1 \cap V_1 \neq \emptyset$. Therefore, $(T(t))_{t\in\mathbb{R}}$ is weakly mixing. The proof of (ix) ⇒ (i) is essentially contained in that of [62, Lemma 2.5]. Let $(x, y) \in HC(T \oplus T)$ and $s \in \mathbb{R}$. Then $x \in HC(T)$ and $y \in HC(T)$. Since $T(s)$ is bijective and $R(T(s)) = X$, the above implies $(x, T(s)y) \in HC(T \oplus T)$. As a consequence, we have that, for every open non-empty subset U of X, there exists $u \in U$ such that $(x, u) \in HC(T \oplus T)$. Put now $Y = Z = Orb(x, T)$ and $U_k = L(0, \frac{1}{k})$, $k \in \mathbb{N}$. Then an induction argument shows that there exist a sequence (u_k) in X and a sequence (t_k) in $\mathbb{R}$ so that:

$$u_k \in U_k,\ T(t_k)x \in U_k \text{ and } T(t_k)u_k \in x + U_k,\ k \in \mathbb{N}. \tag{496}$$

It is evident that (496) implies $\lim_{k\to\infty} T(t_k)x = 0$ and $\lim_{k\to\infty} T(t_k)y = 0$, $y \in Y$. If $z = T(t)x \in Z$ for some $t \in \Delta$, put $\tilde{u}_k = T(t)u_k$, $k \in \mathbb{N}$. Clearly, $\lim_{k\to\infty} \tilde{u}_k = 0$, $\lim_{k\to\infty} T(t_k)\tilde{u}_k = z$ and this yields (i). Suppose $(T(t))_{t\in\Delta}$ fulfills the Hypercyclicity Criterion with Y, Z and (t_n). Then one can deduce, as in the proofs of Theorem 3.1.1 and Lemma 3.1.12, that $(T(t_n))$ is hereditarily hypercyclic; consequently, (a) follows provided that $\Delta = \mathbb{R}$. Suppose $\Delta = \mathbb{C}$. Then the equivalence of (i), (vii), (viii) and (ix) can be proved as before. Let us prove that (ix) implies (ii). To see this, let us suppose that

U, V are open non-empty subsets of X, W is a zero neighborhood in X and $L(0, \varepsilon) \subseteq W$ for some $\varepsilon > 0$. Then the topological transitivity of $(T \oplus T(t))_{t\in\Delta}$ enables one to show that there exists a number $t \in \Delta$ satisfying $T(t)U \cap L(0, \varepsilon) \neq \emptyset$ and $T(t)L(0, \varepsilon) \cap V \neq \emptyset$. This clearly shows that $T(t)U \cap W \neq \emptyset$, $T(t)W \cap V \neq \emptyset$ and that (ii) holds. The implication (ii) $\Rightarrow$ (i) follows similarly as in the proofs of [60, Theorems 3.3-3.4]. First of all, observe that (ii) implies the topological transitivity of $(T(t))_{t\in\Delta}$ (cf. also [197, Corollary 1.3]). Due to Theorem 3.1.1, $(T(t))_{t\in\Delta}$ has a dense G_δ-set of hypercyclic vectors. Let (U_k) be a base of open zero neighborhoods. Designate by P the set of all $x \in X$ such that there exist a sequence (u_n) in X and a sequence (t_n) in Δ satisfying $\lim_{n\to\infty} u_n = 0$, $\lim_{n\to\infty} T(t_n)u_n = x$ and $\lim_{n\to\infty} T(t_n)x = 0$. Proceeding as in the proof of [60, Theorem 3.4], one obtains that

$$P = \bigcap_{k\in\mathbb{N}} \bigcup_{t\in\Delta} \left[T(t)^{-1}(U_k) \cap \{x \in X : T(t)U_k \cap (x + U_k) \neq \emptyset\}\right]$$

and that P is a dense G_δ-set of X. The intersection of two dense G_δ-sets in X is non-empty so that there exist $x \in \mathrm{HC}(T)$, a sequence (u_n) in X and a sequence (t_n) in Δ satisfying $\lim_{n\to\infty} u_n = 0$, $\lim_{n\to\infty} T(t_n)u_n = x$ and $\lim_{n\to\infty} T(t_n)x = 0$. Put now $Y = Z = \mathrm{Orb}(x, T)$. Then one can simply verify that the Hypercyclicity Criterion holds with Y, Z and a sequence (t_n). This proves (i), and consequently (i) $\Leftrightarrow$ (ii) $\Leftrightarrow$ (vii) $\Leftrightarrow$ (viii) $\Leftrightarrow$ (ix). In order to prove the implication (i) $\Rightarrow$ (v), let us suppose that $I \subseteq \Delta$ is syndetic and that U_i, V_i, $i = 1, 2$ are open non-empty subsets of X. The local equicontinuity of $(T(t))_{t\in R}$ implies the existence of a number $\delta > 0$ and open non-empty subsets $\tilde{U}_i$, $i = 1, 2$ such that, for every $z \in R$ with $|z| \leqslant \delta$, $T(z)\tilde{U}_i \subseteq U_i$, $i = 1, 2$. Let $m \in \mathbb{N}$ and $z \in R$ satisfy $m\delta > K$ and $|z| = \delta$. Taken together, Lemma 3.1.12 and the proof of [244, Theorem 2.5], show that there exists $t \in \Delta$ such that:

$$T(t + jz)\tilde{U}_i \cap V_i \neq \emptyset,\ i = 1, 2,\ 0 \leqslant j \leqslant m. \tag{497}$$

Since I is syndetic, one gets $[t, t + mz] \cap I \neq \emptyset$. Hence, there exist $j \in \mathbb{N}_m$ and $s \in \Delta$ such that $s \in [t + (j - 1)z, t + jz] \cap I$. Combined with (497), this inclusion implies $t + jz - s \in R$, $|t + jz - s| \leqslant \delta$ and

$$\emptyset \neq T(t + jz)\tilde{U}_i \cap V_i = T(s)T(t + jz - s)\tilde{U}_i \cap V_i \subseteq T(s)U_i \cap V_i,\ i = 1, 2.$$

Since $s \in I$, the proof of (v) is completed. The implications (ii) $\Rightarrow$ (iii) $\Rightarrow$ (iv) follow as in the proof of [244, Theorem 2.5], and moreover, the implications (v) $\Rightarrow$ (vi) $\Rightarrow$ (ix) are trivial. Therefore, (i) $\Leftrightarrow$ (ii) $\Leftrightarrow$ (v) $\Leftrightarrow$ (vi) $\Leftrightarrow$ (vii) $\Leftrightarrow$ (viii) $\Leftrightarrow$ (ix). Notice also that the denseness of $R(T(\alpha) - I)$ and the arguments given in the proof of Theorem 3.1.13 imply the validity of (iv) $\Rightarrow$ (ix). This completes the proof of (b). Let us examine the case $\Delta = \Delta(\alpha)$, for some $\alpha \in (0, \frac{\pi}{2}]$. On the basis of proofs of (a) and (b), one gets (vii) $\Leftrightarrow$ (ix) $\Rightarrow$ (ii) $\Rightarrow$ (i) $\Rightarrow$ (ix), (i) $\Rightarrow$ (viii) and (i) $\Rightarrow$ (v) $\Rightarrow$ (vi) $\Rightarrow$ (ix). Suppose that (viii) holds. An application of [60, Theorem 3.3] shows that $(T(t_n))$ satisfies the Hypercyclicity Criterion (cf. [60, Definition 1.1]) so that (i) simply follows. Hence, (i) $\Leftrightarrow$ (ii) $\Leftrightarrow$ (v) $\Leftrightarrow$ (vi) $\Leftrightarrow$ (vii) $\Leftrightarrow$ (viii)

$\Leftrightarrow$ (ix). In order to prove (c.2), suppose first that $\alpha \in (0, \frac{\pi}{2})$. Then one can employ Theorem 3.1.2(i) to conclude that, for every $z \in (\partial\Delta)\setminus\{0\}$, $\overline{R(T(z))} = E$ or $\overline{R(T(\bar{z}))} = E$. In the case $\alpha = \frac{\pi}{2}$, notice only that, for every $z \in (\partial\Delta)$, we have $\overline{R(T(z))} = E$ and $\overline{R(T(\bar{z}))} = E$. This follows from the fact that $\mathrm{Orb}(x, T) \subseteq R(T(z))$, $x \in HC(T)$. The proof of [244, Theorem 2.5] works again; in such a way, we obtain that (iv) holds with $s \in \{z, \bar{z}\}$. The rest of the proof of (c) follows by the use of arguments already given in the proofs of (a) and (b). To prove (d), notice that Theorem 3.1.1 implies that $(T(t))_{t\in\Delta}$ has a dense G_δ-set of hypercyclic vectors so that one can repeat literally the arguments given in the proof of [61, Theorem 2.4] to conclude that (ii) holds; (d) follows by applying (a)-(c). The proof of (e) follows instantly from (d), and this completes the proof of theorem. □

The assertion (e) of the previous theorem is slightly generalized in [159], where the authors have proved that every chaotic semigroup $(T(t))_{t\geqslant 0}$ in a Banach space satisfies the Recurrent Hypercyclicity Criterion (cf. [61, Theorem 2.4] and [159, Corollary 6.2]). The next illustrative example shows that the Recurrent Hypercyclicity Criterion is strictly stronger than the Hypercyclicity Criterion.

Example 3.1.15. Define two sequences $(s_n)_{n\in\mathbb{N}_0}$ and $(r_n)_{n\in\mathbb{N}}$ of positive real numbers by $s_0 := 0$, $s_1 := 1$, $r_1 := e$ and, for every $n \in \mathbb{N}$,

$$(s_{n+1}, r_{n+1}) := \begin{cases} (k(1+s_n), er_n), \ n+1 = k^2 \text{ for some } k \in \mathbb{N}, \\ (1+s_n, e^{1+s_n}), \text{ otherwise.} \end{cases}$$

It can be easily verified that $(s_n)_{n\in\mathbb{N}_0}$ and $(r_n)_{n\in\mathbb{N}}$ are strictly increasing sequences as well as that: $\lim_{n\to\infty} s_n = \infty$, $\lim_{n\to\infty} r_n = \infty$ and $\liminf_{n\to\infty} \frac{\ln(r_n)}{s_n} = 0$. Put now $\rho(0) :=$ 1 and $\rho(s) := r_n e^{-s}$, $s \in (s_{n-1}, s_n]$. Since $(r_n)_{n\in\mathbb{N}}$ is increasing and $r_1 = e$, one obtains that $\rho(s) \leqslant e^t \rho(t+s)$, $t, s \geqslant 0$, and therefore, $\rho(t)$ is an admissible weight function. Inductively, we can prove that

$$r_n = \begin{cases} e^{s_n}, \ n \neq k^2 \text{ for every } k \in \mathbb{N}, \\ e^{\frac{s_n}{k}}, \ n = k^2 \text{ for some } k \in \mathbb{N}, \end{cases}$$

and

(498) $$s_{n^2-1} - s_{(n-1)^2} = 2n-2, \ n \in \mathbb{N}.$$

Clearly, $\rho(s) = r_n e^{-s} \geqslant r_n e^{-s_n}$, $s \in (s_{n-1}, s_n]$, $n \in \mathbb{N}$ and this inequality enables one to see that:

(499) $$\rho(s) \geqslant 1, \text{ if } s \in (s_{n-1}, s_n] \text{ and } n \neq k^2 \text{ for every } k \in \mathbb{N}.$$

Notice that

$$\lim_{n\to\infty} \rho(s_{n^2}) = \lim_{n\to\infty} e^{s_n^2(\frac{1}{n}-1)} = 0.$$

With the notion used in [161], let $X = L^1_\rho([0, \infty), \mathbb{C})$. Owing to [159, Proposition 4.4], we have that the translation semigroup $(T(t))_{t\geqslant 0}$ fulfills the Hypercyclicity

Criterion. Suppose $\varepsilon > 0$ and $(T(t))_{t\geqslant 0}$ fulfills the Recurrent Hypercyclicity Criterion. Due to [159, Theorem 4.6], we get the existence of an increasing sequence (t_n) in $(0, \infty)$ and a number $L \in (0, \infty)$ satisfying:

$$\lim_{n\to\infty} t_n = \infty,\ t_{n+1} - t_n \leqslant L \text{ and } \rho(t_n) \leqslant \varepsilon,\ n \in \mathbb{N}. \tag{500}$$

By (499), we have that there exist $k \in \mathbb{N}$ and $n_0 \in \mathbb{N}$ such that, for every $n \geqslant n_0$, $t_n \in \bigcup_{i\geqslant k}(s_{i^2-1}, s_{i^2}]$. Furthermore, by (498), one has $\lim_{n\to\infty}(s_{n^2-1} - s_{(n-1)^2}) = \infty$ and the contradiction is obvious since (t_n) satisfies (500).

Let us observe that the equivalence of assertions (i), (v) and (ix) quoted in the formulation of Theorem 3.1.14 can be slightly strengthened by means of already given arguments.

Theorem 3.1.16. *Suppose $(T(t))_{t\in\Delta}$ is a strongly continuous semigroup in X. Then the following assertions are equivalent:*

(i) *$(T(t))_{t\in\Delta}$ satifies the Hypercyclicity Criterion.*
(ii) *The family $\{T \oplus T(t) : t \in I\}$ is topologically transitive for every subset I of Δ satisfying the next condition: There exist $K > 0$, $n \in \mathbb{N}$ and rays $R_i \subseteq \Delta$, $i = 1, \cdots, n$ starting at 0 so that, for every $t \in \Delta\setminus\{0\}$, there exists $i \in \mathbb{N}_n$ such that, for every $z \in R_i$ with $|z| \geqslant K : [t, t + z] \cap I \neq \emptyset$.*
(iii) *$(T(t))_{t\in\Delta}$ is weakly mixing.*

Finally, we will prove the inheritance law for the Hypercyclicity Criterion.

Proposition 3.1.17. *Suppose $(T(t))_{t\in\Delta}$ is a strongly continuous semigroup in X and $t_0 \in \Delta\setminus\{0\}$. Then the following assertions are equivalent:*

(i) *$(T(t))_{t\in\Delta}$ is weakly mixing.*
(ii) *The family $\{T\oplus T(z) : z \in \Delta, |z| = k|t_0| \text{ for some } k \in \mathbb{N}\}$ is topologically transitive.*

Proof. The implication (ii) $\Rightarrow$ (i) is trivial. Let us prove that (i) $\Rightarrow$ (ii). In case $\Delta \in \{[0, \infty), \mathbb{R}, \Delta(\alpha)\}$, $\alpha \in (0, \frac{\pi}{2})$, the proof follows from Theorem 3.1.14 and the fact that the set $I = \{z \in \Delta : |z| = k|t_0| \text{ for some } k \in \mathbb{N}\}$ is syndetic. We will show this only in the case $\Delta = \Delta(\alpha)$ for some $\alpha \in (0, \frac{\pi}{2})$. Put $R = [0, \infty)$ and $K \geqslant \frac{2|t_0|}{\cos\alpha}$. Suppose, further, $t \in \Delta$, $|t| \in [k|t_0|, (k + 1)|t_0|)$ for an appropriate $k \in \mathbb{N}_0$ as well as $z \in R$ and $|z| \geqslant K$. Clearly, $|\text{Im } t| \leqslant \text{Re } t \tan\alpha$ and $\text{Re } t \geqslant k|t_0|\cos\alpha$. Hence, $|t + z|^2 \geqslant k^2|t_0|^2 + 2 \text{Re } tz + z^2 \geqslant k^2|t_0|^2 + 2k|t_0|z \cos\alpha \geqslant k^2|t_0|^2 + 2k|t_0|\frac{2|t_0|}{\cos\alpha}\cos\alpha \geqslant (k + 1)^2|t_0|^2$. Accordingly, there exists $s \in [t, t + z]$ such that $|s| = (k + 1)|t_0|$. This implies $[t, t + z] \cap I \neq \emptyset$. The rest of the proof is a sophisticated application of Theorem 3.1.16 and we will sketch the proof only in the case $\Delta = \mathbb{C}$. Put $R_j := \{re^{i(2j-1)\frac{\pi}{4}} : r \geqslant 0\}$, $j = 1, 2, 3, 4$. Then the set $I = \{z \in \mathbb{C} : |z| = k|t_0| \text{ for some } k \in \mathbb{N}\}$ fulfills the condition quoted in the item (ii) of the previous theorem with $n = 4$, R_1, R_2, R_3, R_4 and $K > 4|t_0|$. Namely, if $\text{Re } z \geqslant 0$, $\text{Im } z \geqslant 0$ and $|z| \in [k|t_0|, (k + 1)|t_0|)$ for an appropriate $k \in \mathbb{N}$, then we easily infer that, for every $z \in R_1$ with $|z| \geqslant K$, the segment $[t, t + z]$ contains an element of I since $[t, t + z]$ intersects the circle $\{z \in \mathbb{C} : |z| = (k + 1)|t_0|\}$.

The other cases can be considered similarly. □

Now we shall analyze S-hypercyclicity, chaoticity and topologically mixing properties of various kinds of strongly continuous semigroups. We start with the following recollection of the basic structural properties of function spaces used henceforth. A measurable function $\rho : \Delta \to (0, \infty)$ is said to be an admissible weight function if there exist constantas $M \geqslant 1$ and $\omega \in \mathbb{R}$ such that $\rho(t) \leqslant Me^{\omega|t'|}\rho(t+t')$ for all $t, t' \in \Delta$. For such a function $\rho(\cdot)$, we introduce the following Banach spaces:

$$L^p_\rho(\Delta, \mathbb{K}) := \{u : \Delta \to \mathbb{K}\ ;\ u(\cdot) \text{ is measurable and } \|u\|_p < \infty\},$$

where $p \in [1, \infty)$ and $\|u\|_p := (\int_\Delta |u(t)\rho(t)|^p\, dt)^{1/p}$ as well as

$$C_{0,\rho}(\Delta, \mathbb{K}) := \{u : \Delta \to \mathbb{K}\ ;\ u(\cdot) \text{ is continuous and } \lim_{t\to\infty} u(t)\rho(t) = 0\},$$

with $\|u\| := \sup_{t\in\Delta} |u(t)\rho(t)|$.

Definition 3.1.18. ([244]-[245], [311]) Suppose $n \in \mathbb{N}$ and Ω is an open non-empty subset of $\mathbb{R}^n$. A continuous mapping $\varphi : \Delta \times \Omega \to \Omega$ is called a semiflow if $\varphi(0, x) = x$, $x \in \Omega$,

$$\varphi(t + s, x) = \varphi(t, \varphi(s, x)),\ t, s \in \Delta, x \in \Omega \text{ and}$$
$$x \mapsto \varphi(t, x) \text{ is injective for all } t \in \Delta.$$

Designate by $\varphi(t, \cdot)^{-1}$ the inverse mapping of $\varphi(t, \cdot)$, i.e.,

$$y = \varphi(t, x)^{-1} \text{ iff } x = \varphi(t, y),\ t \in \Delta.$$

In what follows, we also deal with the Banach space $L^p_\rho(\Omega, \mathbb{K})$ where $\rho : \Omega \to (0, \infty)$ is a measurable function, $\rho^p : \Omega \to (0, \infty)$ is a locally integrable function and the norm of an element $f \in L^p_\rho(\Omega, \mathbb{K})$ is given by $\|f\|_p := (\int_\Omega |f(x)\rho(x)|^p\, dx)^{1/p}$. The Banach space $C_{0,\rho}(\Omega, \mathbb{K})$ consists of all continuous functions $f : \Omega \to \mathbb{K}$ satisfying that, for every $\varepsilon > 0$, $\{x \in \Omega : |f(x)|\rho(x) \geqslant \varepsilon\}$ is a compact subset of Ω; herein $\rho : \Omega \to (0, \infty)$ is an upper semicontinuous function and the norm of an element $f \in C_{0,\rho}(\Omega, \mathbb{K})$ is given by $\|f\| := \sup_{x\in\Omega} |f(x)|\rho(x)$. Put, by common consent, $\sup_{x\in\emptyset} \rho(x) := 0$ and denote by $C(\Lambda, \mathbb{K})$ the $\mathbb{K}$-vector space consisting of all continuous functions from Λ into $\mathbb{K}$, where Λ is Δ or Ω. The Fréchet topology on $C(\Omega, \mathbb{K})$ is induced by the following family of increasing seminorms: $\|f\|_n =: \sup_{x\in K_n} |f(x)|$, $f \in C(\Omega, \mathbb{K})$, where (K_n) is a sequence of compact subsets of Ω satisfying $K_1 \subseteq K_2 \subseteq \cdots \subseteq K_n \subseteq \cdots$ and $\bigcup_{n\in\mathbb{N}} K_n = \Lambda$. Notice that $C_c(\Lambda, \mathbb{K})$, the subspace of $C(\Lambda, \mathbb{K})$ which consists of all compactly supported functions, is dense in $L^p_\rho(\Lambda, \mathbb{K})$; obviously, $C_c(\Lambda, \mathbb{K})$ is dense in $C_{0,\rho}(\Lambda, \mathbb{K})$, too (cf. [411, Section 13]). The use of symbol ρ in the continuation of this section is clear from the context.

The following characterization of S-hypercyclic translation semigroups essentially follows from the argumentation given in the papers of W. Desch, W. Schappacher, G. F. Webb [161, Theorems 4.7-4.8] and J. A. Conejero, A. Peris [118, Theorems 5.11-5.12]. Notice only that the last equivalence in (ii) is a consequence of the proof of [311, Theorem 2.7].

Theorem 3.1.19. *Suppose* $p \in [1, \infty)$, $\alpha \in (0, \frac{\pi}{2}]$, $\rho : \Delta \to (0, \infty)$ *is an admissible weight function,* $\tilde{X} \in \{L^p_\rho(\Delta, \mathbb{K}), C_{0,\rho}(\Delta, \mathbb{K})\}$ *and the translation semigroup* $(\tilde{T}(t))_{t\in\Delta}$ *is given by*

(501) $$(\tilde{T}(t)f)(x) := f(x + t),\ x, t \in \Delta, f \in \tilde{X}.$$

(i) *Suppose* $\Delta \in \{[0, \infty), \Delta(\alpha)\}$. *The semigroup* $(\tilde{T}(t))_{t\in\Delta}$ *is S-topologically transitive if* $\sup S = \infty$. *In case* $\sup S < \infty$, $(\tilde{T}(t))_{t\in\Delta}$ *is S-topologically transitive iff for every* $\theta \in [0, \infty)$ *there exist a sequence* (t_j) *in* Δ *satisfying* $\lim_{j\to\infty} |t_j| = \infty$ *and a sequence* (a_j) *in* $S\setminus\{0\}$ *such that:*

$$\lim_{j\to\infty} \frac{1}{a_j} \rho(t_j + \theta) = 0$$

iff $(\tilde{T}(t))_{t\in\Delta}$ *is hypercyclic.*

(ii) *Suppose* $\Delta \in \{\mathbb{R}, \mathbb{C}\}$ *and* $S \subseteq [0, \infty)$. *Then* $(\tilde{T}(t))_{t\in\Delta}$ *is S-topologically transitive iff for every* $\theta > 0$ *there exist a sequence* (t_j) *in* Δ *satisfying* $\lim_{j\to\infty} |t_j| = \infty$ *and a sequence* (a_j) *in* $S\setminus\{0\}$ *such that:*

$$\lim_{j\to\infty} a_j \rho(-t_j + \theta) = \lim_{j\to\infty} \frac{1}{a_j} \rho(t_j + \theta) = 0.$$

If $S = [0, \infty)$, *the above is also equivalent to the existence of a sequence* (t_j) *in* Δ *satisfying* $\lim_{j\to\infty} |t_j| = \infty$ *and*

$$\lim_{j\to\infty} \rho(-t_j) \lim_{j\to\infty} \rho(t_j) = 0.$$

Problem. Let $\gamma \in (0, 1)$, let the requirements of Theorem 3.1.19 hold, and let $\Delta = [0, \infty)$. Denote by $(\tilde{T}_\gamma(t))_{t\geqslant 0}$ the subordinated (g_γ, I)-regularized resolvent family (cf. Theorem 2.4.2). Can we find the necessary and sufficient conditions for the S-topological transitivity of $(\tilde{T}_\gamma(t))_{t\geqslant 0}$ (defined in the obvious way) in terms of the weight function $\rho(\cdot)$?

Fix a number $t \in \Delta$, a function $f : \Omega \to \mathbb{K}$, a semiflow $\varphi : \Delta \times \Omega \to \Omega$ and define after that a function $T_\varphi(t)f : \Omega \to \mathbb{K}$ by $(T_\varphi(t)f)(x) := f(\varphi(t, x))$, $x \in \Omega$. Then $T_\varphi(0)f = f$, $T_\varphi(t)T_\varphi(s)f = T_\varphi(s)T_\varphi(t)f = T_\varphi(t + s)f$, $t, s \in \Delta$ and Brouwer's theorem (cf. [140], [245] and Theorem 3.1.30) implies $C_c(\Omega) \subseteq T_\varphi(t)(C_c(\Omega))$. We refer the reader to [245, Theorem 2.1], resp. [245, Theorem 2.2], for the necessary and sufficient conditions stating when the composition operator $T_\varphi(t) : L^p_\rho(\Omega) \to L^p_\rho(\Omega)$, resp. $T_\varphi(t) : C_{0,\rho}(\Omega) \to C_{0,\rho}(\Omega)$, is well defined and continuous. The strong continuity of semigroup $(T_\varphi(t))_{t\in\Delta}$ in $L^p_\rho(\Omega)$, resp. $C_{0,\rho}(\Omega)$, has been discussed in [245, Theorem 3.2, Theorem 3.4] and [311, Theorem 2.5, Theorem 2.6]. It is worthwhile to mention that such a property can be neglected from the formulation of subsequent theorems whose proofs follow by applying [245, Theorem 4.3, Theorem 4.5] (cf. also [311, Theorem 2.7]).

Theorem 3.1.20. *Let $\varphi : \Delta \times \Omega \to \Omega$ be a semiflow, and let $S \subseteq [0, \infty)$.*

(i) *Suppose $(T_\varphi(t))_{t\in\Delta}$ is a strongly continuous semigroup in $L^p_\rho(\Omega)$. Then the following assertions are equivalent.*

(i.1) *$(T_\varphi(t))_{t\in\Delta}$ is S-hypercyclic in $L^p_\rho(\Omega)$.*

(i.2) *For every compact set $K \subseteq \Omega$ there exist a sequence (L_k) of measurable subsets of K, a sequence (t_k) in Δ and a sequence (c_k) in $S\setminus\{0\}$ such that:*

$$(502)\qquad \lim_{k\to\infty} \int_{K\setminus L_k} \rho^p(x)\,dx = 0 \text{ and}$$

$$\lim_{k\to\infty} c_k^p \int_{\varphi(t_k,\cdot)^{-1}(L_k)} \rho^p(x)\,dx = \lim_{k\to\infty} \frac{1}{c_k^p} \int_{\varphi(t_k,L_k)} \rho^p(x)\,dx = 0.$$

In case $S = [0, \infty)$, the above is also equivalent to the condition (i.3), where:

(i.3) *For every compact set $K \subseteq \Omega$ there exist a sequence (L_k) of measurable subsets of K and a sequence (t_k) in Δ such that (502) holds and*

$$(503)\qquad \lim_{k\to\infty} \left[\int_{\varphi(t_k,\cdot)^{-1}(L_k)} \rho^p(x)\,dx * \int_{\varphi(t_k,L_k)} \rho^p(x)\,dx \right] = 0.$$

(ii) *Suppose $(T_\varphi(t))_{t\in\Delta}$ is a strongly continuous semigroup in $C_{0,\rho}(\Omega)$ and, for every compact set $K \subseteq \Omega$, we have $\inf_{x\in K} \rho(x) > 0$. Then the following assertions are equivalent.*

(ii.1) *$(T_\varphi(t))_{t\in\Delta}$ is S-hypercyclic in $C_{0,\rho}(\Omega)$.*

(ii.2) *For every compact set $K \subseteq \Omega$ there exist a sequence (t_k) in Δ and a sequence (c_k) in $S\setminus\{0\}$ such that:*

$$\lim_{k\to\infty} c_k \sup_{x\in\varphi(t_k,\cdot)^{-1}(K)} \rho(x) = \lim_{k\to\infty} \frac{1}{c_k} \sup_{x\in\varphi(t_k,K)} \rho(x) = 0.$$

In case $S = [0, \infty)$, the above is also equivalent to the condition (ii.3), where:

(ii.3) *For every compact set $K \subseteq \Omega$, there exists a sequence (t_k) in Δ such that:*

$$\lim_{k\to\infty} \left[\sup_{x\in\varphi(t_k,\cdot)^{-1}(K)} \rho(x) * \lim_{k\to\infty} \sup_{x\in\varphi(t_k,K)} \rho(x) \right] = 0.$$

Unfortunately, it is not clear whether Theorem 3.1.20 and the assertion (ii) of Theorem 3.1.19 remain true if $S \nsubseteq [0, \infty)$. Nevertheless, reconsidering the proofs of [350, Theorem 1, Corollary 4] once more we can prove the following important relationship between positivity and S-hypercyclicity:

Theorem 3.1.21. *Suppose X is a complex space, $S \subseteq [0, \infty)$, $x \in X$, $(T(t))_{t\in\Delta}$ is a locally equicontinuous semigroup in X, $\{uv : u, v \in S\} \subseteq S$ and there exists $T \in L(X)$ such that $\overline{R(T-\lambda I)} = X$, $\lambda \in \mathbb{C}$ and that $TT(t) = T(t)T$, $t \in \Delta$. Then $\{cT(t)x : c \in S, t \in \Delta\}$ is dense in X if $\{\lambda T(t)x : \lambda \in \mathbb{C}, |\lambda| \in S, t \in \Delta\}$ is dense in X.*

Keeping in mind the analyses of T. Bermúdez, A. Bonilla, A. Peris [56], F. León-Saavedra, V. Müler [350] and F. León-Saavedra, A. Piqueras-Lerena [351], it may seem reasonable to raise the issue ($\mathbb{I}$ stands for the set of irrational numbers):

Problem. Does there exist a supercyclic strongly continuous semigroup $(T(t))_{t\in\Delta}$ in X that is not positively supercyclic and satisfies

$$\emptyset \neq \sigma_p(T(t)^*) \subseteq (0, \infty)e^{2\pi i\mathbb{I}},\ t \in \Delta?$$

Now we shall investigate the basic hypercyclic properties of the above introduced classes of strongly continuous semigroups on $L^p(\Delta, \mathbb{K})$ and $C_0(\Delta, \mathbb{K})$-type spaces. Our study requires some additional technical rearrangements and we shall first state the following analogue of [499, Lemma 1].

Lemma 3.1.22. *Suppose* $p \in [1, \infty)$, $X \in \{L^p(\Delta, \mathbb{K}), C_0(\Delta, \mathbb{K})\}$, $g : \Delta \times \Delta \to \mathbb{K}$ *is continuous and* $(T(t)f)(x) := g(x, t)f(x + t)$, $x, t \in \Delta, f \in X$. *If* $(T(t))_{t\in\Delta}$ *is a strongly continuous semigroup, then:*

(HT1) $g(x, t + s) = g(x, t)g(x + t, s)$, $x, t, s \in \Delta$,

(HT2) $g(x, 0) = 1$, $x \in \Delta$,

(HT3) $g(x, t) \neq 0$, $x, t \in \Delta$ *and*

(HT4) $g(t, s) = \frac{g(0,t+s)}{g(0,t)}$, $t, s \in \Delta$.

Proof. Certainly, $(T(t + s)f)(x) = g(x, t + s)f(x + t + s)$ and $(T(t)T(s)f)(x) = g(x, t)(T(s)f)(x + t) = g(x, t)g(x + t, s)f(x + t + s)$, $x, t, s \in \Delta, f \in X$. This simply implies (HT1) while (HT2) follows from $T(0) = I$. To prove (HT3), suppose $g(x, t) = 0$, $x, t \in \Delta$. Since $g(x, 0) = 1$ and $g(\cdot, \cdot)$ is continuous, we have the existence of a positive real number ε such that, for every $t' \in \Delta_{\varepsilon}$, $|g(x, t')| \geqslant \frac{1}{2}$. Therefore, $0 < \inf\{|t''| : t'' \in \Delta$ and $g(x, t'') = 0\} := r_0$. The continuity of $g(\cdot, \cdot)$ implies that there exists $t_0 \in \Delta$ such that $g(x, t_0) = 0$ and $|t_0| = r_0$. Let $t_1 \in [0, t_0)$. Clearly, $t_0 - t_1 \in \Delta$ and, due to (HT1), $g(x, t_0) = g(x, t_1)g(x + t_1, t_0 - t_1)$. So, $g(x + t_1, t_0 - t_1) = 0$. Letting $t_1 \to t_0$, we obtain $g(x + t_0, 0) = 0$ which contradicts (HT2); (HT4) is a simple consequence of (HT1) and (HT3). □

Lemma 3.1.23. *Suppose* $g : \Delta \times \Delta \to \mathbb{K}$ *is continuous and satisfies (HT1)-(HT4). Put* $\rho(t) := \frac{1}{|g(0,t)|}$, $t \in \Delta$. *Then* $\rho(t)$ *is an admissible weight function iff there exist numbers* $M \geqslant 1$ *and* $\omega \in \mathbb{R}$ *so that* $|g(t, t')| \leqslant Me^{\omega|t'|}$ *for all* $t, t' \in \Delta$.

Proof. Suppose $\rho(t)$ is an admissible weight function. Then the existence of numbers $M \geqslant 1$ and $\omega \in \mathbb{R}$ satisfying $\frac{1}{|g(0,t)|} \leqslant Me^{\omega|t'|}\frac{1}{|g(0,t+t')|}$, $t, t' \in \Delta$ is obvious. This implies $|\frac{|g(0,t+t')|}{g(0,t)}| \leqslant Me^{\omega|t'|}$, $t, t' \in \Delta$, i.e., $|\frac{g(0,t)g(t,t')}{g(0,t)}| \leqslant Me^{\omega|t'|}$, $t, t' \in \Delta$. Hence, $|g(t, t')| \leqslant Me^{\omega|t'|}$, $t, t' \in \Delta$. The converse statement can be proved in a similar way. □

Example 3.1.24. (a) Suppose $\alpha \in (0, \frac{\pi}{2}]$, $\Delta \in \{\Delta(\alpha), \mathbb{C}\}$ as well as the continuous functions $g_1, g_2 : \mathbb{R} \times \mathbb{R} \to \mathbb{K}$ fulfill the conditions (HT1)-(HT2) for $x, t, s \in \mathbb{R}$. If $\Delta = \Delta(\alpha)$, then we assume that the function $g_1(\cdot)$ is defined and continuous on $[0, \infty) \times [0, \infty)$ and that satisfies (HT1)-(HT2) for $x, t, s \geqslant 0$. Put, for $x = x_1 + ix_2 \in \Delta$ and $t = t_1 + it_2 \in \Delta$, $g(x, t) := g_1(x_1, t_1)g_2(x_2, t_2)$. Then $g(\cdot)$ satisfies

(HT1)-(HT2) and the proof of Lemma 3.1.22 implies (HT1)-(HT4) for $g(\cdot, \cdot)$. Suppose, further, that $h : \mathbb{R} \to \mathbb{K}$ is a bounded measurable function. Then the following functions (see [499]) satisfy (HT1)-(HT4):

(1) $g_i(x_i, t_i) = e^{\int_{x_i}^{x_i+t_i} h(s)\,ds}$, $i = 1, 2$,

(2) $g_1(x_1, t_1) = (1 + \frac{t_1}{x_1+a})^b$, $a > 0$, $b \in \mathbb{C}$ and $g_2(x_2, t_2) = (1 + \frac{t_2}{x_2+a})^n$, $a \in \mathbb{C}\backslash\mathbb{R}$, $n \in \mathbb{N}_0$, if $\mathbb{K} = \mathbb{C}$ and $\Delta = \Delta(\alpha)$.

However, for any function $g(x, t) = g_1(x_1, t_1)g_2(x_2, t_2)$, where $g_1(\cdot, \cdot)$, resp., $g_2(\cdot, \cdot)$, is of the form (1)-(2), there exist appropriate constants $M \geqslant 1$ and $\omega \in \mathbb{R}$ so that $|g(t, t')| \leqslant Me^{\omega|t'|}$ for all $t, t' \in \Delta$. Owing to Lemma 3.1.23, the mapping $\rho(t) = \frac{1}{|g(0,t)|}$, $t \in \Delta$ is an admissible weight function.

(b) Suppose $g : \Delta \times \Delta \to \mathbb{K}$ is continuous and satisfies (HT1)-(HT4). Set, for every $p \in [0, \infty)$, $g_p(x, t) := g^p(x, t)$ and $\hat{g}_p(x, t) := |g^p(x, t)|$, $x, t \in \Delta$. Then $\hat{g}_p(\cdot)$ is continuous and satisfies (HT1)-(HT4); the same conclusion holds for $g_p(\cdot)$ provided that $\mathbb{K} = \mathbb{R}$ because, in this case, we have $g(x, t) > 0$, $x, t \in \Delta$. In general, $g_p(\cdot, \cdot)$ does not satisfy (HT1) if $\mathbb{K} = \mathbb{C}$ and $p \notin \mathbb{N}$. A counterexample can be simply constructed; just put $g(x, t) := e^{it}$, $x, t \in \mathbb{C}$ and notice that $g_p(x, -\pi) \neq g_p(x, \frac{\pi}{2})g_p(x - \frac{\pi}{2}, -\frac{\pi}{2})$ if $x \in \mathbb{C}$ and $p \notin \mathbb{N}$. Nevertheless, Lemma 3.1.23 enables one to see that the admissibility of function $t \to \frac{1}{|g(0,t)|}$, $t \in \Delta$ implies the admissibility of function $t \to \frac{1}{|g_p(0,t)|}$, $t \in \Delta$.

(c) Suppose $f \in C([0, \infty) : \mathbb{K})$ and $g : \Delta \times \Delta \to \mathbb{K}$ is defined by $g(x, t) := e^{f(|x|)-f(|x+t|)}$, $x, t \in \Delta$. Then $g(\cdot, \cdot)$ is continuous and satisfies (HT1)-(HT4). Notice also that the Lipschitz continuity of $f(\cdot, \cdot)$ implies that $t \mapsto \frac{1}{|g(0,t)|}$, $t \in \Delta$ is an admissible weight function.

In what follows, we shall assume that the continuous function $g : \Delta \times \Delta \to \mathbb{K}$ satisfies (HT1)-(HT4) and that the mapping $\rho(t) = \frac{1}{|g(0,t)|}$, $t \in \Delta$ is an admissible weight function. Set $g_n(x, t) := g^n(x, t)$, $x, t \in \Delta$, $n \in \mathbb{N}_0$ and $\rho_n(t) := \rho^n(t)$, $t \in \Delta$, $n \in \mathbb{N}_0$. As in the previous example one gets that, for every $n \in \mathbb{N}_0$, the continuous function $g_n(\cdot)$ satisfies (HT1)-(HT4) and that $t \mapsto \rho_n(t)$, $t \in \Delta$ is an admissible weight function.

Theorem 3.1.25. *Suppose $i, j \in \mathbb{N}_0$, $p \in [0, \infty)$ and X_i is either $L^p_{\rho_i}(\Delta, \mathbb{K})$ or $C_{0,\rho_i}(\Delta, \mathbb{K})$. Define $(T_j(t))_{t\in\Delta} \subseteq L(X_i)$ and $(\widetilde{T}_{i+j}(t))_{t\in\Delta} \subseteq L(X_{i+j})$ by:*

$$(T_j(t)f)(x) := g_j(x, t)f(x + t),\ x, t \in \Delta, f \in X_i \text{ and:} \tag{504}$$

$$(\widetilde{T}_{i+j}(t)\tilde{f})(x) := \tilde{f}(x + t),\ x, t \in \Delta, \tilde{f} \in X_{i+j}.$$

Then the operator family $(T_j(t))_{t\in\Delta}$ is a strongly continuous semigroup in X_i and the following holds:

(i) *$(T_j(t))_{t\in\Delta}$ is S-hypercyclic in X_i if $(\widetilde{T}_{i+j}(t))_{t\in\Delta}$ is S-hypercyclic in X_{i+j}.*

(ii) *$(T_j(t))_{t\in\Delta}$ is chaotic in X_i if $(\widetilde{T}_{i+j}(t))_{t\in\Delta}$ is chaotic in X_{i+j}.*

(iii) *$(T_j(t))_{t\in\Delta}$ is topologically mixing in X_i if $(\widetilde{T}_{i+j}(t))_{t\in\Delta}$ is topologically mixing in X_{i+j}.*

Proof. Define the mapping $\varphi_{i,j} : X_{i+j} \to X_i$ by

$$\varphi_{i,j}(\tilde{f})(\tau) := \frac{1}{g_j(0,\tau)} \tilde{f}(\tau),\ \tilde{f} \in X_{i+j},\ \tau \in \Delta.$$

It can be easily seen that $\varphi_{i,j}(\cdot)$ is an isometric isomorphism. Furthermore,

$$\varphi_{i,j} \circ \widetilde{T_{i+j}}(t) = T_j(t) \circ \varphi_{i,j} \text{ and } \varphi_{i,j}^{-1} \circ T_j(t) = \widetilde{T_{i+j}}(t) \circ \varphi_{i,j}^{-1},\ t \in \Delta,$$

the operator family $(\widetilde{T_{i+j}}(t))_{t\in\Delta}$ is a strongly continuous semigroup in X_{i+j} (cf. [118], [161]) and the existence of numbers $M \geqslant 1$ and $\omega \in \mathbb{R}$ satisfying $\|T_j(t)\| \leqslant M^{i+j}e^{\omega(i+j)|t|}$, $t \in \Delta$ follows from Lemma 3.1.23 and the admissibility of $\rho_i(\cdot)$. Since $g_j(\cdot, \cdot)$ satisfies (HT1)-(HT4), one gets that $(T_j(t))_{t\in\Delta}$ is a semigroup in X_i; the strong continuity of $(T_j(t))_{t\in\Delta}$ can be proved as follows. Suppose $f \in X_i$. Then $\lim_{t\to 0, t\in\Delta} T_j(t)f = f$ is equivalent to $\lim_{t\to 0, t\in\Delta} \varphi_{i,j}^{-1} T_j(t)f = \varphi_{i,j}^{-1} f$, i.e., to $\lim_{t\to 0, t\in\Delta} \widetilde{T_{i+j}}(t)\, \varphi_{i,j}^{-1} f = \varphi_{i,j}^{-1} f$. The last statement holds because $(\widetilde{T_{i+j}}(t))_{t\in\Delta}$ is a strongly continuous semigroup. Hence, $(T_j(t))_{t\in\Delta}$ is a strongly continuous semigroup in X_i. The statements (i) and (ii) follow from the following observations: $f \in \mathrm{HC_S}(T_j)$ iff $\varphi_{i,j}^{-1} f \in \mathrm{HC_S}(\widetilde{T_{i+j}})$ and $f \in X_j$ is a periodic point for $(T_j(t))_{t\in\Delta}$ iff $\varphi_{i,j}^{-1} f \in X_{i+j}$ is a periodic point for $(\widetilde{T_{i+j}}(t))_{t\in\Delta}$. The assertion (iii) can be proved similarly. □

Concerning chaoticity of strongly continuous translation semigroups, we would like to observe that the structural results proved by J. A. Conejero and A. Peris in [118] still hold in case $\alpha = \frac{\pi}{2}$, which is also allowed in our research. The only vital change compared with the case $\alpha \in (0, \frac{\pi}{2})$ is the construction of periodic points given in the proof of implication (3) ⇒ (1) of [118, Corollary 1]. This construction has to be adapted by the use of appropriate rectangles, because we must avoid the overlapping of corresponding sectors $kt + \Delta_{|t|}$, $k \in \mathbb{N}$ appearing in the proof of the cited result. Taking into consideration this observation as well as Lemmas 3.1.22-3.1.23, Theorem 3.1.19 and Theorem 3.1.25, we immediately obtain that the following theorem holds.

Theorem 3.1.26. *Suppose* $p \in [1, \infty)$, $\alpha \in (0, \frac{\pi}{2}]$, $i, j \in \mathbb{N}_0$, $i + j > 0$, $\Delta \in \{\Delta(\alpha), \mathbb{C}\}$ *and consider the strongly continuous semigroups* $(T_j(t))_{t\in\Delta}$ *in* $X_i \in \{L^p_{\rho_i}(\Delta, \mathbb{K}), C_{0,\rho_i}(\Delta, \mathbb{K})\}$ *(see Theorem 3.1.25). Then:*

(i) *The semigroup* $(T_j(t))_{t\in\Delta}$*, given by (504), is chaotic in* $L^p_{\rho_i}(\Delta, \mathbb{K})$ *iff for every* $\theta \in [0, \infty)$*, there exists* $t \in \Delta\backslash\{0\}$ *such that:*

$$\sum_{k=0}^{\infty} \frac{1}{|g(0, \theta + kt)|^{p(i+j)}} < \infty \text{ if } \Delta \neq \mathbb{C}, \text{ resp.,}$$

there exists $t \in \Delta\backslash\{0\}$ *such that:*

$$\sum_{k=-\infty}^{\infty} \frac{1}{|g(0, kt)|^{p(i+j)}} < \infty \text{ if } \Delta = \mathbb{C}.$$

(ii) *The semigroup* $(T_j(t))_{t\in\Delta}$ *is chaotic in* $L^p_{\rho_i}(\Delta, \mathbb{K})$ *iff there exists a ray* $R \subseteq \Delta$ *starting at* 0 *such that, for every* $m \in \mathbb{N}$,

$$\int_{F_{R,m}} \frac{dt}{|g(0,t)|^{p(i+j)}} < \infty \text{ if } \Delta \neq \mathbb{C}, \text{ resp.,}$$

$$\int_{F_{\pm R,m}} \frac{dt}{|g(0,t)|^{p(i+j)}} < \infty \text{ if } \Delta = \mathbb{C}.$$

(iii) *Suppose that a ray* $R \subseteq \Delta$ *is not contained in the boundary of* Δ *and* $0 \in R$. *The following conditions are equivalent and any of them implies that the semigroup* $(T_j(t))_{t\in\Delta}$ *is chaotic in* $L^p_{\rho_i}(\Delta, \mathbb{K})$:

(iii.1) *There exists* $t \in R\setminus\{0\}$ *such that*

$$\sum_{k=0}^{\infty} \frac{1}{|g(0,kt)|^{p(i+j)}} < \infty \text{ if } \Delta \neq \mathbb{C}, \text{ resp.,}$$

$$\sum_{k=-\infty}^{\infty} \frac{1}{|g(0,kt)|^{p(i+j)}} < \infty \text{ if } \Delta = \mathbb{C}.$$

(iii.2) $\int_{F_{R,1}} \frac{dt}{|g(0,t)|^{p(i+j)}} < \infty$ *if* $\Delta \neq \mathbb{C}$, *resp.,* $\int_{F_{\pm R,1}} \frac{dt}{|g(0,t)|^{p(i+j)}} < \infty$ *if* $\Delta = \mathbb{C}$.

(iii.3) *The restriction* $(T_j(t))_{t\in R}$ *of the semigroup* $(T_j(t))_{t\in\Delta}$ *to the ray R admits a non-trivial periodic point.*

(iv) *Suppose* $i_1, i_2, j_1, j_2 \in \mathbb{N}_0$, $i_1 + j_1 > 0$ *and* $i_2 + j_2 > 0$. *Then the semigroup* $(T_{j_1}(t))_{t\in\Delta}$ *is chaotic in* $C_{0,\rho_{i_1}}(\Delta, \mathbb{K})$ *iff the semigroup* $(T_{j_2}(t))_{t\in\Delta}$ *is chaotic in* $C_{0,\rho_{i_2}}(\Delta, \mathbb{K})$. *If* $\Delta \neq \mathbb{C}$, *then the semigroup* $(T_{j_1}(t))_{t\in\Delta}$ *is chaotic in* $C_{0,\rho_{i_1}}(\Delta, \mathbb{K})$ *iff for every* $\theta \in [0, \infty)$, *there exists* $t \in \Delta\setminus\{0\}$ *such that:*

$$\lim_{k\to\infty} |g(0, \theta + kt)| = \infty.$$

The semigroup $(T_{j_1}(t))_{t\in\mathbb{C}}$ *is chaotic in* $C_{0,\rho_{i_1}}(\mathbb{C}, \mathbb{K})$ *iff there exists* $t \in \Delta\setminus\{0\}$ *such that:*

$$\lim_{k\to\infty} |g(0, kt)| = \lim_{k\to\infty} |g(0, -kt)| = \infty.$$

(v) *Suppose that a ray* $R \subseteq \Delta$ *is not contained in the boundary of* Δ *and* $0 \in R$. *The following conditions are equivalent and any of them implies that the semigroup* $(T_j(t))_{t\in\Delta}$ *is chaotic in* $C_{0,\rho_i}(\Delta, \mathbb{K})$:

(v.1) $\lim_{z\to\infty, z\in R} |g(0, z)| = \infty$, *if* $\Delta \neq \mathbb{C}$, *resp.,* $\lim_{z\to\infty, z\in\pm R} |g(0, z)| = \infty$, *if* $\Delta = \mathbb{C}$.

(v.2) *The restriction* $(T_j(t))_{t\in R}$ *of the semigroup* $(T_j(t))_{t\in\Delta}$ *to the ray R admits a non-trivial periodic point.*

(vi) *Suppose* $i_1, i_2, j_1, j_2 \in \mathbb{N}_0$, $i_1 + j_1 > 0$ *and* $i_2 + j_2 > 0$. *Then the semigroup* $(T_{j_1}(t))_{t\in\Delta(\alpha)}$, $\alpha \in (0, \frac{\pi}{2}]$ *is always positively supercyclic in* X_{i_1}. *Moreover, the semigroup* $(T_{j_1}(t))_{t\in\mathbb{C}}$ *is positively supercyclic in* X_{i_1} *iff the semigroup* $(T_{j_2}(t))_{t\in\mathbb{C}}$ *is positively supercyclic in* X_{i_2} *iff there exists a sequence* (t_n) *in* $\mathbb{C}$ *such that* $\lim_{n\to\infty} t_n = \infty$ *and*

$$\lim_{n\to\infty} |g(0, t_n)g(0, -t_n)| = \infty.$$

(vii) *Suppose* $i_1 \in \mathbb{N}_0$ *and* $j_1 \in \mathbb{N}_0$. *If* $\Delta = [0, \infty)$ *or* $\Delta = \Delta(\alpha)$ *for an appropriate* $\alpha \in (0, \frac{\pi}{2}]$, *then the semigroup* $(T_{j_1}(t))_{t\in\Delta}$ *is S-hypercyclic in* X_{i_1} *iff for every* $\theta \in [0, \infty)$, *there exist a sequence* (t_n) *in* $\Delta\backslash\{0\}$ *and a sequence* (a_n) *in* $S\backslash\{0\}$ *such that* $\lim_{n\to\infty} |t_n| = \infty$ *and*

$$\lim_{n\to\infty} a_n^{\frac{1}{i_1+j_1}} |g(0, \theta + t_n)| = \infty.$$

Suppose $\Delta \in \{\mathbb{R}, \mathbb{C}\}$ *and* $S \subseteq [0, \infty)$. *Then the semigroup* $(T_{j_1}(t))_{t\in\mathbb{C}}$ *is S-hypercyclic in* X_{i_1} *iff for every* $\theta \in [0, \infty)$, *there exist a sequence* (t_n) *in* $\Delta\backslash\{0\}$ *and a sequence* (a_n) *in* $S\backslash\{0\}$ *such that* $\lim_{n\to\infty} |t_n| = \infty$ *and*

$$\lim_{n\to\infty} a_n^{\frac{(-1)}{i_1+j_1}} |g(0, \theta - t_n)| = \lim_{n\to\infty} a_n^{\frac{1}{i_1+j_1}} |g(0, \theta + t_n)| = \infty.$$

In particular, the hypotheses $i_1 + j_1 > 0$ *and* $i_2 + j_2 > 0$ *imply that* $(T_{j_1}(t))_{t\in\Delta}$ *is hypercyclic in* X_{i_1} *iff* $(T_{j_2}(t))_{t\in\Delta}$ *is hypercyclic in* X_{i_2}.

The proof of following theorem is omitted (see, e.g., [245, Section 4]).

Theorem 3.1.27. *Suppose* $a : \Omega \to \mathbb{K}\backslash\{0\}$ *is continuous,* $g : \Omega \times \Delta \to \mathbb{K}\backslash\{0\}$ *is given by* $g(x, t) := \frac{a(x)}{a(\varphi(t,x))}$, $x \in \Omega$, $t \in \Delta$, $\varphi : \Omega \times \Delta \to \Omega$ *is a semiflow and* $(T_\varphi(t))_{t\in\Delta}$ *is a strongly continuous semigroup in* $C_{0,\rho}(\Omega, \mathbb{K})$, *resp.* $L^p_\rho(\Omega, \mathbb{K})$. *Set, for every* $x \in \Omega$, $t \in \Delta$ *and* $f \in C_{0,\frac{\rho}{|a|}}(\Omega, \mathbb{K})$, *resp.*, $f \in L^p_{\frac{\rho}{|a|}}(\Omega, \mathbb{K})$:

$$(T_{g,\varphi}(t)f)(x) := g(x, t)f(\varphi(t, x)).$$

Then $(T_{g,\varphi}(t))_{t\in\Delta}$ *is a strongly continuous semigroup in* $C_{0,\frac{\rho}{|a|}}(\Omega, \mathbb{K})$, *resp.* $L^p_{\frac{\rho}{|a|}}(\Omega, \mathbb{K})$, *and* $(T_\varphi(t))_{t\in\Delta}$ *is S-hypercyclic, resp. chaotic, topologically mixing, in* $C_{0,\rho}(\Omega, \mathbb{K})$, *resp.* $L^p_\rho(\Omega, \mathbb{K})$ *iff* $(T_{g,\varphi}(t))_{t\in\Delta}$ *is S-hypercyclic, resp. chaotic, topologically mixing, in* $C_{0,\frac{\rho}{|a|}}(\Omega, \mathbb{K})$, *resp.* $L^p_{\frac{\rho}{|a|}}(\Omega, \mathbb{K})$.

Example 3.1.28. (i) Suppose $j \in \mathbb{N}$, $\Delta = \Delta(\alpha)$ for some $\alpha \in (0, \frac{\pi}{2}]$, $g(x, t) = (1 + \frac{t_1}{x_1+a})^b e^{\int_{x_2}^{x_2+t_2} h(s)\, ds}$, where $a > 0$, $b \in \mathbb{C}$ and $h : \mathbb{R} \to \mathbb{C}$ is a bounded measurable function. Due to Theorem 3.1.26, the semigroup $(T_j(t))_{t\in\Delta}$, given by (504), is chaotic in $X = C_0(\Delta, \mathbb{C})$ iff for every $\theta \in [0, \infty)$, there exists $t = t_1 + it_2 \in \Delta\backslash\{0\}$ so that:

$$\lim_{k\to\infty} \left(1 + \frac{\theta + kt_1}{a}\right)^{\operatorname{Re} b} e^{\int_0^{kt_2} \operatorname{Re}(h(s))\, ds} = +\infty.$$

If Re $b > 0$, then one can choose $t = 1$ in order to see that the semigroup $(T_j(t))_{t\in\Delta}$ is chaotic. Suppose now Re $b = 0$; then it can be easily seen that the semigroup $(T_j(t))_{t\in\Delta}$ is chaotic iff $\int_0^{+\infty}$ Re$(h(s))\, ds = +\infty$ or $\int_{-\infty}^0$ Re$(h(s))\, ds = -\infty$. The case Re $b < 0$ is non-trivial. For example, if $h(s) = \frac{d}{ds} ln(s^{2n} + 1)$, $s \in \mathbb{R}$, then $(T_j(t))_{t\in\Delta}$ is chaotic iff Re $b > -2n$. Let us suppose now $p \in [1, \infty)$ and $X = L^p(\Delta, \mathbb{K})$. Then the semigroup $(T_j(t))_{t\in\Delta}$ is chaotic in $L^p(\Delta, \mathbb{K})$ iff for every $\theta \in [0, \infty)$, there exists $t = t_1 + it_2 \in \Delta\setminus\{0\}$ so that:

$$\sum_{k=1}^{\infty} \frac{1}{\left(1+\frac{\theta + kt_1}{a}\right)^{jp\,\mathrm{Re}\, b} e^{jp\int_0^{kt_2}\mathrm{Re}(h(s))ds}} < \infty.$$

Hence, the chaoticity of $(T_j(t))_{t\in\Delta}$ depends on j. Consider, for example, the case $j = 1$. If Re $b > \frac{1}{p}$, then one can choose $t = 1$ in order to conclude that $(T_1(t))_{t\in\Delta}$ is chaotic. The situation is more complicated in case Re $b \leqslant \frac{1}{p}$. To illustrate this, suppose $h(s) = \frac{d}{ds} ln(ln(s^2 + 2))$, $s \in \mathbb{R}$. In case Re $b = \frac{1}{p}$ and $p > 1$, the semigroup $(T_1(t))_{t\in\Delta}$ is chaotic since $\int_2^{\infty} \frac{d\xi}{\xi \ln^p \xi} < \infty$; analogously, the semigroup $(T_1(t))_{t\in\Delta}$ is not chaotic provided that $p = 1$ and Re $b = 1$. Finally, suppose Re $b < \frac{1}{p}$ and $h(s) = \frac{d}{ds} ln(s^{2n} + 1)$, $s \in \mathbb{R}$. Then $(T_1(t))_{t\in\Delta}$ is chaotic iff $p(\mathrm{Re}\, b + 2n) > 1$.

(ii) Suppose $S \subseteq [0, \infty)$, $\Delta = \mathbb{K} = \mathbb{C}$, $p \in [1, \infty)$ and $\rho_i : \mathbb{R} \to (0, \infty)$ is an admissible weight function, $i = 1, 2$. Define $\rho(t_1 + it_2) := \rho_1(t_1)\rho_2(t_2)$, $t_1, t_2 \in \mathbb{R}$. Then it can be easily seen that $\rho : \Delta \to (0, \infty)$ is an admissible weight function. Suppose $\tilde{X} \in \{L^p_\rho(\Delta, \mathbb{K}),\ C_{0,\rho}(\Delta, \mathbb{K})\}$; then Theorem 3.1.19 implies that the semigroup $(\tilde{T}(t))_{t\in\mathbb{C}}$, given by (501), is positively supercyclic in $\tilde{X}$ iff there exist two real sequences (a_n) and (b_n) such that $\lim_{n\to\infty} (a_n^2 + b_n^2) = \infty$ and

$$\lim_{n\to\infty} \rho_1(-a_n)\rho_1(a_n)\rho_2(-b_n)\rho_2(b_n) = 0.$$

The necessary and sufficient conditions for the S-hypercyclicity of $(\tilde{T}(t))_{t\in\mathbb{C}}$ can be simply stated by using Theorem 3.1.19. For example, put $\rho_2(t) := 1$ and $\rho_1(t) := e^{-\int_0^t h(s)\,ds}$, $t \in \mathbb{R}$, where $h : \mathbb{R} \to \mathbb{R}$ is a bounded measurable function which satisfies $\int_0^{+\infty} h(s)\, ds = +\infty$ or $\int_{-\infty}^0 |h(s)|\, ds < \infty$. By Theorem 3.1.19, we have that $(\tilde{T}(t))_{t\in\mathbb{C}}$ is not hypercyclic and that $(\tilde{T}(t))_{t\in\mathbb{C}}$ is S-hypercyclic iff infS $= 0$ or supS $= \infty$. We end (ii) with the following adaptation of [498, Example 1]. Put $\Delta = \Delta(\alpha)$ for some $\alpha \in (0, \frac{\pi}{2}]$, $\tilde{X} = C_{0,\rho}(\Delta, \mathbb{K})$ and $\rho(t_1 + it_2) := e^{-(t_1+1)\cos(ln(t_1+1))+1}$, $t_1 + it_2 \in \Delta$. Notice that $\rho(\cdot)$ is an admissible weight function and that the translation semigroup $(\tilde{T}(t))_{t\in\Delta}$ is hypercyclic but not chaotic.

(iii) Suppose $S \subseteq [0, \infty)$, $\Delta = [0, \infty)$ and $\Omega = [0, \infty)^n$, $n \in \mathbb{N}$. Define a semiflow $\varphi : \Delta \times \Omega \to \Omega$ by $\varphi(t, x_1, \cdots, x_n) := (e^t x_1, \cdots, e^t x_n)$ and a continuous function $\rho : \Omega \to (0, \infty)$ by $\rho(x_1, \cdots, x_n) := \frac{1}{1+x_1^2+\cdots+x_n^2}$, $t \in \Delta$, $(x_1, \cdots, x_n) \in \Omega$. Owing to [245, Theorem 3.7] and Theorem 3.1.20, one gets that $(T_\varphi(t))_{t\geqslant 0}$ is a non-hypercyclic strongly continuous semigroup in $C_{0,\rho}(\Omega, \mathbb{K})$ and that $(T_\varphi(t))_{t\geqslant 0}$ is

S-hypercyclic in $C_{0,\rho}(\Omega, \mathbb{K})$ iff $\inf S = 0$. In particular, the above shows that the concepts of hypercyclicity, resp. positive supercyclicity, and S-hypercyclicity do not coincide if S is bounded, resp. unbounded. Suppose, further, $n = 1$ and $a : \Omega \to \mathbb{R}\setminus\{0\}$ is continuously differentiable. The semigroup solution of the following partial differential equation in $C_{0,\frac{\rho}{|a|}}(\Omega, \mathbb{K})$:

$$u_t = xu_x - x\frac{a'(x)}{a(x)}u,\ t \geqslant 0,\ u(0, x) = f(x),\ x \in \Omega,$$

is given by

$$(T_{g,\varphi}(t)f)(x) := \frac{a(x)}{a(e^t x)} f(e^t x),\ t \geqslant 0,\ x \in \Omega.$$

By Theorem 3.1.27, we have that $(T_{g,\varphi}(t))_{t\geqslant 0}$ is S-hypercyclic in $C_{0,\frac{\rho}{|a|}}(\Omega, \mathbb{K})$ iff $(T_\varphi(t))_{t\geqslant 0}$ is S-hypercyclic in $C_{0,\rho}(\Omega, \mathbb{K})$ iff $\inf S = 0$.

(iv) Suppose $p \in [1, \infty)$, $c \geqslant 0$, $\omega \in \mathbb{R}$, $\Delta = \Delta(\alpha)$ for some $\alpha \in (0, \frac{\pi}{2})$ and $f_n : [n, n + 1] \to \mathbb{R}$ is a function of bounded variation for all $n \in \mathbb{N}_0$; by $\mathrm{V}_n^{n+1}(f_n)$ we denote the total variation of function $f_n(\cdot)$ on $[n, n + 1]$. Suppose, in addition, that for every $m, n \in \mathbb{N}_0$ with $m > n$ and, for every $t \in [n, n + 1]$,

$$f_n(t) \leqslant c + \omega(m - n) + f_m(t + m - n) \text{ and} \tag{505}$$

$$\mathrm{V} := \sup_{n\in\mathbb{N}_0} \mathrm{V}_n^{n+1}(f_n) < \infty. \tag{506}$$

Define $\rho : \Delta \to (0, \infty)$ by $\rho(t) := e^{f_{\lfloor |t| \rfloor}(|t|)}$, $t \in \Delta$. Let $t, x \in \Delta$, $|t| \in [n, n + 1)$ and let $|t + x| \in [m, m + 1)$ for some $m, n \in \mathbb{N}_0$ with $m \geqslant n$. The assumption $m = n$ immediately implies that $\rho(t) \leqslant e^{\mathrm{V}}\rho(t + x)$. Suppose now that $m > n$; then one gets $|x| \geqslant |t + x| - |t| \geqslant m - n - 1$ and:

$$\begin{aligned}\rho(t) = e^{f_n(|t|)} &\leqslant e^{c+\omega(m-n)+f_m(|t|+(m-n))} \leqslant e^{\left(c+\mathrm{V}\right)+\omega(m-n)+f_m(|t+x|)}\\ &\leqslant e^{\left(c+\mathrm{V}+|\omega|\right)+|\omega||x|+f_m(|t+x|)} = e^{\left(c+\mathrm{V}+|\omega|\right)}e^{|\omega||x|}\rho(t + x).\end{aligned}$$

Hence, $\rho(t) \leqslant e^{(c+\mathrm{V}+|\omega|)}e^{|\omega||x|}\rho(t + x)$, $t, x \in \Delta$ and $\rho(\cdot)$ is an admissible weight function. Let us reflect on the next special case:

$$f_n(t) := f_0(t - n) - a_n\omega,\ t \in [n, n + 1],\ n \in \mathbb{N},$$

where $\omega > 0$ and (a_n) is a sequence of real numbers satisfying:

$$1 + a_n \geqslant a_{n+1},\ n \in \mathbb{N}_0. \tag{507}$$

Notice that (507) forces $f_n(t) \leqslant \omega + f_{n+1}(t + 1)$, $n \in \mathbb{N}_0$, $t \in [n, n + 1]$. Now an induction argument shows the validity of (505) with $c = 0$. Furthermore, $\mathrm{V}_n^{n+1}(f_n) = \mathrm{V}_0^1(f_0)$, $n \in \mathbb{N}$, (506) holds and the translation semigroup $(\tilde{T}(t))_{t\in\Delta}$ is hypercyclic in $\tilde{X} \in \{L^p_\rho(\Delta, \mathbb{K}),\ C_{0,\rho}(\Delta, \mathbb{K})\}$ iff $\limsup_{n\to\infty} a_n = +\infty$. Let us prove that $(\tilde{T}(t))_{t\in\Delta}$ is chaotic in $C_{0,\rho}(\Delta, \mathbb{K})$ iff there exists $t > 0$ such that $\lim_{n\to\infty} a_{\lfloor nt\rfloor} = +\infty$ if $\lim_{n\to\infty} a_n = +\infty$. Indeed, suppose that $(\tilde{T}(t))_{t\in\Delta}$ is chaotic. According to [118, Theorem 5], we have the existence of a complex number $t_0 \in \Delta\setminus\{0\}$ satisfying $\lim_{n\to\infty} f_{\lfloor n|t_0|\rfloor}(n|t_0|) = -\infty$. Put $t = |t_0|$ and observe that the boundedness of $f_0(\cdot)$ implies $\lim_{n\to\infty} a_{\lfloor nt\rfloor} = +\infty$. Let us suppose now $\theta \geqslant 0$, $t > 0$, $\lim_{n\to\infty} a_{\lfloor nt\rfloor} = +\infty$ and $\theta \in [kt, (k+1)t)$ for some $k \in \mathbb{N}_0$. Owing to (507),

$$(kt - \theta + 2) + a_{\lfloor nt+\theta\rfloor} \geqslant a_{\lfloor (n+k)t\rfloor}.$$

Therefore, $\lim_{n\to\infty} a_{\lfloor nt+\theta\rfloor} = +\infty$, $\lim_{n\to\infty} \rho(\theta + nt) = 0$ and this shows that the condition 1. given in the formulation of [118, Theorem 5] holds with $R = [0, \infty)$. Hence, $(\tilde{T}(t))_{t\in\Delta}$ is chaotic. Keeping in mind (507), we have that the existence of a number $t > 0$ satisfying $\lim_{n\to\infty} a_{\lfloor nt\rfloor} = +\infty$ is equivalent with $\lim_{n\to\infty} a_n = +\infty$. Analogously, $(\tilde{T}(t))_{t\in\Delta}$ is chaotic in $L^p_\rho(\Delta, \mathbb{K})$ iff there exists $t > 0$ such that $\sum_{n=1}^{\infty} e^{-p\omega a_{\lfloor nt\rfloor}} < \infty$. Suppose, for the time being, $p\omega \leqslant 1$ and $a_n = ln(n+1)$, $n \in \mathbb{N}_0$. Then $(\tilde{T}(t))_{t\in\Delta}$ is chaotic in $C_{0,\rho}(\Delta, \mathbb{K})$ but $(\tilde{T}(t))_{t\in\Delta}$ is not chaotic in $L^p_\rho(\Delta, \mathbb{K})$. Finally, suppose that (a_n) satisfies (507), $\limsup_{n\to\infty} a_n = +\infty$ and $\lim_{n\to\infty} a_n \neq +\infty$. Then $(\tilde{T}(t))_{t\in\Delta}$ is hypercyclic in $\tilde{X}$ but $(\tilde{T}(t))_{t\in\Delta}$ is not chaotic in $\tilde{X}$.

(v) Suppose $\Delta = [0, \infty)$, $\Omega = \{(x, y) \in \mathbb{R}^2 : x^2 + y^2 > 1\}$, $p > 0$, $q \in \mathbb{R}$ and $\varphi(t, x, y) := e^{pt}(x\cos qt - y\sin qt, x\sin qt + y\cos qt)$, $t \geqslant 0$, $(x, y) \in \Omega$.
Then one can simply verify that $\varphi : \Delta \times \Omega \to \Omega$ is a semiflow and $\varphi(t, z) = e^{t(p+iq)}z$, $z = x + iy \in \Omega$, $k \in \mathbb{N}$. Proceeding as in [311, Example 4], one obtains that $(T_\varphi(t))_{t\geqslant 0}$ is topologically mixing in $C(\Omega, \mathbb{K})$. We will prove that $(T_\varphi(t))_{t\geqslant 0}$ is chaotic in $C(\Omega, \mathbb{K})$. Suppose $f \in C_c(\Omega, \mathbb{K})$, $a > a_0 > 1$, $\operatorname{supp}(f) \subseteq \{z \in \Omega : a_0 \leqslant |z| \leqslant a\}$, (a_n) is a strictly increasing sequence in (a, ∞) satisfying $\lim_{n\to\infty} a_n = \infty$ and $t_n := \frac{ln\, a_n}{p}$, $n \in \mathbb{N}$. Define, for every $n \in \mathbb{N}$, a function $f_n : \Omega \to \mathbb{K}$ as follows. Fix a number $z \in \Omega$ and suppose that $|z| \in [e^{kt_n p}, e^{(k+1)t_n p}) = [e^{ka_n}, e^{(k+1)a_n})$ for some $k \in \mathbb{N}_0$. Put now $f_n(z) := f(e^{-kt_n p}ze^{-ikqt_n})$. By construction, $f_n \in C(\Omega, \mathbb{K})$, $f_n(z) = f(z)$, $z \in \Omega$, $|z| \leqslant a_n$ and $f_n(\varphi(t_n, z)) = f_n(z)$, $z \in \Omega$, $n \in \mathbb{N}$.
Thereby, $f_n(\cdot)$ is a t_n-periodic point of $(T_\varphi(t))_{t\geqslant 0}$ and $\lim_{n\to\infty} f_n = f$ in $C(\Omega, \mathbb{K})$.

In the next example, we identify $\mathbb{C}$ and $\Delta(\alpha)$, $\alpha \in (0, \frac{\pi}{2}]$ with the corresponding subsets of $\mathbb{R}^2$.

Example 3.1.29. Let $m \in \mathbb{N}_0$, and let $C^m(\Delta, \mathbb{K})$ denote the vector space of all functions $\varphi : \Delta \to \mathbb{K}$ that are m times continuously differentiable in Δ° and whose partial derivatives $D^\alpha\varphi$ can be extended continuously throughout Δ; if $|\alpha| \leqslant m$ and $\varphi \in C^m(\Delta, \mathbb{K})$, then we also denote by $D^\alpha\varphi$ the extended partial derivative on Δ. Set $C^\infty(\Delta, \mathbb{K}) := \bigcap_{m\in\mathbb{N}} C^m(\Delta, \mathbb{K})$. The Fréchet topology on $C^m(\Delta, \mathbb{K})$, resp. $C^\infty(\Delta, \mathbb{K})$, induces the following system of increasing seminorms:

$$p_n(f) := \sup_{\tau\in\Delta_n} \sup_{|\alpha|\leqslant m} |D^\alpha f(\tau)|, f \in C^m(\Delta, \mathbb{K}), \text{ resp.}$$

$$p_n(f) := \sup_{\tau\in\Delta_n} \sup_{|\alpha|\leqslant n} |D^\alpha f(\tau)|, f \in C^\infty(\Delta, \mathbb{K}), n \in \mathbb{N}.$$

Suppose, further, $X = C^m(\Delta, \mathbb{K})$ for some $m \in \mathbb{N}_0$ or $X = C^\infty(\Delta, \mathbb{K})$. It can be easily verified that the translation semigroup $(T(t))_{t\in\Delta}$ is a locally equicontinuous semigroup in X. We will prove that $(T(t))_{t\in\Delta}$ is chaotic by means of concrete construction of periodic points. Suppose that $\Delta = \Delta(\alpha)$ for some $\alpha \in (0, \frac{\pi}{2}]$ and a C^∞-function $\varphi_\theta : \mathbb{R}^2 \to \mathbb{R}$ satisfies, for every $\theta \in (0, \infty)$,

$$\varphi_\theta(\tau) = \begin{cases} 1, \ |\tau| \leqslant \theta, \\ 0, \ |\tau| \geqslant \theta + 1. \end{cases}$$

Define, for every $f \in X$ and $n \in \mathbb{N}$, the function $f_n : \Delta \to \mathbb{K}$ by $f_n(\tau) := f(\tau)\varphi_n(\tau)$, $\tau \in \Delta$. Clearly, $f_n \in X$ and $\lim_{n\to\infty} f_n = f$ in X. Hence, the set $\{f \in X : \operatorname{supp}(f)$ is a compact subset of $\Delta\}$ is dense in X which implies that X_0 is dense in X. Let us prove that X_∞ is also dense in X. Suppose $g \in X$ and $\operatorname{supp}(g) \subseteq \Delta_\theta$ for some $\theta > 0$. The well known extension type theorems for continuously differentiable functions and the Whitney extension theorem (see [411, p. 350], [456, pp. 305-306] and [496]) imply that there exists a C^m (C^∞) function $\tilde{g} : \mathbb{R}^2 \to \mathbb{K}$ such that $\tilde{g}(\tau) = g(\tau)$, $\tau \in \Delta_{\theta+1}$. Define now, for all $t \in \Delta$ with $|t| > 2\theta + 2$,

$$g_t(\tau) := \begin{cases} \tilde{g}(\tau - t)\varphi_{\theta+1}(\tau - t), \ \tau \in B(t, \theta + 1) \cap \Delta, \\ 0, \text{ otherwise.} \end{cases}$$

It is evident that $g_t \in X$ for all $t \in \Delta$ with $|t| > 2\theta + 2$ and that there exists $n_0 \in \mathbb{N}$ such that $g_n \in X$, $T(n)g_n = g$, $n \geqslant n_0$ and $\lim_{n\to\infty} g_n = 0$ in X. This implies that X_∞ is dense in X. It is well known that the set of all polynomials with rational coefficients is sequentially dense in X; especially, X is separable and Theorem 3.1.4 yields that $(T(t))_{t\in\Delta}$ is topologically transitive. To prove that the set of all periodic points of $(T(t))_{t\in\Delta}$ is dense in X, let us define, for all sufficiently large numbers $n \in \mathbb{N}$, the function $\upsilon_n : \Delta \to \mathbb{K}$ by setting

$$\upsilon_n(\tau) := \begin{cases} g(\tau), \ \tau \in \Delta_\theta, \\ \tilde{g}(\tau - nk)\varphi_{\theta+1}(\tau - nk), \ \tau \in \bigcup_{k\in\mathbb{N}}(nk + B(0, \theta + 1)), \\ 0, \text{ otherwise.} \end{cases}$$

Then $T(n)\upsilon_n = \upsilon_n$, $n \geqslant n_0$ and $\lim_{n\to\infty} \upsilon_n = g$ in X, which completes the proof in case $\Delta = \Delta(\alpha)$. The proof in case $\Delta = [0, \infty)$ follows by making use of E. Borel's theorem (cf. [411, p. 324]) and mollification, while the proof in case $\Delta \in \{\mathbb{R}, \mathbb{C}\}$ is much easier.

It is also worth noting that $(T(t))_{t\in\Delta}$ is topologically mixing. We will prove this provided that $\Delta = \Delta(\alpha)$. So, fix an $\varepsilon > 0$ and a function $g \in X$ with $\operatorname{supp}(g) \subseteq \Delta_\theta$ for some $\theta > 0$. Suppose now that $r \in (2\theta + 2, \infty)$ and $2^{\theta+1-r} < \frac{\varepsilon}{4}$. Then $T(t)g_t = g$ and $d(0, g_t) = \sum_{n=1}^{\infty} \frac{1}{2^n} \frac{p_n(g_t)}{1+p_n(g_t)} = \sum_{n=\lceil r-\theta-1 \rceil}^{\infty} \frac{1}{2^n} \frac{p_n(g_t)}{1+p_n(g_t)} \leqslant \frac{1}{2^{\lceil r-\theta-1 \rceil - 1}} < \varepsilon$ for all $t \in \Delta \setminus \Delta_r$. Since $\lim_{t\to\infty, t\in\Delta} T(t)f = 0$ for all $f \in X$ with compact support, one can proceed as in the proofs of [161, Theorems 2.2.-2.3] in order to see that, for every f, $g \in X$ and $\varepsilon > 0$, there exists $r > 0$ such that for every $t \in \Delta \setminus \Delta_r$ there exists $\upsilon \in X$ so that $d(f, \upsilon) < \varepsilon$ and $d(g, T(t)\upsilon) < \varepsilon$. The last statement simply implies that $(T(t))_{t\in\Delta}$ is topologically mixing. Finally, let us notice that the generator A of $(T(t))_{t\in\Delta}$, where $\Delta \in \{[0, \infty), \mathbb{R}\}$, satisfies $\sigma_p(A) = \mathbb{K}$.

Let X be a separable infinite-dimensional complex Banach space. An uncountable set $\Lambda \subseteq X$ is said to be scrambled if for every pair $x, y \in \Lambda$ of distinct points we have that

$$\liminf_{n\to\infty} \|T(t)x - T(t)y\| = 0 \text{ and } \limsup_{n\to\infty} \|T(t)x - T(t)y\| > 0.$$

Distributional chaos is a very active field of research in the theory of hypercyclicity. A strongly continuous semigroup $(T(t))_{t\geqslant 0}$ on X is said to be distributionally chaotic if there are an uncountable set $S \subseteq X$ and $\sigma > 0$ such that for each $\varepsilon > 0$ and for each pair $x, y \in S$ of distinct points we have that

$$\overline{Dens}(\{s \geqslant 0 : \|T(s)x - T(s)y\| \geqslant \sigma\}) = 1 \text{ and}$$
$$\overline{Dens}(\{s \geqslant 0 : \|T(s)x - T(s)y\| < \varepsilon\}) = 1,$$

where the upper density of a set $D \subseteq [0, \infty)$ is defined by

$$\overline{Dens}(D) := \limsup_{t\to+\infty} \frac{m(D \cap [0,t])}{t},$$

with $m(\cdot)$ being the Lebesgue's measure on $[0, \infty)$. If, moreover, we can choose S to be dense in X, then $(T(t))_{t\geqslant 0}$ is said to be densely distributionally chaotic. The question whether $(T(t))_{t\geqslant 0}$ is distributionally chaotic or not is closely connected with the existence of distributionally irregular vectors, i.e., those elements $x \in X$ such that for each $\sigma > 0$ we have that

$$\overline{Dens}(\{s \geqslant 0 : \|T(s)x\| > \sigma\}) = 1 \text{ and } \overline{Dens}(\{s \geqslant 0 : \|T(s)x\| < \sigma\}) = 1.$$

For further information concerning distributionally chaotic strongly continuous semigroups, as well as distributionally chaotic cosine functions and fractional PDEs, the reader may consult the references [6], [41]-[42] and [122]. The basic facts about frequently hypercyclic semigroups can be found in [401]; concerning hypercyclic strongly continuous semigroups and hypercyclic single operators, mention should be made of the references [125], [208] and [520], too.

3.1.1. Disjoint hypercyclic semigroups. Before we introduce the notion of disjoint hypercyclic semigroups, our first task will be to enquire into the basic structural properties of positively supercyclic strongly continuous semigroups induced by locally Lipschitz continuous semiflows in the setting of weight L^p and C_0-type spaces. Let $\alpha \in (0, \frac{\pi}{2}]$, $\delta > 0$ and $I \neq \emptyset$. Suppose $\Delta \in \{[0, \infty), \mathbb{R}, \mathbb{C}\}$ or $\Delta = \Delta(\alpha)$ for an appropriate angle $\alpha \in (0, \frac{\pi}{2}]$.

Suppose that X is an infinite-dimensional separable Fréchet space over the field $\mathbb{K} \in \{\mathbb{R}, \mathbb{C}\}$. It is said that an operator family $(S(\tau))_{\tau\in I}$ ($S(\tau) \in L(X)$, $\tau \in I$) is:

(i) hypercyclic, if there exists $x \in X$ whose orbit $\{S(\tau)x : \tau \in I\}$ is dense in X,
(ii) topologically transitive, if for every open non-empty subsets U, V of X, there exists $\tau \in I$ such that $S(\tau)U \cap V \neq \emptyset$,

(iii) supercyclic, if there exists $x \in X$ such that its projective orbit $\{cS(\tau)x : c \in \mathbb{K}, \tau \in I\}$ is dense in X,
(iv) positively supercyclic, if there exists $x \in X$ such that its positive projective orbit $\{cS(\tau)x : c \geqslant 0, \tau \in I\}$ is dense in X.

Before we continue any further, observe that the notions of many other hypercyclic properties can be introduced for $(S(\tau))_{\tau \in I}$. Recall that T. Kalmes [244]-[245] has analyzed the hypercyclicity of strongly continuous semigroups induced by semiflows. The state space in his analysis is chosen to be the space $C_{0,\rho}(X_1, \mathbb{K})$, resp. $L^p(X_1, \mu, \mathbb{K})$, where X_1 is a locally compact, Hausdorff space and $\rho : X_1 \to (0, \infty)$ is an upper semicontinuous function, resp. X_1 is a locally compact, σ-compact Hausdorff space, $p \in [1, \infty)$ and μ is a locally finite Borel measure on X_1. Let us recall that the space $C_{0,\rho}(X_1, \mathbb{K})$ consists of all continuous functions $f : X_1 \to \mathbb{K}$ satisfying that, for every $\varepsilon > 0$, $\{x \in X_1 : |f(x)|\rho(x) \geqslant \varepsilon\}$ is a compact subset of X_1; equipped with the norm $\|f\| := \sup_{x \in X_1} |f(x)|\rho(x)$, $C_{0,\rho}(X_1, \mathbb{K})$ becomes a Banach space. Let $L^p(X_1, \mu, \mathbb{K})$ be the classical space of p-integrable function and let $C_c(X_1, \mathbb{K})$ be the space of all continuous functions $f : X_1 \to \mathbb{K}$ whose support is a compact subset of X_1. Then $C_c(X_1, \mathbb{K})$ is dense in $L^p(X_1, \mu, \mathbb{K})$ and in $C_{0,\rho}(X_1, \mathbb{K})$. For the purpose of research of strongly continuous semigroups induced by non-differentiable locally Lipschitz continuous semiflows (cf. Example 3.1.34), we primarily deal with the space $L^p_{\rho_1}(\Omega, \mathbb{K})$, where Ω is an open non-empty subset of $\mathbb{R}^n$, $\rho_1 : \Omega \to (0, \infty)$ is a locally integrable function, m_n is the Lebesgue measure in $\mathbb{R}^n$ and the norm of an element $f \in L^p_{\rho_1}(\Omega, \mathbb{K})$ is given by $\|f\|_p := (\int_\Omega |f(\cdot)|^p \rho_1(\cdot)\, dm_n)^{1/p}$. Let $C(\Omega, \mathbb{K})$ be the $\mathbb{K}$-vector space consisting of all continuous functions from Ω into $\mathbb{K}$. We equip $C(\Omega, \mathbb{K})$ with the usual Fréchet topology. In the sequel, it will not be confusing to write $C_{0,\rho}(X_1)$, $L^p_{\rho_1}(\Omega)$, $C_c(\Omega)$, and m, instead of $C_{0,\rho}(X_1, \mathbb{K})$, $L^p_{\rho_1}(\Omega, \mathbb{K})$, $C_c(\Omega, \mathbb{K})$, and m_n, respectively.

In Theorem 3.1.20, we draw our attention to the study of positive supercyclicity of strongly continuous semigroups induced by semiflows, continuing the research of M. Matsui, M. Yamada and F. Takeo [405]. The full importance of positive supercyclicity of strongly continuous semigroups is vividly exhibited in Example 3.1.37.

On the other hand, disjointness for finitely many operators has been introduced by L. Bernal-Gonzáles [59] and J. Bès, A. Peris [63]. Our objective in this subsection is to extend the notion of disjoint hypercyclicity to strongly continuous semigroups. We establish sufficient conditions for d-hypercyclicity of strongly continuous semigroups on the Fréchet space $C(\Omega)$ and on a class of weighted function spaces. The concrete construction of d-hypercyclic semigroups induced by semiflows, obtained by means of Theorem 3.1.39 and Theorem 3.1.40 given below, is one of the main purposes of this subsection.

The following recollection of well known results from real analysis and measure theory will be helpful in our further work.

Theorem 3.1.30. *Suppose $k, n \in \mathbb{N}$ and Ω is an open non-empty subset of $\mathbb{R}^n$.*

(i) *(Brouwer's theorem, [140]) Suppose that the mapping $f : \Omega \to \mathbb{R}^n$ is continuous and injective. Then $f(\Omega)$ is an open subset of $\mathbb{R}^n$.*

(ii) *(Rademacher's theorem, [64], [183]) Suppose $f : \Omega \to \mathbb{R}^k$ is a locally Lipschitz continuous function. Then $f(\cdot)$ is differentiable at almost every point in Ω.*

(iii) *(The change of variables in Lebesgue's integral, [220], [421]) Suppose $f : \Omega \to \mathbb{R}^n$ is locally Lipschitz continuous and injective. Then, for every measurable subset E of Ω, $f(E)$ is a measurable subset of $\mathbb{R}^n$. Suppose, further, that $g : f(\Omega) \to \mathbb{R}$ is a measurable function and the function $x \mapsto g(x)$ is integrable on $f(E)$. Then the function $x \mapsto g(f(x))|\det Df(x)|$ is integrable on E and the following formula holds:*

$$\int_{f(E)} g(x)\, dx = \int_E g(f(x))|\det Df(x)|\, dx,$$

where $Df(\cdot)$ denotes the Jacobian of the mapping $f(\cdot)$ (which exists for a.e. $x \in \Omega$).

(iv) *([374], [421]) Suppose that the mapping $f : \Omega \to \mathbb{R}^n$ is locally Lipschitz continuous. Then for every measurable set $E \subseteq \Omega$, we have that $m(E) = 0$ implies $m(f(E)) = 0$.*

Given a number $t \in \Delta$, a semiflow $\varphi : \Delta \times \Omega \to \Omega$ and a function $f : \Omega \to \mathbb{K}$, we define $T_\varphi(t)f : \Omega \to \mathbb{K}$ as before $(T_\varphi(t)f)(x) =: f(\varphi(t, x))$, $x \in \Omega$. Recall that $T_\varphi(0) f = f$, $T_\varphi(t)T_\varphi(s)f = T_\varphi(s)T_\varphi(t)f = T_\varphi(t + s)f$, $t, s \in \Delta$, and that Brouwer's theorem implies $C_c(\Omega) \subseteq T_\varphi(t)(C_c(\Omega))$; the necessary and sufficient conditions stating when the composition operator $T_\varphi(t) : L^p_{\rho_1}(\Omega) \to L^p_{\rho_1}(\Omega)$, resp. $T_\varphi(t) : C_{0,\rho}(\Omega) \to C_{0,\rho}(\Omega)$, is well defined and continuous can be found in [245, Theorem 2.1], resp. [245, Theorem 2.2]. In order to see when the semigroup $(T_\varphi(t))_{t\in\Delta}$ is strongly continuous in $L^p_{\rho_1}(\Omega)$, resp. $C_{0,\rho}(\Omega)$, we need the following auxiliary lemma which is inspired by [245, Proposition 3.2].

Lemma 3.1.31. *Suppose $\varphi : \Delta \times \Omega \to \Omega$ is a semiflow. Then for every compact set $K \subseteq \Omega$ and for every $\delta > 0$ with $K + B(0, \delta) \subseteq \Omega$, there exists $n \in \mathbb{N}$ such that:*

$$K \cap \varphi(t, (\Omega \setminus (K + B(0, \delta)))) = \emptyset \textit{ for all } t \in \Delta_{\frac{1}{n}}.$$

Herein $B(0, \delta) = \{x \in \mathbb{R}^n : |x| < \delta\}$ and $K + B(0, \delta) = \{x + y : x \in K, y \in B(0, \delta)\}$.

Proof. We will sketch the proof only in the non-trivial case $\Delta = \Delta(\alpha)$, where $\alpha \in (0, \frac{\pi}{2})$. For the sake of argument suppose that, for every $n \in \mathbb{N}$, there exist $t_n \in \Delta_{\frac{1}{n}}$ and $x_n \in \Omega \setminus (K + B(0, \delta))$ such that $y_n = \varphi(t_n, x_n) \in K$. The continuity of $\varphi(\cdot,\cdot)$ implies that there exist $\tilde{t}_1 \in (\Delta_1)^\circ$ and $x_1 \in \Omega \setminus (K + B(0, \delta))$ such that $\varphi(\tilde{t}_1, x_1) \in K + B(0, \frac{\delta}{2})$. Put $\tilde{x}_1 := x_1$ and choose a natural number $n_1 \geqslant 2$ such that $\tilde{t}_1 - \Delta_{\frac{1}{n_1}} \in \Delta^\circ$. Apply again the continuity of $\varphi(\cdot,\cdot)$ in order to conclude that there exists $t'_{n_1} \in (\Delta_{\frac{1}{n_1}})^\circ$ such that $\varphi(t'_{n_1}, x_{n_1}) \in K + B(0, \frac{\delta}{2})$. Put $\tilde{t}_2 := t'_{n_1}$ and $\tilde{x}_2 := x_{n_1}$. Then $\tilde{t}_2 \in (\Delta_{\frac{1}{2}})^\circ$, $\tilde{x}_2 \in \Omega \setminus (K + B(0,$

$\delta)),\ \varphi(\tilde{t}_2, \tilde{x}_2) \in K + B(0, \frac{\delta}{2})$ and $\tilde{t}_1 - \tilde{t}_2 \in \Delta^\circ$. Inductively, one obtains the existence of a sequence $(\tilde{t}_n)$ in Δ° and a sequence $(\tilde{x}_n)$ in $\Omega\backslash(K + B(0, \delta))$ such that: $\tilde{t}_n \in (\Delta_{\frac{1}{n}})^\circ$, $\tilde{t}_n - \tilde{t}_{n+1} \in \Delta^\circ$ and $\varphi(\tilde{t}_n, \tilde{x}_n) \in K + B(0, \frac{\delta}{2})$, $n \in \mathbb{N}$. Especially, $\tilde{t}_1 - \tilde{t}_n \in \Delta^\circ$, $n \in \mathbb{N}$ and, without loss of generality, we may assume that $\lim_{n\to\infty} \varphi(\tilde{t}_n, \tilde{x}_n) = x \in K + B(0, \frac{\delta}{2})$. Then one gets $\lim_{n\to\infty} \varphi(\tilde{t}_1, \tilde{x}_n) = \lim_{n\to\infty} \varphi(\tilde{t}_1 - \tilde{t}_n, \varphi(\tilde{t}_n, \tilde{x}_n)) = \varphi(\tilde{t}_1, x)$. Since the mapping $\varphi(\tilde{t}_1, \cdot) : \Omega \to \Omega$ is continuous and injective, Brouwer's theorem implies that the inverse mapping $\varphi(\tilde{t}_1, \cdot)^{-1} : \varphi(\tilde{t}_1, \Omega) \to \Omega$ is continuous. Hence, one obtains that $\lim_{n\to\infty} \tilde{x}_n = x$ contradicting $\tilde{x}_n \in \Omega\backslash(K + B(0, \delta))$. $\square$

Lemma 3.1.32. *Let $f : \Omega \to \Omega$ be locally Lipschitz continuous and injective, and let $f^{-1}(\cdot)$ also be locally Lipschitz continuous. Then $Df(x)Df^{-1}(f(x)) = I$ a.e. $x \in \Omega$, where I denotes the identity matrix.*

Proof. Denote $N := \{x \in \Omega : f(\cdot)$ is not differentiable in $x\}$ and $N^- := \{x \in f(\Omega) : f^{-1}(\cdot)$ is not differentiable in $x\}$. Then $m(N) = m(N^-) = 0$, $f^{-1}(N^-) = \{x \in \Omega : f^{-1}$ is not differentiable in $f(x)\}$, and by Theorem 3.1.30(iv), $m(f^{-1}(N^-)) = 0$. This implies $m(N \cup f^{-1}(N^-)) = 0$ and by the chain rule we have $Df(x)Df^{-1}(f(x)) = I$, $x \in \Omega\backslash(N \cup f^{-1}(N^-))$. $\square$

Theorem 3.1.33. *Suppose $\varphi : \Delta \times \Omega \to \Omega$ is a semiflow and $\varphi(t, \cdot)$ is a locally Lipschitz continuous function for all $t \in \Delta$. Then (ii) implies (i), where:*

(i) $(T_\varphi(t))_{t\in\Delta}$ *is a strongly continuous semigroup in* $L^p_{\rho_1}(\Omega)$ *and*
(ii) $\exists M, \omega \in \mathbb{R}\ \forall t \in \Delta : \rho_1(\cdot) \leqslant Me^{\omega|t|}\rho_1(\varphi(t, \cdot))|\det D\varphi(t, \cdot)|$ *a.e.*

If, additionally, $\varphi(t, \cdot)^{-1}$ is locally Lipschitz continuous for all $t \in \Delta$, then the above are equivalent.

Proof. Suppose that (ii) holds. Then Theorem 3.1.30(iii) implies:

$$\begin{aligned}
||T_\varphi(t)f||^p &= \int_\Omega |f(\varphi(t, x))|^p \rho_1(x)\, dx \\
&\leqslant Me^{\omega|t|} \int_\Omega |f(\varphi(t, x))|^p \rho_1(\varphi(t, x))|\det D\varphi(t, x)|\, dx \\
&= Me^{\omega|t|} \int_{\varphi(t,\Omega)} |f(x)|^p \rho_1(x)\, dx \leqslant Me^{\omega|t|}||f||^p,\ t \in \Delta, f \in L^p_{\rho_1}(\Omega).
\end{aligned}$$

Hence, $T_\varphi(t) \in L(L^p_{\rho_1}(\Omega))$ and

$$||T_\varphi(t)|| \leqslant M^{1/p}e^{\omega|t|/p},\ t \in \Delta. \tag{508}$$

Furthermore, the dominated convergence theorem and Lemma 3.1.31 imply that $\lim_{t\to0, t\in\Delta} T_\varphi(t)f = f$ for all $f \in C_c(\Omega)$; then the strong continuity of $(T_\varphi(t))_{t\in\Delta}$ follows easily from the standard limit procedure and (508). Suppose now that $\varphi(t, \cdot)^{-1}$ is locally Lipschitz continuous for all $t \in \Delta$ and (i) holds. The existence of real numbers M and ω satisfying (508) is obvious and, as a simple consequence of Theorem 3.1.30(iii), one obtains:

(509) $$\int_{\varphi(t,\cdot)^{-1}(L\cap\varphi(t,\Omega))} \rho_1(\cdot)\, dm = \int_L \chi_{\varphi(t,\Omega)}(\cdot)\rho_1(\varphi(t,\cdot)^{-1})|\det D\varphi(t,\cdot)^{-1}|\, dm,\ t \in \Delta.$$

Then one can apply [245, Theorem 2.1] and (509) (cf. also [244, Appendix B]) in order to see that, for every $t \in \Delta$, the inequality:

(510) $$\chi_{\varphi(t,\Omega)}(\cdot)\rho_1(\varphi(t,\cdot)^{-1})|\det D\varphi(t,\cdot)^{-1}| \leqslant Me^{\omega|t|}\rho_1(\cdot)$$

holds almost everywhere in Ω. By Lemma 3.1.32, we have

(511) $$\det D\varphi(t,x) \times \det D\varphi(t,\cdot)^{-1}(\varphi(t,x)) = 1 \text{ for a.e. } x \in \Omega.$$

In view of (510)-(511), we get that there exists a measurable subset N of Ω such that $m(N) = 0$ and

(512)

$$\chi_{\varphi(t,\Omega)}(y)\rho_1(\varphi(t,y)^{-1}) \leqslant Me^{\omega|t|}\chi_{\varphi(t,\Omega)}(y)\rho_1(y)|\det D\varphi(t,\varphi(t,y)^{-1})|,\ y \in \Omega\setminus N.$$

By Theorem 3.1.30(iv), we obtain that $m(\varphi(t,\cdot)^{-1}(N)) = 0$ and, thanks to (512), (ii) holds for every $x \in \Omega\setminus(N \cup \varphi(t,\cdot)^{-1}(N))$. This completes the proof of the theorem. □

Suppose $T_\varphi(t) : L^p_{\rho_1}(\Omega) \to L^p_{\rho_1}(\Omega)$ is well defined and continuous for all $t \in \Delta$. Since $R(T_\varphi(t))$, $t \in \Delta$ is dense in $L^p_{\rho_1}(\Omega)$, one can employ [210, Theorem 1, Proposition 1] in order to see that the hypercyclicity of $(T_\varphi(t))_{t\in\Delta}$ is equivalent to its topological transitivity. By [245, Theorem 2.4], $(T_\varphi(t))_{t\in\Delta}$ is hypercyclic in $L^p_{\rho_1}(\Omega)$ if for every compact set $K \subseteq \Omega$ there exist a sequence of measurable subsets (L_k) of K and a sequence (t_k) in Δ such that:

$$\lim_{k\to\infty} \int_{K\setminus L_k} \rho_1(x)\, dx = 0,\ \lim_{k\to\infty} \int_{\varphi(t_k, L_k)} \rho_1(x)\, dx = 0 \text{ and}$$

(513) $$\lim_{k\to\infty} \int_{\varphi(t_k,\cdot)^{-1}(L_k)} \rho_1(x)\, dx = 0.$$

Example 3.1.34. Let $\Delta = [0,\infty)$, $\Omega = (0,\infty)$, $p \in [1,\infty)$ and let (a_n) be a decreasing sequence of positive real numbers satisfying $\Sigma_{n=1}^{\infty} a_n = \infty$. Put, by common consent, $\Sigma_{i=1}^{0} a_i := 0$ and define $f : (0,\infty) \to (0,\infty)$ by $f(x) := a_{n+1}(x-n) + \Sigma_{i=1}^{n} a_i$ if $x \in (n, n+1]$ for some $n \in \mathbb{N}_0$. Then $f(\cdot)$ is a strictly increasing, bijective and locally Lipschitz continuous mapping, and moreover, the inverse mapping $f^{-1} : (0,\infty) \to (0,\infty)$ possesses the same properties. Define $\varphi : \Delta \times \Omega \to \Omega$ and $\rho_1 : \Omega \to (0,\infty)$ by $\varphi(t,x) := f^{-1}(t + f(x))$ and $\rho_1(x) := \frac{1}{f(x)+1}$, $t \in \Delta$, $x \in \Omega$. Clearly, the mapping $\varphi(\cdot,\cdot)$ is a semiflow and the mapping $x \mapsto \varphi(t,x)$, $x \in \Omega$ is locally Lipschitz continuous for every fixed $t \in \Delta$. In general, the mapping $x \mapsto \varphi(t,x)$, $x \in \Omega$ need not be differentiable and one can simply verify that $\frac{d}{dx}f(x) = a_{n+1}$, $x \in (n, n+1)$, $n \in \mathbb{N}_0$ and $\frac{d}{dx}f^{-1}(x) = \frac{1}{a_{n+1}}$, $x \in (\Sigma_{i=1}^{n} a_i, \Sigma_{i=1}^{n+1} a_i)$, $n \in \mathbb{N}_0$. Suppose $t \geqslant 0$, $n \in \mathbb{N}_0$, $k \in \mathbb{N}_0$, $x \in (n, n+1)$ and $t + f(x) \in (\Sigma_{i=1}^{k} a_i, \Sigma_{i=1}^{k+1} a_i)$. Then $k \geqslant n$, $\frac{d}{dx}\varphi(t,x) = \frac{a_{n+1}}{a_{k+1}} \geqslant 1$, $\frac{\rho_1(x)}{\rho_1(\varphi(t,x))} =$

$1 + \frac{1}{f(x)+1} \leqslant 1 + t \leqslant e^t \leqslant e^t|\frac{d}{dx}\varphi(t, x)|$, and Theorem 3.1.33 implies that $(T_\varphi(t))_{t\geqslant 0}$ is a strongly continuous semigroup in $L^p_{\rho_1}(\Omega)$. Let us prove that $(T_\varphi(t))_{t\geqslant 0}$ is hypercyclic whenever the sequence $(\frac{1}{a_n})$ is bounded. Suppose $K = [a, b] \subseteq (0, \infty)$, (t_k) is any sequence of positive real numbers satisfying $\lim_{k\to\infty} t_k = \infty$ and $M := \sup_{n\in\mathbb{N}}\{\frac{1}{a_n}: n \in \mathbb{N}\}$. Notice that, for every $k \in \mathbb{N}$ and $n \in \mathbb{N}$ with $t_k + f(b) < \Sigma_{i=1}^n a_i$,

$$\left|f^{-1}(t_k + f(b)) - f^{-1}(t_k + f(a))\right| \leqslant \max\left(\frac{1}{a_1}, \cdots, \frac{1}{a_{n+1}}\right) |f(b) - f(a)|.$$

This inequality implies

$$\int_{\varphi(t_k, K)} \rho_1(x)\, dx = \int_{f^{-1}(t_k + f(a))}^{f^{-1}(t_k + f(b))} \frac{dx}{f(x)+1}$$

$$\leqslant \frac{f^{-1}(t_k + f(b)) - f^{-1}(t_k + f(a))}{f(a) + t_k + 1} \leqslant M \frac{f(b) - f(a)}{f(a) + t_k + 1}$$

and $\lim_{k\to\infty} \int_{\varphi(t_k, K)} \rho_1(x)\, dx = 0$. Furthermore, it is clear that there exists $k_0 \in \mathbb{N}$ such that $\varphi(t_k, \cdot)^{-1}(K) = \emptyset$, $k \geqslant k_0$; hence, $\lim_{k\to\infty} \int_{\varphi(t_k,\cdot)^{-1}(K)} \rho_1(x)\, dx = 0$, (513) holds and $(T_\varphi(t))_{t\geqslant 0}$ is hypercyclic, as claimed.

Taking into account Lemma 3.1.31 and the proof of [245, Theorem 3.4], one immediately obtains the following theorem which states when $(T_\varphi(t))_{t\in\Delta}$ is a strongly continuous semigroup in $C_{0,\rho}(\Omega)$.

Theorem 3.1.35. *Let $\varphi : \Delta \times \Omega \to \Omega$ be a semiflow. Then $(T_\varphi(t))_{t\in\Delta}$ is a strongly continuous semigroup in $C_{0,\rho}(\Omega)$ iff the following holds:*

(i) *$\exists M, \omega \in \mathbb{R}\ \forall t \in \Delta\ \forall x \in \Omega : \rho(x) \leqslant Me^{\omega|t|}\, \rho(\varphi(t, x))$ and*
(ii) *for every compact set $K \subseteq \Omega$ and, for every $\delta > 0$ and $t \in \Delta$,*

$$\varphi(t, \cdot)^{-1}(K) \cap \{x \in \Omega : \rho(x) \geqslant \delta\} \text{ is a compact subset of } \Omega.$$

Suppose that $T_\varphi(t) : C_{0,\rho}(\Omega) \to C_{0,\rho}(\Omega)$ is well defined and continuous for all $t \in \Delta$ as well as that, for every compact set $K \subseteq \Omega$, we have $\inf_{x\in K} \rho(x) > 0$. Then [245, Corollary 2.11] implies that $(T_\varphi(t))_{t\in\Delta}$ is hypercyclic in $C_{0,\rho}(\Omega)$ iff for every compact set $K \subseteq \Omega$ there exists a sequence (t_k) in Δ such that:

$$\lim_{k\to\infty} \sup_{x\in\varphi(t_k,\cdot)^{-1}(K)} \rho(x) = \lim_{k\to\infty} \sup_{x\in\varphi(t_k,K)} \rho(x) = 0.$$

Theorem 3.1.36. *Let $\varphi : \Delta \times \Omega \to \Omega$ be a semiflow.*

(i) *Suppose $T_\varphi(t) : L^p_{\rho_1}(\Omega) \to L^p_{\rho_1}(\Omega)$ is well defined and continuous for all $t \in \Delta$. Then the following assertions are equivalent.*
   (i.1) *$(T_\varphi(t))_{t\in\Delta}$ is positively supercyclic in $L^p_{\rho_1}(\Omega)$.*
   (i.2) *For every compact set $K \subseteq \Omega$ there exist a sequence (L_k) of measurable subsets of K, a sequence (t_k) in Δ and a sequence (c_k) in $(0, \infty)$ such that:*

$$\lim_{k\to\infty} \int_{K\setminus L_k} \rho_1(x)\, dx = 0 \text{ and} \tag{514}$$

$$\lim_{k\to\infty} c_k \int_{\varphi(t_k,\cdot)^{-1}(L_k)} \rho_1(x)\, dx = \lim_{k\to\infty} \frac{1}{c_k} \int_{\varphi(t_k,L_k)} \rho_1(x)\, dx = 0.$$

(i.3) *For every compact set* $K \subseteq \Omega$ *there exist a sequence* (L_k) *of measurable subsets of* K *and a sequence* (t_k) *in* Δ *such that (514) holds and*

$$\lim_{k\to\infty}\left[\int_{\varphi(t_k,\cdot)^{-1}(L_k)} \rho_1(x)\, dx * \int_{\varphi(t_k,L_k)} \rho_1(x)\, dx\right] = 0.$$

(ii) *Suppose that* $T_\varphi(t) : C_{0,\rho}(\Omega) \to C_{0,\rho}(\Omega)$ *is well defined and continuous for all* $t \in \Delta$ *as well as that, for every compact set* $K \subseteq \Omega$, *we have* $\inf_{x\in K} \rho(x) > 0$. *Then the following assertions are equivalent.*

(ii.1) $(T_\varphi(t))_{t\in\Delta}$ *is positively supercyclic in* $C_{0,\rho}(\Omega)$.

(ii.2) *For every compact set* $K \subseteq \Omega$ *there exist a sequence* (t_k) *in* Δ *and a sequence* (c_k) *in* $(0, \infty)$ *such that:*

$$\lim_{k\to\infty} c_k \sup_{x\in\varphi(t_k,\cdot)^{-1}(K)} \rho(x) = \lim_{k\to\infty} \frac{1}{c_k} \sup_{x\in\varphi(t_k,K)} \rho(x) = 0.$$

(ii.3) *For every compact set* $K \subseteq \Omega$ *there exists a sequence* (t_k) *in* Δ *such that:*

$$\lim_{k\to\infty}\left[\sup_{x\in\varphi(t_k,\cdot)^{-1}(K)} \rho(x) * \lim_{k\to\infty} \sup_{x\in\varphi(t_k,K)} \rho(x)\right] = 0.$$

Proof. Put $I := \{(c, t) : c > 0, t \in \Delta\}$, $T_\varphi(c, t) := cT_\varphi(t)$, $(c, t) \in I$ and notice that the operators $T_\varphi(c, t)$, $(c, t) \in I$ have dense ranges and commute with each other. According to [210, Theorem 1, Proposition 1], one obtains that the positive supercyclicity of $(T_\varphi(t))_{t\in\Delta}$ is equivalent with topological transitivity of $(T_\varphi(c, t))_{(c,t)\in I}$. In view of this, the equivalence of (i.1) and (i.2) follows automatically from an application of [245, Theorem 4.3]. Suppose now K is a compact subset of Ω. Then there exist a sequence (L_k) of measurable subsets of K and a sequence (t_k) in Δ such that (514) and (503) hold. Notice that, for two arbitrary sequences of non-negative real numbers $(\alpha_k)_{k\in\mathbb{N}}$ and $(\beta_k)_{k\in\mathbb{N}}$ with $\lim_{k\to\infty} \alpha_k\beta_k = 0$ there are subsequences $(\alpha_{k_l})_{l\in\mathbb{N}}$ and $(\beta_{k_l})_{l\in\mathbb{N}}$ as well as a sequence $(c_l)_{l\in\mathbb{N}}$ of positive numbers such that $\lim_{l\to\infty} c_l\alpha_{k_l} = \lim_{l\to\infty} c_l^{-1}\beta_{k_l} = 0$ simply by choosing $(k_l)_{l\in\mathbb{N}}$ as a strictly increasing sequence of natural numbers with $k_l > l^2$ and $\alpha_k\beta_k < 1/l^2$ for all $k \geqslant k_l$ and by setting $c_l := l(\beta_{k_l} + k_l^{-1}(1 + \alpha_{k_l})^{-1})$. The proof of implication (i.3) $\Rightarrow$ (i.2) follows by applying this to $\alpha_k = \int_{\varphi(t_k,\cdot)^{-1}(L_{k_l})} \rho_1(x)\, dx$ and $\beta_k = \int_{\varphi(t_{k_l},L_{k_l})} \rho_1(x)\, dx$. The proof of part (ii) is done in exactly the same way as the proof of part (i), so it can be omitted. □

Concerning Theorem 3.1.36(i), let us stress that it is not clear whether, as in the case of hypercyclicity (cf. [245, Example 3.19]), we can get into a situation where one must choose a sequence (L_k) of measurable subsets of K which satisfies $L_k \neq K$, $k \geqslant k_0$.

Now we shall provide an example of a positively supercyclic semigroup that is not hypercyclic.

Example 3.1.37. Suppose $\Delta = \Omega = \mathbb{R}$, $m \in \mathbb{N}$, $p : \mathbb{R} \to \mathbb{R}$, $p(x) = \Sigma_{i=0}^{2m+1} a_i x^i$, $\widetilde{p}(x) = \Sigma_{i=0}^{2m+1} |a_i| x^i$, $x \in \mathbb{R}$, $a_{2m+1} > 0$ and $p'(x) \geqslant c > 0$, $x \in \mathbb{R}$. Then $p(\cdot)$ is bijective and strictly increasing so that we can define a semiflow $\varphi : \Delta \times \Omega \to \Omega$ by $\varphi(t, x) := p^{-1}(t + p(x))$, $t, x \in \mathbb{R}$. Suppose that $f : \mathbb{R} \to (0, \infty)$ is an admissible weight function. Define a locally integrable function $\rho_1 : \mathbb{R} \to (0, \infty)$ by $\rho_1(x) := f(p(x))$, $x \in \mathbb{R}$. We will prove that there exists $c_1 > 0$ such that for every $t, x_0 \in \mathbb{R}$:

$$|\det D\varphi(t, x_0)| \geqslant \frac{1}{c_1\left(1+|t|^{\frac{2m}{2m+1}}\right)}. \tag{515}$$

Notice that $|\det D\varphi(t, x_0)| = \frac{p'(x_0)}{p'(\varphi(t, x_0))}$ and define $q : \mathbb{R} \to \mathbb{R}$ by $q(x) := \frac{p(x)-(t+p(x_0))}{a_{2m+1}}$, $x \in \mathbb{R}$. Then every zero ξ of a real polynomial $r(x) = x^s + \Sigma_{i=0}^{s-1} b_i x^i$, $b_0 \neq 0$, $s \geqslant 2$ satisfies $|\xi| < 2 \max\{|b_i|^{\frac{1}{s-i}} : 0 \leqslant i \leqslant s-1\}$ ([448]). Since $q(\varphi(t, x_0)) = 0$, this assertion enables one to prove that there exists $c > 0$, independent of t and x_0, such that $|\varphi(t, x_0)|$

$$\leqslant 2 \max\left(\left|\frac{a_{2m}}{a_{2m+1}}\right|, \left|\frac{a_{2m-1}}{a_{2m+1}}\right|^{\frac{1}{2}}, \ldots, \left|\frac{a_1}{a_{2m+1}}\right|^{\frac{1}{2m}}, \left|\frac{a_0 - t - p(x_0)}{a_{2m+1}}\right|^{\frac{1}{2m+1}}\right)$$

$$\leqslant 2\left(\sum_{i=0}^{2m}\left|\frac{a_i}{a_{2m+1}}\right|^{\frac{1}{2m+1-i}} + \left|\frac{t}{a_{2m+1}}\right|^{\frac{1}{2m+1}} + \left|\frac{p(x_0)}{a_{2m+1}}\right|^{\frac{1}{2m+1}}\right)$$

$$\leqslant c\left(1 + |t|^{\frac{1}{2m+1}} + |p(x_0)|^{\frac{1}{2m+1}}\right).$$

Taken together, this estimate and the elementary inequalities $|p(x_0)| \leqslant \widetilde{p}|(x_0)|$, $1 + |t|^{\frac{i}{2m+1}} \leqslant 2(1 + |t|^{\frac{2m}{2m+1}})$, $0 \leqslant i \leqslant 2m$, $(a + b + c)^i \leqslant 3^{i-1}(a^i + b^i + c^i)$, $i \in \mathbb{N}$, $a, b, c \geqslant 0$, imply the existence of positive real number $\overline{c}$, independent of t and x_0, such that

$$|p'(\varphi(t, x_0))| \leqslant \sum_{i=1}^{2m}(i + 1)|a_{i+1}|c^i\left(1 + |t|^{\frac{1}{2m+1}} + |p(x_0)|^{\frac{1}{2m+1}}\right)^i + |a_1|$$

$$\leqslant \sum_{i=1}^{2m}(i + 1)|a_{i+1}|c^i 3^{i-1}\left(1 + |t|^{\frac{i}{2m+1}} + |p(x_0)|^{\frac{i}{2m+1}}\right) + |a_1|$$

$$\leqslant \overline{c}\left(1 + |t|^{\frac{2m}{2m+1}} + \sum_{i=0}^{2m} |p(x_0)|^{\frac{i}{2m+1}}\right)$$

$$\leqslant \overline{c}\left(1 + |t|^{\frac{2m}{2m+1}} + \sum_{i=0}^{2m} \widetilde{p}(|x_0|)^{\frac{i}{2m+1}}\right)$$

$$\leqslant 2\overline{c}\left(1 + |t|^{\frac{2m}{2m+1}}\right) \sum_{i=0}^{2m} \widetilde{p}(|x_0|)^{\frac{i}{2m+1}}.$$

Hence,

(516) $$|\det D\varphi(t, x_0)| = \frac{p'(x_0)}{p'(\varphi(t,x_0))} \geqslant \frac{|p'(x_0)|}{2\overline{c}\left(1+|t|^{\frac{2m}{2m+1}}\right)\sum_{i=0}^{2m} \tilde{p}(|x_0|)^{\frac{i}{2m+1}}}.$$

Then, taken together, the positivity of $x \mapsto p'(x) - c, x \in \mathbb{R}$, (516), and the following obvious equality

$$\lim_{x\to\infty} \frac{|p'(x)|}{\sum_{i=0}^{2m} \tilde{p}(|x|)^{\frac{i}{2m+1}}} = (2m+1)a_{2m+1}^{\frac{1}{2m+1}},$$

implies (515) with a suitable positive constant c_1. Now the condition (ii) given in the formulation of Theorem 3.1.33 follows from the admissibility of $f(\cdot)$ and (515); in conclusion, one gets that $(T_\varphi(t))_{t\in\mathbb{R}}$ is a strongly continuous group in $L^p_{\rho_1}(\Omega)$. Since $\varphi(t, x)^{-1} = p^{-1}(p(x) - t)$, $t, x \in \mathbb{R}$, we obtain analogously that there exists $c_2 > 0$ such that:

(517) $$|\det D\varphi(t, x_0)^{-1}| \geqslant \frac{1}{c_2\left(1+|t|^{\frac{2m}{2m+1}}\right)},\ t, x_0 \in \mathbb{R}.$$

Making use of Theorem 3.1.30(iii), (515) and (517), it follows immediately that for every measurable subset E of $\mathbb{R}$:

(518)

$$m(\varphi(t, E)) = \int_{\varphi(t,E)} dx = \int_E |\det D\varphi(t, x)|\, dx \in \left[\frac{m(E)}{c_1\left(1+|t|^{\frac{2m}{2m+1}}\right)}, \frac{1}{c}\int_E p'(x)\, dx\right]$$

and

(519) $$m(\varphi(t, \cdot)^{-1}(E)) \in \left[\frac{m(E)}{c_2\left(1+|t|^{\frac{2m}{2m+1}}\right)}, \frac{1}{c}\int_E p'(x)\, dx\right],\ t \in \Delta.$$

Suppose now that $\beta \geqslant \frac{2m}{2m+1}$ and a bounded measurable function $h : \mathbb{R} \to (0, \infty)$ is defined by:

$$h(s) := \begin{cases} \frac{d}{ds} ln((s+1)^\beta + 1), & s \geqslant 0, \\ 1, & s < 0. \end{cases}$$

Put now $f(x) := \exp(\int_0^x h(s)\, ds)$, $x \in \mathbb{R}$; then

$$\frac{f(x)}{f(x+t)} = \exp\left(\int_x^{x+t} h(s)\, ds\right) \leqslant \exp\left(\sup_{s\in\mathbb{R}} h(s)|t|\right),\ x, t \in \mathbb{R},$$

$f(\cdot)$ is admissible and $\rho_1(x) = \exp(\int_0^{p(x)} h(s)\, ds)$, $x \in \mathbb{R}$. We will prove that $(T_\varphi(t))_{t\in\mathbb{R}}$ is positively supercyclic in $L^p_{\rho_1}(\mathbb{R})$ and that $(T_\varphi(t))_{t\in\mathbb{R}}$ is not hypercyclic in $L^p_{\rho_1}(\mathbb{R})$. To this end, let $-\infty < a < b < \infty$, $K = [a, b]$ and let (t_k) be an arbitrary sequence of positive real numbers such that $\lim_{k\to\infty} t_k = \infty$. It is clear that there exists $k_0 \in \mathbb{N}$ such that, for every $k \in \mathbb{N}$ with $k \geqslant k_0$, $p(a) + t_k \geqslant 0$ and $p(b) - t_k \leqslant 0$. The assumption $x \in \varphi(t_k, \cdot)^{-1}(K)$, resp. $x \in \varphi(t_k, K)$, is equivalent to $p(x) \in [p(a) - t_k, p(b) - t_k]$, resp. $p(x) \in [p(a) + t_k, p(b) + t_k]$. Thus, $\rho_1(x) = \exp(\int_0^{p(x)} h(s)\, ds) = e^{p(x)}$, $k \geqslant k_0$, $x \in \varphi(t_k, \cdot)^{-1}(K)$ and

$$\rho_1(x) = e^{\int_0^{p(x)} h(s)\, ds} \leqslant e^{\int_0^{t_k + p(b)}\left[\frac{d}{ds} ln(s+1)^{\beta}+1\right] ds} = \frac{1}{2}\Big((t_k + p(b) + 1)^{\beta} + 1\Big),$$

provided $k \geqslant k_0$, $x \in \varphi(t_k, K)$. Keeping in mind these inequalities as well as (518)-(519), one gets:

$$\text{(520)} \qquad \int\limits_{\varphi(t_k,\cdot)^{-1}(K)} \rho_1(x)\, dx \leqslant e^{p(b)} e^{-t_k} \Big(\frac{1}{c} \int\limits_K p'(x)\, dx\Big),\ k \geqslant k_0 \text{ and}$$

$$\text{(521)} \qquad \int\limits_{\varphi(t_k,K)} \rho_1(x)\, dx \leqslant \frac{1}{2}\Big((t_k + p(b) + 1)^{\beta} + 1\Big)\Big(\frac{1}{c} \int\limits_K p'(x)\, dx\Big),\ k \geqslant k_0.$$

Now one can employ (520)-(521) and Theorem 3.1.20(a) with $L_k = K$, $k \in \mathbb{N}$ to conclude that $(T_\varphi(t))_{t\in\mathbb{R}}$ is positively supercyclic in $L^p_{\rho_1}(\mathbb{R})$. Suppose that $(T_\varphi(t))_{t\in\mathbb{R}}$ is hypercyclic in $L^p_{\rho_1}(\mathbb{R})$ and K is a compact subset of $\mathbb{R}$ such that $\inf K \geqslant \zeta$, where ζ is a unique real zero of the polynomial $p(\cdot)$. Then we have the existence of a sequence of measurable subsets (L_k) of K and a sequence (t_k) in $\mathbb{R}$ such that (513) holds. It can be proved that (t_k) must be unbounded; we may and shall assume that $\lim_{k\to\infty} t_k = +\infty$. Since $p(x) \geqslant 0$, $x \in K$ one gets $\rho_1(x) = \frac{1}{2}((1 + p(x))^{\beta} + 1) \geqslant 1$, $x \in K$, $\lim_{k\to\infty} m(K \setminus L_k) = 0$, and, a fortiori, there exists $k_1 \in \mathbb{N}$, $k_1 \geqslant k_0$ such that $m(L_k) \geqslant \frac{1}{2} m(K)$, $k \geqslant k_1$. Then (518) implies:

$$\text{(522)} \qquad m(\varphi(t, L_k)) \geqslant \frac{m(L_k)}{c_1\left(1+|t|^{\frac{2m}{2m+1}}\right)} \geqslant \frac{m(K)}{2c_1\left(1+|t|^{\frac{2m}{2m+1}}\right)},\ t \in \mathbb{R},\ k \geqslant k_1.$$

Since $\beta \geqslant \frac{2m}{2m+1}$ and

$$\rho_1(x) = e^{\int_0^{p(x)} h(s)\, ds} \geqslant e^{\int_0^{t_k + p(a)}[\frac{d}{ds} ln((s+1)^{\beta}+1)] ds} = \frac{1}{2}\Big((t_k + p(a) + 1)^{\beta} + 1\Big),$$

provided $k \geqslant k_0$ and $x \in \varphi(t_k, L_k)$, (522) yields:

$$\int\limits_{\varphi(t_k,L_k)} \rho_1(x)\, dx \geqslant \frac{m(K)}{2c_1\left(1+t_k^{\frac{2m}{2m+1}}\right)} \Big(\frac{1}{2}(t_k + p(a) + 1)^{\beta} + \frac{1}{2}\Big) \nrightarrow 0,\ k \to \infty.$$

The last estimate proves that $(T_\varphi(t))_{t\in\mathbb{R}}$ is not hypercyclic in $L^p_{\rho_1}(\mathbb{R})$.

In what follows, our intention will be to clarify the basic facts about disjoint hypercyclic semigroups induced by semiflows.

Definition 3.1.38. Let $n \in \mathbb{N}$, $n \geqslant 2$ and let $(T_i(t))_{t\in\Delta}$ be hypercyclic strongly continuous semigroups in X, $i = 1, 2, \cdot\cdot\cdot, n$. It is said that the semigroups $(T_i(t))_{t\in\Delta}$, $i = 1, 2, \cdot\cdot\cdot, n$ are:

(i) disjoint hypercyclic, in short d-hypercyclic, if there exists $x \in X$ such that

$$\overline{\{(T_1(t)x, \cdot\cdot\cdot, T_n(t)x) : t \in \Delta\}} = X^n. \tag{523}$$

An element $x \in X$ which satisfies (523) is called a d-hypercyclic vector associated to the semigroups $(T_1(t))_{t\in\Delta}, (T_2(t))_{t\in\Delta}, \cdot\cdot\cdot, (T_n(t))_{t\in\Delta}$;

(ii) disjoint topologically transitive, in short d-topologically transitive, if for any open non-empty subsets $V_0, V_1, \cdot\cdot\cdot, V_n$ of X, there exists $t \in \Delta$ such that $V_0 \cap T_1(t)^{-1}(V_1) \cap \cdot\cdot\cdot \cap T_n(t)^{-1}(V_n) \neq \emptyset$.

It immediately follows from Definition 3.1.38 that the d-hypercyclicity of $(T_i(t))_{t\in\Delta}$, $i \in \mathbb{N}_n$ implies that for every $i, j \in \mathbb{N}_n$ with $i \neq j$, there exists $t \in \Delta\setminus\{0\}$ such that $T_i(t) \neq T_j(t)$.

Suppose $(T_i(t))_{t\in\Delta}$, $i \in \mathbb{N}_n$ are strongly continuous semigroups. Arguing as in the proof of [63, Proposition 2.3], one obtains that the d-topological transitivity of $(T_i(t))_{t\in\Delta}$, $i \in \mathbb{N}_n$ implies that $(T_i(t))_{t\in\Delta}$, $i \in \mathbb{N}_n$ are d-hypercyclic and that the set of all d-hypercyclic vectors associated to the semigroups $(T_1(t))_{t\in\Delta}, (T_2(t))_{t\in\Delta}, \cdot\cdot\cdot, (T_n(t))_{t\in\Delta}$ is a dense G_δ-subset of X.

Now we are in a position to state the following theorem which concerns sufficient conditions for d-topological transitivity of strongly continuous semigroups on a class of weighted function spaces.

Theorem 3.1.39. *Suppose $p \in [1, \infty)$, $n \in \mathbb{N}\setminus\{1\}$, $\varphi_i : \Delta \times \Omega \to \Omega$ is a semiflow for all $i = 1, 2, \cdot\cdot\cdot, n$, $\rho : \Omega \to (0, \infty)$ is an upper semicontinuous function and $\rho_1 : \Omega \to (0, \infty)$ is a locally integrable function.*

(i) *Suppose that $X = C_{0,\rho}(\Omega)$ and that $(T_{\varphi_i}(t))_{t\in\Delta}$, $i = 1, 2, \cdot\cdot\cdot, n$ are strongly continuous semigroups in X. If for every compact set $K \subseteq \Omega$ there exists a sequence (t_k) in Δ which satisfies the following conditions:*
(A) $\lim_{k\to\infty} \sup_{\varphi_i(t_k,x)\in\varphi_j(t_k,K)} \rho(x) = 0$, *$i, j \in \mathbb{N}_n$, $i \neq j$ and*
(B) $\lim_{k\to\infty} \sup_{x\in\varphi_i(t_k,\cdot)^{-1}(K)} \rho(x) = \lim_{k\to\infty} \sup_{x\in\varphi_i(t_k,K)} \rho(x) = 0$, $i \in \mathbb{N}_n$,
then the semigroups $(T_{\varphi_i}(t))_{t\in\Delta}$, $i \in \mathbb{N}_n$ are d-topologically transitive.

(ii) *Suppose that $X = L^p_{\rho_1}(\Omega)$ and $(T_{\varphi_i}(t))_{t\in\Delta}$, $i \in \mathbb{N}_n$ are strongly continuous semigroups in X. If, for every compact set $K \subseteq \Omega$, there exist a sequence of measurable subsets (L_k) of K and a sequence (t_k) in Δ which satisfies the following conditions:*
(A1) $\lim_{k\to\infty} \int_{K\setminus L_k} \rho_1(x)\, dx = 0$,
(B1) $\lim_{k\to\infty} \int_{\varphi_i(t_k,\cdot)^{-1}(\varphi_j(t_k,L_k))} \rho_1(x)\, dx = 0$, *$i, j \in \mathbb{N}_n$, $i \neq j$ and*
(C1) $\lim_{k\to\infty} \int_{\varphi_i(t_k,\cdot)^{-1}(L_k)} \rho_1(x)\, dx = \lim_{k\to\infty} \int_{\varphi_i(t_k,L_k)} \rho_1(x)\, dx = 0$, $i \in \mathbb{N}_n$,
then the semigroups $(T_{\varphi_i}(t))_{t\in\Delta}$, $i \in \mathbb{N}_n$ are d-topologically transitive.

Proof. In order to prove (i), let us suppose $\varepsilon > 0$, $u, v_1, \cdots, v_n \in C_c(\Omega)$ and $K = \text{supp}(u) \cup \text{supp}(v_1) \cup \cdots \cup \text{supp}(v_n)$. The prescribed assumption implies that, for this compact set K, one can find a sequence (t_k) in Δ satisfying (A)-(B). Define, for every $k \in \mathbb{N}$, a function $\omega_k : \Omega \to \mathbb{K}$ by setting

$$\omega_k := u + \sum_{i=1}^{n} v_i(\varphi_i(t_k, \cdot)^{-1})\chi_{\varphi_i(t_k, \text{supp}(v_i))}.$$

Clearly, $\text{supp}(\omega_k)$ is a compact set for every $k \in \mathbb{N}$ and Brouwer's theorem implies that $\omega_k \in C_c(\Omega)$, $k \in \mathbb{N}$. Hence, the proof of (i) follows immediately if we prove that there exist $k_0 \in \mathbb{N}$ and $t \in \Delta$ which fulfill the next condition:

$$\max\big(\|\omega_{k_0} - u\|, \|T_{\varphi_1}(t)\omega_{k_0} - v_1\|, \cdots, \|T_{\varphi_n}(t)\omega_{k_0} - v_n\|\big) < \varepsilon. \tag{524}$$

By definition of $\omega_k(\cdot)$, we have the next inequality:

$$\|\omega_k - u\| \leqslant \sum_{i=1}^{n} \|v_i\|_\infty \sum_{i=1}^{n} \sup_{x \in \varphi_i(t_k, \text{supp}(v_i))} \rho(x), \; k \in \mathbb{N}. \tag{525}$$

Due to (A) and (525), there exists $k_{0,0} \in \mathbb{N}$ such that:

$$\|\omega_k - u\| < \varepsilon, \; k \geqslant k_{0,0}. \tag{526}$$

Proceeding in a similar way, one gets that, for every $k \in \mathbb{N}$ and $i \in \mathbb{N}_n$,

$$\|T_{\varphi_i}(t_k)\omega_k - v_i\| \leqslant \|u\|_\infty \sup_{x \in \varphi_i(t_k, \cdot)^{-1}(K)} \rho(x)$$
$$+ \sum_{\substack{1 \leqslant j \leqslant n \\ j \neq i}} \|v_i\|_\infty \sum_{\substack{1 \leqslant j \leqslant n \\ j \neq i}} \sup_{x \in \varphi_i(t_k, \cdot)^{-1}(\varphi_j(t_k, \text{supp}(v_j)))} \rho(x).$$

Now an application of (A)-(B) shows that, for every $i \in \mathbb{N}_n$, there exists $k_{0,i} \in \mathbb{N}$ such that:

$$\|T_{\varphi_i}(t_k)\omega_k - v_i\| < \varepsilon, \; k \geqslant k_{0,i}. \tag{527}$$

Put $k_0 := \max(k_{0,0}, \cdots, k_{0,n})$ and notice that (526)-(527) imply the validity of (524) with $t = t_{k_0}$. To prove (ii), suppose $\varepsilon > 0$, $u, v_1, \cdots, v_n \in C_c(\Omega)$ and $K = \text{supp}(u) \cup \text{supp}(v_1) \cup \cdots \cup \text{supp}(v_n)$. For this compact set $K \subseteq \Omega$, one can find a sequence of measurable subsets (L_k) of K and a sequence (t_k) in Δ satisfying (A1)-(C1). Define, for every $k \in \mathbb{N}$, a function $\omega_k : \Omega \to \mathbb{K}$ as follows:

$$\omega_k := u\chi_{L_k} + \sum_{i=1}^{n} v_i(\varphi_i(t_k, \cdot)^{-1})\chi_{\varphi_i(t_k, L_k)}.$$

It can be simply proved that $\omega_k \in L^p_{\rho_1}(\Omega)$, $k \in \mathbb{N}$. Arguing as in the proof of (i), we have the existence of a positive real number c such that, for every $k \in \mathbb{N}$ and $i \in \mathbb{N}_n$,

$$||\omega_k - u||^p \leqslant c\Big[||u||_\infty^p \int_{K\setminus L_k} \rho_1(x)\,dx + \sum_{i=1}^{n}||v_i||_\infty^p \int_{\varphi_i(t_k,L_k)} \rho_1(x)\,dx\Big] \text{ and}$$

$$||T_{\varphi_i}(t_k)\omega_k - v_i||^p \leqslant c\Big[||v_i||_\infty^p \int_{K\setminus L_k} \rho_1(x)\,dx$$

$$+ ||u||_\infty^p \int_{\varphi_i(t_k,\cdot)^{-1}(L_k)} \rho_1(x)\,dx + \sum_{\substack{1\leqslant j\leqslant n\\ j\neq i}} ||v_j||_\infty^p \int_{\varphi_i(t_k,\cdot)^{-1}(\varphi_j(t_k,L_k))} \rho_1(x)\,dx\Big].$$

By (A1)-(C1), one gets that the semigroups $(T_{\varphi_i}(t))_{t\in\Delta}$, $i = 1, 2, \cdot\cdot\cdot, n$ are d-topologically transitive in $L^p_{\rho_1}(\Omega)$, as required. □

Problem (2008). Suppose K is a compact subset of Ω and the strongly continuous semigroups $(T_{\varphi_i}(t))_{t\in\Delta}$, $i \in \mathbb{N}_n$ are d-topologically transitive in $C_{0,\rho_1}(\Omega)$, resp. $L^p_{\rho_1}(\Omega)$. Is there a sequence (t_k) in Δ satisfying (A)-(B), resp., are there a sequence of measurable subsets (L_k) of K and a sequence (t_k) in Δ satisfying (A1)-(C1)?

Repeating literally the arguments given in the proof of Theorem 3.1.39(i), one can prove the following assertion concerning d-topological transitivity of strongly continuous semigroups on the Fréchet space $C(\Omega)$.

Theorem 3.1.40. *Suppose that $\varphi_i : \Delta \times \Omega \to \Omega$ is a semiflow for all $i \in \mathbb{N}_n$ and, for every compact set $K \subseteq \Omega$, there exists a sequence (t_k) in Δ satisfying the following condition: For every compact set $K' \subseteq \Omega$ there exists $k_0(K') \in \mathbb{N}$ such that:*

(A2) $\varphi_i(t_k, \cdot)^{-1}(\varphi_j(t_k, K)) \cap K' = \emptyset$, $i, j \in \mathbb{N}_n$, $i \neq j$, $k \geqslant k_0(K')$ and

(B2) $\varphi_i(t_k, K) \cap K' = \varphi_i(t_k, \cdot)^{-1}(K) \cap K' = \emptyset$, $i \in \mathbb{N}_n$, $k \geqslant k_0(K')$.

Then $(T_{\varphi_i}(t))_{t\in\Delta}$ is a strongly continuous semigroup in $C(\Omega)$ for every $i \in \mathbb{N}_n$, and $(T_{\varphi_1}(t))_{t\in\Delta}, \cdot\cdot\cdot, (T_{\varphi_n}(t))_{t\in\Delta}$ are d-topologically transitive in $C(\Omega)$.

Example 3.1.41. (i) Suppose $p \in [1, \infty)$, $\alpha \in (0, \frac{\pi}{2}]$, $\Delta \in \{[0, \infty), \Delta(\alpha)\}$, $\Omega = (1, \infty)$, $n \in \mathbb{N}\setminus\{1\}$ and $0 < \alpha_1 < \cdot\cdot\cdot < \alpha_n \leqslant 1$. Define $\varphi_i : \Delta \times \Omega \to \Omega$, $i = 1, 2, \cdot\cdot\cdot, n$ and $\rho_1 : \Omega \to (0, \infty)$ by

$$\varphi_i(t, x) := (\mathrm{Re}\, t + x^{\alpha_i})^{1/\alpha_i} \text{ and } \rho_1(x) := e^{-x^{\alpha_1}},\ t \in \Delta,\ x \in \Omega.$$

It is clear that the mapping $\varphi_i(\cdot, \cdot)$ is a semiflow for all $i = 1, 2, \cdot\cdot\cdot, n$. We will prove that the semigroups $(T_{\varphi_i}(t))_{t\in\Delta}$, $i = 1, 2, \cdot\cdot\cdot, n$ are d-topologically transitive in $L^p_{\rho_1}(\Omega)$; without loss of generality, we may assume that $\Delta = [0, \infty)$. The existence of numbers $M \geqslant 1$ and $\omega \in \mathbb{R}$ satisfying:

$$\rho_1(x) \leqslant Me^{\omega|t|}\rho_1(\varphi_i(t, x)),\ t \geqslant 0,\ x \in \Omega,\ i = 1, 2, \cdot\cdot\cdot, n, \tag{528}$$

is obvious. Furthermore, we have that, for every $t \geqslant 0$, $x \in \Omega$ and $i \in \mathbb{N}_n$,

$$\left|\frac{d}{dx}\varphi_i(t, x)\right| = \left(1 + \frac{t}{x^{\alpha_i}}\right)^{\frac{1-\alpha_i}{\alpha_i}} \in \left[1, (1+t)^{\frac{1-\alpha_i}{\alpha_i}}\right]. \tag{529}$$

We easily infer from (528)-(529) that the condition (ii) of Theorem 3.1.33 is fulfilled so that $(T_{\varphi_i}(t))_{t\in\Delta}$, $i = 1, 2, \cdot\cdot\cdot, n$ are strongly continuous semigroups in $L^p_{\rho_1}(\Omega)$. Suppose $1 < a < b < \infty$, $K = [a, b]$, $L_k = K$, $k \in \mathbb{N}$ and (t_k) is any increasing sequence of positive real numbers satisfying $\lim_{k\to\infty} t_k = \infty$. Then (A1) holds and there exists $k_0 \in \mathbb{N}$ such that for every $k \geqslant k_0$ and $i, j \in \mathbb{N}_n$ with $i < j$:

$$\varphi_i(t_k, \cdot)^{-1}(K) = \varphi_i(t_k, \cdot)^{-1}(\varphi_j(t_k, K)) = \emptyset. \tag{530}$$

Furthermore, one can simply verify that $\lim_{k\to\infty} \int_{\varphi_i(t_k, K)} \rho_1(x)\, dx = 0$ for all $i \in \mathbb{N}_n$. Now one can employ (530) in order to conclude that (C1) holds and that (B1) holds with $i < j$. So, it is enough to prove the validity of (B1) with $i > j$; to this end, define $f : [a, b] \to \mathbb{R}$ by $f(x) := ((t_k + x^{\alpha_j})^{\frac{\alpha_i}{\alpha_j}} - t_k)^{\frac{1}{\alpha_i}}$, $x \in [a, b]$. Then $f'(x) = x^{\alpha_j - 1}((t_k + x^{\alpha_j})^{\frac{\alpha_i}{\alpha_j}} - t_k)^{\frac{1}{\alpha_i}-1}(t_k + x^{\alpha_j})^{\frac{\alpha_i}{\alpha_j}-1} \leqslant a^{\alpha_j - 1}(t_k + b^{\alpha_j})^{\frac{\alpha_i}{\alpha_j}\frac{1-\alpha_i}{\alpha_i}}(t_k + b^{\alpha_j})^{\frac{\alpha_i}{\alpha_j}-1} = a^{\alpha_j - 1}(t_k + b^{\alpha_j})^{\frac{1-\alpha_j}{\alpha_j}}$, $x \in [a, b]$ and the Lagrange mean value theorem implies that, for every $k \in \mathbb{N}$,

$$\frac{\left((t_k + b^{\alpha_j})^{\frac{\alpha_i}{\alpha_j}} - t_k\right)^{\frac{1}{\alpha_i}} - \left((t_k + a^{\alpha_j})^{\frac{\alpha_i}{\alpha_j}} - t_k\right)^{\frac{1}{\alpha_i}}}{(b-a)a^{\alpha_j - 1}} \leqslant (t_k + b^{\alpha_j})^{\frac{1-\alpha_j}{\alpha_i}}.$$

In other words,

$$m\big(\varphi_i(t_k, \cdot)^{-1}(\varphi_j(t_k, K))\big) \leqslant (b-a)a^{\alpha_j - 1}(t_k + b^{\alpha_j})^{\frac{1-\alpha_j}{\alpha_i}}. \tag{531}$$

The existence of an integer $k_{i,j} \in \mathbb{N}$ satisfying $\rho_1(x) \leqslant \exp(-t_k^{\alpha_1/\alpha_i})$ for all $x \in \varphi_i(t_k, \cdot)^{-1}(\varphi_j(t_k, K))$ and $k \geqslant k_{i,j}$ is clear. Thereby, we have the following:

$$\int_{\varphi_i(t_k,\cdot)^{-1}(\varphi_j(t_k,K))} \rho_1(x)\, dx \leqslant m\big(\varphi_i(t_k, \cdot)^{-1}(\varphi_j(t_k, K))\big)e^{-t_k^{\alpha_1/\alpha_i}},\ k \geqslant k_{i,j}. \tag{532}$$

Now (B1) follows from (531)-(532) and Theorem 3.1.39(ii) implies that the semigroups $(T_{\varphi_i}(t))_{t\geqslant 0}$, $i = 1, 2, \cdot\cdot\cdot, n$ are d-topologically transitive in $L^p_{\rho_1}(\Omega)$, as claimed.

(ii) Suppose $\alpha \in (0, \frac{\pi}{2}]$, $m \in \mathbb{N}$, $\Delta \in \{[0, \infty), \Delta(\alpha)\}$, $\Omega = (0, \infty)^m$, $\Theta = [1, \infty)^m$, $n \in \mathbb{N}\setminus\{1\}$, $[\alpha_{ij}]_{1\leqslant i\leqslant n, 1\leqslant j\leqslant m}$ is a matrix whose elements are positive real numbers and $c = \min_{1\leqslant i\leqslant n, 1\leqslant j\leqslant m} \alpha_{ij}$. Suppose that, for every $i, j \in \mathbb{N}_n$ with $i \neq j$, there exists $l \in \mathbb{N}_m$ such that $\alpha_{il} \neq \alpha_{jl}$. Define $\tilde{\varphi}_i : \Delta \times \Omega \to \Omega$, $i = 1, 2, \cdot\cdot\cdot, n$ and $\tilde{\rho} : \Omega \to (0, \infty)$ by setting

$$\tilde{\varphi}_i(t, x) := ((\mathrm{Re}\, t + x_1^{\alpha_{i1}})^{1/\alpha_{i1}}, \cdot\cdot\cdot, (\mathrm{Re}\, t + x_m^{\alpha_{im}})^{1/\alpha_{im}}) \text{ and}$$
$$\tilde{\rho}(x) := e^{-(x_1^c + \cdots + x_m^c)},\ t \in \Delta,\ x = (x_1, \cdot\cdot\cdot, x_m) \in \Omega.$$

Notice that, for every $i \in \mathbb{N}_n$ and $t \in \Delta$, $T_{\tilde{\varphi}_1}(t) \notin L(C_{0,\tilde{\rho}}(\Omega))$ since the condition (ii) given in the formulation of Theorem 3.1.35 does not hold. Define $\rho : \Theta \to (0, \infty)$ and $\varphi_i : \Delta \times \Theta \to \Theta$ by $\rho(x) := \tilde{\rho}(x)$ and $\varphi_i(t, x) := \tilde{\varphi}_i(t, x)$, $t \in \Delta$, $x \in \Theta$. Let us show that, for every fixed $i = 1, 2, \cdot\cdot\cdot, n$ and $t \in \Delta$, the mapping $T_{\varphi_i}(t)$

$: C_{0,\rho}(\Theta) \rightarrow C_{0,\rho}(\Theta)$ is well defined and continuous. The simple calculation $s^c - (\text{Re}\, t + s^{\alpha_{ij}})^{\frac{c}{\alpha_{ij}}} \geqslant s^c - ((\text{Re}\, t)^{\frac{c}{\alpha_{ij}}} + (s^{\alpha_{ij}})^{\frac{c}{\alpha_{ij}}}) = -(\text{Re}\, t)^{\frac{c}{\alpha_{ij}}} \geqslant -1 - \text{Re}\, t,\ s \geqslant 1,\ 1 \leqslant j \leqslant m$ implies $-(x_1^c + \cdots + x_m^c) \leqslant -((\text{Re}\, t + x_1^{\alpha_{i1}})^{c/\alpha_{i1}} + \cdots + (\text{Re}\, t + x_m^{\alpha_{im}})^{c/\alpha_{im}}) + (m + \text{Re}\, t)$, $x \in \Theta$, i.e.,

$$\rho(x) \leqslant e^m e^{\text{Re}\, t} \rho(\varphi_i(t, x)),\ x \in \Theta. \tag{533}$$

Thereby, the condition (ii)(a) quoted in the formulation of [245, Theorem 2.2, p. 1601] holds. On the other hand, for every compact set $K \subseteq \Theta$ and for every $t \in \Delta$ and $\delta > 0$, $\varphi(t, \cdot)^{-1}(K) \cap \{x \in \Theta : \rho(x) \geqslant \delta\}$ is a compact subset of Θ. Therefore, the condition (ii)(b) quoted in the formulation of [245, Theorem 2.2] also holds and the cited theorem implies $T_{\varphi_i}(t) \in L(C_{0,\rho}(\Theta))$. Due to Lemma 3.1.31, (533) and the proof of [245, Theorem 3.4], we get that, for every $i = 1, 2, \cdots, n$, $(T_{\varphi_i}(t))_{t \in \Delta} \subseteq L(C_{0,\rho}(\Theta))$ is a strongly continuous semigroup in $C_{0,\rho}(\Theta)$ and the analysis given in (i) of this example implies that the semigroups $(T_{\varphi_i}(t))_{t \in \Delta}$, $i = 1, 2, \cdots, n$ are d-topologically transitive in $C_{0,\rho}(\Theta)$.

(iii) Suppose that every element of a real matrix $[\alpha_{ij}]_{1 \leqslant i \leqslant n, 1 \leqslant j \leqslant m}$ is a positive real number and that for every $i, j \in \mathbb{N}_n$ with $i \neq j$, there exists $l \in \mathbb{N}_m$ such that $\alpha_{il} \neq \alpha_{jl}$. Let $p \geqslant 1$, $q > \frac{m}{2}$, $\Delta = [0, \infty)$ and let Ω be as in (ii). Define semiflows $\varphi_i : \Delta \times \Omega \rightarrow \Omega$, $i = 1, 2, \cdots, n$ and $\rho_1 : \Omega \rightarrow (0, \infty)$ as follows:

$$\varphi_i(t, x_1, \cdots, x_m) := (e^{a_{i1}t} x_1, \cdots, e^{a_{im}t} x_m) \text{ and} \tag{534}$$

$$\rho_1(x_1, \cdots, x_m) := \frac{1}{(1 + |x|^2)^q},\ t \in \Delta,\ x = (x_1, \cdots, x_m) \in \Omega. \tag{535}$$

Then $(T_{\varphi_i}(t))_{t \geqslant 0}$ is a strongly continuous semigroup in $L^p_{\rho_1}(\Omega)$, $1 \leqslant i \leqslant n$. Suppose $K = [a_1, b_1] \times \cdots \times [a_m, b_m]$ is a compact subset of Ω; set $L_k := K$, $k \in \mathbb{N}$. Let (t_k) be a sequence in Δ such that $\lim_{k \to \infty} t_k = \infty$. It can be simply checked that (A1) and (C1) hold. To see that (B1) also holds, suppose $i, j \in \mathbb{N}_n$, $i \neq j$, $\alpha_{il} \neq \alpha_{jl}$, $x = (x_1, \cdots, x_m) \in \varphi_i(t_k, \cdot)^{-1}(\varphi_j(t_k, K))$ and notice that:

$$\lim_{r \to \infty} \int_{|x| \geqslant r} \frac{dx}{(1 + |x|^2)^q} = 0. \tag{536}$$

Obviously, $x_s \in [e^{(a_{js} - a_{is})t_k} a_s, e^{(a_{js} - a_{is})t_k} b_s]$, $s = 1, \cdots, m$. In case $a_{il} < a_{jl}$, (536) immediately leads us to the following:

$$\int_{\varphi_i(t_k, \cdot)^{-1}(\varphi_j(t_k, K))} \frac{dx}{(1 + |x|^2)^q} \leqslant \int_{|x| \geqslant e^{(a_{jl} - a_{il})t_k a_l}} \frac{dx}{(1 + |x|^2)^q} \rightarrow 0,\ k \rightarrow \infty.$$

Suppose now $a_{il} > a_{jl}$. Then the inequality:

$$(1 + |x|^2)^q \geqslant (1 + x_1^2)^{q/m} \cdots (1 + x_m^2)^{q/m}$$

and (536) imply the existence of an appropriate positive real number c, depending only on K, p, m and $[\alpha_{ij}]_{1 \leqslant i \leqslant n, 1 \leqslant j \leqslant m}$, so that:

$$\int_{\varphi_i(t_k,\cdot)^{-1}(\varphi_j(t_k,K))} \frac{dx}{(1+|x|^2)^q}$$

$$\leqslant \int_{e^{(a_{jl}-a_{il})t_k a_l}}^{e^{(a_{jl}-a_{il})t_k b_l}} \frac{dx}{(1+x_l^2)^{q/m}} \prod_{\substack{1\leqslant s\leqslant m\\ s\neq l}} \int_{e^{(a_{js}-a_{is})t_k a_s}}^{e^{(a_{js}-a_{is})t_k b_s}} \frac{dx}{(1+x_s^2)^{q/m}}$$

$$\leqslant e^{(a_{jl}-a_{il})t_k}(b_l-a_l) \prod_{\substack{1\leqslant s\leqslant m\\ s\neq l}} \int_{e^{(a_{js}-a_{is})t_k a_s}}^{e^{(a_{js}-a_{is})t_k b_s}} \frac{dx}{(1+x_s^2)^{1/2}}$$

$$= e^{(a_{jl}-a_{il})t_k}(b_l-a_l) \prod_{\substack{1\leqslant s\leqslant m\\ s\neq l}} ln \frac{e^{(a_{js}-a_{is})t_k} b_s + \sqrt{e^{2(a_{js}-a_{is})t_k} b_s^2+1}}{e^{(a_{js}-a_{is})t_k} a_s + \sqrt{e^{2(a_{js}-a_{is})t_k} a_s^2+1}}$$

$$\leqslant ce^{(a_{jl}-a_{il})t_k}(b_l-a_l) \rightarrow 0,\ k\rightarrow\infty,$$

whence it follows that (B1) holds and the semigroups $(T_{\varphi_i}(t))_{t\geqslant 0}$, $i = 1, 2, \cdots, n$ are d-topologically transitive in $L^p_{\rho_1}(\Omega)$ (cf. also [245, Example 3.19] and [244, Theorem 6.22]). Define $\widetilde{\varphi}_i : \Delta \times \mathbb{R}^m \rightarrow \mathbb{R}^m$, $i = 1, 2, \cdots, n$ and $\widetilde{\rho}_1 : \mathbb{R}^m \rightarrow (0, \infty)$ through (534) and (535). In this case, the strongly continuous semigroups $(T_{\widetilde{\varphi}_i}(t))_{t\geqslant 0}$, $i = 1, 2, \cdots, n$ are d-topologically transitive in $L^p_{\widetilde{\rho}_1}(\mathbb{R}^m)$. This fact follows from the previous computations and Theorem 3.1.39; notice only that we must use an appropriate sequence (L_k) of measurable subsets of K satisfying $0 \notin L_k^{\circ}$, $k \in \mathbb{N}$. Herein we point out that an employment of [245, Theorem 3.7] implies that, for every $i = 1, 2, \cdots, n$, $(T_{\varphi_i}(t))_{t\geqslant 0}$, resp. $(T_{\widetilde{\varphi}_i}(t))_{t\geqslant 0}$, is a non-hypercyclic strongly continuous semigroup in $C_{0,\rho_1}(\Omega)$, resp. $C_{0,\widetilde{\rho}_1}(\mathbb{R}^m)$.

Example 3.1.42. Suppose $\Delta = [0, \infty)$, $\Omega = \{(x, y) \in \mathbb{R}^2 : x^2 + y^2 > 1\}$, $|(x, y)| = \sqrt{x^2+y^2}$, $(x, y) \in \mathbb{R}^2$, $n \in \mathbb{N}\setminus\{1\}$, $0 < p_1 < \cdots < p_n < \infty$, $q_i \in \mathbb{R}$, $1 \leqslant i \leqslant n$, K is a compact subset of Ω and:

$\varphi_i(t, x, y) = e^{p_i t}(x \cos q_i t - y \sin q_i t, x \sin q_i t + y \cos q_i t)$, $t \geqslant 0$, $(x, y) \in \Omega$, $1 \leqslant i \leqslant n$. Since $|\varphi_i(t, x, y)| = e^{p_i t}|(x, y)|$, $t \geqslant 0$, $(x, y) \in \Omega$, $1 \leqslant i \leqslant n$, one can simply prove that, for every $i \in \mathbb{N}_n$, $\varphi_i : \Delta \times \Omega \rightarrow \Omega$ is a semiflow. Let (t_k) be a sequence in Δ such that $\lim_{k\rightarrow\infty} t_k = \infty$. Then, for an arbitrary compact subset K' of Ω, there exists $k_0(K') \in \mathbb{N}$ such that (A2) and (B2) hold. An application of Theorem 3.1.40 gives that the strongly continuous semigroups $(T_{\varphi_i}(t))_{t\geqslant 0}$, $i = 1, 2, \cdots, n$ are d-topologically transitive in $C(\Omega)$.

Finally, let us notice that Example 3.1.34 can be used for the construction of d-topologically transitive semigroups induced by non-differentiable semiflows.

3.2 Hypercyclic and topologically mixing properties of abstract second order equations

The main purpose of this section is to display the main structural properties of hypercyclic and chaotic integrated C-cosine functions. The notions of hypercyclicity, mixing and chaoticity of an α-times integrated C-cosine functions ($\alpha \geqslant 0$) will be defined by using distributional techniques.

We continue the research of A. Bonilla, P. J. Miana [70] and T. Kalmes [247]. In the second subsection, we introduce and systematically analyze the class of C-distribution cosine functions and slightly improve the results obtained in cooperation with P. J. Miana ([414], [295]-[296]). Motivated by the study of R. deLaubenfels, H. Emamirad and K.-G. Grosse Erdmann [150], we clarify the necessary and sufficient conditions for hypercyclicity, mixing and chaoticity of C-distribution cosine functions. Furthermore, we provide sufficient conditions for mixing and chaoticity of certain classes of C-distribution cosine functions. The last two subsections are devoted to the study of hypercyclic and chaotic cosine functions generated by squares of gradient operators ([244]-[247]) and disjoint hypercyclicity of cosine functions on weighted function spaces. The notion of subspace chaoticity introduced by J. Banasiak and M. Moszyński [35] plays an important role in our research. It is worth noting that several results established in this section (chapter) are obvious modifications of corresponding results from the theory of hypercyclic single valued operators and that we are not primarily concerned with studying new concepts in the theory of hypercyclicity. Our main intention is, in fact, to analyze the basic properties of a new important class of abstract second order (ill-posed) PDEs (cf. [14], [169], [223], [266], [292], [295]-[296], [414], [551] and [554] for further information in this direction). With the exception of Remark 3.2.23, in the remaining part of this chapter we shall work in the setting of complex Banach spaces; the kernel space of an operator A will be denoted by Kern(A). Unless specified otherwise, we shall assume in the sequel that $(E, \|\cdot\|)$ is a complex Banach space, and that $C \in L(E)$ is an injective operator with $CA \subseteq AC$.

In this section, the notations of convolution products might create some confusion; precisely speaking, the convolution product of complex-valued functions $f(\cdot)$ and $g(\cdot)$, defined on $\mathbb{R}$, will be understood in the following sense:

$$(f * g)(t) = \int_{-\infty}^{+\infty} f(t-s)g(s)\,ds, \quad t \in \mathbb{R},$$

(cf. [20, Chapter 1] and [292, Section 1.3] for further information) and the finite convolution product $*_0$ of complex-valued functions $f(\cdot)$ and $g(\cdot)$, defined on $[0, \infty)$, will be understood in the following sense:

$$(f *_0 g)(t) = \int_0^t f(t-s)g(s)\,ds, \quad t \geqslant 0.$$

The convolution of vector-valued distributions is taken in the sense of [331, Proposition 1.1]:

Proposition 3.2.1. *Suppose X, Y and Z are Banach spaces and* $b : X \times Y \to Z$ *is bilinear and continuous. Then there is a unique bilinear, separately continuous mapping* $*_b : \mathcal{D}'_0(X) \times \mathcal{D}'_0(Y) \to \mathcal{D}'_0(Z)$ *such that*

$$(S \otimes x) *_b (T \otimes y) = S * T \otimes b(x, y),$$

for all $S, T \in \mathcal{D}'_0$ *and* $x \in X$, $y \in Y$. *Moreover, this mapping is continuous.*

Definition 3.2.2. ([293]) Let $G \in \mathcal{D}'_0(L(E))$ satisfy $CG(\varphi) = G(\varphi)C$, $\varphi \in \mathcal{D}$. If $G(\varphi *_0 \psi)C = G(\varphi)G(\psi)$, $\varphi, \psi \in \mathcal{D}$, then G is called a pre–$(C-DS)$. If, additionally, $\mathcal{N}(G) = \bigcap_{\varphi\in\mathcal{D}_0} \text{Kern}(G(\varphi)) = \{0\}$, then G is called a C-distribution semigroup, $(C-DS)$ in short. It is said that a pre –$(C-DS)$ is dense if the set $\mathcal{R}(G) = \bigcup_{\varphi\in\mathcal{D}_0} \mathcal{R}(G(\varphi))$ is dense in E. A pre–$(C-DS)$ G is said to be exponential if there exists $\omega \in \mathbb{R}$ such that $e^{-\omega t}G \in \mathcal{S}'_0(L(E))$; the shorthand $(E-CDS)$ is used to denote an exponential $(C-DS)$.

Let G be a $(C-DS)$ and let $T \in \mathcal{E}'_0(\mathbb{C})$, i.e., T is a scalar-valued distribution with compact support in $[0, \infty)$. Then we define $G_1(T)$ on a subset of E by

$$y = G_1(T)x \text{ iff } G(T * \varphi)x = G(\varphi)y \text{ for all } \varphi \in \mathcal{D}_0.$$

Then $G_1(T)$ is a closed linear operator, $G_1(\delta) = I$ and the (infinitesimal) generator of a $(C-DS)$ G is defined by $A := G_1(-\delta')$. In the case $C = I$, there is no danger of ambiguity and we do not distinguish G and G_1.

Put $\mathcal{D}_+ := \{f \in C^\infty([0, \infty)) : f \text{ is compactly supported}\}$ and define $\mathcal{K} : \mathcal{D} \to \mathcal{D}_+$ by $\mathcal{K}(\varphi)(t) := \varphi(t)$, $t \geqslant 0$, $\varphi \in \mathcal{D}$. As is known, $\mathcal{D}_+$ is an (LF) space and there exists a linear continuous operator $\Lambda : \mathcal{D}_+ \to \mathcal{D}$ which satisfies $\mathcal{K}\Lambda = I_{\mathcal{D}_+}$.

3.2.1. *C*-Distribution cosine functions, almost *C*-distribution cosine functions and integrated *C*-cosine functions. We begin this subsection by recalling the following notion ([295]). If $\emptyset \neq \Upsilon \subseteq \mathbb{R}$, then we set $\mathcal{D}_\Upsilon := \{\varphi \in \mathcal{D} : \text{supp}(\varphi) \subseteq \Upsilon\}$. Let $\zeta \in \mathcal{D}_{[-2,-1]}$ be a fixed test function satisfying $\int_{-\infty}^{\infty} \zeta(t)\, dt = 1$. Then, with ζ chosen in this way, we define $I(\varphi)$ $(\varphi \in \mathcal{D})$ as follows

$$I(\varphi)(\cdot) := \int_{-\infty}^{\cdot} \left[\varphi(t) - \zeta(t) \int_{-\infty}^{\infty} \varphi(u)\, du\right] dt.$$

Then $I(\varphi) \in \mathcal{D}$, $I(\varphi') = \varphi$, $\frac{d}{dt}I(\varphi)(t) = \varphi(t) - \zeta(t) \int_{-\infty}^{\infty} \varphi(u)\, du$, $t \in \mathbb{R}$ and, for every $G \in \mathcal{D}'(L(E))$, the primitive G^{-1} of G is defined by setting $G^{-1}(\varphi) := -G(I(\varphi))$, $\varphi \in \mathcal{D}$. It is clear that $G^{-1} \in \mathcal{D}'(L(E))$, $(G^{-1})' = G$, i.e., $-G^{-1}(\varphi') = G(I(\varphi')) = G(\varphi)$, $\varphi \in \mathcal{D}$; moreover, $\text{supp}(G) \subseteq [0, \infty)$ implies $\text{supp}(G^{-1}) \subseteq [0, \infty)$.

Definition 3.2.3. An element $\mathbf{G} \in \mathcal{D}'_0(L(E))$ is called a pre–$(C-DCF)$ iff $\mathbf{G}(\varphi)C = C\mathbf{G}(\varphi)$, $\varphi \in \mathcal{D}$ and

$$(C-DCF_1) : \mathbf{G}^{-1}(\varphi *_0 \psi)C = \mathbf{G}^{-1}(\varphi)\mathbf{G}(\psi) + \mathbf{G}(\varphi)\mathbf{G}^{-1}(\psi),\ \varphi, \psi \in \mathcal{D};$$

if, additionally,

$$(C-DCF_2): x = y = 0 \text{ iff } \mathbf{G}(\varphi)x + \mathbf{G}^{-1}(\varphi)y = 0,\ \varphi \in \mathcal{D}_0,$$

then $\mathbf{G}$ is called a C-distribution cosine function, in short $(C-DCF)$. A pre $-(C-DCF)$ $\mathbf{G}$ is called dense if the set $\mathcal{R}(\mathbf{G}) := \bigcup_{\varphi\in\mathcal{D}_0} \mathcal{R}(\mathbf{G}(\varphi))$ is dense in E.

Notice that (DCF_2) implies $\bigcap_{\varphi\in\mathcal{D}_0} \mathrm{Kern}(\mathbf{G}(\varphi)) = \{0\}$ and $\bigcap_{\varphi\in\mathcal{D}_0} \mathrm{Kern}(\mathbf{G}^{-1}(\varphi)) = \{0\}$, and that the assumption $\mathbf{G} \in \mathcal{D}_0'(L(E))$ implies $\mathbf{G}(\varphi) = 0$, $\varphi \in \mathcal{D}_{(-\infty,0]}$. For $\psi \in \mathcal{D}$, we get $\psi_+(t) := \psi(t)H(t)$, $t \in \mathbb{R}$. Then $\psi_+ \in \mathcal{E}_0'(\mathbb{C})$, $\psi \in \mathcal{D}$, and therefore, $\varphi * \psi_+ \in \mathcal{D}_0$ for any $\varphi \in \mathcal{D}_0$.

The following proposition can be achieved by making use of the arguments given in [295].

Proposition 3.2.4. (i) *Let* $\mathbf{G} \in \mathcal{D}_0'(L(E))$ *and* $\mathbf{G}(\varphi)C = C\mathbf{G}(\varphi)$, $\varphi \in \mathcal{D}$. *Then* $\mathbf{G}$ *is a pre-(C-DCF) in E iff* $\mathcal{G} \equiv \begin{pmatrix} \mathbf{G} & \mathbf{G}^{-1} \\ \mathbf{G}' - \delta\otimes C & \mathbf{G} \end{pmatrix}$ *is a pre-(*$\mathcal{C}$*-DS) in* $E \oplus E$, *where* $\mathcal{C} \equiv \begin{pmatrix} C & 0 \\ 0 & C \end{pmatrix}$. *Moreover,* $\mathcal{G}$ *is a (*$\mathcal{C}$*-DS) iff* $\mathbf{G}$ *is a pre-(C-DCF) which satisfies* $(C-DCF_2)$.

(ii) *Let* $\mathbf{G} \in \mathcal{D}_0'(L(E))$ *and* $\mathbf{G}(\varphi)C = C\mathbf{G}(\varphi)$, $\varphi \in \mathcal{D}$. *Then* $\mathbf{G}$ *is a (C-DCF) iff* (DCF_2) *holds and*

$$\mathbf{G}^{-1}(\varphi * \psi_+)C = \mathbf{G}^{-1}(\varphi)\mathbf{G}(\psi) + \mathbf{G}(\varphi)\mathbf{G}^{-1}(\psi),\ \varphi \in \mathcal{D}_0,\ \psi \in \mathcal{D}.$$

Assume $\mathbf{G}$ is a $(C-DCF)$ and $T \in \mathcal{E}_0'(\mathbb{C})$. Then the (infinitesimal) generator A of $\mathbf{G}$ is defined by

$$A := G(\delta'') := \{(x, y) \in E \oplus E : \mathbf{G}^{-1}(\varphi'')x = \mathbf{G}^{-1}(\varphi)y \text{ for all } \varphi \in \mathcal{D}_0\}.$$

Then A is a closed linear operator and, by the proof of [292, Lemma 3.1.6], we have $C^{-1}AC = A$.

Theorem 3.2.5. *(cf.* [295]*)*

(i) *Let A be the generator of a (C-DCF)* $\mathbf{G}$. *Then* $\mathcal{A} \subseteq \mathcal{B}$, *where* $\mathcal{A} \equiv \begin{pmatrix} 0 & I \\ A & 0 \end{pmatrix}$ *and* $\mathcal{B}$ *is the generator of* $\mathcal{G}$. *Furthermore,* $(x, y) \in A \Leftrightarrow \left(\binom{x}{0}, \binom{0}{y}\right) \in \mathcal{B}$.

(ii) *Let* $\mathbf{G}$ *be a (C-DCF) generated by A. Then the following holds:*

(a) $(\mathbf{G}(\psi)x,\ \mathbf{G}(\psi'')x + \psi'(0)Cx) \in A$, $\psi \in \mathcal{D}$, $x \in E$.

(b) $(\mathbf{G}^{-1}(\psi)x,\ -\mathbf{G}(\psi')x - \psi(0)Cx) \in A$, $\psi \in \mathcal{D}$, $x \in E$.

(c) $\mathbf{G}(\psi)A \subseteq A\mathbf{G}(\psi)$, $\psi \in \mathcal{D}$.

(d) $\mathbf{G}^{-1}(\psi)A \subseteq A\mathbf{G}^{-1}(\psi)$, $\psi \in \mathcal{D}$.

(iii) *A closed linear operator A is the generator of a (C-DCF)* $\mathbf{G}$ *iff for every* $\tau > 0$ *there exist an integer* $n \in \mathbb{N}$ *and a local n-times integrated C-cosine function* $(C_n(t))_{t\in[0,\tau)}$ *with the integral generator A. If this is the case, then the following equality holds:*

$$\mathbf{G}(\varphi)x = (-1)^n \int_0^\tau \varphi^{(n)}(t)C_n(t)x\,dt,\ x \in E,\ \varphi \in \mathcal{D}_{(-\infty,\tau)}.$$

Theorem 3.2.6. *Suppose a, b, α, $M > 0$, $E^2(a, b) \subseteq \rho_C(A)$, the mapping $\lambda \mapsto (\lambda^2 - A)^{-1}C$, $\lambda \in E(a, b)$ is continuous and $\|(\lambda^2 - A)^{-1}C\| \leqslant M(1 + |\lambda|)^{\alpha}$, $\lambda \in E(a, b)$. Put $\widetilde{\varphi}(\lambda) := \int_{-\infty}^{\infty} e^{\lambda t}\varphi(t)\,dt$, $\varphi \in \mathcal{D}$ (in this section, the correspondence $\varphi \mapsto \widetilde{\varphi}$ is the Fourier-Laplace transform actually) and*

$$\mathbf{G}(\varphi)x := \frac{1}{2\pi i}\int_{\Gamma} \lambda\widetilde{\varphi}(\lambda)(\lambda^2 - A)^{-1}Cx\,d\lambda,\ x \in E,\ \varphi \in \mathcal{D},$$

where Γ is the upwards oriented boundary of $E(a, b)$. Then $\mathbf{G}$ is a (C-DCF) generated by $C^{-1}AC$.

Proof. The prescribed assumptions combined with [292, Proposition 2.1.24] imply that there exist $\beta \geqslant 0$ and $M_1 > 0$ such that $E(a, b) \subseteq \rho_{\mathcal{C}}(\mathcal{A})$, $\|(\lambda - \mathcal{A})^{-1}\mathcal{C}\| \leqslant M_1(1 + |\lambda|)^{\beta}$, $\lambda \in E(a, b)$ and that the mapping $\lambda \mapsto (\lambda - \mathcal{A})^{-1}\mathcal{C}$, $\lambda \in E(a, b)$ is continuous. Put $\mathcal{G}(\varphi)\binom{x}{y} := \frac{1}{2\pi i}\int_{\Gamma}\widetilde{\varphi}(\lambda)(\lambda - \mathcal{A})^{-1}\mathcal{C}\binom{x}{y}\,d\lambda$, $x, y \in E$, $\varphi \in \mathcal{D}$. By [301, Theorem 2.1], $\mathcal{G}$ is a $(\mathcal{C} - DS)$ generated by $\mathcal{C}^{-1}\mathcal{A}\mathcal{C}$. Using [292, Proposition 2.1.24] again, one gets that, for every $x, y \in E$ and $\varphi \in \mathcal{D}$, $\mathcal{G}(\varphi) = \begin{pmatrix}\mathbf{G}_1(\varphi) & \mathbf{G}_2(\varphi)\\ \mathbf{G}_3(\varphi) & \mathbf{G}_1(\varphi)\end{pmatrix}$, where $\mathbf{G}_1(\varphi)x = \frac{1}{2\pi i}\int_{\Gamma}\lambda\widetilde{\varphi}(\lambda)(\lambda^2 - A)^{-1}Cx\,d\lambda$, $\mathbf{G}_2(\varphi)x = \frac{1}{2\pi i}\int_{\Gamma}\widetilde{\varphi}(\lambda)(\lambda^2 - A)^{-1}Cx\,d\lambda$ and $\mathbf{G}_3(\varphi)x = \frac{1}{2\pi i}\int_{\Gamma}\widetilde{\varphi}(\lambda)[\lambda^2(\lambda^2 - A)^{-1}C - C]x\,d\lambda$, $x \in E$, $\varphi \in \mathcal{D}$. Obviously, $\operatorname{supp}(\mathbf{G}) \subseteq [0, \infty)$, $\frac{1}{2\pi i}\int_{\Gamma}\widetilde{\varphi}(\lambda)\,d\lambda = \varphi(0)$, $\varphi \in \mathcal{D}$ and $\lambda\widetilde{I(\varphi)}(\lambda) = -\widetilde{I(\varphi)'}(\lambda) = \widetilde{\varphi - \int_{-\infty}^{\infty}\varphi(t)\,dt\,\zeta}(\lambda) = \widetilde{\varphi}(\lambda)$, $\lambda \in \mathbb{C}$. Therefore, $\mathbf{G}_2 = \mathbf{G}_1^{-1}$ and $\mathbf{G}_3 = \mathbf{G}_1' - \delta \otimes C$. This simply implies that $\mathbf{G}_1$ is a $(C - DCF)$. Denote by B the generator of G. Then we finally obtain

$$(x, y) \in B \Leftrightarrow \left(\binom{x}{0}, \binom{0}{y}\right) \in \mathcal{C}^{-1}\mathcal{A}\mathcal{C} \Leftrightarrow (x, y) \in C^{-1}AC.$$

□

Proposition 3.2.7. *Assume that $\pm A$ generate C-distribution semigroups $G_{\pm}$ and A^2 is closed. Then $C^{-1}A^2C$ generates a (C-DCF) $\mathbf{G}$, which is given by $\mathbf{G}(\varphi) := \frac{1}{2}(G_+(\varphi) + G_-(\varphi))$, $\varphi \in \mathcal{D}$.*

Proof. Since $\pm A$ generate C-distribution semigroups, it follows that, for every $\tau > 0$, there exists $n \in \mathbb{N}$ such that $\pm A$ generate local n-times integrated C-semigroups $(S_{n,\pm}(t))_{t\in[0,\tau)}$. The closedness of A^2 taken together with [292, Proposition 2.1.17] imply that, for such a number $\tau > 0$, the operator A^2 is a subgenerator of the local n-times integrated C-cosine function $(\frac{1}{2}(S_{n,-}(t) + S_{n,-}(t)))_{t\in[0,\tau)}$. Keeping in mind Theorem 3.2.5(iii), the above ensures that the operator $C^{-1}A^2C$ is the generator of a $(C - DCF)$ $\mathbf{G}$. □

Theorem 3.2.8. (i) *Let A be the generator of a (C-DCF) $\mathbf{G}$. Then $\mathbf{G} \in \mathcal{D}_0'(L(E,[D(A)]))$,*

(537) $\quad \mathbf{G} * P = \delta' \otimes C_{[D(A)]} \in \mathcal{D}_0'(L([D(A)]))$ *and* $P * \mathbf{G} = \delta' \otimes C \in \mathcal{D}_0'(L(E))$,

where $P := \delta'' \otimes I - \delta \otimes A \in \mathcal{D}_0'(L([D(A)], E))$ and I denotes the inclusion $[D(A)] \hookrightarrow E$.

(ii) *Suppose A is a closed linear operator, $\mathbf{G} \in \mathcal{D}_0'(L(E, [D(A)]))$, $\mathbf{G}(\varphi)C = C\mathbf{G}(\varphi)$, $\varphi \in \mathcal{D}$ and (537) holds. Then $\mathbf{G}$ is a (C-DCF) generated by $C^{-1}AC$.*

(iii) *Let $\mathbf{G} \in \mathcal{D}_0'(L(E))$ and $\mathbf{G}(\varphi)C = C\mathbf{G}(\varphi)$, $\varphi \in \mathcal{D}$. Then $\mathbf{G}$ is a (C-DCF) in E generated by A iff $\mathcal{G}$ is a ($\mathcal{C}$-DS) in $E \oplus E$ generated by $\mathcal{A}$.*

Proof. Let $X = L(E, [D(A)])$, $Y = L([D(A)], E)$, $Z = L([D(A)])$ and let $b : X \times Y \to Z$ be defined by $b(B, D) := BD$, $B \in X$, $D \in Y$. The definition of $\mathbf{G} * P$ is given by Proposition 3.2.1; the convolution $P * \mathbf{G}$ can be understood similarly. Let $x \in D(A)$, $k \in \mathbb{N}_0$ and $\varphi \in \mathcal{D}$. Then it is obvious that $(\mathbf{G} * (\delta^{(k)} \otimes I))(\varphi)x = (-1)^k\mathbf{G}(\varphi^{(k)})x$, $(\mathbf{G} * (\delta^{(k)} \otimes A))(\varphi)x = (-1)^k\mathbf{G}(\varphi^{(k)})Ax$, $((\delta^{(k)} \otimes I) * \mathbf{G})(\varphi)x = (-1)^k\mathbf{G}(\varphi^{(k)})x$ and $((\delta^{(k)} \otimes A) * \mathbf{G})(\varphi)x = (-1)^kA\mathbf{G}(\varphi^{(k)})x$, $\varphi \in \mathcal{D}$, $x \in E$, $k \in \mathbb{N}_0$. Suppose that $\mathbf{G}$ is a $(C-DCF)$ generated by A and $x \in E$. Then $A\mathbf{G}(\varphi)x = \mathbf{G}(\varphi'')x + \varphi'(0)Cx$, which implies $\mathbf{G} \in \mathcal{D}_0'(L(E, [D(A)]))$, $(P * \mathbf{G})(\varphi)x = \mathbf{G}(\varphi'')x - A\mathbf{G}(\varphi)x = -\varphi'(0)Cx$ and $P * \mathbf{G} = \delta' \otimes C$. We obtain $\mathbf{G} * P = \delta' \otimes C_{[D(A)]}$ along the same lines, which completes the proof of (i). In order to prove (ii), let us assume $\mathbf{G} \in \mathcal{D}_0'(L(E, [D(A)]))$, $\mathbf{G} * P = \delta' \otimes C_{[D(A)]}$ and $P * \mathbf{G} = \delta' \otimes C$. Since $\operatorname{supp}(\mathbf{G}) \subseteq [0, \infty)$, it follows that $\operatorname{supp}(\mathbf{G}^{-1}) \subseteq [0, \infty)$ and $\operatorname{supp}(\mathcal{G}) \subseteq [0, \infty)$. If $x \in E$, then the assumptions $\mathbf{G} * P = \delta' \otimes C_{[D(A)]}$ and $P * \mathbf{G} = \delta' \otimes C$ imply $\mathbf{G}(\varphi)Ax = \mathbf{G}(\psi'')x + \psi'(0)Cx$, $\varphi \in \mathcal{D}$, $x \in D(A)$, $A\mathbf{G}^{-1}(\varphi)x = -\mathbf{G}(\varphi')x - \varphi(0)Cx$, $\varphi \in \mathcal{D}$, $x \in E$ and $\mathbf{G}^{-1}(\varphi)Ax = -\mathbf{G}(\varphi')x - \varphi(0)Cx$, $\varphi \in \mathcal{D}$, $x \in E$. It is also clear that $\mathcal{G}$ commutes with $\mathcal{C}$. Then one can repeat literally the proof of [292, Theorem 3.1.7] with a view to obtain that, for every $\tau > 0$, there exists $n_\tau \in \mathbb{N}$ such that $\mathcal{A}$ is a subgenerator of a local $(n_\tau + 1)$-times integrated $\mathcal{C}$-semigroup $(S_{n_\tau+1}(t))_{t\in[0,\tau)}$ whose integral generator is $\mathcal{C}^{-1}\mathcal{A}\mathcal{C}$ and which satisfies $\mathcal{G}(\varphi)\binom{x}{y} = (-1)^{n_\tau+1} \int_0^\tau \varphi^{(n_\tau + 1)}(t)S_{n_\tau+1}(t)\binom{x}{y}\, dt$, $x, y \in E$, $\varphi \in \mathcal{D}_{(-\infty,\tau)}$. By making use of [292, Theorem 2.1.11], we get that, for every $\tau > 0$, there exists $n_\tau \in \mathbb{N}$ such that A is a subgenerator of a local n_τ-times integrated C-cosine function $(C_{n_\tau}(t))_{t\in[0,\tau)}$ whose integral generator is $C^{-1}AC$. Furthermore, the next equality holds $S_{n_\tau+1}(t) =$

$$\begin{pmatrix} \int_0^t C_{n_\tau}(s)\, ds & \int_0^t (t-s)C_{n_\tau}(s)\, ds \\ C_{n_\tau}(t) - \frac{t^{n_\tau}}{n_\tau!}C & \int_0^t C_{n_\tau}(s)\, ds \end{pmatrix}, \quad t \in [0, \tau).$$

This implies that $C^{-1}AC$ is the generator of a $(C-DCF)$ $\mathbf{G}$, finishing the proof of (ii). The proof of (iii) can be deduced as in the case of distribution cosine functions. □

Definition 3.2.9. A $(C-DCF)$ $\mathbf{G}$ is said to be an exponential C-distribution cosine function, $(E-CDCF)$ in short, if $\mathcal{G}$ is an $(E-\mathcal{C}DS)$ in $E \oplus E$.

Theorem 3.2.10. *(cf.* [295], [292]*)*

(i) *Let $\mathbf{G}$ be a (C-DCF). Then $\mathbf{G}$ is exponential iff there exists $\omega \in \mathbb{R}$ such that $e^{-\omega t}\mathbf{G}^{-1} \in \mathcal{S}_0'(L(E))$.*

(ii) *Let A be a closed operator. Then the following assertions are equivalent:*

   (a) *The operator A is the generator of an (E-CDCF) in E.*

(b) *The operator* $\mathcal{A}$ *is the generator of an (E-CDS) in* $E \oplus E$.

(c) *There exists* $n \in \mathbb{N}$ *such that* A *is the generator of an exponentially bounded n-times integrated C-cosine function.*

(d) *There exist* $\omega > 0$, $M > 0$ *and* $k \in \mathbb{N}$ *such that* $\Pi_\omega = \{x + iy \in \mathbb{C} : x > \omega^2 - \frac{y^2}{4\omega^2}\} \subseteq \rho_C(A)$, $\|(\lambda - A)^{-1}C\| \leqslant M|\lambda|^k$, $\lambda \in \Pi_\omega$ *and that the mapping* $\lambda \mapsto (\lambda - A)^{-1}C$, $\lambda \in \Pi_\omega$ *is strongly continuous.*

(iii) *Let* A *be a densely defined operator and let* $R(C)$ *be dense in* E. *If* A *is the generator of an (exponential) (C-DCF) in* E, *then* A^* *is the generator of an (exponential) (*C^**-DCF) in* E^*.

Theorem 3.2.11. *(*[295]*)*

(i) *Let* $\mathbf{G}$ *be a (C-DCF). Then for each* $\binom{x}{y} \in \mathcal{R}(\mathcal{G})$ *there exists a unique function* $u \in C^1([0, \infty) : E)$ *satisfying* $u(0) = Cx$, $u'(0) = Cy$ *and*

$$\mathbf{G}(\psi)x + \mathbf{G}^{-1}(\psi)y = \int_0^\infty \psi(t)u(t)\,dt,\ \psi \in \mathcal{D}.$$

(ii) *Let* $\mathbf{G}$ *be a (C-DCF) generated by* A. *Then, for every* $x, y \in D_\infty(A)$, *there exists a unique function* $u \in C^1([0, \infty) : E)$ *satisfying* $u(0) = Cx$, $u'(0) = Cy$ *and*

$$\mathbf{G}(\varphi)x + \mathbf{G}^{-1}(\varphi)y = \int_0^\infty \varphi(t)u(t)\,dt,\ \varphi \in \mathcal{D}_0.$$

If $\mathbf{G}$ is a (C-DCF) generated by A, then $C(D_\infty(A))$ is contained in the closure of the set $\mathcal{R}(\mathbf{G})$. Moreover, if $R(C)$ is dense in E and $\mathbf{G}$ is a (C-DCF) generated by A, then we have the equivalence of following assertions:

(a) $\mathbf{G}$ is dense.
(b) A is densely defined.
(c) $\mathbf{G}^*$ is a (C^**-DCF*) in E^*.
(d) $\mathcal{G}$ is dense.
(e) $\mathcal{A}$ is densely defined.
(f) $\mathcal{G}^*$ is a (C^*-DS) in $(E \oplus E)^*$.

In order to complete the structural theory of C-distribution cosine functions ([414], [295]-[296]), one has to consider global integrated C-cosine functions with corresponding growth order, cosine convolution products and almost C-distribution cosine functions. Assume that $\tau_0 : [0, \infty) \to [0, \infty)$ is a measurable function such that $\inf_{t>0} \tau_0(t) > 0$ and that there exists $c_0 > 0$ satisfying:

$$\tau_0(t + s) \leqslant c_0\tau_0(t)\tau_0(s),\ t, s \geqslant 0 \text{ and } \tau_0(t - s) \leqslant c_0\tau_0(t)\tau_0(s),\ 0 < s < t.$$

Then $(L^1([0, \infty) : \tau_0), \|\cdot\|_{\tau_0})$ denotes the Banach space which consists of those measurable functions $f: [0, \infty) \to \mathbb{C}$ such that $\|f\|_{\tau_0} := \int_0^\infty |f(t)|\tau_0(t)\,dt < \infty$. If $f, g \in L^1([0, \infty) : \tau_0)$, define $f \circ g(t) := \int_t^\infty f(s - t)g(s)\,ds$, $t \geqslant 0$. Clearly, $f *_0 g \in L^1([0, \infty) : \tau_0)$ and $f \circ g \in L^1([0, \infty) : \tau_0)$. The cosine convolution product $f *_c g$ is defined by

$f *_c g := \frac{1}{2}(f *_0 g + f \circ g + g \circ f)$; the sine convolution product by $f *_s g := \frac{1}{2}(f *_0 g - f \circ g - g \circ f)$ and the sine-cosine convolution product by $f *_{sc} g := \frac{1}{2}(f *_0 g - f \circ g + g \circ f)$. It can be easily proved that $f *_c g, f *_s g, f *_{sc} g \in L^1([0, \infty) : \tau_0)$, resp. $\mathcal{D}_+$, if $f, g \in L^1([0, \infty) : \tau_0)$, resp. $f, g \in \mathcal{D}_+$.

Proposition 3.2.12. *([296])*

(i) *Let* $\mathbf{G}$ *be a (C-DCF) generated by A. Then the following holds:*

$$\mathbf{G}(\varphi *_0 \psi)Cx = \mathbf{G}(\varphi)\mathbf{G}(\psi)x + A\mathbf{G}^{-1}(\varphi)\mathbf{G}^{-1}(\psi)x,\ \varphi,\ \psi \in \mathcal{D},\ x \in E.$$

(ii) *Let* $\mathbf{G} \in \mathcal{D}_0'(L(E))$ *satisfy* $\mathbf{G}(\varphi)\mathbf{G}(\psi) = \mathbf{G}(\psi)\mathbf{G}(\varphi)$, $\varphi, \psi \in \mathcal{D}$. *Then the following assertions are equivalent:*

(a) $\mathbf{G}$ *is a pre-(C-DCF) and* $\mathbf{G}^{-1}(\Lambda(f\circ g - g\circ f))C = \mathbf{G}(\Lambda(f))\mathbf{G}^{-1}(\Lambda(g)) - \mathbf{G}^{-1}(\Lambda(f))\mathbf{G}(\Lambda(g))$, $f, g \in \mathcal{D}_+$.

(b) $\mathbf{G}^{-1}(\Lambda(f *_{sc} g))C = \mathbf{G}^{-1}(\Lambda(f))\mathbf{G}(\Lambda(g))$, $f, g \in \mathcal{D}_+$.

Definition 3.2.13. An element $\mathrm{G} \in L(\mathcal{D}_+, L(E))$ is called an almost C-distribution cosine function, $(A - CDCF)$ in short, iff $\mathrm{G}(f)C = C\mathrm{G}(f)$, $f \in \mathcal{D}_+$,

(i) $\mathrm{G}(f *_c g)C = \mathrm{G}(f)\mathrm{G}(g)$, $f, g \in \mathcal{D}_+$, and

(ii) $\bigcap_{f\in\mathcal{D}_+} \mathrm{Kern}(\mathrm{G}(f)) = \{0\}$.

The (infinitesimal) generator A of G is defined by

$$A := \{(x, y) \in E \oplus E : \mathrm{G}(f)y = \mathrm{G}(f'')x + f'(0)Cx \text{ for all } f \in \mathcal{D}_+\}.$$

It can be simply proved that A is a closed linear operator which satisfies $\mathrm{G}(f)A \subseteq A\mathrm{G}(f)$, $\mathrm{G}(f)x \in D(A)$, $A\mathrm{G}(f)x = \mathrm{G}(f'')x + f'(0)Cx$, $f \in \mathcal{D}_+$, $x \in E$ and $C^{-1}AC = A$.

Theorem 3.2.14. *([296])*

(i) *Let* $\mathbf{G}$ *be a (C-DCF) generated by A. Then* $\mathbf{G}\Lambda$ *is an (A-CDCF) generated by A.*

(ii) *Let* $\mathbf{G}$ *be a (C-DCF) generated by A. Then*

$$G(\Lambda(f *_s g))C = AG^{-1}(\Lambda(f))G^{-1}(\Lambda(g)),\ f, g \in \mathcal{D}_+.$$

(iii) *Let* G *be an (A-CDCF) generated by A. Then A is the generator of a (C-DCF)* $\mathbf{G}$, *which is given by* $\mathbf{G}(\varphi) := \mathrm{G}(\mathcal{K}(\varphi))$, $\varphi \in \mathcal{D}$.

(iv) *Every (almost) C-distribution cosine function is uniquely determined by its generator.*

(v) *Let A be a closed linear operator. Then A is the generator of a (C-DCF) if A is the generator of an (A-CDCF).*

(vi) *Let* $\mathbf{G}$ *be a (C-DCF). Then* $\mathbf{G}(\varphi)\mathbf{G}(\psi) = \mathbf{G}(\psi)\mathbf{G}(\varphi)$, $\varphi, \psi \in \mathcal{D}$.

Let $f \in \mathcal{D}_+$. Then the Weyl fractional integral of order $\alpha > 0$ is defined by $(W_+^{-\alpha} f)(t) := \int_t^\infty g_\alpha(s - t)f(s)\, ds$, $f \in \mathcal{D}_+$, $t \geqslant 0$. It is well known that, for every $\alpha > 0$, the mapping $W_+^{-\alpha} : \mathcal{D}_+ \to \mathcal{D}_+$ is bijective. We will refer to the inverse mapping of $W_+^{-\alpha}(\cdot)$, denoted by $W_+^{\alpha}(\cdot)$, as the Weyl fractional derivative of order $\alpha > 0$. If $\alpha \in$

$\mathbb{N}$, then $W_+^\alpha f = (-1)^n f^{(\alpha)}, f \in \mathcal{D}_+$. Furthermore, $W_+^\alpha W_+^\beta = W_+^{\alpha+\beta}$ for all $\alpha, \beta \in \mathbb{R}$, where we put $W_+^0 := I$. Let us recall [414] that the family of Bochner-Riesz functions (R_t^θ), $\theta > -1$, $t > 0$, is defined by $R_t^\theta(s) := g_{\theta+1}(t-s)\chi_{(0,t)}(s)$, $s > 0$. The Weyl fractional calculus can be applied to the functions lying outside of the space $\mathcal{D}_+$; for example, in the case of Bochner-Riesz functions, one has $W_+^\alpha R_t^\theta = R_t^{\theta-\alpha}$, $\theta + 1 > \alpha \geqslant 0$. Designate by Ω_α, $\alpha > 0$ the set which consists of all nondecreasing continuous functions $\tau_\alpha(\cdot)$ on $(0, \infty)$ such that $\inf_{t>0} t^{-\alpha}u(t) > 0$ and there is a constant $c_\alpha > 0$ satisfying

$$\int_{[0,t]\cap[s,s+t]} u^{\alpha-1}\tau_\alpha(t+s-u)\,du \leqslant c_\alpha\tau_\alpha(t)\tau_\alpha(s),\ 0 < t \leqslant s.$$

The typical functions $\tau_\alpha(t) = t^\alpha$; $t^\beta(1+t)^\gamma$ ($\beta \in [0,\alpha]$, $\beta + \gamma \geqslant \alpha$); $t^\beta e^{\tau t}$ ($\beta \in [0,\alpha]$, $\tau > 0$) belong to Ω_α. Suppose $\tau_\alpha \in \Omega_\alpha$ and $v > \alpha$; then the function $\tau_v = t^{v-\alpha}\tau_\alpha$, $t > 0$ belongs to Ω_v. Denote by Ω_α^h the subset of Ω_α, $\alpha > 0$ which consists of all functions of the form $\tau_\alpha(t) = t^\alpha\omega_0(t)$, $t > 0$, where the continuous nondecreasing function $\omega_0 : [0,\infty) \to [0,\infty)$ satisfies $\inf_{t>0}\omega_0(t) > 0$ and $\omega_0(t+s) \leqslant \omega_0(t)\omega_0(s)$, $t, s > 0$. Suppose $\alpha > 0$, $\tau_\alpha \in \Omega_\alpha$ and define

$$q_{\tau_\alpha}(\varphi) := \int_0^\infty \frac{\tau_\alpha(t)}{\Gamma(\alpha+1)}\left|W_+^\alpha\varphi(t)\right|dt,\ \varphi \in \mathcal{D}_+.$$

Then $q_{\tau_\alpha}(\cdot)$ is a norm on $\mathcal{D}_+$ and there exists a constant $c_\alpha > 0$ such that $q_{\tau_\alpha}(\varphi *_c \phi) \leqslant c_\alpha q_{\tau_\alpha}(\varphi)q_{\tau_\alpha}(\phi)$, $\varphi, \phi \in \mathcal{D}_+$ ([414]). Let $\mathfrak{T}_+^\alpha(\tau_\alpha, *_c)$ denote the completion of the normed space $(\mathcal{D}_+, q_{\tau_\alpha})$; then $\mathfrak{T}_+^\alpha(\tau_\alpha, *_c)$ is invariant under the cosine convolution product $*_c$ and the following holds (cf. [414, Theorem 3]):

(i) $\mathfrak{T}_+^\alpha(\tau_\alpha, *_c) \hookrightarrow \mathfrak{T}_+^\alpha(t^\alpha, *_c) \hookrightarrow L^1([0,\infty), *_c)$, where $\hookrightarrow$ denotes the dense and continuous embedding,
(ii) $\mathfrak{T}_+^\beta(t^\beta, *_c) \hookrightarrow \mathfrak{T}_+^\alpha(t^\alpha, *_c)$, $\beta > \alpha > 0$,
(iii) $R_t^{v-1} \in \mathfrak{T}_+^\alpha(\tau_\alpha, *_c)$, $v > \alpha$, $t > 0$ and there exists $c_{v,\alpha} > 0$ such that $q_{\tau_\alpha}(R_t^{v-1}) \leqslant c_{v,\alpha}t^{v-\alpha}\tau_\alpha(t)$, $t > 0$.

An $(A - CDCF)$ G is said to be of order $\alpha > 0$ and growth $\tau_\alpha \in \Omega_\alpha$ if G can be extended to a continuous linear mapping from $\mathfrak{T}_+^\alpha(\tau_\alpha, *_c)$ into $L(E)$.

Theorem 3.2.15. *([414])*

(i) *Let A be the generator of an α-times integrated C-cosine function $(C_\alpha(t))_{t\geqslant 0}$ and let $\|C_\alpha(t)\| = O(\tau_\alpha(t))$, $t > 0$. Then the mapping* $\mathrm{G} : \mathfrak{T}_+^\alpha(\tau_\alpha, *_c) \to L(E)$, *given by*

$$\mathrm{G}(f)x := \int_0^\infty W_+^\alpha f(t)C_\alpha(t)x\,dt,\ f \in \mathfrak{T}_+^\alpha(\tau_\alpha, *_c),\ x \in E,$$

is a continuous algebra homomorphism satisfying:

$$\int_0^t \frac{(t-s)^{\nu-\alpha-1}}{\Gamma(\nu-\alpha)} C_\alpha(s)x\,ds = \mathrm{G}(R_t^{\nu-1})x,\ \nu > \alpha,\ x \in E$$

and

$$\int_0^\infty W_+^\alpha f(t) C_\alpha(t)x\,dt = \int_0^\infty W_+^\nu f(t) \int_0^t \frac{(t-s)^{\nu-\alpha-1}}{\Gamma(\nu-\alpha)} C_\alpha(s)x\,ds\,dt,$$

for all $f \in \mathfrak{T}_+^\nu(t^{\nu-\alpha}\tau_\alpha, *_c)$ *and* $x \in E$. *Furthermore, the restriction of* G *to* $\mathcal{D}_+$ *is an almost C-distribution cosine function of order* $\alpha > 0$ *and growth* τ_α *with the generator A.*

(ii) *Suppose A is the generator of an (A-CDCF)* G *of order* $\alpha > 0$ *and growth* $\tau_\alpha \in \Omega_\alpha$. *Then, for every* $\nu > \alpha$, *A generates a ν-times integrated C-cosine function* $(C_\nu(t))_{t\geqslant 0}$ *such that* $||C_\nu(t)|| \leqslant c_\nu t^{\nu-\alpha}\tau_\alpha(t)$, $t > 0$ *and*

$$\mathrm{G}(f)x = \int_0^\infty W_+^\nu f(t) \int_0^t \frac{(t-s)^{\nu-\alpha-1}}{\Gamma(\nu-\alpha)} C_\alpha(s)x\,ds\,dt,\ f \in \mathcal{D}_+, x \in E.$$

(iii) *Let* $\alpha > 0$, $\tau_\alpha \in \Omega_\alpha^h$, *and let D(A) and R(C) be dense in E. Then the following assertions are equivalent:*
  (a) *The operator A is the generator of an α-times integrated C-cosine function* $(C_\alpha(t))_{t\geqslant 0}$ *such that* $||C_\alpha(t)|| = O(\tau_\alpha(t))$, $t > 0$.
  (b) *The operator A is the generator of an (A-CDCF)* G *of order* $\alpha > 0$ *and growth* τ_α *such that* $\mathrm{G}(\mathcal{D}_+)$ *is dense in E.*

Remark 3.2.16. One of the most undeveloped subjects in the theory of abstract Volterra integro-differential equation is, undoubtely, the analysis of solutions in the generalized function spaces. This section provides a general information on distributional solutions of abstract second order differential equations, the corresponding results for equations of first order can be found in the third chapter of monograph [292]. On the other hand, distributional solutions of abstract Volterra equations have been considered by G. Da Prato and M. Iannelli in [135]. In connection with this, we would like to address the problem of describing the class of integrated (g_α, C)-regularized resolvent families ($\alpha \in (0, 2)\backslash\{1\}$) in terms of a corresponding class of C-distribution resolvents. Also, it could be of importance to analyze solutions of abstract Volterra equations in the spaces of vector-valued tempered ultradistributions ([327], [455], [292]).

The following theorem will be useful in our further work.

Theorem 3.2.17. *Assume* $\alpha \geqslant 0$ *and A is a subgenerator of a global α-times integrated C-cosine function* $(C_\alpha(t))_{t\geqslant 0}$. *Then, for every* $\beta > \alpha$, *the operator A is a subgenerator of a global β-times integrated C-cosine function* $(C_\beta(t))_{t\geqslant 0}$, *which is given by* $C_\beta(t)x = \int_0^t g_{\beta-\alpha}(t-s)C_\alpha(s)x\,ds$, $x \in E$, $t \geqslant 0$. *Define*

$$\mathbf{G}(\varphi)x := \int_0^\infty W_+^\alpha(\mathcal{K}(\varphi))(t)C_\alpha(t)x\,dt,\ x \in E,\ \varphi \in \mathcal{D}.$$

Then $\mathbf{G}$ *is a (C-DCF) generated by* $C^{-1}AC$ *and the following equality holds:* $\mathbf{G}(\varphi)x = \int_0^\infty W_+^\beta(\mathcal{K}(\varphi))(t)C_\beta(t)x\,dt,\ x \in E,\ \varphi \in \mathcal{D}.$

Let us repeat the following notions. A function $u(t)$ is said to be a mild solution of the abstract Cauchy problem

$$(ACP_1) : u'(t) = Au(t),\ t \geqslant 0,\ u(0) = x, \text{ resp.,}$$
$$(ACP_2) : u''(t) = Au(t),\ t \geqslant 0,\ u(0) = x,\ u'(0) = y,$$

iff the mapping $t \mapsto u(t)$, $t \geqslant 0$ is continuous, $\int_0^t u(s)\,ds \in D(A)$ and $A\int_0^t u(s)\,ds = u(t) - x$, $t \geqslant 0$, resp., iff the mapping $t \mapsto u(t)$, $t \geqslant 0$ is continuous, $\int_0^t (t-s)u(s)\,ds \in D(A)$ and $A\int_0^t (t-s)u(s)\,ds = u(t) - x - ty$, $t \geqslant 0$. We have proved that there exists at most one mild solution of (ACP_1), resp. (ACP_2), provided that there exists $\alpha \geqslant 0$ such that A is a subgenerator of a local α-times integrated C-semigroup, resp., a local α-times integrated C-cosine function. Recall that if mild solutions of (ACP_1) are unique, then the solution space for A, denoted by $Z(A)$, has been defined as the set of those elements $x \in E$ for which there exists a unique mild solution of (ACP_1). In order not to put a strain on the exposition, and to stay consistent with previously given definitions of hypercyclicity and chaos of cosine functions ([70], [247]), we shall primarily consider mild solutions of (ACP_2) with $y = 0$. This, however, may not be the optimal choice and we refer the reader to [120] and Subsection 3.3.2 for further information in this direction. Denote, with a little abuse of notation, by $Z_2(A)$ the set which consists of all $x \in E$ for which there exists such a solution. Let $\pi_1 : E \oplus E \to E$ and $\pi_2 : E \oplus E \to E$ be the projections, and let G be a $(C-DCF)$ generated by A. Then $\mathcal{G}$ is a $(\mathcal{C}-DS)$ generated by $\mathcal{A}$ and the solution space $Z(\mathcal{A})$ can be characterized as follows: Denote by $D(\mathcal{G})$ the set of all $x \in \bigcap_{t>0} D(\mathcal{G}_1(\delta_t))$ satisfying that the mapping $t \mapsto \mathcal{G}_1(\delta_t)x$, $t \geqslant 0$ is continuous; here

$$\mathcal{G}_1(\delta_t) = \left\{\left(\binom{x_1}{y_1}, \binom{x_2}{y_2}\right) : \mathcal{G}_1(\varphi(\cdot - t))\binom{x_1}{y_1} = \mathcal{G}_1(\varphi)\binom{x_2}{y_2},\ \varphi \in \mathcal{D}_0\right\}.$$

Then $Z(\mathcal{A}) = D(\mathcal{G})$ and the mild solution $u(\cdot;\binom{x}{y})$ of (ACP_1) with initial value $\binom{x}{y} \in Z(\mathcal{A})$ is given by $u(t;\binom{x}{y}) = \mathcal{G}_1(\delta_t)\binom{x}{y}$, $t \geqslant 0$. Assume that, for every $\tau > 0$, A is the integral generator of a local n_τ-times integrated C-cosine function $(C_{n_\tau}(t))_{t\in[0,\tau)}$. Then the solution space $Z(\mathcal{A})$ consists of those pairs $\binom{x}{y}$ in $E \oplus E$ which fulfill that, for every $\tau > 0$, $C_{n_\tau}(t)x + \int_0^t C_{n_\tau}(s)y\,ds \in R(C)$, $t \in [0, \tau)$ and that the mapping $t \mapsto C^{-1}(C_{n_\tau}(t)x + \int_0^t C_{n_\tau}(s)y\,ds)$, $t \in [0, \tau)$ is $(n_\tau + 1)$-times continuously differentiable. By prior arguments, we have that $x \in Z_2(A)$ iff $\binom{0}{x} \in Z(\mathcal{A})$ iff $\binom{0}{x} \in D(\mathcal{G})$, and $u(t; x) = \pi_2(\mathcal{G}_1(\delta_t)\binom{0}{x})$, $t \geqslant 0$, where $u(\cdot; x)$ denotes the mild solution of (ACP_2) with $y = 0$. Define $G(\delta_t)x := \pi_2(\mathcal{G}_1(\delta_t)\binom{0}{x})$, $t \geqslant 0$, $x \in Z_2(A)$.

Proposition 3.2.18. *Assume that, for every* $\tau > 0$, *there exists* $n_\tau \in \mathbb{N}$ *such that* A *is a subgenerator of a local* n_τ*-times integrated C-cosine function* $(C_{n_\tau}(t))_{t\in[0,\tau)}$.

Then the solution space $Z_2(A)$ consists exactly of those vectors $x \in E$ such that, for every $\tau > 0$, $C_{n_\tau}(t)x \in R(C)$ and the mapping $t \mapsto C^{-1}C_{n_\tau}(t)x$, $t \in [0, \tau)$ is n_τ-times continuously differentiable. If $x \in Z_2(A)$ and $t \in [0, \tau)$, then $G(\delta_t)x = \frac{d^{n_\tau}}{dt^{n_\tau}}C^{-1}C_{n_\tau}(t)x$.

Proposition 3.2.19. *Let A be the generator of a (C-DCF) $\mathbf{G}$, and let $x \in Z_2(A)$. Then $G(\delta_t)(Z_2(A)) \subseteq Z_2(A)$, $t \geqslant 0$, $2G(\delta_s)G(\delta_t)x = G(\delta_{t+s})x + G(\delta_{|t-s|})x$, $t, s \geqslant 0$ and $\mathbf{G}(\varphi)x = \int_0^\infty \varphi(t)CG(\delta_t)x\, dt$, $\varphi \in \mathcal{D}_0$.*

Proof. We will only prove the d'Alambert formula $2G(\delta_s)G(\delta_t)x = G(\delta_{t+s})x + G(\delta_{|t-s|})x$, $t, s \geqslant 0$. Fix a number $t \geqslant 0$ and define afterwards

$$u(s; G(\delta_t)x) := \frac{1}{2}[G(\delta_{t+s})x + G(\delta_{|t-s|})x],\ s \geqslant 0.$$

Then the mapping $s \mapsto u(s; G(\delta_t)x)$, $s \geqslant 0$ is continuous and, for every $s \in [0, t]$, $\int_0^s (s-r)u(r; G(\delta_t)x)\, dr = \int_0^{t+s}(t+s-r)u(r; G(\delta_t)x)\, dr - \int_0^t (t-r)u(r; G(\delta_t)x)\, dr + \int_0^{t-s}(t-s-r)u(r; G(\delta_t)x)\, dr \in D(A)$ and $A\int_0^s (s-r)u(r; G(\delta_t)x)\, dr = \frac{1}{2}[G(\delta_{t+s})x - x] - \frac{1}{2}[G(\delta_t)x - x] + \frac{1}{2}[G(\delta_{t-s})x - x] = \frac{1}{2}[G(\delta_{t+s})x - x] = u(s; G(\delta_t)x) - G(\delta_t)x$. One can similarly prove that $A\int_0^s (s-r)u(r; G(\delta_t)x)\, dr = u(s; G(\delta_t)x) - G(\delta_t)x$, $s > t$, which completes the proof of theorem. □

Assume $\mathbf{G}$ is a $(C-DCF)$ generated by A and $x \in Z_2(A)$. Then Proposition 3.2.19 implies $C(Z_2(A)) \subseteq \overline{\mathcal{R}(\mathcal{G})}$ and $\mathcal{G}(\varphi)x \in R(C)$, $\varphi \in \mathcal{D}_0$. Further on, $C(Z_2(A)) \subseteq Z_2(A)$ and $G(\delta_t)Cx = CG(\delta_t)x$, $t \geqslant 0$.

Proposition 3.2.20. (i) *Assume $\mathbf{G}$ is a (C-DCF) generated by A. Then $\mathcal{R}(\mathbf{G}) \subseteq Z_2(A)$.*

(ii) *Assume A is a closed linear operator, $x \in Z(A) \cap Z(-A)$, $u_1(\cdot; x)$ and $u_2(\cdot; x)$ are mild solutions of (ACP_1) for A and $-A$, respectively, and $u(t; x) := \frac{1}{2}(u_1(t; x) + u_2(t; x))$, $t \geqslant 0$. If A^2 is closed, then $u(\cdot; x)$ is a mild solution of (ACP_2) for A^2.*

Proof. We will only prove part (i). Assume $x \in \mathcal{R}(\mathbf{G})$ and $x = \mathbf{G}(\varphi)y$ for some $\varphi \in \mathcal{D}_0$ and $y \in E$. Put

$$u(t; x) := \frac{1}{2}[\mathbf{G}(\varphi(\cdot - t))y + \mathbf{G}(\varphi(\cdot + t))y + \mathbf{G}(\varphi(t - \cdot))y],\ t \geqslant 0. \tag{538}$$

Using the continuity of $\mathbf{G}$, one gets that $u(\cdot; x) \in C([0, \infty) : E)$. Denote $f(t) := \mathbf{G}(\varphi(\cdot - t))y$, $g(t) := \mathbf{G}(\varphi(\cdot + t))y$ and $h(t) := \mathbf{G}(\varphi(t - \cdot))y$, $t \geqslant 0$. Then $f, g, h \in C^2([0, \infty) : E)$, $f'(t) = -\mathbf{G}(\varphi'(\cdot - t))y$, $f''(t) = \mathbf{G}(\varphi''(\cdot - t))y$, $g'(t) = \mathbf{G}(\varphi'(\cdot + t))y$, $g''(t) = \mathbf{G}(\varphi''(\cdot + t))y$, $h'(t) = -\mathbf{G}(\varphi'(t - \cdot))y$ and $h''(t) = \mathbf{G}(\varphi''(t - \cdot))y$, $t \geqslant 0$. The above equalities, the partial integration, the representation formula (538) and Theorem 3.2.5(ii)(a) taken together imply:

$$A\int_0^t (t-s)u(s; x)\, ds$$

$$= \frac{1}{2}\int_0^t (t-s)[\mathbf{G}(\varphi''(\cdot - s))y + \mathbf{G}(\varphi''(\cdot + s))y + \varphi'(s)Cy$$

$$+ \mathbf{G}(\varphi''(s - \cdot))y - \varphi'(s)Cy]\, ds$$

$$= \frac{1}{2}\Big[- \int_0^t \mathbf{G}(\varphi'(\cdot - s))y\, ds + \int_0^t \mathbf{G}(\varphi'(\cdot + s))y\, ds - \int_0^t \mathbf{G}(\varphi'(s - \cdot))y\, ds \Big]$$

$$= u(t; x) - x,\ t \geqslant 0.$$

□

3.2.2. Hypercyclicity and chaos for C-distribution cosine functions and integrated C-cosine functions. In the remaining part of this section, we shall assume that E is a separable infinite-dimensional complex Banach space and that S is a non-empty closed subset of $\mathbb{C}$ satisfying $S \setminus \{0\} \neq \emptyset$.

Let $\mathbf{G}$ be a $(C - DCF)$. A closed linear subspace $\tilde{E}$ of E is said to be $\mathbf{G}$-admissible iff $G(\delta_t)(Z_2(A) \cap \tilde{E}) \subseteq Z_2(A) \cap \tilde{E}$, $t \geqslant 0$. Define $\mathbf{G}_{wm}(\varphi)\binom{x}{y} := \binom{\mathbf{G}(\varphi)x}{\mathbf{G}(\varphi)y}$, $x, y \in E$, $\varphi \in \mathcal{D}$. Then $\mathbf{G}_{wm}$ is a $(\mathcal{C} - DCF)$ in $E \oplus E$ generated by $A \oplus A$, $Z_2(A \oplus A) = Z_2(A) \oplus Z_2(A)$, and $\tilde{E} \oplus \tilde{E}$ is $\mathbf{G}_{wm}$-admissible provided that $\tilde{E}$ is $\mathbf{G}$-admissible.

Definition 3.2.21. Let $\mathbf{G}$ be a $(C - DCF)$ and let $\tilde{E}$ be $\mathbf{G}$-admissible. Then it is said that $\mathbf{G}$ is:

(i) $\tilde{E}$-hypercyclic, if there exists $x \in Z_2(A) \cap \tilde{E}$ such that the set $\{G(\delta_t)x : t \geqslant 0\}$ is dense in $\tilde{E}$,
(ii) $\tilde{E}$-chaotic, if $\mathbf{G}$ is $\tilde{E}$-hypercyclic and the set of $\tilde{E}$-periodic points of $\mathbf{G}$, $\mathbf{G}_{\tilde{E},per}$, defined by $\{x \in Z_2(A) \cap \tilde{E} : G(\delta_{t_0})x = x$ for some $t_0 > 0\}$, is dense in $\tilde{E}$,
(iii) $\tilde{E}$-topologically transitive, if for every $y, z \in \tilde{E}$ and $\varepsilon > 0$, there exist $\upsilon \in Z_2(A) \cap \tilde{E}$ and $t \geqslant 0$ such that $\|y - \upsilon\| < \varepsilon$ and $\|z - G(\delta_t)\upsilon\| < \varepsilon$,
(iv) $\tilde{E}$-topologically mixing, if for every $y, z \in \tilde{E}$ and $\varepsilon > 0$, there exists $t_0 \geqslant 0$ such that, for every $t \geqslant t_0$, there exists $\upsilon_t \in Z_2(A) \cap \tilde{E}$ such that $\|y - \upsilon_t\| < \varepsilon$ and $\|z - G(\delta_t)\upsilon_t\| < \varepsilon$, $t \geqslant t_0$,
(v) $\tilde{E}$-weakly mixing, if $\mathbf{G}_{wm}$ is $(\tilde{E} \oplus \tilde{E})$-topologically transitive in $E \oplus E$,
(vi) $\tilde{E}$-supercyclic, if there exists $x \in Z_2(A) \cap \tilde{E}$ such that its projective orbit $\{cG(\delta_t)x : c \in \mathbb{C}, t \geqslant 0\}$ is dense in $\tilde{E}$,
(vii) $\tilde{E}$-positively supercyclic, if there exists $x \in Z_2(A) \cap \tilde{E}$ such that its positive projective orbit $\{cG(\delta_t)x : c \geqslant 0, t \geqslant 0\}$ is dense in $\tilde{E}$,
(viii) $\tilde{E}_S$-hypercyclic, if there exists $x \in Z_2(A) \cap \tilde{E}$ such that its S-projective orbit $\{cG(\delta_t)x : c \in S, t \geqslant 0\}$ is dense in $\tilde{E}$; any element $x \in Z_2(A) \cap \tilde{E}$ which satisfies the above property is called a $\tilde{E}_S$-hypercyclic vector of $\mathbf{G}$,
(ix) $\tilde{E}_S$-topologically transitive, if for every $y, z \in \tilde{E}$ and $\varepsilon > 0$, there exist $\upsilon \in Z_2(A) \cap \tilde{E}$, $t \geqslant 0$ and $c \in S$ such that $\|y - \upsilon\| < \varepsilon$ and $\|z - cG(\delta_t)\upsilon\| < \varepsilon$,
(x) sub-chaotic, if there exists a $\mathbf{G}$-admissible subset $\hat{E}$ such that $\mathbf{G}$ is $\hat{E}$-chaotic.

In what follows, we use the fact that the notion of $\tilde{E}$-periodic points and $\tilde{E}$-topological transitivity ($\tilde{E}_S$-topological transitivity) of a $(C - DCF)$ $\mathbf{G}$ (or a $(C$

$-DS)$ G, cf. [292] and [310] for the notion) can be defined even in the case that $\tilde{E}$ is not $\mathbf{G}$-admissible.

Assume that there exists $\alpha \geqslant 0$ such that A is the integral generator of an α-times integrated C-cosine function $(C_\alpha(t))_{t\geqslant 0}$. Put

$$\mathbf{G}_\alpha(\varphi)x := \int_0^\infty W_+^\alpha(\mathcal{K}(\varphi))(t)C_\alpha(t)x\,dt,\ x \in E,\ \varphi \in \mathcal{D}.$$

Then Theorem 3.2.17 implies that $\mathbf{G}_\alpha$ is a $(C-DCF)$ generated by A.

Definition 3.2.22. Let $\tilde{E}$ be a closed linear subspace of E. Then it is said that $\tilde{E}$ is $(C_\alpha(t))_{t\geqslant 0}$-admissible iff $\tilde{E}$ is $\mathbf{G}_\alpha$-admissible, and that $(C_\alpha(t))_{t\geqslant 0}$ is $\tilde{E}$-hypercyclic iff $\mathbf{G}_\alpha$ is; all other dynamical properties of $(C_\alpha(t))_{t\geqslant 0}$ are understood in the same sense. Let $\tilde{E}$ be $(C_\alpha(t))_{t\geqslant 0}$-admissible; then a point $x \in \tilde{E}$ is said to be a $\tilde{E}$-periodic point ($\tilde{E}_S$-hypercyclic vector) of $(C_\alpha(t))_{t\geqslant 0}$ iff x is a $\tilde{E}$-periodic point ($\tilde{E}_S$-hypercyclic vector) of $\mathbf{G}_\alpha$.

It is clear that the notion of $\tilde{E}_S$-hypercyclicity generalizes the notions of (positive) $\tilde{E}$-supercyclicity and $\tilde{E}$-hypercyclicity. In the case $\tilde{E} = E$, it is also said that $\mathbf{G}$ $((C_\alpha(t))_{t\geqslant 0})$ is hypercyclic, chaotic, · · ·, S-hypercyclic, S-topologically transitive, and we write $\mathbf{G}_{per}$ instead of $\mathbf{G}_{\tilde{E},per}$. Using Theorem 3.2.17 again, we get that a closed linear subspace $\tilde{E}$ of E is $(C_\alpha(t))_{t\geqslant 0}$-admissible iff $\tilde{E}$ is $(C_\beta(t))_{t\geqslant 0}$-admissible, and that $(C_\alpha(t))_{t\geqslant 0}$ is $\tilde{E}$-hypercyclic ($\tilde{E}$-chaotic, · · ·, sub-chaotic) if $(C_\beta(t))_{t\geqslant 0}$ is; this is why we assume in the sequel that $\alpha \in \mathbb{N}_0$. Let $\mathbf{G}_i$ be a $(C_i - DCF)$ generated by A, $i = 1, 2$. Then a closed linear subspace $\tilde{E}$ of E is $\mathbf{G}_1$-admissible iff $\tilde{E}$ is $\mathbf{G}_2$-admissible. Furthermore, it follows from Definition 3.2.21 that $\mathbf{G}_1$ and $\mathbf{G}_2$ share common dynamical properties, which can be simply reformulated for the class of global integrated C-cosine functions.

It is easily seen that $\tilde{E}_S$-hypercyclicity ($\tilde{E}_S$-topological transitivity) of $\mathbf{G}$ implies $\overline{\tilde{E} \cap Z_2(A)} = \tilde{E}$. By Proposition 3.2.19, the assumption $G(\delta_{t_0})x = x$ for some $t_0 > 0$ and $x \in Z_2(A)$ implies by induction $G(\delta_{t_0})^n x = G(\delta_{nt_0})x = x$, $n \in \mathbb{N}$, so that the notion of $\tilde{E}$-periodic points of $\mathbf{G}$ is meaningful in some sense.

Remark 3.2.23. Let $\alpha \geqslant 0$. In the following important remark, we shall explain how one can simply define the above considered dynamical properties for a non-degenerate α-times integrated C-cosine function (the same approach works for non-degenerate α-times integrated C-semigroups) $(C_\alpha(t))_{t\geqslant 0}$ acting on a separable sequentially complete locally convex space X over the field $\mathbb{K} \in \{\mathbb{R}, \mathbb{C}\}$; keeping in mind the observations made in the preceding two paragraphs, we may assume without loss of generality that $\alpha \in \mathbb{N}$. First of all, the definitions of $(C_\alpha(t))_{t\geqslant 0}$ and its integral generator (subgenerators) will be taken in the sense of [292, Definition 2.1.19]. Then we define the space $Z_2(A)$ and the linear operator $G(\delta_t)$ $(t \geqslant 0)$ by using Proposition 3.2.18. Now it becomes apparent that the notion of $(C_\alpha(t))_{t\geqslant 0}$-admissibility of a closed linear subspace $\tilde{E}$ of E can be understood in the sense of Definition 3.2.22, and that the notions of all considered dynamical properties from Definition 3.2.21 can be simply understood for $(C_\alpha(t))_{t\geqslant 0}$. The interested reader may try to construct some examples of hypercyclic integrated C-cosine functions on SCLCSs, as well as to see which ones of the assertions clarified below continue to hold in this setting.

Before going any further, we would like to make a general observation on infinitely regular S-hypercyclic vectors of cosine functions. Let $(C(t))_{t\geqslant 0}$ be an S-topologically transitive cosine function, and let $HC_S(C(\cdot))$ denote the set which consists of all S-hypercyclic vectors of $(C(t))_{t\geqslant 0}$. Then one can prove by means of [244, Lemma 3.1, Theorem 3.2] that $HC_S(C(\cdot)) \cap D_\infty(A)$ is a dense subset of E.

Given $t > 0$ and $\sigma > 0$, set

$$\Phi_{t,\sigma} := \Big\{\varphi \in \mathcal{D}_0 : \operatorname{supp}(\varphi) \subseteq (t-\sigma,\, t+\sigma),\ \varphi \geqslant 0,\ \int \varphi(s)\, ds = 1\Big\}.$$

The following theorem can be deduced by making use of Proposition 3.2.19 and the proof of [150, Theorem 4.6].

Theorem 3.2.24. *([150], [310])*

(i) *Assume $n \in \mathbb{N}_0$, A is the integral generator of an n-times integrated C-cosine function $(C_n(t))_{t\geqslant 0}$, $C(\tilde{E}) = \tilde{E}$ and $\tilde{E}$ is $\mathbf{G}_n$-admissible. Then the following holds:*

(a) *$(C_n(t))_{t\geqslant 0}$ is $\tilde{E}_S$-hypercyclic iff there exists $x \in \tilde{E}$ such that the mapping $t \mapsto C_n(t)x$, $t \geqslant 0$ is n-times continuously differentiable and the set $\{c\frac{d^n}{dt^n} C_n(t)x : c \in S, t \geqslant 0\}$ is dense in $\tilde{E}$.*

(b) *$(C_n(t))_{t\geqslant 0}$ is $\tilde{E}_S$-topologically transitive iff for every $y, z \in \tilde{E}$ and $\varepsilon > 0$, there exist $\upsilon \in \tilde{E}$, $t_0 \geqslant 0$ and $c \in S$ such that the mapping $t \mapsto C_n(t)\upsilon$, $t \geqslant 0$ is n-times continuously differentiable as well as $\|y - \upsilon\| < \varepsilon$ and $\|z - c(\frac{d^n}{dt^n} C_n(t)\upsilon)_{t=t_0}\| < \varepsilon$.*

(c) *$(C_n(t))_{t\geqslant 0}$ is $\tilde{E}$-chaotic iff $(C_n(t))_{t\geqslant 0}$ is $\tilde{E}$-hypercyclic and there exists a dense subset of $\tilde{E}$ consisting of those vectors $x \in \tilde{E}$ for which there exists $t_0 > 0$ such that the mapping $t \mapsto C_n(t)x$, $t \geqslant 0$ is n-times continuously differentiable and $(\frac{d^n}{dt^n} C_n(t)x)_{t=t_0} = Cx$.*

(ii) *Let A be the generator of a (C-DCF) $\mathbf{G}$ and let $\tilde{E}$ be $\mathbf{G}$-admissible. Then:*

(a) *$\mathbf{G}$ is $\tilde{E}_S$-hypercyclic iff there exists $x_0 \in Z_2(A) \cap \tilde{E}$ such that, for every $x \in \tilde{E}$ and $\varepsilon > 0$, there exist $t_0 > 0$, $c \in S$ and $\sigma > 0$ such that*

$$\left\| cC^{-1}\mathbf{G}(\varphi)x_0 - x \right\| < \varepsilon,\ \varphi \in \Phi_{t_0,\sigma}.$$

(b) *$\mathbf{G}$ is $\tilde{E}_S$-topologically transitive iff for every $y, z \in \tilde{E}$ and $\varepsilon > 0$, there exist $t_0 > 0$, $c \in S$, $\sigma > 0$ and $\upsilon \in Z_2(A) \cap \tilde{E}$ such that, for every $\varphi \in \Phi_{t_0,\sigma}$,*

$$\|y - \upsilon\| < \varepsilon \text{ and } \left\| z - cC^{-1}\mathbf{G}(\varphi)\upsilon \right\| < \varepsilon.$$

(c) *$\mathbf{G}$ is $\tilde{E}$-chaotic iff $\mathbf{G}$ is $\tilde{E}$-hypercyclic and if there exists a dense set in $\tilde{E}$ of vectors $x \in Z_2(A) \cap \tilde{E}$ for which there exists $\tau > 0$ such that, for every $\varepsilon > 0$, there exists $\sigma > 0$ satisfying*

$$\left\| C^{-1}\mathbf{G}(\varphi)x - x \right\| < \varepsilon,\ \varphi \in \Phi_{\tau,\sigma}.$$

Corollary 3.2.25. *Let A be the generator of a (C-DCF)* **G**. *Assume $\tilde{E}$ is* **G***-admissible and* **G** *is $\tilde{E}_S$-hypercyclic ($\tilde{E}_S$-topologically transitive). Then $\overline{C(\tilde{E})} \subseteq \overline{\mathcal{R}(\mathbf{G})} \subseteq \overline{D_\infty(A)}$.*

The proof of the subsequent theorem follows from Proposition 3.2.19 and the fact that the continuity of a single operator $C(t)$ ($t \geqslant 0$) of a cosine operator function $(C(t))_{t \geqslant 0}$ is not used in the proofs of [70, Theorem 1.2, Corollary 1.3, Theorem 1.4].

Theorem 3.2.26. *Let* **G** *be a (C-DCF) and let $\tilde{E}$ be* **G***-admissible.*

(i) *Assume that there exists a sequence (t_n) of non-negative real numbers such that*

$$X_{0,\tilde{E}} := \{x \in Z_2(A) \cap \tilde{E} : \lim_{n \to \infty} G(\delta_{t_n})x = 0\}$$

and

$$X_{\infty,\tilde{E}} := \{y \in \tilde{E} : \text{there exists a zero sequence } (u_n) \text{ in } Z_2(A) \cap \tilde{E} \text{ and } c \in S \setminus \{0\} \text{ such that } \lim_{n \to \infty} G(\delta_{t_n})cu_n = y\}$$

are dense subsets of $\tilde{E}$. Then **G** *is $\tilde{E}_S$-topologically transitive.*

(ii) *Assume that there exists a sequence (t_n) of non-negative real numbers such that the set*

$$X_{1,\tilde{E}} := \{x \in Z_2(A) \cap \tilde{E} : \lim_{n \to \infty} G(\delta_{t_n})x = \lim_{n \to \infty} G(\delta_{2t_n})x = 0\}$$

is dense in $\tilde{E}$. Then **G** *is $\tilde{E}$-topologically transitive.*

(iii) *Assume that the set*

$$X_{\tilde{E}} := \{x \in Z_2(A) \cap \tilde{E} : \lim_{t \to \infty} G(\delta_t)x = 0\}$$

is dense in $\tilde{E}$. Then **G** *is $\tilde{E}$-topologically mixing.*

Remark 3.2.27. (i) Assume $x, y \in E$, $\lambda_1, \lambda_2 \in \mathbb{C}$, $Ax = \lambda_1 x$ and $Ay = \lambda_2 y$. Then $x \in Z(A) \cap Z_2(A)$, the mild solution of (ACP_1) is given by $u(t; x) = e^{\lambda_1 t}x$, $t \geqslant 0$ and the mild solution of (ACP_2) is given by $u(t; x, y) = \sum_{n=0}^{\infty} \frac{t^{2n}}{(2n)!} \lambda_1^n x + \sum_{n=0}^{\infty} \frac{t^{2n+1}}{(2n+1)!} \lambda_2^n y$, $t \geqslant 0$. This implies that the condition $f(\lambda) \in Z(A)$, $\lambda \in \Omega$ stated in the formulation of [294, Theorem 11] automatically holds and that the proof of [294, Theorem 13] can be simplified.

(ii) Let $t_0 > 0$. By the proof of Theorem 3.2.28 (cf. also [127]), we obtain that C-distribution semigroups appearing in the formulation of [294, Theorem 11] are (subspace) topologically mixing. With a little abuse of notation, we have that every single operator $G_1(\delta_{t_0})$ in [294, Theorem 11(i)] is topologically mixing and has a dense set of periodic points in E, resp. the part of the operator $G_1(\delta_{t_0})$ in the Banach space $\tilde{E}$ appearing in the formulation of [294, Theorem 11(ii)] is topologically mixing in $\tilde{E}$ and the set of $\tilde{E}$-periodic points of such an operator is dense in $\tilde{E}$.

The following theorem is an important extension of [170, Theorem 2.1], [117, Proposition 2.1], [120, Theorem 1.1] and [294, Theorem 11(i)].

Theorem 3.2.28. *([292])*

(i) *Assume G is a (C-DS) generated by A, $\omega_1, \omega_2 \in \mathbb{R} \cup \{-\infty, \infty\}$, $\omega_1 < \omega_2$ and $t_0 > 0$. If $\sigma_p(A) \cap i\mathbb{R} \supseteq (i\omega_1, i\omega_2) \cap \frac{2\pi i\mathbb{Q}}{t_0}$, $k \in \mathbb{N}$ and $g_j : (\omega_1, \omega_2) \cap \frac{2\pi\mathbb{Q}}{t_0} \to E$ is a function which satisfies that, for every $j = 1, \cdots, k$, $Ag_j(s) = isg_j(s)$, $s \in (\omega_1, \omega_2) \cap \frac{2\pi\mathbb{Q}}{t_0}$, then every point in $span\{g_j(s) : s \in (\omega_1, \omega_2) \cap \frac{2\pi\mathbb{Q}}{t_0}, 1 \leqslant j \leqslant k\}$ is a periodic point of $G_1(\delta_{t_0})$. Assume now that $f_j : (\omega_1, \omega_2) \to E$ is a Bochner integrable function which satisfies that, for every $j = 1, \cdots, k$, $Af_j(s) = isf_j(s)$ for a.e. $s \in (\omega_1, \omega_2)$. Put $\psi_{r,j} := \int_{\omega_1}^{\omega_2} e^{irs} f_j(s)\, ds$, $r \in \mathbb{R}$, $1 \leqslant j \leqslant k$.*
   - (a) *Assume $span\{f_j(s) : s \in (\omega_1, \omega_2)\setminus\Omega, 1 \leqslant j \leqslant k\}$ is dense in E for every subset Ω of (ω_1, ω_2) with zero measure. Then G is topologically mixing and $G_1(\delta_{t_0})$ is topologically mixing.*
   - (b) *Put $\tilde{E} := \overline{span\{\psi_{r,j} : r \in \mathbb{R}, 1 \leqslant j \leqslant k\}}$. Then G is $\tilde{E}$-topologically mixing and the part of $G_1(\delta_{t_0})$ in $\tilde{E}$ is topologically mixing in the Banach space $\tilde{E}$.*

(ii) *Assume G is a (C-DS) generated by A, $t_0 > 0$, $\tilde{E}$ is a closed linear subspace of E, $E_0 := span\{x \in Z(A) : \exists\lambda \in \mathbb{C}, \operatorname{Re}\lambda < 0, G_1(\delta_t)x = e^{\lambda t}x, t \geqslant 0\}$, $E_\infty := span\{x \in Z(A) : \exists\lambda \in \mathbb{C}, \operatorname{Re}\lambda > 0, G_1(\delta_t)x = e^{\lambda t}x, t \geqslant 0\}$ and $E_{per} := span\{x \in Z(A) : \exists\lambda \in \mathbb{Q}, G_1(\delta_t)x = e^{\pi\lambda i t}x, t \geqslant 0\}$. Then the following holds:*
   - (a) *If $E_0 \cap \tilde{E}$ is dense in $\tilde{E}$ and if E_∞ is a dense subspace of $\tilde{E}$, then G is $\tilde{E}$-topologically mixing; if $G_1(\delta_t)(E_0 \cap \tilde{E}) \subseteq \tilde{E}$, $t \geqslant 0$, then the part of $G_1(\delta_{t_0})$ in $\tilde{E}$ is topologically mixing in the Banach space $\tilde{E}$.*
   - (b) *If $E_{per} \cap \tilde{E}$ is dense in $\tilde{E}$, then the set of $\tilde{E}$-periodic points of G is dense in $\tilde{E}$; if, additionally, E_{per} is a dense subspace of $\tilde{E}$, then the set of all periodic points of the part of the operator $G_1(\delta_{t_0})$ in $\tilde{E}$ is dense in $\tilde{E}$.*

Proof. We will prove the assertion (i)(a). By the Riemann-Lebesgue lemma and the dominated convergence theorem, we have that $\lim_{|r|\to\infty} \psi_{r,j} = 0$ and that the mapping $r \mapsto \psi_{r,j}$, $r \in \mathbb{R}$ is continuous ($1 \leqslant j \leqslant k$). By Remark 3.2.27 and [294, Lemma 6(i)], we obtain $G_1(\delta_t)f_j(s) = e^{its}f_j(s)$ for a.e. $s \in (\omega_1, \omega_2)$, $G_1(\delta_t)\psi_{r,j} = \psi_{r+t,j}$, $t \geqslant 0$, $r \in \mathbb{R}$, $1 \leqslant j \leqslant k$ and $span\{\psi_{r,j} : r \in \mathbb{R}, 1 \leqslant j \leqslant k\} \subseteq D(G)$. Keeping in mind the proof of [170, Theorem 2.1], it can be easily seen that $span\{\psi_{r,j} : r \in \mathbb{R}, 1 \leqslant j \leqslant k\}$ is dense in E. So, it suffices to show that, given $y, z \in span\{\psi_{r,j} : r \in \mathbb{R}, 1 \leqslant j \leqslant k\}$ and $\varepsilon > 0$ in advance, there exists $t_0 \geqslant 0$ such that, for every $t \geqslant t_0$, there exists $x_t \in Z(A) = D(G)$ such that:

(539)
$$\|y - x_t\| < \varepsilon \text{ and } \left\|z - G_1(\delta_t)x_t\right\| < \varepsilon.$$

Let $y = \sum_{l=1}^{m} \alpha_l \psi_{r_l,i_l}$ and $z = \sum_{l=1}^{n} \beta_l \psi_{\tilde{r}_l,\tilde{i}_l}$ for some $\alpha_l, \beta_l \in \mathbb{C}$, $r_l, \tilde{r}_l \in \mathbb{R}$ and $1 \leqslant i_l, \tilde{i}_l \leqslant k$. Then there exists $t_0(\varepsilon) > 0$ such that $\|\sum_{l=1}^{n} \beta_l \psi_{\tilde{r}_l - t, \tilde{i}_l}\| < \varepsilon$ and $G_1(\delta_t) \sum_{l=1}^{n} \beta_l \psi_{\tilde{r}_l - t,\tilde{i}_l} = z$, $t \geqslant t_0(\varepsilon)$. Furthermore, there exists $t_1(\varepsilon) > 0$ such that $\|G_1(\delta_t)y\| = \|\sum_{l=1}^{m} \alpha_l \psi_{r_l + t, i_l}\| < \varepsilon$, $t \geqslant$

$t_1(\varepsilon)$. Then (539) holds with $t_0 = \max(t_0(\varepsilon), t_1(\varepsilon))$ and $x_t = \Sigma_{l=1}^{n} \beta_l \psi_{\tilde{r}_{l-t},\tilde{i}_l} + y$, $t \geqslant t_0$. The operator $G_1(\delta_{t_0})$ is obviously topologically mixing, which completes the proof.

Remark 3.2.29. (i) Assume the function $f_j : (\omega_1, \omega_2) \to E$ is weakly continuous for every $j = 1, \cdots, k$, $t_0 > 0$ and Ω is a subset of (ω_1, ω_2) with zero measure.

$\overline{\text{Then span}\{f_j(s) : s \in (\omega_1, \omega_2) \cap \frac{2\pi\mathbb{Q}}{t_0}, 1 \leqslant j \leqslant k\}}$

$= \overline{\text{span}\{f_j(s) : s \in (\omega_1, \omega_2), 1 \leqslant j \leqslant k\}}$

$= \overline{\text{span} \bigcup_{j=1}^{k} \{f_j(s) : s \in (\omega_1, \omega_2)\setminus\Omega\}}$.

(ii) Let Ω be a subset of (ω_1, ω_2) with zero measure, let $r \in \mathbb{R}$ and let $1 \leqslant j \leqslant k$. Then $\psi_{r,j} = \int_{\omega_1}^{\omega_2} e^{irs} f_j(s)\, ds \in \overline{\text{span}\{f_j(s) : s \in (\omega_1, \omega_2)\setminus\Omega\}}$.

(iii) Assume that the mapping $r \mapsto \psi_{r,j}$, $r \in \mathbb{R}$ is an element of the space $L^1(\mathbb{R} : E)$ for every $j = 1, \cdots, k$. Then the inversion theorem for the Fourier transform implies that there exists a subset Ω of (ω_1, ω_2) with zero measure such that $\overline{\text{span}\{f_j(s) : s \in (\omega_1, \omega_2)\setminus\Omega, 1 \leqslant j \leqslant k\}} = \overline{\text{span}\{\psi_{r,j} : r \in \mathbb{R}, 1 \leqslant j \leqslant k\}}$.

(iv) By multiplying with an appropriate scalar-valued function, we may assume that, for every $j = 1, \cdots, k$, the function $f_j(\cdot)$ is strongly measurable (cf. also [170, Remark 2.4]).

The following example illustrates an application of Theorem 3.2.28(i) and can be formulated in a more general setting.

Example 3.2.30. ([292]) Assume $\alpha > 0$, $\tau \in i\mathbb{R}\setminus\{0\}$ and $E := BUC(\mathbb{R})$. After the usual matrix reduction to a first order system, the equation $\tau u_{tt} + u_t = \alpha u_{xx}$ becomes

$$\frac{d}{dx}\vec{u}(t) = P(D)\vec{u}(t),\ t \geqslant 0,$$

where $D \equiv -i\frac{d}{dx}$, $P(x) \equiv \begin{bmatrix} 0 & 1 \\ -\frac{\alpha}{\tau}x^2 & -\frac{1}{\tau} \end{bmatrix}$ and $P(D)$ acts on $E \oplus E$ with its maximal distributional domain. The polynomial matrix $P(x)$ is not Petrovskii correct and [141, Theorem 14.1] implies that there exists an injective operator $C \in L(E \oplus E)$ such that $P(D)$ generates an entire C-regularized group $(T(z))_{z\in\mathbb{C}}$, with $R(C)$ dense (cf. also [292, Example 1.2.9] and [69] for further information and examples of Petrovskii correct matrices). Put $\omega_1 = -\infty$ and $\omega_2 = 0$, resp. $\omega_1 = 0$ and $\omega_2 = +\infty$, if $\operatorname{Im} \tau > 0$, resp. $\operatorname{Im} \tau < 0$. Then $\frac{-\tau s^2+is}{\alpha} \in (-\infty, 0)$, $s \in (\omega_1, \omega_2)$. Let $h_1(s) := \cos(\cdot(\frac{\tau s^2-is}{\alpha})^{1/2})$, $h_2(s) := \sin(\cdot(\frac{\tau s^2-is}{\alpha})^{1/2})$, $s \in (\omega_1, \omega_2)$ and let $f \in C^\infty((0, \infty))$ be such that the mapping $s \mapsto f_j(s) := (f(s)h_j(s), isf(s)h_j(s))^T$, $s > 0$ is Bochner integrable and that the mapping $s \mapsto \begin{cases} f_j(s), & s \in (\omega_1, \omega_2) \\ 0, & s \notin (\omega_1, \omega_2) \end{cases}$ belongs to the space $H^1(\mathbb{R}) \equiv W^{1,2}(\mathbb{R})$ for $j = 1, 2$. Put $\psi_{r,j} = \int_{\omega_1}^{\omega_2} e^{irs} f_j(s)\, ds$, $r \in \mathbb{R}$, $j = 1, 2$ and $\tilde{E} = \overline{\text{span}\{\psi_{r,j} : r \in \mathbb{R}, j = 1, 2\}}$. By Bernstein lemma [20, Lemma 8.2.1, p. 429], Theorem 3.2.28(i)(b) and Remark 3.2.29(i)-(iii), one gets that $(T(t))_{t\geqslant 0}$ is $\tilde{E}$-topologically mixing as well as that for each $t_0 > 0$ the part of the operator $C^{-1}T(t_0)$ in $\tilde{E}$ is topologically mixing in $\tilde{E}$ and that the set of $\tilde{E}$-periodic points of such an operator is dense in $\tilde{E}$.

Theorem 3.2.31. *Let $\pm A$ be the generators of C-distribution semigroups $G_{\pm}$, let A^2 be closed and let* $\mathbf{G}(\varphi) = \frac{1}{2}(G_+(\varphi) + G_-(\varphi))$, $\varphi \in \mathcal{D}$. *Assume* $\omega_1, \omega_2 \in \mathbb{R} \cup \{-\infty, \infty\}$, $\omega_1 < \omega_2$, $t_0 > 0$, $\sigma_p(A) \supseteq (i\omega_1, i\omega_2) \cap \frac{2\pi i\mathbb{Q}}{t_0}$, $k \in \mathbb{N}$ *and* $f_j : (\omega_1, \omega_2) \cap \frac{2\pi\mathbb{Q}}{t_0} \to E$ *satisfies* $Af_j(s) = isf_j(s)$, $s \in (\omega_1, \omega_2) \cap \frac{2\pi\mathbb{Q}}{t_0}$ $(1 \leqslant j \leqslant k)$. *Then* $\mathbf{G}$ *is a (C-DCF) generated by* $C^{-1}A^2C$ *and, for every* $x \in span\{f_j(s) : s \in (\omega_1, \omega_2) \cap \frac{2\pi\mathbb{Q}}{t_0}, 1 \leqslant j \leqslant k\}$, *there exists* $n \in \mathbb{N}$ *such that x is a fixed point of* $G(\delta_{nt_0})$.

Proof. Clearly, $\mathbf{G}$ is a $(C - DCF)$ generated by $C^{-1}A^2C$. By Remark 3.2.27 and [294, Lemma 6(i)], we have $G_{\pm,1}(\delta_t)f_j(s) = e^{\pm ist}f_j(s)$, $t \geqslant 0$, $s \in (\omega_1, \omega_2) \cap \frac{2\pi\mathbb{Q}}{t_0}$. Now it becomes apparent that, for every $x \in \text{span}\{f_j(s) : s \in (\omega_1, \omega_2) \cap \frac{2\pi\mathbb{Q}}{t_0}, 1 \leqslant j \leqslant k\}$, there exists $n \in \mathbb{N}$ such that $G_{\pm,1}(\delta_{t_0})^n x = x$. Then

$$G(\delta_{nt_0})x = \frac{1}{2}\big(G_{+,1}(\delta_{nt_0})x + G_{-,1}(\delta_{nt_0})x\big)$$

$$= \frac{1}{2}\big(G_{+,1}(\delta_{t_0})^n x + G_{-,1}(\delta_{t_0})^n x\big) = \frac{1}{2}(x + x) = x. \qquad \square$$

Remark 3.2.32. Assume Ω is an open connected subset of $\mathbb{C}$, which satisfies $\sigma_p(A) \supseteq \Omega$ and intersects the imaginary axis, $f : \Omega \to E$ is an analytic mapping with $f(\lambda) \in \text{Kern}(A - \lambda)$, $\lambda \in \Omega$, $E_0 = \text{span}\{f(\lambda) : \lambda \in \Omega\}$, $k = 1$ and $f_1(s) = f(is)$, $s \in (\omega_1, \omega_2) \cap \frac{2\pi\mathbb{Q}}{t_0}$, where $\omega_1, \omega_2 \in \mathbb{R}$ and $(i\omega_1, i\omega_2) \subseteq \Omega$. Then [35, Lemma 2.4] implies that $\text{span}\{f_1(s) : s \in (\omega_1, \omega_2) \cap \frac{2\pi\mathbb{Q}}{t_0}\}$ is dense in $\tilde{E}$.

Lemma 3.2.33. *Let* $\lambda \in \mathbb{C}$. *Then* $\lambda \in \sigma_p(\mathcal{A})$ *iff* $\lambda^2 \in \sigma_p(A)$; *if* $f(\lambda^2)$ *an eigenvector of A with the eigenvalue* λ^2, *then* $F(\lambda) = (f(\lambda^2), \lambda f(\lambda^2))^T$ *is an eigenvector of* $\mathcal{A}$ *with the eigenvalue* λ.

The proof of the first part of the following theorem follows immediately from Lemma 3.2.33 and Theorem 3.2.28, while the proof of the second part of the theorem follows from Lemma 3.2.33, [310, Theorem 11(ii)] and Remark 3.2.27.

Theorem 3.2.34. (i) *Assume A is the generator of a (C-DCF)* $\mathbf{G}$, $t_0 > 0$, $\omega_1, \omega_2 \in \mathbb{R} \cup \{-\infty, \infty\}$, $\omega_1 < \omega_2$, $k \in \mathbb{N}$ *and* $\Psi(\omega_1, \omega_2, t_0) := \{-s^2 : s \in (\omega_1, \omega_2) \cap \frac{2\pi\mathbb{Q}}{t_0}\}$. *Then the existence of functions* $g_j : \Psi(\omega_1, \omega_2, t_0) \to E$ *which satisfy that, for every* $j = 1, \cdots, k$, $Ag_j(-s^2) = -s^2 g_j(-s^2)$, $s \in (\omega_1, \omega_2) \cap \frac{2\pi\mathbb{Q}}{t_0}$, *implies that every* $x \in span\{(g_j(-s^2), isg_j(-s^2))^T : s \in (\omega_1, \omega_2) \cap \frac{2\pi\mathbb{Q}}{t_0}, 1 \leqslant j \leqslant k\}$ *is a periodic point of* $\mathcal{G}_1(\delta_{t_0})$. *Let* $f_j : (-\omega_2^2, -\omega_1^2) \to E$ *be a measurable function which satisfies that, for every* $j = 1, \cdots, k$, $Af_j(-s^2) = -s^2 f_j(-s^2)$ *for a.e.* $s \in (\omega_1, \omega_2)$. *Put* $F_j(s) := (f_j(-s^2), is f_j(-s^2))^T$, $s \in (\omega_1, \omega_2)$, $1 \leqslant j \leqslant k$. *Let the mapping* $F_j : (\omega_1, \omega_2) \to E \oplus E$ *be Bochner integrable provided* $1 \leqslant j \leqslant k$ *and let* $\zeta_{r,j} := \int_{\omega_1}^{\omega_2} e^{irs} F_j(s)\, ds$, $r \in \mathbb{R}$, $1 \leqslant j \leqslant k$.

(a) *Assume* $span\{f_j(s) : s \in (\omega_1, \omega_2) \backslash \Omega, 1 \leqslant j \leqslant k\}$ *is dense in* $E \oplus E$ *for every subset* Ω *of* (ω_1, ω_2) *with zero measure. Then* $\mathcal{G}$ *is topologically mixing and* $\mathcal{G}_1(\delta_{t_0})$ *is topologically mixing.*

(b) *Let* $\hat{E} = \overline{span\{\zeta_{r,j} : r \in \mathbb{R}, 1 \leqslant j \leqslant k\}}$. *Then* $\mathcal{G}$ *is* $\hat{E}$*-topologically mixing and the part of* $\mathcal{G}_1(\delta_{t_0})$ *in* $\hat{E}$ *is topologically mixing in the Banach space* $\hat{E}$.

(ii) *Assume A is the generator of a (C-DCF)* **G**, *there exists an open connected subset* Ω *of* $\mathbb{C}$ *which satisfies* $\sigma_p(A) \supseteq \{\lambda^2 : \lambda \in \Omega\}$ *and* $\Omega \cap i\mathbb{R} \neq \emptyset$. *Let* $f : \{\lambda^2 : \lambda \in \Omega\} \to E$ *be an analytic mapping satisfying* $f(\lambda^2) \in Kern(A - \lambda^2) \backslash \{0\}$, $\lambda \in \Omega$, *let* $F(\lambda) := (f(\lambda^2), \lambda f(\lambda^2))^T$, $\lambda \in \Omega$ *and let* $\hat{E} = \overline{span\{F(\lambda) : \lambda \in \Omega\}}$. *Then* $\mathcal{G}$ *is* $\hat{E}$*-topologically mixing, the part of the operator* $\mathcal{G}_1(\delta_{t_0})$ *in* $\hat{E}$ *is topologically mixing in the Banach space* $\hat{E}$, *the set* $\mathcal{G}_{\hat{E},per}$ *is dense in* $\hat{E}$ *and the set of all* $\hat{E}$*-periodic points of the part of the operator* $\mathcal{G}_1(\delta_{t_0})$ *in* $\hat{E}$ *is dense in* $\hat{E}$.

Remark 3.2.35. (i) Assume **G** is a $(C - DCF)$ generated by A. Then one can prove with the help of [292, Theorem 2.1.11], Proposition 3.2.18 and [294, Lemma 6] that x is a periodic point of **G**, resp. a hypercyclic vector of **G**, if $\binom{x}{0}$ $(\binom{0}{x})$ is a periodic point of $\mathcal{G}$, resp. a hypercyclic vector of $\mathcal{G}$. Moreover, the **G**-admissibility of a closed linear subspace $\tilde{E}$ of E implies $\mathcal{G}_1(\delta_t)(\{0\} \oplus \tilde{E}) \subseteq \tilde{E} \oplus \tilde{E}$.

(ii) Assume now $\hat{E}$ is $\mathcal{G}$-admissible and $\binom{x}{y}$ is a $\hat{E}_S$-hypercyclic vector for $\mathcal{G}$. Then $\mathcal{G}_1(\delta_t)\binom{x}{y} = (\pi_1(\mathcal{G}_1(\delta_t)\binom{x}{y}), \frac{d}{dt}\pi_1(\mathcal{G}_1(\delta_t)\binom{x}{y}))^T$, $t \geqslant 0$, and $u(t) = \pi_1(\mathcal{G}_1(\delta_t)\binom{x}{y})$, $t \geqslant 0$ is a mild solution of (ACP_2). Then $\{cu(t) : c \in \mathrm{S}, t \geqslant 0\}$ and $\{cu'(t) : c \in \mathrm{S}, t \geqslant 0\}$ are dense subsets of $\hat{E}$, which can be simply formulated and proved for any of the considered hypercyclic properties.

The following theorem can be reworded by assuming that there exists $\alpha \geqslant 0$ such that $-A$ generates an exponentially bounded, analytic α-times integrated semigroup of angle $\theta \in (0, \frac{\pi}{2})$ and that $\sigma_p(-A)$ strictly lies on the imaginary axis (cf. Theorem 3.2.28).

Theorem 3.2.36. *Let* $\theta \in (0, \frac{\pi}{2})$ *and let* $-A$ *generate an analytic strongly continuous semigroup of angle* θ. *Assume* $n \in \mathbb{N}$, $a_n > 0$, $a_{n-i} \in \mathbb{C}$, $1 \leqslant i \leqslant n$, $D(p(A)) = D(A^n)$, $p(A) = \Sigma_{i=0}^{n} a_i A^i$ *and* $n(\frac{\pi}{2} - \theta) < \frac{\pi}{2}$. *Then there exists* $\omega \in \mathbb{R}$ *such that, for every* $\alpha \in (1, \frac{\pi}{n\pi - 2n\theta})$, $p(A)$ *generates an entire* $C \equiv e^{-(p(A)-\omega)^{\alpha}}$*-regularized group* $(T(t))_{t\in\mathbb{C}}$. *Put* $C(z) := \frac{1}{2}(T(z) + T(-z))$, $z \in \mathbb{C}$. *Then* $(C(t))_{t\geqslant 0}$ *is a C-regularized cosine function generated by* $p^2(A)$ *and the mapping* $z \mapsto C(z)$, $z \in \mathbb{C}$ *is entire.*

(i) *Assume that there exists an open connected subset* Ω *of* $\mathbb{C}$, *which satisfies* $\sigma_p(-A) \supseteq \Omega$, $p(-\Omega) \cap i\mathbb{R} \neq \emptyset$, *and let* $f : \Omega \to E$ *be an analytic mapping satisfying* $f(\lambda) \in Kern(-A - \lambda) \backslash \{0\}$, $\lambda \in \Omega$.

(a) *Assume that* $\langle x^*, f(\lambda)\rangle = 0$, $\lambda \in \Omega$, *for some* $x^* \in E^*$, *implies* $x^* = 0$. *Then there exists a dense subspace* C_{per} *of E which satisfies* $C_{per} \subseteq Z_2(A)$ *and that, for every* $t_0 > 0$ *and* $x \in C_{per}$, *there exists* $n_0 \in \mathbb{N}$ *such that* $C^{-1}C(nn_0t_0)x = x$, $n \in \mathbb{N}$. *In particular, the set of all periodic points of* $(C(t))_{t\geqslant 0}$ *is dense in E.*

(b) *Let the supposition* $\langle x^*, f(\lambda)\rangle + \langle y^*, p(-\lambda)f(\lambda)\rangle = 0$, $\lambda \in \Omega$, *for some* $x^*, y^* \in E^*$ *imply* $x^* = y^* = 0$. *Set*

$$S_0(z) := \begin{pmatrix} C(z) & \int_0^z C(s)\,ds \\ \frac{d}{dz}C(z) & C(z) \end{pmatrix}, \; z \in \mathbb{C}.$$

Then $(S_0(z))_{z\in\mathbb{C}}$ *is an entire C-regularized group generated by the operator* $\begin{pmatrix} 0 & I \\ p^2(A) & 0 \end{pmatrix}$, $(S_0(t))_{t\geqslant 0}$ *is both topologically mixing and chaotic, and for every* $t > 0$, *the operator* $C^{-1}S_0(t) \oplus C^{-1}S_0(t)$ *is chaotic and topologically mixing.*

(ii) *Assume that there exists an open connected subset* Ω *of* $\mathbb{C}$, *which satisfies* $\sigma_p(-A) \supseteq \Omega$ *and* $p(-\Omega) \cap i\mathbb{R} \neq \emptyset$. *Let* $f : \Omega \to E$ *be an analytic mapping satisfying* $f(\lambda) \in Kern(-A - \lambda) \setminus \{0\}$, $\lambda \in \Omega$. *Set* $\hat{E} := \overline{span\{(f(\lambda), p(-\lambda)f(\lambda))^T : \lambda \in \Omega\}}$ *and* $\tilde{E} := \overline{\{f(\lambda) : \lambda \in \Omega\}}$.

(a) *Then there exists a dense subspace* C_{per} *of* $\tilde{E}$ *which satisfies* $C_{per} \subseteq Z_2(A)$ *and that, for every* $t_0 > 0$ *and* $x \in C_{per}$, *there exists* $n_0 \in \mathbb{N}$ *such that* $C^{-1}C(nn_0t_0)x = x$, $n \in \mathbb{N}$. *In particular, the set of all* $\tilde{E}$*-periodic points of* $(C(t))_{t\geqslant 0}$ *is dense in* $\tilde{E}$.

(b) *Let* $(S_0(z))_{z\in\mathbb{C}}$ *be as in (i). Then* $(S_0(z))_{z\in\mathbb{C}}$ *is an entire C-regularized group generated by* $\begin{pmatrix} 0 & I \\ p^2(A) & 0 \end{pmatrix}$, $(S_0(t))_{t\geqslant 0}$ *is* $\hat{E}$*-topologically mixing, the set of* $\hat{E}$*-periodic points of* $(S_0(t))_{t\geqslant 0}$ *is dense in* $\hat{E}$, *and* $R(C_{\hat{E}})$ *is dense in the Banach space* $\hat{E}$. *Let* $t > 0$ *be fixed and let* $T_0(t)$ *be the part of* $C^{-1}S_0(t)$ *in* $\hat{E}$. *Then the operator* $T(t) := T_0(t) \oplus T_0(t)$ *is chaotic and topologically mixing in the Banach space* $\hat{E} \oplus \hat{E}$.

Proof. We will only prove the part (b) of (ii). Notice that the mapping $z \mapsto p(z)$, $z \in \mathbb{C}$ is open and that the set $p(-\Omega)$ is open and connected. Put

$$S_1(s) := \begin{pmatrix} \int_0^s C(r)\,dr & \int_0^s (s-r)C(r)\,dr \\ C(s) - C & \int_0^s C(r)\,dr \end{pmatrix}, \; s \geqslant 0.$$

By the foregoing, $(S_1(s))_{s\geqslant 0}$ is a once integrated C-semigroup generated by the operator $\begin{pmatrix} 0 & I \\ p^2(A) & 0 \end{pmatrix}$. On the other hand, it is clear that the mapping $s \mapsto S_1(s)$, $s \geqslant 0$ can be analytically extended to the whole complex plane, which simply implies that $(S_0(z))_{z\in\mathbb{C}}$ is an entire C-regularized group generated by $\begin{pmatrix} 0 & I \\ p^2(A) & 0 \end{pmatrix}$. In order to prove that $(S_0(s))_{s\geqslant 0}$ is $\hat{E}$-topologically mixing and that the set of $\hat{E}$-periodic points of $(S_0(s))_{s\geqslant 0}$ is dense in $\hat{E}$, one can use the equalities $p^2(A)f(\lambda) = p^2(-\lambda)f(\lambda)$, $\lambda \in \Omega$,

$$\begin{pmatrix} 0 & I \\ p^2(A) & 0 \end{pmatrix} \begin{pmatrix} f(\lambda) \\ p(-\lambda)f(\lambda) \end{pmatrix} = p(-\lambda) \begin{pmatrix} f(\lambda) \\ p(-\lambda)f(\lambda) \end{pmatrix}, \; \lambda \in \Omega$$

as well as Remark 3.2.27 and [294, Theorem 11(ii)]. Arguing in the same manner, we obtain that the single operator $T(t)$, considered as an unbounded linear operator in the Banach space $\hat{E} \oplus \hat{E}$, is topologically mixing and that the set of $(\hat{E} \oplus \hat{E})$-periodic points of $T(t)$ is dense in $\hat{E} \oplus \hat{E}$. By [294, Remark 14(ii)], $R(C_{\hat{E}})$ is dense in $\hat{E}$. Therefore, it remains to be shown that the operator $T(t)$ is hypercyclic in the Banach space $\hat{E} \oplus \hat{E}$. Towards this end, put $X_0 := \operatorname{span}\{(f(\lambda), p(-\lambda)f(\lambda))^T : \lambda \in \Omega, \operatorname{Re}(p(-\lambda)) < 0\}$, $X_\infty := \operatorname{span}\{(f(\lambda), p(-\lambda)f(\lambda))^T : \lambda \in \Omega, \operatorname{Re}(p(-\lambda)) > 0\}$, $Y_1 := X_0 \oplus X_0$, $Y_2 := X_\infty \oplus X_\infty$, and for every $k, l \in \mathbb{N}$ and $\alpha_i \in \mathbb{C}$,

$$S\left(\sum_{i=1}^{k}\alpha_i\begin{pmatrix} f(\lambda_i) \\ p(-\lambda_i)f(\lambda_i)\end{pmatrix}, \sum_{i=1}^{l}\beta_i\begin{pmatrix} f(z_i) \\ p(-z_i)f(z_i)\end{pmatrix}\right)$$
$$:= \left(\sum_{i=1}^{k}\alpha_i e^{-p(-\lambda_i)}\begin{pmatrix} f(\lambda_i) \\ p(-\lambda_i)f(\lambda_i)\end{pmatrix}, \sum_{i=1}^{l}\beta_i e^{-p(-z_i)}\begin{pmatrix} f(z_i) \\ p(-z_i)f(z_i)\end{pmatrix}\right),$$

$\operatorname{Re}(p(-\lambda_i)) < 0$, $1 \leqslant i \leqslant k$, $\beta_i \in \mathbb{C}$, $\operatorname{Re}(p(-z_i)) < 0$, $1 \leqslant i \leqslant l$. Then it follows from [150, Theorem 2.3] (with $C_{\hat{E}}$) that the operator $T(t)$ is hypercyclic in $\hat{E} \oplus \hat{E}$, as claimed. □

In the following instructive example, we consider a class of abstract second order differential equations which cannot be treated by integrated cosine functions. The analysis of this example will be continued in the next section.

Example 3.2.37. (i) ([161, Example 4.12], [152, Example 2.4], [294, Example 15]) Let $a, b, c > 0$ and $c < \frac{b^2}{2a} < 1$. Consider the equation

$$\begin{cases} u_t = au_{xx} + bu_x + cu := -Au, \\ u(0, t) = 0, \ t \geqslant 0, \\ u(x, 0) = u_0(x), \ x \geqslant 0. \end{cases}$$

Then the operator $-A$, with domain $D(-A) = \{f \in W^{2,2}([0, \infty)) : f(0) = 0\}$, generates an analytic strongly continuous semigroup of angle $\frac{\pi}{2}$ in the space $E = L^2([0, \infty))$; the same assertion holds in the case when the operator $-A$ acts on $E = L^1([0, \infty))$ with domain $D(-A) = \{f \in W^{2,1}([0, \infty)) : f(0) = 0\}$. Put

$$\Omega := \left\{\lambda \in \mathbb{C} : \left|\lambda - \left(c - \frac{b^2}{4a}\right)\right| \leqslant \frac{b^2}{4a}, \ \operatorname{Im}\lambda \neq 0 \text{ if } \operatorname{Re}\lambda \leqslant c - \frac{b^2}{4a}\right\}$$

and assume that $p(x) = \Sigma_{i=0}^{n} a_i x^i$ is a nonconstant polynomial such that $a_n > 0$ and $p(-\Omega) \cap i\mathbb{R} \neq \emptyset$ (this, in particular, holds if $a_0 \in i\mathbb{R}$). An application of Theorem 3.2.36(i) gives that there exists an injective operator $C \in L(E)$ such that $p^2(A)$ generates a global C-regularized cosine function $(C(t))_{t\geqslant 0}$ satisfying that the set of periodic points of $(C(t))_{t\geqslant 0}$ is dense in E. Let $\hat{E} := \overline{\{(f_\lambda(\cdot), p(-\lambda)f_\lambda(\cdot))^T : \lambda \in \Omega\}}$, where the function $f_\lambda(\cdot)$ is defined in [161, Example 4.12]. By Theorem 3.2.36(ii), we get that the induced entire C-regularized semigroup $(S_0(t))_{t\geqslant 0}$ generated by

$\begin{pmatrix} 0 & I \\ p^2(A) & 0 \end{pmatrix}$ is $\hat{E}$-topologically mixing and that the set of all $\hat{E}$-periodic points of $(S_0(t))_{t\geqslant 0}$ is dense in $\hat{E}$. Herein it is worth noting that every single operator $T(t)$ (cf. the formulation of Theorem 3.2.36) is chaotic and topologically mixing in the Banach space $\hat{E} \oplus \hat{E}$. Using the composition property of regularized semigroups, it simply follows that there exist $x, y \in \hat{E}$ such that the set $\{C^{-1}S_0(nt)\binom{x}{y} : n \in \mathbb{N}_0\}$ is a dense subset of $\hat{E}$. Since $R(C_{\hat{E}})$ is dense in $\hat{E}$, one gets that $\{S_0(nt)\binom{x}{y} : n \in \mathbb{N}_0\}$ is also a dense subset of $\hat{E}$. This implies that $(S_0(t))_{t\geqslant 0}$ is $\hat{E}$-hypercyclic in the sense of [294, Remark 14(i)], which remains true in examples given in (ii) and (iii).

(ii) ([150]-[151]) Assume that ω_1, ω_2, V_{ω_2,ω_1}, Q, N, h_μ and E possess the same meaning as in [150, Section 5] and $Q(\text{int}(V_{\omega_2,\omega_1})) \cap i\mathbb{R} \neq \emptyset$; the operator $Q(B)$ is defined by means of the $H_{a,b}$ functional calculus developed in the above-mentioned paper. Then $\pm Q(B)h_\mu = \pm Q(\mu)h_\mu$, $e^{-(-B^2)^N}h_\mu = e^{-(-\mu^2)^N}h_\mu$, $\mu \in$ int (V_{ω_2,ω_1}) and $h_\mu \in (\text{Kern}(Q(B))\setminus\{0\})$, provided Re $\mu \in (\omega_2, \omega_1)$. Set

$$\hat{E} := \overline{\text{span}\{(h_\mu, Q(\mu)h_\mu)^T : \mu \in \text{int}(V_{\omega_2,\omega_1})\}}.$$

By Theorem 3.2.31, one obtains that $Q^2(B)$ is the integral generator of a global $(e^{-(-z^2)^N})(B)$-regularized cosine function $((\cosh(tQ(z))e^{-(-z^2)^N})(B))_{t\geqslant 0}$ which has a dense set of periodic points and satisfies that the mapping $t \mapsto (\cosh(tQ(z))e^{-(-z^2)^N})(B)$, $t \geqslant 0$ can be analytically extended to the whole complex plane. It is readily seen that the mapping $\mu \mapsto h_\mu$, $\mu \in \text{int}(V_{\omega_2,\omega_1})$ is analytic. Owing to [294, Theorem 11(ii)], the induced entire $\begin{pmatrix} (e^{-(-z^2)^N})(B) & 0 \\ 0 & (e^{-(-z^2)^N})(B) \end{pmatrix}$-regularized semigroup $(S_0(t))_{t\geqslant 0}$ generated by $\begin{pmatrix} 0 & I \\ Q^2(B) & 0 \end{pmatrix}$ is $\hat{E}$-topologically mixing and the set of all $\hat{E}$-periodic points of $(S_0(t))_{t\geqslant 0}$ is dense in $\hat{E}$. Observe also that the analysis given in [150, Theorem 5.8] can serve one to construct important examples of regular ultradistribution semigroups of Beurling class ([292]).

(iii) ([237]) It is clear that Theorem 3.2.31, Theorem 3.2.34 and Theorem 3.2.36 can be applied to the operators considered by L. Ji and A. Weber in [237, Theorem 3.1(a), Theorem 3.2, Corollary 3.3]. For example, if X is a symmetric space of non-compact type (of rank one) and $p > 2$, then there exist a closed linear subspace $\tilde{X}$ of X (X, if the rank of X is one), a number $c_p > 0$ and an injective operator $C \in L(L^p_\natural(X))$ such that for any $c > c_p$ the operator $(-\Delta^\natural_{X,p} + c)^2$ generates a global C-regularized cosine function $(C(t))_{t\geqslant 0}$ in $L^p_\natural(X)$ which satisfies that the set of $\tilde{X}$-periodic points of $(C(t))_{t\geqslant 0}$ is dense in $\tilde{X}$. By Theorem 3.2.36(ii), we infer that there exists a closed linear subspace $\hat{X}$ of $X \oplus X$ such that the induced entire C-regularized semigroup $(S_0(t))_{t\geqslant 0}$ generated by

$\begin{pmatrix} 0 & I \\ (-\Delta_{X,p}^{\natural} + c)^2 & 0 \end{pmatrix}$ is $\hat{X}$-topologically mixing and that the set of all $\hat{X}$-periodic points of $(S_0(t))_{t\geqslant 0}$ is dense in $\hat{X}$.

Before proceeding further, we would like to observe that the conclusions established in [293, Example 10] (cf. also [292, Example 3.1.35(ii)]) are false. Let $(O_n)_{n\in\mathbb{N}}$ be an open base of the topology of E and let $O_n \neq \emptyset$ for every $n \in \mathbb{N}$. We then need the following proposition.

Proposition 3.2.38. *Suppose A is the integral generator of a C-regularized cosine function $(C(t))_{t\geqslant 0}$ and $R(C)$ is dense in E. Put*

$$\mathcal{T} := \bigcap_{n\in\mathbb{N}} \bigcup_{t\geqslant 0} C(t)^{-1}(O_n).$$

Then

$$\mathcal{T} = \{x \in E : \text{the set } \{C(t)x : t \geqslant 0\} \text{ is dense in } E\} \tag{540}$$

and the following holds:

(i) *Let $(C(t))_{t\geqslant 0}$ be topologically transitive. Then $\mathcal{T}$ is a dense G_δ-subset of E and $C(\mathcal{T}) \subseteq HC(C(\cdot))$. In particular, $(C(t))_{t\geqslant 0}$ is hypercyclic and the set $HC(C(\cdot))$ is dense in E.*

(ii) *Let $(C(t))_{t\geqslant 0}$ be hypercyclic and $x \in HC(C(\cdot))$. Then $x \in \mathcal{T}$.*

Proof. The proof of (540) is trivial and the proof of (ii) follows from the definition of hypercyclic vectors of $(C(t))_{t\geqslant 0}$, the denseness of $R(C)$ in E and (540). Assume now that $(C(t))_{t\geqslant 0}$ is topologically transitive. Let U and V be arbitrary open subsets of E, and let $y, z \in E$ and $\varepsilon > 0$ such that $\{x \in E : \|x - y\| \leqslant \varepsilon\} =: B(y, \varepsilon) \subseteq U$ and $B(Cz, \varepsilon) \subseteq V$. Then there exists $x \in Z_2(A)$ such that $\|y - x\| < \varepsilon$ and $\|z - C^{-1}C(t)x\| < \varepsilon/\|C\|$, which implies $\|y - x\| < \varepsilon$, $\|Cz - C(t)x\| < \varepsilon$, and $C(t)U \cap V \neq \emptyset$. Consequently, $\bigcup_{t\geqslant 0} C(t)^{-1}(O_n)$ is a dense open subset of E for every $n \in \mathbb{N}$ and $\mathcal{T}$ is a dense G_δ-subset of E. The inclusion $C(\mathcal{T}) \subseteq HC(C(\cdot))$ is trivial, which completes the proof of (i). □

Example 3.2.39. ([292]) Let $n \in \mathbb{N}$, $\rho(t) := \frac{1}{t^{2n+1}}$, $t \in \mathbb{R}$, $Af := f'$, $D(A) := \{f \in C_{0,\rho}(\mathbb{R}) : f' \in C_{0,\rho}(\mathbb{R})\}$, $E_n := (C_{0,\rho}(\mathbb{R}))^{n+1}$, $D(A_n) := D(A)^{n+1}$ and $A_n(f_1, \cdots, f_{n+1}) := (Af_1 + Af_2, Af_2 + Af_3, \cdots, Af_n + Af_{n+1}, Af_{n+1})$, $(f_1, \cdots, f_{n+1}) \in D(A_n)$. By the proof of [434, Proposition 2.4] we have that $\pm A_n$ generate global polynomially bounded n-times integrated semigroups $(S_{n,\pm}(t))_{t\geqslant 0}$ and that neither A_n nor $-A_n$ generates a local $(n-1)$-times integrated semigroup. Denote by $G_\pm$ distribution semigroups generated by $\pm A$. Then it can be easily proved that, for every $\varphi_1, \cdots, \varphi_{n+1} \in \mathcal{D}$,

$$G_\pm(\delta_t)(\varphi_1, \cdots, \varphi_{n+1})^T = (\psi_1, \cdots, \psi_{n+1})^T,$$

where $\psi_i(\cdot) = \sum_{j=0}^{n+1-i} \frac{(\pm t)^j}{j!} \varphi_{i+j}^{(j)}(\cdot \pm t)$, $1 \leqslant i \leqslant n+1$. This immediately implies the concrete representation formula for $(S_{n,\pm}(t))_{t\geqslant 0}$. Denote by $\mathbf{G}_n$ and $(C_n(t))_{t\geqslant 0}$ the (DCF) and global polynomially bounded n-times integrated cosine function generated by A_n^2. By Proposition 3.2.20(ii), we get that $\mathbf{G}_n(\delta_t)(\varphi_1, \cdots, \varphi_{n+1})^T = \frac{1}{2}[G_+(\delta_t)(\varphi_1, \cdots, \varphi_{n+1})^T + G_-(\delta_t)(\varphi_1, \cdots, \varphi_{n+1})^T]$, $t \geqslant 0$, $\varphi_1, \cdots, \varphi_{n+1} \in \mathcal{D}$. It is clear that the assumptions $0 \leqslant i \leqslant n$, $\varphi \in \mathcal{D}$ and $\operatorname{supp}(\varphi) \subseteq [a, b]$ imply $t^i \sup_{x\in\mathbb{R}} |\varphi(x \pm t)|\rho(x) \leqslant t^i \sup_{x\in[a\mp t, b\mp t]} \frac{1}{x^{2n}+1} \leqslant t^i \left(\frac{1}{(a-t)^{2n}+1} + \frac{1}{(a+t)^{2n}+1} + \frac{1}{(b-t)^{2n}+1} + \frac{1}{(b+t)^{2n}+1}\right) \to 0$, $|t| \to \infty$. Keeping this and Theorem 3.2.26(iii) in mind, it follows that $\mathbf{G}_n$ and $(C_n(t))_{t\geqslant 0}$ are topologically mixing. Arguing in the same way, we infer that $\mathbf{G}_n \oplus \mathbf{G}_n$ is also topologically mixing, which clearly implies that $\mathbf{G}_n$ and $(C_n(t))_{t\geqslant 0}$ are weakly mixing. Herein it is worthwhile to note that, for every $t > 0$, the operators $G_\pm(\delta_t) \oplus G_\pm(\delta_t)$ are hypercyclic in $\hat{E}_n \equiv E_n \oplus E_n$ ([150], [294]). Before proceeding further, we would like to observe that, for every $\tau > 0$, the mapping $t \mapsto C_n(t)$, $t \in [0, \tau)$ is not strongly differentiable and that A_n^2 cannot be the generator of any local $(n-1)$-times integrated cosine function. The existence of a positive real number λ_0 which belongs to the set $\rho(A_n) \cap \rho(-A_n)$ is obvious and the use of [292, Proposition 2.3.13] gives that $\pm A_n$ are the integral generators of global exponentially bounded $(\lambda_0 \mp A_n)^{-n}$-regularized semigroups $(S_{0,\pm}(t))_{t\geqslant 0}$ satisfying the equalities $S_{n,\pm}(t)x = (\lambda_0 \mp A_n)^n \int_0^t \frac{(t-s)^{n-1}}{(n-1)!} S_{0,\pm}(s)x\, ds$, $t \geqslant 0$, $x \in E_n$. This implies that A_n^2 is the integral generator of a topologically mixing $((\lambda_0 - A_n)^{-n}(\lambda_0 + A_n)^{-n})$-regularized cosine function $(C_0(t))_{t\geqslant 0}$, where $C_0(t) = \frac{1}{2}(S_{0,+}(t)(\lambda_0 + A_n)^{-n} + S_{0,-}(t)(\lambda_0 - A_n)^{-n})$, $t \geqslant 0$. By Proposition 3.2.38, one gets that $\mathbf{G}_n$ and $(C_n(t))_{t\geqslant 0}$ are hypercyclic. Put $C_n := I \oplus (\lambda_0 - A_n)^{-n}(\lambda_0 + A_n)^{-n}$. Then $\frac{d^n}{dt^n}S_{n,\pm}(t)(\varphi_1, \cdots, \varphi_{n+1})^T = (\lambda_0 \mp A_n)^n S_{0,\pm}(t)(\varphi_1, \cdots, \varphi_{n+1})^T$, $t \geqslant 0$, $\varphi_1, \cdots, \varphi_{n+1} \in \mathcal{D}$, and an application of Theorem 3.2.26(iii) yields that $A_n^2 \oplus A_n^2$ is the generator of a global topologically mixing n-times integrated C_n-cosine function $(\overline{C}_n(t) := C_n(t) \oplus \int_0^t \frac{(t-s)^{n-1}}{(n-1)!} C_0(s)\, ds)_{t\geqslant 0}$. By [292, Proposition 2.3.12] and Proposition 3.2.38, $(\overline{C}_n(t))_{t\geqslant 0}$ is also hypercyclic. Finally, $A_n^2 \oplus A_n^2$ cannot be the generator of any local $(n-1)$-times integrated C_n-cosine function in $\hat{E}_n$.

3.2.3. Hypercyclic and chaotic properties of cosine functions. In this subsection, we consider hypercyclic and chaotic properties of various types of cosine functions in the space $L^p(\Omega, \mu, \mathbb{C})$, resp. $C_{0,\rho}(\Omega, \mathbb{C})$, where Ω is an open non-empty subset of $\mathbb{R}^d$, $p \in [1, \infty)$ and μ is a locally finite Borel measure on Ω, resp. $\rho : \Omega \to (0, \infty)$ is an upper semicontinuous function ([244]). Let $\varphi : \mathbb{R} \times \Omega \to \Omega$ be a semiflow, and let $\varphi(t, \cdot)^{-1}$ the inverse mapping of $\varphi(t, \cdot)$, i.e., $y = \varphi(t, x)^{-1}$ iff $x = \varphi(t, y)$, $t \in \mathbb{R}$. We assume that, for every $t \in \mathbb{R}$, the mapping $x \mapsto \varphi(t, x)$, $x \in \Omega$ is a homeomorphism of Ω.

Let $h : \Omega \to \mathbb{R}$ be a continuous function. A locally finite Borel measure μ on Ω is said to be p-admissible for φ and h if the expression

$$(T(t)f)(x) := e^{\int_0^t h(\varphi(r,x))\, dr} f(\varphi(t, x)),\ t \in \mathbb{R},\ x \in \Omega,\ f \in L^p(\Omega, \mu),$$

defines a strongly continuous group on $L^p(\Omega, \mu)$. The C_0-admissibility of $(T(t))_{t\in\mathbb{R}}$ and the integral generator of cosine function $(C(t))_{t\geqslant 0}$, where $C(t) := \frac{1}{2}(T(t)$

$+ T(-t)), t \geqslant 0$, are precisely characterized in [247, Theorem 4(d)-(e)]. Using [210, Theorem 1, Proposition 1] and the proof of [247, Corollary 2], one gets that $(C(t))_{t\geqslant 0}$ is S-topologically transitive iff $(C(t))_{t\geqslant 0}$ is S-hypercyclic. Given a number $t \in \mathbb{R}$, we define $h_t(x) := \exp(\int_0^t h(\varphi(r, x))\, dr)$ and the Borel measures $v_{p,t}(B) := \int_{\varphi(-t,B)} h_t^p\, d\mu$, $t \in \mathbb{R}$, $B \subseteq \Omega$ measurable.

The following theorem slightly improves [247, Theorem 5, Theorem 9].

Theorem 3.2.40. (i) *Let $E = L^p(\Omega, \mu)$ and let μ be p-admissible for φ and h. Then (a) $\Rightarrow$ (b) $\Rightarrow$ (c) $\Rightarrow$ (d) $\Rightarrow$ (e) $\Rightarrow$ (f), where:*

(a) *For every compact set $K \subseteq \Omega$ there exist sequences (L_n^+) and (L_n^-) of Borel measurable subsets of K and a sequence of positive real numbers (t_n) such that for $L_n := L_n^+ \cup L_n^-$ one has*

$$\lim_{n\to\infty} \mu(K \setminus L_n) = \lim_{n\to\infty} \upsilon_{p,t_n}(L_n) = \lim_{n\to\infty} \upsilon_{p,-t_n}(L_n) = 0 \tag{541}$$

and

$$\lim_{n\to\infty} \upsilon_{p,2t_n}(L_n^+) = \lim_{n\to\infty} \upsilon_{p,-2t_n}(L_n^-) = 0. \tag{542}$$

(b) $(C(t))_{t\geqslant 0}$ *is weakly mixing.*

(c) $(C(t))_{t\geqslant 0}$ *is hypercyclic.*

(d) $(C(t))_{t\geqslant 0}$ *is S-hypercyclic for every closed subset S of $\mathbb{C}$ which satisfies $S\setminus\{0\} \neq \emptyset$.*

(e) $(C(t))_{t\geqslant 0}$ *is S-hypercyclic for every (some) bounded closed subset S of $[0, \infty)$ which satisfies* $\inf S > 0$.

Furthermore, if for every compact subset K of Ω one has $\lim_{|t|\to\infty} \varphi(K \cap \varphi(t, K)) = 0$, the above are equivalent.

(ii) *Let ρ be C_0-admissible for φ and h. Then (a) $\Rightarrow$ (b) $\Rightarrow$ (c) $\Rightarrow$ (d) $\Rightarrow$ (e) $\Rightarrow$ (f), where:*

(a) *For every compact set $K \subseteq \Omega$ there exist sequences of positive real numbers (t_n) and open subsets (U_n^+) and (U_n^-) of Ω such that $K \subseteq U_n^+ \cup U_n^-$,*

$$\limsup_{n\to\infty}\sup_{x\in K} \frac{\rho(\varphi(-t_n, x))}{h_{-t_n}(x)} = \limsup_{n\to\infty}\sup_{x\in K} \frac{\rho(\varphi(t_n, x))}{h_{t_n}(x)} = 0$$

and

$$\limsup_{n\to\infty}\sup_{K\cap U_n^-} \frac{\rho(\varphi(-2t_n, x))}{h_{-2t_n}(x)} = \limsup_{n\to\infty}\sup_{K\cap U_n^+} \frac{\rho(\varphi(2t_n, x))}{h_{2t_n}(x)} = 0.$$

(b) $(C(t))_{t\geqslant 0}$ *is weakly mixing on $C_{0,\rho}(\Omega)$.*

(c) $(C(t))_{t\geqslant 0}$ *is hypercyclic on $C_{0,\rho}(\Omega)$.*

(d) $(C(t))_{t\geqslant 0}$ *is S-hypercyclic on $C_{0,\rho}(\Omega)$ for every closed subset S of $\mathbb{C}$ which satisfies $S\setminus\{0\} \neq \emptyset$.*

(e) $(C(t))_{t\geqslant 0}$ *is S-hypercyclic on $C_{0,\rho}(\Omega)$ for every (some) bounded closed subset S of $[0, \infty)$ which satisfies* $\inf S > 0$.

Furthermore, if for every compact subset K of Ω one has

$$\lim_{|t|\to\infty} \sup_{x\in\varphi(K\cap\varphi(t,K))} \rho(x) = 0 \textit{ and } \inf_{x\in K} \rho(x) > 0,$$

the above are equivalent.

Proof. The implications (a) $\Rightarrow$ (b) $\Rightarrow$ (c) in (i) are consequences of [247, Theorem 5] and the implications (c) $\Rightarrow$ (d) $\Rightarrow$ (e) are trivial. Therefore, it suffices to show that the preassumption (e) combined with the additional condition $\lim_{|t|\to\infty} \varphi(K \cap \varphi(t, K)) = 0$ for each compact subset K of Ω implies (541)-(542). This can be obtained by a slight modification of the proof of the aforementioned theorem. Let $\Omega \supseteq K$ be compact and let S be a bounded subset of $[0, \infty)$ satisfying that $\inf S > 0$ and $(C(t))_{t\geqslant 0}$ is S-hypercyclic. We shall consider in the sequel only the non-trivial case $\int_K d\mu > 0$. Assuming this, we get that there do not exist $c \in S$ and $t \geqslant 0$ such that $-\chi_K = cC(t)\chi_K$, which implies by the proof of [309, Lemma 3] that for given $\varepsilon \in (0, 1)$ in advance, there exist $c_\varepsilon \in S\setminus\{0\}$, $t_\varepsilon > 0$ and $\upsilon_\varepsilon \in L^p(\Omega, \mu)$ such that $\|\upsilon_\varepsilon - \chi_K\| < \varepsilon^{2/p}$, $\|c_\varepsilon C(t_\varepsilon)\upsilon_\varepsilon + \chi_K\| < \varepsilon^{2/p}$, $\mu(K \cap \varphi(2t_\varepsilon, K)) < \varepsilon^2$ and $\mu(K \cap \varphi(-2t_\varepsilon, K)) < \varepsilon^2$. Set $L_\varepsilon := K \cap \{|1 - \upsilon_\varepsilon|^p \leqslant \varepsilon\} \cap \{|1 - c_\varepsilon C(t_\varepsilon)\upsilon_\varepsilon|^p \leqslant \varepsilon\}$, $L_\varepsilon^- := \{x \in L : (c_\varepsilon T(t_\varepsilon)\upsilon_\varepsilon)(x) \leqslant \varepsilon^{1/p} - 1\}$ and $L_\varepsilon^+ := L_\varepsilon \setminus L_\varepsilon^-$. Then it is obvious that $\mu(K\setminus L_\varepsilon) < 2\varepsilon$, $\upsilon_{|L_\varepsilon} \geqslant 1 - \varepsilon^{1/p}$ and $(c_\varepsilon C(t_\varepsilon)\upsilon_\varepsilon)_{|L_\varepsilon} \leqslant \varepsilon^{1/p} - 1$. Adopting the same notation as in [247], it follows that for every measurable subsets A, B of Ω we have $\|\upsilon_\varepsilon^- \chi_B\| < \varepsilon^{2/p}$ and

$$\begin{aligned}\|c_\varepsilon(C(t_\varepsilon)(\upsilon_\varepsilon^+ \chi_B))\chi_A\| &\leqslant \|c_\varepsilon(C(t_\varepsilon)\upsilon_\varepsilon - c_\varepsilon^{-1}(-\chi_K) + c_\varepsilon^{-1}(-\chi_K))^+\| \\ &\leqslant \|c_\varepsilon(C(t_\varepsilon)\upsilon_\varepsilon + c_\varepsilon^{-1}\chi_K)\| \leqslant \|c_\varepsilon C(t_\varepsilon)\upsilon_\varepsilon + \chi_K\| < \varepsilon^{2/p}.\end{aligned}$$

This yields $\varepsilon^2 \geqslant 2^{-p}c_\varepsilon^p(v_{p,t_\varepsilon}(L_\varepsilon) + v_{p,-t_\varepsilon}(L_\varepsilon))$, and because ε was arbitrary, we have (541). Furthermore, $|c_\varepsilon| \frac{\upsilon_\varepsilon^-(x)}{h_{-t_\varepsilon}(x)} \geqslant 1 - \varepsilon^{1/p}$, $x \in \varphi(t_\varepsilon, L_\varepsilon^-)$, $|c_\varepsilon| \frac{\upsilon_\varepsilon^-(x)}{h_{t_\varepsilon}(x)} \geqslant 1 - \varepsilon^{1/p}$, $x \in \varphi(-t_\varepsilon, L_\varepsilon^+)$, and the following holds:

$$\begin{aligned}&(1 - \varepsilon^{1/p})^p v_{p,2t_\varepsilon}(L_\varepsilon^+) \\ &\leqslant 2^{p+1} \int_{\varphi(-2t_\varepsilon, L_\varepsilon^+)} (c_\varepsilon C(t_\varepsilon))^p(x)\, d\mu(x) \\ &= 2^{p+1}\left\| c_\varepsilon(C(t_\varepsilon)\upsilon_\varepsilon^+)\chi_{\varphi(-2t_\varepsilon, L_\varepsilon^+)} - (c_\varepsilon C(t_\varepsilon)\upsilon + \chi_K)\chi_{\varphi(-2t_\varepsilon, L_\varepsilon^+)} + \chi_{K\cap\varphi(-2t_\varepsilon, L_\varepsilon^+)}\right\| \\ &\leqslant 2^{3p+1}\left(2\varepsilon^2 + \mu(K \cap \varphi(-2t_\varepsilon, L_\varepsilon^+))\right) \leqslant 2^{3(p+1)}\varepsilon^2.\end{aligned}$$

The estimate $(1 - \varepsilon^{1/p})^p v_{p,-2t_\varepsilon}(L_\varepsilon^-) \leqslant 2^{3(p+1)}\varepsilon^2$ can be proved analogously, which completes the proof of (i). The proof of (ii) is similar and therefore omitted. □

Remark 3.2.41. (i) A careful examination of the proof of [247, Theorem 5] implies that the condition $\lim_{|t|\to\infty} \mu(K \cap \varphi(t, K)) = 0$, for every compact subset K of Ω, can be neglected from the formulation of [247, Corollary 8]. Assume now that, for every compact subset K of Ω, one has $\inf_{x\in K} \rho(x) > 0$; then we get by means of the proofs of [247, Theorem 9] and [245, Theorem 4.11] that the hypercyclicity of cosine function $(C_\varphi(t))_{t\geqslant 0}$ in $C_{0,\rho}(\Omega)$ implies the hypercyclicity of $(T_\varphi(t))_{t\geqslant 0}$ in $C_{0,\rho}(\Omega)$.

(ii) Notice that we have constructed in Subsection 3.1.1 a strongly continuous semigroup induced by semiflow, denoted by $(T_\varphi(t))_{t\geqslant 0}$, which is both non-hypercyclic and positively supercyclic. Therefore, Theorem 3.2.40 might be surprising.

The characterizations of hypercyclicity and mixing can be simplified in the case that $\Omega \subseteq \mathbb{R}$. More precisely, for every $x_0 \in \Omega$, the semiflow $\varphi(\cdot, x_0)$ can be given as the unique solution of the initial value problem $\dot{x} = F(x)$, $x(0) = x_0$, where $F(\cdot)$ is locally Lipschitz continuous vector field on Ω. For the sake of simplicity, we focus our attention to the case when $F(\cdot)$ is continuously differentiable, which implies that the mapping $x \mapsto \varphi(t, x)$, $x \in \Omega$ is continuously differentiable for every fixed $t \in \mathbb{R}$. By the proofs of [247, Theorem 12, Theorem 15], we have the following.

Theorem 3.2.42. *Let $\Omega \subseteq \mathbb{R}$, let $F(\cdot)$ be continuously differentiable and let the locally finite p-admissible measure μ have a positive Lebesgue density $\rho(\cdot)$, resp., let $\rho(\cdot)$ be a positive function C_0-admissible for $F(\cdot)$ and $h(\cdot)$. Then the following assertions are equivalent:*

(a) $(C(t))_{t\geqslant 0}$ *is hypercyclic on $L^p(\Omega, \mu)$, resp. $C_{0,\rho}(\Omega)$.*
(b) $(C(t))_{t\geqslant 0}$ *is S-hypercyclic on $L^p(\Omega, \mu)$, resp. $C_{0,\rho}(\Omega)$, for every closed subset S of $\mathbb{C}$ which satisfies $S\setminus\{0\} \neq \emptyset$.*
(c) $(C(t))_{t\geqslant 0}$ *is S-hypercyclic on $L^p(\Omega, \mu)$, resp. $C_{0,\rho}(\Omega)$, for every (some) bounded closed subset S of $[0, \infty)$ which satisfies* $\inf S > 0$.

In the subsequent theorems we analyze chaoticity of cosine functions on weighted function spaces.

Theorem 3.2.43. *Assume $\Omega \subseteq \mathbb{R}^d$ is open and $\rho(\cdot)$ is a positive function on Ω that is C_0-admissible for $\varphi(\cdot)$ and $h(\cdot)$. Assume further that, for every compact set K of Ω, there exists $t_K > 0$ such that $\varphi(t, K) \cap K = \emptyset$, $t \geqslant t_K$ and* $\inf_{x\in K} \rho(x) > 0$. *Then, the following statements are equivalent:*

(a) $(C(t))_{t\geqslant 0}$ *is chaotic on $C_{0,\rho}(\Omega)$.*
(b) *The set of periodic points of $(C(t))_{t\geqslant 0}$ is dense in $C_{0,\rho}(\Omega)$.*
(c) *For every compact set K there exists $P > 0$ such that*

$$\lim_{n\to\infty} \sup_{x\in\varphi(nP,K)} \frac{\rho(x)}{h_{nP}(\varphi(-nP,x))} = \lim_{n\to\infty} \sup_{x\in\varphi(-nP,K)} h_{nP}(x)\rho(x) = 0.$$

(d) $(T(t))_{t\geqslant 0}$ *is chaotic on $C_{0,\rho}(\Omega)$.*
(e) $(T(-t))_{t\geqslant 0}$ *is chaotic on $C_{0,\rho}(\Omega)$.*

Proof. The equivalence relation (d) $\Leftrightarrow$ (e) follows from [161, Theorem 2.5] and the fact that $(T(t))_{t\geqslant 0}$ and $(T(-t))_{t\geqslant 0}$ have the same set of periodic points, while the equivalence of (c), (d) and (e) follows from [245, Theorem 5.7]; notice also that (c) implies the hypercyclicity of $(C(t))_{t\geqslant 0}$ in (a) since the assertion (i) of [247, Theorem 9] holds with $t_n = nP$ and $U_n^+ = U_n^- = \Omega$. Since every periodic point of $(T(t))_{t\geqslant 0}$ is also a periodic point of $(C(t))_{t\geqslant 0}$, we obtain that (c) and (d) together imply (a). The implication (a) $\Rightarrow$ (b) is trivial and it remains to be proved the implication (b) $\Rightarrow$ (c). For that, assume K is a compact subset of Ω, U_K is a relatively compact, open neighborhood of K, $t > 0$ and, for every $s \geqslant t$, $\varphi(s, \overline{U_K}) \cap \overline{U_K} = \emptyset$. Let $f \in C_c(\Omega)$ be such that $f(x) = 1, x \in K, f(x) \geqslant 0, x \in \Omega$ and $f(x) = 0, x \in \Omega \setminus U_K$. Let $\varepsilon \in (0, \inf_{x\in K} \rho(x)/2)$ and let υ be a real-valued P-periodic point of $(C(t))_{t\geqslant 0}$ with $\varepsilon > \|f - \upsilon\|$. Then $\upsilon(x) \geqslant 1/2$, $x \in K$. Using induction and the composition property of cosine functions, one gets that $C(nP)\upsilon = \upsilon$, $n \in \mathbb{N}$ so that one can assume that $P > t$. Taking into account the equalities

$$2\upsilon(x) = h_{nP}(x)\upsilon(\varphi(nP, x)) + h_{-nP}(x)\upsilon(\varphi(-nP, x)),\ n \in \mathbb{N}, x \in \Omega, \tag{543}$$

and $h_t(x)h_s(\varphi(t, x)) = h_{t+s}(x)$, $x \in \Omega$, $t, s \in \mathbb{R}$, it follows inductively that, for every $x \in \Omega$ and $n \in \mathbb{Z}$:

$$h_{nP}(x)\upsilon(\varphi(nP, x)) = nh_P(x)\upsilon(\varphi(P, x)) - (n-1)\upsilon(x), \tag{544}$$

$$h_{-nP}(x)\upsilon(\varphi(-nP, x)) = -nh_P(x)\upsilon(\varphi(P, x)) + (n+1)\upsilon(x), \tag{545}$$

and

$$h_{-nP}(x)\upsilon(\varphi(-nP, x)) = nh_{-P}(x)\upsilon(\varphi(-P, x)) - (n-1)\upsilon(x). \tag{546}$$

Put $a_n := \sup_{x\in K} h_{nP}(x)|\upsilon(\varphi(nP, x))|$, $n \in \mathbb{Z}$. Without loss of generality, we may assume that $\max(a_1, a_{-1}) = a_1$. Clearly, (543) implies $a_1 \geqslant a_0$. By taking supremum on both sides of (544), we get

$$a_n \geqslant na_1 - (n-1)a_0 \geqslant a_1 \geqslant a_0 \geqslant \frac{1}{2},\ n \in \mathbb{N}. \tag{547}$$

There exist two possibilities. The first one is $a_1 = a_0$, which implies by (546): $a_{-n} \geqslant na_{-1} - (n-1)a_0 \geqslant a_{-1} \geqslant a_0 \geqslant \frac{1}{2}$, $n \in \mathbb{N}$; then

$$\sup_{x\in\varphi(-nP,\overline{U_K})} |\upsilon(x)|\rho(x) \geqslant \sup_{x\in\varphi(-nP,K)} |\upsilon(x)|\rho(x) \geqslant \frac{1}{2}\sup_{x\in K} \frac{\rho(\varphi(-nP,x))}{h_{-nP}(x)}$$

$$= \frac{1}{2}\sup_{x\in K} h_{nP}(\varphi(-nP, x))\rho(\varphi(-nP, x)),$$

$$\sup_{x\in\varphi(nP,\overline{U_K})} |\upsilon(x)|\rho(x) \geqslant \sup_{x\in K} |\upsilon(\varphi(nP, x))|\rho(\varphi(nP, x)) \geqslant \frac{1}{2}\sup_{x\in K} \frac{\rho(\varphi(nP,x))}{h_{nP}(x)},$$

and the proof in this case completes an application of [245, Lemma 5.6]. The second one is $a_1 > a_0$ and the proof in this case is quite similar; as a matter of fact, (547) implies $a_n \geqslant \frac{1}{2}$, $n \in \mathbb{N}$ and we obtain from (545) that $a_{-n} \geqslant na_1 - (n+1)a_0 \to +\infty$, $n \to \infty$ and that there exists $n_0(K) \in \mathbb{N}$ such that $a_{-n} \geqslant \frac{1}{2}$, $n \geqslant n_0(K)$. This completes the proof. □

Regarding the chaoticity of $(C(t))_{t\geqslant 0}$ in $L^p(\Omega, \mu)$, we have the following simple observation. Assume that there exists a closed μ-zero subset N of Ω such that $\varphi(t, N) = N$, $t > 0$ and that, for every compact subset K of $\Omega \setminus N$ and sufficiently large t, one has $\varphi(t, K) \cap K = \emptyset$. By [245, Theorem 5.3, Remark 5.4] and the proof of [247, Theorem 5], it follows that the chaoticity of $(T(t))_{t\geqslant 0}$ implies the chaoticity of $(C(t))_{t\geqslant 0}$. It is not clear whether the converse statement holds.

Let the spaces $L^p_\rho(\mathbb{R})$ and $C_{0,\rho}(\mathbb{R})$ possess the same meaning as in [161]. Arguing in a similar way, we get that the condition $\lim_{|t|\to\infty} \rho(t) = 0$ is equivalent to say that the cosine function $(C(t))_{t\geqslant 0}$, given by $(C(t)f)(x) := \frac{1}{2}(f(x+t) + f(x-t))$, $f \in C_{0,\rho}(\mathbb{R})$, $t \geqslant 0$, $x \in \mathbb{R}$, is chaotic in $C_{0,\rho}(\mathbb{R})$. This enables one to simply construct an example of hypercyclic cosine function $(C(t))_{t\geqslant 0}$ that is not chaotic. Put, for example, $\rho(t) := e^{-(|t|+1)\cos(ln(|t|+1))+1}$, $t \in \mathbb{R}$ and notice that $\rho(\cdot)$ is an admissible weight function which satisfies $\lim_{|t|\to\infty} \rho(t) \neq 0$ ([498]). Hence, $(C(t))_{t\geqslant 0}$ is not chaotic in $C_{0,\rho}(\mathbb{R})$. The hypercyclicity of $(C(t))_{t\geqslant 0}$ follows from the fact that there exists a sequence (t_n) of positive real numbers satisfying

$$\lim_{n\to\infty} \rho(t_n) = \lim_{n\to\infty} \rho(-t_n) = \lim_{n\to\infty} \rho(2t_n) = \lim_{n\to\infty} \rho(-2t_n) = 0.$$

In the following theorem we consider the necessary and sufficient conditions for the chaoticity of cosine function $(C(t))_{t\geqslant 0}$, $(C(t)f)(x) = \frac{1}{2}(f(x+t) + f(x-t))$, $t \geqslant 0$, $x \in \mathbb{R}$, in the space $L^p_\rho(\mathbb{R})$.

Theorem 3.2.44. *Assume that* $\rho : \mathbb{R} \to (0, \infty)$ *is measurable and that there exist* $M \geqslant 1$ *and* $\omega \in \mathbb{R}$ *such that* $\rho(x) \leqslant Me^{\omega|t|}\rho(x+t)$, $x, t \in \mathbb{R}$. *Then* $(T(t))_{t\in\mathbb{R}}$ *is a* C_0*-group in* $L^p_\rho(\mathbb{R})$ *and the following assertions are equivalent.*

(a) $(C(t))_{t\geqslant 0}$ *is chaotic on* $L^p_\rho(\mathbb{R})$.
(b) *The set of periodic points of* $(C(t))_{t\geqslant 0}$ *is dense in* $L^p_\rho(\mathbb{R})$.
(c) *For every* $\varepsilon > 0$ *there exists* $P > 0$ *such that*

(548) $$\sum_{n\in\mathbb{Z}\setminus\{0\}} \rho(nP) < \varepsilon.$$

(d) $(T(t))_{t\geqslant 0}$ *is chaotic on* $L^p_\rho(\mathbb{R})$.
(e) $(T(-t))_{t\geqslant 0}$ *is chaotic on* $L^p_\rho(\mathbb{R})$.

Proof. The implication (a) ⇒ (b) is trivial, the equivalence of (c) and (d) is well known, and the equivalence of (d) and (e) follows from [161, Theorem 2.5] and the fact that $(T(t))_{t\geqslant 0}$ and $(T(-t))_{t\geqslant 0}$ have the same set of periodic points. Since every periodic point of $(T(t))_{t\geqslant 0}$ is also a periodic point of $(C(t))_{t\geqslant 0}$, (c) and (d) taken together imply (a) by [70, Theorem 1.1, 2.2]. Therefore, it remains to be proved the implication (b) $\Rightarrow$ (c). Let $\varepsilon > 0$ be fixed, let $\theta > 0$ and let $z \in C_c(\mathbb{R})$ be such that $\|z\| = (\int_{-\infty}^{\infty} |z(x)|^p \rho(x)\, dx)^{1/p} = 1$, $z \geqslant 0$ and $\operatorname{supp}(z) \subseteq [-\theta, \theta]$. Then there exist $P > 0$ and a real-valued P-periodic point υ of $(C(t))_{t \geqslant 0}$ such that $\|z - \upsilon\| < \varepsilon$. Put $a_n := \int_{-\theta+nP}^{\theta+nP} |2\upsilon(x)|^p\, dx$, $n \in \mathbb{Z}$. By the proof of Theorem 3.2.43, we have $C(nP)\upsilon = \upsilon$ and $2\upsilon(x + nP) = \upsilon(x + (n + 1)P) + \upsilon(x + (n - 1)P)$, $x \in \mathbb{R}$, $n \in \mathbb{Z}$, which implies

$$2\int_{-\theta}^{\theta} |2\upsilon(x + nP)|^p\, dx \leqslant \int_{-\theta}^{\theta} |2\upsilon(x + (n + 1)P)|^p\, dx + \int_{-\theta}^{\theta} |2\upsilon(x + (n - 1)P)|^p\, dx,\ n \in \mathbb{Z},$$

i.e.,

$$2a_n \leqslant a_{n+1} + a_{n-1},\ n \in \mathbb{Z}. \tag{549}$$

We may assume without loss of generality that $P > 2\theta$ and $a_1 = \max(a_1, a_{-1})$. Then $a_1 \geqslant a_0$ and an induction argument combined with (549) shows that:

$$a_{n+1} \geqslant (n + 1)a_1 - na_0,\ a_n - a_0 \geqslant n(a_1 - a_0) \text{ and } a_{n+1} \geqslant a_n,\ n \in \mathbb{N}_0. \tag{550}$$

We first consider the case $a_1 = a_0$. Then $a_{-1} \geqslant a_0$ and, by (549), we have $a_{-(n+1)} \geqslant a_{-n} \geqslant a_{-1}$, $n \in \mathbb{N}$. Since

$$\int_{-\theta}^{\theta} [2^{1-p} |2z(x)|^p - |2\upsilon(x)|^p]\rho(x)\, dx \leqslant \int_{-\theta}^{\theta} |2z(x) - 2\upsilon(x)|^p \rho(x)\, dx < (2\varepsilon)^p,$$

we get from [161, Lemma 4.2] that there exist $m_1 > 0$ and $M_1 > 0$ such that, for every $\sigma \in \mathbb{R}$, $m_1\rho(\sigma - \theta) \leqslant \rho(t) \leqslant M_1\rho(\sigma + \theta)$, and

$$a_0 = \int_{-\theta}^{\theta} |2\upsilon(x)|^p\, dx \geqslant \frac{1}{M_1\rho(\theta)} \int_{-\theta}^{\theta} |2\upsilon(x)|^p\rho(x)\, dx \geqslant \frac{1}{M_1\rho(\theta)} 2(1 - \varepsilon^p).$$

Therefore, the following calculations are correct:

$$\begin{aligned}(2\varepsilon)^p &> \sum_{n\in\mathbb{Z}\setminus\{0\}} \int_{-\theta+nP}^{\theta+nP} |2\upsilon(x)|^p\rho(x)\, dx \geqslant \sum_{n\in\mathbb{Z}\setminus\{0\}} m_1\rho(-\theta + nP)a_n \\ &\geqslant \sum_{n\in\mathbb{Z}\setminus\{0\}} m_1\rho(-\theta + nP)a_0 \geqslant \sum_{n\in\mathbb{Z}\setminus\{0\}} m_1\rho(-\theta + nP)\frac{1}{M_1\rho(\theta)}2(1 - \varepsilon^p).\end{aligned}$$

This immediately implies (548) by a straightforward computation. Assume now $a_1 > a_0$. Then (550) implies $\lim_{n\to+\infty} a_n = +\infty$ and the existence of an integer $n_0 \in \mathbb{N}$ such that $2^{1-p} a_{n_0} > a_0$. Using again the proof of Theorem 3.2.43, we obtain $\upsilon(x - nn_0P) + (n + 1)\upsilon(x) = n\upsilon(x + n_0P)$, $x \in \mathbb{R}$, $n \in \mathbb{Z}$,

$$2^{p-1}(|2\upsilon(x - nn_0P)|^p + (n + 1)^p|2\upsilon(x)|^p) \geqslant n^p|2\upsilon(x + n_0P)|^p$$

and after integration

$$a_{-nn_0} \geqslant 2^{1-p}n^pa_{n_0} - (n + 1)^pa_0 \to +\infty,\ n \to +\infty.$$

This enables one to conclude that there exists $n_1 \in \mathbb{N}$ such that $a_{-nn_0} \geqslant a_0 \geqslant \frac{1}{M_1\rho(\theta)} 2(1-\varepsilon^p)$, $n \geqslant n_1$. Hence,

$$(2\varepsilon)^p > \sum_{n\in\mathbb{Z}\setminus\{0\}} \int_{-\theta+nn_0n_1P}^{\theta+nn_0n_1P} |2v(x)|^p\rho(x)\,dx \geqslant \sum_{n\in\mathbb{Z}\setminus\{0\}} m_1\rho(-\theta+nn_0n_1P)a_{nn_0n_1}$$

$$\geqslant \sum_{n\in\mathbb{Z}\setminus\{0\}} m_1\rho(-\theta+nn_0n_1P)a_0 \geqslant \sum_{n\in\mathbb{Z}\setminus\{0\}} m_1\rho(-\theta+nn_0n_1P)\frac{2(1-\varepsilon^p)}{M_1\rho(\theta)}.$$

By choosing appropriate constants, the above estimate yields (548) with n_0n_1P, finishing the proof of theorem. □

Let $h : \mathbb{R} \to \mathbb{C}$ be bounded and continuous. Then it is well known that the semigroup solution of the equation $u_t = u_x + h(x)u$, $u(0, x) = f(x)$, $t \geqslant 0$ is given by $(T(t)f)(x) := \exp(\int_x^{x+t} h(s)\,ds)f(x+t)$, $t \geqslant 0$, $x \in \mathbb{R}$. If the solution can be extended to the whole real axis, then one can consider hypercyclic properties of the equation

$$u_{tt} = u_{xx} + h(x)u_x + \frac{\partial}{\partial x}(h(x)u) + h^2(x)u,$$
$$u(0, x) = f_1(x),\ u_t(0, x) = f_2(x);\ t \geqslant 0,\ x \in \mathbb{R}.$$

Assume, more generally, that $g : \mathbb{R} \times \mathbb{R} \to \mathbb{C}$ is continuous, as well as that there exist $M \geqslant 1$ and $\omega \in \mathbb{R}$ such that $|g(x, t)| \leqslant Me^{\omega|t|}$, $x, t \in \mathbb{R}$ and the conditions (HT1)-(HT4) hold. Put $\rho(t) := \frac{1}{|g(0,t)|}$, $t \in \mathbb{R}$, $\rho_i(t) := \rho^i(t)$, $t \in \mathbb{R}$, $i \in \mathbb{N}_0$ and $g_j(x, t) := g^j(x, t)$, $x, t \in \mathbb{R}$, $j \in \mathbb{N}_0$. Let E_i be either $L^p_{\rho_i}(\mathbb{R})$ or $C_{0,\rho_i}(\mathbb{R})$, and let $(T_j(t)f)(x) := g_j(x, t)f(x+t)$, $x, t \in \mathbb{R}$, $f \in E_i$. By [309, Lemma 21, Theorem 23] we have that, for every $i \in \mathbb{N}_0$, $\rho_i(\cdot)$ is an admissible weight function and that $(T_j(t))_{t\geqslant 0}$ is a strongly continuous group in E_i. By the proof of [309, Theorem 23], the cosine function $(C_j(t))_{t\geqslant 0}$, given by $(C_j(t)f)(x) := \frac{1}{2}(g_j(x, t)f(x+t) + g_j(x, -t)f(x-t))$, $x, t \in \mathbb{R}$, $f \in E_i$ is chaotic in E_i iff the cosine function $(\widetilde{C}_j(t))_{t\geqslant 0}$, given by $(\widetilde{C}_j(t)f)(x) := \frac{1}{2}(f(x+t) + f(x-t))$, $x, t \in \mathbb{R}$, $f \in E_{i+j}$ is chaotic in E_{i+j}. Assume $i + j > 0$; then Theorem 3.2.43-Theorem 3.2.44 imply that the chaoticity of $(C_j(t))_{t\geqslant 0}$ in $C_{0,\rho_i}(\mathbb{R})$ is equivalent with $\lim_{|t|\to\infty} |g(0, t)| = \infty$, and that the chaoticity of $(C_j(t))_{t\geqslant 0}$ in $L^p_{\rho_i}(\mathbb{R})$ is equivalent to say that, for every $\varepsilon > 0$, there exists $P > 0$ such that $\Sigma_{n\in\mathbb{Z}\setminus\{0\}}|g(0, nP)|^{-i-j} < \varepsilon$.

3.2.4. Disjoint hypercyclicity of *C*-distribution cosine functions. In the following definition we intend to limit ourselves specifically to the analysis of disjoint hypercyclicity and disjoint topological transitivity of *C*-distribution cosine functions.

Definition 3.2.45. Let $n \in \mathbb{N}$, $n \geqslant 2$ and let $\mathbf{G}_i$ be a $(C_i - DCF)$ generated by A_i, $i = 1, 2, \cdots, n$. Then it is said that $\mathbf{G}_i$, $i = 1, 2, \cdots, n$ are:

(i) disjoint hypercyclic, in short d-hypercyclic, if there exists $x \in Z_2(A_1) \cap \cdots \cap Z_2(A_n)$ such that $\overline{\{(G_1(\delta_t)x, \cdots, G_n(\delta_t)x) : t \geqslant 0\}} = E^n$. An element $x \in E$ which satisfies the above equality is called a d-hypercyclic vector associated to $\mathbf{G}_1$, $\mathbf{G}_2, \cdots, \mathbf{G}_n$;

(ii) disjoint topologically transitive, in short d-topologically transitive, if for any open non-empty subsets $V_0, V_1, \cdots, V_n$ of E, there exist $t \geqslant 0$ and $x \in Z_2(A_1) \cap \cdots \cap Z_2(A_n)$ such that $x \in V_0 \cap G_1(\delta_t)^{-1}(V_1) \cap \cdots \cap G_n(\delta_t)^{-1}(V_n)$.

It is clear that the preceding definition can be introduced for the class of fractionally integrated C-cosine functions in Banach spaces and that d-hypercyclicity (d-topological transitivity) of $(C_i - DCF)$'s $\mathbf{G}_i$, $i = 1, 2, \cdots, n$ implies that, for every $i, j \in \mathbb{N}_n$ with $i \neq j$, there exists $t > 0$ such that $G_i(\delta_t) \neq G_j(\delta_t)$. If $(C_i(t))_{t \geqslant 0}$, $i = 1, 2, \cdots, n$ are cosine functions, then the proof of [63, Proposition 2.3] yields that d-topological transitivity of $(C_i(t))_{t \geqslant 0}$, $i = 1, 2, \cdots, n$ implies d-hypercyclicity of $(C_i(t))_{t \geqslant 0}$, $1 \leqslant i \leqslant n$ and that the set of all d-hypercyclic vectors associated to $(C_i(t))_{t \geqslant 0}$, $1 \leqslant i \leqslant n$ is a dense G_δ-subset of E.

The main objective in the subsequent theorem is to clarify sufficient conditions for d-topological transitivity of cosine functions on a class of weighted function spaces. It could be tempting to give an alternative proof of this theorem by using d-Hypercyclicity Criterion from [63].

Theorem 3.2.46. *Suppose* $\Omega \subseteq \mathbb{R}^d$ *is open,* $p \in [1, \infty)$, $n \in \mathbb{N} \setminus \{1\}$, $\varphi_i : [0, \infty) \times \Omega \to \Omega$ *is a semiflow for all* $i = 1, 2, \cdots, n$, $\rho : \Omega \to (0, \infty)$ *is an upper semicontinuous function and* μ *is a locally finite Borel measure on* Ω.

(i) *Suppose* $E = C_{0,\rho}(\Omega)$, ρ *is* C_0*-admissible for* φ_i *and* $h_{\cdot,i}$, $1 \leqslant i \leqslant n$ *and*

$$(C_{\varphi_i}(t)f)(\cdot) = \frac{1}{2}\big(h_{t,i}(\cdot)f(\varphi_i(t, \cdot)) + h_{-t,i}(\cdot)f(\varphi_i(-t, \cdot))\big)$$

for any $t \geqslant 0$, $f \in E$ *and* $1 \leqslant i \leqslant n$. *If for every compact subset* $K \subseteq \Omega$ *there exist a sequence* (t_k) *of non-negative real numbers and sequences* $(U_{k,i}^+)$ *and* $(U_{k,i}^-)$ *of open subsets of* Ω *such that, for every* $i \in \mathbb{N}_n$ *and* $k \in \mathbb{N}$, $K \subseteq U_{k,i}^+ \cup U_{k,i}^-$ *and:*

(a)

$$\limsup_{k\to\infty} \sup_{x\in K} \frac{\rho(\varphi_i(-t_k, x))}{h_{-t_k,i}(x)} = \limsup_{k\to\infty} \sup_{x\in K} \frac{\rho(\varphi_i(t_k, x))}{h_{t_k,i}(x)} = 0,$$

(b)

$$\lim_{k\to\infty} \sup_{x\in K\cap U_{k,i}^-} \frac{\rho(\varphi_i(-2t_k, x))}{h_{-2t_k,i}(x)} = \lim_{k\to\infty} \sup_{x\in K\cap U_{k,i}^+} \frac{\rho(\varphi_i(2t_k, x))}{h_{2t_k,i}(x)} = 0,$$

(c) *for every* $i, j \in \mathbb{N}_n$ *with* $i \neq j$:

$$\lim_{k\to\infty} (A_{ijk} + B_{ijk} + C_{ijk} + D_{ijk}) = 0,$$

where

$$A_{ijk} := \sup_{x \in K \cap U^-_{k,j}} \frac{h_{t_k,i}(\varphi_i(-t_k,\varphi_j(-t_k,x)))\rho(\varphi_i(-t_k,\varphi_j(-t_k,x)))}{h_{-t_k,j}(x)},$$

$$B_{ijk} := \sup_{x \in K \cap U^-_{k,j}} \frac{h_{-t_k,i}(\varphi_i(t_k,\varphi_j(-t_k,x)))\rho(\varphi_i(t_k,\varphi_j(-t_k,x)))}{h_{-t_k,j}(x)},$$

$$C_{ijk} := \sup_{x \in K \cap U^+_{k,j}} \frac{h_{t_k,i}(\varphi_i(-t_k,\varphi_j(t_k,x)))\rho(\varphi_i(-t_k,\varphi_j(t_k,x)))}{h_{t_k,j}(x)},$$

$$D_{ijk} := \sup_{x \in K \cap U^+_{k,j}} \frac{h_{-t_k,i}(\varphi_i(t_k,\varphi_j(t_k,x)))\rho(\varphi_i(t_k,\varphi_j(t_k,x)))}{h_{t_k,j}(x)},$$

then the cosine functions $(C_{\varphi_i}(t))_{t \geqslant 0}$, $i = 1, 2, \cdots, n$ *are d-topologically transitive.*

(ii) *Suppose* $X = L^p(\Omega, \mu)$ *and* μ *is p-admissible for* φ_i *and* $h_{\cdot,i}$, $1 \leqslant i \leqslant n$. *If for every compact subset* $K \subseteq \Omega$ *there exists a sequence* (t_k) *of non-negative real numbers and sequences of Borel measurable subsets* $(L^+_{k,i})$ *and* $(L^-_{k,i})$ *of K such that for* $L_{k,i} := L^+_{k,i} \cup L^-_{k,i}$ *the following holds:*

(a) $\lim_{k\to\infty} \mu(K \setminus L_{k,i}) = \lim_{k\to\infty} v_{p,t_k}(L_{k,i}) = \lim_{k\to\infty} v_{p,-t_k}(L_{k,i}) = 0$, $1 \leqslant i \leqslant n$,

(b) $\lim_{k\to\infty} v_{p,2t_k}(L^+_{k,i}) = \lim_{k\to\infty} v_{p,-2t_k}(L^-_{k,i}) = 0$, $1 \leqslant i \leqslant n$, *and*

(c) *for every* $i, j \in \mathbb{N}_n$ *with* $i \neq j$:

$$\lim_{k\to\infty} \int_{\varphi_i(-t_k,\varphi_j(-t_k,L^+_{k,j}))} h^p_{t_k,i}(x) h^p_{t_k,j}(\varphi_i(t_k, x))\, d\mu = 0,$$

$$\lim_{k\to\infty} \int_{\varphi_i(-t_k,\varphi_j(t_k,L^-_{k,j}))} h^p_{t_k,i}(x) h^p_{-t_k,j}(\varphi_i(t_k, x))\, d\mu = 0,$$

$$\lim_{k\to\infty} \int_{\varphi_i(t_k,\varphi_j(-t_k,L^+_{k,j}))} h^p_{-t_k,i}(x) h^p_{t_k,j}(\varphi_i(-t_k, x))\, d\mu = 0,$$

$$\lim_{k\to\infty} \int_{\varphi_i(t_k,\varphi_j(t_k,L^-_{k,j}))} h^p_{-t_k,i}(x) h^p_{-t_k,j}(\varphi_i(-t_k, x))\, d\mu = 0,$$

then the cosine functions $(C_{\varphi_i}(t))_{t \geqslant 0}$, $i = 1, 2, \cdots, n$ *are d-topologically transitive.*

Proof. We will only prove the first part of the theorem. Let $\varepsilon > 0$, $u, v_1, \cdots, v_n \in C_c(\Omega)$ and $K = \text{supp}(u) \cup \text{supp}(v_1) \cup \cdots \cup \text{supp}(v_n)$. Then there exists a sequence (t_k) of non-negative real numbers and sequences $(U^+_{k,i})$ and $(U^-_{k,i})$ of open subsets of Ω satisfying that, for every $i \in \mathbb{N}_n$, $K \subseteq U^+_{k,i} \cup U^-_{k,i}$ and that (a)-(c) hold. Further on, for every $i \in \mathbb{N}_n$ and $k \in \mathbb{N}$, there exist non-negative C^∞-functions $\psi^\pm_{k,i}(\cdot)$ such that $\text{supp}(\psi^+_{k,i}) \subseteq U^+_{k,i}$, $\text{supp}(\psi^-_{k,i}) \subseteq U^-_{k,i}$ and $\psi^-_{k,i}(x) + \psi^-_{k,i}(x) = 2$, $x \in K$. Define, for every $k \in \mathbb{N}$, a function $\omega_k : \Omega \to \mathbb{C}$ by setting

$$\omega_k(\cdot) := u(\cdot)$$
$$+ \sum_{i=1}^{n} \left[h_{t_k,i}(\cdot) v_i(\varphi_i(t_k, \cdot)) \psi^-_{k,i}(\varphi_i(t_k, \cdot)) + h_{-t_k,i}(\cdot) v_i(\varphi_i(-t_k, \cdot)) \psi^+_{k,i}(\varphi_i(-t_k, \cdot)) \right].$$

Clearly, $\omega_k \in C_c(\Omega)$, $k \in \mathbb{N}$ and it is enough to prove that there exists $k \in \mathbb{N}$ such that:

$$\max\left(\|\omega_k - u\|, \|C_{\varphi_1}(t_k)\omega_k - v_1\|, \cdots, \|C_{\varphi_n}(t_k)\omega_k - v_n\| \right) < \varepsilon.$$

By definition of $\omega_k(\cdot)$, we easily infer that:

$$\|\omega_k - u\| \leqslant \sum_{i=1}^{n} 2\|v_i\|_\infty \left[\sup_{x\in\varphi_i(-t_k,K)} h_{t_k,i}(x)\rho(x) + \sup_{x\in\varphi_i(t_k,K)} h_{-t_k,i}(x)\rho(x) \right]$$

(551) $$= \sum_{i=1}^{n} 2\|v_i\|_\infty \left[\sup_{x\in K} \frac{\rho(\varphi_i(-t_k,x))}{h_{-t_k,i}(x)} + \sup_{x\in K} \frac{\rho(\varphi_i(t_k,x))}{h_{t_k,i}(x)} \right], \quad k\in\mathbb{N}.$$

Set, for every $x\in\Omega$, $k\in\mathbb{N}$ and $1\leqslant i,j\leqslant n$:

$a_{ijk}(x) := \varphi_j(t_k, \varphi_i(t_k, x))$,
$b_{ijk}(x) := \varphi_j(t_k, \varphi_i(-t_k, x))$,
$c_{ijk}(x) := \varphi_j(-t_k, \varphi_i(t_k, x))$,
$d_{ijk}(x) := \varphi_j(-t_k, \varphi_i(-t_k, x))$,
$A_{ik}(x) := \sum_{1\leqslant j\leqslant n, j\neq i} [h_{t_k,i}(x)h_{t_k,j}(\varphi_i(t_k, x))v_j(a_{ijk}(x))\psi^-_{j,i}(a_{ijk}(x))]$,
$B_{ik}(x) := \sum_{1\leqslant j\leqslant n, j\neq i} [h_{-t_k,i}(x)h_{t_k,j}(\varphi_i(-t_k, x))v_j(b_{ijk}(x))\psi^-_{j,i}(b_{ijk}(x))]$,
$C_{ik}(x) := \sum_{1\leqslant j\leqslant n, j\neq i} [h_{t_k,i}(x)h_{-t_k,j}(\varphi_i(t_k, x))v_j(c_{ijk}(x))\psi^+_{j,i}(c_{ijk}(x))]$ and
$D_{ik}(x) := \sum_{1\leqslant j\leqslant n, j\neq i} [h_{-t_k,i}(x)h_{-t_k,j}(\varphi_i(-t_k, x))v_j(d_{ijk}(x))\psi^+_{j,i}(d_{ijk}(x))]$.

A simple computation shows that, for every $x\in\Omega$, $k\in\mathbb{N}$ and $1\leqslant i\leqslant n$,

$2(C_{\varphi_i}(t_k)\omega_k - v_i)(x) = [h_{t_k,i}(x)u(\varphi_i(t_k, x)) + h_{-t_k,i}(x)u(\varphi_i(-t_k, x))]$
$+ [h_{2t_k,i}(x)v_i(\varphi_i(2t_k, x))\psi^-_{k,i}(\varphi_i(2t_k, x))$
$+ h_{-2t_k,i}(x)v_i(\varphi_i(-2t_k, x))\psi^+_{k,i}(\varphi_i(-2t_k, x))]$
$+ A_{ik}(x) + B_{ik}(x) + C_{ik}(x) + D_{ik}(x)$.

By virtue of (a)-(b), we get the following estimates:

$$\sup_{x\in\Omega} |h_{t_k,i}(x)u(\varphi_i(t_k, x)) + h_{-t_k,i}(x)u(\varphi_i(-t_k, x))|\rho(x)$$

(552) $$\leqslant \|u\|_\infty \left[\sup_{x\in K} \frac{\rho(\varphi_i(-t_k,x))}{h_{-t_k,i}(x)} + \sup_{x\in K} \frac{\rho(\varphi_i(-t_k,x))}{h_{-t_k,i}(x)} \right], \quad k\in\mathbb{N},$$

and

(553) $$\left| h_{2t_k,i}(y)v_i(\varphi_i(2t_k, y))\psi^-_{k,i}(\varphi_i(2t_k, y)) + h_{-2t_k,i}(y)v_i(\varphi_i(-2t_k, y))\psi^+_{k,i}(\varphi_i(-2t_k, y)) \right|$$
$$\leqslant \frac{2\|v_i\|_\infty}{\rho(y)} \left[\sup_{x\in K\cap U^-_{k,i}} \frac{\rho(\varphi_i(-2t_k,x))}{h_{-2t_k,i}(x)} + \sup_{x\in K\cap U^+_{k,i}} \frac{\rho(\varphi_i(2t_k,x))}{h_{2t_k,i}(x)} \right], \quad k\in\mathbb{N},\ y\in\Omega.$$

Since $0 \leqslant \psi^{\pm}_{k,i} \leqslant 2$ on K we obtain that for every $x \in \Omega$, $k \in \mathbb{N}$ and $1 \leqslant i \leqslant n$,

$$(554) \qquad |A_{ik}(x)| + |B_{ik}(x)| + |C_{ik}(x)| + |D_{ik}(x)| \leqslant \sum_{\substack{1 \leqslant j \leqslant n \\ j \neq i}} \frac{2\|\upsilon_j\|_\infty}{\rho(x)} (A_{ijk} + B_{ijk} + C_{ijk} + D_{ijk}).$$

Taking into account (554) and (c), we get that for $1 \leqslant i \leqslant n$:

$$(555) \qquad \lim_{k\to\infty} \sup_{x\in\Omega} \big(|A_{ik}(x)| + |B_{ik}(x)| + |C_{ik}(x)| + |D_{ik}(x)|\big)\rho(x) = 0.$$

The proof of the theorem now follows from (551)-(553) and (555). □

Example 3.2.47. Suppose $a_{ij} > 0$, $1 \leqslant i \leqslant n$, $1 \leqslant j \leqslant m$, and for every $i, j \in \mathbb{N}_n$ with $i \neq j$ there exists $l \in \mathbb{N}_m$ such that $a_{il} \neq a_{jl}$. Let $p \geqslant 1$, $q > \frac{m}{2}$, $\Omega = (0, \infty)^m$, resp. $\Omega = \mathbb{R}^m$, and $h_t(x) = 1$, $t \in \mathbb{R}$, $x \in \Omega$. Define $\varphi_i : \mathbb{R} \times \Omega \to \Omega$, $i = 1, 2, \cdot\cdot\cdot, n$ and $\rho : \Omega \to (0, \infty)$ by

$$\varphi_i(t, x_1, \cdot\cdot\cdot, x_m) := (e^{a_{i1}t}x_1, \cdot\cdot\cdot, e^{a_{im}t}x_m) \text{ and}$$

$$\rho(x_1, \cdot\cdot\cdot, x_m) := \frac{1}{(1+|x|^2)^q}, \; t \in \mathbb{R}, \; x = (x_1, \cdot\cdot\cdot, x_m) \in \Omega.$$

Let μ be the measure on Ω with Lebesgue density ρ. Then one can simply ascertain with the help of [247, Theorem 4] that $(T_{\varphi_i}(t))_{t\in\mathbb{R}}$ is a strongly continuous group in $L^p(\Omega, \mu)$ ($C_{0,\rho}(\Omega)$), $1 \leqslant i \leqslant n$. Suppose first $\Omega = (0, \infty)^m$. Let $K = [a_1, b_1] \times \cdot\cdot\cdot \times [a_m, b_m]$ be a compact subset of Ω, let $L^+_{k,i} = L^-_{k,i} = K$, $k \in \mathbb{N}$ and let (t_k) be a sequence of positive real numbers such that $\lim_{k\to\infty} t_k = \infty$. Proceeding as in Example 3.1.41(iii), one can simply check that the conditions (a)-(c) stated in the formulation of Theorem 3.2.46(ii) hold, which implies that the induced cosine functions $(C_{\varphi_1}(t))_{t\geqslant 0}$, $(C_{\varphi_2}(t))_{t\geqslant 0}$, $\cdot\cdot\cdot$, $(C_{\varphi_n}(t))_{t\geqslant 0}$, are d-topologically transitive in $L^p(\Omega, \mu)$. The above assertion remains true in the case that $\Omega = \mathbb{R}^m$, which follows by choosing an appropriate sequence ($L^+_{k,i} = L^-_{k,i} = L_k$) of measurable subsets of K satisfying $0 \notin L^\circ_k$, $k \in \mathbb{N}$. By [245, Theorem 3.7], we have that, for every $i = 1, 2, \cdot\cdot\cdot, n$, $(T_{\varphi_i}(t))_{t\geqslant 0}$ is a non-hypercyclic strongly continuous semigroup in $C_{0,\rho}(\Omega)$. With Remark 3.2.41(i) in view, we obtain that $(C_{\varphi_i}(t))_{t\geqslant 0}$ is a non-hypercyclic cosine function in $C_{0,\rho}(\Omega)$, which implies that $(C_{\varphi_i}(t))_{t\geqslant 0}$, $1 \leqslant i \leqslant n$, cannot be d-hypercyclic in $C_{0,\rho}(\Omega)$.

3.3 Hypercyclic and topologically mixing properties of abstract multi-term fractional differential equations

3.3.1. Hypercyclic and topologically mixing properties of α-times C-regularized resolvent families. This subsection provides information on hypercyclic and topologically mixing properties of (g_α, C)-regularized resolvent families, also called α-times C-regularized resolvent families ($\alpha > 0$). We shall assume that A is a densely defined subgenerator of an α-times C-regularized resolvent family $(R_\alpha(t))_{t\geqslant 0}$.

Recall that $Z_\alpha(A)$ denotes the set which consists of those vectors $x \in E$ such that $R_\alpha(t)x \in R(C)$, $t \geqslant 0$ and the mapping $t \mapsto C^{-1}R_\alpha(t)x$, $t \geqslant 0$ is continuous. Then $R(C) \subseteq Z_\alpha(A)$, and the proof of [463, Proposition 1.1] implies that $x \in Z_\alpha(A)$ iff there exists a unique strong solution of (61) with $f \equiv 0$; if this is the case, the unique strong solution of this problem is given by $u(t; x) = C^{-1}R_\alpha(t)x$, $t \geqslant 0$.

Lemma 3.3.1. *Suppose $\alpha > 0$, $\lambda \in \mathbb{C}$, $x \in E$ and $Ax = \lambda x$. Then $x \in Z_\alpha(A)$ and the unique strong solution of (61) with $f \equiv 0$ is given by $u(t; x) = E_\alpha(\lambda t^\alpha)x$, $t \geqslant 0$.*

Proof. The uniqueness of solutions follows from the fact that A is a subgenerator of $(R_\alpha(t))_{t\geqslant 0}$, so that it suffices to be shown that the function $u(t; x) = E_\alpha(\lambda t^\alpha)x$, $t \geqslant 0$ is a solution of the homogeneous counterpart of (61). This is a consequence of the following simple computation involving the closedness of A and the dominated convergence theorem:

$$A\int_0^t \frac{(t-s)^{\alpha-1}}{\Gamma(\alpha)} u(s; x)\, ds = A\int_0^t \frac{(t-s)^{\alpha-1}}{\Gamma(\alpha)} \sum_{n=0}^{\infty} \frac{\lambda^n s^{\alpha n}}{\Gamma(\alpha n+1)} x\, ds$$

$$= \int_0^t \frac{(t-s)^{\alpha-1}}{\Gamma(\alpha)} \sum_{n=0}^{\infty} \frac{\lambda^{n+1} s^{\alpha n}}{\Gamma(\alpha n+1)} x\, ds = \sum_{n=0}^{\infty} \int_0^t \frac{(t-s)^{\alpha-1}}{\Gamma(\alpha)} \frac{\lambda^{n+1} s^{\alpha n}}{\Gamma(\alpha n+1)} x\, ds$$

$$= \sum_{n=0}^{\infty} \frac{\lambda^{n+1} t^{\alpha(n+1)}}{\Gamma(\alpha(n+1)+1)} x = u(t; x) - x,\ t \geqslant 0.$$

□

Definition 3.3.2. (cf. also Definition 3.3.8 below) Let $\alpha > 0$, let A be a densely defined subgenerator of an α-times C-regularized resolvent family $(R_\alpha(t))_{t\geqslant 0}$ and let $\tilde{E}$ be a closed linear subspace of E. Then it is said that $(R_\alpha(t))_{t\geqslant 0}$ is:

(i) $\tilde{E}$-hypercyclic if there exists $x \in Z_\alpha(A) \cap \tilde{E}$ such that $\{C^{-1}R_\alpha(t)x : t \geqslant 0\}$ is a dense subset of $\tilde{E}$; such an element is called a $\tilde{E}$-hypercyclic vector of $(R_\alpha(t))_{t\geqslant 0}$;

(ii) $\tilde{E}$-topologically transitive if for every $y, z \in \tilde{E}$ and for every $\varepsilon > 0$, there exist $x \in Z_\alpha(A) \cap \tilde{E}$ and $t \geqslant 0$ such that $\|y - x\| < \varepsilon$ and $\|z - C^{-1}R_\alpha(t)x\| < \varepsilon$;

(iii) $\tilde{E}$-topologically mixing if for every $y, z \in \tilde{E}$ and for every $\varepsilon > 0$, there exists $t_0 \geqslant 0$ such that, for every $t \geqslant t_0$, there exists $x_t \in Z_\alpha(A) \cap \tilde{E}$ such that $\|y - x_t\| < \varepsilon$ and $\|z - C^{-1}R_\alpha(t)x_t\| < \varepsilon$.

In the case $\tilde{E} = E$, it is also said that a $\tilde{E}$-hypercyclic vector of $(R_\alpha(t))_{t\geqslant 0}$ is a hypercyclic vector of $(R_\alpha(t))_{t\geqslant 0}$ and that $(R_\alpha(t))_{t\geqslant 0}$ is topologically transitive, resp. topologically mixing.

The following facts are generally enough for our purpose. Let $C = I$, let $\tilde{E} = E$ and let $(R_\alpha(t))_{t\geqslant 0}$ be topologically transitive. Then $(R_\alpha(t))_{t\geqslant 0}$ is hypercyclic and the set of all hypercyclic vectors of $(R_\alpha(t))_{t\geqslant 0}$, denoted by HC($R_\alpha$), is a dense G_δ-subset of E ([210]). Furthermore, the integral generator $\hat{A}$ of $(R_\alpha(t))_{t\geqslant 0}$ is the unique subgenerator of $(R_\alpha(t))_{t\geq 0}$ and the condition $\rho(\hat{A}) \neq \emptyset$ combined with the proofs of [244, Lemma 3.1, Theorem 3.2] implies that $HC(R_\alpha) \cap D_\infty(\hat{A})$ is a dense subset of E.

The following theorem is the kind of Desch-Schappacher-Webb and Banasiak-Moszyński criteria for chaos of strongly continuous semigroups. For a general result regarding the integer case $\alpha = 1$, we refer the reader to [292, Theorem 3.1.36].

Theorem 3.3.3. *Assume* $\alpha \in (0, 2)\backslash\{1\}$, *A is a densely defined subgenerator of an α-times C-regularized resolvent family* $(R_\alpha(t))_{t\geq 0}$ *and there exists an open connected subset* Ω *of* $\mathbb{C}$ *which satisfies* $\Omega \cap (-\infty, 0] = \emptyset$, $\Omega^\alpha := \{\lambda^\alpha : \lambda \in \Omega\} \subseteq \sigma_p(A)$ *and* $\Omega \cap i\mathbb{R} \neq \emptyset$. *Let* $f : \Omega^\alpha \to E$ *be an analytic mapping such that* $f(\lambda^\alpha) \in Kern(A - \lambda^\alpha)\backslash\{0\}$, $\lambda \in \Omega$, *and let* $\tilde{E} := \overline{span\{f(\lambda^\alpha) : \lambda \in \Omega\}}$. *Then* $(R_\alpha(t))_{t\geq 0}$ *is* $\tilde{E}$*-topologically mixing.*

Proof. Let Ω_0 be an arbitrary open connected subset of Ω which admits a cluster point in Ω. Then it follows from the (weak) analyticity of the mapping $\lambda \mapsto f(\lambda^\alpha) \in \tilde{E}$, $\lambda \in \Omega$ that $\Psi(\Omega_0) := \text{span}\{f(\lambda^\alpha) : \lambda \in \Omega_0\}$ is dense in the Banach space $\tilde{E}$. Further on, it follows from Lemma 3.3.1 that

$$C^{-1}R_\alpha(t)f(\lambda^\alpha) = E_\alpha(\lambda^\alpha t^\alpha)f(\lambda^\alpha),\ t \geq 0,\ \lambda \in \Omega. \tag{556}$$

Without loss of generality, we may assume that $\Omega \cap i(0, \infty) \neq \emptyset$. Then it is clear that there exist $\lambda_0 \in \Omega \cap i(0, \infty)$ and $\delta > 0$ such that any of the sets $\Omega_{0,+} := \{\lambda \in \Omega : |\lambda - \lambda_0| < \delta, \arg(\lambda) \in (\frac{\pi}{2} - \delta, \frac{\pi}{2})\}$ and $\Omega_{0,-} := \{\lambda \in \Omega : |\lambda - \lambda_0| < \delta, \arg(\lambda) \in (\frac{\pi}{2}, \frac{\pi}{2} + \delta)\}$ admits a cluster point in Ω as well as that $\arg(\lambda^\alpha t^\alpha) < \alpha\pi/2$, $\lambda \in \Omega_{0,+}$ and $\arg(-\lambda^\alpha t^\alpha) \in \pi - \alpha\pi/2$, $\lambda \in \Omega_{0,-}$. By (17)-(19), one gets:

$$E_\alpha(\lambda^\alpha t^\alpha) \to \infty,\ t \to \infty,\ \lambda \in \Omega_{0,+} \text{ and } E_\alpha(\lambda^\alpha t^\alpha) \to 0,\ t \to \infty,\ \lambda \in \Omega_{0,-}. \tag{557}$$

Assume $y \in \Psi(\Omega_{0,-})$, $z \in \Psi(\Omega_{0,+})$, $\varepsilon > 0$, $y = \sum_{i=1}^n \beta_i f(\lambda_i^\alpha)$, $z = \sum_{j=1}^m \gamma_j f(\tilde{\lambda}_j^\alpha)$, $\alpha_i, \beta_j \in \mathbb{C}$, $\lambda_i \in \Omega_{0,-}$ and $\tilde{\lambda}_j \in \Omega_{0,+}$ for $1 \leq i \leq n$ and $1 \leq j \leq m$. By (557), we get that there exists $t_0(z) > 0$ such that $E_\alpha(\tilde{\lambda}_j^\alpha t^\alpha) \neq 0$, $t \geq t_0(z)$. Put $z_t := \sum_{j=1}^m \frac{\beta_j}{E_\alpha(\tilde{\lambda}_j^\alpha t^\alpha)} f(\tilde{\lambda}_j^\alpha)$ and $x_t := y + z_t$, $t \geq t_0(z)$. Clearly, $x_t \in Z_\alpha(A) \cap \tilde{E}$, $t \geq t_0(z)$. Keeping in mind (556)-(557), we obtain that there exists $t(y, z, \varepsilon) > t_0(z)$ such that: $\|C^{-1}R_\alpha(t)y\| < \varepsilon$, $C^{-1}R_\alpha(t)z_t = z$ and $\|z_t\| < \varepsilon$, $t \geq t(y, z, \varepsilon)$. This implies $\|x_t - y\| = \|z_t\| < \varepsilon$ and $\|C^{-1}R_\alpha(t)x_t - z\| = \|C^{-1}R_\alpha(t)y\| < \varepsilon$, $t \geq t(y, z, \varepsilon)$, which completes the proof. □

Remark 3.3.4. (i) Assume that $\langle x^*, f(\lambda^\alpha)\rangle = 0$, $\lambda \in \Omega$ for some $x^* \in E^*$ implies $x^* = 0$. Then $\tilde{E} = E$.

(ii) It is not clear how one can prove an extension of [170, Theorem 2.1] for time-fractional evolution equations.

(iii) It is worth noting that Theorem 3.3.3 can be slightly improved by assuming that there exist $n \in \mathbb{N}$, open connected subsets Ω_i of $\mathbb{C}$ and analytic mappings $f_i : \Omega_i^\alpha \to E$ which satisfy, for every $i = 1, \cdots, n$: $\Omega_i \cap (-\infty, 0] = \emptyset$, $\Omega_i^\alpha \subseteq \sigma_p(A)$, $\Omega_i \cap i\mathbb{R}$

$\neq \emptyset$ and $f_i(\lambda^\alpha) \in Kern(A - \lambda^\alpha) \setminus \{0\}$, $\lambda \in \Omega_i$. Set $\widetilde{E} := \overline{span\{f_i(\lambda^\alpha) : \lambda \in \Omega_i, 1 \leqslant i \leqslant n\}}$ and assume that Ω_i' is an open connected subset of Ω_i which admits a cluster point in Ω_i for $1 \leqslant i \leqslant n$. Then

$$\widetilde{E} = \overline{span\{f_i(\lambda^\alpha) : \lambda \in \Omega_i', 1 \leqslant i \leqslant n\}}$$

and one can repeat literally the proof of Theorem 3.3.3 in order to see that $(R_\alpha(t))_{t \geqslant 0}$ is $\widetilde{E}$-topologically mixing. Notice that the idea used here has appeared for the first time in [117].

(iv) Let $Cf(\lambda^\alpha) \in \widetilde{E}$, $\lambda \in \Omega$. Then $A_{|\widetilde{E}}$ is a densely defined subgenerator of the α-times $C_{|\widetilde{E}}$-regularized resolvent family $(R_\alpha(t)_{|\widetilde{E}})_{t \geqslant 0}$ in the Banach space $\widetilde{E}$ and the proof of Theorem 3.3.3 implies that $(R_\alpha(t)_{|\widetilde{E}})_{t \geqslant 0}$ is topologically mixing in $\widetilde{E}$. The additional assumption $\overline{C(\widetilde{E})} = \widetilde{E}$ implies that $(R_\alpha(t)_{|\widetilde{E}})_{t \geqslant 0}$ is hypercyclic and that the set of all hypercyclic vectors of $(R_\alpha(t)_{|\widetilde{E}})_{t \geqslant 0}$ is dense in $\widetilde{E}$.

Before stating the following extension of [161, Theorem 3.3], we would like to observe that the notion of chaoticity of an α-times C-regularized resolvent family makes no sense if $\alpha \in (0, 2) \setminus \{1\}$; more precisely, it is not clear whether the assumption $C^{-1}R_\alpha(t_0)x = x$ for some $t_0 > 0$ and $x \in E$ (here $(R_\alpha(t))_{t \geqslant 0}$ denotes an α-times C-regularized resolvent family subgenerated by A) implies that $C^{-1}R_\alpha(nt_0)$ $x = x$ for all $n \in \mathbb{N}$ (observe, however, that the affirmative answer maybe depend on some particular choices of number α).

Theorem 3.3.5. *Suppose* $0 < \alpha \leqslant 2$, *A is a densely defined subgenerator of a hypercyclic α-times C-regularized resolvent family* $(R_\alpha(t))_{t \geqslant 0}$ *and $R(C)$ is dense in E. Then* $\sigma_p(A^*) = \emptyset$.

Proof. We will prove the theorem in the case $\alpha \in (0, 2) \setminus \{1\}$. Since $R(C)$ is dense in E and $(R_\alpha(t))_{t \geqslant 0}$ is hypercyclic, we obtain that there exists $x \in E$ such that the set $\{R_\alpha(t)x : t \geqslant 0\}$ is dense in E. Proceeding as in the proof of [161, Theorem 3.3(i)], one gets that the set $\{R_\alpha(t)^*\phi : t \geqslant 0\}$ is unbounded in E^*, provided $\phi \in E^*$ and $\phi \neq 0$. Assume to the contrary $\lambda \in \sigma_p(A^*)$ and $A^*\phi = \lambda\phi$ for some $\phi \in E^*$ with $\phi \neq 0$. Then it follows from Theorem 2.1.12(ii) that $A^*_{|\overline{D(A^*)}}$ is a subgenerator of the α-times $C^*_{|\overline{D(A^*)}}$-regularized resolvent family $(R_\alpha(t)^*_{|\overline{D(A^*)}})_{t \geqslant 0}$ in $\overline{D(A^*)}$, which implies $R_\alpha(t)^*\phi = E_\alpha(\lambda t^\alpha)C^*\phi$, $t \geqslant 0$, $\lambda \neq 0$ and $\langle \phi, Cx \rangle \neq 0$. On the other hand, (17)-(19) and the unboundedness of the set $\{R_\alpha(t)^*\phi : t \geqslant 0\}$ imply that $\arg(\lambda) < \alpha\frac{\pi}{2}$. Using (17) again, we get that there exists $t_0 > 0$ such that $|E_\alpha(\lambda t^\alpha)| \geqslant 1$, $t \geqslant t_0$. Take any $y \neq 0$ with $\langle \phi, y \rangle = 0$. Then there exist $n \in \mathbb{N}$ and $t > t_0$ such that $\|R_\alpha(t)x - ny\| < \frac{|\langle \phi, Cx \rangle|}{2\|\phi\|}$ and

$$|\langle \phi, Cx \rangle| < |\langle \phi, R_\alpha(t)x \rangle| \leqslant |\langle \phi, ny \rangle| + \|\phi\| * \|R_\alpha(t)x - ny\| < \frac{|\langle \phi, Cx \rangle|}{2},$$

which is a contradiction. □

We close this subsection with the observation that Theorem 3.3.3 can be also applied to the operators considered in [35]-[37], [151], [172], [407] and [498]-[499].

3.3.2. Hypercyclic and topologically mixing properties of solutions of (2) with $A_j = c_jI$; $c_j \in \mathbb{C}$, $j \in \mathbb{N}_{n-1}$. In this subsection, we shall assume that there exist complex constants $c_1, \cdots, c_{n-1}$ such that $A_j = c_jI$, $j \in \mathbb{N}_{n-1}$. We analyze only global C-resolvent propagation families, i.e., global k-regularized C-resolvent propagation families, for (2) with $k(t) \equiv 1$; in the case $C = I$, such a resolvent family is also called a resolvent propagation family for (2), or simply a resolvent propagation family, if there is no risk for confusion. As before, it will be assumed that every single operator family $(R_i(t))_{t\geqslant 0}$ of the tuple $((R_0(t))_{t\geqslant 0}, \cdots, (R_{m_n-1}(t))_{t\geqslant 0})$ is non-degenerate, i.e., that the supposition $R_i(t)x = 0$, $t \geqslant 0$ implies $x = 0$. Then it is also said that the operator A is a subgenerator of $((R_0(t))_{t\geqslant 0}, \cdots, (R_{m_n-1}(t))_{t\geqslant 0})$. Recall that the integral generator $\hat{A}$ of $((R_0(t))_{t\geqslant 0}, \cdots, (R_{m_n-1}(t))_{t\geqslant 0})$ is defined as the set of all pairs $(x, y) \in E \times E$ such that, for every $i = 0, \cdots, m_n-1$ and $t \geqslant 0$, the following holds:

$$[R_i(\cdot)x - (k * g_i)(\cdot)Cx] + \sum_{j=1}^{n-1} c_j g_{\alpha_n-\alpha_j} * [R_i(\cdot)x - (k * g_i)(\cdot)Cx]$$

$$+ \sum_{j\in\mathbb{N}_{n-1}\setminus D_i} c_j[g_{\alpha_n-\alpha_j+i} * k](\cdot)Cx$$

$$= \begin{cases} [g_{\alpha_n-\alpha} * R_i](\cdot)y, & m - 1 < i, \\ g_{\alpha_n-\alpha} * [R_i(\cdot)y - (k * g_i)(\cdot)Cy], & m - 1 \geqslant i. \end{cases}$$

The standing hypothesis in the sequel of this subsection will be that, for every $i \in \mathbb{N}^0_{m_n-1}$ with $m - 1 \geqslant i$, one has $\mathbb{N}_{n-1}\setminus D_i \neq \emptyset$ and $\Sigma_{j\in\mathbb{N}_{n-1}\setminus D_i} |c_j|^2 > 0$. Then the problem (397) has at most one mild (strong) solution.

The proof of following auxiliary lemma is simple and therefore omitted.

Lemma 3.3.6. (i) *Suppose A generates an exponentially bounded, analytic C-regularized semigroup of angle $\beta \in (0, \pi/2]$ and A is densely defined. Then A is the integral generator of an exponentially bounded, analytic C-regularized propagation family $((R_0(t))_{t\geqslant 0}, \cdots, (R_{m_n-1}(t))_{t\geqslant 0})$ of angle* $\min(\frac{\frac{\pi}{2}+\beta}{\alpha_n-\alpha} - \frac{\pi}{2}, \frac{\pi}{2})$, *provided that* $\frac{\pi}{2} + \beta > \frac{\pi}{2}(\alpha_n - \alpha)$.

(ii) *Suppose A generates an exponentially bounded C-regularized semigroup and A is densely defined. Then A is the integral generator of an exponentially bounded, analytic C-regularized propagation family $((R_0(t))_{t\geqslant 0}, \cdots, (R_{m_n-1}(t))_{t\geqslant 0})$ of angle* $\min(\frac{\pi}{2(\alpha_n-\alpha)} - \frac{\pi}{2}, \frac{\pi}{2})$, *provided that* $\frac{\pi}{2} < \frac{\pi}{2(\alpha_n-\alpha)}$.

Before we go any further, the following facts should be emphasized. If $n = 2$, $c_1 = 0$, $\alpha = 0$, and A is a subgenerator of a global C-regularized propagation family $((R_0(t))_{t\geqslant 0}, \cdots, (R_{m_2-1}(t))_{t\geqslant 0})$, then it is obvious that $(R_0(t))_{t\geqslant 0}$ is a global α_2-times C-regularized resolvent family with A as a subgenerator. In the previous subsection, we have considered hypercyclic and topologically mixing properties of fractional C-regularized resolvent families. Therefore, the results of this subsection can be viewed as slight generalizations of corresponding results from the previous one.

We recall the basic notations used: E is a separable infinite-dimensional complex Banach space, A is a closed linear operator on E, $n \in \mathbb{N}\setminus\{1\}$, $0 \leqslant \alpha_1 < \cdots < \alpha_n$, $0 \leqslant \alpha < \alpha_n$, $A_j = c_j I$ for certain complex constants $c_1, \cdots, c_{n-1}$, $m_j = \lceil \alpha_j \rceil$, $1 \leqslant j \leqslant n$, $m = m_0 = \lceil \alpha \rceil$, $A_0 = A$ and $\alpha_0 = \alpha$. We assume, in addition, that $C^{-1}AC = A$ is densely defined and subgenerates a global C-resolvent propagation family $((R_0(t))_{t\geqslant 0}, \cdots, (R_{m_n-1}(t))_{t\geqslant 0})$. Then we know that A is, in fact, the integral generator of $((R_0(t))_{t\geqslant 0}, \cdots, (R_{m_n-1}(t))_{t\geqslant 0})$.

Let $i \in \mathbb{N}^0_{m_n-1}$. Then we denote by $Z_i(A)$ (again with a little abuse of notation) the set which consists of those vectors $x \in E$ such that $R_i(t)x \in R(C)$, $t \geqslant 0$ and the mapping $t \mapsto C^{-1}R_i(t)x$, $t \geqslant 0$ is continuous. Then $R(C) \subseteq Z_i(A)$, and it can be simply proved that $x \in Z_i(A)$ iff there exists a unique mild solution of (397) with $u_k = \delta_{k,i}x$, $k \in \mathbb{N}^0_{m_n-1}$; if this is the case, the unique mild solution of (397) is given by $u(t; x) := u_i(t; x) := C^{-1}R_i(t)x$, $t \geqslant 0$. Set $\mathcal{D}_i := \{j \in \mathbb{N}^0_{n-1} : m_j - 1 \geqslant i\}$.

The Laplace transform can be used to prove the following extension of Lemma 3.3.1.

Lemma 3.3.7. *Suppose $\lambda \in \mathbb{C}$, $x \in E$ and $Ax = \lambda x$. Then $x \in Z_i(A)$ and the unique strong solution of* (397) *is given by*

$$u_i(t; x) = \mathcal{L}^{-1}\left(\frac{z^{-i-1} + \Sigma_{j\in D_i} c_j z^{-\alpha_n - i - 1 + \alpha_j} - \chi_{\mathcal{D}_i}(0)\lambda z^{-\alpha_n - i - 1 + \alpha}}{1 + \Sigma_{j=1}^{n-1} c_j z^{\alpha_j - \alpha_n} - \lambda z^{\alpha - \alpha_n}}\right)(t)x,$$

for any $t > 0$ and $i \in \mathbb{N}^0_{m_n-1}$.

Set $P_\lambda := \lambda^{\alpha_n - \alpha} + \Sigma_{j=1}^{n-1} c_j \lambda^{\alpha_j - \alpha}$, $\lambda \in \mathbb{C}\setminus\{0\}$ and

$$F_i(\lambda, t) := \mathcal{L}^{-1}\left(\frac{z^{-i-1} + \Sigma_{j\in D_i} c_j z^{-\alpha_n - i - 1 + \alpha_j} - \chi_{\mathcal{D}_i}(0)P_\lambda z^{-\alpha_n - i - 1 + \alpha}}{1 + \Sigma_{j=1}^{n-1} c_j z^{\alpha_j - \alpha_n} - P_\lambda z^{\alpha - \alpha_n}}\right)(t),$$

for any $t > 0$, $i \in \mathbb{N}^0_{m_n-1}$ and $\lambda \in \mathbb{C}\setminus\{0\}$.

Definition 3.3.8. Let $i \in \mathbb{N}^0_{m_n-1}$, and let $\tilde{E}$ be a closed linear subspace of E. Then it is said that $(R_i(t))_{t\geqslant 0}$ is:

(i) $\tilde{E}$-hypercyclic if there exists $x \in Z_i(A) \cap \tilde{E}$ such that $\{C^{-1}R_i(t)x : t \geqslant 0\}$ is a dense subset of $\tilde{E}$; such an element is called a $\tilde{E}$-hypercyclic vector of $(R_i(t))_{t\geqslant 0}$;

(ii) $\tilde{E}$-topologically transitive if for every $y, z \in \tilde{E}$ and for every $\varepsilon > 0$, there exist $x \in Z_i(A) \cap \tilde{E}$ and $t \geqslant 0$ such that $\|y - x\| < \varepsilon$ and $\|z - C^{-1}R_i(t)x\| < \varepsilon$;

(iii) $\tilde{E}$-topologically mixing if for every $y, z \in \tilde{E}$ and for every $\varepsilon > 0$, there exists $t_0 \geqslant 0$ such that, for every $t \geqslant t_0$, there exists $x_t \in Z_i(A) \cap \tilde{E}$ such that $\|y - x_t\| < \varepsilon$ and $\|z - C^{-1}R_i(t)x_t\| < \varepsilon$.

In the case $\widetilde{E} = E$, it is also said that a $\widetilde{E}$-hypercyclic vector of $(R_i(t))_{t\geq 0}$ is a hypercyclic vector of $(R_i(t))_{t\geq 0}$ and that $(R_i(t))_{t\geq 0}$ is topologically transitive, resp. topologically mixing.

Suppose $C = I$, $\widetilde{E} = E$ and $(R_i(t))_{t\geq 0}$ is topologically transitive for some $i \in \mathbb{N}^0_{m_n-1}$. Then $(R_i(t))_{t\geq 0}$ is hypercyclic and the set of all hypercyclic vectors of $(R_i(t))_{t\geq 0}$, denoted by $HC(R_i)$, is a dense G_δ-subset of E ([210]). Furthermore, the condition $\rho(A) \neq \emptyset$ combined with the proofs of [245, Lemma 3.1, Theorem 3.2] implies that $HC(R_i) \cap D_\infty(A)$ is a dense subset of E.

The proof of following theorem can be derived by using Lemma 3.3.7 and the proof of Theorem 3.3.3.

Theorem 3.3.9. *Suppose* $i \in \mathbb{N}^0_{m_n-1}$, Ω *is an open connected subset of* $\mathbb{C}$, $\Omega \cap (-\infty, 0] = \emptyset$ *and* $P_\Omega := \{P_\lambda : \lambda \in \Omega\} \subseteq \sigma_p(A)$. *Let* $f : P_\Omega \to E$ *be an analytic mapping such that* $f(P_\lambda) \in Kern(P_\lambda - A)\setminus\{0\}$, $\lambda \in \Omega$ *and let* $\widetilde{E} := \overline{span\{f(P_\lambda) : \lambda \in \Omega\}}$. *Suppose* Ω_+ *and* Ω_- *are two open connected subsets of* Ω, *and each of them admits a cluster point in* Ω. *If*

$$\lim_{t\to+\infty} |F_i(\lambda, t)| = +\infty,\ \lambda \in \Omega_+ \text{ and } \lim_{t\to+\infty} F_i(\lambda, t) = 0,\ \lambda \in \Omega_-, \tag{558}$$

then $(R_i(t))_{t\geq 0}$ *is* $\widetilde{E}$*-topologically mixing.*

Remark 3.3.10. (i) Assume that $\langle x^*, f(P_\lambda)\rangle = 0$, $\lambda \in \Omega$ for some $x^* \in E^*$ implies $x^* = 0$. Then $\widetilde{E} = E$.

(ii) The previous theorem can be slightly improved in the following manner. Suppose $l \in \mathbb{N}$, $\Omega_1, \cdots, \Omega_l$ are open connected subsets of $\mathbb{C}$, as well as $\Omega_{j,+}$ and $\Omega_{j,-}$ are open connected subsets of Ω_j which admits a cluster point in Ω_j, and satisfy (558) with Ω_+ and Ω_- replaced respectively by $\Omega_{j,+}$ and $\Omega_{j,-}$ $(1 \leq j \leq l)$. Assume, additionally, that $f_j : P_{\Omega_j} \to E$ is an analytic mapping, $\Omega_j \cap (-\infty, 0] = \emptyset$, $P_{\Omega_j} \subseteq \sigma_p(A)$, and $f_j(P_\lambda) \in \text{Kern}(A - P_\lambda)\setminus\{0\}$, $\lambda \in \Omega_j$ $(1 \leq j \leq l)$. Set $\widetilde{E} := \overline{span\{f_j(P_\lambda) : \lambda \in \Omega_j, 1 \leq j \leq n\}}$ and assume that Ω'_j is an open connected subset of Ω_j which admits a cluster point in Ω_j for $1 \leq j \leq l$. Then

$$\widetilde{E} = \overline{span\{f_j(P_\lambda) : \lambda \in \Omega'_j, 1 \leq j \leq l\}}$$

and $(R_i(t))_{t\geq 0}$ is $\widetilde{E}$-topologically mixing.

(iii) Let $Cf(P_\lambda) \in \widetilde{E}$, $\lambda \in \Omega$. Then $A_{|\widetilde{E}}$ is the densely defined integral generator of the $C_{|\widetilde{E}}$-resolvent propagation family $((R_0(t)_{|\widetilde{E}})_{t\geq 0}, \cdots, (R_{m_n-1}(t)_{|\widetilde{E}})_{t\geq 0})$ in the Banach space $\widetilde{E}$, $C^{-1}_{|\widetilde{E}} A_{|\widetilde{E}} C_{|\widetilde{E}} = A_{|\widetilde{E}}$ and each single operator family of the tuple $((R_0(t)_{|\widetilde{E}})_{t\geq 0}, \cdots, (R_{m_n-1}(t)_{|\widetilde{E}})_{t\geq 0})$ is topologically mixing in $\widetilde{E}$. The additional assumption $\overline{C(\widetilde{E})} = \widetilde{E}$ implies that each single operator family of the tuple $((R_0(t)_{|\widetilde{E}})_{t\geq 0}, \cdots, (R_{m_n-1}(t)_{|\widetilde{E}})_{t\geq 0})$ is hypercyclic in $\widetilde{E}$.

(iv) In Subsection 3.3.1, we have seen that the assumptions of Theorem 3.3.9 hold provided that $n = 2$, $c_1 = 0$, $\alpha_2 > 0$, $\alpha = 0$, $i = 0$ and $\Omega \cap i\mathbb{R} \neq \emptyset$. In this case,

$F_0(\lambda, t) = E_{\alpha_2}(\lambda^{\alpha_2} t^{\alpha_2})$, $t \geqslant 0$, and there exist $\lambda_0 \in \Omega$ and $\delta > 0$ such that (558) holds with $\Omega_+ = \{\lambda \in \Omega : |\lambda - \lambda_0| < \delta, \arg(\lambda) \in (\frac{\pi}{2} - \delta, \frac{\pi}{2})\}$ and $\Omega_- = \{\lambda \in \Omega : |\lambda - \lambda_0| < \delta, \arg(\lambda) \in (\frac{\pi}{2}, \frac{\pi}{2} + \delta)\}$.

(v) It is worth noting that the condition (558) of Theorem 3.3.9 does not hold in general. In order to illustrate this, suppose that $n = 4$, $\alpha_j = j - 1, j \in \mathbb{N}_4$, $\alpha = 1$, $i = 2$ and $c_1 \in \mathbb{C} \setminus \{0\}$. Then $\mathcal{D}_2 = \emptyset$ and, for every $t \geqslant 0$,

$$F_2(\lambda, t) = \frac{e^{\lambda t}}{(\lambda - \lambda_1)(\lambda - \lambda_2)} + \frac{e^{\lambda_1 t}}{(\lambda_1 - \lambda)(\lambda_1 - \lambda_2)} + \frac{e^{\lambda_2 t}}{(\lambda_2 - \lambda)(\lambda_2 - \lambda_1)},$$

where $\lambda_{1,2} := (-\lambda^2 \pm \sqrt{\lambda^4 + 4c_1\lambda})/(2\lambda)$. It is not difficult to prove that, for every $\lambda \in \mathbb{C} \setminus \{0\}$, the following relation holds: $\operatorname{Re} \lambda \neq \operatorname{Re} \lambda_1$. This implies that, for every $\lambda \in \mathbb{C}$ with $\operatorname{Re} \lambda > 0$, one has $\lim_{t \to +\infty} |F_2(\lambda, t)| = +\infty$. Regrettably, there does not exist an open connected subset Ω_- of $\mathbb{C}$ such that $\lim_{t \to +\infty} F_2(\lambda, t) = 0$, $\lambda \in \Omega_-$.

(vi) As far as we know, in the handbooks containing tables of Laplace transforms, the explicit forms of functions like $F_i(\lambda, t)$ have not been presented as known images, except for some very special cases of the coefficients α_j, c_j. In this place, we would like to point out the following fact. Suppose $\alpha_n - \alpha_j \in \mathbb{Q}$, $j \in \mathbb{N}^0_{n-1}$. By (20), we obtain that there exists a number $\zeta \in (0, 1)$, independent of λ, such that the function $F_i(\lambda, t)$ can be represented as the finite convolution products of functions like $E_{1,\zeta}(p_\lambda t)$.

(vii) The assertion of Theorem 3.3.9 has been recently generalized in the author's recent study of hypercyclic and topologically mixing properties of degenerate multi-term fractional differential equations (the paper will appear soon in Diff. Eqn. Dyn. Sys.). As an application, we have considered topologically mixing properties of fractional analogues of strongly damped Klein-Gordon equation and vibrating beam type equation.

Theorem 3.3.11. ([313]) *Suppose $R(C)$ is dense in E and there exists $i \in \mathbb{N}^0_{m_n - 1}$ such that $(R_i(t))_{t \geqslant 0}$ is hypercyclic. Then $\sigma_p(A^*) = \emptyset$.*

Example 3.3.12. (i) ([161], [152], [315]) Let $a, b, c > 0$, $\zeta \in (0, 2)$, $c < \frac{b^2}{2a} < 1$ and

$$\Lambda := \left\{\lambda \in \mathbb{C} : \left|\lambda - \left(c - \frac{b^2}{4a}\right)\right| \leqslant \frac{b^2}{4a}, \operatorname{Im} \lambda \neq 0 \text{ if } \operatorname{Re} \lambda \leqslant c - \frac{b^2}{4a}\right\}.$$

Consider the following abstract time-fractional equation:

$$\begin{cases} \mathbf{D}_t^{\alpha} u(t) = a u_{xx} + b u_x + c u := -Au, \\ u(0, t) = 0, \ t \geqslant 0, \\ u(x, 0) = u_0(x), \ x \geqslant 0, \text{ and } u_t(x, 0) = 0, \text{ if } \alpha \in (1, 2). \end{cases}$$

As is known, the operator $-A$ with domain $D(-A) = \{f \in W^{2,2}([0, \infty)) : f(0) = 0\}$, generates an analytic strongly continuous semigroup of angle $\frac{\pi}{2}$ in the space $E = L^2([0, \infty))$; the same assertion holds in the case that the operator $-A$ acts on

$E = L^1([0, \infty))$ with domain $D(-A) = \{f \in W^{2,1}([0, \infty)) : f(0) = 0\}$. Assume first $\zeta \in [1, 2)$, $\theta \in (\zeta\frac{\pi}{2} - \pi, \pi - \zeta\frac{\pi}{2})$ and $P(z) = \Sigma_{j=0}^{n} a_j z^j$ is a non-constant complex polynomial such that $a_n > 0$ and

$$(559) \qquad -e^{i\theta} P(-\Lambda) \cap \{te^{\pm i\zeta\frac{\pi}{2}} : t \geqslant 0\} \neq \emptyset.$$

Then it is not difficult to prove that $-e^{i\theta}P(A)$ generates an analytic C_0-semigroup of angle $\frac{\pi}{2} - |\theta|$. Taking into account Theorem 2.2.5, one gets that the operator $-e^{i\theta} P(A)$ is the integral generator of an exponentially bounded, analytic ζ-times regularized resolvent family $(R_{\zeta,\theta,P}(t))_{t\geqslant 0}$ of angle $\frac{\pi-|\theta|}{\zeta} - \frac{\pi}{2}$. Moreover, the conditions of Theorem 3.3.9 are satisfied with $\tilde{E} = E$, which implies that $(R_{\zeta,\theta,P}(t))_{t\geqslant 0}$ is topologically mixing. Suppose now $\zeta \in (0, 1)$, $\theta \in (-\frac{\pi}{2}, \frac{\pi}{2})$ and $P(z) = \Sigma_{j=0}^{n} a_j z^j$ is a non-constant complex polynomial such that $a_n >$ 0 and (559) holds. Then $-e^{i\theta} P(A)$ is the integral generator of an exponentially bounded, analytic ζ-times regularized resolvent family $(R_{\zeta,\theta,P}(t))_{t\geqslant 0}$ of angle $\min((\frac{1}{\zeta} - 1)\frac{\pi}{2}, \frac{\pi}{2})$. By the above arguments, we easily infer that $(R_{\zeta,\theta,P}(t))_{t\geqslant 0}$ is topologically mixing. Notice that (559) holds if $c < \frac{b^2}{4a}$; in the case $c \geqslant \frac{b^2}{4a}$, one can prove that (559) holds provided $a_0 = 0$ or $P(z) = \Sigma_{j=0}^{n} a_j(z + d)^j$, $z \in \mathbb{C}$, where $d \in \mathbb{C}$ and $0 \in \operatorname{int}(d - \Lambda)$. Consider now the equation (2) with $n = 2$, $\alpha_2 = 2$, $\alpha_1 = 0$, $\alpha = 1$, $c_1 > 0$, and A replaced by $-e^{i\theta} P(A)$ therein, where $P(z) = \Sigma_{j=0}^{n} a_j z^j$ is a non-constant complex polynomial such that $a_n > 0$. Using Lemma 3.3.6(1), one gets that $-e^{i\theta} P(A)$ is the integral generator of an exponentially bounded, analytic resolvent propagation family $((R_0(t))_{t\geqslant 0}, (R_1(t))_{t\geqslant 0})$ of angle $\frac{\pi}{2} - |\theta|$. Moreover, $F_0(\lambda, t) = (\lambda^2 + \lambda)(\lambda^2 - c_1)^{-1}e^{\lambda t} - (c_1 + \lambda)(\lambda^2 - c_1)^{-1}e^{c_1 t/\lambda}$, $t \geqslant 0$. By Theorem 3.3.9, we easily infer that the condition

$$(560) \qquad e^{i\theta} P(-\Lambda) \cap i\mathbb{R} \neq \emptyset$$

implies that $(R_0(t))_{t\geqslant 0}$ is topologically mixing. Finally, suppose that $n = 2$, $\alpha_2 - \alpha = 1$, $\alpha_1 - \alpha = -1$, $i = 1$, $c_1 > 0$ and $2 < \alpha_2 \leqslant 3$. Then $m_2 = 3$, $\mathcal{D}_1 = \emptyset$ and $F_1(\lambda, t) = \lambda^{-1}(1 + c_1(\lambda^2 - c_1)^{-1})e^{\lambda t} - \lambda(\lambda^2 - c_1)^{-1}e^{c_1 t/\lambda}$, $t \geqslant 0$. By Lemma 3.3.6(1), we get that $-e^{i\theta}P(A)$ is the integral generator of an exponentially bounded, analytic resolvent propagation family $((R_0(t))_{t\geqslant 0}, (R_1(t))_{t\geqslant 0}, (R_2(t))_{t\geqslant 0})$ of angle $\frac{\pi}{2} - |\theta|$. If the condition (560) is satisfied, then one can apply Theorem 3.3.9 in order to see that $(R_1(t))_{t\geqslant 0}$ is topologically mixing.

(ii) ([237], [315]) Theorem 3.3.9 can be applied in the analysis of (subspace) topologically mixing properties of time-fractional wave equation and time-fractional heat equation on symmetric spaces of non-compact type (cf. [237, Theorem 3.1(a), Theorem 3.2, Corollary 3.3]); here we shall also provide some applications of the aforementioned theorem to the abstract Cauchy problem (2). Consider, for example, the situation of [237, Theorem 3.1(a)]. Let X be a symmetric space of non-compact type and rank one, let $p > 2$, let the parabolic domain P_p and the positive real number c_p possess the same meaning as in

[237], and let $P(z) = \Sigma_{j=0}^{n} a_j z^j$, $z \in \mathbb{C}$ be a non-constant complex polynomial with $a_n > 0$. Assume first $\zeta \in (1, 2)$, $\pi - n \arctan \frac{|p-2|}{2\sqrt{p-1}} - \zeta\frac{\pi}{2} > 0$ and

$$\theta \in \left(n \arctan\frac{|p-2|}{2\sqrt{p-1}} + \zeta\frac{\pi}{2} - \pi,\ \pi - n \arctan\frac{|p-2|}{2\sqrt{p-1}} - \zeta\frac{\pi}{2}\right).$$

Then it can be simply proved that $-e^{i\theta} P(\Delta_{X,p}^{\natural})$ is the integral generator of an exponentially bounded, analytic ζ-times regularized resolvent family $(R_{\zeta,\theta,P}(t))_{t\geqslant 0}$ of angle $\frac{1}{\zeta}(\pi - n \arctan \frac{|p-2|}{2\sqrt{p-1}} - \zeta\frac{\pi}{2} - |\theta|)$. Keeping in mind that $\operatorname{int}(P_p) \subseteq \sigma_p(\Delta_{X,p}^{\natural})$, the condition

$$-e^{i\theta} P(\operatorname{int}(P_p)) \cap \{te^{\pm i\zeta\frac{\pi}{2}} : t \geqslant 0\} \neq \emptyset \tag{561}$$

implies that $(R_{\zeta,\theta,P}(t))_{t\geqslant 0}$ is topologically mixing. Suppose now $n = 2$, $0 < a < 2$, $\alpha_2 = 2a$, $\alpha_1 = 0$, $\alpha = a$, $c_1 > 0$, $i = 0$ and $|\theta| < \min(\frac{\pi}{2} - n \arctan \frac{|p-2|}{2\sqrt{p-1}}, \frac{\pi}{2} - n \arctan \frac{|p-2|}{2\sqrt{p-1}} - \frac{\pi}{2}a)$. Then $\mathcal{D}_0 = \{0\}$ and, by Lemma 3.3.6(i), $-e^{i\theta} P(\Delta_{X,p}^{\natural})$ is the integral generator of an exponentially bounded, analytic resolvent propagation family $((R_{\theta,P,0}(t))_{t\geqslant 0}, \cdots, (R_{\theta,P,\lceil 2a\rceil -1}(t))_{t\geqslant 0})$ of angle $\min(\frac{\pi - n \arctan \frac{|p-2|}{2\sqrt{p-1}} - |\theta|}{a} - \frac{\pi}{2}, \frac{\pi}{2})$. Furthermore, the equality (20) can serve one to simply verify that:

$$\begin{aligned}
F_0(\lambda, t) = {} & \frac{\lambda^a t^{-a}}{\lambda^{2a} - c_1}(E_{a,2-a}(\lambda^a t^a) - E_{a,2-a}(c_1\lambda^{-a}t^a)) \\
& + \frac{\lambda^a}{\lambda^{2a} - c_1}[\lambda^a E_a(\lambda^a t^a) + (a-1)\lambda^a E_{a,2}(\lambda^a t^a) \\
& - c_1\lambda^{-a}E_a(c_1\lambda^{-a}t^a) - (a-1)c_1\lambda^{-a}E_{a,2}(c_1\lambda^{-a}t^a)] \\
& + (\lambda^a + c_1\lambda^{-a})\frac{\lambda^a}{\lambda^{2a} - c_1}[E_a(\lambda^a t^a) - E_a(c_1\lambda^{-a}t^a)],\ t > 0.
\end{aligned}$$

Invoking the asymptotic expansion formulae (17)-(19) and the above expression, it can be shown that the condition

$$-e^{i\theta} P(\operatorname{int}(P_p)) \cap \{(it)^a + c_1(it)^{-a} : t \in \mathbb{R}\setminus\{0\}\} \neq \emptyset$$

implies that $(R_{\theta,P,0}(t))_{t\geqslant 0}$ is topologically mixing. Finally, let $\zeta \in (0, 1)$ and

$$\theta \in \left(n \arctan\frac{|p-2|}{2\sqrt{p-1}} - \frac{\pi}{2}, \frac{\pi}{2} - n \arctan\frac{|p-2|}{2\sqrt{p-1}}\right).$$

Then the validity of (561) provides that $-e^{i\theta} P(\Delta_{X,p}^{\natural})$ is the integral generator of a topologically mixing ζ-times regularized resolvent family $(R_{\zeta,\theta,P}(t))_{t\geqslant 0}$ of angle $\min((\frac{1}{\zeta} - 1)\frac{\pi}{2}, \frac{\pi}{2})$. It is clear that (561) holds if $P(z)$ is of the form $P(z) = \Sigma_{j=0}^{n} a_j(z - c)^j$, $z \in \mathbb{C}$, where $c > c_p$.

(iii) ([117], [413], [315]) Suppose $\zeta \in (0, 1)$, $E := L^2(\mathbb{R})$, $c > \frac{b}{2} > 0$, $\Omega := \{\lambda \in \mathbb{C} : \operatorname{Re} \lambda < c - \frac{b}{2}\}$, $\phi \in E^* = E$ and $\mathcal{A}_c u := u'' + 2bxu' + cu$ is the bounded perturbation of the one-dimensional Ornstein-Uhlenbeck operator acting with domain $D(\mathcal{A}_c) := \{u \in L^2(\mathbb{R}) \cap W^{2,2}_{loc}(\mathbb{R}) : \mathcal{A}_c u \in L^2(\mathbb{R})\}$. Then $\mathcal{A}_c$ is the integral generator of a topologically mixing ζ-times regularized resolvent family $(R_\zeta(t))_{t\geqslant 0}$ which cannot be hypercyclic provided $b < 0$ or $c \leqslant \frac{b}{2}$ ([117], [315]). Notice also that the above assertions continue to hold in the case of ζ-times regularized resolvent families generated by bounded perturbations of multi-dimensional Ornstein-Uhlenbeck operators [117, Proposition 4.1, Theorem 4.2]; for the sake of simplicity, in the sequel of this example we shall consider only the hypercyclic and topologically mixing properties of resolvent propagation families generated by the operator $\mathcal{A}_c$ defined above. Suppose $\alpha_n - \alpha < 1$. Then an application of Lemma 3.3.6(2) shows that $\mathcal{A}_c$ is the integral generator of an exponentially bounded, analytic resolvent propagation family $((R_0(t))_{t\geqslant 0}, \cdots, (R_{m_n-1}(t))_{t\geqslant 0})$ of angle $\min(\frac{\pi}{2(\alpha_n-\alpha)} - \frac{\pi}{2}, \frac{\pi}{2})$. If $b < 0$, then $\sigma_p(\mathcal{A}_c^*) \neq \emptyset$ (cf. [117]) and, by Theorem 3.3.11, there does not exist $i \in \mathbb{N}^0_{m_n-1}$ such that $(R_i(t))_{t\geqslant 0}$ is hypercyclic (the case $c \leqslant \frac{b}{2}$ is more complicated in the newly arisen situation since it is not clear how one can prove the boundedness of $(R_i(t))_{t\geqslant 0}$, in general). Consider now the following case: $n = 3$, $\frac{1}{3} < a < \frac{1}{2}$, $\alpha_3 = 3a$, $\alpha_2 = 2a$, $\alpha_1 = 0$, $\alpha = a$, $c_1 < 0$, $c_2 > 0$ and $i = 1$. Then $\mathcal{D}_1 = \emptyset$ and

$$\mathcal{L}(F_1(\lambda, t))(z) = \frac{z^{3a-2}}{z^{3a} + c_2 z^{2a} - z^a(\lambda^{2a} + c_1\lambda^{-a} + c_2\lambda^a) + c_1}.$$

Set $\lambda_{1,2} := \frac{-c_2 - \lambda^a \pm \sqrt{(c_2+\lambda^a)^2 + 4c_1\lambda^{-a}}}{2}$. Then one can simply prove that the set $\Upsilon = \{\lambda \in \mathbb{C} : (\lambda^a - \lambda_1)(\lambda^a - \lambda_2)(\lambda_1 - \lambda_2) \neq 0\}$ is finite and that, for every $z \in \mathbb{C}\setminus\{0\}$ and $\lambda \in \mathbb{C}\setminus\Upsilon$,

$$z^{3a} + c_2 z^{2a} - z^a(\lambda^{2a} + c_1\lambda^{-a} + c_2\lambda^a) + c_1 = (z^a - \lambda^a)(z^a - \lambda_1)\,(z^a - \lambda_2).$$

Then the equality (20) implies that, for every $\lambda \in \mathbb{C}\setminus\Upsilon$,

$$\text{(562)} \qquad F_1(\lambda, t) = \frac{t^{1-2a}E_{a,2-2a}(\lambda^a t^a)}{(\lambda^a - \lambda_1)(\lambda^a - \lambda_2)} + \frac{t^{1-2a}E_{a,2-2a}(\lambda_1 t^a)}{(\lambda_1 - \lambda_2)(\lambda_1 - \lambda^a)} + \frac{t^{1-2a}E_{a,2-2a}(\lambda_2 t^a)}{(\lambda_2 - \lambda_1)(\lambda_2 - \lambda^a)}.$$

Clearly, $P_\lambda = \lambda^{2a} + c_2\lambda^a + c_1\lambda^{-a}$, $\lambda \in \mathbb{C}\setminus\{0\}$, $\lim_{\lambda\to 0}(\lambda_1 - (-\frac{c_2}{2} + \sqrt{c_1\lambda^{-a}})) = 0$ and $\lim_{\lambda\to 0}(\lambda_2 - (-\frac{c_2}{2} - \sqrt{c_1\lambda^{-a}})) = 0$. This implies that there exists a sufficiently small number $\varepsilon_1 > 0$ such that, for every $\lambda \in \mathbb{C}$ with $\operatorname{Re}\lambda > 0$ and $|\lambda| \leqslant \varepsilon_1$, the following holds: $\operatorname{Re}\lambda_2 \leqslant -\frac{c_2}{4}$ and

$$\text{(563)} \qquad \operatorname{dist}\Big(\lambda_1, \Big\{z \in \mathbb{C} : \arg\Big(z + \frac{c_2}{2}\Big) \in \Big[\frac{\pi}{2} - \frac{\pi a}{4}, \frac{\pi}{2}\Big]\Big\}\Big) < \min\Big(\frac{c_2}{4}, \frac{c_2}{2}\cot\frac{\pi a}{4}\Big).$$

Arguing similarly, we obtain that there exists a sufficiently small number $\varepsilon_2 > 0$ such that, for every $\lambda \in \mathbb{C}$ with $\arg(\lambda) \in (\frac{\pi}{2}, \frac{\pi}{2a})$ and $|\lambda| \leqslant \varepsilon_2$, the following holds: $\operatorname{Re} \lambda_2 \leqslant -\frac{c_2}{4}$ and

(564) $$\operatorname{dist}\left(\lambda_1, \left\{z \in \mathbb{C} : \arg\left(z + \frac{c_2}{2}\right) \in \left[\frac{\pi}{4}, \frac{\pi}{2} - \frac{\pi a}{4}\right]\right\}\right) < \frac{c_2}{4}.$$

Furthermore, our assumption $c_1 < 0$ implies that there exists a sufficiently small number $\varepsilon_3 > 0$ such that, for every $\lambda \in \mathbb{C}\setminus\{0\}$ with $|\arg(\lambda)| \leqslant \frac{\pi}{2a}$ and $|\lambda| \leqslant \varepsilon_3$, we have $\operatorname{Re}(P_\lambda) = \operatorname{Re}(\lambda^{2a} + c_2\lambda^a + c_1\lambda^{-a}) \leqslant \varepsilon_3^{2a} + |c_2|\varepsilon_3^a < c - \frac{b}{2}$. Let $\varepsilon_4 > 0$ satisfy that, for every $\lambda \in \mathbb{C}\setminus\{0\}$ and $|\lambda| \leqslant \varepsilon_4$, one has $\lambda \in \Upsilon$. Put $\varepsilon := \min(\varepsilon_1, \varepsilon_2, \varepsilon_3, \varepsilon_4)$, $\Omega := \Omega_1 := \Omega_2 := \{z \in \mathbb{C}\setminus\{0\} : |\arg(z)| \leqslant \frac{\pi}{2a}, |z| < \varepsilon\}$, $\Omega_+ := \Omega_{1,+} := \Omega_{2,+} := \{z \in \mathbb{C} : \operatorname{Re} z > 0, |z| < \varepsilon\}$ and $\Omega_- := \Omega_{1,-} := \Omega_{2,-} := \{z \in \mathbb{C}\setminus\{0\} : \arg(z) \in (\frac{\pi}{2}, \frac{\pi}{2a}), |z| < \varepsilon\}$. Then it is obvious that $P_\Omega \subseteq \sigma_p(\mathcal{A}_c)$. Define $f_1 : P_\Omega \to E$ and $f_2 : P_\Omega \to E$ by $f_1(z) := \mathcal{F}^{-1}(e^{-\frac{\xi^2}{2b}} \xi|\xi|^{-(2+\frac{z-c}{b})})(\cdot)$, $z \in P_\Omega$ and $f_2(z) := \mathcal{F}^{-1}(e^{-\frac{\xi^2}{2b}} |\xi|^{-(1+\frac{z-c}{b})})(\cdot)$, $z \in P_\Omega$. Exploiting (562)-(564) and (17)-(19), we easily infer that:

$$\lim_{t\to+\infty} |F_1(\lambda, t)| = +\infty,\ \lambda \in \Omega_+ \text{ and } \lim_{t\to+\infty} F_1(\lambda, t) = 0,\ \lambda \in \Omega_-.$$

By Remark 3.3.10(iii) and the consideration given in [315, Example 2.5(iii)], we reveal that $(R_1(t))_{t\geqslant 0}$ is topologically mixing. On the other hand, by performing the Laplace transform it readily follows from Theorem 2.1.5 that the operator $\mathcal{A}_c$ is the integral generator of an exponentially bounded (a, k)-regularized resolvent family $(S(t))_{t\geqslant 0}$ with

$$a(t) := \mathcal{L}^{-1}\left(\frac{\lambda^a}{\lambda^{3a} + c_2\lambda^{2a} + c_1}\right)(t),\ t \geqslant 0$$

and

$$k(t) := \mathcal{L}^{-1}\left(\frac{\lambda^{3a-2}}{\lambda^{3a} + c_2\lambda^{2a} + c_1}\right)(t),\ t \geqslant 0.$$

Therefore, the abstract Volterra equation

$$u(t) = k(t)\cdot + \int_0^t a(t-s)\mathcal{A}_c u(s)\, ds,\ t \geqslant 0,$$

possesses certain topologically mixing properties.

(iv) Our intention here is to clarify the most striking facts about hypercyclic and topologically mixing properties of once integrated solutions of the abstract Basset-Boussinesq-Oseen equation (3) with $f(t) \equiv 0$. Clearly, $n = 2$, $\alpha_2 = 1$, $\alpha_1 = 0$, $c_1 = 1$, $\mathcal{D}_0 = \emptyset$ and the analysis is quite complicated in the general case since

$$\mathcal{L}(F_0(\lambda, t))(z) = \frac{1 - (\lambda^{1-\alpha} + \lambda^{-\alpha})z^{\alpha-1}}{z + 1 - z^\alpha(\lambda^{1-\alpha} + \lambda^{-\alpha})}.$$

The cases $\alpha = \frac{1}{2}$ and $\alpha = \frac{1}{3}$ can be considered similarly as in the previous examples. Suppose now $\alpha = \frac{2}{3}$, $A \equiv \mathcal{A}_c$ and $c - \frac{b}{2} > 2^{1/3} + 2^{2/3}$. Then $\mathcal{A}_c$ is the integral generator of an exponentially bounded, analytic resolvent propagation family $(R_0(t))_{t\geqslant 0}$ of angle $\frac{\pi}{2}$. Put $\lambda_{1,2} := \frac{-\lambda^{1/3} \pm \sqrt{\lambda^{2/3} + 4\lambda^{(-1)/3}}}{2}$. Then the sets $\Upsilon_1 := \{\lambda \in \mathbb{C}\setminus\{0, -4\} : (\lambda - \lambda_1)(\lambda - \lambda_2) = 0\}$ and $\Upsilon_2 := \{\lambda \in \mathbb{C}\setminus\{0\} : \operatorname{Re}\lambda = \operatorname{Re}(\lambda_1^3)\}$ are finite. Furthermore, for every $\lambda \in \mathbb{C}\setminus((-\infty, 0] \cup \Upsilon_1)$, we have:

$$
\begin{aligned}
F_0(\lambda, t) = & \frac{E_{1/3,1/3}(\lambda^{1/3}t^{1/3})}{(\lambda^{1/3} - \lambda_1)(\lambda^{1/3} - \lambda_2)} \\
& + \frac{E_{1/3,1/3}(\lambda_1 t^{1/3})}{(\lambda_1 - \lambda^{1/3})(\lambda_1 - \lambda_2)} + \frac{E_{1/3,1/3}(\lambda_2 t^{1/3})}{(\lambda_2 - \lambda^{1/3})(\lambda_2 - \lambda_1)} \\
& - (\lambda^{1/3} + \lambda^{(-2)/3}) \Bigg[\frac{E_{1/3,2/3}(\lambda^{1/3}t^{1/3})}{t^{1/3}(\lambda^{1/3} - \lambda_1)(\lambda^{1/3} - \lambda_2)} \\
& + \frac{E_{1/3,2/3}(\lambda_1 t^{1/3})}{t^{1/3}(\lambda_1 - \lambda^{1/3})(\lambda_1 - \lambda_2)} + \frac{E_{1/3,2/3}(\lambda_2 t^{1/3})}{t^{1/3}(\lambda_2 - \lambda^{1/3})(\lambda_2 - \lambda_1)} \Bigg].
\end{aligned}
$$

Since the function $s \mapsto s^{1/3} + s^{(-2)/3}$, $s > 0$ attains its global minimum $2^{1/3} + 2^{2/3}$ for $s = 2$, we obtain that there exist positive real numbers ε_1 and ε_2 such that $\varepsilon_1 < 2 < \varepsilon_2$ and $\operatorname{Re}(P_\lambda) = \operatorname{Re}(\lambda^{1/3} + \lambda^{(-2)/3}) < c - \frac{b}{2}$, provided $\varepsilon_1 < |\lambda| < \varepsilon_2$. Set $\Omega := \Omega_1 := \Omega_2 := \{\lambda \in \mathbb{C} : \varepsilon_1 < |\lambda| < \varepsilon_2\}$ and $\Omega_+ := \Omega_{1,+} := \Omega_{2,+} := \{\lambda \in \mathbb{C} : \operatorname{Re}\lambda > 0, \varepsilon_1 < |\lambda| < \varepsilon_2, \lambda \notin \Upsilon_2\}$. It is clear that $\operatorname{Re}\lambda_2 < 0$ for $\lambda \in \mathbb{C}\setminus\{0\}$, and that $\lim_{\lambda\to -2, \operatorname{Im}\lambda > 0} \lambda_1 = \lim_{\lambda\to -2, \operatorname{Im}\lambda > 0} \frac{-\lambda^{1/3} + \sqrt{\lambda^{2/3} + 4\lambda^{(-1)/3}}}{2} = \frac{-(-2)^{1/3} + \sqrt{(-2)^{2/3} + 4(-2)^{(-1)/3}}}{2}$. Direct calculation shows that the argument of the last written number belongs to the set $(-\frac{2\pi}{3}, -\frac{\pi}{6})$, which implies that there exists a sufficiently small number $\varepsilon > 0$ such that the set $\Omega_- := \Omega_{1,-} := \Omega_{2,-} := \{\lambda \in \mathbb{C} : \operatorname{Im}\lambda > 0, |\lambda + 2| < \varepsilon\}$ is a subset of Ω, and that $\arg(\lambda_1) \in (-\frac{2\pi}{3}, -\frac{\pi}{6})$ for $\lambda \in \Omega_-$. Owing to (17)-(19), we obtain that $(R_0(t))_{t\geqslant 0}$ is topologically mixing.

(v) ([150]) Let B, ω_1, ω_2, V_{ω_2,ω_1}, E, a and b possess the same meaning as in [150, Section 5] and let $Q(z)$ be a non-constant complex polynomial of degree n. Assume $0 < \alpha < 2$, $N \in \mathbb{N}$, $N > \frac{n}{2\alpha}$ and

$$R_\alpha(t) = \left(E_\alpha(t^\alpha Q(z))e^{-(-z^2)^N}\right)(B),\ t \geqslant 0. \tag{565}$$

Then the operator $(e^{-(-z^2)^N})(B)$ has a dense range, and it is not difficult to prove that $(R_\alpha(t))_{t\geqslant 0}$ is an α-times $(e^{-(-z^2)^N})(B)$-regularized resolvent family generated by $Q(B)$, and that the condition

$$Q(\operatorname{int}(V_{\omega_2,\omega_1})) \cap \{te^{\pm i\alpha\frac{\pi}{2}} : t \geqslant 0\} \neq \emptyset$$

implies that $(R_\alpha(t))_{t\geqslant 0}$ is both topologically mixing and hypercyclic.

3.3.3. Topological dynamics of certain classes of abstract time-fractional PDEs with unilateral backward shifts. The blank hypothesis in the previous subsection was that there exist complex constants $c_1, \cdots, c_{n-1}$ such that, for every $j \in \mathbb{N}_{n-1}$, the operator A_j satisfies the equality $A_j = c_j I$. In what follows, we shall consider topologically mixing solutions of the equation (2) with $A_0, A_1, \cdots, A_{n-1}$ being functions of unilateral backward shift operators. Here we would like to observe that various types of hypercyclic and topologically mixing properties of backward shift operators on Banach or Fréchet sequence spaces have been widely studied (cf. [84], [161], [184], [193], [475]-[476] and [486] for further information in this direction). We shall work only with non-degenerate operator families. Recall that the Wright function $\Phi_\gamma(t)$ can be extended to an entire function, and that there exists a finite constant $M > 0$ such that $0 \leqslant \Phi_\gamma(t) \leqslant M$, $t \geqslant 0$.

By E we denote a separable infinite-dimensional Banach space over the field of complex numbers. We use the hypotheses that $A_0, A_1, \cdots, A_{n-1}$ are bounded linear operators acting on E as well as that $n \in \mathbb{N} \setminus \{1\}$, $0 \leqslant \alpha_1 < \cdots < \alpha_n$, $0 \leqslant \alpha < \alpha_n$, $m_j = \lceil \alpha_j \rceil$, $1 \leqslant j \leqslant n$, $m = m_0 = \lceil \alpha \rceil$, $\alpha_0 = \alpha$, and that there exists a global resolvent propagation family $((R_0(t))_{t\geqslant 0}, \cdots, (R_{m_n-1}(t))_{t\geqslant 0})$ for (2). Then we know that the unique mild solution of (341) is given by $u(t) = \Sigma_{i=0}^{m_n-1} R_i(t)x_i$, $t \geqslant 0$. The definitions of $\tilde{E}$-hypercyclicity, $\tilde{E}$-topological transitivity and $\tilde{E}$-topological mixing of $(R_i(t))_{t\geqslant 0}$ will be understood in the sense of Definition 3.3.8 (with $Z_i(A) = E$); observe also that the notions of $\tilde{E}$-hypercyclicity and $\tilde{E}$-topological transitivity (mixing) can be introduced for an arbitrary strongly continuous operator family $(R(t))_{t\geqslant 0} \subseteq L(E)$. In the sequel, we shall always assume that $\tilde{E} = E$; then the topological transitivity of $(R_i(t))_{t\geqslant 0}$ for some $i \in \mathbb{N}^0_{m_n-1}$ implies that $(R_i(t))_{t\geqslant 0}$ is hypercyclic and that the set of all hypercyclic vectors of $(R_i(t))_{t\geqslant 0}$ is a dense G_δ-subset of E ([313]).

Let $\zeta > 0$, and let $(r_k)_{k\in\mathbb{N}}$ be a sequence of positive real numbers satisfying that there exists $M > 0$ such that $r_k r_{k+1}^{-1} \leqslant M$ for all $k \in \mathbb{N}$. Consider the weighted l^1-space

$$l^1_r := \left\{ (x_k)_{k\in\mathbb{N}} : x_k \in \mathbb{C}, \sum_{k=1}^{\infty} r_k |x_k| < \infty \right\},$$

normed by

$$\left\| (x_k)_{k\in\mathbb{N}} \right\| := \sum_{k=1}^{\infty} r_k |x_k|, \quad (x_k)_{k\in\mathbb{N}} \in l^1_r.$$

Define now the unilateral backward shift $A : l^1_r \to l^1_r$ by $A(x_k)_{k\in\mathbb{N}} := (x_{k+1})_{k\in\mathbb{N}}$, $(x_k)_{k\in\mathbb{N}} \in l^1_r$. Clearly, $A \in L(l^1_r)$ and the norm of A can be majorized by the constant M mentioned above. Recall that H. R. Salas [475] has proved that the operator $I + A$ is hypercyclic. Details of his proof have been essentially used by W. Desch, W. Schappacher and G. F. Webb [161], where it has been shown that the strongly continuous semigroup $(T(t))_{t\geqslant 0}$, generated by A, is hypercyclic. Observe further that [49, Theorem 2.5] and its proof imply that the operator A is the integral generator of a global exponentially bounded ζ-times regularized resolvent family

$$\left(R^{\zeta}(t) \equiv \sum_{k=0}^{\infty} \frac{t^{\zeta k}}{\Gamma(\zeta k+1)} A^k\right)_{t \geqslant 0}.$$

A slight modification of the arguments given in the proof of [161, Theorem 5.2] implies that the following theorem holds.

Theorem 3.3.13. *Let $\zeta > 0$, and let A be defined as above. Denote by $(R^{\zeta}(t))_{t \geqslant 0}$ the ζ-times regularized resolvent family generated by A. Then $(R^{\zeta}(t))_{t \geqslant 0}$ is topologically mixing.*

The importance of Theorem 3.3.13 lies in the fact that, for any arbitrarily large finite number $\zeta > 0$, we have the existence of a topologically mixing ζ-times regularized resolvent family on a Banach space, here concretely on l^1_r.

Now we would like to mention the following problem connected with the existence of topologically mixing solutions of the abstract Cauchy problem (ACP_n) : if there exist at least two indices $i, j \in \mathbb{N}^0_{n-1}$ such that the operators A_i and A_j are not scalar multiples of the identity operator, then we would be unable to find in the existing literature an example of the abstract Cauchy problem (ACP_n) with topologically mixing solution $u(t)$. The main goal of the following theorem is to show that, for every $n \geqslant 2$, there exists an example of the abstract Cauchy problem (ACP_n) with such properties. In order to help one to better understand the proof, we will consider separately the cases $n = 2$ and $n > 2$.

Theorem 3.3.14. *Let $A \in L(l^1_r)$ be as in the formulation of Theorem 3.3.13, and let $(T(t))_{t \geqslant 0}$ be the strongly continuous semigroup generated by A.*

(i) *Consider the abstract Cauchy problem*

$$(P_2): \begin{cases} u''(t) - (2A - I)u'(t) + A(A - I)u(t) = 0, \; t \geqslant 0, \\ u(0) = x, \; u'(0) = y. \end{cases}$$

Then there exists a resolvent propagation family $((R_0(t))_{t \geqslant 0}, (R_1(t))_{t \geqslant 0})$ for (P_2), given by $R_0(t)x = T(t)(x - Ax) + e^{-t}T(t)Ax$ and $R_1(t)x = T(t)(1 - e^{-t})x$ ($t \geqslant 0$, $x \in l^1_r$). Furthermore, $(R_0(t))_{t \geqslant 0}$ and $(R_1(t))_{t \geqslant 0}$ are topologically mixing, and the operator family $(R_0(t) + R_1(t))_{t \geqslant 0}$ is also topologically mixing.

(ii) *Suppose $n > 2$ and $0 < c_1 < \cdots < c_{n-1}$. Consider the abstract Cauchy problem*

$$(P_n): \begin{cases} \prod_{i=0}^{n-1}\left(\frac{d}{dt} - (-c_i + A)\right)u(t) = 0, \; t \geqslant 0, \\ u^{(k)}(0) = x_k, \; k = 0, \cdots, n-1, \end{cases}$$

with $c_0 = 0$. Then there exists a global exponentially bounded resolvent propagation family $((R_0(t))_{t \geqslant 0}, \cdots, (R_{n-1}(t))_{t \geqslant 0})$ for (P_n). Furthermore, $(R_i(t))_{t \geqslant 0}$ is topologically mixing for any $i \in \mathbb{N}^0_{n-1}$, and the operator family $(R_0(t) + \cdots + R_{n-1}(t))_{t \geqslant 0}$ is also topologically mixing.

Proof. Notice that the problem (P_2) is a special case of the problem (P_n) with $n = 2$ and $c_1 = 1$, so that the first statement in (i) is an almost immediate consequence of [141, Theorem 25.6]. Suppose now that $y = (y_k)_{k\in\mathbb{N}}$ and $z = (z_k)_{k\in\mathbb{N}}$ belong to the dense subset

$$D := \{(x_k)_{k\in\mathbb{N}} : \exists L \in \mathbb{N}\ \forall k > L\ x_k = 0\}$$

of l^1_r. Let $y_k = z_k = 0$ for $k > L$. For any sufficiently large number $t > 0$, we will construct the vector $\upsilon(t) = (\upsilon_k(t))_{k\in\mathbb{N}} \in l^1_r$ such that
(566)

$$\|y - \upsilon(t)\| = O(t^{-1}) \text{ as } t\to+\infty, \text{ and } \|z - R_0(t)\upsilon(t)\| = O(t^{-1}) \text{ as } t\to+\infty.$$

Towards this end, observe that

$$T(t)(x_k)_{k\in\mathbb{N}} = \left(\sum_{j=k}^{\infty}\frac{t^{j-k}}{(j-k)!}x_j\right)_{k\in\mathbb{N}}, \ t \geqslant 0,\ (x_k)_{k\in\mathbb{N}} \in l^1_r,$$

and
(567)

$$R_0(t)(x_k)_{k\in\mathbb{N}} = \left(\sum_{j=k}^{\infty}\frac{t^{j-k}}{(j-k)!}(x_j + (e^{-t} - 1)x_{j+1})\right)_{k\in\mathbb{N}}, \ t \geqslant 0,\ (x_k)_{k\in\mathbb{N}} \in l^1_r.$$

Define $\upsilon_k(t) := y_k$ for $1 \leqslant k \leqslant L$, and $\upsilon_k(t) := 0$ for $L + 1 \leqslant k \leqslant 2L$ and $k \geqslant 3L + 1$. The numbers $\upsilon_{2L+1}(t), \cdots, \upsilon_{3L}(t)$ are defined as the unique solutions of system (cf. the first L elements of sequence appearing on the right hand side of (567), with x_j replaced by $\upsilon_j(t)$)

$$(S):\quad \begin{cases} \sum_{j=1}^{\infty}\frac{t^{j-1}}{(j-1)!}(\upsilon_j(t) + (e^{-t} - 1)\upsilon_{j+1}(t)) = z_1 \\ \sum_{j=2}^{\infty}\frac{t^{j-2}}{(j-2)!}(\upsilon_j(t) + (e^{-t} - 1)\upsilon_{j+1}(t)) = z_2 \\ \quad\cdots \\ \sum_{j=L}^{\infty}\frac{t^{j-L}}{(j-L)!}(\upsilon_j(t) + (e^{-t} - 1)\upsilon_{j+1}(t)) = z_L \end{cases}$$

i.e., $\upsilon_{2L+1}(t), \cdots, \upsilon_{3L}(t)$ satisfy the following matrix equality:

$$A(t)[\upsilon_{2L+1}(t) \cdots \upsilon_{3L}(t)]^T = [z_1 \cdots z_L]^T - B(t)[y_1 \cdots y_L]^T,$$

where

$$A(t) = [a_{ij}(t)]_{L\times L} = \left[(e^{-t} - 1)\frac{t^{2L-i+j-1}}{(2L-i+j-1)!} + \frac{t^{2L-i+j}}{(2L-i+j)!}\right]_{L\times L}$$

and $B(t) = [b_{ij}(t)]_{L\times L}$ with

$$b_{ij}(t) = \begin{cases} (e^{-t} - 1)\frac{t^{j-i-1}}{(j-i-1)!} + \frac{t^{j-i}}{(j-i)!}, \text{ for } j > i, \\ 1, \text{ for } i = j, \\ 0, \text{ for } i > j. \end{cases}$$

Notice that any element $a_{ij}(t)[b_{ij}(t)]$ of the matrix $A(t)$ $[B(t)]$ asymptotically behave as $t \to +\infty$ like the corresponding element of the matrix $\widetilde{A}(t)$ $[\widetilde{B}(t)]$, where

$$\widetilde{A}(t) = [\widetilde{a_{ij}}(t)]_{L \times L} = \left[\frac{t^{2L-i+j}}{(2L-i+j)!}\right]_{L \times L}$$

and $\widetilde{B}(t) = [\widetilde{b_{ij}}(t)]_{L \times L}$ with

$$\widetilde{b_{ij}}(t) = \begin{cases} \frac{t^{j-i}}{(j-i)!}, \text{ for } j > i, \\ 1, \text{ for } i = j, \\ 0, \text{ for } i > j. \end{cases}$$

The matrices $\widetilde{A}(t)$ and $\widetilde{B}(t)$ play an important role in the proof of [161, Theorem 5.2], which in combination with the above given arguments also shows that there exists an absolute constant $C_2 > 0$ such that, for every $k \in \{2L + 1, \cdots, 3L\}$,

$$|\upsilon_k(t)| \leqslant C_2 t^{L-k}. \quad (568)$$

Now it is not difficult to prove that (566) holds as well as that, for every $y_1, z_1 \in l_r^1$ and for every $\varepsilon > 0$, there exists $t_0 \geqslant 0$ such that, for every $t \geqslant t_0$, there exists $\upsilon_1(t) \in l_r^1$ such that $\|y_1 - \upsilon_1(t)\| < \varepsilon$ and $\|z_1 - R_0(t)\upsilon_1(t)\| < \varepsilon$. Hence, $(R_0(t))_{t \geqslant 0}$ is topologically mixing. The proof of topologically mixing property of $(R_1(t))_{t \geqslant 0}$ and $(R_0(t) + R_1(t))_{t \geqslant 0}$ is quite similar and as such will not be given. Consider now the assertion (ii). Denote by X the operator Van der Monde matrix $X = [x_{kl}]_{n \times n} = [(-c_{l-1} + A)^{k-1}]_{n \times n}$. Let $i \in \mathbb{N}_{n-1}^0$, let $x \in l_r^1$, and let

$$[y_{0,i}(x)\, y_{1,i}(x) \cdots y_{n-1,i}(x)]^T = X^{-1}[0 \cdots x \cdots 0]^T,$$

where x appears in the i-th place of the last vector column, starting from 0. The existence of a global exponentially bounded resolvent propagation family $((R_0(t))_{t \geqslant 0}, \cdots, (R_{n-1}(t))_{t \geqslant 0})$ for (P_n) follows again from an application of [141, Theorem 25.6]. This theorem yields that, for every $i \in \mathbb{N}_{n-1}^0$, we have:

$$R_i(t)x = \sum_{l=0}^{n-1} e^{-c_l t} T(t) y_{l,i}(x), \ t \geqslant 0, \ x \in l_r^1.$$

Using the analysis given on page 15 of [461] (cf. the problems 245-246), one can simply prove that there exist $m \in \mathbb{N}$ and complex polynomials $P_{l,i}(z) \equiv a_{l,i}^m z^m + \cdots + a_{l,i}^0$ $(0 \leqslant l \leqslant n-1, 0 \leqslant i \leqslant n-1)$ such that the following holds:

(a) $y_{l,i}(x) = P_{l,i}(A)x$ $(0 \leqslant l \leqslant n-1, 0 \leqslant i \leqslant n-1)$, where the operator $P_{l,i}(A)$ is defined in the obvious way,
(b) $a_{0,i}^0 \neq 0$ $(0 \leqslant i \leqslant n-1)$,
(c) $\Sigma_{i=0}^{n-1} a_{0,i}^0 \neq 0$.

This implies that, for every $x = (x_k)_{k \in \mathbb{N}} \in l_r^1$,

$$R_i(t)(x_k)_{k\in\mathbb{N}} = \Bigg(\sum_{j=k}\frac{t^{j-k}}{(j-k)!}\bigg\{[a^m_{0,i}x_{j+m}+\cdots+a^0_{0,i}x_j] + \sum_{l=1}^{n-1} e^{-c_l t}[a^m_{l,i}x_{j+m}+\cdots+a^0_{l,i}x_j]\bigg\}\Bigg)_{k\in\mathbb{N}}. \tag{569}$$

Suppose now that $y = (y_k)_{k\in\mathbb{N}}$ and $z = (z_k)_{k\in\mathbb{N}}$ belong to D, and $y_k = z_k = 0$ for $k > L$. Now we will construct the vector $w(t) = (w_k(t))_{k\in\mathbb{N}} \in l^1_r$ such that (566) holds with $R_0(\cdot)$ and $\upsilon(\cdot)$ replaced respectively with $R_i(\cdot)$ and $w(\cdot)$. The sequence $w(t)$ is defined by $w_k(t) := y_k$ for $1 \leqslant k \leqslant L$, and $w_k(t) := 0$ for $L+1 \leqslant k \leqslant 2L$ and $k \geqslant 3L+1$; similarly as in the first part of the proof, the numbers $w_{2L+1}(t), \cdots, w_{3L}(t)$ satisfy the following system of equations (cf. (569) and the first part of the proof):

$$(S'): \begin{cases} \sum_{j=1}\frac{t^{j-1}}{(j-1)!}\Big\{[a^m_{0,i}w_{j+m}(t)+\cdots+a^0_{0,i}w_j(t)] \\ \quad + \sum_{l=1}^{n-1} e^{-c_l t}[a^m_{l,i}w_{j+m}(t)+\cdots+a^0_{l,i}w_j(t)]\Big\} = z_1 \\ \sum_{j=2}\frac{t^{j-2}}{(j-2)!}\Big\{[a^m_{0,i}w_{j+m}(t)+\cdots+a^0_{0,i}w_j(t)] \\ \quad + \sum_{l=1}^{n-1} e^{-c_l t}[a^m_{l,i}w_{j+m}(t)+\cdots+a^0_{l,i}w_j(t)]\Big\} = z_2 \\ \cdots \\ \sum_{j=L}\frac{t^{j-L}}{(j-L)!}\Big\{[a^m_{0,i}w_{j+m}(t)+\cdots+a^0_{0,i}w_j(t)] \\ \quad + \sum_{l=1}^{n-1} e^{-c_l t}[a^m_{l,i}w_{j+m}(t)+\cdots+a^0_{l,i}w_j(t)]\Big\} = z_L. \end{cases}$$

It is clear that the matricial form of system (S') looks like:

$$A_1(t)[\upsilon_{2L+1}(t)\cdots\upsilon_{3L}(t)]^T = [z_1\cdots z_L]^T - B_1(t)[y_1\cdots y_L]^T,$$

where any element $a^1_{kl}(t)[b^1_{kl}(t)]$ of the matrix $A_1(t)$ $[B_1(t)]$ asymptotically behave as $t\to+\infty$ like $a^0_{0,i}a^{\sim}_{kl}(t)$ $[a^0_{0,i}b^{\sim}_{kl}(t)]$; cf. also (b). Arguing as in the proof of (i), we get that there exists $C_2 > 0$ such that, for every $k \in \{2L+1, \cdots, 3L\}$, the estimate (568) holds. Now one can simply prove that $(R_i(t))_{t\geqslant 0}$ is topologically mixing. Using (c) instead of (b), we obtain similarly that the operator family $(R_0(t)+\cdots+R_{n-1}(t))_{t\geqslant 0}$ is topologically mixing. □

Let A, l^1_r, $(T(t))_{t\geqslant 0}$ and $(R^{\zeta}(t))_{t\geqslant 0}$ be defined as before ($\zeta > 0$), let $n \geqslant 2$, and let $0 < c_1 < \cdots < c_{n-1}$. Then it is clear that the problem (P_n) is a special case of the abstract Cauchy problem (ACP_n), with the operators $A_0, A_1, \cdots, A_{n-1}$ being certain functions of A; for example, $A_{n-1} = \Sigma_{j=1}^{n-1} c_j I - nA$ and $A_0 = (-1)^n \Pi_{j=0}^{n-1}(-c_j + A)$. Consider now the problem

$$(P^{\gamma}_n): \begin{cases} \mathbf{D}^{n\gamma}_t u(t) + A_{n-1}\mathbf{D}^{(n-1)\gamma}_t u(t) + \cdots + A_1\mathbf{D}^{\gamma}_t u(t) + A_0 u(t) = 0,\ t > 0, \\ u^{(k)}(0) = u_k,\ k = 0, \cdots, \lceil n\gamma\rceil - 1, \end{cases}$$

where $\gamma \in (0, 1)$. Then it is not difficult to prove that there exists a global exponentially bounded resolvent propagation family $((R_0^\gamma(t))_{t\geqslant 0}, \cdots, (R_{\lceil n\gamma\rceil-1}^\gamma(t))_{t\geqslant 0})$ for (P_n^γ), given by

$$(570) \qquad R_i^\gamma(t)x = \mathcal{L}^{-1}\left(\left(\lambda^{n\gamma} + \sum_{j=0}^{n-1} \lambda^{j\gamma} A_j\right)^{-1}\left(\lambda^{n\gamma-i-1}x + \sum_{j\in D_i'} \lambda^{j\gamma-i-1} A_j x\right)\right)(t),$$

for any $t \geqslant 0$, $x \in l_r^1$ and $i \in \mathbb{N}^0_{\lceil n\gamma\rceil-1}$, where $D_i' = \{j \in \mathbb{N}^0_{n-1} : \lceil j\gamma\rceil - 1 \geqslant i\}$; speaking matter-of-factly, we have that, for every $t \geqslant 0$, $x \in l_r^1$ and $i \in \mathbb{N}^0_{\lceil n\gamma\rceil-1}$,

$$R_i^\gamma(t)x = g_i * \left[P_0(A)R_{c_0,\gamma}(\cdot)x + \cdots + P_{n-1}(A)R_{c_{n-1},\gamma}(\cdot)x\right](t)$$
$$+ \sum_{j\in D_i'} A_j\left\{g_{i+(n-j)\gamma} * \left[P_0(A)R_{c_0,\gamma}(\cdot)x + \cdots + P_{n-1}(A)R_{c_{n-1},\gamma}(\cdot)x\right](t)\right\},$$

where

$$P_j(A) = \prod_{\substack{0\leqslant l\leqslant n-1\\ j\neq l}} (-c_j + c_l)(-c_j + A)^{n-1}, j \in \mathbb{N}^0_{n-1},$$

and $(R_{c_j,\gamma}(t))_{t\geqslant 0}$ denotes the γ-times regularized resolvent family generated by $-c_j + A$ $(j \in \mathbb{N}^0_{n-1})$. By the foregoing, we have that:

$$(571) \qquad R_0^\gamma(t)x = t^{-\gamma}\int_0^\infty \Phi_\gamma(st^{-\gamma})R_0(s)x\, ds,\ t > 0,\ x \in l_r^1,$$

where $((R_0(t))_{t\geqslant 0}, \cdots, (R_{n-1}(t))_{t\geqslant 0})$ is the resolvent propagation family for (P_n). With the same notation as before, we obtain from (569)-(571) that, for every $x = (x_k)_{k\in\mathbb{N}} \in l_r^1$ and $t > 0$,

$$R_0^\gamma(t)(x_k)_{k\in\mathbb{N}} = t^{-\gamma}\left(\int_0^\infty \Phi_\gamma(st^{-\gamma})\left[\sum_{j=k}\frac{s^{j-k}}{(j-k)!}\left\{[a_{0,0}^m x_{j+m} + \cdots + a_{0,0}^0 x_j]\right.\right.\right.$$
$$\left.\left.\left.+ \sum_{l=1}^{n-1} e^{-c_l s}[a_{l,0}^m x_{j+m} + \cdots + a_{l,0}^0 x_j]\right\}\right] ds\right)_{k\in\mathbb{N}}.$$

Applying Theorem 2.4.2, we get that:

$$t^{-\gamma}\int_0^\infty \Phi_\gamma(st^{-\gamma})\frac{s^l}{l!}\, ds = \frac{t^{\gamma l}}{\Gamma(\gamma l+1)},\ t > 0,\ l \in \mathbb{N}_0,$$

which further implies that, for every $x = (x_k)_{k\in\mathbb{N}} \in l_r^1$ and $t > 0$,

$$R_0^\gamma(t)(x_k)_{k\in\mathbb{N}} = \left(\sum_{j=k}\frac{t^{\gamma(j-k)}}{\Gamma(\gamma(j-k)+1)}[a_{0,0}^m x_{j+m} + \cdots + a_{0,0}^0 x_j]\right.$$
$$\left.+ t^{-\gamma}\sum_{j=k}^\infty\sum_{l=1}^{n-1}\int_0^\infty \Phi_\gamma(st^{-\gamma})\frac{s^{j-k}}{(j-k)!}e^{-c_l s}[a_{l,0}^m x_{j+m} + \cdots + a_{l,0}^0 x_j]\, ds\right)_{k\in\mathbb{N}}.$$

Observe also that

$$t^{-\gamma}\int_0^{\infty}\Phi_{\gamma}(st^{-\gamma})\sum_{l=1}^{n-1}\frac{s^l}{l!}e^{-c_l s}\,ds \leqslant Mt^{-\gamma}\sum_{l=1}^{n-1}\int_0^{\infty}\frac{s^l}{l!}e^{-c_l s}\,ds \to 0 \text{ as } t\to+\infty,$$

and that, for every $t \geqslant 0$ and $(x_k)_{k\in\mathbb{N}} \in l_r^1$,

$$R^{\gamma}(t)(x_k)_{k\in\mathbb{N}} = \left(\sum_{j=k}^{\infty}\frac{t^{\gamma(j-k)}}{\Gamma(\gamma(j-k)+1)}x_j\right)_{k\in\mathbb{N}},\ t\geqslant 0,\ (x_k)_{k\in\mathbb{N}}\in l_r^1.$$

Proceeding as in the proofs of Theorem 3.3.13 (in this case, there exists $C_2 > 0$ such that, for every $k \in \{2L + 1, \cdots, 3L\}$, the corresponding vector $(v_k^{\gamma}(t))_{k\in\mathbb{N}} \in l_r^1$ satisfies $|v_k^{\gamma}(t)| \leqslant C_2 t^{\gamma(L-k)}$) and Theorem 3.3.14, we obtain that the following theorem is true.

Theorem 3.3.15. *The operator family* $(R_0^{\gamma}(t))_{t\geqslant 0}$ *is topologically mixing.*

Remark 3.3.16. (i) If $0 < \gamma \leqslant 1/n$, then it makes no sense to define $(R_i(t))_{t\geqslant 0}$ for $i \geqslant 1$; if this is not the case, then it is not clear whether there exists an index $i \in \mathbb{N}_{\lceil n\gamma\rceil - 1}$ such that the operator family $(R_i(t))_{t\geqslant 0}$ is topologically mixing.

(ii) Concerning the invariance of hypercyclic and topologically mixing properties under the action of subordination principles, it should be noted that the unilateral backward shifts have some advantages over other operators used in the theory of hypercyclicity (cf. Theorem 3.3.13, Theorem 3.3.15, Theorem 2.4.2 and Subsection 3.3.1 for further information in this direction).

We would like to propose the following problem.

Problem. Suppose $n \in \mathbb{N}\setminus\{1\}$, $0 \leqslant \alpha_1 < \cdots < \alpha_n$ and $0 \leqslant \alpha < \alpha_n$. Is it possible to construct a separable infinite-dimensional complex Banach space E and closed linear operators $A_0, A_1, \cdots, A_{n-1}$ on E such that there exists a global resolvent propagation family $((R_0(t))_{t\geqslant 0}, \cdots, (R_{m_n-1}(t))_{t\geqslant 0})$ for (2) satisfying that some (every) single operator family $(R_i(t))_{t\geqslant 0}$ of this tuple is topologically mixing?

ADDENDUM: ERRORS AND CORRECTIONS

Assume that (M_p) is a sequence of positive real numbers satisfying $M_0 = 1$, (M.1), (M.2) and (M.3)'; cf. Preliminaries. Let $M(\cdot)$ denote the associated function of sequence (M_p). It is worth noting that there exist some places in this book where we have used the following entire function of exponential type zero: $\omega_l(\lambda) = \Pi_{p=1}^{\infty}(1 + \frac{l\lambda}{m_p})$, $\lambda \in \mathbb{C}$, where $m_p = M_{p-1}/M_p$, $p \in \mathbb{N}$ $(l > 0)$. Recall that an old result of C. Roumieu says that $|\omega_l(\lambda)| \geq e^{M(l|\lambda|)}$, $\Re\lambda \geq 0$ (cf. [283, pp. 88-89], as well as [101] and [106] for some other estimates of this type). In a series of our recent papers, whose results have been included in the monograph [292] and this book, we have used the wrong estimate

$$(1) \qquad |\omega_l(\lambda)| \geq e^{M(l(1+\tan\alpha)^{-1}|\lambda|)},\ \alpha \in (0, \pi/2),\ l > 0,\ \lambda \in \Sigma_{\frac{\pi}{2}+\alpha};$$

see, e.g., [302, (36)], [292, (274)], [301, (1.8)] and the equation (122) contained in Example 2.4.6 (ii). It is checked at once that the validity of (1) implies $|\omega_l(\lambda)| \geq 1$, $\lambda \in \mathbb{C}$, which contradicts Liouville's theorem in a drastic manner. Furthermore, the following holds: Let $\alpha \in (0, \pi/2)$, let $l > 0$, and let a sequence (M_p) of positive real numbers satisfy $M_0 = 1$, (M.1), (M.2) and (M.3)'. Then we cannot find, in general, two finite constants $a_{\alpha,l} > 0$ and $b_{\alpha,l} > 0$ such that

$$(2) \qquad |\omega_l(\lambda)| \geq a_{\alpha,l} e^{M(b_{\alpha,l}|\lambda|)},\quad \lambda \in \Sigma_{\frac{\pi}{2}+\alpha}.$$

Consider, for example, the case in which $M_p = p!^s$, $s > 1$, $\alpha \in (0, \pi/2)$, $l > 0$ and

$$\frac{\pi}{2} + \alpha \geq \frac{\pi}{2}s.$$

Since $M(\lambda) \sim \lambda^{1/s}$ as $\lambda \to +\infty$, the validity of (2) would imply that the function

$$f_l(\lambda) := \frac{1}{\omega_l\left((\lambda + 1)^{\frac{(\pi/2+\alpha)}{\pi/2}}\right)},\quad \Re\lambda > -\frac{1}{2},$$

is analytic and

$$|f_l(\lambda)| = O\left(e^{-|\lambda|^{(\frac{\pi}{2}+\alpha)/(\frac{\pi}{2}s)}}\right),\quad \Re\lambda \geq 0.$$

This is in contradiction with some well-known consequences of the Phragmén-Lindelöf theorem (see, e.g., [297, Example 6.2(2)] and the page 40 of B. Ya. Levin's monograph: *Lectures on Entire Functions,* Translations of Mathematical Monographs, Volume 150, American Mathematical Society, 1996).

Although the inaccuracy of the estimate (1) inevitably causes some other inaccuracies from our analyses, we can simply debug the errors we have made and formulate the corresponding results in a proper way. The following is a list of corrections that should be made:

1. Example 2.4.6(ii), cf. also [304, Example 3.1(ii)]: The family $\{\lambda e^{-\zeta|\lambda|^{1/s}}(\lambda - A)^{-1} : \lambda \in \Sigma_{\pi-\epsilon}\} \subseteq L(E)$ is equicontinuous, provided $\epsilon \in (0, \pi)$, $s > 1$ and $\zeta > 0$. This implies that for each $s \in (1, 2)$ and $v > 0$ the operator A generates an exponentially equicontinuous, analytic $\mathcal{L}^{-1}(\exp(-v\lambda^{1/s}))$-convoluted semigroup $(R(t))_{t\geq 0}$ of angle $\pi/2$. In the final part of Example 2.4.6(ii), the operator A generates an exponentially equicontinuous, analytic $(a, \mathcal{L}^{-1}(\exp(-v\lambda^{1/s})))$-regularized resolvent family $(S(t))_{t\geq 0}$ of angle $\pi/2$ ($v > 0$, $1 < s < 1/\sigma$); $(R(t))_{t\geq 0}$ and $(S(t))_{t\geq 0}$ satisfy the properties stated in the above-mentioned example.
2. Example 2.10.34: The equation (122) has been unnecessary quoted here; the final conclusions are completely true.
3. The references [301]-[302]: The sentence directly before [301, Proposition 3.12] should be deleted. The final conclusion stated in Example after [301, Remark 2.23] should be stated as follows: For every $\alpha \in (0, \pi/2)$, there exist $a_\alpha \in (1/2, 1)$ and $b_\alpha > 0$ such that the operator $-\Delta$ generates an exponentially bounded analytic $(K*_0 \mathcal{L}^{-1}(\exp(-b_\alpha\lambda^{a_\alpha})))$-convoluted semi-group of angle α. Similar modifications are to be made in a part of [302, Example 2.31(iii)] (cf. also [292, Example 2.8.1]). Concerning [302, Example 2.31(ii)] (cf. also the third part of [292, Example 2.8.3]), one has to replace the function $k_l(\lambda)$ with $\mathcal{L}^{-1}(\exp(-c\lambda^{\beta/s}))$, for a sufficiently large number $c > 0$, and the angle of analyticity with $\min(\pi/2(-1 + 1/\beta), \pi/2)$.
4. Line 4 above Theorem 2.9.48 on p. 207 should be preceded with:
   We define the modified Liouville right-sided fractional derivative of non-integer order $\beta > 0$, $D^\beta_- u(s)$ for short, for those continuously differentiable functions $u : (0, \infty) \to E$ for which $\lim_{T\to\infty} \int_s^T g_{\lceil\beta\rceil-\beta}(t-s)u'(t)\,dt = \int_s^\infty g_{\lceil\beta\rceil-\beta}(t-s)u'(t)\,dt$ exists and defines an $\lceil\beta - 1\rceil$-times continuously differentiable function on $(0, \infty)$, by
   $$D^\beta_- u(s) := (-1)^{\lceil\beta\rceil} \frac{d^{\lceil\beta-1\rceil}}{ds^{\lceil\beta-1\rceil}} \int_s^\infty g_{\lceil\beta\rceil-\beta}(t-s)u'(t)\,dt,\ t > 0.$$

If $\beta = n \in \mathbb{N}$, then $D^n_- u$ is defined for all n-times continuously differentiable functions $u(\cdot)$ on $(0, \infty)$, by $D^n_- u := (-1)^n d/d^n$, where d/d^n denotes the usual derivative operator of order n.

BIBLIOGRAPHY

1. T. Abdeljawad, D. Baleanu, F. Jarad, O. G. Mustafa, J. J. Trujillo, *A finite type result for sequential fractional differential equations*, Dynam. Systems Appl. **19** (2010), 383–394.
2. S. Agmon, A. Douglis, L. Nirenberg, *Estimates near the boundary for solutions of elliptic partial differential equations satisfying general boundary conditions*, Comm. Pure Applied Math. **12** (1959), 623–727.
3. F. Alabau-Boussouira, P. Cannarsa, D. Sforza, *Decay estimates for second order evolution equations with memory*, J. Funct. Anal. **254** (2008), 1342–1372.
4. A. A. Albanese, F. Kühnemund, *Trotter-Kato approximation theorems for locally equicontinuous semigroups*, Riv. Mat. Univ. Parma **1** (2002), 19–53.
5. A. A. Albanese, E. Mangino, *Trotter-Kato approximation theorems for bi-continuous semigroups and applications to Feller semigroups*, J. Math. Anal. Appl. **289** (2004), 477–492.
6. A. A. Albanese, X. Barrachina, E. M. Mangino, A. Peris, *Distributional chaos for strongly continuous semigroups of operators*, Commun. Pure Appl. Anal. **12** (2013), 2069–2082.
7. E. Alvarez-Pardo, C. Lizama, *The maximal subspace for generation of (a, k)-regularized families*, Abstr. Appl. Anal. 2012, Art. ID 683021, 14 pp. 47A16.
8. H. Amann, *Linear and Quasilinear Parabolic Problems*, Vol. I, Birkhäuser, Basel, 1995.
9. M. H. Annaby, Z. S. Mansour, *q-Fractional Calculus and Equations*, Lecture Notes Math. **2056**, Springer-Verlag, 2012.
10. W. Arendt, *Vector-valued Laplace transforms and Cauchy problems*, Israel J. Math. **59** (1987), 327–352.
11. W. Arendt, O. El–Mennaoui, V. Keyantuo, *Local integrated semigroups: evolution with jumps of regularity*, J. Math. Anal. Appl. **186** (1994), 572–595.
12. W. Arendt, *Sobolev imbeddings and integrated semigroups*, 2nd International Conference on Trends in Semigroup Theory and Evolution Equations, (Ph. Clément, E. Mitidieri and B. de Pagter, eds.), Lecture Notes in Pure and Appl. Math. vol. **135**, Marcel Dekker, New York, 1991.
13. W. Arendt, F. Neubrander, U. Schlotterbeck, *Interpolation of semigroups and integrated semigroups*, Semigroup Forum **45** (1992), 26–37.
14. W. Arendt, H. Kellermann, *Integrated solutions of Volterra integrodifferential equations and applications*, Volterra integrodierential equations in Banach spaces and applications, Proc. Conf. Trento/Italy 1987, Pitman Res. Notes Math. Ser. **190** (1989), 21–51.
15. W. Arendt, O. El–Mennaoui, M. Hieber, *Boundary values of holomorphic semigroups*, Proc. Amer. Math. Soc. **125** (1997), 635–647.
16. W. Arendt, C. J. K. Batty, *Rank-1 perturbations of cosine functions and semigroups*, J. Funct. Anal. **238** (2006), 340–352.
17. W. Arendt, N. Nikolski, *Vector-valued holomorphic functions revisited*, Math. Z. **234** (2000), 777–805.
18. W. Arendt, A. Rhandi, *Perturbations of positive semigroups*, Arch. Math. (Basel) **56** (1991), 107–119.
19. W. Arendt, C. J. K. Batty, *Rank-1 perturbations of cosine functions and semigroups*, J. Funct. Anal. **238** (2006), 340–352.

20. W. Arendt, C. J. K. Batty, M. Hieber, F. Neubrander, *Vector-valued Laplace Transforms and Cauchy Problems*, Monographs in Mathematics **96**, Birkhäuser/Springer Basel AG, Basel, 2001.
21. J. Aroza, T. Kalmes, E. Magino, *Chaotic C_0-semigroups induced by semiflows in Lebesgue and Sobolev spaces*, J. Math. Anal. Appl. **412** (2014), 77–98.
22. T. M. Atanacković, S. Pilipović, D. Zorica, *A diffusion wave equation with two fractional derivatives of different order*, J. Phys. A **40** (2007), 5319–5333.
23. L. Autret, H. A. Emamirad, *Entire propagator*, Proc. Amer. Math. Soc. **120** (1994), 1151–1158.
24. Kh. K. Avad, A. V. Glushak, *Perturbation of an abstract differential equation containing fractional Riemann-Liouville derivatives*, Differential Equations **46** (2010), 1–15.
25. V. A. Babalola, *Semigroups of operators on locally convex spaces*, Trans. Amer. Math. Soc. **199** (1974), 163–179.
26. B. Baeumer, M. M. Meerschaert, J. Mortensen, *Space-time fractional derivative operators*, Proc. Amer. Math. Soc. **133** (2005), 2273–2282.
27. B. Baeumer, M. Haase, M. Kovacs, *Unbounded functional calculus for bounded groups with applications*, J. Evol. Equ. **9** (2009), 171–195.
28. B. Baeumer, M. Haase, M. Kovacs, *Fractional reproduction-dispersal equations and heavy tail dispersal kernels*, Bull. Math. Biol. **69** (2007), 2281–2297.
29. K. Balachandran, S. Kiruthika, *Existence of solutions of abstract fractional integrodifferential equations of Sobolev type*, Comput. Math. Appl. **64** (2012), 3406–3413.
30. M. K. Balaev, *Solvability of the Cauchy problem for linear operator-differential equations with variable operator coefficients*, Math. Notes **90** (2011), 651–665.
31. A. V. Balakrishnan, *Fractional powers of closed operators and the semigroups generated by them*, Pacific J. Math. **10** (1960), 419–437.
32. A. V. Balakrishnan, *An operation calculus for infinitesimal generators of semigroups*, Trans. Amer. Math. Soc. **91** (1959), 330–353.
33. D. Baleanu, K. Diethelm, E. Scalas, J. J. Trujillo, *Fractional Calculus Models and Numerical Methods*, Series on Complexity, Nonlinearity and Chaos, World Scientific, Singapore, 2012.
34. D. Baleanu, J. J. Trujillo, *On exact solutions of a class of fractional Euler-Lagrange equations*, Nonlinear Dynam. **52** (2008), 331–335.
35. J. Banasiak, M. Moszynski, *A generalization of Desch–Schappacher–Webb criteria for chaos*, Discrete Contin. Dyn. Syst. **12** (2005), 959–972.
36. J. Banasiak, M. Lachowicz, *Topological chaos for birth-and-death-type-models with proliferation*, Math. Models Methods Appl. Sci. **12** (2002), 755–775.
37. J. Banasiak, M. Lachowicz, M. Moszyński, *Chaotic behaviour of semigroups related to process of gene amplification-deamplification with cell proliferation*, Math. Biosci. **206** (2007), 200–215.
38. V. Barbu, *Nonlinear Semigroups and Differential Equations in Banach Spaces*, Noordho Int. Publ. Leyde the Netherlands, 1976.
39. V. Barbu, *Differentiable distribution semi-groups*, Ann. Sc. Norm. Super. Pisa Cl. Sci. **23** (1969), 413–429.
40. E. Barletta, S. Dragomir, *Vector valued holomorphic functions*, Bull. Math. Soc. Sci. Math. Roumanie **52** (2009), 211–226.
41. X. Barrachina, A. Peris, *Distributionally chaotic translation semigroups*, J. Difference Equ. Appl. **18** (2012), 751–761.
42. X. Barrachina, J. A. Conejero, *Devaney chaos and distributional chaos in the solution of certain partial differential equations*, Abstr. Appl. Anal. 2012, Art. ID 457019, 11 pp.
43. T. Bárta, *Smooth solutions of Volterra equations via semigroups*, Bull. Austral. Math. Soc. **78** (2008), 249–260.
44. B. Bäumer, *Approximate solutions to the abstract Cauchy problem*, Evolution Equations and Their Applications in Physical and Life Sciences (Bad Herrenalb, 1998), 33–41. Lect. Notes Pure Appl. Math. **215**, Dekker, New York, 2001.
45. C. J. K. Batty, *Differentiability of perturbed semigroups and delay semigroups*, Banach Center Publications, vol. **75**, pp. 39–53, Polish Academy of Science, Warszawa, 2007.

46. C. J. K. Batty, *On a perturbation theorem of Kaiser and Weis*, Semigroup Forum **70** (2005), 471–474.
47. F. Bayart, *Hypercyclic operators failing the Hypercyclicity Criterion on classical Banach spaces*, J. Funct. Anal. **250** (2007), 426–441.
48. F. Bayart, T. Bermúdez, *Semigroups of chaotic operators*, Bull. London Math. Soc. **41** (2009), 823–830.
49. E. Bazhlekova, *Fractional Evolution Equations in Banach Spaces*, Ph.D. Thesis, Eindhoven University of Technology, Eindhoven, 2001.
50. B. Bäumer, *A Vector-valued Operational Calculus and Abstract Cauchy Problems*, Ph.D. Thesis, Louisiana State University, Baton Rouge, 1997.
51. R. Beals, *On the abstract Cauchy problem*, J. Funct. Anal. **10** (1972), 281–299.
52. R. Beals, *Semigroups and abstract Gevrey spaces*, J. Funct. Anal. **10** (1972), 300–308.
53. H. Berens, P. L. Butzer, U. Westphal, *Representation of fractional powers of infinitesimal generators of semigroups*, Bull. Amer. Math. Soc. **74** (1968), 191–196.
54. T. Bermúdez, A. Bonilla, J. A. Conejero, A. Peris, *Hypercyclic, topologically mixing and chaotic semigroups on Banach spaces*, Studia Math. **170** (2005), 57–75.
55. T. Bermúdez, A. Bonilla, A. Peris, *On hypercyclicity and supercyclicity criteria*, Bull. Austral. Math. Soc. **70** (2004), 45–54.
56. T. Bermúdez, A. Bonilla, A. Peris, *$\mathbb{C}$-supercyclic versus $\mathbb{R}^+$-supercyclic vectors*, Arch. Math. (Basel) **79** (2002), 125–130.
57. T. Bermúdez, A. Bonilla, A. Martinón, *On the existence of chaotic and hypercyclic semigroups in Banach spaces*, Proc. Amer. Math. Soc. **131** (2003), 2435–2441.
58. S. Bermudo, A. Montes-Rodríguez, S. Shkarin, *Orbits of operators commuting with the Volterra operator*, J. Math. Pures Appl. **89** (2008), 145–173.
59. L. Bernal-Gonzáles, *Disjoint hypercyclic operators*, Studia Math. **182** (2007), 113–131.
60. L. Bernal-Gonzáles, K.-G. Grosse Erdmann, *The Hypercyclicity Criterion for sequences of operators*, Studia Math. **157** (2003), 17–32.
61. L. Bernal-Gonzáles, K.-G. Grosse Erdmann, *Existence and nonexistence of hypercyclic semigroups*, Proc. Amer. Math. Soc. **135** (2007), 755–766.
62. J. Bès, A. Peris, *Hereditarily hypercyclic operators*, J. Funct. Anal. **167** (1999), 94–112.
63. J. Bès, A. Peris, *Disjointness in hypercyclicity*, J. Math. Anal. Appl. **336** (2007), 297–315.
64. D. N. Bessis, F. H. Clarke, *Partial subdifferentials, derivatives and Rademacher's theorem*, Trans. Amer. Math. Soc. **351** (1999), 2899–2926.
65. C. Blondia, *A Radon-Nikodym Theorem for vector valued measures*, Bull. Soc. Math. Belg. Sér. B **33** (1981), 231–249.
66. G. Bluman, V. Shtelen, *Nonlocal transformations of Kolmogorov equations into the backward heat equation*, J. Math. Anal. Appl. **291** (2004), 419–437.
67. S. Bochner, *Diffusion equation and stochastic processes*, Proc. Natl. Acad. Sci. USA **35** (1949), 368–370.
68. Yu. V. Bogacheva, *Resolution's Problems of Initial Problems for Abstract Differential Equations with Fractional Derivatives*, Ph.D. Thesis, Belgorod, 2006.
69. J. L. Boldrini, *Asymptotic behaviour of traveling wave solutions of the equations for the flow of a fluid with small viscosity and capillarity*, Quart. Appl. Math. **44** (1987), 697–708.
70. A. Bonilla, P. J. Miana, *Hypercyclic and topologically mixing cosine functions on Banach spaces*, Proc. Amer. Math. Soc. **136** (2008), 519–528.
71. A. Borichev, Y. Tomilov, *Optimal polynomial decay of functions and operator semigroups*, Math. Ann. **347** (2010), 455–478.
72. J. Bourgain, *A Hausdorff–Young inequality for B-convex Banach spaces*, Pacific J. Math. **101** (1982), 255–262.
73. S. Busenberg, B. H. Wu, *Convergence theorems for integrated semigroups*, Differential Integral Equations **5** (1992), 509–520.
74. S. Calzadillas, C. Lizama, *Bounded mild solutions of perturbed Volterra equations with infinite delay*, Nonlinear Anal. **72** (2010), 3976–3983.

75. P. Cannarsa, D. Sforza, *Integro-differential equations of hyperbolic type with positive definite kernels*, J. Differential Equations **250** (2011), 4289–4335.
76. P. Cannarsa, D. Sforza, *Semilinear integrodifferential equations of hyperbolic type: existence in the large*, Mediterr. J. Math. **1** (2004), 151–174.
77. P. Cannarsa, D. Sforza, *Global solutions of abstract semilinear parabolic equations with memory terms*, NoDEA **10** (2003), 399–430.
78. R. Carmicheal, A. Kamiński, S. Pilipović, *Notes on Boundary Values in Ultradistribution Spaces*, Lect. Notes Ser. **49**, Seoul, 1999.
79. A. N. Carvalho, T. Dlotko, M. J. D. Nascimento, *Non-autonomous semilinear evolution equations with almost sectorial operators*, J. Evol. Equ. **8** (2008), 631–659.
80. J. A. van Casteren, *Generators of Strongly Continuous Semigroups*, Pitman Res. Notes Math. **115**, Longman, Harlow, 1985.
81. N. D. Chakraborty, SK. J. Ali, *On strongly Pettis integrable functions in locally convex spaces*, Revista Math. **6** (1993), 241–262.
82. C. Chalk, *Nonlinear Evolutionary Equations in Banach Spaces with Fractional Time Derivative*, Ph.D. Thesis, University of Hull, 2006.
83. J.-C. Chang, S.-Y. Shaw, *Perturbation theory of abstract Cauchy problems and Volterra equations*, Proceedings of the Second World Congress of Nonlinear Analysts, Part 6 (Athens, 1996). Nonlinear Anal. **30** (1997), 3521–3528.
84. S.-J. Chang, C.-C. Chen, *Topologically mixing for cosine operator functions generated by shifts*, Topology Appl. **160** (2013), 382–386.
85. Y.-C. Chang, S.-Y. Shaw, *Optimal and non-optimal rates of approximation for integrated semigroups and cosine functions*, J. Approx. Theory **90** (1997), 200–223.
86. Y.-H. Chang, C.-H. Hong, *The chaos of the solution semigroup for the quasi-linear Lasota equation*, Taiwanese J. Math. **16** (2012), 1707–1717.
87. J. Chazarain, *Problémes de Cauchy abstraites et applicationsá quelques problémes mixtes*, J. Funct. Anal. 7 (1971), 386–446.
88. C. Chen, M. Li, F.-B. Li, *On boundary values of fractional resolvent families*, J. Math. Anal. Appl. **384** (2011), 453–467.
89. C. Chen, M. Li, *On fractional resolvent operator functions*, Semigroup Forum **80** (2010), 121–142.
90. C. Chen, M. Kostić, M. Li, M. Žigić, *Complex powers of C-sectorial operators*. Part I, Taiwanese J. Math. **17** (2013), 465–499.
91. C. Chen, X.-Q. Song, H.-M. Li, *The reduction square root and perturbation for a class of strongly continuous operator families*, Acta Math. Sinica (English Ser.) **26** (2010), 1993–2002.
92. C. Chen, M. Kostić and M. Li, *Complex powers of almost C-nonnegative operators*. Contemporary Analysis and Applied Mathematics **2** (1) (2014), 1–77.
93. C. Chen, M. Kostić and M. Li, *Representation of complex powers of C-sectorial operators*. Fract. Calc. Appl. Anal. **17** (3), 827–854.
94. G. Chen, R. Grimmer, *Semigroups and integral equations*, J. Integral Equations **2** (1980), 133–54.
95. G. Chen, R. Grimmer, *Integral equations as evolution equations*, J. Differential Equations **45** (1982), 53–74.
96. J.-C. Chen, S.-Y. Shaw, *Topological mixing and hypercyclicity criterion for sequences of operators*, Proc. Amer. Math. Soc. **134** (2006), 3171–3179.
97. P. R. Chernoff, *Semigroup product formulas and addition of unbounded operators*, Bull. Amer. Math. Soc. **76** (1970), 395–398.
98. R. Chill, V. Keyantuo, M. Warma, *Generation of cosine families on $L^p(0, 1)$ by elliptic operators with Robin boundary conditions*, Functional Analysis and Evolution Equations. Basel, Birkhäuser Verlag, 2008, pp. 113–130.
99. R. Chill, J. Prüss, *Asymptotic behaviour of linear evolutionary integral equations*, Integral Equations Operator Theory **39** (2001), 193–213.
100. Y. H. Choe, *C_0-semigroups on a locally convex space*, J. Math. Anal. Appl. **106** (1985), 293–320.

101. I. Ciorănescu, *Beurling spaces of class* (M_p) *and ultradistribution semi-groups*, Bull. Sci. Math. **102** (1978), 167–192.
102. I. Ciorănescu, *Local convoluted semigroups*, Evolution Equations (Baton Rauge, LA, 1992), 107–122, Dekker, New York, 1995.
103. I. Ciorănescu, G. Lumer, *Problèmes d'évolution régularisés par un noyan général* $K(t)$, *Formule de Duhamel, prolongements, théorèmes de génération*, C. R. Acad. Sci. Paris Sér. I Math. **319** (1995), 1273–1278.
104. I. Cioranescu, G. Lumer, *On* $K(t)$*-convoluted semigroups*, Recent Developments in Evolution Equations (Glasgow, 1994), 86–93. Longman Sci. Tech., Harlow, 1995.
105. I. Ciorănescu, G. Lumer, *Regularization of evolution equations via kernels* $K(t)$, *K-evolution operators and convoluted semigroups, generation theorems*, Seminar Notes in Functional Analysis and PDE, Louisiana State Univ., Baton Rouge, 1994, pp. 45–52.
106. I. Ciorănescu, L. Zsidó, *ω-Ultradistributions and Their Applications to Operator Theory*, Spectral Theory, Banach Center Publications **8**, Warszawa 1982, 77–220.
107. I. Ciorănescu, L. Zsidó, *ω-Ultradistributions in the abstract Cauchy problem*, An. Ştiinţ. Univ. "Al. I. Cuza" Iaşi Secţ. I a Mat. (N. S.) 25 (1979), 79–94.
108. I. Cioranescu, H. Henriquez, *Interpolation properties associated with second order differential equations*, Aequationes Math. **47** (1994), 150–163.
109. I. Cioranescu, V. Keyantuo, *On operator cosine functions in UMD spaces*, Semigroup Forum **63** (2001), 429–440.
110. P. Clément, R. C. MacCamy, J. A. Nohel, *Asymptotic properties of solutions of nonlinear abstract Volterra equations*, J. Integral Equations **3** (1981), 185–216.
111. P. Clément, J. A. Nohel, *Asymptotic behavior of solutions of nonlinear Volterra equations with completely positive kernels*, SIAM J. Math. Anal. **10** (1981), 514–535.
112. P. Clément, J. A. Nohel, *Abstract linear and nonlinear Volterra equations preserving positivity*, SIAM J. Math. Anal. **10** (1979), 365–388.
113. B. D. Coleman, M. E. Gurtin, *Equipresence and constitutive equations for rigid heat conductors*, Z. Angew. Math. Phys. **18** (1967), 199–208.
114. B. D. Coleman, V. J. Mizel, *Norms and semigroups in the theory of fading memory*, Arch. Rational Mech. Anal. **28** (1966), 87–123.
115. J. A. Conejero Casares, *Operadores y Semigrupos de Operadores Espacios de Fréchet y Espacios Localmente Convexos*, Ph.D. Thesis, Univ. Politècnica de València, 2004.
116. J. A. Conejero, *On the existence of transitive and topologically mixing semigroups*, Bull. Belg. Math. Soc. Simon Stevin **14** (2007), 463–471.
117. J. A. Conejero, E. Mangino, *Hypercyclic semigroups generated by Ornstein-Uhlenbeck operators*, Mediterr. J. Math. 7 (2010), 101–109.
118. J. A. Conejero, A. Peris, *Chaotic translation semigroups*, Discrete Contin. Dyn. Syst., Supplement, 269–276 (2007).
119. J. A. Conejero, A. Peris, *Linear transitivity criteria*, Topology Appl. **153** (2005), 767–773.
120. J. A. Conejero, A. Peris, M. Trujillo, *Chaotic asymptotic behaviour of the hyperbolic heat transfer equation solutions*, Internat. J. Bifur. Chaos Appl. Sci. Engrg. **20** (2010), 2943–2947.
121. J. A. Conejero, V. Müler, A. Peris, *Hypercyclic behavior of operators in a hypercyclic* C_0*-semigroup*, J. Funct. Anal. **244** (2007), 342–348.
122. J. A. Conejero, M. Kostić, P. J. Miana, *Distributionally chaotic cosine functions and fractional PDEs*, preprint.
123. C. Corduneanu, *Integral Equations and Applications*, Monographs in Mathematics **87**, Cambridge University Press, 1991.
124. G. Costakis, M. Sambarino, *Topologically mixing hypercyclic operators*, Proc. Amer. Math. Soc. **132** (2004), 385–389.
125. G. Costakis, A. Peris, *Hypercyclic semigroups and somewhere dense orbits*, C. R. Acad. Sci. Paris Ser. I **335** (2002), 895–898.
126. M. G. Crandall, S.-O. Londen, J. A. Nohel, *An abstract nonlinear Volterra integrodifferential equation*, J. Math. Anal. Appl. **64** (1978), 701–735.

127. R. A. A. Cuello, *Traslaciones y Semigrupos C_0 Topológicamente Mezclantes*, Ph.D. Thesis, Universidad de Puerto Rico, Recinto Universitario de Mayagüez, 2008.
128. C. Cuevas, C. Lizama, *Almost automorphic solutions to integral equations on the line*, Semigroup Forum **79** (2009), 461–472.
129. C. M. Dafermos, *An abstract Volterra equation with applications to linear viscoelasticity*, J. Differential Equations **7** (1970), 554–569.
130. C. M. Dafermos, *Asymptotic stability in viscoelasticity*, Arch. Rational Mech. Anal. **37** (1970), 297–308.
131. G. Da Prato, E. Sinestrari, *Differential operators with nondense domain*, Ann. Scuola Norm. Sup. Pisa Cl. Sci. **14** (1987), 285–344.
132. G. Da Prato, *Semigruppi regolarizzabilli*, Ricerche Mat. **15** (1966), 223–248.
133. G. Da Prato, *Semigruppi di crescenza n*, Ann. Scuola Norm. Sup. Pisa Cl. Sci. **20** (1966), 753–782.
134. G. Da Prato, M. Iannelli, *Linear integro-differential equations in Banach spaces*, Rend. Sem. Mat. Padova **62** (1980), 207–219.
135. G. Da Prato, M. Iannelli, *Distribution resolvents for Volterra equations in a Banach space*, J. Integral Equations **6** (1984), 93–103.
136. G. Da Prato, A. Lunardi, *Stabilizability of integrodifferential parabolic equations*, J. Integral Equations **2** (1990), 281–304.
137. R. Dautray, J.-L. Lions, *Mathematical Analysis and Numerical Methods for Science and Technology, Vol. 5: Evolution Problems I*, Springer-Verlag, Berlin, 2000.
138. E. B. Davies, *One–Parameter Semigroups*, Academic Press, London, 1980.
139. E. B. Davies, M. M. Pang, *The Cauchy problem and a generalization of the Hille–Yosida theorem*, Proc. London Math. Soc. **55** (1987), 181–208.
140. K. Deimling, *Nonlinear Functional Analysis*, Springer-Verlag, Berlin, 1985.
141. R. deLaubenfels, *Existence Families, Functional Calculi and Evolution Equations*, Lect. Notes Math. **1570**, Springer-Verlag, New York, 1994.
142. R. deLaubenfels, *Holomorphic C-existence families*, Tokyo J. Math. **15** (1992), 17–38.
143. R. deLaubenfels, *Polynomials of generators of integrated semigroups*, Proc. Amer. Math. Soc. **107** (1989), 197–204.
144. R. deLaubenfels, F. Yao, S. W. Wang, *Fractional powers of operators of regularized type*, J. Math. Anal. Appl. **199** (1996), 910–933.
145. R. deLaubenfels, J. Pastor, *Semigroup approach to the theory of fractional powers of operators*, Semigroup Forum **199** (2008), 910–933.
146. R. deLaubenfels, G. Sun, S. W. Wang, *Regularized semigroups, existence families and the abstract Cauchy problem*, Differential Integral Equations **8** (1995), 1477–1496.
147. R. deLaubenfels, *Integrated semigroups and integrodifferential equations*, Math. Z. **204** (1990), 501–514.
148. R. deLaubenfels, *Matrices of operators and regularized semigroups*, Math. Z. **212** (1993), 619–629.
149. R. deLaubenfels, Y. Lei, *Regularized functional calculi, semigroups, and cosine functions for pseudodifferential operators*, Abstr. Appl. Anal. **2** (1997), 121–136.
150. R. deLaubenfels, H. Emamirad, K.-G. Grosse–Erdmann, *Chaos for semigroups of unbounded operators*, Math. Nachr. **261/262** (2003), 47–59.
151. R. deLaubenfels, H. Emamirad, *Chaos for functions of discrete and continuous weighted shift operators*, Ergodic Theory Dynam. Systems **21** (2001), 1411–1427.
152. R. deLaubenfels, H. Emamirad, V. Protopopescu, *Linear chaos and approximation*, J. Approx. Theory **105** (2000), 176–187.
153. R. deLaubenfels, *Existence and uniqueness families for the abstract Cauchy problem*, J. London Math. Soc. **s2-44** (1991), 310–338.
154. B. Dembart, *On the theory of semigroups on locally convex spaces*, J. Funct. Anal. **16** (1974), 123–160.
155. W. Desch, R. Grimmer, *Initial-boundary value problems for integrodifferential equations*, J. Integral Equations **10** (1985), 73–97.

156. W. Desch, R. Grimmer, *Propagation of singularities for integro-differential equations*, J. Differential Equations **65** (1986), 411–426.
157. W. Desch, R. Grimmer, W. Schappacher, *Some considerations for linear integrodifferential equations*, J. Math. Anal. Appl. **104** (1984), 219–234.
158. W. Desch, W. Schappacher, *Some perturbation results for analytic semigroups*, Math. Ann. **281** (1988), 157–162.
159. W. Desch, W. Schappacher, *On products of hypercyclic semigroups*, Semigroup Forum **71** (2005), 301–311.
160. W. Desch, G. Schappacher, W. Schappacher, *Relatively bounded rank one perturbations of non-analytic semigroups can generate large point spectrum*, Semigroup Forum **75** (2007), 470–476.
161. W. Desch, W. Schappacher, G. F. Webb, *Hypercyclic and chaotic semigroups of linear operators*, Ergodic Theory Dynam. Systems **17** (1997), 1–27.
162. W. Desch, K. B. Hannsgen, R. L. Wheeler, *Passive boundary damping of viscoelastic structures*, J. Integral Equations Appl. **8** (1996), 125–171.
163. K. Diethelm, *The Analysis of Fractional Differential Equations*, Springer-Verlag, Berlin, 2010.
164. E. Di Nezza, G. Palatucci, E. Valdinoci, *Hitchhiker's guide to the fractional Sobolev spaces*, Bull. Sci. Math. **136** (2012), 521–573.
165. I. Dobrakov, T. V. Panchapagesan, *A generalized Pettis measurability criterion and integration of vector functions*, Studia Math. **164** (2004), 205–229.
166. S. Dugowson, *Les Différentielles Métaphysiques: Historie et Philosophie de la Généralisation de l'Ordre de Dérivation*, Ph.D. Thesis, University of Paris, 1994.
167. O. El–Mennaoui, *Trace des Semi–Goupes Holomorphes Singuliersà l'Origine et Comportement Asymptotique*, Ph.D. Thesis, Besançon, 1992.
168. O. El–Mennaoui, V. Keyantuo, *On the Schrödinger equation in L^p spaces*, Math. Ann. **304** (1996), 293–302.
169. O. El–Mennaoui, V. Keyantuo, *Trace theorems for holomorphic semigroups and the second order Cauchy problem*, Proc. Amer. Math. Soc. **124** (1996), 1445–1458.
170. S. El Mourchid, *The imaginary point spectrum and hypercyclicity*, Semigroup Forum **76** (2006), 313–316.
171. S. El Mourchid, *On a hypercyclicity criterion for strongly continuous semigroups*, Discrete Contin. Dyn. Syst. **13** (2005), 271–275.
172. S. El Mourchid, G. Metafune, A. Rhandi, J. Voigt, *On the chaotic behaviour of size structured cell population*, J. Math. Anal. Appl. **339** (2008), 918–924.
173. H. Emamirad, *Hypercyclicity in the scattering theory for linear transport equation*, Trans. Amer. Math. Soc. **350** (1998), 3707–3716.
174. H. Emamirad, G. R. Goldstein, J. A. Goldstein, *Chaotic solution for the Black-Scholes equation*, Proc. Amer. Math. Soc. **140** (2012), 2043–2052.
175. K. J. Engel, R. Nagel, *One–Parameter Semigroups for Linear Evolution Equations*, Springer–Verlag, Berlin, 2000.
176. A. Es-Sarhir, B. Farkas, *Perturbation for a class of transition semigroups on the Hölder space $C^{\theta}_{b,\mathrm{loc}}(H)$*, J. Math. Anal. Appl. **315** (2006), 666–685.
177. K. J. Engel, *On singular perturbations of second order Cauchy problems*, Pacific J. Math. **152** (1992), 79–91.
178. E. Fašangová, J. Prüss, *Asymptotic behaviour of a semilinear viscoelastic beam model*, Arch. Math. (Basel) **77** (2001), 488–497.
179. H. O. Fattorini, *The Cauchy Problem*, Addison–Wesley, 1983. MR84g:34003.
180. H. O. Fattorini, *Second Order Linear Differential Equations in Banach Spaces*, North–Holland Math. Stud. **108**, North–Holland, Amsterdam, 1985.
181. H. O. Fattorini, *A note on fractional derivatives of semigroups and cosine functions*, Pacific J. Math. **109** (1983), 335–347.
182. A. Favini, R. Labbas, K. Lemrabet, S. Keddour, B.-K. Sadallah, *Study of a complete abstract differential equation of elliptic type with variable operator coefficients*, Rev. Mat. Complut. **21** (2008), 89–133.

183. H. Federer, *Geometric Measure Theory*, Springer-Verlag, New York, 1969.
184. N. S. Feldman, *Hypercyclicity and supercyclicity for invertible bilateral weighted shifts*, Proc. Amer. Math. Soc. **131** (2003), 479–485.
185. W. Feller, *On a generalization of Marcel Riesz' potentials and the semigroups generated by them*, Comm. Sém. Math. Univ. Lund [Medd. Lunds Univ. Mat. Sem.], Tome Supplementaire (1952), 72–81.
186. C. Fernández, C. Lizama, V. Poblete, *Regularity of solutions for a third order differential equation in Hilbert spaces*, Appl. Math. Comput. **217** (2011), 8522–8533.
187. K. Floret, J. Wloka, *Einführung in die Theorie der Lokalkonvexen Räume*, Lect. Notes Math. **56**, Springer-Verlag, 1968.
188. L. S. Frank, H. W. Norde, *On a singular perturbation in the linear soliton theory*, Asymptotic Anal. **4** (1991), 17–59.
189. A. Friedman, *Generalized Functions and Partial Differential Equations*, Prentice Hall, New York, 1963.
190. A. Friedman, M. Shinbrot, *Volterra integral equations in Banach spaces*, Trans. Amer. Math. Soc. **126** (1967), 131–179.
191. J. E. Galé, P. J. Miana, P. R. Stinga, *Extension problem and fractional operators: semigroups and wave equation*, J. Evol. Equ. **13** (2013), 343–368.
192. R. M. Gethner, J. H. Shapiro, *Universal vectors for operators on spaces of holomorphic functions*, Proc. Amer. Math. Soc. **100** (1987), 281–288.
193. F. Martinez-Giménez, P. Oprocha, A. Peris, *Distributional chaos for backward shifts*, J. Math. Anal. Appl. **351** (2009), 607–615.
194. C. Giorgi, V. Patta, E. Vuk, *On the extensible viscoelastic beam*, Nonlinearity **21** (2008), 713–733.
195. A. V. Glushak, *On the relationship between the integrated cosine functions and the operator Bessel function*, Dierentsial'nye Uravneniya **42** (2006), 583–589.
196. A. V. Glushak, *Cauchy-type problem for an abstract differential equation with fractional derivatives*, Math. Notes **77** (2005), 26–38.
197. G. Godefroy, J. H. Shapiro, *Operators with dense, invariant, cyclic, vector manifolds*, J. Funct. Anal. **98** (1991), 229–269.
198. J. A. Goldstein, *Semigroups of Linear Operators and Applications*, Oxford Univ. Press, 1985.
199. J. A. Goldstein, *On the convergence and approximations of cosine functions*, Aequationes Math. **10** (1974), 201–205.
200. R. Gorenflo, F. Mainardi, *Fractals and Fractional Calculus in Continuum Mechanics*, Springer-Verlag, Vienna and New York, 1997, pp. 223–276.
201. R. Gorenflo, Y. Luchko, F. Mainardi, *Analytical properties and applications of the Wright function*, Fract. Calculus Appl. Anal. **2** (1999), 383–414.
202. E. A. Gorin, *On the square-integrable solutions of partial differential equations with constant coefficients*, Siberian Math. J. **11** (1961), 221–232 (in Russian).
203. R. Grimmer, J. Liu, *Singular perturbations in linear viscoelasticity*, Rocky Mountain J. Math. **24** (1994), 61–75.
204. R. Grimmer, J. Liu, *Integrated semigroups and integro-differential equations*, Semigroup Forum **48** (1994), 79–95.
205. R. Grimmer, J. Prüss, *On linear Volterra equations in Banach spaces. Hyperbolic partial differential equations, II*. Comput. Math. Appl. **11** (1985), 189–205.
206. G. Gripenberg, *On a nonlinear Volterra integral equation in a Banach space*, J. Math. Anal. Appl. **66** (1978), 207–219.
207. S. Grivaux, *Hypercyclic operators, mixing operators, and the Bounded Steps problem*, J. Operator Theory **54** (2005), 147–168.
208. S. Grivaux, *Sums of hypercyclic operators*, J. Funct. Anal. **202** (2003), 486–503.
209. K.-G. Grosse-Erdmann, *A weak criterion for weak holomorphy*, Math. Proc. Camb. Phil. Soc. **136** (2004), 399–411.
210. K. G. Grosse-Erdmann, *Universal families and hypercyclic operators*, Bull. Amer. Math. Soc. **36** (1999), 345–381.

211. K.-G. Grosse-Erdmann, A. Peris, *Linear Chaos*, Springer-Verlag, London, 2011.
212. G. M. N'Guérékata, *Topics in Almost Automorphy*, Springer-Verlag, New York, 2005.
213. D. Guidetti, B. Karasrözen, S. Piskarev, *Approximations of abstract differential equations*, J. Math. Sci. (N. Y.) **122** (2004), 3013–3054.
214. M. Haase, *A general framework for holomorphic functional calculi*, Proc. Edinburgh Math. Soc. **48** (2005), 423–444.
215. M. Haase, *The Functional Calculi for Sectorial Operators*, Number **169** in Operator Theory: Advances and Applications. Birkhäuser Verlag, Basel, 2006.
216. M. Haase, *The group reduction for bounded cosine functions on UMD spaces*, Math. Z. **262** (2009), 281–299.
217. M. Haase, *A functional calculus description of real interpolation spaces for sectorial operators*, Studia Math. **171** (2005), 177–195.
218. J. Haluška, *On integration in complete bornological locally convex spaces*, Czechoslovak Math. J. **47** (1997), 205–218.
219. J. Haluška, O. Hutník, *On integrable functions in complete bornological locally convex spaces*, Mediterr. J. Math. **9** (2012), 165–186.
220. P. Hajłasz, *Change of variables formula under minimal assumptions*, Colloq. Math. **64** (1993), 93–101.
221. H. J. Haubold, A. M. Mathai, R. K. Saxena, *Mittag-Leffler functions and their applications*, J. Appl. Math., vol. 2011 (2011), Article ID 298628, 51 pages.
222. B. Hennig, F. Neubrander, *On representations, inversion and approximations of Laplace transform in Banach spaces*, Appl. Anal. **49** (1993), 151–170.
223. M. Hieber, *Integrated semigroups and differential operators on L^p spaces*, Math. Ann. **291** (1995), 1–16.
224. M. Hieber, *L^p spectra of pseudodifferential operators generating integrated semigroups*, Trans. Amer. Math. Soc. **347** (1995), 4023–4035.
225. M. Hieber, *Laplace transforms and α-times integrated semigroups*, Forum Math. **3** (1991), 595–612.
226. M. Hieber, *Integrated semigroups and the Cauchy problem for systems in Lp spaces*, J. Math. Anal. Appl. **162** (1991), 300–308.
227. M. Hieber, A. Holderrieth, F. Neubrander, *Regularized semigroups and systems of linear partial differential equations*, Ann. Scuola Norm. Sup. Pisa Cl. Sci. **19** (1992), 363–379.
228. R. Hilfer, *Applications of Fractional Calculus in Physics*, World Scientic Publ. Co., Singapore, 2000.
229. E. Hille, R. S. Phillips, *Functional Analysis and Semigroups*, AMS, Providence, 1957.
230. R. H. W. Hoppe, *Interpolation of cosine operator functions*, Ann. Mat. Pura Appl. **136** (1984), 183–212.
231. R. H. W. Hoppe, *Representations of fractional powers of infinitesimal generators of cosine operator functions*, Results Math. **7** (1984), 65–70.
232. L. Hörmander, *Estimates for translation invariant operators in L^p spaces*, Acta Math. **104** (1960), 93–140.
233. H. W. Hövel, U. Westphal, *Fractional powers of closed operators*, Studia Math. **42** (1972),177–194.
234. J. W. Hrusa, J. A. Nohel, *The Cauchy problem in one-dimensional nonlinear viscoelasticity*, J. Differential Equations **59** (1985), 388–412.
235. F. Huang, F. Liu, *The space-time fractional diffusion equation with Caputo derivatives*, J. Appl. Math. Comp. **19** (2005), 179–190.
236. P. Iley, *Perturbations of differentiable semigroups*, J. Evol. Equ. **7** (2007), 765–781.
237. L. Ji, A. Weber, *Dynamics of the heat semigroup on symmetric spaces*, Ergodic Theory Dynam. Systems **30** (2010), 457–468.
238. H. Jiang, F. Liu, I. Turner, K. Burrage, *Analytical solutions for the multi-term timespace Caputo-Riesz fractional advection-diffusion equations on a finite domain*, J. Math. Anal. Appl. **389** (2012), 1117–1127.

239. E. Jordá, *Vitali's and Harnack's type results for vector-valued functions*, J. Math. Anal. Appl. **327** (2007), 739–743.
240. M. Jung, *Duality theory for solutions to Volterra integral equations*, J. Math. Anal. Appl. **230** (1999), 112–134.
241. C. Kaiser, L. Weis, *Perturbation theorems for α-times integrated semigroups*, Arch. Math. (Basel) **81** (2003), 215–228.
242. C. Kaiser, L. Weis, *A perturbation theorem for operator semigroups in a Hilbert space*, Semigroup Forum **67** (2003), 63–75.
243. C. Kaiser, *Integrated semigroups and linear partial dierential equations with delay*, J. Math. Anal. Appl. **292** (2004), 328–339.
244. T. Kalmes, *Hypercyclic, Mixing, and Chaotic C_0-semigroups*, Ph.D. Thesis, Universität Trier, 2006.
245. T. Kalmes, *Hypercyclic, mixing, and chaotic C_0-semigroups induced by semiflows*, Ergodic Theory Dynam. Systems **27** (2007), 1599–1631.
246. T. Kalmes, *Hypercyclic C_0-semigroups and evolution families generated by first order differential operators*, Proc. Amer. Math. Soc. **137** (2009), 3833–3848.
247. T. Kalmes, *Hypercyclicity and mixing for cosine operator functions generated by second order partial differential operators*, J. Math. Anal. Appl. **365** (2010), 363–375.
248. A. D. Kandilakis, N. S. Papageorgiou, *Periodic solutions for nonlinear Volterra integrodifferential equations in Banach spaces*, Comment. Math. Univ. Carolin. **38** (1997), 283–296.
249. S. Kantorovitz, *The Hille-Yosida space of an arbitrary operator*, J. Math. Anal. Appl. **136** (1988), 107–111.
250. A. Karczewska, *Stochastic Volterra equations of nonscalar type in Hilbert space*, Transactions XXV International Seminar on Stability Problems for Stochastic Models, ed. C. D'Apice et al., University of Salerno (2005), 78–83.
251. A. Karczewska, C. Lizama, *Solutions to stochastic fractional oscillation equations*, Appl. Math. Lett. **23** (2010), 1361–1366.
252. T. Kato, *Perturbation Theory for Linear Operators*, 2nd. ed., Springer-Verlag, Berlin, 1976.
253. T. Kato, *Note on fractional powers of linear operators*, Proc. Japan Acad. **36** (1960), 94–96.
254. T. Kato, *Remarks on pseudo-resolvents and infinitesimal generators of semi-groups*, Proc. Japan Acad. **35** (1959), 467–468.
255. H. Kellermann, M. Hieber, *Integrated semigroups*, J. Funct. Anal. **84** (1989), 160–180.
256. L. Kexue, P. Jigen, *Fractional abstract Cauchy problems*, Integral Equations Operator Theory **70** (2011), 333–361.
257. L. Kexue, P. Jigen, J. Junxiong, *Cauchy problems for fractional differential equations with Riemann-Liouville fractional derivatives*, J. Funct. Anal. **263** (2012), 476–510.
258. V. Keyantuo, *The Laplace transform and the ascent method for abstract wave equations*, J. Differential Equations **122** (1995), 27–47.
259. V. Keyantuo, *A note on interpolation of semigroups*, Proc. Amer. Math. Soc. **123** (1995), 2123–2132.
260. V. Keyantuo, *Integrated semigroups and related partial differential equations*, J. Math. Anal. Appl. **212** (1997), 135–153.
261. V. Keyantuo, C. Lizama, *Hölder continuous solutions for integro-differential equations and maximal regularity*, J. Differential Equations **230** (2006), 634–660.
262. V. Keyantuo, C. Lizama, *Periodic solutions of second order differential equations in Banach spaces*, Math. Z. **253** (2006), 489–514.
263. V. Keyantuo, C. Lizama, *Fourier multipliers and integro-differential equations in Banach spaces*, J. Lond. Math. Soc. **69** (2004), 737–750.
264. V. Keyantuo, C. Lizama, *A characterization of periodic solutions for time-fractional differential equations in UMD spaces and applications*, Math. Nachr. **284** (2011), 494–506.
265. V. Keyantuo and C. Lizama, *On a connection between powers of operators and fractional Cauchy problems*, J. Evol. Equ. **12** (2012), 245–265.
266. V. Keyantuo, M. Warma, *The wave equation in L^p-spaces*, Semigroup Forum **71** (2005), 73–92.

267. V. Keyantuo, M. Warma, *The wave equation with Wentzell-Robin boundary conditions on L^p-spaces*, J. Differential Equations **229** (2006), 680–697.
268. V. Keyantuo, C. Lizama, P. J. Miana, *Algebra homomorphisms defined via convoluted semigroups and cosine functions*, J. Funct. Anal. **257** (2009), 3454–3487.
269. V. Keyantuo, P. J. Miana, L. Sánchez-Lajusticia, *Sharp extensions for convoluted solutions of abstract Cauchy problems*, Integral Equations Operator Theory **77** (2013), 211–241.
270. S. S. Khurana, *Strong measurability in Fréchet spaces*, Indian J. Pure Appl. Math. **7** (1979), 810–814.
271. A. A. Kilbas, H. M. Srivastava, J. J. Trujillo, *Theory and Applications of Fractional Differential Equations*, Elsevier Science B.V., Amsterdam, 2006.
272. M. Kim, *Abstract Volterra Equations*, Ph.D. Thesis, Louisiana State University, Baton Rouge, 1995.
273. M. Kim, *Trotter–Kato theorems for convoluted resolvent families*, Commun. Korean Math. Soc. **19** (2004), 293–305.
274. V. Kiryakova, *Generalized Fractional Calculus and Applications*, Longman Scientific &Technical, Harlow, 1994, copublished in the United States with John Wiley & Sons, Inc.,New York.
275. J. Kisyński, *The Petrovskiĭ correctness and semigroups of operators*, preprint, 2009.
276. J. Kisyński, *On cosine operator functions and one-parameter groups of operators*, Studia Math. **44** (1972), 93–105.
277. C. Kitai, *Invariant Closed Sets for Linear Operators*, Ph.D. Thesis, University of Toronto, 1982.
278. J. Klafter, S. C. Lim, R. Metzler (Eds.), *Fractional Dynamics in Physics: Recent Advances*, World Scientific, Singapore, 2011.
279. Ch. Klein, S. Rolowicz, *On Riemann integration of functions with values in topologically vector spaces*, Studia Math. **80** (1984), 109–118.
280. A. N. Kochubei, *Hyperfunction solutions of differential-operator equations*, Siberian Math. J. **20** (1979), 778–791.
281. A. N. Kochubei, *Fractional-parabolic systems*, Potential Anal. **37** (2012), 1–30.
282. A. N. Kochubei, *General fractional calculus, evolution equations, and renewal processes*, Integral Equations Operator Theory **71** (2011), 583–600.
283. H. Komatsu, *Ultradistributions, I. Structure theorems and a characterization*, J. Fac. Sci. Univ. Tokyo Sect. IA Math. **20** (1973), 25–105.
284. H. Komatsu, *Ultradistributions, II. The kernel theorem and ultradistributions with support in a manifold*, J. Fac. Sci. Univ. Tokyo Sect. IA Math. **24** (1977), 607–628.
285. H. Komatsu, *Ultradistributions, III. Vector valued ultradistributions. The theory of kernels*, J. Fac. Sci. Univ. Tokyo Sect. IA Math. **29** (1982), 653–718.
286. H. Komatsu, *Operational calculus and semi-groups of operators*, in: Functional Analysis and Related topics (Kyoto), Springer, Berlin, 213–234, 1991.
287. H. Komatsu, *Fractional powers of operators*, Pac. J. Math. **19** (1966), 285–346.
288. H. Komatsu, *Fractional powers of operators, II. Interpolation spaces*, Pacific J. Math. **21** (1967), 89–111.
289. H. Komatsu, *Fractional powers of operators, III. Negative powers*, J. Math. Soc. Japan **21** (1969), 205–220.
290. H. Komatsu, *Semi-groups of operators in locally convex spaces*, J. Math. Soc. Japan **16** (1964), 230–262.
291. T. Kōmura, *Semigroups of operators in locally convex spaces*, J. Funct. Anal. **2** (1968), 258–296.
292. M. Kostić, *Generalized Semigroups and Cosine Functions*, Mathematical Institute Belgrade, 2011.
293. M. Kostić, *C-Distribution semigroups*, Studia Math. **185** (2008), 201–217.
294. M. Kostić, *Chaotic C-distribution semigroups*, Filomat **23** (2009), 51–65.
295. M. Kostić, *Distribution cosine functions*, Taiwanese J. Math. **10** (2006), 739–775.

296. M. Kostić, P. J. Miana, *Relations between distribution cosine functions and almost distribution cosine functions*, Taiwanese J. Math. **11** (2007), 531–543.
297. M. Kostić, S. Pilipović, *Global convoluted semigroups*, Math. Nachr. **280** (2007), 1727–1743.
298. M. Kostić, S. Pilipović, *Convoluted C-cosine functions and semigroups. Relations with ultradistribution and hyperfunction sines*, J. Math. Anal. Appl. **338** (2008), 1224–1242.
299. M. Kostić, *Complex powers of operators*, Publ. Inst. Math., Nouv. Sér. **83** (2008), 15–25.
300. M. Kostić, *Complex powers of non-densely dened operators*, Publ. Inst. Math., Nouv. Sér. **90** (2011), 47–64.
301. M. Kostić, *Differential and analytical properties of semigroups of operators*, Integral Equations Operator Theory **67** (2010), 499–557.
302. M. Kostić, *(a, k)-regularized C-resolvent families: regularity and local properties*, Abstr. Appl. Anal., vol. 2009, Article ID 858242, 27 pages, 2009.
303. M. Kostić, *Abstract Volterra equations in locally convex spaces*, Sci. China Math. **55** (2012), 1797–1825.
304. M. Kostić, *On a class of (a, k)-regularized C-resolvent families*, Electron. J. Qual. Theory Differ. Equ. 2012, No. 94, 27 pp.
305. M. Kostić, *Time-dependent perturbations of abstract Volterra equations*, Bull. Cl. Sci. Math. Nat. Sci. Math. **36** (2011), 89–104.
306. M. Kostić, *Ill-posed abstract Volterra equations*, Publ. Inst. Math., Nouv. Sér. **93** (2013), 49–63.
307. M. Kostić, *Systems of abstract time-fractional equations*, Publ. Inst. Math., Nouv. Sér., **95** (109) (2014), 119–132.
308. M. Kostić, *(a, k)-Regularized* (C_1,C_2)*-existence and uniqueness families*, Bull. Cl. Sci. Math. Nat. Sci. Math. **38** (2013), 9–26.
309. M. Kostić, *S-hypercyclic, mixing and chaotic strongly continuous semigroups*, Funct. Anal. Appr. Comp. **1:1** (2009), 49–86.
310. M. Kostić, *Hypercyclic and chaotic integrated C-cosine functions*, Filomat **26** (2012), 1–44.
311. M. Kostić, *On hypercyclicity and supercyclicity of strongly continuous semigroups induced by semiflows. Disjoint hypercyclic semigroups*, Discrete Contin. Dyn. Syst., accepted.
312. M. Kostić, *Abstract time-fractional equations: existence and growth of solutions*, Fract. Calc. Appl. Anal. **14** (2011), 301–316.
313. M. Kostić, *Hypercyclic and topologically mixing properties of certain classes of abstract time-fractional equations*, Springer Proceedings in Mathematics, Discrete dynamical systems and applications, edited by Lluis Alseda, J. Cushing, S. Elaydi and A. Pinto, forthcoming paper (2015).
314. M. Kostić, *Perturbation theory for abstract Volterra equations*, Abstr. Appl. Anal. 2013, Art. ID 307684, 26 pp.
315. M. Kostić, *Hypercyclicity and mixing for abstract time-fractional equations*, Dyn. Syst. **27** (2012), 213–221.
316. M. Kostić, *Some contributions to the theory of abstract Volterra equations*, Int. J. Math. Anal. (Russe) **5** (2011), 1529–1551.
317. M. Kostić, *Generalized well-posedness of hyperbolic Volterra equations of non-scalar type*, Ann. Acad. Rom. Sci. Ser. Math. Appl. **6** (2014), 21–49.
318. M. Kostić, C.-G. Li, M. Li, *On a class of abstract time-fractional equations on locally convex spaces*, Abstr. Appl. Anal., Volume 2012, Article ID 131652, 41 pages.
319. M. Kostić, C.-G. Li, M. Li, S. Piskarev, *On a class of time-fractional differential equations*, Fract. Calc. Appl. Anal. **27** (2012), 639–668.
320. M. Kostić, *Hypercyclic and topologically mixing properties of abstract time-fractional equations with discrete shifts*, Sarajevo J. Math. **9** (2013), 1–13.
321. M. Kostić, *Abstract differential operators generating fractional resolvent families*, Acta Math. Sinica, Sinica (English Ser.), **30** (2014), 1989–1998.
322. M. Kostić, *Approximations and convergence of (a, k)-regularized C-resolvent families*, Numerical Funct. Anal. Appl. **35** (2014), 1579–1604.

323. M. Kostić, *On the existence and uniqueness of solutions of certain classes of abstract multi-term fractional differential equations*, Funct. Anal. Appr. Comp. **6** (2014), 13–33.
324. M. Kostić, *A note on the existence and growth of mild solutions of abstract Cauchy problems for generators of integrated C-semigroups and cosine functions*, Novi Sad J. Math. **43** (2013), 157–165.
325. M. Kostić, C.-G. Li, M. Li, *Abstract multi-term fractional differential equations*, Krag. J. Math. **38** (2014), 51–71.
326. M. Kostić, *Bi-continuous (a, k)-regularized C-resolvent families*, preprint.
327. D. Kovačević, *On Integral Transforms and Convolution Equations on the Spaces of Tempered Ultradistributions*, Ph.D. Thesis, University of Novi Sad, 1992.
328. M. A. Krasnoselskii, P. E. Sobolevskii, *Fractional powers of operators defined on Banach space*, Dokl. Acad. Nauk SSSR **129** (1959), 499–502.
329. A. M. Krägeloh, *Two families of functions related to the fractional powers of generators of strongly continuous contraction semigroups*, J. Math. Anal. Appl. **283** (2003), 459–467.
330. S. Krol, *Perturbation theorems for holomorphic semigroups*, J. Evol. Equ. **9** (2009), 449–468.
331. P. C. Kunstmann, *Distribution semigroups and abstract Cauchy problems*, Trans. Amer. Math. Soc. **351** (1999), 837–856.
332. P. C. Kunstmann, *Stationary dense operators and generation of non-dense distribution semigroups*, J. Operator Theory **37** (1997), 111–120.
333. P. C. Kunstmann, *Nonuniqueness and wellposedness of abstract Cauchy problems in a Fréchet space*, Bull. Austral. Math. Soc. **63** (2001), 123–131.
334. P. C. Kunstmann, *Banach space valued ultradistributions and applications to abstract Cauchy problems*, preprint.
335. C.-C. Kuo, *Perturbation theorems for local integrated semigroups*, Studia Math. **197** (2010), 13–26.
336. C.-C. Kuo, *On perturbation of local integrated cosine functions*, Taiwanese J. Math. **90** (2012), 1613–1628.
337. C.-C. Kuo, *On existence and approximation of solutions of abstract Cauchy problem*, Taiwanese J. Math. **13** (2009), 137–155.
338. C.-C. Kuo, *On existence and approximation of solutions of second order abstract Cauchy problem*, Taiwanese J. Math. **14** (2010), 1093–1109.
339. C.-C. Kuo, *Local K-convoluted C-semigroups and abstract Cauchy problems*. Filomat, accepted.
340. C.-C. Kuo, *Local K-convoluted C-cosine functions and abstract Cauchy problems*, preprint.
341. F. Kühnemund, *A Hille-Yosida Theorem for Bi-continuous semigroups*, Semigroup Forum **67** (2003), 205–225.
342. T. A. M. Langlands, B. I. Henry, S. L. Wearne, *Fractional cable equation models for anomalous electrodiffusion in nerve cells: finite domain solutions*, SIAM J. Appl. Math. **71** (2011), 1168–1203.
343. T. A. M. Langlands, B. I. Henry, S. L. Wearne, *Fractional cable equation models for anomalous electrodiffusion in nerve cells: infinite domain solutions*, J. Math. Biol. **59** (2009), 761–808.
344. T. A. M. Langlands, B. I. Henry, S. L. Wearne, *Anomalous subdiffusion with multispecies linear reaction dynamics*, Phys. Rev. E **(3)** **77** (2008), no. 2, 021111, 9 pp.
345. T. A. M. Langlands, B. I. Henry, S. L. Wearne, *Anomalous diffusion with linear reaction dynamics: from continuous time random walks to fractional reaction-diffusion equations*, Phys. Rev. E **(3)** **74** (2006), no. 3, 031116, 15 pp.
346. O. E. Lanford III, D. W. Robinson, *Fractional powers of generators of equicontinuous semigroups and fractional derivatives*, J. Australian Math. Soc. **46** (1989), 473–504.
347. Y. Lei, W. Yi, Q. Zheng, *Semigroups of operators and polynomials of generators of bounded strongly continuous groups*, Proc. London Math. Soc. **69** (1994), 144–170.
348. L. D. Lemle, *$L^1(\mathbb{R}^d, dx)$-uniqueness of weak solutions for the Fokker-Planck equation associated with a class of Dirichlet operators*, Electron. Res. Announc. Math. Sci. **15** (2008), 65–70.
349. F. León-Saavedra, *Notes about the hypercyclicity criterion*, Math. Slovaca **53** (2003), 313–319.

350. F. León-Saavedra, V. Müler, *Rotations of hypercyclic and supercyclic operators*, Integral Equations Operator Theory **50** (2004), 385–391.
351. F. León-Saavedra, A. Piqueras-Lerena, *Positivity in the theory of supercyclic operators*, Perspectives in operator theory, Banach Center Publications **75** (2007), 221–232.
352. F. León-Saavedra, A. Piqueras-Lerena, *Cyclic properties of Volterra operators*, Pacific J. Math. **211** (2003), 157–162.
353. F. Li, *Multiplicative perturbations of incomplete second order abstract differential equations*, Kybernetes **37** (2008), 1431–1437.
354. F. Li, J. H. Liu, *Note on multiplicative perturbation of local C-regularized cosine functions with nondensely defined generators*, Electr. J. Qual. Theory Diff. Equ. **57** (2010), 1–12.
355. F. Li, J. Liang, T.-J. Xiao, J. Zhang, *On perturbation of convoluted C-regularized operator families*, J. Funct. Spaces Appl., Vol. 2013, Article ID 579326, 8 pp.
356. F. Li, H. Wang, J. Zhang, *Multiplicative perturbations of convoluted C-cosine functions and convoluted C-semigroups*, J. Funct. Spaces Appl., Vol. 2013, Article ID 426459, 9 pp.
357. F.-B. Li, M. Li, Q. Zheng, *Fractional evolution equations governed by coercive differential operators*, Abstr. Appl. Anal. 2009, Art. ID 438690, 14 pp. 34G10.
358. K. Li, J. Peng, J. Jia, *Cauchy problems for fractional differential equations with Riemann-Liouville fractional derivatives*, J. Funct Anal. **263** (2012), 476–510.
359. M. Li, S. Piskarev, *On approximations of integrated semigroups*, Taiwanese J. Math. **14** (2010), 2137–2161.
360. M. Li, Q. Zheng, *On product formulas for C-semigroups*, Semigroup Forum **78** (2009), 536–546.
361. M. Li, Q. Zheng, J. Zhang, *Regularized resolvent families*, Taiwanese J. Math. **11** (2007), 117–133.
362. M. Li, Q. Zheng, *α-times integrated semigroups: local and global*, Studia Math. **154** (2003), 243–252.
363. M. Li, F.-B. Li, Q. Zheng, *Elliptic operators with variable coefficients generating fractional resolvent families*, Int. J. Evol. Equ. **2** (2007), 195–204.
364. M. Li, Q. Zheng, *On spectral inclusions and approximations of α-times resolvent families*, Semigroup Forum **69** (2004), 356–368.
365. M. Li, C. Chen, F.-B. Li, *On fractional powers of generators of fractional resolvent families*, J. Funct Anal. **259** (2010), 2702–2726.
366. X.-M. Li, X.-Q. Song, Y.-Y. Zhao, *The Trotter-Kato approximation theorems of N–times integrated C cosine functions*, J. Huaiyin Institute of Technology **01** (in Chinese).
367. Y.-C. Li, S.-Y. Shaw, *N-times integrated C-semigroups and the abstract Cauchy problem*, Taiwanese J. Math. **1** (1997), 75–102.
368. Y.-C. Li, S.-Y. Shaw, *Perturbation of non-exponentially-bounded α-times integrated C-semigroups*, J. Math. Soc. Japan **55** (2003), 1115–1136.
369. J. Liang, J. Liu, T.-J. Xiao, *Hyperbolic singular perturbations for integrodifferential equations*, Appl. Math. Comput. **163** (2005), 609–620.
370. Y. Lin, *Time-dependent perturbation theory for abstract evolution equations of second order*, Studia Math. **130** (1998), 263–274.
371. J. L. Lions, *Semi-groupes distributions*, Portugalie Math. **19** (1960), 141–164.
372. J. L. Lions, J. Peetre, *Sur une classe d'espaces d'interpolation*, Inst. Hautes Études Sci. Publ. Math. **19** (1964), 5–68.
373. W. Littman, *The wave operator and L_p norms*, J. Math. Mech. **12** (1963), 55–68.
374. J. Lindenstrauss, J. E. Matoušková, D. E. Preiss, *Lipschitz image of a measure-null set can have a null complement*, Israel J. Math. **118** (2000), 207–219.
375. J. Liu, J. Sochacki, P. Dostert, *Singular perturbations and approximations for integrodifferential equations*, Differential equations and control theory (Athens, OH, 2000), 233–244, Lecture Notes in Pure and Appl. Math., **225**, Dekker, New York, 2002.
376. J. H. Liu, *Singular perturbations in a nonlinear viscoelasticity*, J. Integral Equations Appl. **9** (1997), 99–112.

377. J. H. Liu, *A singular perturbation problem in integrodifferential equations*, Electronic J. Diff. Equ. **02** (1993), 1–10.
378. L. Liu, F. Guo, C. Wu, Y. Wu, *Existence theorems of global solutions for nonlinear Volterra type integral equations in Banach spaces*, J. Math. Anal. Appl. **309** (2005), 638–649.
379. C. Lizama, *A characterization of uniform continuity for Volterra equations in Hilbert spaces*, Proc. Amer. Math. Soc. **126** (1998), 3581–3587.
380. C. Lizama, *On Volterra equations associated with a linear operator*, Proc. Amer. Math. Soc. **118** (1993), 1159–1166.
381. C. Lizama, *On the convergence and approximations of integrated semigroups*, J. Math. Anal. Appl. **181** (1994), 89–103.
382. C. Lizama, *Regularized solutions for abstract Volterra equations*, J. Math. Anal. Appl. **243** (2000), 278–292.
383. C. Lizama, *On approximation and representation of K–regularized resolvent families*, Integral Equations Operator Theory **41** (2001), 223–229.
384. C. Lizama, J. Sánchez, *On perturbation of K–regularized resolvent families*, Taiwanese J. Math. 7 (2003), 217–227.
385. C. Lizama, V. Poblete, *On multiplicative perturbation of integral resolvent families*, J. Math. Anal. Appl. **327** (2007), 1335–1359.
386. C. Lizama, H. Prado, *Rates of approximations and ergodic limits of regularized operator families*, J. Approx. Theory **122** (2003), 42–61.
387. C. Lizama, H. Prado, *Singular perurbation for Volterra equations of convolution type*, Appl. Math. Comp. **181** (2006), 1624–1634.
388. C. Lizama, H. Prado, *On duality and spectral properties of* (a, k)*-regularized resolvents*, Proc. R. Soc. Edinb., Sect. A, Math. **139** (2009), 505–517.
389. C. Lizama, P. J. Miana, *A Landau-Kolmogorov inequality for generators of families of bounded operators*, J. Math. Anal. Appl. **371** (2010), 614–623.
390. C. Lizama, F. Poblete, *On a functional equation associated with* (a, k)*-regularized resolvent families*, Abstr. Appl. Anal., volume 2012 (2012), Article ID 495487, 23 pages.
391. C. Lizama, G. N'Guérékata, *Mild solutions for abstract fractional differential equations*, Appl. Anal. **92** (2013), 1731–1754.
392. C. Lizama, H. Prado, *Fractional relaxation equations on Banach spaces*, Appl. Math. Lett. **23** (2010), 137–142.
393. P. Loreti, D. Sforza, *Exact reachibility problems for second-order integro-differential equations*, C. R. Acad. Sci. Paris, Ser. I **347** (2009), 1153–1158.
394. P. Loreti, D. Sforza, *Reachibility problems for a class of integro-differential equations*, J. Differential Equ. **248** (2010), 1711–1755.
395. Y. Luchko, R. Gorenflo, *An operational method for solving fractional differential equations with the Caputo derivatives*, Acta Math. Vietnam. **24** (1999), 207–233.
396. A. Lunardi, *Analytic Semigroups and Optimal Regularity in Parabolic Problems*, Birkhäuser, Basel, 1995.
397. Yu. I. Lyubich, *Investigation of the deficiency of the abstract Cauchy problem*, Sov Math., Dokl. 7 (1966), 166–169.
398. R. C. MacCamy, *A model for one-dimensional nonlinear viscoelasticity*, Q. Appl. Math. **35** (1997), 21–33.
399. C. R. MacCluer, *Chaos in linear distributed spaces*, J. Dynam. Systems Measurement Control **114** (1992), 322–324.
400. F. Mainardi, *Fractional Calculus and Waves in Linear Viscoelasticity. An Introduction to Mathematical Models*, Imperial College Press, London, 2010.
401. E. Mangino, A. Peris, *Frequently hypercyclic semigroups*, Studia Math. **202** (2011), 227–242.
402. V. Marraffa, *Riemann type integrals for functions taking values in locally convex spaces*, Czechoslovak Math. J. **56** (2006), 475–490.
403. C. Martinez, M. Sanz, *The Theory of Fractional Powers of Operators*, North–Holland Math. Stud. **187**, Elseiver, Amsterdam, 2001.

404. C. Martinez, M. Sanz, A. Redondo, *Fractional powers of almost non-negative operators*, Fract. Calc. Appl. Anal. **8** (2005), 201–230.
405. M. Matsui, M. Yamada, F. Takeo, *Supercyclic and chaotic translation semigroups*, Proc. Amer. Math. Soc. **131** (2003), 3535–3546.
406. M. Matsui, M. Yamada, F. Takeo, *Erratum to "Supercyclic and chaotic translation semigroups"*, Proc. Amer. Math. Soc. **132** (2004), 3751–3752.
407. M. Matsui, F. Takeo, *Chaotic semigroups generated by certain differential operators of order 1*, SUT J. Math. **37** (2001), 51–67.
408. T. Matsumoto, S. Oharu, H. R. Thieme, *Nonlinear perturbations of a class of integrated semigroups*, Hiroshima Math. J. **26** (1996), 433–473.
409. A. McIntosh, *Operators which have an H^∞ functional calculus*, Miniconference on Operator Theory and PDE, Proc. of the Center for Math. Analysis, Austral. Nat. Univ., Canberra **14** (1986), 210–231.
410. M. M. Meerschaert, E. Nane, P. Vellaisamy, *Distributed-order fractional diffusions on bounded domains*, J. Math. Anal. Appl. **379** (2011), 216–228.
411. R. Meise, D. Vogt, *Introduction to Functional Analysis*, Translated from the German by M. S. Ramanujan and revised by the authors. Oxf. Grad. Texts Math., Clarendon Press, New York, 1997.
412. I. V. Melnikova, A. I. Filinkov, *Abstract Cauchy Problems: Three Approaches*, Chapman Hall/CRC, 2001.
413. G. Metafune, *L^p-Spectrum of Ornstein-Uhlenbeck operators*, Ann. Scuola Norm. Sup. Pisa Cl. Sci. **30** (2001), 97–124.
414. P. J. Miana, *Almost-distribution cosine functions and integrated cosine functions*, Studia Math. **166** (2005), 171–180.
415. P. J. Miana, V. Poblete, *Sharp extensions for convoluted solutions of wave equations*, preprint. arXiv:1303.7346 [math.FA].
416. K. S. Miller, B. Ross, *An Introduction to Fractional Calculus and Fractional Differential Equations*, Wiley, New York, 1993.
417. M. Mijatović, S. Pilipović, F. Vajzović, *α-times integrated semigroups* ($\alpha \in \mathbb{R}^+$), J. Math. Anal. Appl. **210** (1997), 790–803.
418. L. Miller, *Non-structural controllability of linear elastic systems with structural damping*, J. Funct. Anal. **236** (2006), 592–608.
419. R. K. Miller, *Well-posedness of abstract Volterra problems*, Volterra equations (Proc. Helsinki Sympos. Integral Equations, Otaniemi, 1978), pp. 192–205, Lecture Notes in Math., **737**, Springer, Berlin, 1979.
420. R. K. Miller, *Resolvent operators for integral equations in a Banach space*, Trans. Amer. Math. Soc. **273** (1982), 333–349.
421. B. Mirković, *Teorija Mera i Integrala, Naučna knjiga*, Beograd, 1990 (in Serbian).
422. I. Miyadera, M. Okubo, N. Tanaka, *On integrated semigroups which are not exponentially bounded*, Proc. Japan Acad. **69** (1993), 199–204.
423. I. Miyadera, *Semi-groups of operators in Fréchet spaces and applications to partial differential equations*, Tôhoku Math. J. **11** (1959), 162–183.
424. A. Montes-Rodríguez, A. Rodríguez-Martínez, S. Shkarin, *Cyclic behaviour of Volterra composition operators*, Proc. Lond. Math. Soc. **103** (2011), 535–562.
425. A. Montes-Rodríguez, A. Rodríguez-Martínez, S. Shkarin, *Spectral theory of Volterra composition operators*, Math. Z. **261** (2009), 431–472.
426. R. T. Moore, *Generation of equicontinuous semigroups by hermitian and sectorial operators*, I. Bull. Amer. Math. Soc. **77** (1971), 224–229.
427. G. M. Mophou, G. N'Guérékata, *On a class of fractional differential equations in a Sobolev space*, Appl. Anal. **91** (2012), 15–34.
428. D. Mügnolo, *Damped wave equations with dynamic boundary conditions*, J. Appl. Anal. **17** (2011), 241–275.

429. C. Muller, *Approximation of local convoluted semigroups*, J. Math. Anal. Appl. **269** (2002), 401–420.
430. J.M.A.M. van Neerven, *Some recent results on adjoint semigroups*, Centrum voor Wiskunde en Informatica. Centre for Mathematics and Computer Science. CWI Quarterly **6** (1993), 139–153.
431. J.M.A.M. van Neerven, B. Straub, *On the existence and growth of mild solutions of the abstract Cauchy problem for operators with polynomially bounded resolvent*, Houston J. Math. **24** (1998), 137–171.
432. E. Nelson, *A functional calculus using singular Laplace integrals*, Trans. Amer. Math. Soc. **88** (1958), 400–413.
433. F. Neubrander, *Wellposedness of higher order abstract Cauchy problems*, Trans. Amer. Math. Soc. **295** (1986), 257–290.
434. F. Neubrander, *Integrated semigroups and their applications to the abstract Cauchy problem*, Pacific J. Math. **135** (1998), 111–155.
435. F. Neubrander, *Well-posedness of higher order abstract Cauchy problems*, Trans. Amer. Math. Soc. **295** (1986), 257–290.
436. F. Neubrander, *Integrated semigroups and their application to complete second order Cauchy problems*, Semigroup Forum **38** (1989), 233–251.
437. L. T. P. Ngoc, N. T. Long, *On a nonlinear Volterra-Hammerstein integral equation in two variables*, Acta Math. Sci. Ser. B Engl. Ed. **33B(2)** (2013), 484–494.
438. L. Nguyen, *Periodicity of mild solutions to higher order differential equations in Banach spaces*, Electronic J. Diff. Equ. **79** (2004), 1–12.
439. L. Nguyen, *On the periodic mild solutions to complete higher order differential equations on Banach spaces*, Surv. Math. Appl. **6** (2011), 23–41.
440. Nguyen Thanh Lan, *On the mild solutions of higher-order differential equations in Banach spaces*, Abstr. Appl. Anal. 2003, no. 15, 865–880.
441. S. Nicaise, *The Hille–Yosida and Trotter–Kato theorems for integrated semigroups*, J. Math. Anal. Appl. **180** (1993), 303–316.
442. J. A. Nohel, M. J. Renardy, *Development of singularities in nonlinear viscoelasticity. Amorphous polymers and non-Newtonian fluids*, (Minneapolis, Minn., 1985), 139–152, IMA Vol. Math. Appl., 6, Springer, New York, 1987.
443. J. W. Nunziato, *On heat conduction in materials with memory*, Quarterly Appl. Math. **29** (1971), 187–204.
444. H. Oka, *A class of complete second order linear differential equations*, Proc. Amer. Math. Soc. **124** (1996), 3143–3150.
445. H. Oka, *Linear Volterra equations and integrated solution families*, Semigroup Forum **53** (1996), 278–297.
446. N. Okazawa, *A generation theorem for semigroups of growth order*, Tôhoku Math. J. **26** (1974), 39–51.
447. K. B. Oldham, J. Spanier, *The Fractional Calculus*, Academic Press, New-York-London, 1974.
448. A. Ostrowski, *Solutions of Equations and Systems of Equations*, Academic Press, New York, 1966.
449. S. Ōuchi, *Hyperfunction solutions of the abstract Cauchy problems*, Proc. Japan Acad. **47** (1971), 541–544.
450. A. Pazy, *Semigroups of Linear Operators and Applications to Partial Differential Equations*, Springer-Verlag, Berlin, 1983.
451. J. Peng, K. Li, *A novel characteristic of solution operator for the fractional abstract Cauchy problem*, J. Math. Anal. Appl. **385** (2012), 786–796.
452. F. Periago, B. Straub, *On the existence and uniqueness of solutions for an incomplete second-order abstract Cauchy problem*, Studia Math. **155** (2003), 183–193.
453. F. Periago, B. Straub, *A functional calculus for almost sectorial operators and applications to abstract evolution equations*, J. Evol. Equ. **2** (2002), 41–68.

454. A. Peris, L. Saldivia, *Syndetically hypercyclic operators*, Integral Equations Operator Theory **51** (2005), 275–281.
455. S. Pilipović, *Tempered ultradistributions*, Boll. Un. Mat. Ital. 7 (1998), 235–251.
456. S. Pilipović, B. Stanković, *Prostori Distribucija*, SANU, Novi Sad, 2000 (in Serbian).
457. S. Piskarev, S.-Y. Shaw, *Multiplicative perturbations of C_0-semigroups and some applications to step responses and cumulative outputs*, J. Funct. Anal. **128** (1995), 315–340.
458. I. Podlubny, *Fractional Differential Equations*, Academic Press, New York, 1999.
459. H. Prado, *On the asymptotic behavior of regularized operator families*, Semigroup Forum **73** (2006), 243–252.
460. H. Prado, *Stability properties for solution operators*, Semigroup Forum **77** (2008), 243–252.
461. I. V. Proskuryakov, *Problems in Linear Algebra*, Savremena administracija, Beograd, 1988 (in Serbian).
462. V. Protopopescu, Y. Azmy, *Topological chaos for a class of linear models*, Math. Models Methods Appl. Sci. **2** (1992), 79–90.
463. J. Prüss, *Evolutionary Integral Equations and Applications*, Monogr. Math. **87**, Birkhäuser, Basel, Boston, Berlin, 1993.
464. J. Prüss, *Decay properties for the solutions of a partial differential equation with memory*, Arch. Math. (Basel) **92** (2009), 158–173.
465. J. Prüss, H. Sohr, *Imaginary powers of elliptic second order differential operators in Lp-spaces*, Hiroshima Math. J. **23** (1993), 161–192.
466. M. M. Rao, *Integration with vector valued measures*, Discrete Cont. Dyn. Sys. **33** (2013), 5429–5440.
467. M.-L. Raynal, *On some nonlinear problems of diffusion*, Volterra equations (Proc. Helsinki Sympos. Integral Equations, Otaniemi, 1978), pp. 251–266, Lecture Notes in Math., **737**, Springer, Berlin, 1979.
468. M. Renardy, W. J. Hrusa, J. A. Nohel, *Mathematical Problems in Viscoelasticity*, Longman Science Technology, New York, 1987.
469. A. Rhandi, *Multiplicative perturbations of linear Volterra equations*, Proc. Amer. Math. Soc. **119** (1993), 493–501.
470. A. Rhandi, *Positive perturbations of linear Volterra equations and sine functions of operators*, J. Integral Equations Appl. **4** (1992), 409–420.
471. S. Rolewicz, *On orbits of elements*, Studia Math. **32** (1969), 17–22.
472. W. V. Smith, D. H. Tucker, *Weak integral convergence theorems and operator measures*, Pacific J. Math. **111** (1984), 243–256.
473. A. Rodríguez-Martínez, *Residuality of sets of hypercylic operators*, Integral Equations Operator Theory **72** (2012), 301–308.
474. J. Rodríguez, *Absolutely summing operators and integration of vector-valued functions*, J. Math. Anal. Appl. **316** (2006), 579–600.
475. H. R. Salas, *Hypercyclic weighted shifts*, Trans. Amer. Math. Soc. **347** (1995), 993–1004.
476. H. R. Salas, *Supercyclicity and weighted shifts*, Studia Math. **135** (1999), 55–74.
477. S. Samko, *Approximative approach to fractional powers of operators*, Proc. Second ISAAC Congress, Kluwer Academic Publishers 2000, Vol. 2, Chapter 131, pp. 1163–1170.
478. S. G. Samko, A. A. Kilbas, O. I. Marichev, *Fractional Derivatives and Integrals: Theory and Applications*, Gordon and Breach, New York, 1993.
479. J. T. Sandefur, *Higher Order Abstract Cauchy Problems*, Ph.D. Thesis, Tulane University, New Orleans, 1974.
480. R. K. Saxena, A. M. Mathai, H. J. Haubold, *Reaction-diffusion systems and nonlinear waves*, Astrophys. Space Sci. **305** (2006), 297–303.
481. L. Schwartz, *Theorie des Distributions*, 2 vols., Hermann, Paris, 1950–1951.
482. H. Serizawa, M. Watanabe, *Time-dependent perturbation for cosine families in Banach spaces*, Houston J. Math. **12** (1986), 579–586.
483. D. Sforza, *Maximal regularity results for a second order integro-differential equation*, J. Math. Anal. Appl. **191** (1995), 203–228.

484. S.-Y. Shaw, *Cosine operator functions and Cauchy problems*, Conf. Semin. Mat. Univ. Bari **287** (2002), 1–75.
485. S.-Y. Shaw, *Ergodic theorems and approximation theorems with rates*, Taiwanese J. Math. **4** (2010), 365–383.
486. S. Shkarin, *The Kitai criterion and backwards shifts*, Proc. Amer. Math. Soc. **136** (2008), 1659–1670.
487. S. Shkarin, *On supercyclicity of operators from an supercyclic semigroup*, J. Math. Anal. Appl. **382** (2011), 516–522.
488. S. Shkarin, *Operators commuting with the Volterra operator are not weakly supercyclic*, Integral Equations Operator Theory **68** (2010), 229–241.
489. S. Shkarin, *Orbits of semigroups of truncated convolution operators*, Glasgow Math. J. **54** (2012), 399–414.
490. S. Shkarin, *Antisupercyclic operators and orbits of the Volterra operator*, J. London Math. Soc. **73** (2006), 506–528.
491. D. Smith, *Singular Perturbation Theory*, Cambridge University Press, Cambridge, 1985.
492. I. N. Sneddon, *The Use of Integral Transforms*, McGraw–Hill, New York, 1972.
493. P. E. Sobolevskiĭ, *On semigroups of growth order α*, Sov. Math., Dokl. **12** (1971), 202–205.
494. O. J. Staffans, *Systems of nonlinear Volterra equations with positive definite kernels*, Trans. Amer. Math. Soc. **228** (1977), 99–116.
495. B. Stanković, *On the function of E. M. Wright*, Publ. Inst. Math., Nouv. Sér. **10** (1970), 113–124.
496. E. M. Stein, *Singular Integrals and Differentiability Properties of Functions*, Princeton University Press, New Jersey, 1970.
497. B. Straub, *Fractional powers of operators with polynomially bounded resolvent and the semigroups generated by them*, Hiroshima Math. J. **24** (1994), 529–548.
498. F. Takeo, *Chaos and hypercyclicity for solution semigroups to partial differential equations*, Nonlinear Anal. **63** (2005), 1943–1953.
499. F. Takeo, *Chaotic or hypercyclic semigroups on a function space $C_0(I, \mathbb{C})$ or $L^p(I, \mathbb{C})$*, SUT J. Math. **41** (2005), 43–61.
500. N. Tanaka, *Holomorphic C-semigroups and holomorphic semigroups*, Semigroup Forum **38** (1989), 253–261.
501. G. Teschl, *Mathematical Methods in Quantum Mechanics: with Applications to Schrödinger Operators*, AMS, Providence, Rhode Island, 2009.
502. G. E. F. Thomas, *Integration of functions with values in locally convex spaces*, Trans. Amer. Math. Soc. **212** (1975), 61–81.
503. C. C. Travis, G. F. Webb, *Cosine families and abstract nonlinear second order differential equations*, Acta Math. Acad. Sci. Hungar. **32** (1978), 75–96.
504. H. Triebel, *Analysis and Mathematical Physics*, Translated from the German by Bernhard Simon and Hedwig Simon. BSB B. G. Teubner Verlagsgesellschaft, Leipzig, 1986.
505. H. F. Trotter, *Approximation of operator-semigroups*, Pacific J. Math. **8** (1958), 887–919.
506. K. Tsuruta, *Regularity of solutions to an abstract inhomogeneous linear integro-differential equation*, Math. Japon. **26** (1981), 65–76.
507. K. Tsuruta, *Bounded linear operators satisfying second-order integro-differential equations in a Banach space*, J. Integral Equations **6** (1984), 231–268.
508. T. Ushijima, *On the generation and smoothness of semi-groups of linear operators*, J. Fac. Sci., Univ. Tokyo, Sect. I A **19** (1972), 65–126.
509. T. Ushijima, *On the abstract Cauchy problem and semi-groups of linear operators in locally convex spaces*, Sci. Pap. Coll. Gen. Educ., Univ. Tokyo **21** (1971), 93–122.
510. R. G. Venter, *Measures and Functions in Locally Convex Spaces*, Ph.D. Thesis, University of Pretoria, 2010.
511. W. von Wahl, *Gebrochene Potenzen eines elliptischen Operators und parabolische Differentialgleichungen in Raumen hölderstetiger Funktionen*, Nachr. Akad. Wiss. Gottingen Math.-Phys. Kl. **11** (1972), 231–258.
512. J. L. Walsh, *On the location of the roots of certain types of polynomials*, Trans. Amer. Math. Soc. **24** (1922), 163–180.

513. J.-R. Wang, X.-W. Dong, W. Wei, *On the existence of solutions for a class of fractional differential equations*, Stud. Univ. Babes-Bolyai Math. **57** (2012), 15–24.
514. R.-N. Wang, D.-H. Chen, T.-J. Xiao, *Abstract fractional Cauchy problems with almost sectorial operators*, J. Differential Equations **252** (2012), 202–235.
515. S. W. Wang, *Mild integrated C-existence families*, Studia Math. **112** (1995), 251–266.
516. S. W. Wang, M. Y. Wang, Y. Shen, *Perturbation theorems for local integrated semigroups and their applications*, Studia Math. **170** (2005), 121–146.
517. G. F. Webb, *An abstract semilinear Volterra integrodifferential equation*, Proc. Amer. Math. Soc. **69** (1978), 255–260.
518. G. F. Webb, *Periodic and chaotic behaviour in structured models of popolation dynamics*, Recent Developments in Evolution Equations, Pitman Research Notes in Math. **324** (1995), 40–49.
519. L. Weis, V. Wrobel, *Asymptotic behaviour of C_0-semigroups in Banach spaces*, Proc. Amer. Math. Soc. **124** (1996), 3663–3671.
520. J. Wengenroth, *Hypercyclic operators on non-locally convex spaces*, Proc. Amer. Math. Soc. **131** (2002), 1759–1761.
521. U. Westphal, *An approach to fractional powers of operators via fractional differences,* Proc. Amer. Math. Soc. **29** (1974), 557–576.
522. R. Wong, Y.-Q. Zhao, *Exponential asymptotics of the Mittag-Leffler function*, Constr. Approx. **18** (2002), 355–385.
523. E. M. Wright, *The asymptotic expansion of integral functions defined by Taylor series*, Philos. Trans. Roy. Soc. London, Ser. A. **238** (1940), 423–451.
524. E. M. Wright, *The generalized Bessel functions of order greater than one*, Quarterly J. Math. (Oxford Ser.) **11** (1940), 36–48.
525. L. Wu, Y. Zhang, *A new topological approach to the L^∞-uniqueness of operators and L^1-uniqueness of Fokker-Planck equations*, J. Funct. Anal. **241** (2006), 557–610.
526. T.-J. Xiao, J. Liang, *Laplace transforms and integrated, regularized semigroups in locally convex spaces*, J. Funct. Anal. **148** (1997), 448–479.
527. T.-J. Xiao, J. Liang, *Widder-Arendt theorem and integrated semigroups in locally convex space*, Sci. China Math. **39** (1996), 1121–1130.
528. T.-J. Xiao, J. Liang, *Approximations of Laplace transforms and integrated semigroups*, J. Funct. Anal. **172** (2000), 202–220.
529. T.-J. Xiao, J. Liang, *Higher order abstract Cauchy problems: their existence and uniqueness families*, J. Lond. Math. Soc. **67** (2003), 149–164.
530. T.-J. Xiao, J. Liang, *Differential operators and C-wellposedness of complete second order abstract Cauchy problems*, Pacific J. Math. **186** (1998), 167–191.
531. T.-J. Xiao, J. Liang, *The Cauchy Problem for Higher-Order Abstract Differential Equations*, Springer-Verlag, Berlin, 1998.
532. T.-J. Xiao, J. Liang, *Schrödinger-type evolution equations in $L^p(\Omega)$*, J. Math. Anal. Appl. **260** (2001), 55–69.
533. T.-J. Xiao, J. Liang, *Perturbations of existence families for abstract Cauchy problems*, Proc. Amer. Math. Soc. **130** (2002), 2275–2285.
534. T.-J. Xiao, J. Liang, *Abstract degenerate Cauchy problems in locally convex spaces*, J. Math. Anal. Appl. **259** (2001), 398–412.
535. T.-J. Xiao, R. Nagel, J. Liang, *Approximation theorems for the propagators of higher order abstract Cauchy problems*, Trans. Amer. Math. Soc. **360** (2007), 1723–1739.
536. T.-J. Xiao, J. Liang, J. V. Casteren, *Time dependent Desch-Schappacher type perturbations of Volterra integral equations*, Integral Equations Operator Theory **44** (2002), 494–506.
537. T.-J. Xiao, J. Liang, F. Li, *Multiplicative perturbations of local C-regularized semigroups*, Semigroup Forum **72** (2006), 375–386.
538. T.-J. Xiao, J. Liang, *Wellposedness results for certain classes of higher order abstract Cauchy problems connected with integrated semigroups*, Semigroup Forum **56** (1998), 84–103.

539. T.-J. Xiao, J. Liang, *Integrated semigroups and higher order abstract equations*, J. Math. Anal. Appl. **222** (1998), 110–125.
540. T.-J. Xiao, J. Liang, *Entire solutions of higher order abstract Cauchy problems*, J. Math. Anal. Appl. **208** (1997), 298–310.
541. Y. Xin, C. Liang, *Multiplicative perturbations of C-regularized resolvent families*, J. Zheijang Univ. SCI **5(5)** (2004), 528–532.
542. A.Yagi, *Abstract Parabolic Evolution Equations and Their Applications*, Springer-Verlag, Berlin, 2010.
543. K. Yoshinaga, *Fractional powers of an infinitesimal generator in the theory of semi-group distributions*, Bull. Kyushu Inst. Tech. Math. Natur. Sci. **18** (1971), 1–15.
544. K. Yosida, *On the differentiability of semi-groups of linear operators*, J. Math. Soc. Japan **1** (1948), 15–21.
545. K. Yosida, *On the differentiability of semi-groups of linear operators*, Proc. Japan Acad. **34** (1958), 337–340.
546. A. V. Zafievskii, *On semigroups with singularites summable with a power-weight at zero*, Sov. Math. Dokl. **11** (1970), 1408–1411.
547. J. Z. Zhang, *Regularized cosine existence and uniqueness families for second order abstract Cauchy problems*, Studia Math. **152** (2002), 131–145.
548. J. Zhang, Q. Zheng, *Pseudodifferential operators and regularized semigroups*, China J. Contemp. Math. **19** (1998), 387–394.
549. J. Zhang, Q. Zheng, *On α-times integrated cosine functions*, Math. Japon. **50** (1999), 401–408.
550. Q. Zheng, *Cauchy problems for polynomials of generators of bounded C_0-semigroups and for differential operators*, Tubinger Berichte zur Funktionalanalysis **4**, Tübingen, (1995).
551. Q. Zheng, *Matrices of operators and regularized cosine functions*, J. Math. Anal. Appl. **315** (2006), 68–75.
552. Q. Zheng, Y. Li, *Abstract parabolic systems and regularized semigroups*, Pacific J. Math. **182** (1998), 183–199.
553. Q. Zheng, *Integrated cosine functions*, Int. J. Math. Math. Sci. **19** (1996), 575–580.
554. Q. Zheng, *Coercive differential operators and fractionally integrated cosine functions*, Taiwanese J. Math. **6** (2002), 59–65.
555. Q. Zheng, *The analyticity of abstract parabolic and correct systems*, Science in China (Ser. A) **45** (2002), 859–865.
556. Q. Zheng, Y.-S. Lei, *Exponentially bounded C-semigroup and integrated semigroup with nondensely dened generators I: approximation*, Acta Math. Sinica **13** (1993), 251–260.

INDEX